Incredible Anaerobes

From Physiology to Genomics to Fuels

LARS G. LJUNGDAHL

ANNALS OF THE NEW YORK ACADEMY OF SCIENCES
Volume 1125

Incredible Anaerobes
From Physiology to Genomics to Fuels

Edited by
JUERGEN WIEGEL, ROBERT J. MAIER, AND MICHAEL W. W. ADAMS

Published by Blackwell Publishing on behalf of the New York Academy of Sciences
Boston, Massachusetts
2008

Library of Congress Cataloging-in-Publication Data

Incredible anaerobes : from physiology to genomics to fuels/editors, Juergen Wiegel, Robert J. Maier, and Michael W. W. Adams.
p.; cm. – (Annals of the New York Academy of Sciences, ISSN 0077-8923)
Includes bibliographical references.
ISBN-13: 978-1-57331-705-4 (paper : alk. paper)
ISBN-10: 1-57331-705-5 (paper : alk. paper)
1. Anaerobic bacteria–Congresses. 2. Ljungdahl, Lars G.–Congresses. I. Wiegel, Juergen. II. Adams, Michael W. W., 1954- III. Maier, Robert J. IV. New York Academy of Sciences. V. Series.
[DNLM: 1. Bacteria, Anaerobic–physiology–Congresses. 2. Bacteria, Anaerobic–physiology–Festschrift. 3. Bacteria, Anaerobic–genetics–Congresses. 4. Bacteria, Anaerobic–genetics–Festschrift. W1 AN626YL 2007/QW 52 I37 2007]

QR89.5.I53 2007
579.3'149–dc22

2007050339

The *Annals of the New York Academy of Sciences* (ISSN: 0077-8923 [print]; ISSN: 1749-6632 [online]) is published 28 times a year on behalf of the New York Academy of Sciences by Blackwell Publishing with offices at (US) 350 Main St., Malden, MA 02148-5020, (UK) 9600 Garsington Road, Oxford, OX4 2ZG, and (Asia) 165 Cremorne St., Richmond VIC 3121, Australia. Blackwell Publishing was acquired by John Wiley & Sons in February 2007. Blackwell's program has been merged with Wiley's global Scientific, Technical, and Medical business to form Wiley-Blackwell.

MAILING: *Annals* is mailed Standard Rate. Mailing to rest of world by IMEX (International Mail Express). Canadian mail is sent by Canadian publications mail agreement number 40573520. POSTMASTER: Send all address changes to *Annals of the New York Academy of Sciences*, Blackwell Publishing Inc., Journals Subscription Department, 350 Main St., Malden, MA 02148-5020.

Blackwell Publishing is now part of Wiley-Blackwell.

Information for subscribers: For ordering information, claims, and any inquiry concerning your subscription please contact your nearest office:

UK: Tel: +44 (0)1865 778315; Fax: +44 (0) 1865 471775
USA: Tel: +1 781 388 8599 or 1 800 835 6770 (toll free in the USA & Canada); Fax: +1 781 388 8232 or Fax: +44 (0) 1865 471775
Asia: Tel: +65 6511 8000; Fax: +44 (0)1865 471775,
Email: customerservices@blackwellpublishing.com

Subscription prices for 2008 are: Premium Institutional: US$4265 (The Americas), £2370 (Rest of World). Customers in the UK should add VAT at 7%; customers in the EU should also add VAT at 7%, or provide a VAT registration number or evidence of entitlement to exemption. Customers in Canada should add 5% GST or provide evidence of entitlement to exemption. The Premium institutional price also includes online access to the current and all online back files to January 1, 1997, where available. For other pricing options, including access information and terms and conditions, please visit www.blackwellpublishing.com/nyas.

Delivery Terms and Legal Title: Prices include delivery of print publications to the recipient's address. Delivery terms are Delivered Duty Unpaid (DDU); the recipient is responsible for paying any import duty or taxes. Legal title passes to the customer on despatch by our distributors.

Membership information: Members may order copies of *Annals* volumes directly from the Academy by visiting www.nyas.org/annals, emailing membership@nyas.org, faxing +1 212 298 3650, or calling 1 800 843 6927 (toll free in the USA), or +1 212 298 8640. For more information on becoming a member of the New York Academy of Sciences, please visit www.nyas.org/membership. Claims and inquiries on member orders should be directed to the Academy at email: membership@nyas.org or Tel: 1 800 843 6927 (toll free in the USA) or +1 212 298 8640.

Printed in the USA. Printed on acid-free paper.

Annals is available to subscribers online at Blackwell Synergy and the New York Academy of Sciences Web site. Visit www.blackwell-synergy.com or www.annalsnyas.org to search the articles and register for table of contents e-mail alerts.

The paper used in this publication meets the minimum requirements of the National Standard for Information Sciences Permanence of Paper for Printed Library Materials, ANSI Z39.48 1984.

ISSN: 0077-8923 (print); 1749-6632 (online)
ISBN-10: 1-57331-705-5 (paper); ISBN-13: 978-1-57331-705-4 (paper)

A catalogue record for this title is available from the British Library.

A Note of Thanks

In March 2007 colleagues Michael W.W. Adams, Robert J. Maier, and Juergen K. Wiegel organized the symposium Incredible Anaerobes: From Physiology to Genomics to Fuels. The symposium was timely, considering the interest in using anaerobic microorganisms for the production of biofuels and other feed stock chemicals from renewable biomass. The symposium was also to honor me on the occasion of my retirement and 80th birthday. The symposium was a great success, and I hereby thank the organizers from the bottom of my heart. It was a fantastic tribute, much more grandiose than I could ever have imagined. My thanks also go to all participants of the symposium and especially to those who gave outstanding talks and poster presentations, as well as to those who gave me a "hard time" with tough, but funny and enjoyable comments at the banquet. I treasured every minute at the symposium. Thanks are also extended to those who remembered me in letters, which were so very much appreciated.

I joined the newly established Department of Biochemistry at the University of Georgia (UGA) in 1967, a time when the university, including the biological sciences, was being strengthened. New laboratories and a fermentation plant were built, and Harry D. Peck was the chairman. I also joined the Department of Microbiology led by William Jackson Payne. It was clear to me that UGA was the place for me, especially with these two leaders and with all the productive activity. Subsequent years have proven this correct; it became a fantastic place for research and teaching of microbial physiology and biochemistry. I would like to thank all my colleagues at UGA, and especially David Puett, my last department head, for support and friendship. The list of international conferences that preceded the Incredible Anaerobes symposium (most of them leading to published books or special journal issues) is evidence that UGA is a great place for microbial biochemistry.

1990. D.E. Akin, L.G. Ljungdahl, J.K. Wilson, and R.J. Harris, Eds. *Microbial and Plant Opportunities to Improve Lignocellulose Utilization by Ruminants.* Elsevier. New York, NY.

1991. J.E. Rogers and W.B. Whitman, Eds. *Microbial Production and Consumption of Greenhouse Gases: Methane, Nitrogen Oxides, and Halomethanes.* American Society for Microbiology, Washington, D.C.

1992. J.E. Rogers and J. Wiegel. Anaerobic Dehalogenation and Its Environmental Implications. Athens, GA.

1996. Harold Drake, Stephen W. Ragsdale, and Juergen Wiegel. *The Arts of Anaerobes.* Bio-Factors Special Issue, Vol. 6, No. 1, 1997.

1998. J. Wiegel and M.W.W. Adams, Eds. Thermophiles: *The Keys to Molecular Evolution and the Origin of Life.* Taylor & Francis Ltd., Philadelphia, PA.

2003. L.G. Ljungdahl, M.W. Adams, L, Barton, J.G. Ferry, and M. Johnson, Eds. *Physiology and Biochemistry of Anaerobic Bacteria.* Springer Verlag. New York, NY.

It seems to me, considering the present interest in microbiology and its potential for use in solving the world's ecological, environmental, and energy problems, that studies of microbiology will continue as an important subject of study at the University of Georgia, as well as worldwide.

LARS G. LJUNGDAHL

ANNALS OF THE NEW YORK ACADEMY OF SCIENCES

Volume 1125
March 2008

Incredible Anaerobes

From Physiology to Genomics to Fuels

Editors
JUERGEN WIEGEL, ROBERT J. MAIER, AND MICHAEL W. W. ADAMS

This volume is the result of a conference entitled **Incredible Anaerobes: From Physiology to Genomics to Fuels** held on March 2–3, 2007 in Athens, Georgia.

CONTENTS

Part III. Methanogens and Methanogenesis

Part IV. Metal Reductions and Metal Enzymes

Part V. Cellulolytic Anaerobes and Their Cellulolytic Enzymes

Part VI. Applied Aspects and Fuel Production

Foreword

The first microorganisms observed by humans were aerobes, but all of the evidence suggests that life first originated on this planet under anoxic conditions. This subsequently led to the evolution of facultative aerobes, and aerobes when molecular oxygen became available billions of years later. Anaerobes, therefore, took a relatively long time to develop diverse metabolic pathways to enable them to thrive under the conditions of the early Earth. Consequently, anaerobes as a group can be found in virtually all ecosystems, which range from deep sea hydrothermal vents with temperatures far exceeding the normal boiling point to permafrost soil below the normal freezing point, from the intestines of insects and mammals to the oxygen-exposed skin of humans, and from oligotrophic marine environments to salt-saturated freshwater ponds. This extensive microbial, physiological, and biochemical diversity is assumed to arise, at least in part, due to the lower free-energy yield of fermentation and of virtually all anaerobic respiratory systems, in comparison to aerobic respiration where oxygen serves as the terminal electron acceptor. It appears that most anaerobes are more specialized and have a range of metabolic processes that are unique to them as a group, and not observed with many, if any, aerobes. Indeed, the biochemical pathways that are evident from recent genome sequences indicate that many anaerobes have even more capacities than anticipated.

This book stemmed from a two-day conference held in March 2007, in Athens, Georgia, to honor the biochemist and microbiologist Lars G. Ljungdahl, on the occasion of his 80th birthday and of his retirement from the University of Georgia. Lars began his career in 1943 as a technician at the Karolinska Institute in Sweden. In 1947 he joined the Stockholm Brewery Company, where he first became interested in anaerobes and their ability to degrade biomass. This interest developed into a career under the guidance of Harland Wood at Case Western Reserve University, where Lars earned his Ph.D. in 1964, working on a bacterium able to convert glucose nearly stoichiometrically to acetate. Elucidating the biochemistry of this once-obscure microorganism, named *Clostridium thermoaceticum* (now *Moorella thermoacetica*), continued to be the major thrust of his research, and in 1967 he joined the Departments of Biochemistry and Microbiology of the University of Georgia in Athens (UGA).

Many of the chapters in this book were written by former students and postdoctoral candidates of Lars and also by many of his longtime colleagues at UGA and from all over the world, now leading authorities in their field. Three chapters are devoted to the homoacetogenic fermentation process by such anaerobes: a chapter on the historical aspects and microbial diversity; a chapter on the biochemistry of the Wood–Ljungdahl pathway, the central part of homoacetogenic fermentation; and one on bioenergetic aspects. Lars's group characterized many of the enzymes in the pathway, and one of the many surprises with *M. thermoacetica* was the discovery of tungsten (W) as a biologically relevant metal; one chapter deals with the events that led to this discovery and the current status of this field. This section of the book on metals also includes two chapters on nickel, which is an essential component of the enzymes, hydrogenase, as well the carbon monoxide dehydrogenase/acetyl-CoA synthase, the latter one described in the chapter on the biochemistry of the acetogens. The role of nickel in the hydrogen metabolism of mesophilic intestinal bacteria and in hyperthermophilic microorganisms is discussed by two current colleagues of Lars. A comparison of the homoacetogenic and methanogenic fermentation with the homoacetogenic pathway and the pecularities of the biochemistry of the acetoclastic and CO_2/H_2-using methanogens are discussed in two adjacent chapters, followed by chapters on the biochemistry and ecology of methanogens.

In addition to enzyme systems involved in fermentative pathways, a significant part of this book is devoted to another area where Lars has made significant contributions, the anaerobic degradation of cellulosic biomass. His focus was the cellulosome, a multienzyme complex found in many anaerobes, including *Clostridium thermocellum* JW20, which was isolated by his group in the late 1970s and shown on the cover of this volume. Chapters include a comparison of cellulolytic systems and descriptions of the various strategies that anaerobes use to break down plant material, a chapter on how intestinal anaerobes degrade plant cell walls, and a chapter on the cellulosomes. At

Ann. N.Y. Acad. Sci. 1125: xi–xiii (2008). © 2008 New York Academy of Sciences.
doi: 10.1196/annals.1419.031

From left to right, front row: Michael W.W. Adams (co-organizer), Lars G. Ljungdahl, Juergen Wiegel (co-organizer); *second row*: Edward A. Bayer (*behind Lars Ljungdahl*), Harald L. Drake, Douglas E. Eveleigh, Harry J. Flint, Thomas W. Jeffries; *third row*: Tairo Oshima, Jan R. Andreesen, Roy H. Doi, David B. Wilson, Judy D. Wall, Harry J. Gilbert, Lonnie O. Ingram, Stephen W. Ragsdale; *fourth row*: Michael J. McInerney, Georg Fuchs, Gerhard Gottschalk, Wolfgang H. Schwarz; *fifth row*: Rudolf K. Thauer, Gerrit Voordouw, Volker Müller; (*not shown*: Robert Maier (co-organizer), David Lee, C. Michael Cassidy, Thomas W. Johnson, and Gregory Dilworth).

Conference attendees for Incredible Anaerobes: From Physiology to Genomics to Fuels

the end of this section, Lars describes the highly efficient cellulolytic system of the little-known anaerobic fungi.

The diversity of anaerobes is described in the first five chapters, which begins with an overview of anaerobic thermophiles followed by a description of anaerobes under the multiple extremes of high salt, alkaline pH, and elevated temperatures. The following chapter, authored by several leaders in the field, describes the impact that the study of genomic sequences have on our understanding of the intriguing group of anaerobic syntrophs; a separate chapter presents new insights into the capabilities of the clostridia as elucidated through comparative genomics. The final chapter in this section provides an overview of the anaerobic metabolism of aromatic compounds, which involves a range of unique enzymes. These compounds are important constituents of lignin, the third major component of plant biomass.

There is much current interest in the production of alternative fuels from plant-derived biomass. This could lead to a degree of independence from fossil fuels, a goal that has the attention of political leaders. The rising worldwide demand for energy, coupled with various supply-side instability factors regarding the delivery of crude or refined oil and the overall petroleum market, make it imperative that we develop processes to produce alternative fuels. Biofuels, such as ethanol and biodiesel, produced from renewable resources, have been promoted by President Bush and the Department of Energy (DOE). The DOE has designated this approach as a primary alternative for immediate and long-term replacement of fossil fuels. Consequently, our volume closes with the industrial applications of thermophiles on fuel production by anaerobes, including methane and butanol. The final chapter describes processes that are being considered for the industrial production of ethanol by an engineered strain of *E. coli*.

The topics covered in this volume, therefore, range from the diversity and physiology of major types of anaerobes to industrial fuel production, topics that mirror the major research interests of Lars Ljungdahl throughout his career. It is hoped that this volume is a fitting tribute to his accomplishments. His career began with work on an obscure bacterium, *Clostridium thermoaceticum*, progressed dramatically with the elucidation of the Wood–Ljungdahl pathway, merged in the 1980s into the applied world with the production of Ca-Mg-acetate from cornstarch-derived glucose using thermophilic acetogens, and latterly has focused on the use of thermophilic anaerobes to efficiently degrade cellulose and produce ethanol. These studies included the discovery in the 1980s, in a hot spring in Yellowstone National Park, of *Thermoanaerobacter ethanolicus*, an organism with a fermentation balance similar to that of yeast. In addition, a variety of anaerobic fungi with remarkably efficient cellulolytic enzyme systems have been described. Lars's work has also resulted in several patents, demonstrating that he not only excelled in basic research but that he also kept his "eye on the ball" and could readily appreciate the applied aspects of his progress. In a career spanning over 50 years of academic research, Lars has influenced many scientists, as demonstrated by the many researchers from a wide variety of backgrounds that traveled to Athens, Georgia to celebrate and honor Lars, both as a scientist and a gentleman.

The editors are indebted to the Office of Biological and Environmental Research of the U.S. Department of Energy; to New England Biolabs, Inc.; and to the Department of Biochemistry and Molecular Biology, and the Department of Microbiology of the University of Georgia for their generous support of the conference. We also gratefully acknowledge the offices of the provost and of the president of the University of Georgia for their support of the conference program, and the helpful staff of the *Annals of the New York Academy of Sciences*, who made it possible to publish this volume.

Michael W.W. Adams
Robert J. Maier
Juergen Wiegel

University of Georgia
Athens, Georgia

Diversity of Thermophilic Anaerobes

ISAAC D. WAGNER AND JUERGEN WIEGEL

Department of Microbiology, University of Georgia, Athens, Georgia, USA

Thermophilic anaerobes are Archaea and Bacteria that grow optimally at temperatures of 50°C or higher and do not require the use of O_2 as a terminal electron acceptor for growth. The prokaryotes with this type of physiology are studied for a variety of reasons, including (a) to understand how life can thrive under extreme conditions, (b) for their biotechnological potential, and (c) because anaerobic thermophiles are thought to share characteristics with the early evolutionary life forms on Earth. Over 300 species of thermophilic anaerobes have been described; most have been isolated from thermal environments, but some are from mesobiotic environments, and others are from environments with temperatures below 0°C. In this overview, the authors outline the phylogenetic and physiological diversity of thermophilic anaerobes as currently known. The purpose of this overview is to convey the incredible diversity and breadth of metabolism within this subset of anaerobic microorganisms.

Key words: **thermophile; anaerobe; thermobiotic; diversity of anaerobic thermophiles; Bacteria; Archaea**

Introduction

Bacteria and Archaea that grow optimally at elevated temperatures and do not require oxygen for growth are described as thermophilic anaerobes, and the taxa having this physiology are of interest from basic and applied scientific perspectives. Because these prokaryotes grow optimally at elevated temperatures, thermophilic anaerobes are designated *extremophiles* and are studied to understand how life can thrive in environments previously considered inhospitable to life. Such environments include volcanic solfatares and hot springs high in sulfur and toxic metals, as well as abyssal hydrothermal vents with extremely high pressure and temperature.[1,2] Isolated species of thermophilic anaerobes include astonishing forms of life: for example, the mothercell of the alkalithermophile *Clostridium paradoxum* becomes highly motile when sporulating,[3] and *Moorella thermacetica*–like strains have exceptionally heat-resistant spores with D_{10} times of nearly 2 h at 121°C.[4] Also, *Pyrolobus fumarii* grows optimally at 106°C,[5] and a recently isolated *Methanopyrus kandleri*–like strain grows at 122°C under increased pressure.[6] *Thermobrachium celere* strains have doubling times of about 10 min under optimal conditions,[7] and the triple extremophile *Natranaerobius thermophilus* grows optimally at high temperature (53°C), high pH (9.5), and high salt concentration (3.3–3.9 M Na^+) simultaneously.[8]

The analyses of the biodiversity and patterns of biodiversity within thermal environments is an area of active research that continually expands as technology allows for novel approaches and more detailed analyses. Additionally, their thermostable enzymes, among other characteristics, make thermophilic anaerobes of significant interest for their biotechnological potential.[9–11]

Thermophilic anaerobes also attract research attention because it is assumed that they have properties similar, in various aspects, to those of the early evolutionary life forms on Earth.[a] There is little doubt that the first forms of life on Earth occurred at a time when significantly less oxygen was present. Apparent biogenic signatures have been dated to 3.85–3.8 billion years ago (Ga) and complex microfossil communities are dated to 3.5–3.4 Ga, whereas the accumulation of oxygen happened later, approximately 2.1–2.3 Ga.[12,13] The early forms of life from which everything else then evolved (i.e., the progenotic life forms), were therefore anaerobes.[14] Considering Earth history and these progenotic life forms, the authors believe that present-day life has several roots as proposed by Kandler.[15] In addition to having an anoxic origin, the current mainstream opinion is that life began at elevated temperatures, and was

Address for correspondence: Juergen Wiegel, 212 Biological Sciences Building, 1000 Cedar Street, University of Georgia, Athens, GA 30602-2605. Voice: +1-706-542-2651; fax: +1-706-542-2651.
jwiegel@uga.edu

[a]Further discussion of thermophiles and the origin of life can be found in the conference proceedings book: Wiegel, J. and M.W.W. Adams. 1998. Thermophiles: the Keys to Molecular Evolution and the Origin of Life? Taylor & Francis Ltd. London.

Ann. N.Y. Acad. Sci. 1125: 1–43 (2008). © 2008 New York Academy of Sciences.
doi: 10.1196/annals.1419.029

consequently thermophilic. Although some contest this view of a thermophilic origin of life and postulate that prebiotic chemistry implies the emergence of living systems at a low temperature or a rapid selection for hyperthermophiles during the late bombardment,[16–18] many believe that evolution from mesophily to hyperthermophily is improbable.[19] Thus, the authors and most others posit that life began around 80°C on clay or iron-sulfur mineral surfaces in shallow pools.[20]

Although the first forms of life no longer exist, natural thermal environments do still exist and some have properties similar to those environments in which life presumably first began. Many of these environments are characteristically anaerobic or have low levels of oxygen. The anaerobic feature can stem from a number of factors: remoteness of the environment from the atmosphere; low solubility of oxygen in water at elevated temperatures; hypersalinity; inputs of reducing gasses, such as H_2 and H_2S; or the consumption of oxygen by aerobic microorganisms on or near the water surface.[21,22] Broadly, natural thermobiotic environments are of terrestrial, marine, or subsurface nature. Terrestrial or continental geothermally heated features include hot springs, geysers, solfatares, mud pools ("mud pots"), and some solar-heated environments. One example, Yellowstone National Park, USA, contains the highest concentration of terrestrial geothermal features on Earth.[23] Other locales of notable terrestrial thermal activity include Japan, New Zealand, Iceland, the Hawaiian volcanoes, various South Pacific Islands located at The Ring of Fire, and the Kamchatka Peninsula in the Russian Far East. Thermobiotic marine environments include geothermally heated beaches, shallow hydrothermal vents, and abyssal hydrothermal vents, where water escapes at temperatures over 300°C.[24] Two examples of geothermally heated beaches are those on the North Island of New Zealand and on SavuSavu (Fiji Islands); the geothermally heated spots are exposed at low tide and covered with water at high tide. Some shallow thermal marine systems where thermophilic anaerobes have been isolated include Volcano Shore of Sicily and Lucrino Beach near Naples, Italy; Palaeochori Bay of Milos, Greece; the coastal hot springs of Ibuski, Kangoshima Prefecture, Japan; the marine solfataric fields of Kraternaya Cove, Ushishir Archipelago, Northern Kurils; a hydrothermal field at a depth of about 100 m at the Eyjafjörður Fjord, Northern Iceland; and the Kolbeinsey Ridge located at a depth about 105 m, north of Iceland. First described in the 1970s on the Galapagos Rift,[25] deep-sea hydrothermal vent regions have been found and studied in the Pacific and Atlantic Oceans, including: the "Rainbow," "Snakepit," and Logatchev hydrothermal vent regions of the Mid-Atlantic Ridge; the 9°N, 13°N, and 21°N deep-sea hydrothermal vent systems of the East Pacific Rise; the Iheya Ridge and Yonaguni Knoll IV of the Okinawa Trough; the Guaymas Basin, Gulf of California; and recently, hydrothermal vents were discovered as far north as the Mohns Ridge, Norway.[26] Oil reservoirs, mines, and geothermal aquifers are examples of subsurface environments thermophiles populate. Species of the genera *Geotoga* and *Petrotoga* (family Thermotogaceae $\{B34\}$[b]) have thus far only been found in deep subsurface oil reservoirs; on this basis, it has been proposed that these taxa represent typical indigenous *Bacteria* in this particular ecosystem.[27] Geothermal aquifers, such as the Great Artesian Basin of Australia, are considered to be markedly different from volcanically related hot springs in that they have low flow rates and long recharge times (around 1000 years) that affect the microbial populations therein.[28] Besides natural thermal environments, thermophilic anaerobes are also found within anthropogenically heated environments, including coal refuse piles; compost heaps, which contain not only spore-forming species, but also methanogenic Archaea[29]; and nuclear power plant effluent channels. Contrary to expectations, thermophilic anaerobes have also been isolated from mesobiotic and even psychrobiotic environments: two *Thermosediminibacter* species were isolated from ocean sediments of the Peru Margin at temperatures at or below 12°C,[30] uncharacterized *Thermoanaerobacter* species have been isolated from melted snow from Antarctica (J. Wiegel, unpublished results), alkalithermophiles have been isolated from many river sediments and wet meadows,[7] and *Methanothermobacter thermoautotrophicus* and other thermophilic methanogens can readily be found in lake sediments (e.g., Lake Mendota, Wisconsin, USA) and rivers, streams, and ponds (e.g., in Northern Germany).[31] Possible reasons for the presence of thermophilic anaerobes in environments where they ought not to grow, considering their physiological properties, include (a) that the microorganisms are present but not growing in these environments, (b) that they dispersed only transiently from other thermal environments, or (c) as the authors propose, that they are surviving by taking advantage of temporary thermal microniches that become available when proteinaceous biomass is degraded.[7]

[b]The designation within the brackets corresponds to the particular thermophilic anaerobe-containing family as shown in FIG. 1., a 16S rRNA gene sequence-based phylogenetic tree, and as listed in TABLE 1, which lists the validly described thermophilic anaerobes.

Undoubtedly, the wide spectrum of properties of the environments thermophilic anaerobes inhabit has been, and continues to be, of significance in the evolution and diversification of these microorganisms. The evolutionary implications are a fascinating topic alone. Nevertheless, the result of this diversification—the diversity of thermophilic anaerobes—is examined in this chapter. The goal herein is to provide an overview of the diversity of thermophilic anaerobes from phylogenetic and physiological perspectives and, where appropriate, to highlight unique and noteworthy taxa.

Measuring the Diversity of Thermophilic Anaerobes

Although biological diversity is usually considered to be a combination of two qualities of a population, species richness (the number of different kinds) and species evenness (the relative numbers of those present),[32] the focus herein will primarily be species richness. As such, this overview of the diversity of thermophilic anaerobes will be primarily limited to describing diversity as a function of what is known of these Bacteria and Archaea through studies on axenic cultures. The caveat is that most of the prokaryotes from any environment, including the thermal environments previously discussed, are currently uncultured.[33] Consequently, our understanding of prokaryotic biodiversity and physiological properties is based on very limited knowledge. Although culture-independent studies are not the focus of this chapter, such studies, including analyses of environmental 16S ribosomal RNA gene sequences, metagenomics (i.e., the sequencing of DNA isolated from an environment), and sequences of genes encoding particular functional proteins, help address this major limitation.

Culture-independent studies of terrestrial thermobiotic sites include sites in Yellowstone National Park,[34–38] hot springs of Japan,[39] hot springs of New Zealand,[40] hot springs of Greenland,[41] subterranean and terrestrial Icelandic hot springs.[42,43] the thermal subterranean Great Artesian Basin of Australia,[44] and sun-heated salt lakes (Wadi An Natrun) in Egypt,[45] to name a few. Examples of deep-sea hydrothermal sites studied by culture-independent means include the Manus Basin near Papua New Guinea,[46] the Suiyo Sea Mount and Myojin Knoll in the Ogasawara area of Japan, and the hydrothermal fields at the Iheya Basin in the Okinawa area of Japan.[47] Besides strictly culture-independent techniques, additional methods have been developed or employed to study the thermophilic anaerobes inhabiting thermobiotic environments. For example, an *in situ* growth chamber deployed at a hydrothermal vent field revealed novel lineages.[48] 16S rRNA hybridization probes developed for members of the *Thermoanaerobacter*/*Caldanaerobacter* clade detected *Thermoanaerobacter* members from deep-sea hydrothermal regions, an environment not previously known to harbor *Thermoanaerobacter* species.[49] Similarly, *Desulfotomaculum* 16S rRNA hybridization probes were employed to measure the change in abundance in thermophilic anaerobic digesters among other sites.[50]

Thermophilic anaerobes in pure culture are characterized through the polyphasic approach, in which phenotypic and genotypic/phylogenetic properties are examined.[51,52] Phenotypic characteristics of particular interest for this discussion include oxygen relationships and metabolic properties, such as energy production and carbon assimilation. Group-defining properties, such as temperature growth range (e.g., T_{min}, T_{opt}, and T_{max}) and pH growth range (e.g., pH_{min}, pH_{opt}, and pH_{max}), are particularly important. These values should be determined by measuring the doubling times over the range for growth and specifically noting where growth was obtained and where growth was not obtained. (For example, in the authors' laboratory, a shaking gradient incubator with 1°C–3°C intervals is used to determine the temperature growth profile for a strain.) Other properties, such as salt tolerance and response to pressure, are important when considering thermophilic anaerobes from habitats such as hypersaline lakes and deep-sea hydrothermal regions. Although genotypic characteristics such as G+C mol% and DNA–DNA relatedness between strains have been studied since the 1960s, in the past 20 years, analysis of the 16S rRNA gene sequence has become standard, and the analysis of housekeeping genes and whole-genome sequencing of prokaryotes is becoming increasingly common.[53] Known thermophilic anaerobes with available sequenced genomes are designated within TABLE 1 with the § symbol, but the continually increasing number can be obtained from the National Center for Biotechnology Information Taxonomy Database: http://www.ncbi.nlm.nih.gov/genomes/MICROBES/microbial_taxtree.html. Analysis of the 16S rRNA gene sequence-based phylogenetic tree (FIG. 1), reveals that, at present (November 2007), thermophilic anaerobes reside in 51 known prokaryotic families (TABLE 1). Still, this will undoubtedly change as novel thermophilic anaerobes are isolated and the subsequent phylogenetic reorganization of taxa proceeds.

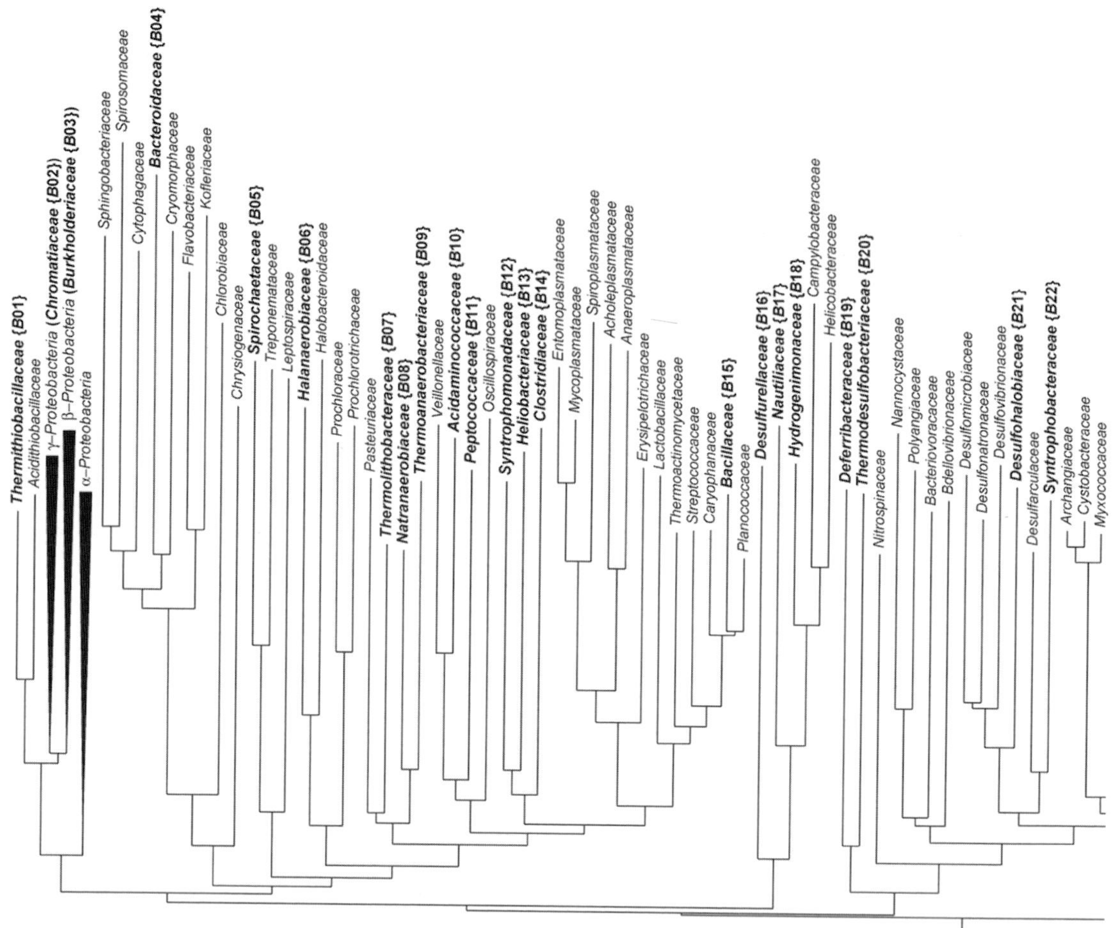

FIGURE 1. Prokaryotic phylogenetic tree highlighting lineages with anaerobic thermophilic taxa. 16S rRNA gene sequence-based phylogenetic tree, highlighting lineages with anaerobic thermophilic taxa. Lineages with thermophilic or hyperthermophilic ($T_{opt} \geq 80^{\circ}C$) anaerobic taxa are indicated with a bold text. Hyperthermophilic lineages are further differentiated with extra-bold branches. Sequences were aligned using ClustalW and the tree was constructed using the neighbor-joining method with the Jukes and Cantor distance corrections through the PHYLIP program.[147–150] Scale bar = 5 nucleotide changes per 100 base pairs. When possible, the 16S rRNA gene sequence of the type species of the type genus of the family was used as a proxy for the prokaryotic family. If the nucleotide sequence of the type species was unavailable or of poor quality, a sequence from another member of the type genus was used. To our knowledge, no thermophilic anaerobes are present within the *Actinobacteria* or *α-Proteobacteria* phylogenetic groups. Within the condensed *β-Proteobacteria* clade, thermophilic anaerobes have been found only within the *Burkholderiaceae* {B03}, and within the *γ-Proteobacteria*, thermophilic anaeorbes have been found only within the *Thermithiobacillaceae* {B01} and the *Chromatiaceae* {B02}. Designations within the brackets correspond to the families as listed within TABLE 1 and as presented throughout the text. Family, species, and corresponding GenBank accession numbers for the 16S rRNA gene sequences used for the construction of the tree are as follows: *Acholeplasmataceae, Acholeplasma laidlawii* U14905; *Acidaminococcaceae* {B10}, *Acidaminococcus fermentans* X65935; *Acidimicrobiaceae, Acidimicrobium ferrooxidans* U75647; *Acidithiobacillaceae, Acidithiobacillus thiooxidans* Y11596; *Anaerolinaceae* {B27}, *Anaerolinea thermophila* AB046413; *Anaeroplasmataceae, Anaeroplasma abactoclasticum* M25050; *Aquificaceae* {B32}, *Aquifex pyrophilus* M83548; *Archaeoglobaceae* {A09}, *Archaeoglobus fulgidus* X05567; *Archangiaceae, Archangium gephyra* DQ768106; *Bacillaceae* {B15}, *Bacillus subtilis* AJ276351; *Bacteriovoracaceae, Bacteriovorax stolpii* AJ288899; *Bacteroidaceae* {B04}, *Bacteroides fragilis* X83935; *Bdellovibrionaceae, Bdellovibrio bacteriovorus* AJ292759; *Bifidobacteriaceae, Bifidobacterium bifidum* EF589113; *Burkholderiaceae* {B03}, *Burkholderia cepacia* U96927; *Caldilineaceae* {B28}, *Caldilinea aerophila* AB067647; *Campylobacteraceae, Campylobacter fetus* L04314; *Caryophanaceae, Caryophanon latum* AJ491302; *Chlamydiaceae, Chlamydia trachomatis* D89067; *Chlorobiaceae, Chlorobium limicola* NZ_AAHJ01000048 (region 4935–6399); *Chloroflexaceae* {B29}, *Chloroflexus aggregans* D32255; *Chromatiaceae* {B02}, *Chromatium okenii* AJ223234; *Chrysiogenaceae, Chrysiogenes arsenates* X81319; *Clostridiaceae* {B14}, *Clostridium butyricum* AB075768; *Conexibacteraceae,*

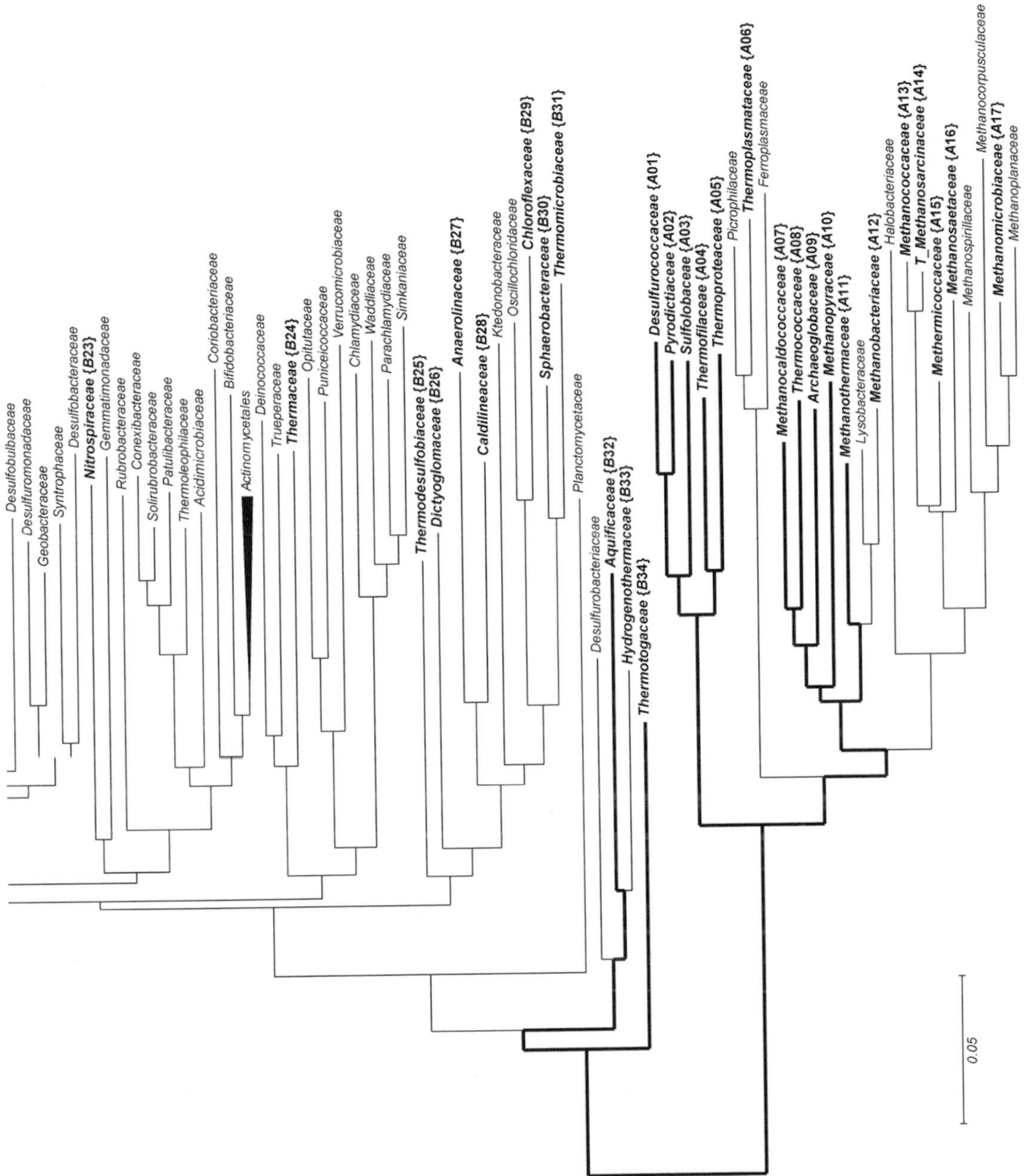

FIGURE 1 (*Continued*). *Conexibacter woesei* AJ440237; *Coriobacteriaceae, Coriobacterium glomerans* X79048; *Cryomorphaceae, Cryomorpha ignava* AF170738; *Cystobacteraceae, Cystobacter fuscus* M94276; *Cytophagaceae, Cytophaga hutchinsonii* CP000383 (region 120118–121499); *Deferribacteraceae* {B19}, *Deferribacter thermophilus* U75602; *Deinococcaceae, Deinococcus radiodurans* Y11332; *Desulfarculaceae, Desulfarculus baarsii* M34403; *Desulfobacteraceae, Desulfobacter postgatei* AF418180; *Desulfobulbaceae, Desulfobulbus propionicus* AY548789; *Desulfohalobiaceae* {B21}, *Desulfohalobium retbaense* U48244; *Desulfomicrobiaceae, Desulfomicrobium baculatum* AF030438; *Desulfonatronaceae, Desulfonatronum lacustre* Y14594; *Desulfovibrionaceae, Desulfovibrio desulfuricans* AF192153; *Desulfurellaceae* {B16}, *Desulfurella acetivorans* X72768; *Desulfurobacteriaceae, Desulfurobacterium thermolithotrophum* AJ001049; *Desulfurococcaceae* {A01}, *Desulfurococcus mobilis* M36474; *Desulfuromonadaceae, Desulfuromonas acetoxidans* NZ_AAEW02000008 (region 189409–190966); *Dictyoglomaceae* {B26}, *Dictyoglomus thermophilum* X69194; *Entomoplasmataceae, Entomoplasma freundtii* AF036954; *Erysipelotrichaceae, Erysipelothrix rhusiopathiae* M23728; *Ferroplasmaceae, Ferroplasma acidiphilum* AJ224936; *Flavobacteriaceae, Flavobacterium anhuiense* EU046269; *Gemmatimonadaceae, Gemmatimonas aurantiaca* AB072735;

(Continued)

FIGURE 1 (*Continued*). *Geobacteraceae, Geobacter metallireducens* L07834; *Halanaerobiaceae* {*B06*}, *Halanaerobium praevalens* AB022034; *Halobacteriaceae, Halobacterium salinarum* AJ496185; *Halobacteroidaceae, Halobacteroides halobius* U32595; *Helicobacteraceae, Helicobacter pylori* Z25741; *Heliobacteriaceae* {*B13*}, *Heliobacterium sulfidophilum* AF249678; *Hydrogenimonaceae* {*B18*}, *Hydrogenimonas thermophila* AB105048; *Hydrogenothermaceae* {*B33*}, *Hydrogenothermus marinus* AJ292525; *Kofleriaceae, Kaistella flava* AM421015; *Ktedonobacteraceae, Ktedobacter racemifer* AM180156; *Lactobacillaceae, Lactobacillus delbrueckii* M58814; *Leptospiraceae, Leptospira interrogans* Z12817; *Lysobacteraceae, Lysobacter enzymogenes* AY947529; *Methanobacteriaceae* {*A12*}, *Methanobacterium thermaggregans* AF095264; *Methanocaldococcaceae* {*A07*}, *Methanocaldococcus jannaschii* NC_000909 (region 157985–159459); *Methanococcaceae* {*A13*}, *Methanococcus vannielii* NC_009634 (region 155–1619); *Methanocorpusculaceae, Methanocorpusculum parvum* AY260435; *Methanomicrobiaceae* {*A17*}, *Methanomicrobium mobile* AY196679; *Methanoplanaceae, Methanoplanus limicola* M59143; *Methanopyraceae* {*A10*}, *Methanopyrus kandleri* AE010349 (region 6963–8474); *Methanosaetaceae* {*A16*}, *Methanosaeta thermoacetophila* AB071701; *Methanosarcinaceae* {*A14*}, *Methanosarcina thermophila* M59140; *Methanospirillaceae, Methanospirillum hungatei* AY196683; *Methanothermaceae* {*A11*}, *Methanothermus fervidus* M32222; *Methermicoccaceae* {*A15*}, *Methermicoccus shengliensis* EF026570; *Mycoplasmataceae, Mycoplasma mycoides* EU040177; *Myxococcaceae, Myxococcus fulvus* AB218224; *Nannocystaceae, Nannocystis exedens* AB084253; *Natranaerobiaceae* {*B08*}, *Natranaerobius thermophilus* DQ417202; *Nautiliaceae* {*B17*}, *Nautilia lithotrophica* AJ404370; *Nitrospinaceae, Nitrospina gracilis* L35504; *Nitrospiraceae* {*B23*}, *Nitrospira moscoviensis* X82558; *Opitutaceae, Opitutus terrae* AJ229235; *Oscillochloridaceae, Oscillochloris trichoides* AF093427; *Oscillospiraceae, Oscillospira guilliermondii* AB040495; *Parachlamydiaceae, Parachlamydia acanthamoebae* YO7556; *Pasteuriaceae, Pasteuria hartismerei* AJ878853; *Patulibacteraceae, Patulibacter minatonensis* AB193261; *Peptococcaceae* {*B11*}, *Peptococcus niger* X55797; *Picrophilaceae, Picrophilus oshimae* X84901; *Planctomycetaceae, Planctomyces brasiliensis* AJ231190; *Planococcaceae, Planococcus citreus* X62172; *Polyangiaceae, Polyangium vitellinum* AJ233944; *Prochloraceae, Prochloron* sp. X63141; *Prochlorotrichaceae, Prochlorothrix hollandica* AJ007907; *Puniceicoccaceae, Puniceicoccus vermicola* DQ539046; *Pyrodictiaceae* {*A02*}, *Pyrodictium occultum* M21087; *Rubrobacteraceae, Rubrobacter radiotolerans* X87134; *Simkaniaceae, Simkania negevensis* U68460; *Solirubrobacteraceae, Solirubrobacter pauli* AY039806; *Sphaerobacteraceae* {*B30*}, *Sphaerobacter thermophilus* AJ420142; *Sphingobacteriaceae, Sphingomonas paucimobilis* U20776; *Spirochaetaceae* {*B05*}, *Spirochaeta bajacaliforniensis* AJ698859; *Spiroplasmataceae, Spiroplasma citri* AM157769; *Spirosomaceae, Spirosoma lingual* AM000023; *Streptococcaceae,* Streptococcus *pyogenes* AB002521; *Sulfolobaceae* {*A03*}, *Sulfolobus acidocaldarius* NC_007181 (region 1107140–1108619); *Syntrophaceae, Syntrophus buswellii* X85131; *Syntrophobacteraceae* {*B22*}, *Syntrophobacter wolinii* X70905; *Syntrophomonadaceae* {*B12*}, *Syntrophomonas wolfei* NC_008346 (region 43738–45267); *Thermaceae* {*B24*}, *Thermus aquaticus* L09663; *Thermithiobacillaceae* {*B01*}, *Thermithiobacillus tepidarius* AJ459801; *Thermoactinomycetaceae, Thermoactinomyces vulgaris* AF138739; *Thermoanaerobacteriaceae* {*B09*}, *Thermoanaerobacter ethanolicus* L09162; *Thermococcaceae* {*A08*}, *Thermococcus celer* M21529; *Thermodesulfobacteriaceae* {*B20*}, *Thermodesulfobacterium commune* AF418169; *Thermodesulfobiaceae* {*B25*}, *Thermodesulfobium narugense* AB077817; *Thermofilaceae* {*A04*}, *Thermofilum pendens* CP000505 (region 366243–367743); *Thermoleophilaceae, Thermoleophilum album* AJ458462; *Thermolithobacteraceae* {*B07*}, *Thermolithobacter ferrireducens* AF282252; *Thermomicrobiaceae* {*B31*}, *Thermomicrobium roseum* M34115; *Thermoplasmataceae* {*A06*}, *Thermoplasma acidophilum* NC_002578 (region 1474300–1475770); *Thermoproteaceae* {*A05*}, *Thermoproteus tenax* M35966; *Thermotogaceae* {*B34*}, *Thermotoga maritime* M21774; *Treponemataceae, Treponema pallidum* M88726; *Trueperaceae, Truepera radiovictrix* DQ022076; *Veillonellaceae, Veillonella parvula* X84005; *Verrucomicrobiaceae, Verrucomicrobium spinosum* X90515; *Waddliaceae, Waddlia chondrophila* AF042496.

Oxygen Relationship

As noted previously, many thermobiotic environments are either anaerobic or low in oxygen. Therefore, one would expect that thermophiles are predominantly anaerobic and, indeed, this is what is seen. Most axenic thermophiles are anaerobes or facultative aerobes.[21,22] By the authors' definition, anaerobes are unable to use O_2 as the terminal electron acceptor, even if they can grow in the presence of O_2 (i.e., an O_2-tolerant anaerobe). Facultative aerobes are able to use oxygen as a terminal electron acceptor and some obligately anaerobic thermophiles can survive exposure to oxygenic atmospheres, especially if they are metabolically inactive (e.g., at suboptimal temperatures, or with the absence of metabolizable substrates). In the absence of oxygen, respiring anaerobic or facultative aerobic thermophiles can use, via energy production through electron transport phosphorylation, a variety of compounds as electron acceptors, including: CO_2, CO, NO_3^-, NO_2^-, NO, N_2O, SO_4^{-2}, SO_3^{-2}, $S_2O_3^{-2}$, S^0, Fe(III), Mn(IV), and Mo(VI).[54,128] The energy gleaned from these respiratory pathways is in addition to the energy produced anaerobically through substrate-level phosphorylation.

An examination of families of thermophilic anaerobes reveals that most (38 of 51) are currently composed solely of anaerobic taxa. Although a majority

TABLE 1. Validly described thermophilic anaerobes[c]

Species	O_2 relationship and metabolism	Temperature range [optimum]	pH Range [optimum]	Originally isolated from
Bacteria; Proteobacteria; Gammaproteobacteria; Acidithiobacillales; Thermithiobacillaceae {*B*01};				
Genus: *Thermithiobacillus*				
Thermithiobacillus tepidarius[151,152]	AN CLA	37–50 [43–45]	5.5–7.7 [6–7.5]	The Roman Bath, Avon, UK
Bacteria; Proteobacteria; Gammaproteobacteria; Chromatiales; Chromatiaceae {*B*02};				
Genus: *Thermochromatium*				
Thermochromatium tepidum[113,114]§	AN PA/PH	34–57 [48–50]	[7]	Mammoth Hot Spring, Yellowstone National Park, USA
Bacteria; Proteobacteria; Betaproteobacteria; Burkholderiales; Burkholderiaceae {*B*03};				
Genera: *Thermothrix, Thiobacter* (unclassified *Betaproteobacteria*)				
Thermothrix thiopara[153]	FAE F-CLA	55–85 [70–73]	[7]	Jemez Spring, New Mexico, USA
Thiobacter subterraneus[154]	AN CLA	35–62 [50–55]	5.2–7.7 [6.5–7]	Subsurface geothermal aquifer, Hishikari gold mine, Japan
Bacteria; Bacteroidetes/Chlorobi group; Bacteroidetes; Bacteroidetes; Bacteroidales; Bacteroidaceae {*B*04};				
Genera: *Acetomicrobium, Anaerophaga*				
Acetomicrobium flavidum[155]	AN COH	35–65 [58]	6.2–8	60°C biogas sewage fermentor
Acetomicrobium faecale[156]	AN COH	55–74 [70–73]	5.5–9 [6.5–7]	Mesophilically digested sewage sludge
Anaerophaga thermohalophila[157]	AN COH	37–55 [50]	NR	Blackish-oily sedimentary residues of an oil separation tank near Hannover, Germany
Bacteria; Spirochaetes; Spirochaetes; Spirochaetales; Spirochaetaceae {*B*05};				
Genus: *Spirochaeta*				
Spirochaeta caldaria[158]	AN COH	[48–52]	5.8–8.5 [7.2]	Cyanobacterial mat samples from Oregon and Utah, USA
Spirochaeta thermophila[159]	AN COH	40–73 [66–68]	5.9–7.7 [7.5]	Marine hot spring on the beach of an island from Kamchatka, Russia; also a hot spring, Raoul Island, New Zealand
Bacteria; Firmicutes; Clostridia; Halanaerobiales; Halanaerobiaceae {*B*06};				
Genus: *Halothermothrix*				
Halothermothrix orenii[160]§	AN COH	45–68 [60]	5.5–8.2 [6.5–7]	Chott El Guettar hypersaline lake, Tunisia
Bacteria; Firmicutes; Thermolithobacteria; Thermolithobacterales; Thermolithobacteraceae {*B*07};				
Genus: *Thermolithobacter*				
Thermolithobacter ferrireducens[99]	AN F-CLA	50–75 [73]	6.5–8.5 [7.1–7.3]	Calcite Spring, Yellowstone National Park, USA
Thermolithobacter carboxydivorans[99]	AN CLA	40–78 [70]	6.6–7.6 [6.8–7.0]	Terrestrial hot spring at Raoul Island, Archipelago Kermadeck, New Zealand
Bacteria; Firmicutes; Clostridia; Natranaerobiales; Natranaerobiaceae {*B*08};				
Genus: *Natranaerobius*				
Natranaerobius thermophilus[8]§	AN COH	35–56 [53]	8.5–10.6 [9.5]	Sediment of alkaline, hypersaline lakes of the Wadi An Natrun
Bacteria; Firmicutes; Clostridia; Thermoanaerobacteriales; Thermoanaerobacteriaceae {*B*09};				
Genera: *Coprothermobacter, Gelria, Moorella, Thermacetogenium, Mahella, Thermoanaerobacterium, Thermoanaerobacter, Thermosediminibacter, Caldanaerobacter, Thermovenabulum, Tepidanaerobacter, Ammonifex, Thermanaeromonas, Thermhydrogenium*				
Coprothermobacter platensis[161]	AN COH	35–65 [55]	4.3–8.3 [7]	Methanogenic mesophilic reactor treating protein-rich wastewater

Continued

TABLE 1. *Continued*

Species	O_2 relationship and metabolism	Temperature range [optimum]	pH Range [optimum]	Originally isolated from
Coprothermobacter proteolyticus[162,163]§	AN COH	35–70 [63]	5–8.5 [7.5]	Thermophilic digester fermenting tannery wastes and cattle manure
Gelria glutamica[164]	AN COH	37–60 [50–55]	5.5–8 [7]	Thermophilic, syntrophic, propionate-oxidizing enrichment culture
Moorella glycerini[165]	AN COH	43–65 [58]	5.9–7.8 [6.3]	Calcite Spring area hot spring, Yellowstone National Park, USA
Moorella thermoautotrophica[31,166]	AN F-CLA	36–70 [55–58]	4.5–7.6 [5.7]	Yellowstone National Park, USA; also Georgia and Hawaii, USA; Zaire, Africa; and Germany
Moorella mulderi[167]	AN F-CLA	40–70 [65]	5.5–8.5 [7]	A methanol-degrading enrichment culture obtained from a thermophilic anaerobic reactor
Moorella thermoacetica[166,168] §	AN COH	45–65 [55–60]	NR	Horse manure
Thermacetogenium phaeum[169]	AN F-CLA	40–65 [58]	5.9–8.4 [6.8]	Thermophilic anaerobic methanogenic reactor treating kraft-pulp production plant wastewater, Japan
Mahella australiensis[170]	AN COH	30–60 [50–60]	5.5–8.8 [7.5]	Riverslea oilfield, Bown-Surat Basin, Queensland, Australia
Thermoanaerobacterium thermosulfurigenes[171]	AN COH	55–75 [60]	4–7.6 [5.5–6.5]	Octopus Spring, Yellowstone National Park, USA
Thermoanaerobacterium saccharolyticum[171]	AN COH	45–70 [60]	5–7.5 [6]	Yellowstone National Park, USA
Thermoanaerobacterium xylanolyticum[171]	AN COH	45–70 [60]	5–7.5 [6]	Geothermal areas, Wyoming and Nevada, USA
Thermoanaerobacterium aotearoense[68]	AN COH	35–66 [60–63]	3.8–6.8 [5.2]	Geothermal hot springs, New Zealand
Thermoanaerobacterium polysaccharolyticum[172]	AN COH	45–70 [65–68]	5–8 [6.8–7]	Canning factory waste, Illinois, USA
Thermoanaerobacterium zeae[172]	AN COH	55–72 [65–70]	3.9–7.9	Canning factory waste, Illinois, USA
Thermoanaerobacterium aciditolerans[69]	AN COH	37–68 [55]	3.2–7.1 [5.7]	Hydrothermal vent in the Orange Field, Uzon Caldera, Kamchatka, Far Eastern Russia
Thermoanaerobacterium thermosaccharolyticum[166,173]	AN COH	[55–60]	NR	Soil
Thermoanaerobacter brockii subsp. *lactiethylicus*[174]	AN COH	37–75 [55–60]	[7.3]	Oilfields of France; and Cameroon, Africa
Thermoanaerobacter thermocopriae[166,175]	AN COH	47–74 [60]	6–8 [6.5–7.3]	Compost of cattle feces and grasses, at the University of Tokyo; other strains from camel feces, compost, soil, and a hot spring in Japan
Thermoanaerobacter brockii subsp. *finnii*[174]	AN COH	40–75 [65]	[6.5–6.8]	Sediment sludge, Lake Kivu, Africa
Thermoanaerobacter acetoethylicus[163,176]	AN COH	[65]	5.5–8.5	Hot springs, Yellowstone National Park, USA
Thermoanaerobacter kivui[166,177]	AN F-CLA	50–72 [66]	5.3–7.3 [6.4]	Lake Kivu sediment, Africa
Thermoanaerobacter wiegelii[178]	AN COH	38–78 [65–68]	5.5–7.2 [6.8]	Anthropogenically heated freshwater pool, Rotorua, New Zealand

TABLE 1. *Continued*

Species	O_2 relationship and metabolism	Temperature range [optimum]	pH Range [optimum]	Originally isolated from
Thermoanaerobacter thermohydrosulfuricus[171]	AN COH	37–78 [67–69]	5.5–9.2 [6.9–7.5]	Extraction juices from beet sugar factories; mud and soil; hot springs in Utah and Wyoming, and a sewage plant in Georgia, USA
Thermohydrogenium kirishiense[179]	AN COH	45–75 [65]	5–8 [7–7.4]	Industrial yeast biomass at the stages of thermal treatment
Thermoanaerobacter brockii subsp. *brockii*[174]	AN COH	40–80 [65–70]	5.5–9.5 [7.5]	Washburn thermal spring, Yellowstone National Park, USA
Thermoanaerobacter italicus[180]	AN COH	45–78 [70]	[7]	Thermal spas, water and mud samples, northern Italy
Thermoanaerobacter siderophilus[133]	AN F-CLA	39–78 [69–71]	4.8–8.2 [6.3–6.5]	Hydrothermal vents near the Karymsky volcano, Kamchatka, Russia
Thermoanaerobacter mathranii[181]	AN COH	50–75 [70–75]	4.7–8.8 [7]	Alkaline hot spring at Hveragerdi-Hengil, Iceland
Thermoanaerobacter ethanolicus[84]§	AN COH	37–78 [69]	4.4–9.9 [5.8–8.5]	Hot springs, Yellowstone National Park, USA
Thermoanaerobacter pseudoethanolicus[171,182,183]§	AN COH	[65]	NR	Hot Spring, Yellowstone National Park, USA
Thermoanaerobacter sulfurigignens[184]	AN COH	34–72 [65]	4–8 [5.0–6.5]	Acidic volcanic steam outlet on White Island, New Zealand
Thermoanaerobacter sulfurophilus[185]	AN COH	44–75 [55–60]	4.5–8.0 [6.9–7.2]	Cyanobacterial mat from a hot spring, Uzon Caldera, Kamchatka, Far Eastern Russia
Thermosediminibacter litoriperuensis[30]	AN COH	43–76 [64]	5–9.5 [7.9–8.4]	Nonhydrothermal deep sea sediments of Peru Margin
Thermosediminibacter oceani[30]	AN COH	52–76 [68]	6.3–9.3 [7.5]	Nonhydrothermal deep sea sediments of Peru Margin
Caldanaerobacter subterraneus subsp. *subterraneus*[103,186]	AN COH	45–75 [65]	6–8.5 [7.5]	Oilfield reservoir, southwest France
Caldanaerobacter subterraneus subsp. *pacificus*[103,187]	AN F-CLA	50–80 [70]	5.8–7.6 [6.8–7.2]	Submarine hot vent in the Okinawa Trough
Caldanaerobacter subterraneus subsp. *tengcongensis*[103,188,189]§	AN COH	50–80 [75]	5.5–9 [7–7.5]	Hot spring, Tengcong, China
Caldanaerobacter subterraneus subsp. *yonseiensis*[103,190]	AN COH	50–85 [75]	4.5–9 [6.5]	Hot water, mud, and soil from hot streams at Silcri, Java Island
Thermovenabulum ferriorganovorum[134]	AN COH	45–76 [63–65]	4.8–8.2 [6.7–6.9]	Hydrothermal spring, Uzon Caldera, Kamchatka, Russia
Tepidanaerobacter syntrophicus[191]	AN COH	25–60 [45–50]	5.5–8.5 [6–7]	Sludges of thermophilic (55°C) digesters that decomposed either municipal solid wastes or sewage sludge
Ammonifex degensii[192]§	AN F-CLA	57–77 [70]	5–8 [7.5]	Kawah Candradimuka Crater, Dieng Plateau, Java, Indonesia
Thermanaeromonas toyohensis[193]	AN COH	55–73 [70]	5.5–8.5 [7.5]	Geothermal water at Toyoho Mine, Hokkkaido, Japan
Bacteria; *Firmicutes*; *Clostridia*; *Clostridiales*; *Acidaminococcaceae* {*B*10}; Genus: *Thermosinus*				
Thermosinus carboxydivorans[104]§	AN CLA	40–68 [60]	6.5–7.6 [6.8–7]	Norris Basin hot spring, Yellowstone National Park, USA
Bacteria; *Firmicutes*; *Clostridia*; *Clostridiales*; *Peptococcaceae* {*B*11}; Genera: *Desulfotomaculum*, *Pelotomaculum*, *Carboxydothermus*, *Thermincola*				
Desulfotomaculum thermosapovorans[194]	AN COH	35–60 [50]	[7.2–7.5]	Mixed compost containing rice hulls and peanut shells

Continued

TABLE 1. *Continued*

Species	O_2 relationship and metabolism	Temperature range [optimum]	pH Range [optimum]	Originally isolated from
Desulfotomaculum alkaliphilum[107]	AN CLH	30–58 [50–55]	8–9.15 [8.6–8.7]	Mixed cow/pig manure
Desulfotomaculum solfataricum[124]	AN COH	48–65 [60]	6.4–7.9 [7.3]	Solfataric mud pools, Krafla, northeast Iceland
Desulfotomaculum thermoacetoxidans[195]	AN COH	45–65 [55–60]	6–7.5 [6.5]	A thermophilic anaerobic digestor converting cellulosic waste to methane
Desulfotomaculum thermobenzoicum subsp. *thermobenzoicum*[196]	AN F-CLA	40–70 [62]	6–8 [7.2]	Thermophilic methane fermentation reactor treating kraft-pulp wastewater
Desulfotomaculum thermobenzoicum subsp. *thermosyntrophicum*[196]	AN F-CLA	45–62 [55]	6–8 [7]	Thermophilic granular methanogenic sludge
Desulfotomaculum putei[197]	AN COH	22–65 [64]	6–7.8	Deep terrestrial rock, Taylorsville Triassic Basin, Virginia, USA
Desulfotomaculum australicum[198]	AN F-CLA	40–74 [68]	5.5–8.5 [7–7.4]	Bore wells of the nonvolcanically heated waters of the Great Artesian Basin, Australia
Desulfotomaculum geothermicum[199]	AN F-CLA	37–57 [54]	6–8 [7.2–7.3]	Geothermally heated ground water, at a depth of 2500 m, Creil production well, France
Desulfotomaculum luciae[197,200]	AN CLA	50–70 [60–65]	6.3–7.8	Hot spring, St. Lucia
Desulfotomaculum thermocisternum[201]	AN F-CLA	41–75 [62]	6.2–8.9 [6.7]	Brent group formation water originating 2.6 km below the sea floor, Norwegian sector, North Sea
Desulfotomaculum thermosubterraneum[202]	AN F-CLA	50–72 [61–66]	6.4–7.8 [7.2–7.4]	Underground mine in a geothermally active region, Japan.
Desulfotomaculum carboxydivorans[97]	AN CLH	30–68 [55]	6.8–8 [7.2]	Sludge from an anaerobic bioreactor treating paper mill wastewater
Desulfotomaculum kuznetsovii[203,204]	AN F-CLA	50–85 [60–65]	NR	Thermal water sample from a spontaneous effusion from a rift in the Sukhumsk deposit
Desulfotomaculum nigrificans[205]	AN COH	[55]	NR	Spoiled food
Pelotomaculum thermopropionicum[85]§	AN COH	45–65 [55]	6.7–7.5 [7]	Thermophilic upflow anaerobic sludge blanket reactor
Carboxydothermus ferrireducens[206,207]	AN COH	50–74 [65]	5.5–7.6 [6–6.2]	Hot springs of Yellowstone National Park, USA; and New Zealand
Carboxydothermus hydrogenoformans[98,208]§	AN CLA	40–78 [70–72]	6.4–7.7 [6.8–7]	Freshwater hydrothermal springs, Kunashir Island, Kamchatka, Russia
Thermincola carboxydiphila[101]	AN CLH	37–68 [55]	6.7–9.5 [8]	Hot spring of the Baikal Lake region, Russia
Thermincola ferriacetica[102]	AN F-CLA	45–70 [57–60]	5.9–8.0 [7.0–7.2]	Terrestrial hydrothermal spring, Kunashir Island, Kuril Island, Russian Far East
Bacteria; *Firmicutes*; *Clostridia*; *Clostridiales*; *Syntrophomonadaceae* {*B*12}; Genera: *Anaerobaculum*, *Syntrophothermus*, *Thermanaerovibrio*, *Carboxydocella*, *Anaerobranca*, *Thermosyntropha*, *Caldicellulosiruptor*, *Thermovirga*				
Anaerobaculum thermoterrenum[209]	AN COH	28–60 [55]	5.5–8.6 [7–7.6]	Redwash oilfield production fluids, Utah, USA
Anaerobaculum mobile[210]	AN COH	35–65 [55–60]	5.4–8.7 [6.6–7.3]	Anaerobic wool-scouring wastewater treatment lagoon sludge, Trinidad, Uraguay
Syntrophothermus lipocalidus[211]	AN COH	45–60 [55]	5.8–7.5 [6.5–7]	Granular sludge of a thermophilic upflow anaerobic sludge blanket

TABLE 1. *Continued*

Species	O_2 relationship and metabolism	Temperature range [optimum]	pH Range [optimum]	Originally isolated from
Thermanaerovibrio velox[212]	AN COH	45–70 [60–65]	4.5–8 [7.3]	Cyanobacterial mat, Uzon Caldera, Kamchatka, Russia
Thermanaerovibrio acidaminovorans[213]	AN COH	40–58 [55]	[6.5–8.1]	Granular methanogemic sludge from a sugar refinery, Breda, the Netherlands
Carboxydocella thermautotrophica[100]	AN CLA	40–68 [58]	6.5–7.6 [7]	Hot spring, Gyzer Valley, Kamchatka, Russia
Carboxydocella sporoproducens[214]	AN F-CLA	50–70 [60]	6.2–8 [6.8]	Hot spring of Karymskoe Lake, Kamchatka Peninsula
Anaerobranca gottschalkii[215]	AN COH	30–65 [50–55]	6–10 [9.5]	Hot inlet of Lake Bogoria, Kenya
Anaerobranca horikoshii[216]	AN COH	34–66 [57]	6.9–10.3 [8.5]	Thermal pools, Yellowstone National Park, USA
Anaerobranca californiensis[217]	AN COH	45–70 [58]	8.6–10.4 [9–9.5]	Paoho Island hot springs Mono Lake, California, USA
Thermosyntropha lipolytica[95]	AN COH	52–70 [60–66]	7.15–9.5 [8.1–8.9]	Alkaline hot springs, Lake Bogoria, Kenya
Caldicellulosiruptor lactoaceticus[218]	AN COH	50–78 [68]	5.8–8.2 [7]	Hveragerdi alkaline hot spring, Iceland
Caldicellulosiruptor owensensis[219]	AN COH	50–80 [75]	5.5–9 [7.5]	Freshwater pond within the dry Owens Lake bed, California, USA
Caldicellulosiruptor kristjanssonii[220]	AN COH	45–82 [78]	5.8–8 [7]	Hot spring, Iceland
Caldicellulosiruptor saccharolyticus[221]§	AN COH	45–80 [70]	5.5–8.0 [7.0]	Geothermal spring, Taupo, New Zealand
Caldicellulosiruptor acetigenus[222,223]	AN COH	50–78 [65–68]	5.2–8.6 [7.0]	Combined biomat and sediment from a slightly alkaline hot spring, Hveragerdi, Iceland.
Thermovirga lienii[224]	AN COH	37–68 [58]	6.2–8.0 [6.5–7]	Production water obtained from an oil reservoir in the North Sea
Bacteria; *Firmicutes*; *Clostridia*; *Clostridiales*; *Heliobacteriaceae* {*B*13}; Genus: *Heliobacterium*				
Heliobacterium modesticaldum[120]§	AN PH & COH	25–56 [52]	[6–7]	Iceland; and Yellowstone National Park, USA
Bacteria; *Firmicutes*; *Clostridia*; *Clostridiales*; *Clostridiaceae* {*B*14}; Genera: *Alkaliphilus*, *Clostridium*, *Tepidibacter*, *Caloramator*, *Garciella*, *Caminicella*, *Caloranaerobacter*, *Thermobrachium*, *Thermohalobacter*, *Tepidimicrobium*				
Clostridium isatidis[225]	AN COH	30–55 [50]	5.9–9.9 [7.2]	Fermenting woad vat
Clostridium thermoalcaliphilum[80]	AN COH	27–57.5 [48–51]	7–11 [9.6–10.1]	Anaerobic digestor and aerobic oxidation basin, Municipal Sewage Plant, Atlanta, Georgia, USA
Clostridium straminisolvens[226]	AN COH	46–64 [50–55]	6–8.5 [7.5]	Cellulose-degrading bacterial community
Clostridium thermobutyricum[227]	AN COH	26–61.5 [55]	5.8–9 [6.8–7.1]	Horse manure
Clostridium paradoxum[3]	AN COH	30–63 [55–56]	7–11.1 [10.1]	Municipal Sewage Plants, Athens and Atlanta, Georgia, USA
Clostridium thermopapyrolyticum[91]	AN COH	45–66 [59]	NR	Sediment from a river bank, Buenos Aires, Argentina
Clostridium cellulosi[228]	AN COH	40–65 [55–60]	6.2–8.5 [7.3–7.5]	Cow manure compost
Clostridium stercorarium subsp. *stercorarium*[229,230]	AN COH	[65]	[7.3]	Compost heap

Continued

TABLE 1. *Continued*

Species	O_2 relationship and metabolism	Temperature range [optimum]	pH Range [optimum]	Originally isolated from
Clostridium stercorarium subsp. *leptospartum*[230,231]	AN COH	45–71 [60]	6.7–8.9 [7.5]	Cattle manure compost, Ehime Prefecture, Japan
Clostridium stercorarium subsp. *thermolacticum*[230,232]	AN COH	50–70 [60–65]	6.8–7.4 [7.0]	Sludge from a mesophilic digester fed ground duck weed
Tepidibacter thalassicus[233]	AN COH	33–60 [50]	4.8–8.5 [6.5–6.8]	Black smoker chimney, hydrothermal vent field, East Pacific Rise
Tepidibacter formicigenes[234]	AN COH	35–55 [45]	5.0–8.5 [6.0]	Menez-Gwen hydrothermal site on the Mid-Atlantic Ridge
Caloramator proteoclasticus[235]	AN COH	30–68 [55]	6–9.5 [7–7.5]	Mesophilic granular methanogenic sludge
Caloramator coolhaasii[236]	AN COH	37–65 [50–55]	6–8.5 [7]	Thermophilic methanogenic granular sludge
Caloramator viterbiensis[237]	AN COH	33–64 [58]	6–7.8 [6.5–7]	Hot spring at the Bagnaccio Spring, Viterbo, Italy
Caloramator indicus[238]	AN COH	[60–65]	6.2–9.2 [8.1]	Natural well of the artesian aquifer, Surat District, Gujarat State, India
Caloramator fervidus[166,239]	AN COH	37–80 [68]	5.5–9 [7–7.5]	Hot spring, New Zealand
Garciella nitratireducens[240]	AN COH	25–60 [55]	5.5–9 [7.5]	A water separator collecting fluids, the SAMIII oilfield, Gulf of Mexico
Caminicella sporogenes[241]	AN COH	45–64 [55–60]	4.5–8 [7.5–8]	Deep-sea vent, East Pacific Rise
Caloranaerobacter azorensis[242]	AN COH	45–65 [65]	5.5–9 [7]	Deep-sea hydrothermal chimney rocks, Mid-Atlantic Ridge
Thermobrachium celere[7]	AN COH	37–75 [62–65]	5–9.7 [8–8.5]	Geothermally and anthroprogenically heated environments on three continents
Thermohalobacter berrensis[243]	AN COH	45–70 [65]	5.2–8.8 [7]	Solar saltern canal near Berre Lagoon, southern France
Tepidimicrobium ferriphilum[146]	AN COH	26–62 [50]	5.5–9.5 [7.8–8]	Freshwater hot spring at Barguzin Valley, Buryatiya, Russia
Bacteria; *Firmicutes*; *Bacilli*; *Bacillales*; *Bacillaceae* {*B*15}; Genera: *Anoxybacillus*, *Bacillus*, *Geobacillus*, *Vulcanibacillus*				
Anoxybacillus pushchinoensis[244]	AN COH	37–66 [62]	8–10.5 [9.5–9.7]	Manure from farms near Moscow, Russia
Anoxybacillus ayderensis[247]	FAE COH	30–70 [50]	6–11 [7.5–8.5]	Ayder hot spring, Rize province, Turkey
Anoxybacillus voinovskiensis[248]	FAE COH	30–64 [54]	7–8	Voinovskie Hot Springs, Kamchatka, Russia
Anoxybacillus kestanbolensis[247]	FAE COH	40–70 [50–55]	6–10.5 [7.5–8.5]	Kestanbol hot spring, Canakkale province, Turkey
Anoxybacillus gonensis[249]	FAE COH	40–70 [55–60]	6–10 [7.5–8]	Gonen hot spring, Balikesir province, Turkey
Anoxybacillus flavithermus[244]	FAE COH	30–72 [60–65]	5.5–9 [7]	A hot spring, New Zealand
Bacillus infernus[245]	AN COH	[61]	[7.3–7.8]	Deep terrestrial rock, Taylorsville Triassic Basin, Virginia, USA
Bacillus thermoamylovorans[246]	FAE COH	[50]	5.4–8.5 [7]	Palm wine collected in Rafisque, Senegal
Geobacillus thermocatenulatus[250,251]	FAE COH	42–69 [55–60]	6.5–8.5	Thermal zone of the Tangan-Tau Mountain, Southern Urals, Russia
Geobacillus thermodenitrificans[250,252,253§]	FAE COH	45–70	6–8	Sugar beet juice from extraction installations, Austria

TABLE 1. *Continued*

Species	O_2 relationship and metabolism	Temperature range [optimum]	pH Range [optimum]	Originally isolated from
Geobacillus thermoleovorans[250,254,255]	FAE COH	35–78 [55–65]	[6.2–6.8]	Soil near hot water effluent, Bethlehem, Pennsylvania, USA
Geobacillus uzenensis[250]	FAE COH	45–65	6.2–7.8	The Uzen oilfield, Kazakhstan
Vulcanibacillus modesticaldus[256]	AN COH	37–60 [55]	6–8.5 [7]	"Rainbow" deep-sea hydrothermal vent field, Mid-Atlantic Ridge (36° 14′ N 33° 54′ W)
Bacteria; *Proteobacteria*; *delta/epsilon subdivisions*; *Deltaproteobacteria*; *Desulfurellales*; *Desulfurellaceae* {*B*16}; Genera: *Desulfurella*, *Hippea*				
Desulfurella kamchatkensis[257]	AN F-CLA	40–70 [54]	[6.9–7.2]	The Pauzhetka hot spring, Kamchatka, Far East Russia
Desulfurella propionica[257]	AN F-CLA	33–63 [55]	[6.9–7.2]	Cyanobacterial mat from sulfide-rich hot pond, Uzon Caldera, Kamchatka, Russia
Desulfurella acetivorans[258]	AN COH	44–70 [52–57]	4.3–7.5 [6.8–7]	Hot water pool, Uzon Caldera, Kamchatka, Russia
Desulfurella multipotens[126]	AN F-CLA	42–77 [58–60]	6.0–7.2 [6.4–6.8]	Green Lake, Raoul Island, Kermadec Archipelago, New Zealand
Hippea maritima[259]	AN COH	40–65 [52–54]	5.4–6.5 [5.8–6.2]	Shallow water hot vents, Bay of Plenty, New Zealand; and Matupi Harbour, Papua New Guinea
Bacteria; *Proteobacteria*; *delta/epsilon subdivisions*; *Epsilonproteobacteria*; *Nautiliales*; *Nautiliaceae* {*B*17}; Genera: *Nautilia*, *Lebetimonas*, *Caminibacter*				
Nautilia lithotrophica[260]	AN CLA	37–68 [53]	6.4–7.4 [6.8–7]	13°N hydrothermal vent field, East Pacific Rise
Lebetimonas acidiphila[70]	AN CLA	30–68 [50]	4.2–7 [5.2]	TOTO caldera of the Mariana arc
Caminibacter mediatlanticus[261] §	AN CLA	45–70 [55]	4.5–7.5 [5.5]	"Rainbow" deep-sea vent field, Mid-Atlantic Ridge
Caminibacter profundus[262]	AN CLA	45–65 [55]	6.5–7.4 [6.9–7.1]	"Rainbow" deep-sea vent field, Mid-Atlantic Ridge
Caminibacter hydrogeniphilus[263]	AN CLA	50–70 [60]	5–7.5 [5.5–6]	Deep-sea vent region, East Pacific Rise
Bacteria; *Proteobacteria*; *delta/epsilon subdivisions*; *Epsilonproteobacteria*; *Campylobacterales*; *Hydrogenimonaceae* {*B*18}; Genus: *Hydrogenimonas*				
Hydrogenimonas thermophila[264]	FAE CLA	35–65 [55]	4.9–7.2 [5.9]	Kairei deep-sea hydrothermal field, Central Indian Ridge
Bacteria; *Deferribacteres*; *Deferribacteres*; *Deferribacterales*; *Deferribacteraceae* {*B*19}; Genera: *Deferribacter*, *Flexistipes*, *Caldithrix* (unclassified *Deferribacteres*)				
Caldithrix abyssi[108]	AN CLH	40–70 [60]	5.8–7.8 [6.8–7]	Logatchev hydrothermal field, Mid-Atlantic Ridge
Deferribacter thermophilus[130]	AN COH	50–65 [60]	5–8 [6.5]	Produced formation water collected from a well in the Beatrice oilfield
Deferribacter abyssi[131]	AN CLA	45–65 [60]	6–7.2 [6.5–6.7]	"Rainbow" vent field, Mid-Atlantic Ridge
Deferribacter desulfuricans[265]	AN COH	40–70 [60–65]	5.0–7.5 [6.5]	A black smoker vent from the hydrothermal fileds at the Suiyo Seamount in the Izu-Bonin Arc, Japan
Flexistipes sinusarabici[266]	AN COH	30–53 [45–50]	6–8	Brine water samples of the Atlantis II Deep of the Red Sea, depth of 2000 m

Continued

TABLE 1. *Continued*

Species	O_2 relationship and metabolism	Temperature range [optimum]	pH Range [optimum]	Originally isolated from
Bacteria; *Thermodesulfobacteria*; *Thermodesulfobacteria*; *Thermodesulfobacteriales*; *Thermodesulfobacteriaceae* {*B*20}; Genera: *Thermodesulfatator*, *Thermodesulfobacterium*				
Thermodesulfatator indicus[86]	AN CLA	55–80 [70]	6–6.7 [6.25]	The Kairei deep-sea hydrothermal vent field, Central Indian Ridge
Thermodesulfobacterium hydrogeniphilum[87]	AN CLA	50–80 [75]	6.3–6.8 [6.5]	Deep-sea hydrothermal vent site, Guaymas Basin
Thermodesulfobacterium commune[267]§	AN COH	50–85 [70]	6.0–8.0	Ink Pot Spring, Yellowstone National Park, USA
Thermodesulfobacterium hveragerdense[268]	AN COH	55–74 [70]	NR	Icelandic hot springs
Thermodesulfobacterium thermophilum[269]	AN COH	45–85 [65]	NR	Strata water of the oil deposit on the Apsheron Peninsula, Caspian Sea
Bacteria; *Proteobacteria*; *delta/epsilon subdivisions*; *Deltaproteobacteria*; *Desulfovibrionales*; *Desulfohalobiaceae* {*B*21}; Genus: *Desulfothermus*				
Desulfothermus naphthae[270]	AN COH	50–69 [60–65]	6.1–7.1 [6.5–6.8]	Guaymas Basin, Gulf of California
Desulfothermus okinawensis[271]	AN COH	35–60 [50]	5.4–7.9 [5.9–6.4]	Black smoker chimney, deep-sea hydrothermal field, Southern Okinawa Trough
Bacteria; *Proteobacteria*; *delta/epsilon subdivisions*; *Deltaproteobacteria*; *Syntrophobacterales*; *Syntrophobacteraceae* {*B*22}; Genera: *Desulfacinum*, *Thermodesulforhabdus*				
Desulfacinum infernum[272]	AN F-CLA	40–65 [60]	6.6–8.4 [7.1–7.5]	Oil well of the Beatrice field platform, the North Sea near the coast of Scotland
Desulfacinum hydrothermale[273]	AN F-CLA	37–64 [60]	6–7.5 [7]	Shallow, submarine hydrothermal vent, Palaeochori Bay, Milos in the Aegean Sea
Thermodesulforhabdus norvegica[274]	AN COH	44–74 [60]	6.1–7.9 [6.9]	Oilfield water (Norwegian oil platform) from the North Sea
Bacteria; *Nitrospirae*; *Nitrospira*; *Nitrospirales*; *Nitrospiraceae* {*B*23}; Genus: *Thermodesulfovibrio*				
Thermodesulfovibrio islandicus[268]	AN COH	45–70 [65]	NR	Icelandic hot springs
Thermodesulfovibrio yellowstonii[275]§	AN COH	40–70 [65]	[6.8–7]	Thermal vent, Yellowstone National Park, USA
Bacteria; *Deinococcus-Thermus*; *Deinococci*; *Thermales*; *Thermaceae* {*B*24}; Genera: *Oceanithermus*, *Vulcanithermus*				
Oceanithermus profundus[110]	FAE F-CLH	40–68 [60]	5.5–8.4 [7.5]	13° North hydrothermal vent field, East Pacific Rise, depth of 2600 m
Oceanithermus desulfurans[276]	FAE COH	30–65 [60]	6–8 [6.5]	Submarine hydrothermal field at the Suiyo Seamount in Izu-Bonin Arc, Western Pacific
Vulcanithermus mediatlanticus[109]	FAE F-CLH	37–80 [70]	5.5–8.4 [6.7]	Hydrothermal vent, Mid-Atlantic Ridge
Bacteria; *Firmicutes*; *Clostridia*; *Thermoanaerobacteriales*; *Thermodesulfobiaceae* {*B*25}; Genus: *Thermodesulfobium*				
Thermodesulfobium narugense[277]	AN CLA	37–65 [50–55]	4–6.5 [5.5–6]	Nogo hot spring, Miyagi Prefecture, Japan

TABLE 1. *Continued*

Species	O_2 relationship and metabolism	Temperature range [optimum]	pH Range [optimum]	Originally isolated from
Bacteria; *Dictyoglomi*; *Dictyoglomi*; *Dictyoglomales*; *Dictyoglomaceae* {*B26*}; Genus: *Dictyoglomus*				
Dictyoglomus thermophilum[278]§	AN COH	50–80 [73–78]	5.9–8.3 [7]	Hot spring, Kumamoto Prefecture, Japan
Dictyoglomus turgidus[279]	AN COH	48–86 [72]	5.2–9.0 [7.0–7.1]	Oranzhevoe pole hot spring, Uzon Caldera, Kamchatka, Far East Russia
Bacteria; *Chloroflexi*; *Anaerolineae*; *Anaerolinaeles*; *Anaerolinaceae* {*B27*}; Genus: *Anaerolinea*				
Anaerolinea thermolimosa[59]	AN COH	42–55 [50]	6–7.5 [7]	Sludge from a thermophilic reactor in which wastewater from manufacture of a Japanese distilled alcohol is treated
Anaerolinea thermophila[60]	AN COH	50–60 [55]	6.0–8.0 [7.0]	Thermophilic upflow anaerobic sludge blanket reactor treating soybean-curd manufacturing wastewater
Bacteria; *Chloroflexi*; *Caldilineae*; *Caldilineales*; *Caldilineaceae* {*B28*}; Genus: *Caldilinea*				
Caldilinea aerophila[60]	FAE COH	37–65 [55]	7.0–9.0 [7.5–8.5]	Hot spring sulfur-turf, Japan.
Bacteria; *Chloroflexi*; *Chloroflexi*; *Chloroflexales*; *Chloroflexaceae* {*B29*}; Genera: *Roseiflexus*, *Chloroflexus*, *Heliothrix*				
Roseiflexus castenholzii[115]§	FAE PH (anaerobic)	45–55 [50]	7–9 [7.5–8]	Hot spring, Nakabusa, Japan
Chloroflexus aggregans[116]§	FAE PH (anaerobic)	[50–60]	7.0–9.0	Hot spring of the Okukinu Meotobuchi hot spring in Tochigi Perfecture, Japan
Chloroflexus aurantiacus[117]§	FAE PH (anaerobic)	[52–60]	[8]	Hot spring in the canyon at Sokokura, Hakone district, Japan
Heliothrix oregonensis[118,119]§	FAE PH	[40–55]	NR	Hot spring near Warm Springs River, Oregon, USA
Bacteria; *Chloroflexi*; *Thermomicrobia*; *Sphaerobacterales*; *Sphaerobacteraceae* {*B30*}; Genus: *Sphaerobacter*				
Sphaerobacter thermophilus[280]	AN COH	[55]	[8.5]	A laboratory-scale 60°C fermentor
Bacteria; *Chloroflexi*; *Thermomicrobia*; *Thermomicrobiales*; *Thermomicrobiaceae* {*B31*}; Genus: *Thermomicrobium*				
Thermomicrobium roseum[281]§	AN COH	[70–75]	6–9.4 [8.2–8.5]	Hot spring, Yellowstone National Park, USA
Bacteria; *Aquificae*; *Aquificae*; *Aquificales*; *Aquificaceae* {*B32*}; Genera: *Hydrogenivirga*, *Aquifex*, *Desulfurobacterium* (unclassified *Aquificales*), *Balnearium* (unclassified *Aquificales*), *Thermovibrio* (unclassified *Aquificales*)				
Hydrogenivirga caldilitoris[282]	AN CLA	55–77.5 [75]	5.5–8.3 [6.5–7]	Coastal hot spring, Ibuski, Kangoshima Prefecture, Japan
Aquifex pyrophilus[283]	FAE CLA	67–95 [85]	5.4–7.5 [6.8]	Hot marine sediments at the Kolbeinsey Ridge, Iceland, 106 m deep
Desulfurobacterium thermolithotrophum[284]	AN CLA	40–75 [70]	4.4–7.5 [6]	"Snake Pit" vent field, Mid-Atlantic ridge

Continued

TABLE 1. *Continued*

Species	O_2 relationship and metabolism	Temperature range [optimum]	pH Range [optimum]	Originally isolated from
Desulfurobacterium crinifex[285]	AN CLA	50–70 [60–65]	5.0–7.5 [6.0–6.5]	A Juan de Fuca Ridge hydrothermal vent sample (tubes of the annelid polychaete *Paralvinella sulfincola* attached to small pieces of hydrothermal chimney), depth of 1581 m
Desulfurobacterium atlanticum[286]	AN CLA	50–80 [70–75]	5–7.5 [6.0–6.2]	A deep-sea hydrothermal vent chimney at the Mid-Atlantic Ridge (23°N)
Desulfurobacterium pacificum[286]	AN CLA	55–85 [75]	5.5–7.5 [6–6.2]	A deep-sea hydrothermal vent chimney at the East Pacific Rise (13°N)
Balnearium lithotrophicum[287]	AN CLA	45–80 [70–75]	5–7 [5.4]	Deep-sea hydrothermal system, Suiyo Seamount
Thermovibrio ruber[288]	AN CLA	50–80 [75]	5–6.5 [6]	Submarine hydrothermal vents in the Papua New Guinea region, type strain from sandy sediments taken from the beach off Lihir Island
Thermovibrio ammonificans[289]	AN CLA	60–80 [75]	5–7 [5.5]	Deep sea hydrothermal vent area, East Pacific Rise
Thermovibrio guaymasensis[286]	AN CLA	50–88 [75–80]	5.5–7.5 [6–6.2]	A deep-sea hydrothermal vent chimney at Guaymas Basin
Bacteria; *Aquificae*; *Aquificae*; *Aquificales*; *Hydrogenothermaceae* {*B*33}; Genera: *Hydrogenothermus*, *Sulfurihydrogenibium*, *Persephonella*				
Hydrogenothermus marinus[290]	FAE CLA	45–80 [65]	5–7	Vulcano Beach, Vulcano Island, Italy
Sulfurihydrogenibium subterraneum[291,292]	FAE CLA	40–70 [60–65]	6.4–8.8 [7.5]	Subsurface hot aquafier water in the Hishikari Japanese gold mine, Kagoshima Prefecture, Japan
Sulfurihydrogenibium azorense[291,293] §	FAE CLA	50–73 [68]	5.5–7 [6]	Near the Água do Caldeirão, Furnas, on São Miguel Island, Azores
Persephonella hydrogeniphila[294]	FAE CLA	50–72.5 [70]	5.5–7.6 [7.2]	The hydrothermal field at the Suiyo Seamount in the Izu-Bonin Arc, Japan
Persephonella guaymasensis[295]	FAE CLA	55–75 [70]	4.7–7.5 [6]	Deep-sea hydrothermal vent chimney in Guaymas Basin, Mexico
Persephonella marina[295]	FAE CLA	55–80 [73]	4.7–7.5 [6]	9°N deep-sea hydrothermal vent, East Pacific Rise
Bacteria; *Thermotogae*; *Thermotogae*; *Thermotogales*; *Thermotogaceae* {*B*34}; Genera: *Geotoga*, *Marinitoga*, *Petrotoga*, *Thermosipho*, *Thermotoga*, *Fervidobacterium*				
Geotoga petraea[296]	AN COH	30–60 [45]	5.5–9 [6.5]	Oilfield brines, Texas and Oklahoma, USA
Geotoga subterranea[296]	AN COH	30–55 [50]	5.5–9 [6.5]	Oilfield brines, Texas and Oklahoma, USA
Marinitoga camini[297] §	AN COH	25–65 [55]	5–9 [7]	Deep sea vent fields, Mid-Atlantic Ridge
Marinitoga piezophila[298]	AN COH	45–70 [65]	5–8 [6]	"Grandbonum" deep-sea-vent field site, East-Pacific Rise
Marinitoga hydrogenitolerans[299]	FAE COH	35–65 [60]	4.5–8.5 [6]	Deep-sea hydrothermal chimney collected at the Rainbow field on the Mid-Atlantic Ridge
Marinitoga okinawensis[300]	AN COH	30–70 [55–60]	5.0–7.4 [5.5–5.8]	Deep-sea hydrothermal field in Yonaguni Knoll IV, Southern Okinawa Trough

TABLE 1. *Continued*

Species	O_2 relationship and metabolism	Temperature range [optimum]	pH Range [optimum]	Originally isolated from
Petrotoga mexicana[301]	AN COH	25–65 [55]	5.8–8.5 [6.6]	Oil/water mixtures of production well heads in Tabasco, Gulf of Mexico, an offshore reservoir
Petrotoga miotherma[296]	AN COH	35–65 [55]	5.5–9 [6.5]	Oilfield brines, Texas and Oklahoma, USA
Petrotoga olearia[413]	AN COH	37–60 [55]	6.5–8.5 [7.5]	Oil/water mixtures of the Samotlor oilfields, western Siberia, Russia
Petrotoga sibirica[413]	AN COH	37–55 [55]	6.5–9.4 [8]	Oil/water mixtures of the Samotlor oilfields, western Siberia, Russia
Petrotoga mobilis[302]§	AN COH	40–65 [58–60]	5.5–8.5 [6.5–7]	Oil reservoir production water from offshore oil platforms, North Sea
Petrotoga halophila[303]	AN COH	45–65 [55–60]	5.6–7.8 [6.7–7.2]	Oil-producing well of the Tchibouella offshore oilfield in Congo, West Africa
Thermosipho atlanticus[414]	AN COH	45–80 [65]	5–9 [6]	Deep sea hydrothermal vent, Mid-Atlantic ridge
Thermosipho melanesiensis[304]§	AN COH	45–80 [70]	3.5–9.5 [6.5–9.5]	Deep sea hydrothermal area, Lau Basin, southwest Pacific Ocean
Thermosipho geolei[305]	AN COH	45–75 [70]	6–9.4 [7.5]	Production well-head oil/water mixture of a deep continental petroleum reservoir, western Siberia, Russia
Thermosipho japonicus[306]	AN COH	45–80 [72]	5.3–9.3 [7.2]	Hydrothermal field, Iheya Basin of the Okinawa Trough, Japan
Thermosipho africanus[307,308]	AN COH	35–77 [75]	6–8 [7.2]	Hydrothermal springs, Gulf of Tadjoura, Republic of Djibouti, Africa
Fervidobacterium gondwanense[309]	AN COH	[65–68]	[7]	Great Artesian Basin, Australia
Fervidobacterium islandicum[310]	AN COH	50–80 [65]	6–8 [7]	Hot spring in Hveragerdi, Iceland
Fervidobacterium nodosum[311]§	AN COH	41–79 [70]	6–8 [7]	Hot spring in New Zealand
Fervidobacterium pennavorans[312]	AN COH	50–80 [70]	5.5–8.0 [6.5]	Hot springs of São Miguel Island, Azores, Portugal
Fervidobacterium changbaicum[313]	AN COH	55–90 [75–80]	6.3–8.5 [7.5]	Hot spring, Changbai Mountains, China
Thermotoga lettingae[314]§	AN COH	50–75 [65]	6–8.5 [7]	Thermophilic, sulfate-reducing, slightly saline bioreactor
Thermotoga elfii[315]	AN COH	50–72 [66]	5.5–8.7 [7.5]	African oil well
Thermotoga hypogea[316]	AN COH	56–90 [70]	6.1–9.1 [7.3–7.4]	Oil-producing well, Cameroon, central Africa
Thermotoga subterranea[317]	AN COH	50–75 [70]	6–8.5 [7]	Deep continental oil reservoir, East Paris Basin, France
Thermotoga thermarum[318]	AN COH	55–84 [70]	5.5–9 [7]	Hot water and mud from the shore of Lac Abbe, Republic of Djibouti, Africa
Thermotoga maritima[319,320]	AN COH	55–90 [80]	5.5–9 [6.5]	Geothermally heated sea floors, Italy and the Azores
Thermotoga petrophila[321]§	AN COH	47–88 [80]	5.2–9 [7]	Production fluid of the Kubiki oil reservoir in Niigata, Japan

Continued

TABLE 1. *Continued*

Species	O_2 relationship and metabolism	Temperature range [optimum]	pH Range [optimum]	Originally isolated from
Thermotoga naphthophila[321]	AN COH	48–86 [80]	5.4–9 [7]	Production fluid of the Kubiki oil reservoir in Niigata, Japan
Thermotoga neapolitana[322]§	AN COH	55–90 [80]	5.5–9 [7]	Shallow submarine hot springs, Lucrino Bay, Naples, Italy
Archaea; *Crenarchaeota*; *Thermoprotei*; *Desulfurococcales*; *Desulfurococcaceae* {A01}; Genera: *Acidilobus*, *Staphylothermus*, *Ignicoccus*, *Desulfurococcus*, *Thermosphaera*, *Sulfophobococcus*, *Stetteria*, *Thermodiscus* (also *Ignishpaera* of the *Ignisphaera* group)				
Thermosphaera aggregans[323]	AN COH	65–90 [85]	5–7 [6.5]	"Obsidian Pool," Yellowstone National Park, USA
Acidilobus aceticus[324]	AN COH	60–92 [85]	2–6 [3.8]	A hot spring of the Mountnovski volcano, Kamchatka, Far East Russia
Staphylothermus marinus[325]§	AN COH	65–98 [92]	4.5–8.5 [6.5]	Vulcano Island, Italy; also a deep-sea black smoker of the East Pacific Rise
Staphylothermus hellenicus[326]	AN COH	70–90 [85]	4.5–7 [6]	Palaeochori Bay, Milos, Greece
Ignicoccus islandicus[327]	AN CLA	70–98 [90]	3.8–6.5 [5.8]	Submarine hydrothermal system, Atlantic Ocean, at the Kolbeinsey Ridge north of Iceland
Ignicoccus hospitalis[328]	AN CLA	73–98 [90]	4.5–7.0 [5.5]	Kolbeinsey Ridge, north of Iceland
Ignicoccus pacificus[327]	AN CLA	75–98 [90]	4.5–7 [6]	Submarine hydrothermal system, black smoker samples, 9°N, 104°W, Pacific Ocean
Desulfurococcus fermentans[92]	AN COH	63–89 [80–82]	4.8–6.8 [6]	Freshwater hot spring of the Uzon Caldera, Kamchatka, Far East Russia
Desulfurococcus amylolyticus[329,330]	AN COH	68–97 [90–92]	5.7–7.5 [6.4]	Thermal springs of Kamchatka and Kunashir Islands, Far East Russia
Desulfurococcus mucosus[331]	AN COH	[85]	[6]	Solfataric hot springs, Iceland
Desulfurococcus mobilis[331]	AN COH	[85]	[6]	Solfataric hot springs, Iceland
Sulfophobococcus zilligii[332]	AN COH	70–90 [85]	6.5–8.5 [7.5]	Hot spring near Hveragerdi, Iceland
Stetteria hydrogenophila[106]	AN CLH	68–102 [95]	4.5–7.0 [6]	Marine hydrothermal sediment, Paleohori Bay, Milos, Greece
Thermodiscus maritimus[333–335]	AN COH	75–98 [90]	5–7 [5.5]	Submarine solfatara field close to Vulcano, Italy
Ignisphaera aggregans[336]	AN COH	85–98 [92–95]	5.4–7 [6.4]	Geothermal Rotorua, Tokaanu, New Zealand
Archaea; *Crenarchaeota*; *Thermoprotei*; *Desulfurococcales*; *Pyrodictiaceae* {A02}; Genera: *Pyrodictium*, *Hyperthermus*, *Pyrolobus*				
Pyrodictium abyssi[62]	AN COH	80–110 [97]	4.7–7.1 [5.5]	Deep-sea hydrothermal areas of Guaymas Basin, Gulf of California and Kolbeinsey Ridge, north of Iceland
Pyrodictium brockii[63]	AN CLA	[105]	5–7 [5.5]	Shallow submarine solfataric field at Porto di Levante, Vulcano Island, Italy
Pyrodictium occultum[63]	AN CLA	[105]	5–7 [5.5]	Shallow submarine solfataric field at Porto di Levante, Vulcano Island, Italy

TABLE 1. *Continued*

Species	O_2 relationship and metabolism	Temperature range [optimum]	pH Range [optimum]	Originally isolated from
Hyperthermus butylicus[64]§	AN COH	[95–107]	[7]	Hydrothermally heated flat-sea sediments off the coast of São Miguel Island, Azores
Pyrolobus fumarii[5]	FAE CLA	90–113 [106]	4.0–6.5 [5.5]	TAG site, Mid-Atlantic Ridge,
Archaea; *Crenarchaeota*; *Thermoprotei*; *Sulfolobales*; *Sulfolobaceae* {*A*03}; Genera: *Stygiolobus*, *Sulfurisphaera*, *Acidianus*				
Stygiolobus azoricus[72]	AN CLA	57–89 [80]	1–5.5 [2.5–3]	Solfataric fields of São Miguel Island, Azores
Sulfurisphaera ohwakuensis[337]	FAE COH	63–92 [84]	1–5 [2]	Hot springs of Ohwaku Valley, Hakone, Japan
Acidianus brierleyi[132]	FAE CLA	45–75 [70]	1–6 [1.5–2]	Distant acidic geo- and hydrothermal heated areas
Acidianus ambivalens[338,339]	FAE CLA	[80]	1–3.5	Leirhnukur fissure, Iceland
Acidianus infernus[132]	FAE CLA	65–96 [90]	1–5.5 [2]	Solfatara Crater and Pisciarelli Solfatara, Naples, Italy
Acidianus sulfidivorans[71]	FAE CLA	45–83 [74]	0.35–3.0 [0.8–1.4]	Solfatara on Lihir Island, Papua New Guinea
Archaea; *Crenarchaeota*; *Thermoprotei*; *Thermoproteales*; *Thermofilaceae* {*A*04}; Genus: *Thermofilum*				
Thermofilum pendens[340]§	AN COH	[85–90]	[5]	Icelandic solfataras
Archaea; *Crenarchaeota*; *Thermoprotei*; *Thermoproteales*; *Thermoproteaceae* {*A*05}; Genera: *Thermoproteus*, *Pyrobaculum*, *Thermocladium*, *Caldvirga*				
Thermoproteus uzoniensis[341]	AN COH	74–102 [90]	4.6–6.8 [5.6]	Uzon Caldera, Kamchatka, Far East Russia
Thermoproteus tenax[342]	AN F-CLA	[85]	[3.7–4.2]	Icelandic solfataric hot springs
Thermoproteus neutrophilus[342,343]§	AN F-CLA	[85]	[6.8]	Hot spring, Iceland
Pyrobaculum oguniense[344]	AN COH	70–97 [90–94]	5.4–7.4 [6.3–7]	Terrestrial hot spring at Oguni-cho, Kumamoto Prefecture, Japan
Pyrobaculum arsenaticum[345]§	AN F-CLA	68–100 [81]	NR	Hot water pond, Pisciarelli Solfatara, Naples, Italy
Pyrobaculum organotrophum[346]	AN COH	74–102 [100]	5–7 [6]	Boiling solfataric waters in Iceland, Italy, and the Azores
Pyrobaculum islandicum[346]§	AN F-CLA	74–102 [100]	5–7 [6]	Boiling solfataras and geothermal waters, Iceland
Pyrobaculum calidifontis[347]§	FAE COH	75–100 [90–95]	5.5–8.0 [7.0]	Terrestrial hot spring Calamba, Laguna, the Philippines
Pyrobaculum aerophilum[348,349]§	FAE F-CLA	75–104 [100]	5.8–9 [7]	Boiling marine water hole, Maronti Beach, Ischia, Italy
Thermocladium modestius[350]	FAE COH	45–82 [72]	2.6–5.9 [4]	Sites of volcanic activity, Japan
Vulcanisaeta souniana[351]	FAE COH	65–89 [85]	3.5–5.0 [4.5]	Hot spring, Sounzan, Kanagawa, Japan
Vulcanisaeta distributa[351]	FAE COH	70–92 [90]	3.5–5.6 [4.5]	Hot spring, Ohwakudani, Kanagawa, Japan
Caldivirga maquilingensis[352]§	FAE COH	62–92 [85]	2.3–6.4 [3.7–4.2]	Acidic hot spring, the Philippines
Archaea; *Euryarchaeota*; *Thermoplasmata*; *Thermoplasmatales*; *Thermoplasmataceae* {*A*06}; Genus: *Thermoplasma*				
Thermoplasma acidophilum[353,354]§	FAE COH	45–63 [59]	0.5–4 [1–2]	Solfatara fields and self-heated coal refuse piles

Continued

TABLE 1. *Continued*

Species	O_2 relationship and metabolism	Temperature range [optimum]	pH Range [optimum]	Originally isolated from
Thermoplasma volcanium[353,355]§	FAE COH	33–67 [60]	1–4 [2]	Submarine and continental solfataras at Vulcano Island, Italy; also from Java, Iceland; and Yellowstone National Park, USA
Archaea; Euryarchaeota; Methanococci; Methanococcales; Methanocaldococcaceae {*A*07}; Genera: *Methanocaldococcus*, *Methanotorris*				
Methanocaldococcus fervens[356]	AN CLA	48–92 [85]	5.5–7.6 [6.5]	Deep-sea hydrothermal vent, Guaymas Basin, Gulf of California
Methanocaldococcus indicus[357]	AN CLA	50–86 [85]	[6.5–6.6]	Deep-sea hydrothermal vent, Central Indian Ridge
Methanocaldococcus infernus[358,359]	AN CLA	55–91 [85]	5.25–7.0 [6.5]	Deep-sea hydrothermal chimney sample collected on the Mid-Atlantic Ridge at a depth of 3000 m
Methanocaldococcus jannaschii[358,360,361]§	AN CLA	50–86 [85]	5.2–7.0 [6.0]	"White smoker" chimney on the 20°N East Pacific Rise
Methanocaldococcus vulcanius[356]	AN CLA	49–89 [80]	5.2–7 [6.5]	Deep-sea vent, 13°N thermal field, East Pacific Rise
Methanotorris formicicus[362]	AN CLA	55–83 [75]	6–8.5 [6.7]	Deep-sea vent, Kairei field, Central Indian Ridge
Methanotorris igneus[363,364]	AN CLA	45–91 [88]	5.0–7.5 [5.7]	Kolbeinsey Ridge shallow (103 and 106 m) submarine hydrothermal system off Iceland
Archaea; Euryarchaeota; Thermococci; Thermococcales; Thermococcaceae {*A*08}; Genera: *Thermococcus*, *Pyrococcus*, *Palaeococcus*				
Thermococcus sibiricus[415]	AN COH	40–88 [78]	5.8–9 [7.5]	Samotlor oil reservoir, Western Siberia, Russia
Thermococcus zilligii[365]	AN COH	55–85 [75–80]	5.4–9.2 [7.4]	Terrestrial fresh water hot pool, New Zealand
Thermococcus profundus[366]	AN COH	50–90 [80]	4.5–8.5 [7.5]	Deep-sea hydrothermal vent system, Middle Okinawa Trough
Thermococcus barossii[367]	AN COH	60–94 [82.5]	4–9 [6.5–7.5]	Hydrothermal vent flange formation, East Pacific Rise of the Juan de Fuca Ridge
Thermococcus siculi[368]	AN COH	50–93 [85]	5–9 [7]	Deep-sea hydrothermal vent, the Mid-Okinawa Trough
Thermococcus celericrescens[369]	AN COH	50–85 [80]	5.6–8.3 [7.0]	Hydrothermal vent at Suiyo Seamount, Izu-Bonin Arc, Western Pacific Ocean
Thermococcus acidaminovorans[75]	AN COH	56–93 [85]	5–9.5 [9]	Vulcano Island, Italy
Thermococcus aegaeus[326]	AN COH	50–95 [85]	4.5–7.5 [6]	Palaeochori Bay, Milos, Greece
Thermococcus alcaliphilus[78]	AN COH	56–90 [85]	6.5–10.5 [9]	Vulcano Island, Italy
Thermococcus atlanticus[370]	AN COH	70–90 [85]	5–8 [7]	"Snakepit" hydrothermal vent region of the Mid-Atlantic Ridge
Thermococcus barophilus[140]§	AN COH	48–100 [85]	[7]	"Snakepit" hydrothermal vent region of the Mid-Atlantic Ridge
Thermococcus chitonophagus[371]	AN COH	60–93 [85]	3.5–9 [6.7]	Guaymas Basin hydrothermal vents, Gulf of California
Thermococcus fumicolans[76]	AN COH	73–103 [85]	4.5–9.5 [8.5]	Deep-sea hydrothermal vent, North Fiji Basin
Thermococcus litoralis[372]	AN COH	50–96 [85]	4–8.5 [6]	Vulcano Island, Italy; also submarine thermal spring, Lucrino Beach, Naples

TABLE 1. *Continued*

Species	O_2 relationship and metabolism	Temperature range [optimum]	pH Range [optimum]	Originally isolated from
Thermococcus thioreducens[373]	AN COH	55–94 [83–85]	5.0–8.5 [7.0]	"Black smoker" chimney material from the "Rainbow" hydrothermal vent site on the Mid-Atlantic Ridge (36.2°N, 33.9°W)
Thermococcus waiotapuensis[374]	AN COH	60–90 [85]	5–8 [7]	Terrestrial freshwater hot spring, New Zealand
Thermococcus stetteri[375]	AN COH	55–95 [75–88]	5.7–7.2 [6.5]	Marine solfataric fields of Kraternaya Cove, Ushishir Archipelago, Northern Kurils
Thermococcus pacificus[412]	AN COH	70–95 [80–88]	6–8 [6.5]	Bay of Plenty, New Zealand
Thermococcus aggregans[376]	AN COH	60–94 [88]	5.6–7.9 [7]	Guaymas Basin hydrothermal vents, Gulf of California
Thermococcus celer[77]	AN COH	[88]	[8.5]	Vulcano Island, Italy
Thermococcus gammatolerans[143]	AN COH	55–95 [88]	[6]	Guaymas Basin, Gulf of California
Thermococcus gorgonarius[412]	AN COH	68–95 [80–88]	5.8–8.5 [6.5–7.2]	Shore of Whale Island, New Zealand
Thermococcus guaymasensis[376]	AN COH	56–90 [88]	5.6–8.1 [7.2]	Guaymas Basin hydrothermal vent site, Gulf of California
Thermococcus hydrothermalis[82]	AN COH	55–100 [80–90]	3.5–9.5 [5.5–6.5]	21°N deep-sea hydrothermal vent region, East Pacific Rise
Thermococcus peptonophilus[377]	AN COH	60–100 [85–90]	4–8 [6]	Izu-Borin forearc; also from the southern Mariana Trough
Thermococcus kodakarensis[378,379,380]§	AN COH	60–100 [85]	5–9 [6.5]	Solfatara on Kodakara Island, Kagoshima, Japan
Thermococcus coalescens[381]	AN COH	57–90 [87]	5.2–8.7 [6.5]	Hydrothermal fluid obtained from Suiyo Seamount of the Izu-Bonin Arc
Pyrococcus furiosus[382,383]§	AN COH	70–103 [100]	5–9 [7]	Shallow marine hydrothermal system at Vulcano Island, Italy
Pyrococcus horikoshii[383,384,385]§	AN COH	80–102 [98]	5–8 [7]	Hydrothermal fluid samples obtained at the Okinawa Trough vents in the NE Pacific Ocean, at a depth of 1395 m
Pyrococcus glycovorans[83]	AN COH	75–104 [95]	2.5–9 [7.5]	Deep-sea hydrothermal vent on the East Pacific Rise
Pyrococcus woesei[386,d]	AN COH	[100–103]	[6–6.5]	Marine solfataras at the northern beach of Porto di Levante, Vulcano Island, Italy
Palaeococcus ferrophilus[387]	AN COH	60–88 [83]	4–8 [6]	Hydrothermal vent chimney, Myojin Knoll, Ogasawara-Bonin Arc, Japan
Palaeococcus helgesonii[388]	FAE COH	45–85 [80]	5–8 [6.5]	Geothermal well on Vulcano Island, Italy
Archaea; *Euryarchaeota*; *Archaeoglobi*; *Archaeoglobales*; *Archaeoglobaceae* {*A09*}; Genera: *Archeoglobus*, *Geoglobus*, *Ferroglobus*				
Archaeoglobus veneficus[389]	AN F-CLA	65–85 [75–80]	6.5–8 [7]	"Snake Pit" hydrothermal vents of the Mid-Atlantic Ridge
Archaeoglobus profundus[105]	AN CLH	65–90 [82]	4.5–7.5 [6]	The Guaymas hot vent area; cores of hot sediment and active smoker chimneys
Archaeoglobus fulgidus[390,391,392]§	AN F-CLA	64–92 [83]	5.5–7.5	Marine hydrothermal systems at Vulcano island and at Stufe di Nerone, Naples, Italy
Geoglobus ahangari[129]	AN F-CLA	65–90 [88]	5–7.6 [7.0]	Guaymas Basin hydrothermal system, Gulf of California, at a depth of 2000 m
Ferroglobus placidus[393]	AN F-CLA	65–95 [85]	6.0–8.5 [7.0]	Shallow submarine hydrothermal systems at Vulcano Island, Italy

Continued

TABLE 1. *Continued*

Species	O_2 relationship and metabolism	Temperature range [optimum]	pH Range [optimum]	Originally isolated from
Archaea; Euryarchaeota; Methanopyri; Methanopyrales; Methanopyraceae {*A10*};				
Genus: *Methanopyrus*				
Methanopyrus kandleri[394,395] §	AN CLA	84–110 [98]	5.5–7 [6.5]	Deep-sea sediment from the Guaymas Basin, Gulf of California; and from the shallow marine hydrothermal system of the Kolbeinsey Ridge, Iceland
Archaea; Euryarchaeota; Methanobacteria; Methanobacteriales; Methanothermaceae {*A11*};				
Genus: *Methanothermus*				
Methanothermus sociabilis[396]	AN CLA	55–97 [88]	5.5–7.5 [6.5]	Hot waters and muds of Icelandic continental geothermal areas
Methanothermus fervidus[397]	AN CLA	55–97 [80–85]	NR	Icelandic solfatara fields at Keringarfjoll and Hveragerdi
Archaea; Euryarchaeota; Methanobacteria; Methanobacteriales; Methanobacteriaceae {*A12*};				
Genera: *Methanobacterium, Methanothermobacter*				
Methanobacterium thermaggregans[398]	AN CLA	40–75 [65]	6.5–9 [7–7.5]	Mud from a cow pasture
Methanothermobacter defluvii[73,74]	AN CLA	[60–65]	[6.5–7.0]	Anaerobic sludge obtained from the pilot-scale plant digesting methacrylic wastewater
Methanothermobacter marburgensis[399]	AN CLA	45–70 [65]	5.0–8.0 [6.8–7.4]	Mesophilic sewage sludge and hot springs
Methanothermobacter thermautotrophicus[399,400] §	AN CLA	40–75 [65–70]	6.0–8.8 [7.2–7.6]	Anaerobic sewage sludge digestor
Methanothermobacter thermoflexus[73,74]	AN CLA	[55]	[7.9–8.2]	Anaerobic sludge obtained from the pilot-scale plant digesting methacrylic wastewater
Methanothermobacter thermophilus[74,401]	AN CLA	45–65 [57]	7.0–8.5 [7.5]	Uzon Caldera, Kamchatka, Far East Russia
Methanothermobacter wolfei[399,402]	AN CLA	37–74 [55–65]	6.0–8.2 [7.0–7.5]	Mixture of sewage sludge and river sediment
Archaea; Euryarchaeota; Methanococci; Methanococcales; Methanococcaceae {*A13*};				
Genus: *Methanothermococcus*				
Methanothermococcus thermolithotrophicus[403,404] §	AN CLA	30–70 [65]	6–8 [7]	Heated sea sediments near Naples, Italy
Methanothermococcus okinawensis[405]	AN CLA	40–75 [60–65]	4.5–8.5 [6–7]	Deep-sea vent at Iheya Ridge in the Okinawa Trough, Japan
Archaea; Euryarchaeota; Methanomicrobia; Methanosarcinales; Methanosarcinaceae {*A14*};				
Genera: *Methanosarcina, Methanomethylovorans*				
Methanosarcina thermophila[406] §	AN F-CLA	[50]	[6–7]	55°C anaerobic sludge digestor
Methanomethylovorans thermophila[407]	AN CLA	42–58 [50]	5–7 [6.5]	Methanol-fed thermophilic anaerobic reactor
Archaea; Euryarchaeota; Methanomicrobia; Methanosarcinales; Methermicoccaceae {*A15*};				
Genus: *Methermicoccus*				
Methermicoccus shengliensis[408]	AN CLA	50–70 [65]	5.5–8.0 [6.0–6.5]	Oil-production water of the Shengli oilfield, China
Archaea; Euryarchaeota; Methanomicrobia; Methanosarcinales; Methanosaetaceae {*A16*};				
Genus: *Methanothrix*				
Methanothrix thermophila[409] §	AN COH	[55]	6.1–7.5 [6.7]	Mesophilic anaerobic sludge digestors

TABLE 1. *Continued*

Species	O_2 relationship and metabolism	Temperature range [optimum]	pH Range [optimum]	Originally isolated from
Archaea; *Euryarchaeota*; *Methanomicrobia*; *Methanomicrobiales*; *Methanomicrobiaceae* {*A*17};				
Genus: *Methanoculleus*				
Methanoculleus thermophilus[410,411]	AN F-CLA	37–65 [55]	6.18–7.82 [7]	Nuclear power plant effluent channel sediment

[c]Symbols and abbreviations: §, sequenced genome; AN, anaerobic; FAE, facultative aerobe; COH, chemoorganoheterotroph; CLA, chemolithoautotroph; F-CLA, facultative chemolithoautotroph; PA, photoautotroph; PH, photoheterotroph; CLH, chemolithoheterotroph; F-CLH, facultative chemolithoheterotroph; NR, not reported.

[d]Note that Kanoksilapatham *et al.* suggest reclassifying *Pyrococcus woesei* as *Pyrococcus furiosus* subsp. *woesei*: Kanoksilapatham, W., J.M. Gonzalez, D.L. Maeder, *et al.* 2004. A proposal to rename the hyperthermophile *Pyrococcus woesei* as *Pyrococcus furiosus* subsp. *woesei*. Archaea. **1:** 277–283.

of the cultured thermophilic Bacteria and Archaea are anaerobic, it is reasonable to ask whether this observation accurately reflects the ecological situation in thermobiotic environments. This question is largely beyond the scope of the present chapter; however, culture-independent analyses have revealed that members of the Aquificales, (Aquificaceae {*B*32}), appear to be ubiquitous in the terrestrial hot springs of Yellowstone National Park, as well as in Japan, Iceland, and Kamchatka. It has been postulated that these Aquificales, particularly the obligately aerobic *Thermocrinis ruber*–like microorganisms, are the primary producers (via chemoautotrophic hydrogen oxidation) in these ecosystems.[34,55]

Temperature Relationship

Etymologically, thermophiles "love" heat, and Bacteria and Archaea with this physiology are further categorized according to their optimal (T_{opt}) and maximal (T_{max}) growth temperatures. The authors classify Bacteria and Archaea with T_{opt} 50°C–64°C as (moderate) thermophiles, those with T_{opt} 65°C–79°C as extreme thermophiles, and prokaryotes with $T_{opt} \geq 80°C$ as hyperthermophiles.[22,56–58] Although not considered true thermophiles, Bacteria and Archaea that grow optimally at mesophilic temperatures but have a $T_{max} > 50°C$ are described as thermotolerant.

Thermophilic prokaryotes able to grow over a 35°C–40°C temperature span are considered temperature-tolerant thermophiles.[57] To the authors' knowledge, the record widest temperature growth range is 22°C–75°C by a *Methanothermobacter thermautotrophicus*–like strain isolated from river sediment (J. Wiegel, unpublished results).By comparison, some thermophilic anaerobes are reported to have especially narrow temperature growth ranges, such as 42°C–55°C for *Anaerolinea thermolimosa*[59] and 50°C–60°C for *Anaerolinea thermophila*[60]—two species from the family *Anaerolinaceae* {*B*27}.

Hyperthermophiles with $T_{opt} \geq 80°C$ and $T_{max} > 100°C$, were first isolated by Stetter from the hot vents at Vulcano Island off the coast of Sicily, Italy, in 1981.[24] Isolates growing optimally above 100°C are often found at deep-sea vents,[61] or deep in terrestrial hot spring channels and sediments. This "deep" requirement is ascribed to the increased pressure, which allows water to remain in liquid form at temperatures above 100°C. Within the *Bacteria*, only the *Thermotogaceae* {*B*34} and *Aquificaceae* {*B*32} lineages contain hyperthermophilic members. Within the *Archaea*, taxonomic families containing hyperthermophilic taxa include: *Methanopyraceae* {*A*10}, *Methanothermaceae* {*A*11}, *Methanocaldococcaceae* {*A*07}, *Thermococcaceae* {*A*08}, *Archaeoglobaceae* {*A*09}, *Sulfolobaceae* {*A*03}, *Desulfurococcaceae* {*A*01}, *Pyrodictiaceae* {*A*02}, *Thermofilaceae* {*A*04}, and *Thermoproteaceae* {*A*05} (see TABLE 1). Many species with extremely high temperature growth ranges and T_{opt} belong to the *Pyrodictiaceae* {*A*02}: *Pyrodictium abyssi* with a growth range of 80°C–110°C, T_{opt} 97°C[62]; *Pyrodictium brockii* and *Pyrodictium occultum*, both with T_{opt} 105°C[63]; *Hyperthermus butylicus* with a reported T_{opt} 95°C–107°C[64]; and *P. fumarii* with a growth range of 90°C–113°C and T_{opt} 106°C.[5] Very recently, Takai provided convincing evidence of a *M. kandleri* strain isolated from a deep-sea hydrothermal region capable of growing under increased pressure at 122°C.[6]

The base of the inferred 16S rRNA gene sequence phylogenetic tree contains hyperthermophilic taxa (FIG. 1). This is one line of evidence for the hypothesis that life evolved at a time of elevated temperature on early Earth. However, thermophilic anaerobes are also found among branches of the phylogenetic tree containing predominantly mesophiles. These taxa having either evolved after a speciation event, during subsequent phyletic evolution, or they are remnants of the thermophilic origins of the clades. Because of the many specialized properties required for thermophilic life, thermophily is not a trait transferrable via horizontal gene transfer. In cases where thermophilic and mesophilic species exist within a genus, the ability to grow at elevated temperatures occurs via combinations of different properties, including the stabilization of intracellular compounds through binding of cations, such as Ca^{2+}; the insertion of additional hydrogen bridges in protein structures via the presence of increased acidic amino acids (e.g., by substituting glutamine with glutamic acid); the formation of more globular protein structures with additional hydrophobic regions inside; the stabilization of membrane fluidity by changes in fatty acid lengths and branching[65]; and changes in proton permeability of the cytoplasmic membranes.[66] In cases where the addition of genes conferred higher growth temperatures, it is assumed that the enzymes encoded by these genes removed the bottleneck of otherwise cryptic thermophiles.[57,67]

pH Relationship

The pH growth minimum (pH_{min}), maximum (pH_{max}), and optimum (pH_{opt}) determine whether a prokaryote is characterized as acidophilic ($pH_{opt} < 5.0$), acidotolerant ($pH_{min} < 4$, pH_{opt} circumneutral), neutrophilic ($pH_{opt} \sim 7.0$), alkalitolerant ($pH_{opt} < 8.5$, $pH_{max} > 8.5$), or alkaliphilic ($pH_{opt} \geq 8.5$). The majority of known thermophilic anaerobes are neutrophilic. None of the known thermophilic anaerobic *Bacteria* grow below pH 3.0.[58] To the authors' current knowledge, the most acidophilic anaerobic thermophilic *Bacteria* are: *Thermoanaerobacterium aotearoense*, pH^{60C} range 3.8–6.8 and pH^{60C}_{opt} 5.2[68]; *Thermoanaerobacterium aciditolerans*, pH range 3.2–7.1, pH_{opt} 5.7[69]; and *Lebetimonas acidiphila*, pH range 4.2–7.0 and pH_{opt} 5.2.[70] The most acidophilic thermophilic anaerobic *Archaea* belong to the *Thermoplasmataceae* {*A*06} and *Sulfolobaceae* {*A*03} families and nearly all are facultative aerobes; the most acidophilic is *Acidianus sulfidivorans*, with pH growth range of 0.35–3.0, and pH_{opt} 0.8–1.4.[71] One exception is the obligately anaerobic *Stygiolobus azoricus* of the *Sulfolobaceae* {*A*03}, which has a pH growth range of 1–5.5 and pH_{opt} 2.5–3.[72]

In contrast to the aerobic and truly acidophilic Archaea, and to the alkalithermophilic Bacteria (Table 1), only a few known anaerobic thermophilic Archaea grow optimally at high pH. All are obligately anaerobic and reside within the *Euryarchaeota* clade: species from the *Methanobacteriaceae* {*A*12}, including *Methanothermobacter thermoflexus*, with pH_{opt} 7.9–8.1[73,74]; and species from the *Thermococcaceae* {*A*08}, including *Thermococcus acidaminovorans* with a pH range of 5–9.5 and pH_{opt} 9.0,[75] *Thermococcus fumicolans*, with a growth range of pH 4.5–9.5 and pH_{opt} 8.5,[76] *Thermococcus celer*, with pH_{opt} 8.5,[77] and *Thermococcus alcaliphilus*, with a pH range of 6.5–10.5 and pH_{opt} 9.0.[78]

Alkalithermophilic ($pH_{opt}^{55C} > 8.5$; $T_{opt} > 55°C$) anaerobic Bacteria are particularly intriguing because of the phylogenetic position of most known taxa and an observed physiological peculiarity. Although anaerobic alkalithermophilic Bacteria taxa differ from one another physiologically, almost all belong to the phylum Firmicutes (formerly called the Gram-type positive Bacillus–Clostridium group) phylogenetic subbranch {*B*07–14}.[58] The one known Gram-type negative exception is the not validly published and poorly characterized "*Thermopallium natrophilum*."[79] Anaerobic alkalithermophilic *Bacteria* include: *C. paradoxum*, the most alkaliphilic thermophile with a pH^{60C} growth range of 7–11.1, pH_{opt} 10.1[3]; *Clostridium thermoalcaliphilum*, pH^{60C} range 7–11, pH_{opt} 9.6–10.1[80]; and members of the genus *Anaerobranca* of the family *Syntrophomonadaceae* {*B*12}. Phylogenetically similar *T. celere*–like strains have been isolated from a variety of niches, including thermobiotic and mesobiotic, and from alkaline and neutrophilic soils and sediments. Isolates from mesobiotic and neutrophilic niches exhibit shorter doubling times (10–15 min) compared with isolates from thermobiotic niches (25–40 min).[7,81] *N. thermophilus* (*Natranaerobiaceae* {*B*08}), isolated from an alkaline, hypersaline lake of the Wadi An Natrun, Egypt, classifies within a novel order of the class *Clostridia* in the phylum *Firmicutes*. It is alkalithermophilic (T_{opt} 53°C, pH_{opt}^{55C} 9.5) and additionally halophilic, growing optimally between 3.3 and 3.9 M Na^+.[8] It is important to note that the determination of the pH of alkalithermophiles (as well acidothermophiles) has to be done with temperature-equilibrated electrodes and the use of a pH meter standardized with buffers at the elevated temperature range.[58]

As with temperature growth ranges, there are thermophilic anaerobes with wide pH growth ranges spanning six or more pH units; for example, *Thermococcus hydrothermalis* with a pH range of 3.5–9.5[82] and *Pyrococcus glycovorans* with a pH growth range of 2.5–9.0.[83] *Thermoanaerobacter ethanolicus* strain JW200 is another interesting example as it has an unusual pH profile with a broad and flat pH optimum from pH 5.5 to 8.5.[84] Thermophilic anaerobes have also been discovered with growth ranges of less than one pH unit; for example, *Pelotomaculum thermopropionicum* with a pH range of 6.7–7.5,[85] *Thermodesulfatator indicus* with a pH range of 6.0–6.7,[86] and *Thermodesulfobacterium hydrogeniphilum* with a pH range of 6.3–6.8.[87]

Metabolic Diversity of Thermophilic Anaerobes

A majority of the axenic thermophilic anaerobic Archaea and Bacteria are chemoorganoheterotrophic, using organic compounds for carbon and energy. However, a variety of other metabolic strategies are seen within the set of thermophilic anaerobic prokaryotes, including chemolithoautotrophy, chemolithoheterotrophy, photoheterotrophy, and photoautotrophy. Considering these metabolic strategies and the variation that can be found within each type, an astonishing diversity of metabolism is observed within the thermophilic anaerobes. In this next section, an overview of commonly observed and unique physiologies of thermophilic anaerobes is provided. Amend and Shock have previously described thermophilic and hyperthermophilic energenic reactions in depth, and their work is a key resource for the study of thermophilic metabolisms.[54]

Chemoorganoheterotrophic metabolisms are further categorized, and include glycolytic, cellulolytic, lipolytic, and peptidolytic metabolisms. The Emden-Meyerhof and Entner-Doudoroff pathways are employed by glycolytic thermophilic anaerobes, but a variety of modifications have been discovered, particularly within the *Archaea*.[88] Principal fermentation products formed by glycolytic thermophilic anaerobes include acetate, lactate, ethanol, CO_2, and H_2. The production of ethanol by glycolytic and cellulolytic taxa has previously been studied.[89] Cellulose is the most abundant renewable natural plant fiber, and its degradation, coupled with the production of "biofuels," such as ethanol, by thermophilic anaerobes is an intensely studied and timely research area. This metabolism, involving many unusual enzymes, has been studied with axenic cultures as well as in co-culture. An example of the latter is the cellulolytic *Clostridium thermocellum* in culture with the glycolytic *T. ethanolicus*.[90] Besides *C. thermocellum*, other thermophilic species within the genus *Clostridium* are cellulolytic, these are: *C. sterocorium*, *C. thermolacticum*, *C. thermocopriae*, and *C. thermopapyrolyticum*.[91] Relatively recently, the first hyperthermophilic cellulosic materia degrading archeaon, *Desulfurococcus fermentans*, was isolated.[92] As with cellulose-degrading thermophilic anaerobes, xylanolytic thermophilic anaerobes generate interest because the conversion of xylan—a component of plant hemicellulose and the second-most abundant renewable polysaccharide in biomass—to useful products might be coupled to increasing the efficiency of processing lignocellulose and to the production of energy from renewable resources.[93] Xylan is widely used among thermophilic anaerobic Bacteria, especially among members of the Firmicutes. Lipolytic chemoorganotrophic thermophilic anaerobes include *Thermosyntropha lipolytica* which, in co-culture, syntrophically grows on saturated and unsaturated fatty acids with 4 to 18 carbon atoms,[95] and *Desulfurothermus naphthae*, which can use long-chain fatty acids with 6 to 18 carbon atoms as well as alkanes having 6 to 14 carbon atoms.[94] Although chemoorganoheterotrophic metabolisms appear to be common metabolic strategies for the axenic thermophilic anaerobes studied in the lab, the natural substrates for these microorganisms are largely unknown[54]; furthermore, culture-independent analyses indicate that lithotrophic prokaryotes are the primary producers in certain thermobiotic environments, such as hot springs.[35]

Among chemolithotrophic pathways, the methanogenic reaction, $4H_2 + CO_2 \rightarrow CH_4 + 2H_2O$, is well characterized and used by thermophilic taxa within the *Methanobacteriaceae* {*A*12}, *Methanothermaceae* {*A*11}, *Methanocaldococcaceae* {*A*07}, and *Methanococcaceae* {*A*13}.[54] Another interesting chemolithotrophic metabolism of anaerobic thermophiles described relatively recently makes use of CO, which occurs as a normal component of escaping volcanic gas of terrestrial and deep-sea hydrothermal origin.[96] Several thermophilic anaerobes have indeed been isolated that grow lithotrophically on CO, performing the metabolic reaction $CO + H_2O \rightarrow CO_2 + H_2$. Thermophilic anaerobes known to employ this strategy include: *Desulfotomaculum carboxydivorans*,[97] *Carboxydothermus hydrogenoformans*,[98] *Thermolithobacter carboxydivorans*,[99] *Carboxydocella thermautotrophica*,[100] *Thermincola carboxydiphila*,[101] *Thermincola ferriacetica*,[102] *Caldanaerobacter subterraneus* subsp. *pacificus*,[103] and *Thermosinus carboxydivorans*.[104] This same CO-using reaction has also been observed within the *Archaea* in an isolate belonging to the genus *Thermococcus* (family *Thermococcaceae* {*A*08}).[96] Another interesting chemolithotrophic strategy is employed by the acetogens using the Wood-Ljungdahl pathway (from the reaction: $3H_2 + CO_2 \rightarrow$ acetate). Both mesophilic and thermophilic taxa (e.g., *Moorella* species) are known to perform this reaction (for additional discussion, see the corresponding chapter within this book).

Chemolithoheterotrophs generate energy chemolithotrophically and assimilate carbon heterotrophically. Thermophilic anaerobes with this metabolism include Archaea, *Archaeoglobus profundus*[105] and *Stetteria hydrogenophila,*[106] and Bacteria, *Desulfotomaculum alkaliphilum,*[107] *Desulfotomaculum carboxydivorans,*[97] *T. carboxydiphila,*[101] *T. ferriacetica* (which can also grow chemolithoautotrophically),[102] *Caldithrix abyssi,*[108] *Vulcanithermus mediatlanticus,*[109] and *Oceanithermus profundus.*[110]

Two mechanisms for collecting light energy and converting it into chemical energy are known—one depends on photochemical reaction centers containing (bacterio)-chlorophyll and the other employs rhodopsins.[111,112] However, to the authors' knowledge, there are no rhodopsin-using thermophilic anaerobes. The question of whether there is a biological explanation for this (e.g., that rhodopsin proteins do not function well at high temperature) or whether it is the result of insufficient searching for this particular physiology, appears unanswered at this time. However, there are phototrophic moderately thermophilic anaerobes (TABLE 1). Presumably because of the temperature sensitivity of the photosystem, there are no known hyperthermophilic phototrophs.[54] Bacteria with (bacterio)-chlorophyll containing photochemical reaction centers have been found within the Cyanobacteria, Chlorobi, Proteobacteria, Chloroflexi and Firmicutes;[111] the few currently known thermophilic anaerobic phototrophs are found within the latter three of the above phyla. *Thermochromatium tepidum*, of the *γ-Proteobacteria* clade (*Chromatiaceae* {*B*02}), is an anaerobe with T_{max} 57°C and T_{opt} 48°C–50°C and is one of the very few thermophiles capable of anaerobic photoautotrophic growth.[113,114] *T. tepidum* is also capable of assimilating compounds such as pyruvate and is, therefore, also a photoheterotroph.[113,114] Four filamentous, gliding, moderately photoheterotrophic thermophilic facultative aerobes, *Roseiflexus castenholzii,*[115] *Chloroflexus aggregans,*[116] *Chloroflexus aurantiacus,*[117] and *Heliothrix oregonensis,*[118,119] are found within the *Chloroflexaceae* {*B*29}. *R. castenholzii, C. aggregans*, and *C. aurantiacus* have a unique metabolic strategy as they grow phototoheterotrophically under anaerobic conditions with available light, but also grow chemoorganotrophically under aerobic light or dark conditions.[115–117] Within the phylum Firmicutes (family Heliobacteriaceae {*B*13}), *Heliobacterium modesticaldum* is an obligately anaerobic photoheterotroph that is also capable of growing chemoorganoheterotrophically.[120] *H. modesicaldum* is among the most recently discovered taxa containing (bacterio)-chlorophyll photochemical reaction centers; however, at present, it is poorly characterized.[111]

Although thermophilic anaerobes are most often studied as axenic cultures—as this is a requirement for the valid publication of any prokaryotic taxon—some intriguing relationships have been observed for microorganisms with this physiology growing in co-culture. For example, in pure culture, the previously mentioned taxon *T. lipolytica* cannot use triacylglycerols or short- and long-chain fatty acids; but when grown in syntrophic co-culture with *Methanobacterium* strain JW/VS-M29, *T. lipolytica* grows on triacylglycerols and linear saturated and unsaturated fatty acids with 4 to 18 carbon atoms.[95] Many Archaea were initially described as being obligately dependent on S^0 reduction for the production of energy.[61] However, Bonch-Osmolovskaya and Stetter showed that some so-called "sulfur-dependent" Archaea grow well in co-culture with hydrogen-using thermophilic methanogens in the absence of sulfur. This is possible through interspecies hydrogen transfer, whereby growth-inhibiting hydrogen (from H^+ used as an electron acceptor) is removed without sulfur serving as the electron acceptor.[121] The *Ignicoccus*–"*Nanoarchaeum*" system has been described as a symbiotic relationship. It was discovered that small cocci were attached to the larger cells of a strain of *Ignococcus* isolated from the Kolbeinsey Ridge, north of Iceland. These tiny cocci could be isolated from the larger cells and subsequently studied, but grew only when attached to their host.[122] The genome sequence analyis of "*Nanoarchaeum*" showed that it is missing most of the enzymes required for nonparasitic growth.[123]

The importance of sulfur in the metabolism of thermophilic anaerobes becomes evident when one considers that a majority of thermophiles (chemolithotrophs, as well as chemoheterotrophs) take advantage of the sulfur redox system. Amend and Shock posit that the most common energy-yielding reaction may be the reduction of elemental sulfur: $H_2 + S° \rightarrow H_2S$.[54] Indeed, the diversity of known thermophilic anaerobic taxa that use this strategy is notable: the sulfur-reducing reaction has been reported within the *Pyrodictiaceae* {*A*02}, *Sulfolobaceae* {*A*03}, *Thermoanaerobacteriaceae* {*B*09}, *Thermoproteaceae* {*A*05}, *Aquificaceae* {*B*032}, *Desulfurellaceae* {*B*16}, *Desulfurococcaceae* {*A*01}, *Thermococcaceae* {*A*08}, *Thermoplasmataceae* {*A*06}, *Thermofilaceae* {*A*04}, and *Thermotogaceae* {*B*34}. Thermophilic, sulfate-reducing *Bacteria* have been isolated from a wide range of environments and many of these thermophiles belong to a phylogenetically coherent cluster of Gram-type positive, spore-forming *Desulfotomaculum* species of the *Firmicutes* (*Peptococcaceae* {*B*11}).[124,125] Thus, the role of sulfur in the metabolisms of thermophilic anaerobes can vary for different groups:

it can be reduced, it can serve as an electron sink during fermentation, and it can function as a terminal electron acceptor to allow sulfur respiration.[126]

Thermophilic anaerobic Fe(III)-reducing Bacteria and Archaea are found within nearly all thermobiotic environments and are usually respirationally diverse, capable of growing chemoorganotrophically with fermentable substrates or chemolithoautotrophically with molecular hydrogen.[102,127,128] Although only relatively recently described, a diverse set of thermophilic anaerobes are known to reduce Fe(III). Families of the Bacteria with taxa known to reduce Fe(III) include the *Bacillaceae* {*B*15}, *Peptococcaceae* {*B*11}, *Thermoanaerobacteriaceae* {*B*09}, *Acidaminococcaceae* {*B*10}, *Syntrophomonadaceae* {*B*12}, *Deferribacteraceae* {*B*19}, *Hydrogenothermaceae* {*B*33}, *Thermotogaceae* {*B*34}, and the *Thermodesulfobacteriaceae* {*B*20}. Families of the Archaea with taxa known to reduce Fe(III) include the *Thermoproteaceae* {*A*05}, *Archaeoglobaceae* {*A*09}, and the *Thermococcaceae* {*A*08}.[128] *Geoglobus ahangari*, of the *Archaeoglobaceae* {*A*09} was reported as the first dissimilatory Fe(III)-reducing prokaryote obligately growing autotrophically on hydrogen.[129] In some genera, such as *Thermoanaerobacter*, *Thermotoga*, and *Anaerobranca*, many of the species tested have been found to be capable of dissimilatory reduction of Fe(III), but overall it appears as though the ability to reduce Fe(III) does not correlate to an affiliation at the genus, or sometimes even at the species, level.[128] For example, although *Deferribacter abyssi* and *Deferribacter thermophilus* are closely related, having 98.1% 16S rRNA gene sequence similarity, *D. abyssi* is unable to reduce Fe(III) whereas it is a primary electron acceptor for *D. thermophilus.*[130,131] Also, *Thermolithobacter ferrireducens* and *Thermolithobacter carboxydivorans* have 99% 16S rRNA gene sequence similarity to each other; however, *T. ferrireducens* is able to reduce Fe(III) but cannot assimilate CO, whereas *T. carboxydivorans* is able to assimilate CO but cannot reduce Fe(III).[99]

Besides Fe(III), the reduction of other metals coupled to the generation of energy by thermophilic anaerobes includes Mn(IV)—which is known to be used by *Acidianus infernus*,[132] *Thermoanaerobacter siderophilus*,[133] *Thermovenabulum ferriorganovorum*,[134] and *D. thermophilus*[130]—as well as Mo(VI), which is used by *Acidianus brierleyi.*[135] *Pyrobaculum arsenaticum* has the ability to grow chemolithoautotrophically by arsenate reduction, and both *P. arsenaticum* and *Pyrobaculum aerophilum* can use selenate, selenite, or arsenate chemolithoorganotrophically.[137] Thermophilic anaerobes are also known to reduce a number of other metals, possibly as a detoxification mechanism; for example, *Thermoanaerobacter* strains isolated from Piceance Basin in Colorado were able to reduce Co(III), Cr(VI), and U(VI), in addition to Mn(IV) and Fe(III).[136]

Unique Physiological Adaptations

In addition to these described characteristics—O_2-relationship, temperature and pH profiles, and metabolic strategies—a number of additional physiological properties of thermophilic anaerobes should be examined and should, thereby, add to what is known about the diversity of thermophilic anaerobes. Nakagawa and Takai recently gave a detailed overview of the metabolic types isolated from deep-sea hydrothermal vents; therefore, these metabolic types are not further discussed here.[138] The NaCl optimum and tolerance of a prokaryote is often assessed. Thermophilic anaerobes of marine origin, for example, would be expected to grow best at marine salinity—around 3.5% (wt/vol) NaCl. Prokaryotes that grow optimally with high salinity are referred to as halophiles, and halophilic thermophilic anaerobes are known, as are halophilic alkalithermophiles, which are further discussed in the chapter by Mesbah and Wiegel within this volume.

Thermophilic anaerobes living at deep-sea hydrothermal vent sites must cope with the additional pressure exerted by the water column and are, therefore, piezotolerant or perhaps even piezophilic. Both *Methanocaldococcus* (basonym *Methanococcus*) *jannaschii,* isolated from the 21°N East Pacific Rise deep-sea hydrothermal vent site, and *Thermococcus barophilus*, obtained from the Snakepit region of the Mid-Atlantic Ridge, grow faster under increased hydrostatic pressure.[139,140] At its optimal growth temperature, the growth rate of *T. barophilus* was more than doubled at elevated hydrostatic pressure (40 MPa) compared with the growth rate at low pressure (0.3 MPa).[140] Furthermore, *T. barophilus*, as well as "*Pyrococcus abyssi*" and *Pyrococcus* strain ES4, isolated from deep-sea hydrothermal vent sites, show an extension of their T_{max} with elevated hydrostatic pressure.[140–142] Takai reported at the Thermophiles 2007 Meeting that under hydrostatic conditions, an H_2/CO_2-utilizing, methanogenic, *Methanopyrus kandleri*–like isolate was able to grow at 122°C, thus representing the new maximum temperature for sustaining life (i.e, for growing and multiplying).[6]

Some of the isolated thermophilic anaerobes also possess ionizing radiation resistance; for example, this characteristic is found in *Tepidimicrobium ferriphilum* (family *Clostridiaceae* {*B*14}), which was isolated from a freshwater hot spring within the Barguzin Valley,

Buryatiya, Russia.[146] The level of natural radioactivity at hydrothermal vents can be 100 times greater than that at Earth's surface because of the increased occurrence of elements, such as ^{210}Pb, ^{210}Po, and ^{222}Rn.[143,144] Indeed, Archaea of the family Thermococcaceae {*A08*}, *Thermococcus gammatolerans* and "*Thermococcus radiotolerans*" isolated from the Guaymas Basin, of the Gulf of California, and "*Thermococcus marinus*," isolated from the Snakepit hydrothermal site of the Mid-Atlantic Ridge have γ-irradiation resistance.[143,145]

Concluding Remarks

More than 300 species of thermophilic anaerobes have been described (TABLE 1). Many of the prokaryotes with this physiology were isolated from natural thermobiotic environments, such as terrestrial hot springs, deep-sea hydrothermal vents, and subsurface environments. Thermophilic anaerobes have also been isolated from anthropogenically heated environments, as well as from mesobiotic environments, such as rivers and lake sediments, and even psychrobiotic environments. The thermophilic anaerobes are metabolically diverse; a majority of isolated taxa are chemoorganoheterotrophic, although it is doubtful that this reflects the ecological situation in most thermobiotic environments. The examples of the phylogenetic and physiological diversity of thermophilic anaerobes given herein are certainly not exhaustive. Besides the diversity found so far, there are physiological types among mesophiles and potentially as-yet unknown or only superficially described physiologies for which no thermophiles have been isolated, probably because of a lack of serious isolation attempts. Thus, the authors believe that additional novel thermophilic anaerobes with unique and exciting properties will undoubtedly be isolated, especially considering the small percentage of assumed existing prokaryotes on Earth that have been described.[416]

Acknowledgments

Research in the laboratory of J.W. was supported by NSF grants MCB 0238407 (Kamchatka Microbial Observatory) and MCB 0348180 (Nevada Microbial Interactions and Processes). J.W. dedicates this chapter especially to Lars Ljungdahl, who introduced him to the world of thermophiles 30 years ago.

Conflict of Interest

The authors declare no conflicts of interest.

References

1. STETTER, K.O. 1999. Extremophiles and their adaptation to hot environments. FEBS Lett. **452:** 22–25.
2. STETTER, K.O. 2006. History of discovery of the first hyperthermophiles. Extremophiles **10:** 357–362.
3. LI, Y., L. MANDELCO & J. WIEGEL. 1993. Isolation and characterization of a moderately thermophilic anaerobic alkaliphile, *Clostridium paradoxum* sp. nov. Int. J. Syst. Bacteriol. **43:** 450–460.
4. BYRER, D.E., F.A. RAINEY & J. WIEGEL. 2000. Novel strains of *Moorella thermoacetica* form unusually heat-resistant spores. Arch. Microbiol. **174:** 334–339.
5. BLOCHL, E., R. RACHEL, S. BURGGRAF, *et al.* 1997. *Pyrolobus fumarii*, gen. and sp. nov., represents a novel group of archaea, extending the upper temperature limit for life to 113 degrees C. Extremophiles **1:** 14–21.
6. TAKAI, K., K.T. NAKAMURA, U. TSUNOGAI, *et al.* 2007. Methanogenesis at 122°C under high pressure produces isotopically 'abnormal' methane. *In* Thermophiles 2007: 9th International Conference on Thermophiles Research. University of Bergen, Bergen, Norway.
7. ENGLE, M., Y. LI, F. RAINEY, *et al.* 1996. *Thermobrachium celere* gen. nov., sp. nov., a rapidly growing thermophilic, alkalitolerant, and proteolytic obligate anaerobe. Int. J. Syst. Bacteriol. **46:** 1025–1033.
8. MESBAH, N.M., D.B. HEDRICK, A.D. PEACOCK, *et al.* 2007. *Natranaerobius thermophilus* gen. nov., sp. nov., a halophilic, alkalithermophilic bacterium from soda lakes of the Wadi An Natrun, Egypt, and proposal of *Natranaerobiaceae* fam. nov. and *Natranaerobiales* ord. nov. Int. J. Syst. Evol. Microbiol. **57:** 2507–2512.
9. VIEILLE, C. & G.J. ZEIKUS. 2001. Hyperthermophilic enzymes: sources, uses, and molecular mechanisms for thermostability. Microbiol. Mol. Biol. R. **65:** 1–43.
10. WIEGEL, J. & L.G. LJUNGDAHL. 1986. The importance of thermophilic bacteria in biotechnology. Crit. Rev. Biotechnol. **3:** 39–107.
11. UNSWORTH, L.D., J. VAN DER OOST & S. KOUTSOPOULOS. 2007. Hyperthermophilic enzymes—stability, activity and implementation strategies for high temperature applications. FEBS J. **274:** 4044–4056.
12. BAROSS, J.A. 1998. Do the geological and geochemical records of the early Earth support the prediction from global phylogenetic models of a thermophilic cenancestor? *In* Thermophiles: The Keys to Molecular Evolution and the Origin of Life? J. Wiegel & M.W.W. Adams, Eds.: 3–18. Taylor & Francis Ltd. London, UK.
13. KASTING, J.F. 1993. Earth's early atmosphere. Science **259:** 920–926.
14. CANFIELD, D.E., M.T. ROSING & C. BJERRUM. 2006. Early anaerobic metabolisms. Philos. T. R. Soc. B. **361:** 1819–1836.
15. KANDLER, O. 1998. The early diversification of life and the origin of the three domains: a proposal. *In* Thermophiles: The Keys to Molecular Evolution and the Origin of Life? J. Wiegel & M.W.W. Adams, Eds.: 19–31. Taylor & Francis Ltd. London, UK.
16. FORTERRE, P. 1998. Were our ancestors actually hyperthermophiles? Viewpoint of a devil's advocate. *In*

Thermophiles: The Keys to Molecular Evolution and the Origin of Life? J. Wiegel & M.W.W. Adams, Eds.: 137–146. Taylor & Francis Ltd. London, UK.

17. MILLER, S.L. & A. LAZCANO. 1998. Facing up to chemical realities: life did not begin at the growth temperatures of hyperthermophiles. *In* Thermophiles: The Keys to Molecular Evolution and the Origin of Life? J. Wiegel & M.W.W. Adams, Eds.: 127–133. Taylor & Francis Ltd. London, UK.
18. MILLER, S.L. & A. LAZCANO. 1995. The origin of life—did it occur at high temperatures? J. Mol. Evol. **41:** 689–692.
19. WÄCHTERSHÄUSER, G. 1998. The case for a hyperthermophilic, chemolithotrophic orign of life in an iron-sulfur world. *In* Thermophiles: The Keys to Molecular Evolution and the Origin of Life? J. Wiegel & M.W.W. Adams, Eds.: 47–57. Taylor & Francis Ltd. London, UK.
20. RUSSELL, M.J., D.E. DAIA & A.J. HALL. 1998. The emergence of life from FeS bubbles at alkaline hot springs in an acid ocean. *In* Thermophiles: The Keys to Molecular Evolution and the Origin of Life? J. Wiegel & M.W.W. Adams, Eds.: 77–126. Taylor & Francis Ltd. London, UK.
21. BROCK, T.D. 1970. High temperature systems. Annu. Rev. Ecol. Syst. **1:** 191–220.
22. STETTER, K.O. 1996. Hyperthermophilic procaryotes. FEMS Microbiol. Rev. **18:** 149–158.
23. SMITH, R.B. & L.J. SIEGEL. 2000. Windows Into the Earth: The Geologic Story of Yellowstone and Grand Teton National Parks. Oxford University Press. New York, New York.
24. STETTER, K.O. 2006. Hyperthermophiles in the history of life. Philos. T. R. Soc. B. **361:** 1837–1843.
25. CORLISS, J.B., J. DYMOND, L.I. GORDON, *et al.* 1979. Submarine thermal springs on the Galapagos Rift. Science **203:** 1073–1083.
26. REED, C. 2006. Boiling points. Nature **439:** 905–907.
27. OLLIVIER, B. & J.L. CAYOL. 2005. The fermentative, iron-reducing, and nitrate-reducing microorganisms. *In* Petroleum Microbiology. B. Ollivier & M. Magot, Eds.: 71–88. American Society for Microbiology. Washington, DC.
28. KIMURA, H., M. SUGIHARA, H. YAMAMOTO, *et al.* 2005. Microbial community in a geothermal aquifer associated with the subsurface of the Great Artesian Basin, Australia. Extremophiles **9:** 407–414.
29. THUMMES, K., J. SCHÄFER, P. KÄMPFER, *et al.* 2007. Thermophilic methanogenic *Archaea* in compost material: occurrence, persistence and possible mechanisms for their distribution to other environments. Syst. Appl. Microbiol. **30:** 634–643.
30. LEE, Y.J., I.D. WAGNER, M.E. BRICE, *et al.* 2005. *Thermosediminibacter oceani* gen. nov., sp. nov. and *Thermosediminibacter litoriperuensis* sp. nov., new anaerobic thermophilic bacteria isolated from Peru Margin. Extremophiles **9:** 375–383.
31. WIEGEL, J., M. BRAUN & G. GOTTSCHALK. 1981. *Clostridium thermoautotrophicum* species nova, a thermophile producing acetate from molecular hydrogen and carbon dioxide. Curr. Microbiol. **5:** 255–260.
32. MAGURRAN, A.E. 2004. Measuring Biological Diversity. Blackwell Science Ltd. Malden, MA.
33. HUGENHOLTZ, P. 2002. Exploring prokaryotic diversity in the genomic era. Genome Biol. **3:** 1–8.
34. BLANK, C.E., S.L. CADY & N.R. PACE. 2002. Microbial composition of near-boiling silica-depositing thermal springs throughout Yellowstone National Park. Appl. Environ. Microbiol. **68:** 5123–5135.
35. HUGENHOLTZ, P., C. PITULLE, K.L. HERSHBERGER, *et al.* 1998. Novel division level bacterial diversity in a Yellowstone hot spring. J. Bacteriol. **180:** 366–376.
36. BARNS, S.M., R.E. FUNDYGA, M.W. JEFFRIES, *et al.* 1994. Remarkable archaeal diversity detected in a Yellowstone National Park hot spring environment. Proc. Natl. Acad. Sci. USA **91:** 1609–1613.
37. REYSENBACH, A.L., M. EHRINGER & K. HERSHBERGER. 2000. Microbial diversity at 83°C in calcite springs, yellowstone national park: another environment where the *Aquificales* and "Korarchaeota" coexist. Extremophiles **4:** 61–67.
38. REYSENBACH, A.L., G.S. WICKHAM & N.R. PACE. Phylogenetic analysis of the hyperthermophilic pink filament community in Octopus Spring, Yellowstone National Park. Appl. Environ. Microbiol. **60:** 2113–2119.
39. YAMAMOTO, H., A. HIRAISHI, K. KATO, *et al.* 1998. Phylogenetic evidence for the existence of novel thermophilic bacteria in hot spring sulfur-turf microbial mats in Japan. Appl. Environ. Microbiol. **64:** 1680–1687.
40. HETZER, A., H.W. MORGAN, I.R. MCDONALD, *et al.* 2007. Microbial life in Champagne Pool, a geothermal spring in Waiotapu, New Zealand. Extremophiles **11:** 605–614.
41. ROESELERS, G., T.B. NORRIS, R.W. CASTENHOLZ, *et al.* 2007. Diversity of phototrophic bacteria in microbial mats from Arctic hot springs (Greenland). Environ. Microbiol. **9:** 26–38.
42. MARTEINSSON, V.T., S. HAUKSDOTTIR, C.F. HOBEL, *et al.* 2001. Phylogenetic diversity analysis of subterranean hot springs in Iceland. Appl. Environ. Microbiol. **67:** 4242–4248.
43. SKIRNISDOTTIR, S., G.O. HREGGVIDSSON, S. HJORLEIFSDOTTIR, *et al.* 2000. Influence of sulfide and temperature on species composition and community structure of hot spring microbial mats. Appl. Environ. Microbiol. **66:** 2835–2841.
44. KIMURA, H., M. SUGIHARA, H. YAMAMOTO, *et al.* 2005. Microbial community in a geothermal aquifer associated with the subsurface of the Great Artesian Basin, Australia. Extremophiles **9:** 407–414.
45. MESBAH, N.M., S.H. ABOU-EL-ELA & J. WIEGEL. 2007. Novel and unexpected prokaryotic diversity in water and sediments of the alkaline, hypersaline lakes of the Wadi An Natrun, Egypt. Microbial. Ecol. **54:** 598–617.
46. TAKAI, K., T. KOMATSU, F. INAGAKI, *et al.* 2001. Distribution of Archaea in a black smoker chimney structure. Appl. Environ. Microbiol. **67:** 3618–3629.
47. TAKAI, K. & K. HORIKOSHI. 1999. Genetic diversity of *Archaea* in deep-sea hydrothermal vent environments. Genetics **152:** 1285–1297.
48. REYSENBACH, A.L., K. LONGNECKER & J. KIRSHTEIN. 2000. Novel Bacterial and Archaeal lineages from an in

situ growth chamber deployed at a Mid-Atlantic Ridge hydrothermal vent. Appl. Environ. Microbiol. **66:** 3798–3806.

49. SUBBOTINA, I.V., N.A. CHERNYH, T.G. SOKOLOVA, *et al.* 2003. Oligonucleotide probes for the detection of representatives of the genus *Thermoanaerobacter*. Mikrobiologiya **72:** 331–339.
50. HRISTOVA, K.R., M. MAU, D. ZHENG, *et al.* 2000. *Desulfotomaculum* genus- and subgenus-specific 16S rRNA hybridization probes for environmental studies. Environ. Microbiol. **2:** 143–159.
51. VANDAMME, P., B. POT, M. GILLIS, *et al.* 1996. Polyphasic taxonomy, a consensus approach to bacterial systematics. Microbiol. Rev. **60:** 407–438.
52. STACKEBRANDT, E., W. FREDERIKSEN, G.M. GARRITY, *et al.* 2002. Report of the ad hoc committee for the re-evaluation of the species definition in bacteriology. Int. J. Syst. Evol. Microbiol. **52:** 1043–1047.
53. SANTOS, S.R. & H. OCHMAN. 2004. Identification and phylogenetic sorting of bacterial lineages with universally conserved genes and proteins. Environ. Microbiol. **6:** 754–759.
54. AMEND, J.P. & E.L. SHOCK. 2001. Energetics of overall metabolic reactions of thermophilic and hyperthermophilic Archaea and Bacteria. FEMS Microbiol Rev. **25:** 175–243.
55. HUBER, R., W. EDER, S. HELDWEIN, *et al.* 1998. *Thermocrinis ruber* gen. nov., sp. nov., A pink-filament-forming hyperthermophilic bacterium isolated from Yellowstone National Park. Appl. Environ. Microbiol. **64:** 3576–3583.
56. CAVICCHIOLI, R. & T. THOMAS. 2000. Extremophiles. *In* Encylopedia of Microbiology, Vol. 2. J. Lederberg, Ed.: 317–337. Academic Press. San Diego, CA.
57. WIEGEL, J. 1990. Temperature spans for growth: hypothesis and discussion. FEMS Microbiol. Rev. **75:** 155–170.
58. WIEGEL, J. 1998. Anaerobic alkalithermophiles, a novel group of extremophiles. Extremophiles **2:** 257–267.
59. YAMADA, T., Y. SEKIGUCHI, S. HANADA, *et al.* 2006. *Anaerolinea thermolimosa* sp. nov., *Levilinea saccharolytica* gen. nov., sp. nov. and *Leptolinea tardivitalis* gen. nov., sp. nov., novel filamentous anaerobes, and description of the new classes *Anaerolineae* classis nov. and *Caldilineae* classis nov. in the bacterial phylum *Chloroflexi*. Int. J. Syst. Evol. Microbiol. **56:** 1331–1340.
60. SEKIGUCHI, Y., T. YAMADA, S. HANADA, *et al.* 2003. *Anaerolinea thermophila* gen. nov., sp. nov. and *Caldilinea aerophila* gen. nov., sp. nov., novel filamentous thermophiles that represent a previously uncultured lineage of the domain *Bacteria* at the subphylum level. Int. J. Syst. Evol. Microbiol. **53:** 1843–1851.
61. ADAMS, M.W.W. 1994. Biochemical diversity among sulfur-dependent, hyperthermophilic microorganisms. FEMS Microbiol. Rev. **15:** 261–277.
62. PLEY, U., J. SCHIPKA, A. GAMBACORTA, *et al.* 1991. *Pyrodictium abyssi* sp. nov. represents a novel heterotrophic marine archaeal hyperthermophile growing at 110°C. Syst. Appl. Bacteriol. **14:** 245–253.
63. STETTER, K.O., H. KONIG & E. STACKEBRANDT. 1983. *Pyrodictium* gen. nov., a new genus of submarine disc-shaped sulphur reducing archaebacteria growing optimally at 105°C. Syst. Appl. Microbiol. **4:** 535–551.
64. ZILLIG, W., I. HOLZ & S. WUNDERL. 1991. *Hyperthermus butylicus* gen. nov., sp. nov., a hyperthermophilic, anaerobic, peptide-fermenting, facultatively H_2S-generating archaebacterium. Int. J. Syst. Bacteriol. **41:** 169–170.
65. POND, J.L. & T.A. LANGWORTHY. 1987. Effect of growth temperature on the long-chain diols and fatty acids of *Thermomicrobium roseum*. J. Bacteriol. **169:** 1328–1330.
66. VAN DE VOSSENBERG, J.L., T. UBBINK-KOK, M.G. ELFERINK, *et al.* 1995. Ion permeability of the cytoplasmic membrane limits the maximum growth temperature of bacteria and archaea. Mol. Microbiol. **18:** 925–932.
67. WIEGEL, J. 1998. Lateral gene exchange, an evolutionary mechanism for extending the upper or lower temperature limits for growth of microorganisms? A hypothesis. *In* Annu. Rev. Ecol. Syst. J. Wiegel & M.W.W. Adams, Eds.: 177–185. Taylor & Francis Ltd. London, UK.
68. LIU, S.Y., F.A. RAINEY, H.W. MORGAN, *et al.* 1996. *Thermoanaerobacterium aotearoense* sp. nov., a slightly acidophilic, anaerobic thermophile isolated from various hot springs in New Zealand, and emendation of the genus *Thermoanaerobacterium*. Int. J. Syst. Bacteriol. **46:** 388–396.
69. KUBLANOV, I.V., M.I. PROKOFEVA, N.A. KOSTRIKINA, *et al.* 2007. *Thermoanaerobacterium aciditolerans* sp. nov., a moderate thermoacidophile from a Kamchatka hot spring. Int. J. Syst. Evol. Microbiol. **57:** 260–264.
70. TAKAI, K., H. HIRAYAMA, T. NAKAGAWA, *et al.* 2005. *Lebetimonas acidiphila* gen. nov., sp. nov., a novel thermophilic, acidophilic, hydrogen-oxidizing chemolithoautotroph within the '*Epsilonproteobacteria*', isolated from a deep-sea hydrothermal fumarole in the Mariana Arc. Int. J. Syst. Evol. Microbiol. **55:** 183–189.
71. PLUMB, J.J., C.M. HADDAD, J.A. GIBSON, *et al.* 2007. *Acidianus sulfidivorans* sp. nov., an extremely acidophilic, thermophilic archaeon isolated from a solfatara on Lihir Island, Papua New Guinea, and emendation of the genus description. Int. J. Syst. Evol. Microbiol. **57:** 1418–1423.
72. SEGERER, A.H., A. TRINCONE, M. GAHRTZ, *et al.* 1991. *Stygiolobus azoricus* gen. nov., sp. nov. represents a novel genus of anaerobic, extremely thermoacidophilic archaebacteria of the order *Sulfolobales*. Int. J. Syst. Bacteriol. **41:** 495–501.
73. KOTELNIKOVA, S.V., A.Y. OBRAZTSOVA, G.M. GONGADZE, *et al.* 1993. *Methanobacterium thermoflexum* sp. nov. and *Methanobacterium defluvii* sp. nov., thermophilic rodshaped methanogens isolated from anaerobic digester sludge. Syst. Appl. Microbiol. **16:** 427–435.
74. BOONE, D.R. 2001. Genus IV. *Methanothermobacter* Wasserfallen, Nölling, Pfister, Reeve and Conway de Macario 2000. *In* Bergey's Manual of Systematic Bacteriology, Vol. 1. D.R. Boone, R.W. Castenholz & G.M. Garrity, Eds.: 230–233. Springer-Verlag. New York, NY.
75. DIRMEIER, R., M. KELLER, D. HAFENBRADL, *et al.* 1998. *Thermococcus acidaminovorans* sp. nov., a new hyperthermophilic alkalophilic archaeon growing on amino acids. Extremophiles **2:** 109–114.
76. GODFROY, A., J.R. MEUNIER, J. GUEZENNEC, *et al.* 1996. *Thermococcus fumicolans* sp. nov., a new hyperthermophilic archaeon isolated from a deep-sea hydrothermal vent in

the north Fiji Basin. Int. J. Syst. Bacteriol. **46:** 1113–1119.

77. ZILLIG, W., L. HOLZ, D. JANEKOVIC, *et al.* 1983. The archaebacterium *Thermococcus celer* represents a novel genus within the thermophilic branch of the archaebacteria. Syst. Appl. Microbiol. **4:** 88–94.
78. KELLER, M., F.J. BRAUN, R. DIRMEIER, *et al.* 1995. *Thermococcus alcaliphilus* sp. nov., a new hyperthermophilic archaeum growing on polysulfide at alkaline pH. Arch. Microbiol. **164:** 390–395.
79. DUCKWORTH, A.W., W.D. GRANT, B.E. JONES, *et al.* 1996. Phylogenetic diversity of soda lake alkaliphiles. FEMS Microbiol. Ecol. **19:** 181–191.
80. LI, Y., M. ENGLE, N. WEISS, *et al.* 1994. *Clostridium thermoalcaliphilum* sp. nov., an anaerobic and thermotolerant facultative alkaliphile. Int. J. Syst. Bacteriol. **44:** 111–118.
81. WIEGEL, J. & V.V. KEVBRIN. 2004. Alkalithermophiles. Biochem. Soc. Trans. **32:** 193–198.
82. GODFROY, A., F. LESONGEUR, G. RAGUENES, *et al.* 1997. *Thermococcus hydrothermalis* sp. nov., a new hyperthermophilic archaeon isolated from a deep-sea hydrothermal vent. Int. J. Syst. Bacteriol. **47:** 622–626.
83. BARBIER, G., A. GODFROY, J.R. MEUNIER, *et al.* 1999. *Pyrococcus glycovorans* sp. nov., a hyperthermophilic archaeon isolated from the East Pacific Rise. Int. J. Syst. Bacteriol. **49:** 1829–1837.
84. WIEGEL, J. & L.G. LJUNGDAHL. 1981. *Thermoanaerobacter ethanolicus* gen. nov., spec. nov., a new, extreme thermophilic, anaerobic bacterium. Arch. Microbiol. **128:** 343–348.
85. IMACHI, H., Y. SEKIGUCHI, Y. KAMAGATA, *et al.* 2002. *Pelotomaculum thermopropionicum* gen. nov., sp. nov., an anaerobic, thermophilic, syntrophic propionate-oxidizing bacterium. Int. J. Syst. Evol. Microbiol. **52:** 1729–1735.
86. MOUSSARD, H., S. L'HARIDON, B.J. TINDALL, *et al.* 2004. *Thermodesulfatator indicus* gen. nov., sp. nov., a novel thermophilic chemolithoautotrophic sulfate-reducing bacterium isolated from the Central Indian Ridge. Int. J. Syst. Evol. Microbiol. **54:** 227–233.
87. JEANTHON, C., S. L'HARIDON, V. CUEFF, *et al.* 2002. *Thermodesulfobacterium hydrogeniphilum* sp. nov., a thermophilic, chemolithoautotrophic, sulfate-reducing bacterium isolated from a deep-sea hydrothermal vent at Guaymas Basin, and emendation of the genus *Thermodesulfobacterium.* Int. J. Syst. Evol. Microbiol. **52:** 765–772.
88. SIEBERS, B. & P. SCHÖNHEIT. 2005. Unusual pathways and enzymes of central carbohydrate metabolism in Archaea. Curr. Opin. Microbiol. **8:** 695–705.
89. WIEGEL, J. 1980. Formation of ethanol by bacteria: a pledge for the use of extreme thermophilic anaerobic bacteria in industrial ethanol fermentation processes. Cell. Mol. Life Sci. **36:** 1434–1446.
90. FREIER, D., C.P. MOTHERSHED & J. WIEGEL. 1988. Characterization of *Clostridium thermocellum* JW20. Appl. Environ. Microbiol. **54:** 204–211.
91. MENDEZ, B.S., M.J. PETTINARI, S.E. IVANIER, *et al.* 1991. *Clostridium thermopapyrolyticum* sp. nov., a cellulolytic thermophile. Int. J. Syst. Bacteriol. **41:** 281–283.
92. PEREVALOVA, A.A., V.A. SVETLICHNY, I.V. KUBLANOV, *et al.* 2005. *Desulfurococcus fermentans* sp. nov., a novel hyperthermophilic archaeon from a Kamchatka hot spring, and emended description of the genus *Desulfurococcus*. Int. J. Syst. Evol. Microbiol. **55:** 995–999.
93. BIELY, P. 1985. Microbial xylanolytic systems. Trends Biotechnol. **3:** 286–290.
94. ALAZARD, D., S. DUKAN, A. URIOS, *et al.* 2003. *Desulfovibrio hydrothermalis* sp. nov., a novel sulfate-reducing bacterium isolated from hydrothermal vents. Int. J. Syst. Evol. Microbiol. **53:** 173–178.
95. SVETLITSHNYI, V., F. RAINEY & J. WIEGEL. 1996. *Thermosyntropha lipolytica* gen. nov., sp. nov., a lipolytic, anaerobic, alkalitolerant, thermophilic bacterium utilizing short- and long-chain fatty acids in syntrophic coculture with a methanogenic archaeum. Int. J. Syst. Bacteriol. **46:** 1131–1137.
96. SOKOLOVA, T.G., C. JEANTHON, N.A. KOSTRIKINA, *et al.* 2004. The first evidence of anaerobic CO oxidation coupled with H_2 production by a hyperthermophilic archaeon isolated from a deep-sea hydrothermal vent. Extremophiles **8:** 317–323.
97. PARSHINA, S.N., J. SIPMA, Y. NAKASHIMADA, *et al.* 2005. *Desulfotomaculum carboxydivorans* sp. nov., a novel sulfate-reducing bacterium capable of growth at 100% CO. Int. J. Syst. Evol. Microbiol. **55:** 2159–2165.
98. SVETLICHNY, V.A., T.G. SOKOLOVA, M. GERHARDT, *et al.* 1991. *Carboxydothermus hydrogenoformans* gen. nov., sp. nov., a CO-utilizing thermophilic anaerobic bacterium from hydrothermal environments of Kunashir Island. Syst. Appl. Microbiol. **14:** 254–260.
99. SOKOLOVA, T., J. HANEL, R.U. ONYENWOKE, *et al.* 2007. Novel chemolithotrophic, thermophilic, anaerobic bacteria *Thermolithobacter ferrireducens* gen. nov., sp. nov. and *Thermolithobacter carboxydivorans* sp. nov. Extremophiles **11:** 145–157.
100. SOKOLOVA, T.G., N.A. KOSTRIKINA, N.A. CHERNYH, *et al.* 2002. *Carboxydocella thermautotrophica* gen. nov., sp. nov., a novel anaerobic, CO-utilizing thermophile from a Kamchatkan hot spring. Int. J. Syst. Evol. Microbiol. **52:** 1961–1967.
101. SOKOLOVA, T.G., N.A. KOSTRIKINA, N.A. CHERNYH, *et al.* 2005. *Thermincola carboxydiphila* gen. nov., sp. nov., a novel anaerobic, carboxydotrophic, hydrogenogenic bacterium from a hot spring of the Lake Baikal area. Int. J. Syst. Evol. Microbiol. **55:** 2069–2073.
102. ZAVARZINA, D.G., T.G. SOKOLOVA, T.P. TOUROVA, *et al.* 2007. *Thermincola ferriacetica* sp. nov., a new anaerobic, thermophilic, facultatively chemolithoautotrophic bacterium capable of dissimilatory Fe(III) reduction. Extremophiles **11:** 1–7.
103. FARDEAU, M.L., M. BONILLA SALINAS, S. L'HARIDON, *et al.* 2004. Isolation from oil reservoirs of novel thermophilic anaerobes phylogenetically related to *Thermoanaerobacter subterraneus*: reassignment of *T. subterraneus, Thermoanaerobacter yonseiensis, Thermoanaerobacter tengcongensis* and *Carboxydibrachium pacificum* to *Caldanaerobacter subterraneus* gen. nov., sp. nov., comb. nov. as four novel subspecies. Int. J. Syst. Evol. Microbiol. **54:** 467–474.
104. SOKOLOVA, T.G., J.M. GONZALEZ, N.A. KOSTRIKINA, *et al.* 2004. *Thermosinus carboxydivorans* gen. nov., sp. nov., a new anaerobic, thermophilic, carbon-monoxide-

oxidizing, hydrogenogenic bacterium from a hot pool of Yellowstone National Park. Int. J. Syst. Evol. Microbiol. **54:** 2353–2359.

105. Burggraf, S., H.W. Jannasch, B. Nicolaus, *et al.* 1990. *Archaeoglobus profundus* sp. nov., represents a new species within the sulfur-reducing *Archaebacteria*. Syst. Appl. Microbiol. **13:** 24–28.

106. Jochimsen, B., S. Peinemann-Simon, H. Volker, *et al.* 1997. *Stetteria hydrogenophila*, gen. nov. and sp. nov., a novel mixotrophic sulfur-dependent crenarchaeote isolated from Milos, Greece. Extremophiles **1:** 67–73.

107. Pikuta, E., A. Lysenko, N. Suzina, *et al.* 2000. *Desulfotomaculum alkaliphilum* sp. nov., a new alkaliphilic, moderately thermophilic, sulfate-reducing bacterium. Int. J. Syst. Evol. Microbiol. **50:** 25–33.

108. Miroshnichenko, M.L., N.A. Kostrikina, N.A. Chernyh, *et al.* 2003. *Caldithrix abyssi* gen. nov., sp. nov., a nitrate-reducing, thermophilic, anaerobic bacterium isolated from a Mid-Atlantic Ridge hydrothermal vent, represents a novel bacterial lineage. Int. J. Syst. Evol. Microbiol. **53:** 323–329.

109. Miroshnichenko, M.L., S. L'Haridon, O. Nercessian, *et al.* 2003. *Vulcanithermus mediatlanticus* gen. nov., sp. nov., a novel member of the family *Thermaceae* from a deep-sea hot vent. Int. J. Syst. Evol. Microbiol. **53:** 1143–1148.

110. Miroshnichenko, M.L., S. L'Haridon, C. Jeanthon, *et al.* 2003. *Oceanithermus profundus* gen. nov., sp. nov., a thermophilic, microaerophilic, facultatively chemolithoheterotrophic bacterium from a deep-sea hydrothermal vent. Int. J. Syst. Evol. Microbiol. **53:** 747–752.

111. Bryant, D.A. & N.U. Frigaard. 2006. Prokaryotic photosynthesis and phototrophy illuminated. Trends Microbiol. **14:** 488–496.

112. Xiong, J. & C.E. Bauer. 2002. Complex evolution of photosynthesis. Annu. Rev. Plant Biol. **53:** 503–521.

113. Imhoff, J.F., J. Suling & R. Petri. 1998. Phylogenetic relationships among the *Chromatiaceae*, their taxonomic reclassification and description of the new genera *Allochromatium, Halochromatium, Isochromatium, Marichromatium, Thiococcus, Thiohalocapsa* and *Thermochromatium*. Int. J. Syst. Bacteriol. **48:** 1129–1143.

114. Madigan, M.T. 1986. *Chromatium tepidum* sp. nov., a thermophilic photosynthetic bacterium of the family *Chromatiaceae*. Int. J. Syst. Bacteriol. **36:** 222–227.

115. Hanada, S., S. Takaichi, K. Matsuura, *et al.* 2002. *Roseiflexus castenholzii* gen. nov., sp. nov., a thermophilic, filamentous, photosynthetic bacterium that lacks chlorosomes. Int. J. Syst. Evol. Microbiol. **52:** 187–193.

116. Hanada, S., A. Hiraishi, K. Shimada, *et al.* 1995. *Chloroflexus aggregans* sp. nov., a filamentous phototrophic bacterium which forms dense cell aggregates by active gliding movement. Int. J. Syst. Evol. Microbiol. **45:** 676–681.

117. Pierson, B.K. & R.W. Castenholz. 1974. A phototrophic gliding filamentous bacterium of hot springs, *Chloroflexus aurantiacus*, gen. and sp. nov. Arch. Microbiol. **100:** 5–24.

118. Pierson, B.K., S.J. Giovannoni & R.W. Castenholz. 1984. Physiological ecology of a gliding bacterium containing bacteriochlorophyll *a*. Appl. Environ. Microbiol. **47:** 576–584.

119. Pierson, B.K., S.J. Giovannoni, D.A. Stahl, *et al.* 1985. *Heliothrix oregonensis*, gen. nov., sp. nov., a phototrophic filamentous gliding bacterium containing bacteriochlorophyll *a*. Arch. Microbiol. **142:** 164–167.

120. Kimble, L.K., L. Mandelco, C.R. Woese, *et al.* 1995. *Heliobacterium modesticaldum*, sp. nov., a thermophilic heliobacterium of hot springs and volcanic soils. Arch. Microbiol. **163:** 259–267.

121. Bonch-Osmolovskaya, E.A. & K.O. Stetter. 1991. Interspecies hydrogen transfer in cocultures of thermophilic Archaea. Syst. Appl. Microbiol. **14:** 205–208.

122. Huber, H., M.J. Hohn, R. Rachel, *et al.* 2002. A new phylum of *Archaea* represented by a nanosized hyperthermophilic symbiont. Nature **417:** 63–67.

123. Waters, E., M.J. Hohn, I. Ahel, *et al.* 2003. The genome of *Nanoarchaeum equitans*: Insights into early archaeal evolution and derived parasitism. Proc. Natl. Acad. Sci. USA **100:** 12984–12988.

124. Goorissen, H.P., H.T. Boschker, A.J. Stams, *et al.* 2003. Isolation of thermophilic *Desulfotomaculum* strains with methanol and sulfite from solfataric mud pools, and characterization of *Desulfotomaculum solfataricum* sp. nov. Int. J. Syst. Evol. Microbiol. **53:** 1223–1229.

125. Stackebrandt, E., C. Sproer, F.A. Rainey, *et al.* 1997. Phylogenetic analysis of the genus *Desulfotomaculum*: evidence for the misclassification of *Desulfotomaculum guttoideum* and description of *Desulfotomaculum orientis* as *Desulfosporosinus orientis* gen. nov., comb. nov. Int. J. Syst. Bacteriol. **47:** 1134–1139.

126. Miroshnichenko, M.L., G.M. Gongadze, A.M. Lysenko, *et al.* 1994. *Desulfurella multipotens* sp. nov., a new sulfur-respiring thermophilic eubacterium from Raoul Island (Kermadec archipelago, New Zealand). Arch. Microbiol. **161:** 88–93.

127. Burgess, E.A., I.D. Wagner & J. Wiegel. 2007. Thermal environments and biodiversity. *In* Physiology and Biochemistry of Extremophiles. C. Gerday & N. Glansdorff, Eds.: 13–29. ASM Press. Washington, DC.

128. Slobodkin, A.I. 2005. Thermophilic microbial metal reduction. Mikrobiologiya **74:** 581–595.

129. Kashefi, K., J.M. Tor, D.E. Holmes, *et al.* 2002. *Geoglobus ahangari* gen. nov., sp. nov., a novel hyperthermophilic archaeon capable of oxidizing organic acids and growing autotrophically on hydrogen with Fe(III) serving as the sole electron acceptor. Int. J. Syst. Evol. Microbiol. **52:** 719–728.

130. Greene, A.C., B.K. Patel & A.J. Sheehy. 1997. *Deferribacter thermophilus* gen. nov., sp. nov., a novel thermophilic manganese- and iron-reducing bacterium isolated from a petroleum reservoir. Int. J. Syst. Bacteriol. **47:** 505–509.

131. Miroshnichenko, M.L., A.I. Slobodkin, N.A. Kostrikina, *et al.* 2003. *Deferribacter abyssi* sp. nov., an anaerobic thermophile from deep-sea hydrothermal vents of the Mid-Atlantic Ridge. Int. J. Syst. Evol. Microbiol. **53:** 1637–1641.

132. Segerer, A., A.M. Neuner, J.K. Kristjansson, *et al.* 1986. *Acidianus infernus* gen. nov., sp. nov., and *Acidianus brierleyi* comb. nov.: facultatively aerobic, extremely acidophilic thermophilic sulfur-metabolizing archaebacteria. Int. J. Syst. Bacteriol. **36:** 559–564.

133. SLOBODKIN, A.I., T.P. TOUROVA, B.B. KUZNETSOV, *et al.* 1999. *Thermoanaerobacter siderophilus* sp. nov., a novel dissimilatory Fe(III)-reducing, anaerobic, thermophilic bacterium. Int. J. Syst. Bacteriol. **49:** 1471–1478.

134. ZAVARZINA, D.G., T.P. TOUROVA, B.B. KUZNETSOV, *et al.* 2002. *Thermovenabulum ferriorganovorum* gen. nov., sp. nov., a novel thermophilic, anaerobic, endospore-forming bacterium. Int. J. Syst. Evol. Microbiol. **52:** 1737–1743.

135. BRIERLEY, C.L. & J.A. BRIERLEY. 1982. Anaerobic reduction of molybdenum by *Sulfolobus* species. Zbl. Bakt. Mik. Hyg. I. C. **3:** 289–294.

136. ROH, Y., S.V. LIU, G. LI, *et al.* 2002. Isolation and characterization of metal-reducing *Thermoanaerobacter* strains from deep subsurface nnvironments of the Piceance Basin, Colorado. Appl. Environ. Microbiol. **68:** 6013–6020.

137. HUBER, R., M. SACHER, A. VOLLMANN, *et al.* 2000. Respiration of arsenate and selenate by hyperthermophilic archaea. Syst. Appl. Microbiol. **23:** 305–314.

138. NAKAGAWA, S. & K. TAKAI. 2006. Methods for the isolation of thermophiles from deep-sea hydrothermal environments. Methods Microbiol. **35:** 55–92.

139. MILLER, J.F., N.N. SHAH, C.M. NELSON, *et al.* 1988. Pressure and temperature effects on growth and methane production of the extreme thermophile *Methanococcus jannaschii*. Appl. Environ. Microbiol. **54:** 3039–3042.

140. MARTEINSSON, V.T., J.L. BIRRIEN, A.L. REYSENBACH, *et al.* 1999. *Thermococcus barophilus* sp. nov., a new barophilic and hyperthermophilic archaeon isolated under high hydrostatic pressure from a deep-sea hydrothermal vent. Int. J. Syst. Bacteriol. **49:** 351–359.

141. HOLDEN, J.F. & J.A. BAROSS. 1995. Enhanced thermotolerance by hydrostatic pressure in the deep-sea hyperthermophile *Pyrococcus* strain ES 4. FEMS Microbiol. Ecol. **18:** 27–34.

142. ERAUSO, G., A.L. REYSENBACH, A. GODFROY, *et al.* 1993. *Pyrococcus abyssi* sp. nov., a new hyperthermophilic archaeon isolated from a deep-sea hydrothermal vent. Arch. Microbiol. **160:** 338–349.

143. JOLIVET, E., S. L'HARIDON, E. CORRE, *et al.* 2003. *Thermococcus gammatolerans* sp. nov., a hyperthermophilic archaeon from a deep-sea hydrothermal vent that resists ionizing radiation. Int. J. Syst. Evol. Microbiol. **53:** 847–851.

144. CHERRY, R., D. DESBRUYERES, M. HEYRAUD, *et al.* 1992. High levels of natural radioactivity in hydrothermal vent polychaetes. C. R. Acad, Sci, III-Vie. **315:** 21–26.

145. JOLIVET, E., E. CORRE, S. L'HARIDON, *et al.* 2004. *Thermococcus marinus* sp. nov. and *Thermococcus radiotolerans* sp. nov., two hyperthermophilic archaea from deep-sea hydrothermal vents that resist ionizing radiation. Extremophiles **8:** 219–227.

146. SLOBODKIN, A.I., T.P. TOUROVA, N.A. KOSTRIKINA, *et al.* 2006. *Tepidimicrobium ferriphilum* gen. nov., sp. nov., a novel moderately thermophilic, Fe(III)-reducing bacterium of the order *Clostridiales*. Int. J. Syst. Evol. Microbiol. **56:** 369–372.

147. THOMPSON, J.D., D.G. HIGGINS & T.J. GIBSON. 1994. CLUSTAL W: improving the sensitivity of progressive multiple sequence alignment through sequence weighting, position-specific gap penalties and weight matrix choice. Nucleic Acids Res. **22:** 4673–4680.

148. SAITOU, N. & M. NEI. 1987. The neighbor-joining method: a new method for reconstructing phylogenetic trees. Mol. Biol. Evol. **4:** 406–425.

149. JUKES, T.H. & C.R. CANTOR. 1969. Evolution of protein molecules. *In* Mammalian Protein Metabolism. H.N. Munro, Ed.: 21–132. Academic Press. New York, NY.

150. FELSENSTEIN, J. 2001. PHYLIP (Phylogenetic Inference Package) Version 3.6a2.1. Department of Genome Sciences, University of Washington, Seattle, WA.

151. KELLY, D.P. & A.P. WOOD. 2000. Reclassification of some species of *Thiobacillus* to the newly designated genera *Acidithiobacillus* gen. nov., *Halothiobacillus* gen. nov. and *Thermithiobacillus* gen. nov. Int. J. Syst. Evol. Microbiol. **50:** 511–516.

152. WOOD, A.P. & D.P. KELLY. 1985. Physiological characteristics of a new thermophilic obligately chemolithotrophic *Thiobacillus* species, *Thiobacillus tepidarius*. Int. J. Syst. Bacteriol. **35:** 434–437.

153. CADWELL, D.E., S.J. CADWELL & J.P. LAYLOCK. 1976. *Thermothrix thiopara* gen. et sp. nov., a facultatively anaerobic facultative chemolithotroph living at neutral pH and high temperature. Can. J. Microbiol. **22:** 1509–1517.

154. HIRAYAMA, H., K. TAKAI, F. INAGAKI, *et al.* 2005. *Thiobacter subterraneus* gen. nov., sp. nov., an obligately chemolithoautotrophic, thermophilic, sulfur-oxidizing bacterium from a subsurface hot aquifer. Int. J. Syst. Evol. Microbiol. **55:** 467–472.

155. STOUTSCHEK, E., J. WINTER, F. SCHINDLER, *et al.* 1984. *Acetomicrobium flavidum*, gen. nov., sp. nov., a thermophilic, anaerobic bacterium from sewage sludge, forming acetate, CO_2 and H_2 from glucose. Syst. Appl. Microbiol. **5:** 377–390.

156. WINTER, J., E. BRAUN & H.P. ZABEL. 1987. *Acetomicrobium faecalis* spec. nov., a strictly anaerobic bacterium from sewage sludge, producing ethanol from pentoses. Syst. Appl. Microbiol. **9:** 71–76.

157. DENGER, K., R. WARTHMANN, W. LUDWIG, *et al.* 2002. *Anaerophaga thermohalophila* gen. nov., sp. nov., a moderately thermohalophilic, strictly anaerobic fermentative bacterium. Int. J. Syst. Evol. Microbiol. **52:** 173–178.

158. POHLSCHROEDER, M., S.B. LESCHINE & E. CANALE-PAROLA. 1994. *Spirochaeta caldaria* sp. nov., a thermophilic bacterium that enhances cellulose degradation by *Clostridium thermocellum*. Arch. Microbiol. **161:** 17–24.

159. AKSENOVA, H.Y., F.A. RAINEY, P.H. JANSSEN, *et al.* 1992. *Spirochaeta thermophila* sp. nov., an obligately anaerobic, polysaccharolytic, extremely thermophilic bacterium. Int. J. Syst. Bacteriol. **42:** 175–177.

160. CAYOL, J.L., B. OLLIVIER, B.K. PATEL, *et al.* 1994. Isolation and characterization of *Halothermothrix orenii* gen. nov., sp. nov., a halophilic, thermophilic, fermentative, strictly anaerobic bacterium. Int. J. Syst. Bacteriol. **44:** 534–540.

161. ETCHEBEHERE, C., M.E. PAVAN, J. ZORZOPULOS, *et al.* 1998. *Coprothermobacter platensis* sp. nov., a new anaerobic proteolytic thermophilic bacterium isolated from an anaerobic mesophilic sludge. Int. J. Syst. Bacteriol. **48:** 1297–1304.

162. OLLIVIER, B.M., R.A. MAH, T.J. FERGUSON, *et al.* 1985. Emendation of the genus *Thermobacteroides*: *Thermobacteroides proteolyticus* sp. nov., a proteolytic acetogen from a methanogenic enrichment. Int. J. Syst. Bacteriol. **35:** 425–428.

163. RAINEY, F.A. & E. STACKEBRANDT. 1993. Transfer of the type species of the genus *Thermobacteroides* to the genus *Thermoanaerobacter* as *Thermoanaerobacter acetoethylicus* (Ben-Bassat and Zeikus 1981) comb. nov., description of *Coprothermobacter* gen. nov., and reclassification of *Thermobacteroides proteolyticus* as *Coprothermobacter proteolyticus* (Ollivier et al. 1985) comb. nov. Int. J. Syst. Bacteriol. **43:** 857–859.

164. PLUGGE, C.M., M. BALK, E.G. ZOETENDAL, *et al.* 2002. *Gelria glutamica* gen. nov., sp. nov., a thermophilic, obligately syntrophic, glutamate-degrading anaerobe. Int. J. Syst. Evol. Microbiol. **52:** 401–407.

165. SLOBODKIN, A., A.L. REYSENBACH, F. MAYER, *et al.* 1997. Isolation and characterization of the homoacetogenic thermophilic bacterium *Moorella glycerini* sp. nov. Int. J. Syst. Bacteriol. **47:** 969–974.

166. COLLINS, M.D., P.A. LAWSON, A. WILLEMS, *et al.* 1994. The phylogeny of the genus *Clostridium*: proposal of five new genera and eleven new species combinations. Int. J. Syst. Bacteriol. **44:** 812–826.

167. BALK, M., J. WEIJMA, M.W. FRIEDRICH, *et al.* 2003. Methanol utilization by a novel thermophilic homoacetogenic bacterium, *Moorella mulderi* sp. nov., isolated from a bioreactor. Arch. Microbiol. **179:** 315–320.

168. FONTAINE, F.E., W.H. PETERSON, E. MCCOY, *et al.* 1942. A new type of glucose fermentation by *Clostridium thermoaceticum* n. sp. J. Bacteriol. **43:** 701–715.

169. HATTORI, S., Y. KAMAGATA, S. HANADA, *et al.* 2000. *Thermacetogenium phaeum* gen. nov., sp. nov., a strictly anaerobic, thermophilic, syntrophic acetate-oxidizing bacterium. Int. J. Syst. Evol. Microbiol. 50: 1601–1609.

170. SALINAS, M.B., M.L. FARDEAU, P. THOMAS, *et al.* 2004. *Mahella australiensis* gen. nov., sp. nov., a moderately thermophilic anaerobic bacterium isolated from an Australian oil well. Int. J. Syst. Evol. Microbiol. **54:** 2169–2173.

171. LEE, Y.E., M.K. JAIN, C. LEE, *et al.* 1993. Taxonomic distinction of saccharolytic thermophilic anaerobes: description of *Thermoanaerobacterium xylanolyticum* gen. nov., sp. nov., and *Thermoanaerobacterium saccharolyticum* gen. nov., sp. nov.; reclassification of *Thermoanaerobium brockii*, *Clostridium thermosulfurogenes*, and *Clostridium thermohydrosulfuricum* E100-69 as *Thermoanaerobacter brockii* comb. nov., *Thermoanaerobacterium thermosulfurigenes* comb. nov., and *Thermoanaerobacter thermohydrosulfuricus* comb. nov., respectively; and transfer of *Clostridium thermohydrosulfuricum* 39E to *Thermoanaerobacter ethanolicus*. Int. J. Syst. Bacteriol. **43:** 41–51.

172. CANN, I.K., P.G. STROOT, K.R. MACKIE, *et al.* 2001. Characterization of two novel saccharolytic, anaerobic thermophiles, *Thermoanaerobacterium polysaccharolyticum* sp. nov. and *Thermoanaerobacterium zeae* sp. nov., and emendation of the genus *Thermoanaerobacterium*. Int. J. Syst. Evol. Microbiol. **51:** 293–302.

173. McClung, L.S. 1935. Studies on anaerobic Bacteria: IV. Taxonomy of cultures of a thermophilic species causing "swells" of canned foods. J. Bacteriol. **29:** 189–203.

174. CAYOL, J.L., B. OLLIVIER, B.K. PATEL, *et al.* 1995. Description of *Thermoanaerobacter brockii* subsp. *lactiethylicus* subsp. nov., isolated from a deep subsurface French oil well, a proposal to reclassify *Thermoanaerobacter finnii* as *Thermoanaerobacter brockii* subsp. *finnii* comb. nov., and an emended description of *Thermoanaerobacter brockii*. Int. J. Syst. Bacteriol. **45:** 783–789.

175. JIN, F., K. YAMASATO & K. TODA. 1988. *Clostridium thermocopriae* sp. nov., a cellulolytic thermophile from animal feces, compost, soil, and a hot spring in Japan. Int. J. Syst. Bacteriol. **38:** 279–281.

176. BEN-BASSAT, A. & J.G. ZEIKUS. 1981. *Thermobacteroides acetoethylicus* gen. nov. and spec. nov., a new chemoorganotrophic, anaerobic, thermophilic bacterium. Arch. Microbiol. **128:** 365–370.

177. LEIGH, J.A. & R.S. WOLFE. 1983. *Acetogenium kivui* gen. nov., sp. nov., a thermophilic acetogenic bacterium. Int. J. Syst. Bacteriol. **33:** 886.

178. COOK, G.M., F.A. RAINEY, B.K. PATEL, *et al.* 1996. Characterization of a new obligately anaerobic thermophile, *Thermoanaerobacter wiegelii* sp. nov. Int. J. Syst. Bacteriol. **46:** 123–127.

179. ZACHAROVA, E.V., T.I. MITROFANOVA, E.N. KRASILNIKOVA, *et al.* 1993. *Thermohydrogenium kirishiense* gen. nov. and sp. nov., a new anaerobic thermophilic bacterium. Arch. Microbiol. **160:** 492–497.

180. KOZIANOWSKI, G., F. CANGANELLA, F.A. RAINEY, *et al.* 1997. Purification and characterization of thermostable pectate-lyases from a newly isolated thermophilic bacterium, *Thermoanaerobacter italicus* sp. nov. Extremophiles **1:** 171–182.

181. LARSEN, L., P. NIELSEN & B.K. AHRING. 1997. *Thermoanaerobacter mathranii* sp. nov., an ethanol-producing, extremely thermophilic anaerobic bacterium from a hot spring in Iceland. Arch. Microbiol. **168:** 114–119.

182. ZEIKUS, J.G., A. BEN-BASSAT & P.W. HEGGE. 1980. Microbiology of methanogenesis in thermal, volcanic environments. J. Bacteriol. **143:** 432–440.

183. ONYENWOKE, R.U., V.V. KEVBRIN & A. LYSENKO. 2007. *Thermoanaerobacter pseudethanolicus* sp. nov., a thermophilic heterotrophic anaerobe from Yellowstone National Park. Int. J. Syst. Evol. Microbiol. **57:** 2191–2193.

184. LEE, Y.J., M. DASHTI, A. PRANGE, *et al.* 2007. *Thermoanaerobacter sulfurigignens* sp. nov., an anaerobic thermophilic bacterium that reduces 1 M thiosulfate to elemental sulfur and tolerates 90 mM sulfite. Int. J. Syst. Evol. Microbiol. **57:** 1429–1434.

185. BONCH-OSMOLOVSKAYA, E.A., M.L. MIROSHNICHENKO, N.A. CHERNYKH, *et al.* 1997. Reduction of elemental sulfur by moderately thermophilic organotrophic bacteria and the description of *Thermoanaerobacter sulfurophilus* sp. nov. Mikrobiologiya **66:** 581–587.

186. FARDEAU, M.L., M. MAGOT, B.K. PATEL, *et al.* 2000. *Thermoanaerobacter subterraneus* sp. nov., a novel thermophile isolated from oilfield water. Int. J. Syst. Evol. Microbiol. **50:** 2141–2149.

187. SOKOLOVA, T.G., J.M. GONZALEZ, N.A. KOSTRIKINA, *et al.* 2001. *Carboxydobrachium pacificum* gen. nov., sp. nov., a new

anaerobic, thermophilic, CO-utilizing marine bacterium from Okinawa Trough. Int. J. Syst. Evol. Microbiol. **51:** 141–149.

188. Xue, Y., Y. Xu, Y. Liu, *et al*. 2001. *Thermoanaerobacter tengcongensis* sp. nov., a novel anaerobic, saccharolytic, thermophilic bacterium isolated from a hot spring in Tengcong, China. Int. J. Syst. Evol. Microbiol. **51:** 1335–1341.
189. Bao, Q., Y. Tian, W. Li, *et al*. 2002. A complete sequence of the *T. tengcongensis* genome. Genome Res. **12:** 689–700.
190. Kim, B.C., R. Grote, D.W. Lee, *et al*. 2001. *Thermoanaerobacter yonseiensis* sp. nov., a novel extremely thermophilic, xylose-utilizing bacterium that grows at up to 85°C. Int. J. Syst. Evol. Microbiol. **51:** 1539–1548.
191. Sekiguchi, Y., H. Imachi, A. Susilorukmi, *et al*. 2006. *Tepidanaerobacter syntrophicus* gen. nov., sp. nov., an anaerobic, moderately thermophilic, syntrophic alcohol- and lactate-degrading bacterium isolated from thermophilic digested sludges. Int. J. Syst. Evol. Microbiol. **56:** 1621–1629.
192. Huber, R., P. Rossnagel, C.R. Woese, *et al*. 1996. Formation of ammonium from nitrate during chemolithoautotrophic growth of the extremely thermophilic bacterium *Ammonifex degensii* gen. nov. sp. nov. Syst. Appl. Microbiol. **19:** 40–49.
193. Mori, K., S. Hanada, A. Maruyama, *et al*. 2002. *Thermanaeromonas toyohensis* gen. nov., sp. nov., a novel thermophilic anaerobe isolated from a subterranean vein in the Toyoha Mines. Int. J. Syst. Evol. Microbiol. **52:** 1675–1680.
194. Fardeau, M.L., B. Ollivier, B.K. Patel, *et al*. 1995. Isolation and characterization of a thermophilic sulfate-reducing bacterium, *Desulfotomaculum thermosapovorans* sp. nov. Int. J. Syst. Bacteriol. **45:** 218–221.
195. Min, H. & S.H. Zinder. 1990. Isolation and characterization of a thermophilic sulfate-reducing bacterium *Desulfotomaculum thermoacetoxidans* sp. nov. Arch. Microbiol. **153:** 399–404.
196. Plugge, C.M., M. Balk & A.J. Stams. 2002. *Desulfotomaculum thermobenzoicum* subsp. *thermosyntrophicum* subsp. nov., a thermophilic, syntrophic, propionate-oxidizing, spore-forming bacterium. Int. J. Syst. Evol. Microbiol. **52:** 391–399.
197. Liu, Y., T.M. Karnauchow, K.F. Jarrell, *et al*. 1997. Description of two new thermophilic *Desulfotomaculum* spp., *Desulfotomaculum putei* sp. nov., from a deep terrestrial subsurface, and *Desulfotomaculum luciae* sp. nov., from a hot spring. Int. J. Syst. Bacteriol. **47:** 615–621.
198. Love, A.C., B.K. Patel, P.D. Nichols, *et al*. 1993. *Desulfotomaculum australicum,* sp. nov., a thermophilic sulfate-reducing bacterium isolated from the Great Artesian Basin in Australia. Syst. Appl. Bacteriol. **16:** 244–251.
199. Daumas, S., R. Cord-Ruwisch & J.L. Garcia. 1988. *Desulfotomaculum geothermicum* sp. nov., a thermophilic, fatty acid-degrading, sulfate-reducing bacterium isolated with H_2 from geothermal ground water. Anton. Leeuw. Int. J. G. **54:** 165–178.
200. Karnauchow, T.M., S.F. Koval & K.F. Jarrell. 1992. Isolation and characterization of three thermophilic anaerobes from a St. Lucia hot spring. Syst. Appl. Bacteriol. **15:** 296–310.
201. Nilsen, R.K., T. Torsvik & T. Lein. 1996. *Desulfotomaculum thermocisternum* sp. nov., a sulfate reducer isolated from a hot North Sea oil reservoir. Int. J. Syst. Bacteriol. **46:** 397–402.
202. Kaksonen, A.H., S. Spring, P. Schumann, *et al*. 2006. *Desulfotomaculum thermosubterraneum* sp. nov., a thermophilic sulfate-reducer isolated from an underground mine located in a geothermally active area. Int. J. Syst. Evol. Microbiol. **56:** 2603–2608.
203. Nazina, T.N., A.E. Ivanova, L.P. Kanchaveli, *et al*. 1988. A new sporeforming thermophilic methylotrophic sulfate-reducing bacterium, *Desulfotomaculum kuznetsovii* sp. nov. Mikrobiologiya **57:** 823–827.
204. Turova, T.P., B.B. Kuznetsov, E.V. Novikova, *et al*. 2001. Heterogeneity of nucleotide sequences of 16S ribosomal RNA genes from the *Desulfotomaculum kuznetsovii* type strain. Mikrobiologiya **70:** 788–795.
205. Campbell, L.L. & J.R. Postgate. 1965. Classification of the spore-forming sulfate-reducing bacteria. Microbiol. Mol. Biol. R. **29:** 359–363.
206. Slobodkin, A., A.L. Reysenbach, N. Strutz, *et al*. 1997. *Thermoterrabacterium ferrireducens* gen. nov., sp. nov., a thermophilic anaerobic dissimilatory Fe(III)-reducing bacterium from a continental hot spring. Int. J. Syst. Bacteriol. **47:** 541–547.
207. Slobodkin, A.I., T.G. Sokolova, A.M. Lysenko, *et al*. 2006. Reclassification of *Thermoterrabacterium ferrireducens* as *Carboxydothermus ferrireducens* comb. nov., and emended description of the genus *Carboxydothermus*. Int. J. Syst. Evol. Microbiol. **56:** 2349–2351.
208. Wu, M., Q. Ren, A.S. Durkin, *et al*. 2005. Life in hot carbon monoxide: the complete genome sequence of *Carboxydothermus hydrogenoformans* Z-2901. PLoS Genetics **1:** 563–574.
209. Rees, G.N., B.K. Patel, G.S. Grassia, *et al*. 1997. *Anaerobaculum thermoterrenum* gen. nov., sp. nov., a novel, thermophilic bacterium which ferments citrate. Int. J. Syst. Bacteriol. **47:** 150–154.
210. Menes, R.J. & L. Muxi. 2002. *Anaerobaculum mobile* sp. nov., a novel anaerobic, moderately thermophilic, peptide-fermenting bacterium that uses crotonate as an electron acceptor, and emended description of the genus *Anaerobaculum*. Int. J. Syst. Evol. Microbiol. **52:** 157–164.
211. Sekiguchi, Y., Y. Kamagata, K. Nakamura, *et al*. 2000. *Syntrophothermus lipocalidus* gen. nov., sp. nov., a novel thermophilic, syntrophic, fatty-acid-oxidizing anaerobe which utilizes isobutyrate. Int. J. Syst. Evol. Microbiol. **50:** 771–779.
212. Zavarzina, D.G., T.N. Zhilina, T.P. Tourova, *et al*. 2000. *Thermanaerovibrio velox* sp. nov., a new anaerobic, thermophilic, organotrophic bacterium that reduces elemental sulfur, and emended description of the genus *Thermanaerovibrio*. Int. J. Syst. Evol. Microbiol. **50:** 1287–1295.
213. Baena, S., M.L. Fardeau, T.H. Woo, *et al*. 1999. Phylogenetic relationships of three amino-acid-utilizing anaerobes, *Selenomonas acidaminovorans, 'Selenomonas acidaminophila'* and *Eubacterium acidaminophilum*, as inferred from partial 16S rDNA nucleotide sequences and

proposal of *Thermanaerovibrio acidaminovorans* gen. nov., comb. nov. and *Anaeromusa acidaminophila* gen. nov., comb. nov. Int. J. Syst. Bacteriol. **49:** 969–974.

214. SLEPOVA, T.V., T.G. SOKOLOVA, A.M. LYSENKO, *et al.* 2006. *Carboxydocella sporoproducens* sp. nov., a novel anaerobic CO-utilizing/H2-producing thermophilic bacterium from a Kamchatka hot spring. Int. J. Syst. Evol. Microbiol. **56:** 797–800.

215. PROWE, S.G. & G. ANTRANIKIAN. 2001. *Anaerobranca gottschalkii* sp. nov., a novel thermoalkaliphilic bacterium that grows anaerobically at high pH and temperature. Int. J. Syst. Evol. Microbiol. **51:** 457–465.

216. ENGLE, M., Y. LI, C. WOESE, *et al.* 1995. Isolation and characterization of a novel alkalitolerant thermophile, *Anaerobranca horikoshii* gen. nov., sp. nov. Int. J. Syst. Bacteriol. **45:** 454–461.

217. GORLENKO, V., A. TSAPIN, Z. NAMSARAEV, *et al.* 2004. *Anaerobranca californiensis* sp. nov., an anaerobic, alkalithermophilic, fermentative bacterium isolated from a hot spring on Mono Lake. Int. J. Syst. Evol. Microbiol. **54:** 739–743.

218. MLADENOVSKA, Z., I.M. MATHRANI & B.K. AHRING. 1995. Isolation and characterization of *Caldicellulosiruptor lactoaceticus* sp. nov., an extremely thermophilic, cellulolytic, anaerobic bacterium. Arch. Microbiol. **163:** 223–230.

219. HUANG, C.Y., B.K. PATEL, R.A. MAH, *et al.* 1998. *Caldicellulosiruptor owensensis* sp. nov., an anaerobic, extremely thermophilic, xylanolytic bacterium. Int. J. Syst. Bacteriol. **48:** 91–97.

220. BREDHOLT, S., J. SONNE-HANSEN, P. NIELSEN, *et al.* 1999. *Caldicellulosiruptor kristjanssonii* sp. nov., a cellulolytic, extremely thermophilic, anaerobic bacterium. Int. J. Syst. Bacteriol. **49:** 991–996.

221. RAINEY, F.A., A.M. DONNISON, P.H. JANSSEN, *et al.* 1994. Description of *Caldicellulosiruptor saccharolyticus* gen. nov., sp. nov: an obligately anaerobic, extremely thermophilic, cellulolytic bacterium. FEMS Microbiol. Lett. **120:** 263–266.

222. NIELSEN, P., I.M. MATHRANI & B.K. AHRING. 1993. *Thermoanaerobium acetigenum* spec. nov., a new anaerobic, extremely thermophilic, xylanolytic non-spore-forming bacterium isolated from an Icelandic hot spring. Arch. Microbiol. **159:** 460–464.

223. ONYENWOKE, R.U., Y.J. LEE, S. DABROWSKI, *et al.* 2006. Reclassification of *Thermoanaerobium acetigenum* as *Caldicellulosiruptor acetigenus* comb. nov. and emendation of the genus description. Int. J. Syst. Evol. Microbiol. **56:** 1391–1395.

224. DAHLE, H. & N.K. BIRKELAND. 2006. *Thermovirga lienii* gen. nov., sp. nov., a novel moderately thermophilic, anaerobic, amino-acid-degrading bacterium isolated from a North Sea oil well. Int. J. Syst. Evol. Microbiol. **56:** 1539–1545.

225. PADDEN, A.N., V.M. DILLON, J. EDMONDS, *et al.* 1999. An indigo-reducing moderate thermophile from a woad vat, *Clostridium isatidis* sp. nov. Int. J. Syst. Bacteriol. **49:** 1025–1031.

226. KATO, S., S. HARUTA, Z.J. CUI, *et al.* 2004. *Clostridium straminisolvens* sp. nov., a moderately thermophilic, aerotolerant and cellulolytic bacterium isolated from a cellulose-degrading bacterial community. Int. J. Syst. Evol. Microbiol. **54:** 2043–2047.

227. WIEGEL, J., S.U. KUK & G.W. KOHRING. 1989. *Clostridium thermobutyricum* sp. nov., a moderate thermophile isolated from a cellulolytic culture, that produces butyrate as the major product. Int. J. Syst. Bacteriol. **39:** 199–204.

228. HE, Y.L., Y.F. DING & Y.Q. LONG. 1991. Two cellulolytic *Clostridium* species: *Clostridium cellulosi* sp. nov. and *Clostridium cellulofermentans* sp. nov. Int. J. Syst. Bacteriol. **41:** 306–309.

229. MADDEN, R.H. 1983. Isolation and characterization of *Clostridium stercorarium* sp. nov., cellulolytic thermophile. Int. J. Syst. Bacteriol. **33:** 837–840.

230. FARDEAU, M.L., B. OLLIVIER, J.L. GARCIA, *et al.* 2001. Transfer of *Thermobacteroides leptospartum* and *Clostridium thermolacticum* as *Clostridium stercorarium* subsp. *leptospartum* subsp. *thermolacticum* subsp. nov., comb. nov. and *C. stercorarium* subsp. *thermolacticum* subsp. nov., comb. nov. Int. J. Syst. Evol. Microbiol. **51:** 1127–1131.

231. TODA, Y., T. SAIKI, T. UOZUMI, *et al.* 1988. Isolation and characterization of a protease-producing, thermophilic, anaerobic bacterium, *Thermobacteroides leptospartum* sp. nov. Agric. Biol. Chem. **52:** 1339–1344.

232. LE RUYET, P., H.C. DUBOURGUIER, G. ALBAGNAC, *et al.* 1985. Characterization of *Clostridium thermolacticum* sp. nov., a hydrolytic thermophilic anaerobe producing high amounts of lactate. Syst. Appl. Microbiol. **6:** 196–202.

233. SLOBODKIN, A.I., T.P. TOUROVA, N.A. KOSTRIKINA, *et al.* 2003. *Tepidibacter thalassicus* gen. nov., sp. nov., a novel moderately thermophilic, anaerobic, fermentative bacterium from a deep-sea hydrothermal vent. Int. J. Syst. Evol. Microbiol. **53:** 1131–1134.

234. URIOS, L., V. CUEFF, P. PIGNET, *et al.* 2004. *Tepidibacter formicigenes* sp. nov., a novel spore-forming bacterium isolated from a Mid-Atlantic Ridge hydrothermal vent. Int. J. Syst. Evol. Microbiol. **54:** 439–443.

235. TARLERA, S., L. MUXI, M. SOUBES, *et al.* 1997. *Caloramator proteoclasticus* sp. nov., a new moderately thermophilic anaerobic proteolytic bacterium. Int. J. Syst. Bacteriol. **47:** 651–656.

236. PLUGGE, C.M., E.G. ZOETENDAL & A.J. STAMS. 2000. *Caloramator coolhaasii* sp. nov., a glutamate-degrading, moderately thermophilic anaerobe. Int. J. Syst. Evol. Microbiol. **50:** 1155–1162.

237. SEYFRIED, M., D. LYON, F.A. RAINEY, *et al.* 2002. *Caloramator viterbensis* sp. nov., a novel thermophilic, glycerol-fermenting bacterium isolated from a hot spring in Italy. Int. J. Syst. Evol. Microbiol. **52:** 1177–1184.

238. CHRISOSTOMOS, S., B.K.C. PATEL, P.P. DWIVEDI, *et al.* 1996. *Caloramator indicus* sp. nov., a new thermophilic anaerobic bacterium isolated from the deep-seated non-volcanically heated waters of an Indian artesian aquifer. Int. J. Syst. Bacteriol. **46:** 497–501.

239. PATEL, B.K.C., C. MONK, H. LITTLEWORTH, *et al.* 1987. *Clostridium fervidus* sp. nov., a new chemoorganotrophic acetogenic thermophile. Int. J. Syst. Bacteriol. **37:** 123–126.

240. MIRANDA-TELLO, E., M.L. FARDEAU, J. SEPULVEDA, *et al.* 2003. *Garciella nitratireducens* gen. nov., sp. nov.,

an anaerobic, thermophilic, nitrate- and thiosulfate-reducing bacterium isolated from an oilfield separator in the Gulf of Mexico. Int. J. Syst. Evol. Microbiol. **53:** 1509–1514.

241. ALAIN, K., P. PIGNET, M. ZBINDEN, *et al.* 2002. *Caminicella sporogenes* gen. nov., sp. nov., a novel thermophilic spore-forming bacterium isolated from an East-Pacific Rise hydrothermal vent. Int. J. Syst. Evol. Microbiol. **52:** 1621–1628.
242. WERY, N., J.M. MORICET, V. CUEFF, *et al.* 2001. *Caloranaerobacter azorensis* gen. nov., sp. nov., an anaerobic thermophilic bacterium isolated from a deep-sea hydrothermal vent. Int. J. Syst. Evol. Microbiol. **51:** 1789–1796.
243. CAYOL, J.L., S. DUCERF, B.K. PATEL, *et al.* 2000. *Thermohalobacter berrensis* gen. nov., sp. nov., a thermophilic, strictly halophilic bacterium from a solar saltern. Int. J. Syst. Evol. Microbiol. **50:** 559–564.
244. PIKUTA, E., A. LYSENKO, N. CHUVILSKAYA, *et al.* 2000. *Anoxybacillus pushchinensis* gen. nov., sp. nov., a novel anaerobic, alkaliphilic, moderately thermophilic bacterium from manure, and description of *Anoxybacillus flavithermus* comb. nov. Int. J. Syst. Evol. Microbiol. **50:** 2109–2117.
245. BOONE, D.R., Y. LIU, Z.J. ZHAO, *et al.* 1995. *Bacillus infernus* sp. nov., an Fe(III)- and Mn(IV)-reducing anaerobe from the deep terrestrial subsurface. Int. J. Syst. Bacteriol. **45:** 441–448.
246. COMBET-BLANC, Y., B. OLLIVIER, C. STREICHER, *et al.* 1995. *Bacillus thermoamylovorans* sp. nov., a moderately thermophilic and amylolytic bacterium. Int. J. Syst. Bacteriol. **45:** 9–16.
247. DULGER, S., Z. DEMIRBAG & A.O. BELDUZ. 2004. *Anoxybacillus ayderensis* sp. nov. and *Anoxybacillus kestanbolensis* sp. nov. Int. J. Syst. Evol. Microbiol. **54:** 1499–1503.
248. YUMOTO, I., K. HIROTA, T. KAWAHARA, *et al.* 2004. *Anoxybacillus voinovskiensis* sp. nov., a moderately thermophilic bacterium from a hot spring in Kamchatka. Int. J. Syst. Evol. Microbiol. **54:** 1239–1242.
249. BELDUZ, A.O., S. DULGER & Z. DEMIRBAG. 2003. *Anoxybacillus gonensis* sp. nov., a moderately thermophilic, xylose-utilizing, endospore-forming bacterium. Int. J. Syst. Evol. Microbiol. **53:** 1315–1320.
250. NAZINA, T.N., T.P. TOUROVA, A.B. POLTARAUS, *et al.* 2001. Taxonomic study of aerobic thermophilic bacilli: descriptions of *Geobacillus subterraneus* gen. nov., sp. nov. and *Geobacillus uzenensis* sp. nov. from petroleum reservoirs and transfer of *Bacillus stearothermophilus*, *Bacillus thermocatenulatus*, *Bacillus thermoleovorans*, *Bacillus kaustophilus*, *Bacillus thermoglucosidasius* and *Bacillus thermodenitrificans* to *Geobacillus* as the new combinations *G. stearothermophilus*, *G. thermocatenulatus*, *G. thermoleovorans*, *G. kaustophilus*, *G. thermoglucosidasius* and *G. thermodenitrificans*. Int. J. Syst. Evol. Microbiol. **51:** 433–446.
251. GOLOVACHEVA, R.S., L.G. LOGINOVA, T.A. SALIKHOV, *et al.* 1975. A new thermophilic species *Bacillus thermocatenulatus* nov. sp. Mikrobiologiya **44:** 230–233.
252. MANACHINI, P.L., D. MORA, G. NICASTRO, *et al.* 2000. *Bacillus thermodenitrificans* sp. nov., nom. rev. Int. J. Syst. Evol. Microbiol. **50:** 1331–1337.
253. FENG, L., W. WANG, J. CHENG, *et al.* 2007. Genome and proteome of long-chain alkane degrading *Geobacillus thermodenitrificans* NG80-2 isolated from a deep-subsurface oil reservoir. Proc. Natl. Acad. Sci. USA **104:** 5602–5607.
254. ZARILLA, K.A. & J.J. PERRY. 1987. *Bacillus thermoleovorans*, sp. nov., a species of obligately thermophilic hydrocarbon utilizing endospore-forming bacteria. Syst. Appl. Microbiol. **9:** 258–264.
255. MERKEL, G.J., S.S. STAPLETON & J.J. PERRY. 1978. Isolation and peptidoglycan of Gram-negative hydrocarbon-utilizing thermophilic bacteria. J. Gen. Microbiol. **109:** 141–148.
256. L'HARIDON, S., M.L. MIROSHNICHENKO, N.A. KOSTRIKINA, *et al.* 2006. *Vulcanibacillus modesticaldus* gen. nov., sp. nov., a strictly anaerobic, nitrate-reducing bacterium from deep-sea hydrothermal vents. Int. J. Syst. Evol. Microbiol. **56:** 1047–1053.
257. MIROSHNICHENKO, M.L., F.A. RAINEY, H. HIPPE, *et al.* 1998. *Desulfurella kamchatkensis* sp. nov. and *Desulfurella propionica* sp. nov., new sulfur-respiring thermophilic bacteria from Kamchatka thermal environments. Int. J. Syst. Evol. Microbiol. **48:** 475–479.
258. BONCH-OSMOLOVSKAYA, E.A., T.G. SOKOLOVA, N.A. KOSTRIKINA, *et al.* 1990. *Desulfurella acetivorans* gen. nov. and sp. nov. -a new thermophilic sulfur-reducing eubacterium. Arch. Microbiol. **153:** 151–155.
259. MIROSHNICHENKO, M.L., F.A. RAINEY, M. RHODE, *et al.* 1999. *Hippea maritima* gen. nov., sp. nov., a new genus of thermophilic, sulfur-reducing bacterium from submarine hot vents. Int. J. Syst. Bacteriol. **49:** 1033–1038.
260. MIROSHNICHENKO, M.L., N.A. KOSTRIKINA, S. L'HARIDON, *et al.* 2002. Nautilia lithotrophica gen. nov., sp. nov., a thermophilic sulfur-reducing epsilon-proteobacterium isolated from a deep-sea hydrothermal vent. Int. J. Syst. Evol. Microbiol. **52:** 1299–1304.
261. VOORDECKERS, J.W., V. STAROVOYTOV & C. VETRIANI. 2005. *Caminibacter mediatlanticus* sp. nov., a thermophilic, chemolithoautotrophic, nitrate-ammonifying bacterium isolated from a deep-sea hydrothermal vent on the Mid-Atlantic Ridge. Int. J. Syst. Evol. Microbiol. **55:** 773–779.
262. MIROSHNICHENKO, M.L., S. L'HARIDON, P. SCHUMANN, *et al.* 2004. *Caminibacter profundus* sp. nov., a novel thermophile of *Nautiliales* ord. nov. within the class *'Epsilonproteobacteria'*, isolated from a deep-sea hydrothermal vent. Int. J. Syst. Evol. Microbiol. **54:** 41–45.
263. ALAIN, K., J. QUERELLOU, F. LESONGEUR, *et al.* 2002. *Caminibacter hydrogeniphilus* gen. nov., sp. nov., a novel thermophilic, hydrogen-oxidizing bacterium isolated from an East Pacific Rise hydrothermal vent. Int. J. Syst. Evol. Microbiol. **52:** 1317–1323.
264. TAKAI, K., K.H. NEALSON & K. HORIKOSHI. 2004. *Hydrogenimonas thermophila* gen. nov., sp. nov., a novel thermophilic, hydrogen-oxidizing chemolithoautotroph within the epsilon-Proteobacteria, isolated from a black smoker in a Central Indian Ridge hydrothermal field. Int. J. Syst. Evol. Microbiol. **54:** 25–32.
265. TAKAI, K., H. KOBAYASHI, K.H. NEALSON, *et al.* 2003. *Deferribacter desulfuricans* sp. nov., a novel sulfur-, nitrate- and arsenate-reducing thermophile isolated from a deep-sea hydrothermal vent. Int. J. Syst. Evol. Microbiol. **53:** 839–846.

266. Fiala, G., C.R. Woese, T.A. Langworthy, *et al.* 1990. *Flexistipes sinusarabici*, a novel genus and species of eubacteria occurring in the Atlantis II Deep brines of the Red Sea. Arch. Microbiol. **154:** 120–126.

267. Zeikus, J.G., M.A. Dawson, T.E. Thompson, *et al.* 1983. Microbial ecology of volcanic sulphidogenesis: isolation and characterization of *Thermodesulfobacterium commune* gen. nov. and sp. nov. J. Gen. Microbiol. **129:** 1159–1169.

268. Sonne-Hansen, J. & B.K. Ahring. 1999. *Thermodesulfobacterium hveragerdense* sp. nov., and *Thermodesulfovibrio islandicus* sp. nov., two thermophilic sulfate reducing bacteria isolated from a Icelandic hot spring. Syst. Appl. Microbiol. **22:** 559–564.

269. Rozanova, E.P. & T.A. Pivovarova. 1988. Reclassification of *Desulfovibrio thermophilus* (Rozanova, Khudyakova, 1974). Mikrobiologiya **57:** 102–106.

270. Kuever, J., F.A. Rainey & F. Widdel. 2005. Genus III. *Desulfothermus* gen. nov. *In* Bergey's Manual of Systematic Bacteriology, Vol. 2. D.J. Brenner, N.R. Krieg, J.T. Staley *et al.*, Eds.: 955–956. Springer-Verlag. New York, NY.

271. Nunoura, T., H. Oida, M. Miyazaki, *et al.* 2007. *Desulfothermus okinawensis* sp. nov., a thermophilic and heterotrophic sulfate-reducing bacterium isolated from a deep-sea hydrothermal field. Int. J. Syst. Evol. Microbiol. **57:** 2360–2364.

272. Rees, G.N., G.S. Grassia, A.J. Sheehy, *et al.* 1995. *Desulfacinum infernum* gen. nov., sp. nov., a thermophilic sulfate-reducing bacterium from a petroleum reservoir. Int. J. Syst. Bacteriol. **45:** 85–89.

273. Sievert, S.M. & J. Kuever. 2000. *Desulfacinum hydrothermale* sp. nov., a thermophilic, sulfate-reducing bacterium from geothermally heated sediments near Milos Island (Greece). Int. J. Syst. Evol. Microbiol. **50:** 1239–1246.

274. Beeder, J., T. Torsvik & T. Lien. 1995. *Thermodesulforhabdus norvegicus* gen. nov., sp. nov., a novel thermophilic sulfate-reducing bacterium from oil field water. Arch. Microbiol. **164:** 331–336.

275. Henry, E.A., R. Devereux, J.S. Maki, *et al.* 1994. Characterization of a new thermophilic sulfate-reducing bacterium *Thermodesulfovibrio yellowstonii*, gen. nov. and sp. nov.: its phylogenetic relationship to *Thermodesulfobacterium commune* and their origins deep within the bacterial domain. Arch. Microbiol. **161:** 62–69.

276. Mori, K., T. Kakegawa, Y. Higashi, *et al.* 2004. *Oceanithermus desulfurans* sp. nov., a novel thermophilic, sulfur-reducing bacterium isolated from a sulfide chimney in Suiyo Seamount. Int. J. Syst. Evol. Microbiol. **54:** 1561–1566.

277. Mori, K., H. Kim, T. Kakegawa, *et al.* 2003. A novel lineage of sulfate-reducing microorganisms: *Thermodesulfobiaceae* fam. nov., *Thermodesulfobium narugense*, gen. nov., sp. nov., a new thermophilic isolate from a hot spring. Extremophiles **7:** 283–290.

278. Saiki, T., Y. Kobayashi, K. Kawagoe, *et al.* 1985. *Dictyoglomus thermophilum* gen. nov., sp. nov., a chemoorganotrophic, anaerobic, thermophilic bacterium. Int. J. Syst. Bacteriol. **35:** 253–259.

279. Svetlichny, V.A. & T.P. Svetlichnaya. 1988. *Dictyoglomus turgidus* sp. nov., a new extremely thermophilic eubacterium isolated from hot springs of the Uzon volcano caldera. Mikrobiologiya **57:** 435–441.

280. Demharter, W., R. Hensel, J. Smida, *et al.* 1989. *Sphaerobacter thermophilus* gen. nov., sp. nov. A deeply rooting member of the actinomycetes subdivision isolated from thermophilically treated sewage sludge. Syst. Appl. Microbiol. **11:** 261–266.

281. Jackson, T.J., R.F. Ramaley & W.G. Meinschein. 1973. *Thermomicrobium*, a new genus of extremely thermophilic bacteria. Int. J. Syst. Bacteriol. **23:** 28–36.

282. Nakagawa, S., S. Nakamura, F. Inagaki, *et al.* 2004. *Hydrogenivirga caldilitoris* gen. nov., sp. nov., a novel extremely thermophilic, hydrogen- and sulfur-oxidizing bacterium from a coastal hydrothermal field. Int. J. Syst. Evol. Microbiol. **54:** 2079–2084.

283. Huber, R., T. Wilharm, D. Huber, *et al.* 1992. *Aquifex pyrophilus* gen. nov. sp. nov., represents a novel group of marine hyperthermophilic hydrogen-oxidizing bacteria. Syst. Appl. Microbiol. **15:** 340–351.

284. L'Haridon, S., V. Cilia, P. Messner, *et al.* 1998. *Desulfurobacterium thermolithotrophum* gen. nov., sp. nov., a novel autotrophic, sulphur-reducing bacterium isolated from a deep-sea hydrothermal vent. Int. J. Syst. Bacteriol. **48:** 701–711.

285. Alain, K., S. Rolland, P. Crassous, *et al.* 2003. *Desulfurobacterium crinifex* sp. nov., a novel thermophilic, pinkish-streamer forming, chemolithoautotrophic bacterium isolated from a Juan de Fuca Ridge hydrothermal vent and amendment of the genus *Desulfurobacterium*. Extremophiles **7:** 361–370.

286. L'Haridon, S., A.L. Reysenbach, B.J. Tindall, *et al.* 2006. *Desulfurobacterium atlanticum* sp. nov., *Desulfurobacterium pacificum* sp. nov. and *Thermovibrio guaymasensis* sp. nov., three thermophilic members of the *Desulfurobacteriaceae* fam. nov., a deep branching lineage within the *Bacteria*. Int. J. Syst. Evol. Microbiol. **56:** 2843–2852.

287. Takai, K., S. Nakagawa, Y. Sako, *et al.* 2003. *Balnearium lithotrophicum* gen. nov., sp. nov., a novel thermophilic, strictly anaerobic, hydrogen-oxidizing chemolithoautotroph isolated from a black smoker chimney in the Suiyo Seamount hydrothermal system. Int. J. Syst. Evol. Microbiol. **53:** 1947–1954.

288. Huber, H., S. Diller, C. Horn, *et al.* 2002. *Thermovibrio ruber* gen. nov., sp. nov., an extremely thermophilic, chemolithoautotrophic, nitrate-reducing bacterium that forms a deep branch within the phylum *Aquificae*. Int. J. Syst. Evol. Microbiol. **52:** 1859–1865.

289. Vetriani, C., M.D. Speck, S.V. Ellor, *et al.* 2004. *Thermovibrio ammonificans* sp. nov., a thermophilic, chemolithotrophic, nitrate-ammonifying bacterium from deep-sea hydrothermal vents. Int. J. Syst. Evol. Microbiol. **54:** 175–181.

290. Stohr, R., A. Waberski, H. Volker, *et al.* 2001. *Hydrogenothermus marinus* gen. nov., sp. nov., a novel thermophilic hydrogen-oxidizing bacterium, recognition of *Calderobacterium hydrogenophilum* as a member of the genus *Hydrogenobacter* and proposal of the reclassification of *Hydrogenobacter acidophilus* as *Hydrogenobaculum acidophilum* gen. nov., comb. nov., in the phylum *'Hydrogenobacter/Aquifex'*. Int. J. Syst. Evol. Microbiol. **51:** 1853–1862.

291. Nakagawa, S., Z. Shtaih, A. Banta, *et al.* 2005. *Sulfurihydrogenibium yellowstonense* sp. nov., an extremely thermophilic, facultatively heterotrophic, sulfur-oxidizing bacterium from Yellowstone National Park, and emended descriptions of the genus *Sulfurihydrogenibium*, *Sulfurihydrogenibium subterraneum* and *Sulfurihydrogenibium azorense*. Int. J. Syst. Evol. Microbiol. **55:** 2263–2268.
292. Takai, K., H. Kobayashi, K.H. Nealson, *et al.* 2003. *Sulfurihydrogenibium subterraneum* gen. nov., sp. nov., from a subsurface hot aquifer. Int. J. Syst. Evol. Microbiol. **53:** 823–827.
293. Aguiar, P., T.J. Beveridge & A.L. Reysenbach. 2004. *Sulfurihydrogenibium azorense*, sp. nov., a thermophilic hydrogen-oxidizing microaerophile from terrestrial hot springs in the Azores. Int. J. Syst. Evol. Microbiol. **54:** 33–39.
294. Nakagawa, S., K. Takai, K. Horikoshi, *et al.* 2003. *Persephonella hydrogeniphila* sp. nov., a novel thermophilic, hydrogen-oxidizing bacterium from a deep-sea hydrothermal vent chimney. Int. J. Syst. Evol. Microbiol. **53:** 863–869.
295. Gotz, D., A. Banta, T.J. Beveridge, *et al.* 2002. *Persephonella marina* gen. nov., sp. nov. and *Persephonella guaymasensis* sp. nov., two novel, thermophilic, hydrogen-oxidizing microaerophiles from deep-sea hydrothermal vents. Int. J. Syst. Evol. Microbiol. **52:** 1349–1359.
296. Davey, M.E., W.A. Wood, R. Key, *et al.* 1993. Isolation of three species of *Geotoga* and *Petrotoga*: two new genera, representing a new lineage in the bacterial line of descent distantly related to the *"Thermotogales"*. Syst. Appl. Microbiol. **16:** 191–200.
297. Wery, N., F. Lesongeur, P. Pignet, *et al.* 2001. *Marinitoga camini* gen. nov., sp. nov., a rod-shaped bacterium belonging to the order *Thermotogales*, isolated from a deep-sea hydrothermal vent. Int. J. Syst. Evol. Microbiol. **51:** 495–504.
298. Alain, K., V.T. Marteinsson, M.L. Miroshnichenko, *et al.* 2002. *Marinitoga piezophila* sp. nov., a rod-shaped, thermo-piezophilic bacterium isolated under high hydrostatic pressure from a deep-sea hydrothermal vent. Int. J. Syst. Evol. Microbiol. **52:** 1331–1339.
299. Postec, A., C.L. Breton, M.L. Fardeau, *et al.* 2005. *Marinitoga hydrogenitolerans* sp. nov., a novel member of the order *Thermotogales* isolated from a black smoker chimney on the Mid-Atlantic Ridge. Int. J. Syst. Evol. Microbiol. **55:** 1217–1221.
300. Nunoura, T., H. Oida, M. Miyazaki, *et al.* 2007. *Marinitoga okinawensis* sp. nov., a novel thermophilic and anaerobic heterotroph isolated from a deep-sea hydrothermal field, Southern Okinawa Trough. Int. J. Syst. Evol. Microbiol. **57:** 467–471.
301. Miranda-Tello, E., M.L. Fardeau, P. Thomas, *et al.* 2004. *Petrotoga mexicana* sp. nov., a novel thermophilic, anaerobic and xylanolytic bacterium isolated from an oil-producing well in the Gulf of Mexico. Int. J. Syst. Evol. Microbiol. **54:** 169–174.
302. Lien, T., M. Madsen, F.A. Rainey, *et al.* 1998. *Petrotoga mobilis* sp. nov., from a North Sea oil-production well. Int. J. Syst. Bacteriol. **48:** 1007–1013.
303. Miranda-Tello, E., M.L. Fardeau, C. Joulian, *et al.* 2007. *Petrotoga halophila* sp. nov., a thermophilic, moderately halophilic, fermentative bacterium isolated from an offshore oil well in Congo. Int. J. Syst. Evol. Microbiol. **57:** 40–44.
304. Antoine, E., V. Cilia, J.R. Meunier, *et al.* 1997. *Thermosipho melanesiensis* sp. nov., a new thermophilic anaerobic bacterium belonging to the order *Thermotogales*, isolated from deep-sea hydrothermal vents in the southwestern Pacific Ocean. Int. J. Syst. Bacteriol. **47:** 1118–1123.
305. Haridon, S.L., M.L. Miroshnichenko, H. Hippe, *et al.* 2001. *Thermosipho geolei* sp. nov., a thermophilic bacterium isolated from a continental petroleum reservoir in Western Siberia. Int. J. Syst. Evol. Microbiol. **51:** 1327–1334.
306. Takai, K. & K. Horikoshi. 2000. *Thermosipho japonicus* sp. nov., an extremely thermophilic bacterium isolated from a deep-sea hydrothermal vent in Japan. Extremophiles **4:** 9–17.
307. Huber, R., C.R. Woese, T.A. Langworthy, *et al.* 1989. *Thermosipho africanus* gen. nov., represents a new genus of thermophilic eubacteria within the *"Thermotogales"*. Syst. Appl. Microbiol. **12:** 32–37.
308. Ravot, G., B. Ollivier, B.K.C. Patel, *et al.* 1996. Emended description of *Thermosipho africanus* as a carbohydrate-fermenting species using thiosulfate as an electron acceptor. Int. J. Syst. Bacteriol. **46:** 321–323.
309. Andrews, K.T. & B.K. Patel. 1996. *Fervidobacterium gondwanense* sp. nov., a new thermophilic anaerobic bacterium isolated from nonvolcanically heated geothermal waters of the Great Artesian Basin of Australia. Int. J. Syst. Bacteriol. **46:** 265–269.
310. Huber, R., C.R. Woese, T.A. Langworthy, *et al.* 1990. *Fervidobacterium islandicum* sp. nov., a new extremely thermophilic eubacterium belonging to the *"Thermotogales"*. Arch. Microbiol. **154:** 105–111.
311. Patel, B.K.C., H.W. Morgan & R.M. Daniel. 1985. *Fervidobacterium nodosum* gen. nov. and spec. nov., a new chemoorganotrophic, caldoactive, anaerobic bacterium. Arch. Microbiol. **141:** 63–69.
312. Friedrich, A.B. & G. Antranikian. 1996. Keratin degradation by *Fervidobacterium pennavorans*, a novel thermophilic anaerobic species of the order *Thermotogales*. Appl. Environ. Microbiol. **62:** 2875–2882.
313. Cai, J., Y. Wang, D. Liu, *et al.* 2007. *Fervidobacterium changbaicum* sp. nov., a novel thermophilic anaerobic bacterium isolated from a hot spring of the Changbai Mountains, China. Int. J. Syst. Evol. Microbiol. **57:** 2333–2336.
314. Balk, M., J. Weijma & A.J. Stams. 2002. *Thermotoga lettingae* sp. nov., a novel thermophilic, methanol-degrading bacterium isolated from a thermophilic anaerobic reactor. Int. J. Syst. Evol. Microbiol. **52:** 1361–1368.
315. Ravot, G., M. Magot, M.L. Fardeau, *et al.* 1995. *Thermotoga elfii* sp. nov., a novel thermophilic bacterium from an African oil-producing well. Int. J. Syst. Bacteriol. **45:** 308–314.
316. Fardeau, M.L., B. Ollivier, B.K. Patel, *et al.* 1997. *Thermotoga hypogea* sp. nov., a xylanolytic, thermophilic bacterium from an oil-producing well. Int. J. Syst. Bacteriol. **47:** 1013–1019.

317. JEANTHON, C., A.L. REYSENBACH, S. L'HARIDON, *et al.* 1995. *Thermotoga subterranea* sp. nov., a new thermophilic bacterium isolated from a continental oil reservoir. Arch. Microbiol. **164:** 91–97.

318. WINDBERGER, E., R. HUBER, A. TRINCONE, *et al.* 1989. *Thermotoga thermarum* sp. nov. and *Thermotoga neapolitana* occurring in African continental solfataric springs. Arch. Microbiol. **151:** 506–512.

319. HUBER, R., T.A. LANGWORTHY, H. KONIG, *et al.* 1986. *Thermotoga maritima* sp. nov. represents a new genus of unique extremely thermophilic eubacteria growing up to 90°C. Arch. Microbiol. **144:** 324–333.

320. NELSON, K.E., R.A. CLAYTON, S.R. GILL, *et al.* 1999. Evidence for lateral gene transfer between *Archaea* and *Bacteria* from genome sequence of *Thermotoga maritima*. Nature **399:** 323–329.

321. TAKAHATA, Y., M. NISHIJIMA, T. HOAKI, *et al.* 2001. *Thermotoga petrophila* sp. nov. and *Thermotoga naphthophila* sp. nov., two hyperthermophilic bacteria from the Kubiki oil reservoir in Niigata, Japan. Int. J. Syst. Evol. Microbiol. **51:** 1901–1909.

322. JANNASCH, H.W., R. HUBER, S. BELKIN, *et al.* 1988. *Thermotoga neapolitana* sp. nov. of the extremely thermophilic, eubacterial genus *Thermotoga*. Arch. Microbiol. **150:** 103–104.

323. HUBER, R., D. DYBA, H. HUBER, *et al.* 1998. Sulfur-inhibited *Thermosphaera aggregans* sp. nov., a new genus of hyperthermophilic archaea isolated after its prediction from environmentally derived 16S rRNA sequences. Int. J. Syst. Bacteriol. **48:** 31–38.

324. PROKOFEVA, M.I., M.L. MIROSHNICHENKO, N.A. KOSTRIKINA, *et al.* 2000. *Acidilobus aceticus* gen. nov., sp. nov., a novel anaerobic thermoacidophilic archaeon from continental hot vents in Kamchatka. Int. J. Syst. Evol. Microbiol. **50:** 2001–2008.

325. FIALA, G., K.O. STETTER, H.W. JANNASCH, *et al.* 1986. *Staphylothermus marinus* sp. nov. represents a novel genus of extremely thermophilic submarine heterotrophic archaebacteria growing up to 98°C. Syst. Appl. Microbiol. **8:** 106–113.

326. ARAB, H., H. VOLKER & M. THOMM. 2000. *Thermococcus aegaeicus* sp. nov. and *Staphylothermus hellenicus* sp. nov., two novel hyperthermophilic archaea isolated from geothermally heated vents off Palaeochori Bay, Milos, Greece. Int. J. Syst. Evol. Microbiol. **50:** 2101–2108.

327. HUBER, H., S. BURGGRAF, T. MAYER, *et al.* 2000. *Ignicoccus* gen. nov., a novel genus of hyperthermophilic, chemolithoautotrophic *Archaea*, represented by two new species, *Ignicoccus islandicus* sp. nov. and *Ignicoccus pacificus* sp. nov. Int. J. Syst. Evol. Microbiol. **50:** 2093–2100.

328. PAPER, W., U. JAHN, M.J. HOHN, *et al.* 2007. *Ignicoccus hospitalis* sp. nov., the host of '*Nanoarchaeum equitans*'. Int. J. Syst. Evol. Microbiol. **57:** 803–808.

329. TUROVA, T.P., B.B. KUZNETSOV, T.V. KALGANOVA, *et al.* 2000. Phylogenetic position of *Desulfurococcus amylolyticus*. Mikrobiologiya **69:** 447–448.

330. BONCH-OSMOLOVSKAYA, E.A., A.I. SLESAREV, M.L. MIROSHNICHENKO, *et al.* 1988. Characteristics of *Desulfurococcus amylolyticus* n. sp., a new extreme thermophilic archaebacterium from hot volcanic vents of Kamchatka and Kunashir. Mikrobiologiya **57:** 78–85.

331. ZILLIG, W., K.O. STETTER, D. PRANGISHVILLI, *et al.* 1982. *Desulfurococcaceae*, the second family of the extremely thermophilic, anaerobic, sulfur-respiring *Thermoproteales*. Zbl. Bakt. Mik. Hyg. I. C. **3:** 304–317.

332. HENSEL, R., K. MATUSSEK, K. MICHALKE, *et al.* 1997. *Sulfophobococcus zilligii* gen. nov., spec. nov. a novel hyperthermophilic archaeum isolated from hot alkaline springs of Iceland. Syst. Appl. Microbiol. **20:** 102–110.

333. BURGGRAF, S., H. HUBER & K.O. STETTER. 1997. Reclassification of the crenarchael orders and families in accordance with 16S rRNA sequence data. Int. J. Syst. Bacteriol. **47:** 657–660.

334. STETTER, K.O. 1986. Diversity of extremely thermophilic archaebacteria. *In* Thermophlies: General, Molecular, and Applied Microbiology. T.D. Brock, Ed. John Wiley & Sons, Inc., New York, NY.

335. STETTER, K.O. 2001. Genus VII. *Thermodiscus* gen. nov. *In* Bergey's Manual of Systematic Bacteriology. Vol. 1. D.R. Boone, R.W. Castenholz & G.M. Garrity, Eds.: 189–190. Springer-Verlag. New York, NY.

336. NIEDERBERGER, T.D., D.K. GOTZ, I.R. MCDONALD, *et al.* 2006. *Ignisphaera aggregans* gen. nov., sp. nov., a novel hyperthermophilic crenarchaeote isolated from hot springs in Rotorua and Tokaanu, New Zealand. Int. J. Syst. Evol. Microbiol. **56:** 965–971.

337. KUROSAWA, N., Y.H. ITOH, T. IWAI, *et al.* 1998. *Sulfurisphaera ohwakuensis* gen. nov., sp. nov., a novel extremely thermophilic acidophile of the order *Sulfolobales*. Int. J. Syst. Bacteriol. **48:** 451–456.

338. FUCHS, T., H. HUBER, S. BURGGRAF, *et al.* 1996. 16S rDNA-based phylogeny of the archeal order *Sulfolobales* and reclassification of *Desulfurolobus ambivalens* as *Acidianus ambivalens* comb. nov. Syst. Appl. Microbiol. **19:** 46–60.

339. ZILLIG, W., S. YEATS, I. HOLZ, *et al.* 1986. *Desulfurolobus ambivalens*, gen. nov., sp. nov., an autotrophic archaebacterium facultatively oxidizing or reducing sulfur. Syst. Appl. Microbiol. **8:** 197–203.

340. ZILLIG, W., A. GIERL, G. SCHREIBER, *et al.* 1983. The archaebacterium *Thermofilum pendens* represents a novel genus of the thermophilic, anaerobic sulfur respiring *Thermoproteales*. Syst. Appl. Microbiol. **4:** 79–87.

341. BONCH-OSMOLOVSKAYA, E.A., M.L. MIROSHNICHENKO, N.A. KOSTRIKINA, *et al.* 1990. *Thermoproteus uzoniensis* sp. nov., a new extremely thermophilic archaebacterium from Kamchatka continental hot springs. Arch. Microbiol. **154:** 556–559.

342. ZILLIG, W. & A. REYSENBACH. 2001. Genus I. *Thermoproteus* Zillig and Stetter 1982. *In* Bergey's Manual of Systematic Bacteriology, Vol. 1. D.R. Boone, R.W. Castenholz & G.M. Garrity, Eds.: 171–173. Springer-Verlag. New York, NY.

343. ZILLIG, W. 1989. Genus I. *Thermoproteus* Zillig and Stetter 1982. *In* Bergey's Manual of Systematic Bacteriology. Vol. 3. J.T. Staley, M.P. Bryant, N. Pfenning & J.G. Holt, Eds.: 2241. Williams and Wilkins, Baltimore, MD.

344. SAKO, Y., T. NUNOURA & A. UCHIDA. 2001. *Pyrobaculum oguniense* sp. nov., a novel facultatively aerobic and

hyperthermophilic archaeon growing at up to 97°C. Int. J. Syst. Evol. Microbiol. **51:** 303–309.

345. HUBER, R., M. SACHER, A. VOLLMANN, *et al.* 2000. Respiration of arsenate and selenate by hyperthermophilic archaea. Syst. Appl. Microbiol. **23:** 305–314.
346. HUBER, R., J.K. KRISTJANSSON & K.O. STETTER. 1987. *Pyrobaculum* gen. nov., a new genus of neutrophilic, rod-shaped archaebacteria from continental solfataras growing optimally at 100°C. Arch. Microbiol. **149:** 95–101.
347. AMO, T., M.L. PAJE, A. INAGAKI, *et al.* 2002. *Pyrobaculum calidifontis* sp. nov., a novel hyperthermophilic archaeon that grows in atmospheric air. Archaea **1:** 113–121.
348. VOLKL, P., R. HUBER, E. DROBNER, *et al.* 1993. *Pyrobaculum aerophilum* sp. nov., a novel nitrate-reducing hyperthermophilic archaeum. Appl. Environ. Microbiol. **59:** 2918–2926.
349. FITZ-GIBBON, S.T., H. LADNER, U.J. KIM, *et al.* 2002. Genome sequence of the hyperthermophilic crenarchaeon *Pyrobaculum aerophilum*. Proc. Natl. Acad. Sci. USA **99:** 984–989.
350. ITOH, T., K. SUZUKI & T. NAKASE. 1998. *Thermocladium modestius* gen. nov., sp. nov., a new genus of rod-shaped, extremely thermophilic crenarchaeote. Int. J. Syst. Bacteriol. **48:** 879–887.
351. ITOH, T., K. SUZUKI & T. NAKASE. 2002. *Vulcanisaeta distributa* gen. nov., sp. nov., and *Vulcanisaeta souniana* sp. nov., novel hyperthermophilic, rod-shaped crenarchaeotes isolated from hot springs in Japan. Int. J. Syst. Evol. Microbiol. **52:** 1097–1104.
352. ITOH, T., K. SUZUKI, P.C. SANCHEZ, *et al.* 1999. *Caldivirga maquilingensis* gen. nov., sp. nov., a new genus of rod-shaped crenarchaeote isolated from a hot spring in the Philippines. Int. J. Syst. Bacteriol. **49:** 1157–1163.
353. SEGERER, A., T.A. LANGWORTHY & K.O. STETTER. 1988. *Thermoplasma acidophilum* and *Thermoplasma volcanium* sp. nov. from solfatara fields. Syst. Appl. Microbiol. **10:** 161–171.
354. RUEPP, A., W. GRAML, M.L. SANTOS-MARTINEZ, *et al.* 2000. The genome sequence of the thermoacidophilic scavenger *Thermoplasma acidophilum*. Nature **407:** 508–513.
355. KAWASHIMA, T., N. AMANO, H. KOIKE, *et al.* 2000. Archaeal adaptation to higher temperatures revealed by genomic sequence of *Thermoplasma volcanium*. Proc. Natl. Acad. Sci. USA **97:** 14257–14262.
356. JEANTHON, C., S. L'HARIDON, A.L. REYSENBACH, *et al.* 1999. *Methanococcus vulcanius* sp. nov., a novel hyperthermophilic methanogen isolated from East Pacific Rise, and identification of *Methanococcus* sp. DSM 4213T as *Methanococcus fervens* sp. nov. Int. J. Syst. Bacteriol. **49:** 583–589.
357. L'HARIDON, S., A.L. REYSENBACH, A. BANTA, *et al.* 2003. *Methanocaldococcus indicus* sp. nov., a novel hyperthermophilic methanogen isolated from the Central Indian Ridge. Int. J. Syst. Evol. Microbiol. **53:** 1931–1935.
358. WHITMAN, W.B. 2001. Genus I. *Methanocaldococcus* gen. nov. *In* Bergey's Manual of Systematic Bacteriology, Vol. 1. D.R. Boone, R.W. Castenholz & G.M. Garrity, Eds.: 685–690. Springer-Verlag, New York, NY.
359. JEANTHON, C., S. L'HARIDON, A.L. REYSENBACH, *et al.* 1998. *Methanococcus infernus* sp. nov., a novel hyperthermophilic lithotrophic methanogen isolated from a deep-sea hydrothermal vent. Int. J. Syst. Bacteriol. **48:** 913–919.
360. JONES, W.J., J.A. LEIGH, F. MAYER, *et al.* 1983. *Methanococcus jannaschii* sp. nov., an extremely thermophilic methanogen from a submarine hydrothermal vent. Arch. Microbiol. **136:** 254–261.
361. BULT, C.J., O. WHITE, G.J. OLSEN, *et al.* 1996. Complete genome sequence of the methanogenic archaeon, *Methanococcus jannaschii*. Science **273:** 1058–1073.
362. TAKAI, K., K.H. NEALSON & K. HORIKOSHI. 2004. *Methanotorris formicicus* sp. nov., a novel extremely thermophilic, methane-producing archaeon isolated from a black smoker chimney in the Central Indian Ridge. Int. J. Syst. Evol. Microbiol. **54:** 1095–1100.
363. BURGGRAF, S., H. FRICKE, A. NEUNER, *et al.* 1990. *Methanococcus igneus* sp. nov., a novel hyperthermophilic methanogen from a shallow submarine hydrothermal system. Syst. Appl. Microbiol. **13:** 263–269.
364. WHITMAN, W.B. 2001. Genus II. *Methanotorris* gen. nov. *In* Bergey's Manual of Systematic Bacteriology, Vol. 1, D.R. Boone, R.W. Castenholz & G.M. Garrity, Eds.: 245–246. Springer-Verlag. New York, NY.
365. RONIMUS, R.S., A.L. REYSENBACH, D.R. MUSGRAVE, *et al.* 1997. The phylogenetic position of the *Thermococcus* isolate AN_1 based on 16S rRNA gene sequence analysis: a proposal that AN_1 represents a new species, *Thermococcus zilligii* sp. nov. Arch. Microbiol. **168:** 245–248.
366. KOBAYASHI, T., Y.S. KWAK, T. AKIBA, *et al.* 1994. *Thermococcus profundus* sp. nov., a new hyperthermophilic archaeon isolated from deep-sea hydrothermal vent. Syst. Appl. Microbiol. **17:** 232–236.
367. DUFFAUD, G.D., O.B. D'HENNEZEL, A.S. PEEK, *et al.* 1998. Isolation and characterization of *Thermococcus barossii*, sp. nov., a hyperthermophilic archaeon isolated from a hydrothermal vent flange formation. Syst. Appl. Microbiol. **21:** 40–49.
368. GROTE, R., L. LI, J. TAMAOKA, *et al.* 1999. *Thermococcus siculi* sp. nov., a novel hyperthermophilic archaeon isolated from a deep-sea hydrothermal vent at the Mid-Okinawa Trough. Extremophiles **3:** 55–62.
369. KUWABARA, T., M. MINABA, N. OGI, *et al.* 2007. *Thermococcus celericrescens* sp. nov., a fast-growing and cell-fusing hyperthermophilic archaeon from a deep-sea hydrothermal vent. Int. J. Syst. Evol. Microbiol. **57:** 437–443.
370. CAMBON-BONAVITA, M.A., F. LESONGEUR, P. PIGNET, *et al.* 2003. *Thermococcus atlanticus* sp nov., a hyperthermophilic Archaeon isolated from a deep-sea hydrothermal vent in the Mid-Atlantic Ridge. Extremophiles **7:** 101–109.
371. HUBER, R., J. STOHR, S. HOHENHAUS, *et al.* 1995. *Thermococcus chitonophagus* sp. nov., a novel, chitin-degrading, hyperthermophilic archaeum from a deep-sea hydrothermal vent environment. Arch. Microbiol. **164:** 255–264.
372. NEUNER, A., H.W. JANNASCH, S. BELKIN, *et al.* 1990. *Thermococcus litoralis* sp. nov.: a new species of extremely thermophilic marine archaebacteria. Arch. Microbiol. **153:** 205–207.
373. PIKUTA, E.V., D. MARSIC, T. ITOH, *et al.* 2007. *Thermococcus thioreducens* sp. nov., a novel hyperthermophilic, obligately

sulfur-reducing archaeon from a deep-sea hydrothermal vent. Int. J. Syst. Evol. Microbiol. **57:** 1612–1618.

374. GONZALEZ, J.M., D. SHECKELLS, M. VIEBAHN, *et al.* 1999. *Thermococcus waiotapuensis* sp. nov., an extremely thermophilic archaeon isolated from a freshwater hot spring. Arch. Microbiol. **172:** 95–101.

375. MIROSHNICHENKO, M.L., E.A. BONCH-OSMOLOVSKAYA, A. NEUNER, *et al.* 1989. *Thermococcus stetteri* sp. nov., a new extremely thermophilic marine sulfur-metabolizing *Archaebacterium*. Syst. Appl. Microbiol. **12:** 257–262.

376. CANGANELLA, F., W.J. JONES, A. GAMBACORTA, *et al.* 1998. *Thermococcus guaymasensis* sp. nov. and *Thermococcus aggregans* sp. nov., two novel thermophilic archaea isolated from the Guaymas Basin hydrothermal vent site. Int. J. Syst. Bacteriol. **48:** 1181–1185.

377. GONZALEZ, J.M., C. KATO & K. HORIKOSHI. 1995. *Thermococcus peptonophilus* sp. nov., a fast-growing, extremely thermophilic archaebacterium isolated from deep-sea hydrothermal vents. Arch. Microbiol. **164:** 159–164.

378. ATOMI, H., T. FUKUI, T. KANAI, *et al.* 2004. Description of *Thermococcus kadakaraensis* sp. nov., a well studied hyperthermophilic archaeon previously reported as *Pyrococcus* sp. KOD1. Archaea **1:** 263–267.

379. FUJIWARA, S., M. TAKAGI & T. IMANAKA. 1998. Archaeon *Pyrococcus kodakaraensis* KOD1: application and evolution. Biotechnol. Annu. Rev. **4:** 259–284.

380. FUKUI, T., H. ATOMI, T. KANAI, *et al.* 2005. Complete genome sequence of the hyperthermophilic archaeon *Thermococcus kodakaraensis* KOD_1 and comparison with *Pyrococcus* genomes. Genome Res. **15:** 352–363.

381. KUWABARA, T., M. MINABA, Y. IWAYAMA, *et al.* 2005. *Thermococcus coalescens* sp. nov., a cell-fusing hyperthermophilic archaeon from Suiyo Seamount. Int. J. Syst. Evol. Microbiol. **55:** 2507–2514.

382. FIALA, G. & K.O. STETTER. 1986. *Pyrococcus furiosus* sp. nov. represents a novel genus of marine heterotrophic archaebacteria growing optimally at 100°C. Arch. Microbiol. **145:** 56–61.

383. MAEDER, D.L., R.B. WEISS, D.M. DUNN, *et al.* 1999. Divergence of the hyperthermophilic *Archaea Pyrococcus furiosus* and *P. horikoshii* inferred from complete genomic sequences. Genetics **152:** 1299–1305.

384. GONZALEZ, J.M., Y. MASUCHI, F.T. ROBB, *et al.* 1998. *Pyrococcus horikoshii* sp. nov., a hyperthermophilic archaeon isolated from a hydrothermal vent at the Okinawa Trough. Extremophiles **2:** 123–130.

385. KAWARABAYASI, Y., M. SAWADA, H. HORIKAWA, *et al.* 1998. Complete sequence and gene organization of the genome of a hyper-thermophilic archaebacterium, *Pyrococcus horikoshii* OT3. DNA Res. **5:** 55–76.

386. ZILLIG, W., I. HOLZ, H.P. KLENK, *et al.* 1987. *Pyrococcus woesei*, sp. nov., an ultra-thermophilic marine Archaebacterium, representing a novel order, *Thermococcales*. Syst. Appl. Microbiol. **9:** 62–70.

387. TAKAI, K., A. SUGAI, T. ITOH, *et al.* 2000. *Palaeococcus ferrophilus* gen. nov., sp. nov., a barophilic, hyperthermophilic archaeon from a deep-sea hydrothermal vent chimney. Int. J. Syst. Evol. Microbiol. **50:** 489–500.

388. AMEND, J.P., D.R. MEYER-DOMBARD, S.N. SHETH, *et al.* 2003. *Palaeococcus helgesonii* sp. nov., a facultatively anaerobic, hyperthermophilic archaeon from a geothermal well on Vulcano Island, Italy. Arch. Microbiol. **179:** 394–401.

389. HUBER, H., H. JANNASCH, R. RACHEL, *et al.* 1997. *Archaeoglobus veneficus* sp. nov., a novel facultative chemolithoautotrophic hyperthermophilic sulfite reducer, isolated from abyssal black smokers. Syst. Appl. Microbiol. **20:** 374–380.

390. STETTER, K.O. 1988. *Archaeoglobus fulgidus* gen. nov., sp. nov. a new taxon of extremely thermophilic *Archaebacteria*. Syst. Appl. Microbiol. **10:** 172–173.

391. STETTER, K.O., G. LAUERER, M. THOMM, *et al.* 1987. Isolation of extremely thermophilic sulfate reducers: evidence for a novel branch of Archaebacteria. Science **236:** 822–824.

392. KLENK, H.P., R.A. CLAYTON, J.F. TOMB, *et al.* 1997. The complete genome sequence of the hyperthermophilic, sulphate-reducing archaeon *Archaeoglobus fulgidus*. Nature **390:** 364–370.

393. HAFENBRADL, D., M. KELLER, R. DIRMEIER, *et al.* 1996. *Ferroglobus placidus* gen. nov., sp. nov., a novel hyperthermophilic archaeum that oxidizes $Fe2+$ at neutral pH under anoxic conditions. Arch. Microbiol. **166:** 308–314.

394. KURR, M., R. HUBER, H. KONIG, *et al.* 1991. *Methanopyrus kandleri*, gen. and sp. nov. represents a novel group of hyperthermophilic methanogens, growing at 110°C. Arch. Microbiol. **156:** 239–247.

395. SLESAREV, A.I., K.V. MEZHEVAYA, K.S. MAKAROVA, *et al.* 2002. The complete genome of hyperthermophile *Methanopyrus kandleri* AV19 and monophyly of archaeal methanogens. Proc. Natl. Acad. Sci. USA **99:** 4644.

396. LAUERER, G., J.K. KRISTJANSSON, T.A. LANGWORTHY, *et al.* 1986. *Methanothermus sociabilis* sp. nov., a second species within the *Methanothermaceae* growing at 97°C. Syst. Appl. Microbiol. **8:** 100–105.

397. STETTER, K.O. 2001. Genus I. *Methanothermus* Stetter 1982b. *In* Bergey's Manual of Systematic Bacteriology, Vol. 1. D.R. Boone, R.W. Castenholz & G.M. Garrity, Eds.: 243–245. Springer-Verlag. New York, NY.

398. BLOTEVOGEL, K. & U. FISCHER. 1985. Isolation and characterization of a new thermophilic and autotrophic methane producing bacterium: *Methanobacterium thermoaggregans* spec. nov. Arch. Microbiol. **142:** 218–222.

399. WASSERFALLEN, A., J. NOLLING, P. PFISTER, *et al.* 2000. Phylogenetic analysis of 18 thermophilic *Methanobacterium* isolates supports the proposals to create a new genus, *Methanothermobacter* gen. nov., and to reclassify several isolates in three species, *Methanothermobacter thermautotrophicus* comb. nov., *Methanothermobacter wolfeii* comb. nov., and *Methanothermobacter marburgensis* sp. nov. Int. J. Syst. Evol. Microbiol. **50:** 43–53.

400. SMITH, D.R., L.A. DOUCETTE-STAMM, C. DELOUGHERY, *et al.* 1997. Complete genome sequence of *Methanobacterium thermoautotrophicum* ΔH: functional analysis and comparative genomics. J. Bacteriol. **179:** 7135–7155.

401. LAURINAVICHIUS, K.S., S.V. KOTELNIKOVA & A.Y. OBRAZTSOVA. 1988. A new species of the thermophilic methane-forming bacterium *Methanobacterium thermophilum*. Mikrobiologiya **57:** 1035–1041.

402. Winter, J., C. Lerp, H.P. Zabel, *et al.* 1985. *Methanobacterium wolfei*, sp. nov., a new tungsten-requiring, thermophilic, autotrophic methanogen. Syst. Appl. Microbiol. **5:** 457–466.

403. Huber, H., M. Thomm, H. Konig, *et al.* 1982. *Methanococcus thermolithotrophicus*, a novel thermophilic lithotrophic methanogen. Arch. Microbiol. **132:** 47–50.

404. Whitman, W.B. 2001. Genus II. *Methanothermococcus* gen. nov. *In* Bergey's Manual of Systematic Bacteriology, Vol. 1. D.R. Boone, R.W. Castenholz & G.M. Garrity, Eds.: 241–242. Springer-Verlag. New York, NY.

405. Takai, K., A. Inoue & K. Horikoshi. 2002. *Methanothermococcus okinawensis* sp. nov., a thermophilic, methane-producing archaeon isolated from a Western Pacific deep-sea hydrothermal vent system. Int. J. Syst. Evol. Microbiol. **52:** 1089–1095.

406. Zinder, S.H., K.R. Sower & J.G. Ferry. 1985. *Methanosarcina thermophila* sp. nov., a thermophilic, acetotrophic, methane-producing bacterium. Int. J. Syst. Bacteriol. **35:** 522–523.

407. Jiang, B., S.N. Parshina, W. van Doesburg, *et al.* 2005. *Methanomethylovorans thermophila* sp. nov., a thermophilic, methylotrophic methanogen from an anaerobic reactor fed with methanol. Int. J. Syst. Evol. Microbiol. **55:** 2465–2470.

408. Cheng, L., T.L. Qiu, X.B. Yin, *et al.* 2007. *Methermicoccus shengliensis* gen. nov., sp. nov., a thermophilic, methylotrophic methanogen isolated from oil-production water, and proposal of *Methermicoccaceae* fam. nov. Int. J. Syst. Evol. Microbiol. **57:** 2964–2969.

409. Kamagata, Y., H. Kawasaki, H. Oyaizu, *et al.* 1992. Characterization of three thermophilic strains of *Methanothrix ("Methanosaeta") thermophila* sp. nov. and rejection of *Methanothrix ("Methanosaeta") thermoacetophila*. Int. J. Syst. Bacteriol. **42:** 463–468.

410. Maestojuan, G.M., D.R. Boone, L. Xun, *et al.* 1990. Transfer of *Methanogenium bourgense, Methanogenium marisnigri, Methanogenium olentangyi*, and *Methanogenium thermophilicum* to the genus *Methanoculleus* gen. nov., emendation of *Methanoculleus marisnigri* and *Methanogenium* and description of new strains of *Methanoculleus bourgense* and *Methanoculleus marisnigri*. Int. J. Syst. Bacteriol. **40:** 117–122.

411. Rivard, C.J. & P.M. Smith. 1982. Isolation and characterization of a thermophilic marine methanogenic bacterium, *Methanogenium thermophilicum* sp. nov. Int. J. Syst. Bacteriol. **32:** 430–436.

412. Miroshnichenko, M.L., G.M. Gongadze, F.A. Rainey, *et al.* 1998. *Thermococcus gorgonarius* sp. nov. and *Thermococcus pacificus* sp. nov.: heterotrophic extremely thermophilic archaea from New Zealand submarine hot vents. Int. J. Syst. Bacteriol. **48:** 23–29.

413. L'Haridon, S., M.L. Miroshnichenko, H. Hippe, *et al.* 2002. *Petrotoga olearia* sp. nov. and *Petrotoga sibirica* sp. nov., two thermophilic bacteria isolated from a continental petroleum reservoir in Western Siberia. Int. J. Syst. Evol. Microbiol. **52:** 1715–1722.

414. Urios, L., V. Cueff-Gauchard, P. Pignet, *et al.* 2004. *Thermosipho atlanticus* sp. nov., a novel member of the *Thermotogales* isolated from a Mid-Atlantic Ridge hydrothermal vent. Int. J. Syst. Evol. Microbiol. **54:** 1953–1957.

415. Miroshnichenko, M.L., H. Hippe, E. Stackebrandt, *et al.* 2001. Isolation and characterization of *Thermococcus sibiricus* sp. nov. from a Western Siberia high-temperature oil reservoir. Extremophiles **5:** 85–91.

416. Whitman, W.B., D.C. Coleman & W.J. Wiebe. 1998. Prokaryotes: the unseen majority. Proc. Natl. Acad. Sci. USA **95:** 6578–6583.

Life at Extreme Limits

The Anaerobic Halophilic Alkalithermophiles

NOHA M. MESBAH AND JUERGEN WIEGEL

Department of Microbiology, University of Georgia, Athens, Georgia, USA

The ability of anaerobic microorganisms to proliferate under extreme conditions is of widespread importance for microbial physiology, remediation, industry, and evolution. The halophilic alkalithermophiles are a novel group of polyextremophiles. Tolerance to alkaline pH, elevated NaCl concentrations, and high temperatures necessitates mechanisms for cytoplasmic pH acidification; permeability control of the cell membrane; and stability of proteins, the cell wall, and other cellular constituents to multiple extreme conditions. Although it is generally assumed that extremophiles growing at more than one extreme combine adaptive mechanisms for each individual extreme, adaptations for individual extremes often counteract each other. However, in alkaline, hypersaline niches heated via intense solar irradiation, culture-independent analyses have revealed the presence of an extensive diversity of aerobic and anaerobic microorganisms belonging to Bacteria and Archaea that survive and grow under multiple harsh conditions. Thus, polyextremophiles must have developed novel adaptive strategies enabling them to grow and proliferate under multiple extreme conditions. The recent isolation of two novel anaerobic, halophilic alkalithermophiles, *Natranaerobius thermophilus* and *Halonatronum saccharophilum*, will provide a platform for detailed biochemical, genomic, and proteomic experiments, allowing a greater understanding of the novel adaptive mechanisms undoubtedly employed by polyextremophiles. In this review, we highlight growth characteristics, ecology, and phylogeny of the anaerobic halophilic alkalithermophiles isolated. We also describe the bioenergetic and physiological problems posed by growth at the multiple extreme conditions of alkaline pH, high NaCl concentration, and elevated temperature under anoxic conditions and highlight recent findings and unresolved problems regarding adaptation to multiple extreme conditions.

Key words: **extremophile; anaerobe; halophile; alkalithermophile**

Introduction

We usually associate extreme conditions with properties outside of the ordinary. This leads to the question "What is biologically normal?" From an anthropogenic point of view, "normal" is what is comfortable for us as human beings. Thus, what we regard as normal includes ambient temperatures (20–25°C), neutral or near-neutral pH values between 6 and 8, atmospheric pressure, and an atmosphere containing 20% wt/vol. oxygen. Everything outside this comfort zone, such as temperatures above 40°C, acidic or alkaline pH values, low atmospheric pressures (e.g., at high altitudes), or high pressure (e.g., in the deep ocean) are regarded as "extreme." However, these extreme conditions are normal conditions for extremophiles: microorganisms that not only survive, but grow and thrive under harsh environmental conditions, such as elevated (≥50°C) or low (≤10°C) temperatures, acidic (≤5) or alkaline (≥8.5) pH values, and very high pressure as occurs in deep ocean sediments. In fact, many extremophiles cannot grow, divide, or even survive under environmental conditions that humans consider normal. Thus, in this article, we will define *normal conditions* as those under which the microorganism will (a) grow optimally at the shortest measured doubling time, (b) attain reasonable cell densities (10^5 cells/mL), (c) remain viable indefinitely in a vegetative state, and (d) form a major component of the prokaryotic community composition in its natural "extreme" habitat. An excellent example is *Thermoplasma*, which lives in burning and smoldering coal piles, and can only grow in acidic (pH 0.5–3.0) environments at elevated temperatures (optimally around 60°C). It cannot survive at neutral pH values or ambient temperatures.[1] We also describe facultative extremophiles, which can survive and grow under extreme conditions but grow and multiply faster under human-defined "normal" conditions.

Address for correspondence: Noha Mesbah, Department of Microbiology, 211 Biological Sciences Building, Athens, GA 30602-2605. Voice: +1-706-542-9275; fax: +1-706-542-2674.
nmesbah@uga.edu

Ann. N.Y. Acad. Sci. 1125: 44–57 (2008).
doi: 10.1196/annals.1419.028

TABLE 1. Definitions of different extremophiles

Growth Characteristic	Minimum	Optimum	Maximum
Thermotolerant	T_{min} –	$T_{opt} < 50°C$	$T_{max} < 60°C$
Thermophile	T_{min} –	$T_{opt} \geq 50°C$	$T_{max} \geq 60°C$
Extreme Thermophile	$T_{min} \geq 35°C$	$T_{opt} \geq 65°C$	$T_{max} < 85°C$
Hyperthermophile	$T_{min} \geq 60°C$	$T_{opt} \geq 80°C$	$T_{max} \geq 85°C$
Alkalitolerant	$pH_{min} \geq 6.0$	$pH_{opt} < 8.5$	$pH_{max} > 9.0$
Alkaliphile[a]			
Facultative	$pH_{min} < 7.5$	$pH_{opt} \geq 8.5$	$pH_{max} \geq 10.0$
Obligate	$pH_{min} \geq 7.5$	$pH_{opt} \geq 8.5$	$pH_{max} \geq 10.0$
Non-halophile	$NaCl_{min}$ –	$NaCl_{opt} \leq 0.5$ M NaCl[b]	$NaCl_{max} \leq 1$ M
Halotolerant	$NaCl_{min}$ –	$NaCl_{opt}$ 0.25–1.5 M	$NaCl_{max} \leq 2.5$ M
Halophile	$NaCl_{min}$ 1 M	$NaCl_{opt} \geq 1.5$ M.	$NaCl_{max}$ –
Extreme Halophile	$NaCl_{min} \geq 1.5$ M	$NaCl_{opt} \geq 2.5$ M	$NaCl_{max}$ –

[a]For thermophiles and psychrophiles, the pH must be measured at the growth temperature for the microorganism.
[b]1 M NaCl = 6% NaCl wt/vol.

What Are Extremophiles?

The term *extremophile* collectively applies to a great variety of Bacteria and Archaea that grow optimally under "extreme" or "abnormal" conditions; such as acid or alkaline pH, high or low temperature, extremes of atmospheric pressure, and extremes of salt or organic ion concentrations. Extremophiles are best characterized according to their growth profiles, using marginal data, under certain culture or environmental conditions, such as NaCl range ($NaCl_{opt}$, $NaCl_{min}$, $NaCl_{max}$) or temperature profile (T_{opt}, T_{min}, T_{max}). Examples of extremophiles include, but are not limited to the following categories (and extremes): thermophiles (high temperature), psychrophiles (low temperature), acidophiles (low pH), alkaliphiles (high pH), piezophiles (high pressure, formerly called barophiles), halophiles (high salt concentration), osmophiles (high concentration of organic solutes), oligotrophs (low concentration of solutes and/or nutrients), and xerophiles (very dry environment). Extremophiles also include microorganisms able to grow in the presence of high metal concentrations or microorganisms that grow with unusually short doubling times, such as *Vibrio natriegeans*, whichgrows with a doubling time of 5–6 min in a continuous culture.[2] TABLE 1 contains definitions for different categories of extremophiles. It is important to note that the criteria defining an extremophile differ for prokaryotes and eukaryotes. In general, a eukaryote is considered to be extreme at far less harsh conditions than a prokaryote.

Oxygen as an Extreme

It is hypothesized that life evolved under anoxic conditions—in other words, an absence of molecular oxygen.[3] It thus follows that the first forms of life on Earth were anaerobic and gained energy by an anaerobic metabolism. There are many anoxic niches in the environment, even inside the human body, where obligate anaerobes predominate. In this article, we will refer to microorganisms that can grow both in the presence and absence of oxygen as *facultative aerobes* to stress the anaerobic origin of life and the later evolution of the ability to use oxygen as an electron acceptor. We also stress that anaerobes as a group are able to survive in nearly all environmental conditions found on Earth.

Multiple Extremophiles

Whereas extremophiles are usually defined by one extreme, many natural environments pose two or more extremes, such as acidic hot springs, alkaline hypersaline lakes, and dry sandy deserts. These environments harbor acidothermophiles, halophilic alkaliphiles, and UV radiation–resistant oligotrophs, respectively. Acidothermophiles are widely distributed because of the frequent occurrence of hot springs with acidic pH values ($\leq$3.0). Acidic hot springs arise because of the presence of sulfuric acid formed by microbial and chemical oxidation of sulfur compounds. Many halophiles are also alkaliphiles as salt lakes, whether they are of marine origin (thalassic) or terrestrial origin (athalassic), are alkaline because of the accumulation of carbonate salts.

When dealing with multiple extremophiles, problems arise in descriptions of their optimal and marginal growth data as the value of one of the extreme growth conditions could be affected by the other. For example, the measured pH value of a medium is dependent on

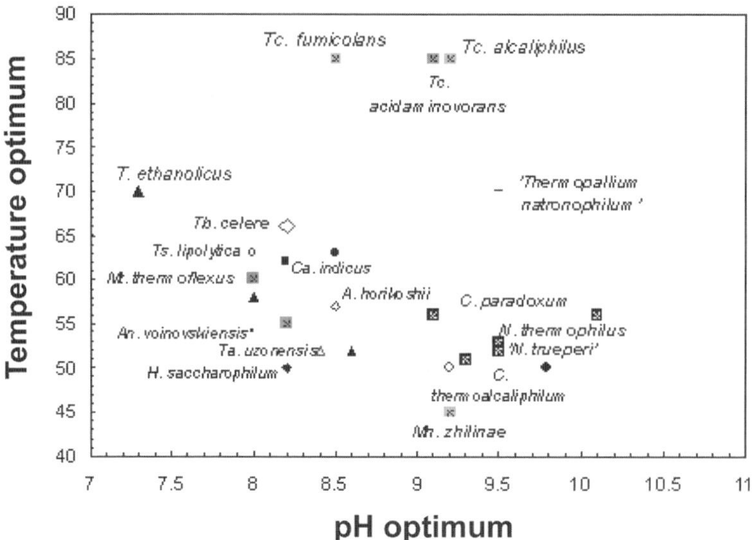

FIGURE 1. Anaerobic alkalithermophiles and halophilic alkalithermophiles graphed according to their temperature and pH optima. Solid symbols represent pH values measured at room temperature, open symbols represent pH values measured at the corresponding growth temperature. Genera have been abbreviated for legibility: A, *Anaerobranca*; An, *Anoxybacillus*; C, *Clostridium*; Ca, *Caloramator*; H, *Halonatronum*; Mh, *Methanosalsum*; Mt, *Methanothermobacter*; N, *Natranaerobius*; T, *Thermoanaerobacter*; Ta, *Thermalkalibacillus*; Tc, *Thermococcus*; Ts, *Thermosyntropha*.

temperature because of changing pKa values of different medium components at different temperatures.[4] Thus, the pH of the medium when measured at room temperature will be different from its pH when it is measured at the elevated growth temperature using temperature-calibrated electrodes and pH meters. For neutral pH, the difference in pH is small, usually less than 0.3 pH units. However, at acidic or alkaline pH values, the difference can be larger than 1 pH unit. Thus, to facilitate comparison of published data, it is important to know the conditions under which the pH was determined. It has been proposed[4] that authors indicate the temperature at which the pH is measured and the pH meter calibrated with a superscript (e.g., $pH^{55^{\circ}C}$).

Similarly, for halophilic alkaliphiles, the elevated salt concentration has an effect on measured pH values. In solutions with high Na^+ concentrations, Na^+ can be read as H^+ by the pH electrode, and this is particularly pronounced at high pH values (>10). This phenomenon, known as the "Na^+ error," can be overcome to some extent with the use of a specific glass combination electrode. The glass membrane in this electrode is specifically constructed to reduce the Na^+ error at high-pH values, and it may also allow use of the electrode over a wider range of temperatures.

Extremophiles adapted to more than two extremes, termed *polyextremophiles*, are much less common than those adapted to two extremes, with the exception of the aerobic thermacidophilic archaea. This most likely reflects a lack of exploration of these novel groups of extremophiles and/or the improper selection of media and culture conditions. Two examples of polyextremophiles, the anaerobic alkalithermophiles and anaerobic halophilic alkalithermophiles, have been isolated, cultivated, and characterized in our laboratory; we discuss these polyextremophiles below.

Anaerobic Alkalithermophiles

The anaerobic alkalithermophiles isolated thus far are neither the most alkaliphilic nor the most thermophilic of the extremophiles. From the presently characterized anaerobic alkalithermophiles, it appears that the pH and temperature optima of the extremophiles are correlated: the larger the temperature optimum, the lower the pH optimum, and vice versa (FIG. 1). This is not only true for species among different genera, but also for different strains of the same species.[5] The scarcity of alkalithermophiles with pH optima ≥9.5 and temperature optima ≥65°C could be attributable to physiological causes. Growth under both of these extreme conditions requires specific adaptations of the cell wall and membrane compositions to minimize permeability to protons and cations. Alkalithermophiles are also faced with the burden of acidifying cytoplasmic pH while growing in a dearth of protons and various other bioenergetic problems, such as suboptimal proton motive force and phosphorylation potential. However, the existence of such isolates cannot be ruled out, as isolates have been reported at such temperatures on the acidic side of the pH scale. Obligately aerobic archaeal *Picrophilus* species have been isolated from solfataric areas in Japan. These isolates grow at pH values of 0–4, and temperatures of 55°C–65°C.[6] The genome of *Picrophilus torridus* contained genes encoding for putative proton-pumping NADH dehydrogenase (complex I) and quinol and cytochrome oxidation.[7] Genes encoding an A_0A_1-adenosine 5′-triphosphatase (ATPase) were also detected. This effective respiratory chain allows rapid extrusion of protons from the cell, thus preventing cytoplasmic acidification. In addition, genome analysis has shown evidence of various pathways for organic acid degradation. Organic acids are lethal for acidophiles as they behave as uncouplers of oxidative phosphorylation. Interestingly, the genes that enhance the ability to *P. torridus* to cope with extremely acidic conditions were acquired by horizontal gene transfer.[7]

Thus, it follows that the lack of "hyper-alkalithermophiles" could be attributable to a lack of exploration of this group and/or improper selection of medium and culture conditions. The acquisition of the genes required for surviving extremely acidic conditions by horizontal gene transfer is intriguing as this implies that hyper-alkalithermophiles could have evolved by obtaining the necessary genes from extreme alkaliphiles and extreme thermophiles.

Anaerobic Halophilic Alkalithermophiles

Anaerobic halophilic alkalithermophiles require, in addition to high temperature and alkaline pH, an elevated salt concentration (>1.5 M NaCl) for growth. Definitions for halophiles are shown in TABLE 1. The first anaerobic, moderately halophilic alkalithermophilic bacterium to be described was *Halonatronum saccharophilum*, a spore-forming bacterium that belongs to the order *Halanaerobiales*.[8] The optima for growth are 1.1–1.5 M NaCl, pH^{RT} 8.5, and 36°C–55°C, which represents a very broad temperature optimum. *H. saccharophilum* has a fermentative metabolism and grows on mono- and disaccharides, starch, and glycogen. *H. saccharophilum* is the first alkaliphilic representative of the order *Halanaerobiales*, which is dominated by extremely halophilic anaerobes.

With a larger Na^+ ion requirement and more alkaline pH optimum, *Natranaerobius thermophilus*, isolated in our laboratory, represents the first true anaerobic halophilic alkalithermophilic bacterium described thus far.[9] *N. thermophilus* grows (at $pH^{55°C}$ 9.5) between 35°C and 56°C, with an optimum at 53°C. The $pH^{55°C}$ range for growth is 8.5–10.6, with an optimum at $pH^{55°C}$ 9.5 and no detectable growth at $\leq pH^{55°C}$ 8.2 or $\geq pH^{55°C}$ 10.8. At the optimum pH and temperature, *N. thermophilus* grows in the Na^+ range of 3.1–4.9 M (1.5–3.3 M of added NaCl, the remainder from added Na_2CO_3 and $NaHCO_3$) and optimally between 3.3 and 3.9 M Na^+ (1.7–2.3 M added NaCl). We have also isolated three additional species of the genus *Natranaerobius*, all of which are extremely halophilic, growing optimally between 1.5 and 2 M NaCl, and obligately alkaliphilic, growing optimally between $pH^{55°C}$ 9.3 and 9.5 and not below $pH^{55°C}$ 8.2. They are also moderately thermophilic, growing optimally between 51°C and 54°C. These isolates are currently undergoing further characterization.

A number of other aerobic and facultatively anaerobic halotolerant alkalithermophiles have been isolated, but not validly published and, in some instances, not characterized beyond the genus level. "*Caloramator halophilus*" is an obligately proteolytic, facultatively aerobic *Firmicute* isolated in our laboratory from salt flats located in northern Nevada. It is thermophilic, growing optimally at 64°C ($temp._{range}$ 42°C–75°C), and alkaliphilic, growing optimally at $pH^{60°C}$ 9.2, and not growing below $pH^{60°C}$ 8.1 or above 10.8. "*Caloramator halophilus*" is unusual in having a large NaCl range, between 0 and 3 M NaCl, for growth, with an optimum at 1.5 M (N.M. Mesbah, P. Maurizio, and J. Wiegel, unpublished results).

"*Bacillus thermoalcaliphilus*" is a chemoorganotrophic, facultatively anaerobic bacterium isolated from mound soil infested with the termite *Odontotermes obesus*.[10] The optima for growth are 1.5 M NaCl, pH^{RT} 8.5–9.0, and 60°C. *Bacillus* STS1, isolated from the same soil termite mounds as "*Bacillus thermoalcaliphilus*," grows optimally at 60°C, pH 9.0, and 1 M NaCl.[10] Another example is *Bacillus* sp. BG-11, isolated from an alkaline thermal environment in India; it is a facultatively aerobic, nonmotile, spore-forming rod that is capable of growth at temperatures greater than 55°C, NaCl concentrations greater than 1.5 M, and pH values 7.5–9.0.[5]

Habitats and Isolation of Anaerobic Halophilic Alkalithermophiles

Hypersaline environments are ubiquitous and arise as a result of evaporative concentration driven by sunlight. The existence of halophiles with high-temperature optima is probably an adaptation to the elevated temperatures reached in these solar-heated environments, even though a few may also contain hot springs, such as saline Mono Lake, California, or the south African Rift lakes. Two types of hypersaline environments occur—thalassohaline and athalassohaline. Thalassohaline environments arise as a result of seawater evaporation; their ionic composition is similar to that of seawater, with sodium chloride as the predominant salt. The Great Salt Lake in Utah and solar salterns, which are found worldwide next to seashores, are examples of thalassohaline environments.[11] Athalassohaline environments arise from non-seawater sources and are usually fed by surface or river water. They are dominated by potassium, magnesium, sodium, and carbonate ions. Some examples of athalassohaline environments include the alkaline soda lakes of Egypt (Wadi An Natrun), the Dead Sea, the East African soda lakes of the Kenyan-Tanzanian Rift Valley, and Mono Lake, California.

The only two anaerobic extremely halophilic alkalithermophiles isolated and characterized thus far, *N. thermophilus*[9] and *H. saccharophilum*,[8] were isolated from sediments of soda lakes in the Wadi An Natrun, Egypt, and Lake Magadi in the Kenyan-Tanzanian Rift Valley.

Wadi An Natrun, Egypt

The Wadi An Natrun is a depression in the Sahara Desert located 90 km northwest of Cairo. The bottom of the valley is 23 m below sea level and 38 m below the Rosette branch of the river Nile. Along the valley stretches a chain of seven large alkaline, hypersaline lakes in addition to a number of ephemeral pools. Water is supplied by underground seepage from the river Nile and occasional winter precipitation. The depth of the lakes ranges between 0.5 and 2 m, and is regulated by seasonal changes in influx seepage and evaporation. High evaporation rates and arid climatic conditions during the summer months cause the salinity to rise above 5 M.[12]

Leachate in seepage water infiltrating the valley is the source of salts in the lakes. The geochemical properties of these lakes were first reported more than 100 years ago[13] and later by Imhoff *et al.*[14] and Taher.[15] The Wadi An Natrun lakes are extreme in several aspects: total dissolved salt concentrations are between 91.0 and 394 g/L in all seven of the lakes, and the water in all lakes has pH values between 9 and 11. The lakes are solar-heated, and salinity and temperature remain the same throughout the water column. Temperatures ranging between 30 and 60°C were measured in the water below the salt crusts formed on the surface (N.M. Mesbah and J. Wiegel, unpublished results). Salinity in most of the lakes approaches saturation, with measured values between 2.5 and 5 M. The water in the most hypersaline lakes is anoxic, with dissolved oxygen concentrations below 0.2 ppm. Consistent with the low measured dissolved oxygen concentrations, the lake waters have high sulfide concentrations, ranging between 45 and 50 μM.[16] It was not possible to measure exact concentrations of cations and anions in the lake waters because of the high concentrations of Na^+ and carbonate, but semiquantitative results showed that Na^+ was present at concentrations of 300–350 ppt (parts per thousand), K^+ at 10–50 ppt, and Ca^{2+} and Rb^{2+} at 1–5 ppm. Other elements, such as Li^+, Sr^{2+}, Zn^{2+} and Cd^{2+}, were present in the low ppb range.[16] The major anions present were ${CO_3}^{2-}$ and Cl^-.

The Wadi An Natrun lakes are populated by dense communities of halophilic alkaliphilic microorganisms. The water displays different shades of red, purple, and green according to the content of halophilic Archaea, photosynthetic purple bacteria, and *Cyanobacteria*. Microbial mats occur along the lake floors and margins. The mats grow under a thin (1–2 mm) layer of sand. They consist of thin pink layers of purple photosynthetic bacteria and red halophilic Archaea followed by thick black layers containing decayed organic matter and sulfide minerals. Molecular analysis of the prokaryotic community of three large lakes of the Wadi An Natrun revealed a diverse range of prokaryotes, a high proportion of which had less than 90% 16S rRNA sequence identity to any sequences deposited in GenBank. The majority of the sequences retrieved were affiliated with the *Firmicutes*, α-Proteobacteria, *Bacteroidetes*, and *Halobacteriales*.[16]

Kenyan-Tanzanian Rift Valley

The Kenyan-Tanzanian Rift Valley harbors a series of six highly alkaline lakes. The salinities of these lakes vary from 1 M in the northerly lakes to 5 M in the lakes in the south, with roughly equal proportions of sodium carbonate and chloride as the major salts.[17] The Rift Valley is a volcanically active region and many of the soda lakes are fed by hot springs at temperatures of 45°C–96°C. A combination of intense solar irradiation, high ambient temperature (30°C–40°C), and continuous supply of CO_2 causes these soda lakes to be among the most productive natural environments in the world in terms of biomass.

The microbial community of these soda lakes contains alkaliphilic representatives of all major trophic groups of Bacteria and Archaea. It is hypothesized

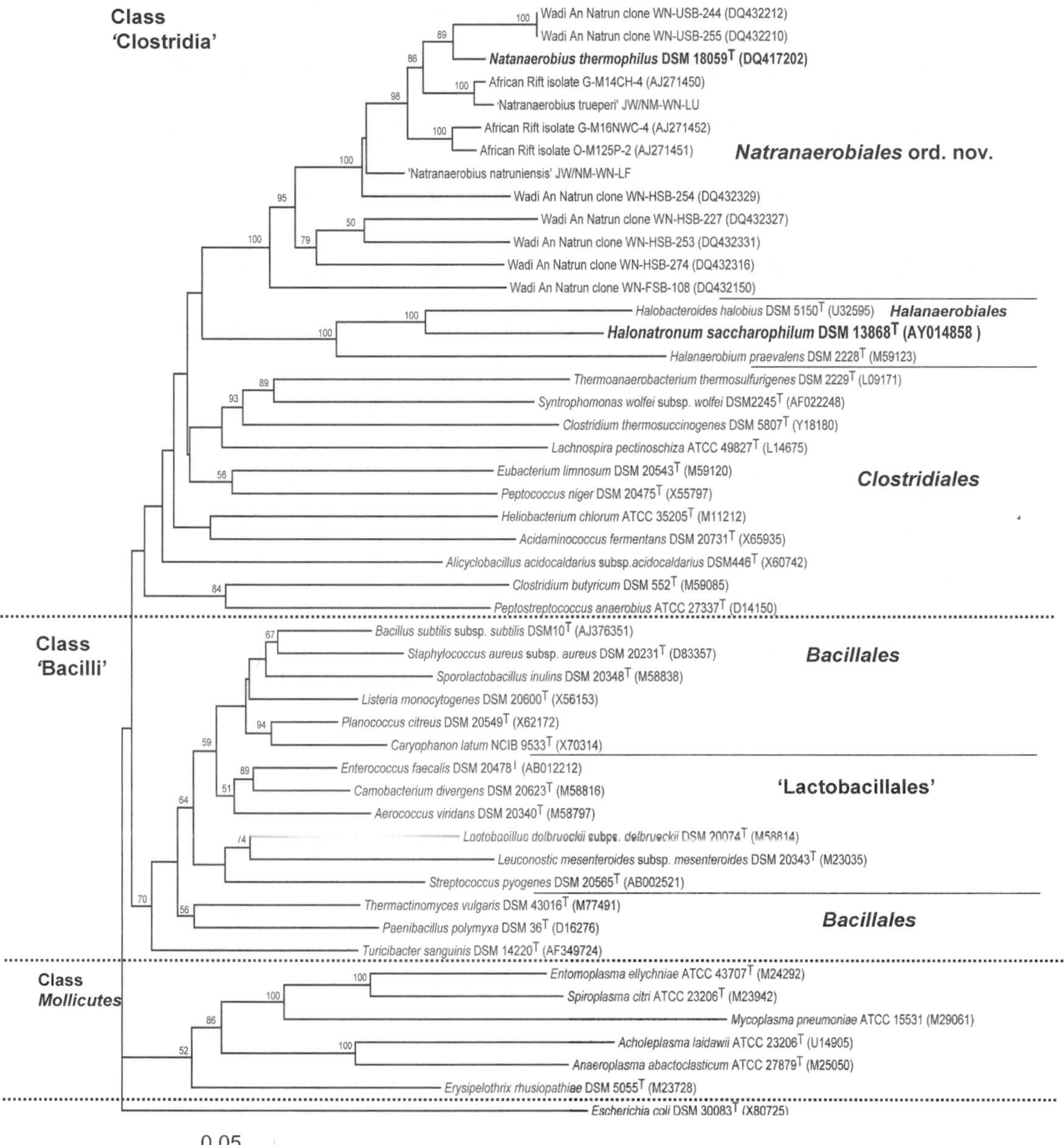

FIGURE 2. Neighbor-joining tree based on 16S rRNA sequences showing position of anaerobic halophilic alkalithermophiles in relation to environmental clones and type species of the type genera within the *Firmicutes*. GenBank accession numbers for the sequences are in parentheses. The tree was rooted with the 16S rRNA gene of *E. coli* DSM 30083^{T} as the outgroup. Numbers at nodes are bootstrap values based on 100 replicates; only values greater than 50 are shown. Bar 5 nucleotide substitution per 100 nt.

that carbon, sulfur, and nitrogen are actively cycled among these groups under both aerobic and anaerobic conditions. The aerobic environment of these soda lakes is dominated by *Cyanobacteria*, which contribute to photosynthetic primary production in the lakes,[18] and haloalkaliphilic archaea.[17] Phototrophic proteobacteria of the genera *Ectothiorhodospira* and *Halorhodospira* also make an unquantified contribution to primary productivity.[19] The soda lake anaerobic environment is dominated by chemoorganotrophic and sulfate-reducing members of the *Firmicutes*. Lacustrine muds at the bottom of the lakes are black and anoxic, and the characteristic odor of hydrogen sulfide associated with sulfate-reducing activity has been reported.[19] The sediments in the lakes also have a sufficiently low redox potential (between −310 and 470 mV) to allow for the activity of sulfate-reducing bacteria. In addition to *H. saccharophilum*, a fermentative gram-negative microorganism belonging to the order *Thermotogales* was isolated. This microorganism is capable of growth at temperatures up to 78°C and pH values above 10.5, but cannot tolerate large sodium chloride concentrations.[19] It has not been characterized beyond the genus stage.

Phylogeny of the Anaerobic Halophilic Alkalithermophiles

Phylogenetically, all halophilic alkalithermophiles isolated, both aerobic and anaerobic, are members of the *Firmicutes*. The two anaerobic halophilic alkalithermophiles described, *H. saccharophilum* and *N. thermophilus*, belong to class *Clostridia* of the *Firmicutes*, but fall into different orders, order *Halanaerobiales* and order *Natranaerobiales* (FIG. 2). Additional anaerobic halophilic alkalithermophiles isolated in our laboratory also cluster within the *Natranaerobiales*. We hypothesize that the narrow phylogeny observed for this group of extremophiles is not attributable to restricted evolution and distribution, but rather to lack of exploration and/or a restricted isolation scheme. Culture-independent analyses have been conducted in alkaline, hypersaline lakes from around the world, including the Wadi An Natrun lakes, Egypt; Mono Lake, California; soda lakes in the Kenyan-Tanzanian Rift Valley; soda lakes in Mongolia and Inner Mongolia, China; athalassohaline lakes of the Atacama desert, Chile; saline meromictic Lake Kaiike, Japan; saline Qinghai Lake, China; and athalassohaline Lake Chaka, China. These studies have revealed an extensive prokaryotic diversity with clones representing bacterial and archaeal groups including *Firmicutes*; α-, γ-, and δ-proteobacteria; *Bacteroidetes*; *Actinobacteria*; *Verrucomicrobia*; *Cyanobacteria*; *Halobacteria*; and *Methanosarcinales*.[16,18,20–26] Intense solar irradiation and anoxic conditions commonly encountered in hypersaline environments must promote the existence of microorganisms capable of not only surviving, but growing and proliferating under harsh environmental conditions. Future isolation and cultivation schemes should be directed toward isolation of specific groups, such as sulfate reducers, methanogens, and/or methane oxidizers, and phototrophic microorganisms.

Adaptive Mechanisms of Anaerobic Alkalithermophiles and Halophilic Alkalithermophiles

Life at high salt concentrations, alkaline pH values, and high temperatures undoubtedly requires special adaptive physiological mechanisms. Each extreme growth condition, whether high salt concentration, alkaline pH, or high temperature, poses a number of physiological and bioenergetic problems outlined below.

Adaptive Mechanisms of Halophiles

Halophilic microorganisms must keep their cytoplasm isoosmotic or hyperosmotic with their surroundings to prevent loss of water to the environment. Maintenance of a turgor pressure requires a hyperosmotic cytoplasm. All halophilic microorganisms, with the possible exception of the halophilic archaea of the family Halobacteriaceae, maintain a turgor pressure.[27]

Two different strategies enable halophilic and halotolerant prokaryotes to withstand the high osmotic stress associated with their hypersaline habitats. In the first strategy, cells accumulate and maintain high concentrations of inorganic ions in the cytoplasm (the "salt-in" strategy). In this case, all the intracellular enzymatic machinery is adapted to the presence of high salt concentrations. In the second strategy, cells maintain low cytoplasmic salt concentrations but accumulate high concentrations of low molecular weight compatible solutes (the "salt-out" or "compatible solute" strategy). No adaptation of intracellular machinery is required.

High Salt + High Temperature Implications for Halophilic Thermophiles

The salt-in strategy is used by three phylogenetically unrelated groups; the extremely halophilic aerobic archaea of the family Halobacteriaceae, the anaerobic bacteria of the order Halanaerobiales, and the extremely halophilic *Salinibacter ruber* of the Bacteroidetes.[11] No organic osmotic solutes have been found in the members of these groups. The cytoplasm in these microorganisms is characterized by the presence of molar concentrations of K^+ cations, and Cl^- is the predominant anion. Microorganisms using the salt-in strategy are obligate halophiles, high concentrations of inorganic ions, both intra- and extra-cellularly, are necessary for proper maintenance of cell shape, membrane structure, and enzyme activity.

The compatible-solute strategy is used by most moderately halophilic bacteria with the exception of the members of the order Haloanaerobiales and *S. ruber*.

The use of organic compatible solutes allows a great deal of flexibility and adaptability to a wide range of NaCl concentrations.[28]

The Salt-In Strategy

Analysis of intracellular ionic concentrations in the halothermophile *Halobacterium salinarum* showed that it contained an extremely high intracellular salt concentration and that the ionic composition of the intracellular milieu was different from that of the outside medium. For *H. salinarum* cells in early exponential growth phase, the intracellular concentration of K^+ was 5.3 M, whereas the medium concentration was 0.05 M. The intracellular concentration of Na^+ was 0.8 M, but the medium concentration was 3.3 M.[28] High intracellular concentrations of Na^+, K^+, and Cl^- were also measured inside cells of mesophilic members of the order *Halanaerobiales*, and K^+ was the predominant cation. The concentrations were high enough to be isotonic with the medium.[29] Apparent intracellular concentrations of K^+ measured in *S. ruber* grown in medium with 3.3 M NaCl was 0.6 M as determined by X-ray microprobe analysis under the electron microscope. Insignificant amounts of compatible solutes were detected in *S. ruber*, so it appears that it also uses the salt-in strategy.[30]

For microorganisms using the salt in strategy for osmotic adaptation, all structural cell components and enzymes must be adapted to the presence of molar concentrations of inorganic salts. High salt concentrations reduce the availability of free water due to hydration of ions and lead to aggregation and collapse of protein structure attributable to alteration of electrostatic and hydrophobic interactions on the solvent-exposed surfaces of proteins.[11] Evolutionary modifications required to engineer proteins so that they become halophilic have included the incorporation of a large number of acidic amino acids (glutamate and aspartate) into the surface of proteins. The pI values of enzymes from extreme halophiles are typically around 4. The carboxylic groups of the acidic amino acids sequester and organize a tight network of water and hydrated K^+ ions around the protein and form internal salt bridges with basic amino acid residues to provide internal structural rigidity to the protein.[31] Another characteristic feature of hydrophilic proteins is their low hydrophobic amino acid residue content, which is offset by a high content of the borderline hydrophobic amino acids serine and threonine. High salt in the surrounding environment of the proteins allows them to maintain the weak hydrophobic interactions and preserve protein structure.[27] Most proteins of microorganisms using the salt-in strategy require high salt concentrations for maintenance of proper conformation and activity. When suspended in hypoosmotic solutions, repulsion of the negatively charged acidic side groups on the surface of the protein results in structural collapse.

The Salt-Out (Compatible Solute) Strategy

In most other halophilic and halotolerant bacteria and the halophilic methanogenic archaea, osmotic balance is provided by small organic molecules that are either synthesized by the cell or taken up from the medium when available. The compatible solute strategy does not require specifically adapted proteins. Compatible solutes are polar, highly soluble molecules that are uncharged or zwitterionic at physiological pH. Compatible solutes include amino acids and derivatives; quaternary amines, such as glycine betaine; sugars or sugar derivatives, such as sucrose, trehalose, and 2-sulfotrehalose; and polyols, such as glycerol and arabitol.[32,33]

Intracellular concentrations of compatible solutes are regulated according to the salinity of the medium, allowing this strategy to provide a high degree of adaptability of cells to changes in the salt concentration of the environment. However, this strategy is only highly effective in environments with up to 1.5 M salt. Above this concentration it is less effective and energetically unfavorable.[27] However, compatible solutes have been observed in the extremely halophilic black yeast *Hortaea werneckii*, in which glycerol accumulates intracellularly when *H. werneckii* grows at salinities up to 1.5 M, but then remains unchanged at NaCl concentrations between 1.5 and 3 M. This indicates that this halophilic black yeast may contain another compatible solute(s).[34] The intracellular cation concentration of *H. werneckii* was far below the extracellular cation concentration, and thus unlikely to be involved in osmotic balance.[34] In the case of halophilic alkalithermophiles with broad NaCl ranges between 1.5 and 4 M, it would be expected that they either combine both strategies to cope with high salt concentrations or use the salt-in strategy alone.

Maintenance of a turgor pressure by a halophile, whether by the salt-in or salt-out strategy, requires a highly impermeable cell wall and cell membrane to maintain the large extracellular sodium gradient and to prevent diffusion of the intracellular K^+ and/or accumulated compatible solute. However, at high temperatures ($\geq$ 45°C), the rate of Na^+ diffusion through the membrane increases as it is a temperature-dependent process.[35] In addition, a study by Vossenberg *et al.* showed a 10-fold increase in the Na^+ permeability of liposomes prepared from haloarchaeal cell membranes as NaCl concentration increased from

0.5 to 4 M.[36] Nevertheless, numerous halophilic thermophiles have been isolated and brought into culture, thus it follows that halophilic thermophiles must have developed unique mechanisms, particularly novel membrane lipid compositions, that remain impermeable to ions, both at elevated salt concentrations and at elevated temperatures.

Adaptive Mechanisms of Alkaliphiles

Microorganisms must maintain a cytoplasmic pH that is compatible with optimal functional and structural integrity of the cytoplasmic proteins that support growth. Many nonextremophilic microorganisms grow over a broad range of external pH values, from 5.5 to 9.0, and maintain a cytoplasmic pH that lies within the narrow range of pH 7.4–7.8.[37] The consequences of not being able to do so are profound. The anaerobic *Caloramator fervidus*, previously named *Clostridium fervidus*,[38] has bioenergetic processes that are entirely Na^+-coupled and lacks active H^+ extrusion or uptake systems that can support pH homeostasis. As a result, it grows only within the narrow pH range of 6.3–7.7 that corresponds to an optimal cytoplasmic pH.[39]

The key bioenergetic difficulty faced by all alkaliphiles is that of cytoplasmic acidification and/or pH homeostasis. A large number of adaptive strategies are used for intracellular pH homeostasis, including (a) increased ATP synthase activity that couples H^+ entry to ATP generation, (b) increased expression and activity of monovalent cation/proton antiporters, (c) changes in cell surface properties, and (d) increased metabolic acid production through amino acid deaminases and sugar fermentation. Among these strategies, monovalent cation/proton antiporters play an essential and dominant role in cytoplasmic pH regulation and also have a role in Na^+ and volume homeostasis.[40,41]

Bacteria have multiple monovalent cation/proton antiporters: there are at least four in the facultative aerobe *Escherichia coli*,[41] five in the aerobic *Bacillus subtilis*,[42,43] and four in the alkaliphilic *Bacillus* strains.[44,45] Detailed information on individual Na^+/H^+ and $Na^+(K^+)/H^+$ antiporters is beginning to provide insights into the basis for the dominant role of specific antiporters in cytoplasmic pH homeostasis of individual bacterial strains.

Although microorganisms have multiple monovalent cation/proton antiporters forming complex H^+ and Na^+ cycles within the cell, these different antiporters do not contribute equally to cytoplasmic pH regulation and homeostasis. Properties of an antiporter that control its impact on cytoplasmic pH acidification include the stoichiometry and kinetics of the exchange, the affinity (K_D) of substrate binding, positive and negative effectors of activity, and reaction mechanisms. Very little work on antiporters and their characteristics has been done in anaerobic alkaliphiles, alkalithermophiles, and halophilic alkalithermophiles.

Importance of Electrogenicity for Na^+/H^+ Antiport-dependent pH Acidification

If Na^+/H^+ antiporters are to be effective at cytoplasmic acidification, transport must be electrogenic rather than electroneutral.[37] An electroneutral exchange of cytoplasmic Na^+ for extracellular H^+ is a 1:1 exchange involving no net flux of charge. Thus, the membrane potential ($\Delta\psi$) component of the proton motive force does not contribute to energizing the exchange. Microorganisms extrude Na^+ from their cytoplasm by diverse expulsion mechanisms because it is cytotoxic.[46] As a result, there is no outwardly directed Na^+ gradient to support electroneutral Na^+/H^+ antiport. Electroneutral Na^+/H^+ antiport can only equilibrate the ΔpH component of the proton motive force, but cannot drive the accumulation of H^+ necessary for cytoplasmic acidification. Electrogenic Na^+/H^+ antiport exchange a larger number of H^+ for each extruded Na^+ during a turnover, so that a net positive charge moves inward. Thus the $\Delta\psi$ component of the proton motive force, which is negative inside relative to outside the cell, is able to drive the inward H^+ proton movement, and cytoplasmic acidification can be achieved. The coupling stoichiometry of the Na^+/H^+ antiporter NhaA of *E. coli* was determined to be 2 $H^+/1\ Na^+$.[47] For other Na^+/H^+ and $Na^+(K^+)/H^+$ antiporters that function in cytoplasmic acidification, precise coupling stoichiometries have not yet been determined. No work has been done to identify and characterize antiporters in anaerobic alkaliphiles and alkalithermophiles. Initial unpublished data from our laboratory on membrane vesicles prepared from the anaerobic halophilic alkalithermophile *N. thermophilus* shows the presence of strong Na^+/H^+ antiporter activity, but no stoichiometry has been determined yet.

Substrate and Cation Specificity of Antiporters

Although studies on antiporter-mediated cytoplasmic acidification have been focused so far on the $Na^+(Li^+)/H^+$-specific antiporters, other types of antiporters have been described. For example, the electroneutral Bs-MleN antiporter catalyzes the 2 H^+-Malate^{2-}/Na^+-Lactate^{1-} exchange.[48] Driven by inwardly directed malate and outwardly directed lactate gradients, external malate enters with H^+ and cytoplasmic lactate exits with Na^+, thus supporting cytoplasmic accumulation of H^+. Theoretically, an electoneutral K^+/H^+ antiporter can support H^+ uptake as alkaliphiles usually maintain an outwardly directed K^+ gradient. Involvement of K^+/H^+ antiporter activity in cytoplasmic acidification has been reported in *E. coli*, *Enterococcus hirae*, *Vibrio alginolyticus*, and most recently *Alkalimonas amylolytica*.[49–52] NhaK, a *B. subtilis* transporter belonging to the CPA1 antiporter family, catalyzes the K^+/H^+ antiport at alkaline pH.[43] The anaerobic halophilic alkalithermophile *N. thermophilus* has also been shown in our laboratory to use K^+ for cytoplasmic acidification. However, whether this activity is attributable to a specific K^+/H^+ antiporter or an $Na^+(K^+)/H^+$ antiporter remains to be determined (N.M. Mesbah and J. Wiegel, unpublished results).

Antiporters and the Alkaliphile Na^+ Cycle

In line with the essential role of Na^+/H^+ antiporters in alkaliphily, alkaliphile pH homeostasis is strictly dependent on the presence of Na^+.[37,53] Alkaliphiles must use Na^+-specific antiporters to avoid the risk of lowering the cytoplasmic K^+ concentration. However, for optimal cytoplasmic acidification, re-entry routes for Na^+ are crucial to provide a substrate for continuous antiporter activity. One such route is Na^+-coupled transport of solutes and metabolites into the cell. Such a route has been postulated for the anaerobic alkalithermophiles *Clostridium paradoxum* and *Anaerobranca gottschalkii*.[54,55] Initial data from our laboratory also showed that solute uptake in the anaerobic halophilic alkalithermophile *N. thermophilus* is Na^+-coupled. An investigation by Pitryuk *et al.*[56] on the effects of metabolic inhibitors on energy-generating processes in the anaerobic haloalkalithermophile *H. saccharophilum* showed that its metabolism depends on both Na^+ and H^+ gradients. Another route for Na^+ re-entry to the cell is via an Na^+-translocating ATPase. This ATPase will catalyze the synthesis of ATP in conjunction with the pumping of Na^+ ions into the cell. Cloning and characterization of the ATPase in the anaerobic alkalithermophile *C. paradoxum* showed that it is Na^+-coupled and incapable of ATP synthesis; it is used solely as an Na^+ pump.[54] ATPase activity in inverted membrane vesicles prepared from cells of *A. gottschalkii* was also found to be stimulated by both Na^+ and Li^+ cations.[55]

Adaptive Strategies of Alkaliphiles Apart from the Na^+ Cycle

Adaptive mechanisms of anaerobic alkaliphiles and alkalithermophiles have yet to be investigated in detail. Studies on aerobic alkaliphilic *Bacillus* species revealed the presence of specific secondary cell wall polymers associated with the peptidoglycan that are crucial for survival at pH 10.5 but not at 7.5.[37] Alkaliphilic *Bacillus* species also have large proportions of cardiolipin and squalene in the cell membrane. It is hypothesized that both these membrane components are involved in trapping protons at the membrane surface. The phospholipid fatty acid (PLFA) profile of *N. thermophilus*, when grown at its optimal growth conditions of $pH^{55°C}$ 9.5, 1.5 M NaCl, and 53°C, revealed a unique pattern of branched chain dimethyl acetals.[9] It is unclear at this stage what role these unique fatty acids are playing in cellular bioenergetics. Future research should focus on identifying the changes in PLFA profiles of anaerobic extremophiles in response to different pH values (and NaCl concentrations) in the growth medium to identify the specific fatty acids involved in alkaline adaptation.

Adaptive Mechanisms of Thermophiles

Thermophiles are faced with the challenge of controlling cytoplasmic membrane permeability at high temperatures. Increased intramolecular motion of lipids at elevated temperatures results in increased permeability to protons. Because of this motion, water molecules become trapped in the lipid core of the membranes, allowing protons to hop from one molecule to the other. Other ions, unlike protons, can diffuse through the membrane. Diffusion is a temperature-dependent process, thus membrane permeability to ions will increase as well.[35] In addition, thermophiles must have mechanisms to preserve protein structure at elevated temperatures.

Studies on permeabilities of bacterial and archaeal cytoplasmic membranes have shown that Bacteria and Archaea adjust the permeability of the cytoplasmic membranes to the growth temperature of the microorganism.[57] In psychrophilic and mesophilic bacteria and archaea, the proton permeability of the cytoplasmic membrane is kept constant by regulation of the phospholipid composition of the membrane as a function of temperature.[58] However, thermophilic bacteria and archaea growing at temperatures greater than 50°C encounter increased membrane proton permeability because they are no longer able to compensate by adjusting the lipid composition. The membrane proton permeability of the facultatively aerobic *Geobacillus stearothermophilus*, anaerobic *Thermotoga maritima*, and aerobic *Sulfolobus acidocaldarius* increases exponentially with temperature. The sodium ion permeability is several orders of magnitude lower than the proton permeability, thus many thermophiles use Na^+-coupled bioenergetics.[35]

The mechanisms underlying thermal stability of proteins are yet to be fully explained. The "equilibrium model," introduced by Daniel *et al.*[59] was formulated to describe genuine temperature optima of enzymes and correlates thermal stability with the temperature optimum of an enzyme in cases where temperature is a major factor in enzyme stability. In general, extreme conditions require either the adaptation of the amino acid sequence of a protein by mutations, the optimization of weak interactions within the protein at the protein–solvent boundary, and the influence of extrinsic factors, such as metabolites and cofactors. With regard to thermal adaptations, instead of being a consequence of one dominant type of interaction or of a general stabilization strategy, adaptation to high temperatures reflects a number of subtle interactions, often characteristic for each protein species. Comparison of protein sequences of thermophilic and mesophilic *Methanococcus* spp. revealed amino acid substitutions involved in adapting proteins to higher temperatures. The main change was an increase in hydrophobic residues.[60] This adaptation minimizes surface energy and the hydration of apolar surface groups while burying hydrophobic residues and maximizing packing of the core.[61] An increase in hydrophobic residues seems to interfere with the mechanism of adaptation to high salt concentrations, as it was shown that the borderline hydrophobic residues serine and threonine are replaced with the more hydrophobic alanine/proline and valine/isoleucine, respectively.[60] This could explain why all obligate extreme anaerobic halophiles are moderate thermophiles, with T_{max} not exceeding 70°C. Above this temperature, minimizing hydration and polar groups at the surface of the protein are necessary to maintain stability. The presence of high concentrations of intracellular ions would interfere with protein folding and decrease stability.

Interactions among Salt, Alkaline, and Temperature Stresses: Implications for the Anaerobic Halophilic Alkalithermophiles

Alkaline, salt, and temperature stresses show significant interplay with one another. First, alkaline stress and salt stress overlap: as pH rises, Na^+ cytotoxicity greatly increases.[46] The toxicity of Na^+ is also dependent on the status of cytoplasmic K^+ concentration, such that the larger the K^+ concentration in the cytoplasm, the more tolerant the cell becomes to Na^+. Second and third, interplay of an alkaline challenge with that of high temperature alters the H^+ permeability more than the Na^+ permeability in many bacteria,[35] and this in turn interplays with changes in osmolarity since some monovalent cation/proton antiporters have roles in osmoadaptation.[37] Fourth, alkaline pH, high salt concentrations, and temperatures all interplay with cell membrane composition. Although increased temperatures cause an increase in membrane fluidity and permeability,[35] alkaline pH and high salt concentrations cause an increase in the degree of unsaturation of the fatty acids of the cytoplasmic membrane with a concomitant increase in rigidity. An additional stress intersection includes the adaptation of intracellular proteins to high temperature, which involves increasing hydrophobic packing, and the adaptation to high intracellular ionic concentrations, which involves increasing the protein content of hydrophilic proteins.

Studies of the adaptive mechanisms of polyextremophiles are scarce and many do not take into account the effect of elevated salt concentration. Studies on the alkalithermophilic anaerobes *C. paradoxum* and *A. gottschalkii* showed that, as in mesophilic alkaliphiles, the magnitude of the proton motive force is suboptimal for H^+-coupled processes.[53,55] *C. paradoxum* is capable of cytoplasm acidification; it increases its ΔpH across the cell membrane (pH_{in}–pH_{out}) by as much of 1.3 units.[53] At pH values greater than 10.0, near its maximum pH range, the ΔpH and $\Delta\psi$ gradually decline, and the intracellular pH significantly increases until it is almost equal that of the extracellular medium. *C. paradoxum* has an absolute requirement for 50–200 mM of Na^+ for growth, and growth is inhibited by the sodium ionophore monensin, and amiloride, an inhibitor of Na^+/H^+ antiporters. This indicates the importance of the Na^+ cycle in this anaerobic alkalithermophile. *C. paradoxum* is also sensitive to the F-type ATPase inhibitor N,N′-dicyclohexylcarbodiimide, indicating the presence of an F-type ATPase in this anaerobe.[54] The F_1F_0-ATPase of *C. paradoxum* is an Na^+-translocating ATPase, which is used to generate an electrochemical gradient of Na^+ that could be used to drive other membrane-related bioenergetic processes, such as solute transport. Cloning of the *atp* operon revealed an absence of the C-terminal region of the ε subunit that is essential for coupled ATP synthesis.[62] Thus, the F-type ATPase of *C. paradoxum* is used solely as a pump for generation of an Na^+ gradient and is not involved in ATP synthesis.

It is unclear how the additional stress of elevated NaCl concentration will interplay with the adaptive mechanisms discussed above to cope with elevated pH and temperature. The presence of an Na^+-translocating ATPase would play a role in extrusion of excess Na^+ from the cytoplasm. However, continuous extrusion of Na^+—and possibly K^+—from the cytoplasm would compromise osmolarity, which is necessary for maintenance of a turgor pressure at elevated salt concentrations.

In addition, the challenge of acidifying the cytoplasm while growing in alkaline conditions, i.e., a dearth of protons, remains to be investigated. Although a high H^+/Na^+ stoichiometry makes it possible for an antiporter to use the $\Delta\psi$ to acidify the cytoplasm at alkaline pH, it does not solve the problem of H^+ capture by the antiporter when the external concentration of H^+ is low. The efficiency of an Na^+/H^+ antiporter is also compromised by the increased membrane permeability to both H^+ and Na^+ posed by elevated temperatures and salt concentrations, respectively. No cellular features have yet been described that could trap protons on the surface of the cell, similar to those suggested for membrane-embedded electron transport or light-driven proton pumps.

Finally, one of the major challenges encountered by the anaerobic alkalithermophiles and halophilic alkalithermophiles concerns their anaerobic, and thus fermentative, metabolism. The anaerobic alkalithermophile *C. paradoxum* and the anaerobic halophilic alkalithermophiles *N. thermophilus* and *H. saccharophilum* are all fermentative chemoorganotrophs that produce their energy (ATP) in the cytoplasm by substrate-level phosphorylation. Thus, compared with the aerobic extremophiles that produce ATP via respiratory electron transport, ATP production in the fermentative anaerobes is significantly less. Many of the key adaptive mechanisms for the halophilic alkalithermophiles and alkalithermophiles, such as maintenance of the crucial Na^+ cycle, solute uptake, compatible solute synthesis, and synthesis of impermeable cell wall and cell membrane constituents, are processes that require ATP. Because the anaerobic extremophiles mentioned above are capable of rapid growth under extreme conditions (doubling times 16 min to 3.5 h), it follows that the anaerobic extremophiles must have developed highly efficient methods for energy production and turnover that allow them not only to survive, but to grow rapidly under multiple extreme conditions.

Concluding Remarks

The ability of microorganisms not only to survive, but to grow under alkaline pH and high temperature and salt concentration is of widespread ecological and industrial importance. The total repertoire of proteins that contribute to structural, metabolic and bioenergetic adaptations are yet to be investigated. Active alkaline pH homeostasis that is crucial to both growth and survival depends on monovalent cation/proton antiporters, which are identified through both genomic and biochemical approaches. An isoosmotic cytoplasm and maintenance of a turgor pressure necessitate accumulation of molar concentrations of either organic solutes, monovalent cations, or both; these features require adaptation of the whole intracellular enzymatic machinery to high-osmotic conditions. Combined extremes must result in unique membrane lipid profiles and cell wall structures that are impermeable to cations. An intricate interplay among alkaline pH, high salt concentration, and elevated temperature regulates the levels of expression, activity, and synthesis of adaptive proteins and cellular structures. The isolation of novel anaerobic halophilic alkalithermophiles will enable biochemical, genomic, and proteomic studies to define the specific mechanisms that underpin the efficacy of physiologically important antiporters, solute symporters, membrane lipids, and cell wall components in adaptation to multiple extreme conditions.

However, one question still remains: "Have the extreme boundaries for life been identified?" More specifically, what are the highest growth temperatures for microorganisms living at a 5-M NaCl concentration? How will the maximum pH allowing growth be affected at a high-NaCl concentration? Is it possible for microorganisms to grow optimally at 80°C, 90°C, and even above 100°C and at NaCl concentrations approaching saturation? The use of novel microbiological and molecular techniques and sampling of additional alkaline, hypersaline environments will lead to the isolation and identification of more novel anaerobic haloalkalithermophiles with exciting new properties, and the known boundaries for life on earth will be further extended.

Acknowledgments

Work in our laboratory is supported by grants INT-021100 and MCB-0604224 from the National Science Foundation and by grant AFOSR 033835-01 from the Air Force Office of Scientific Research Extremophile Core Program to J. Wiegel.

Conflict of Interest

The authors declare no conflicts of interest.

References

1. Segerer, A. *et al*. 1988. *Thermoplasma acidophilum* and *Thermoplasma volcanium* sp. nov. from solfatara fields. Syst. Appl. Microbiol. **10:** 161–171.

2. Eagon, R.G. 1962. *Pseudomonas natriegens*, a marine bacterium with a generation time of less than 10 minutes. J. Bacteriol. **83:** 736–737.
3. Wiegel, J. 1998. Lateral gene exchange, an evolutionary mechanism for extending the upper and lower temperature limits for growth of microorganisms? A hypothesis. *In* Thermophiles: The Keys to Molecular Evolution and the Origin of Life? J. Wiegel & M.W. Adams, Eds.: 177–185. Taylor & Francis Inc. Philadelphia, PA.
4. Wiegel, J. 1998. Anaerobic alkalithermophiles, a novel group of extremophiles. Extremophiles **2:** 257–267.
5. Kevbrin, V.V. *et al.* 2004. Alkalithermophiles: a double challenge from extreme environments. *In* Cellular Origins, Life in Extreme Habitats and Astrobiology. J. Seckback, Ed.: 395–412. Kluwer Academic Publishers. Dordrecht, the Netherlands.
6. Schleper, C. *et al.* 1995. Picrophilus gen. nov., fam. nov.: a novel aerobic, heterotrophic, thermoacidophilic genus and family comprising archaea capable of growth around pH 0. J. Bacteriol. **177:** 7050–7059.
7. Futterer, O. *et al.* 2004. Genome sequence of Picrophilus torridus and its implications for life around pH 0. 10.1073/pnas.0401356101. PNAS **101:** 9091–9096.
8. Zhilina, T.N. *et al.* 2004. *Halonatronum saccharophilum* gen. nov. sp. nov.: a new haloalkaliphilic bacterium of the order *Halaanaerobiales* from Lake Magadi. Microbiology **70:** 64–72.
9. Mesbah, N.M. *et al.* 2007. *Natranaerobius thermophilus* gen. nov. sp. nov., a halophilic, alkalithermophilic bacterium from soda lakes of the Wadi An Natrun, Egypt, and proposal of *Natranaerobiaceae* fam. nov. and *Natranaerobiales* ord. nov. Int. J. Syst. Evol. Microbiol. **57:** 2507–2512.
10. Sarkar, A. 1991. Isolation and characterization of thermophilic, alkaliphilic, cellulose-degrading *Bacillus thermoalcaliphilus* sp. nov. from termite *(Odontotermes obesus)* mound soil of a semiarid area. Geomicrobiol. J. **9:** 225–232.
11. Oren, A. 2002. Halophilic Microorganisms and Their Environments. Kluwer Academic Publishers. Dordrecht, the Netherlands.
12. Abd-el-Malek, Y. & S.G. Rizk. 1963. Bacterial sulfate reduction and development of alkalinity. III. Experiments under natural conditions in the Wadi An Natrun. J. Appl. Microbiol. **26:** 20–26.
13. Schweinfurth, G. & L. Lewin. 1898. Beitrage zur topographie und geochemie des agyptischen Natron-Thals. Zeitschr. d. Ges. f. Erdk. **33:** 1–25.
14. Imhoff, J.F. *et al.* 1979. The Wadi Natrun: chemical composition and microbial mass developments in alkaline brines of eutrophic desert lakes. Geomicrobiol. J. **1:** 219–234.
15. Taher, A.G. 1999. Inland saline lakes of Wadi El Natrun depression, Egypt. Int. J. Salt. Lake Res. **8:** 149–170.
16. Mesbah, N.M. *et al.* 2007. Novel and unexpected prokaryotic diversity in water and sediments of the alkaline, hypersaline lakes of the Wadi An Natrun, Egypt. Microbial. Eco. **54:** 598–617.
17. Grant, S. *et al.* 1999. Novel archaeal phylotypes from an East African alkaline saltern. Extremophiles **3:** 139–145.
18. Rees, H.C. *et al.* 2004. Diversity of Kenyan soda lake alkaliphiles assessed by molecular methods. Extremophiles **8:** 63–71.
19. Jones, B.E. *et al.* 1998. Microbial diversity of soda lakes. Extremophiles **2:** 191–200.
20. Humayoun, S.B. *et al.* 2003. Depth distribution of microbial diversity in Mono Lake, a meromictic soda lake in California. Appl. Environ. Microbiol. **69:** 1030–1042.
21. Sorokin, D.Y. *et al.* 2004. Prokaryotic communities of northeastern Mongolian soda lakes. Hydrobiologia **522:** 235–248.
22. Ma, Y. *et al.* 2004. Bacterial diversity of the Inner Mongolian Baer Soda Lake as revealed by 16S rRNA gene sequence analyses. Extremophiles **8:** 45–51.
23. Demergasso, C. *et al.* 2004. Distribution of prokaryotic genetic diversity in athalassohaline lakes of the Atacama Desert, Northern Chile. FEMS Microbiol. Ecol. **48:** 57–69.
24. Koizumi, Y. *et al.* 2004. Vertical and temporal shifts in microbial communities in the water column and sediment of saline meromictic Lake Kaiike (Japan), as determined by a 16S rDNA-based analysis, and related to physicochemical gradients. Environ. Microbiol. **6:** 622–637.
25. Dong, H. *et al.* 2006. Microbial diversity in sediments of saline Qinghai Lake, China: linking geochemical controls to microbial ecology. Microb. Ecol. **51:** 65–82.
26. Jiang, H. *et al.* 2006. Microbial diversity in water and sediment of Lake Chaka, an athalassohaline lake in northwestern China. Appl. Environ. Microbiol. **72:** 3832–3845.
27. Oren, A. 1999. Bioenergetic aspects of halophilism. Microbiol. Mol. Biol. Rev. **63:** 334–348.
28. Oren, A. 2000. Life at High Salt Concentrations. Springer-Verlag. New York, NY.
29. Oren, A. *et al.* 1997. X-ray microanalysis of intracellular ions in the anaerobic halophilic eubacterium *Haloanaerobium praevalens*. Can. J. Microbiol. **43:** 588–592.
30. Oren, A. *et al.* 2002. Intracellular ion and organic solute concentrations of the extremely halophilic bacterium *Salinibacter ruber*. Extremophiles **6:** 491–498.
31. Dennis, P. & L. Shimmin. 1997. Evolutionary divergence and salinity-mediated selection in halophilic archaea. Microbiol. Mol. Biol. Rev. **61:** 90–104.
32. Roberts, M. 2005. Organic compatible solutes of halotolerant and halophilic microorganisms. Saline Systems **1:** 5.
33. Ventosa, A. *et al.* 1998. Biology of moderately halophilic aerobic bacteria. Microbiol. Mol. Biol. Rev. **62:** 504–544.
34. Plemenitas, A. & N. Gunde-Cimerman. 2005. Cellular responses in the halophilic black yeast *Hortaea weneckii* to high environmental salinity. *In* Adaptation to Life at High Salt Concentrations in Archaea, Bacteria and Eukarya. N. Gunde-Cimerman, A. Oren & A. Plemenitas, Eds.: 455–470. Springer. Dordrecht, the Netherlands.
35. Konings, W.N. *et al.* 2002. The cell membrane plays a crucial role in survival of bacteria and archaea in extreme environments. Antonie van Leeuwenhoek **81:** 61–72.
36. Vossenberg, J.L.C.M. *et al.* 1999. Lipid membranes from halophilic and alkali-halophilic Archaea have a low H+ and Na+ permeability at high salt concentration. Extremophiles **3:** 253–257.
37. Padan, E. *et al.* 2005. Alkaline pH homeostasis in bacteria: new insights. Biochim. Biophys. Acta **1717:** 67–88.
38. Collins, M.D. *et al.* 1994. The phylogeny of the genus *Clostridium*: proposal of five new genera and eleven new

species combinations. Int. J. Syst. Bacteriol. **44:** 812–826.

39. SPEELMANS, G. *et al.* 1993. Energy transduction in the thermophilic anaerobic bacterium *Clostridium fervidus* is exclusively coupled to sodium ions. PNAS **90:** 7975–7979.
40. KRULWICH, T.A. *et al.* 1998. Energetics of alkaliphilic Bacillus species: physiology and molecules. Adv. Microb. Physiol. **40:** 401–438.
41. PADAN, E. *et al.* 2001. Na^+/H^+ antiporters. Biochim. Biophys. Acta **1505:** 144–157.
42. CHENG, J. *et al.* 1994. The chromosomal tetracyline resistance locus of *Bacillus subtilis* encodes a Na^+/H^+ antiporter that is physiologically important at elevated pH. J. Biol. Chem. **269:** 27365–27371.
43. FUJISAWA, M. *et al.* 2005. NhaK, a novel monovalent cation/H+ antiporter of *Bacillus subtilis*. Arch. Microbiol. **183:** 411–420.
44. ITO, M. *et al.* 1997. Role of the *nhaC*-encoded Na+/H+ antiporter of alkaliphilic *Bacillus firmus* OF4. J. Bacteriol. **179:** 3851–3857.
45. TAKAMI, H. *et al.* 2000. Complete genome sequence of the alkaliphilic bacterium *Bacillus halodurans* and genomic sequence comparision with *Bacillus subtilis*. Nucleic Acids Res. **28:** 4317–4331.
46. PADAN, E. & T.A. KRULWICH. 2000. Sodium stress. *In* Bacterial Stress Response. G. Storz & R. Hengge-Aronis, Eds.: 117–130. ASM Press. Washington, DC.
47. TAGLICHT, D. *et al.* 1993. Proton-sodium stoichiometry of NhaA, an electrogenic antiporter from *Escherichia coli*. J. Biol. Chem. **268:** 5382–5387.
48. WEI, Y. *et al.* 2000. *Bacillus subtilis* YqkI is a novel malic/Na+ -lactate antiporter that enhances growth on malate at low proton motive force. J. Biol. Chem. **275:** 30287–30292.
49. PLACK, JR., R.H. & B.P. ROSEN. 1980. Cation/proton antiport systems in *Escherichia coli*. Absence of potassium/proton antiport activity in a pH-sensitive mutant. J. Biol. Chem. **255:** 3824–3825.
50. KAKINUMA, Y. & K. IGARASHI. 1995. Potassium/proton antiport system of growing Enterococcus hirae at high pH. J. Bacteriol. **177:** 2227–2229.
51. NAKAMURA, T. *et al.* 1992. Roles of K^+ and Na^+ in pH homeostasis and growth of the marine bacterium *Vibrio alginolyticus*. J. Gen. Microbiol. **138:** 1271–1276.
52. WEI, Y. *et al.* 2007. Three putative cation/proton antiporters from the soda lake alkaliphile *Alkalimonas amylolytica* N10 complement an alkali-sensitive *Escherichia coli* mutant. Microbiology **153:** 2168–2179.
53. COOK, G. *et al.* 1996. The intracellular pH of *Clostridium paradoxum*, an anaerobic, alkaliphilic, and thermophilic bacterium. Appl. Environ. Microbiol. **62:** 4576–4579.
54. FERGUSON, S.A. *et al.* 2006. Biochemical and molecular characterization of a Na^+-translocating F_1F_o-ATPase from the thermoalkaliphilic bacterium *Clostridium paradoxum*. J. Bacteriol. **188:** 5045–5054.
55. PROWE, S. *et al.* 1996. Sodium-coupled energy transduction in the newly isolated thermoalkaliphilic strain LBS3. J. Bacteriol. **178:** 4099–4104.
56. PITRYUK, A.V. *et al.* 2004. Comparative study of the energy metabolism of anaerobic alkaliphiles from soda lakes. Mikrobiologiia **73:** 293–299.
57. VOSSENBERG, J.L.C.M. *et al.* 1995. Ion permeability of the cytoplasmic membrane limits the maximum growth temperature of bacteria and archaea. Mol. Microbiol. **18:** 925–932.
58. VOSSENBERG, J.L.C.M. *et al.* 1999. Homeostasis of the membrane proton permeability in *Bacillus subtilis* grown at different temperatures. Biochim. Biophys. Acta **1419:** 97–104.
59. DANIEL, R. *et al.* 2001. The temperature optima of enzymes: a new perspective on an old phenomenon. Trends Biochem. Sci. **26:** 223.
60. HANEY, P.J. *et al.* 1999. Thermal adaptation analyzed by comparison of protein sequences from mesophilic and extremely thermophilic *Methanococcus* species. PNAS **96:** 3578–3583.
61. YIP, K.S. *et al.* 1995. The structure of *Pyrococcus furiosus* glutamate dehydrogenase reveals a key role for ion-pair networks in maintaining enzyme stability at extreme temperatures. Structure **3:** 1147–1158.
62. CIPRIANO, D.J. & S.D. DUNN. 2006. The role of the epsilon subunit in the *Escherichia coli* ATP synthase. The C-terminal domain is required for efficient energy coupling. J. Biol. Chem. **281:** 501–507.

Physiology, Ecology, Phylogeny, and Genomics of Microorganisms Capable of Syntrophic Metabolism

MICHAEL J. MCINERNEY,[a] CHRISTOPHER G. STRUCHTEMEYER,[a] JESSICA SIEBER,[a] HOUSNA MOUTTAKI,[a] ALFONS J. M. STAMS,[b] BERNHARD SCHINK,[c] LARS ROHLIN,[d] AND ROBERT P. GUNSALUS[d]

[a]*Department of Botany and Microbiology, University of Oklahoma, Norman, Oklahoma, USA*

[b]*Laboratory of Microbiology, Department of Agrotechnology and Food Sciences, Wageningen University, Wageningen, the Netherlands*

[c]*Department of Biology and Microbial Ecology, University of Konstanz, Konstanz, Germany*

[d]*Department of Microbiology, Immunology, and Molecular Genetics, University of California, Los Angeles, California, USA*

Syntrophic metabolism is diverse in two respects: phylogenetically with microorganisms capable of syntrophic metabolism found in the Deltaproteobacteria and in the low G+C gram-positive bacteria, and metabolically given the wide variety of compounds that can be syntrophically metabolized. The latter includes saturated fatty acids, unsaturated fatty acids, alcohols, and hydrocarbons. Besides residing in freshwater and marine anoxic sediments and soils, microbes capable of syntrophic metabolism also have been observed in more extreme habitats, including acidic soils, alkaline soils, thermal springs, and permanently cold soils, demonstrating that syntrophy is a widely distributed metabolic process in nature. Recent ecological and physiological studies show that syntrophy plays a far larger role in carbon cycling than was previously thought. The availability of the first complete genome sequences for four model microorganisms capable of syntrophic metabolism provides the genetic framework to begin dissecting the biochemistry of the marginal energy economies and interspecies interactions that are characteristic of the syntrophic lifestyle.

Key words: **syntrophy; methanogenesis; hydrogen; formate; acetogenesis; *Syntrophus*; *Syntrophomonas***

Introduction

The complete mineralization of complex organic matter to CO_2 and CH_4 occurs in anoxic environments where electron acceptors, other than CO_2, are limiting.[1–5] Examples of such environments include freshwater sediments, flooded soils, wet wood of trees, tundra, landfills, and sewage digestors. Syntrophic metabolism plays an essential role in the recycling of organic matter to methane and carbon dioxide in these environments. The degradation of natural polymers, such as polysaccharides, proteins, nucleic acids, and lipids, to CO_2 and CH_4 involves a complex microbial community. Fermentative bacteria hydrolyze the polymeric substrates, such as polysaccharides, proteins, and lipids, and ferment the hydrolysis products to acetate and longer-chain fatty acids, CO_2, formate, H_2. Acetogenic bacteria are likely involved in the fermentation of methanol derived from the demethylation of pectin and in the *O*-demethylation of low molecular-weight ligneous materials, and ferment hydroxylated and methoxylated aromatic compounds with the production of acetate.[6] Propionate and longer-chain fatty acids, alcohols, and some amino acids and aromatic compounds are syntrophically degraded to the methanogenic substrates, H_2, formate, and acetate.[2,4] The syntrophic degradation of fatty

Address for correspondence: Michael J. McInerney, Department of Botany and Microbiology, University of Oklahoma, 770 Van Vleet Oval, Norman, OK 73019. Voice: 405-325-6050; fax: 405-325-7619. mcinerney@ou.edu

Ann. N.Y. Acad. Sci. 1125: 58–72 (2008). © 2008 New York Academy of Sciences.
doi: 10.1196/annals.1419.005

acids is often the rate-limiting step, so it is essential that waste treatment reactors be operated under conditions that favor the retention of bacteria capable of syntrophic metabolism. Last, two different groups of methanogens, the hydrogenotrophic methanogens and the acetotrophic methanogens, complete the process by converting the acetate, formate, and hydrogen made by other microorganisms to methane and carbon dioxide. The syntrophic metabolism and methanogenesis must be tightly coupled to accomplish anaerobic degradation.

Large amounts of organic matter are microbially degraded, making methanogenesis an integral part of the global carbon cycle. Methanogenesis also occurs in the gastrointestinal tract of animals; however, organic matter is incompletely degraded to acetate and longer-chain fatty acids, which accumulate and are absorbed and used by the host animal as energy sources.[7] Syntrophic bacteria and acetoclastic methanogens grow too slowly to be maintained in the gastrointestinal tract. The amount of energy released and harvested per unit of biomass degraded during methanogenesis is very low. For this reason, methanogenesis is a treatment of choice for complex waste digestion, because sludge yields are low and most of the energy in the original substrates is retained in the energy-rich product, methane.

The methanogenic fermentation of complex polymeric materials involves a number of diverse, interacting microbial species. The mutual dependence between interacting species can be so extreme that neither species can function without the activity of its partner, and together the partners perform functions that neither species can do alone. Syntrophy is a specialized case of tightly coupled mutualistic interactions. The term syntrophy was first used to describe the interaction between phototrophic green sulfur bacteria and chemolithotrophic, sulfur-reducing bacteria[8] and fatty acid–oxidizing microorganisms and hydrogen/formate-using microorganisms.[3] In both cases, the pool size of intermediates that are exchanged between the partners (sulfur during anaerobic photosynthesis or hydrogen/formate during fatty-acid metabolism) must be kept very low for efficient cooperation among the partners to occur. We will focus on syntrophic interactions active in methanogenic environments where hydrogen and formate are exchanged between the two partners. In these syntrophic interactions, the degradation of the parent compound, for example, the fatty acid, is thermodynamically unfavorable unless the hydrogen and formate produced by the fatty-acid degrader is kept at low levels by a second microorganism, in this case, a hydrogen/formate-consuming methanogen.[4] The thermodynamic basis for these interactions is discussed later in the chapter.

Under optimal conditions, the free energy changes involved in syntrophic metabolism are close to equilibrium[9–12] and the little free energy that is released in these reactions must be shared among partners.[4] Growth rates and growth yields are low, 7-month doubling time and 0.6 g (dry weight) per mole of methane for anaerobic methane–oxidizing consortium,[13] making biochemical investigations very difficult. Thus, it is appropriate to describe syntrophy as an extreme existence, a lifestyle that involves a marginal energy economy.

In this review, we discuss syntrophic interactions operative in natural and man-made environments from the perspective of the organisms involved, their phylogenetic relationships, the range of syntrophic substrates metabolized, and the new perspectives offered by the emerging genome sequencing information for several model syntrophic microorganisms.

Historical Origins

The first example of a thermodynamically based syntrophic interaction was ethanol metabolism performed by members of the "*Methanobacillus omelianskii*" culture.[14] Subsequently, Bryant *et al.*[15] showed that the "*Methanobacillus omelianskii*" culture was in fact a coculture of two organisms, the S organism and *Methanobacterium bryantii* strain M.O.H. (FIG. 1). The S organism fermented ethanol to acetate and hydrogen:

$$2CH_3CH_2OH + 2H_2O \rightarrow 2CH_3COOH + 4H_2$$
$$\Delta G^{\circ\prime} = 19\text{ kJ (per 2 mol ethanol)} \quad (1)$$

The methanogen did not use ethanol, but used the H_2 made by the S organism to reduce CO_2 to CH_4:

$$4H_2 + CO_2 \rightarrow CH_4 + 2H_2O$$
$$\Delta G^{\circ\prime} = -131\text{ kJ (per mole of } CH_4) \quad (2)$$

When the two reactions are combined, the degradation of ethanol becomes favorable:

$$2CH_3CH_2OH + CO_2 \rightarrow 2CH_3COOH + CH_4$$
$$\Delta G^{\circ\prime} = -112\text{ kJ (per mole of } CH_4) \quad (3)$$

The importance of end-product removal on the thermodynamics of syntrophic ethanol, propionate, and butyrate degradation is illustrated in TABLE 1. Under standard conditions, the degradation of the substrates listed in TABLE 1 is endergonic. However, if the hydrogen partial pressure is low, then the degradation

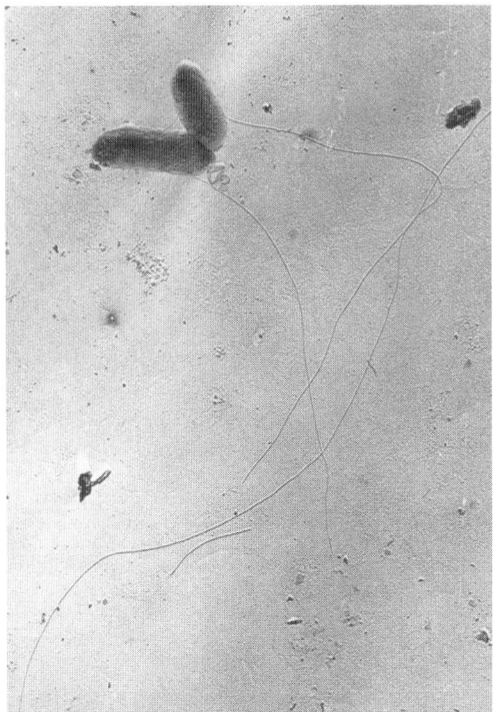

FIGURE 1. Electron micrograph of the ethanol-utilizing, rod-shaped syntrophic "S organism" isolated by Bryant and co-workers.[15] It was the first bacterium capable of syntrophic metabolism to be isolated in pure culture. The S organism possesses a single polar flagellum whose length extends approximately 10-fold the length of the cell body. (Photograph courtesy of Professor R. S. Wolfe.)

TABLE 1. Reactions involved in syntrophic metabolism

Reactions	$\Delta G^{\circ\prime}$ (footnote [a])	pH_2 for $-\Delta G'$ (footnote [b])
Ethanol + $H_2O \rightarrow$ Acetate$^-$ + H^+ + $2H_2$	+9.6	$<10^{-1}$
Propionate$^-$ + $3H_2O \rightarrow$ Acetate$^-$ + HCO_3^- + H^+ + $3H_2$	+76.1	$<10^{-4}$
Butyrate$^-$ + $2H_2O \rightarrow$ 2 Acetate$^-$ + H^+ + $2H_2$	+48.3	$<10^{-4}$

[a]From Reference 104.

[b]The partial pressure of hydrogen needed for the reaction to be thermodynamically favorable ($-\Delta G'$), which was calculated with concentrations of substrate and acetate of 0.1 mM and a bicarbonate concentration of 100 mM.

of these compounds is exergonic. Consistent with the thermodynamic predictions, small increases in H_2 partial pressure inhibit the degradation of butyrate and benzoate by syntrophic cocultures[4,16–20] and propionate degradation in methanogenic mixed cultures.[21]

Since this first description of syntrophic metabolism, numerous studies have led to the isolation of many novel genera and species that are capable of syntrophic metabolism. Syntrophically metabolizing bacteria have been most commonly isolated from freshwater sediments and anaerobic digesters used to treat various types of wastewater.[22–26] Several molecular studies have shown that sequences related to those of bacteria capable of syntrophic metabolism are found in a wide variety of anoxic environments.[22,24,27–30]

The theoretical basis of syntrophic metabolism was based first on the transfer of H_2 between the two partners.[2] However, we now know that interspecies transfer of formate is essential. Zindel *et al.*[31] showed that syntrophic metabolism can occur solely by interspecies formate transfer by growing an amino acid–fermenting bacterium with a sulfate reducer that used formate but not H_2. Syntrophic propionate degradation by *Syntrophobacter fumaroxidans*[32,33] and butyrate degradation by *Syntrophomonas* (*Syntrophospora*) *bryantii*[34] occurred only if the partner used both hydrogen and formate. Additionally, formate dehydrogenase levels were very high in both members of the syntrophic propionate–degrading association consistent with electron flow being coupled to interspecies formate transfer.[35]

Initially, bacteria that syntrophically oxidized fatty and aromatic acids were believed to be obligately dependent on the hydrogen/formate-using partner, since other substrates or electron donor/acceptor combinations could not be found that allowed the growth of the syntrophic metabolizer in pure culture.[3] Microbes capable of the syntrophic metabolism of fatty and aromatic acids were first called obligately proton-reducing acetogenic bacteria.[3] As discussed elsewhere in this volume, the term acetogenic is best reserved for those bacteria that synthesize acetate from CO_2 rather than from compounds with carbon–carbon bonds. Additionally, almost all of the known bacteria capable of syntrophic metabolism can be grown fermentatively in pure culture with a more oxidized derivative of their parent substrate, such as crotonate for fatty-acid degraders or fumarate for propionate degraders.[36,37] Some, like the genus *Syntrophobacter*, have diverse metabolisms and grow fermentatively with several different substrates or by anaerobic respiration using electron acceptors, such as sulfate.[18,23,38] It is noteworthy that there are a few species (*Pelotomaculum schinkii*, *Syntrophomonas zehnderi*, and *Pelotomaculum isophthalicicum*) that appear to be obligately syntrophic microorganisms.[39–41] Verification of this prediction awaits future genome sequencing studies.

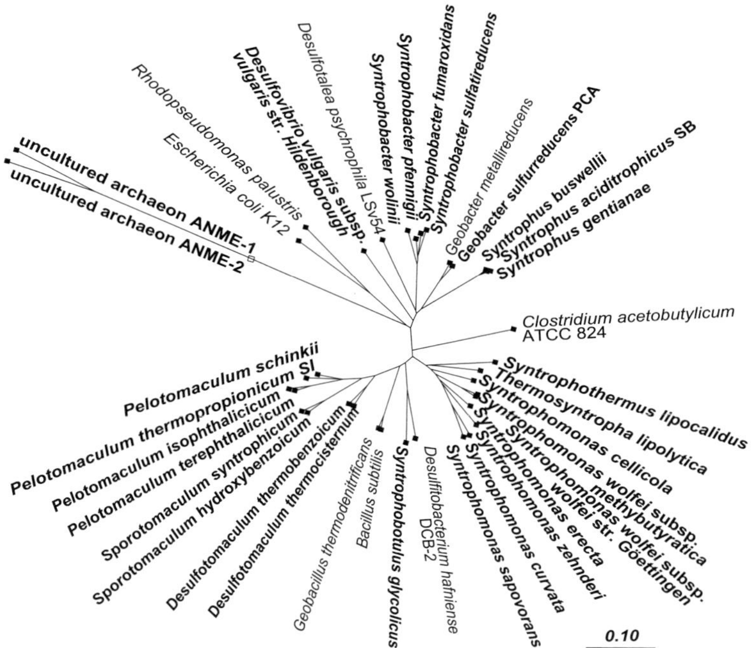

FIGURE 2. A phylogenetic tree containing representative syntrophic bacterial species. The organisms capable of syntrophy are in boldface type. The radial Neighbor-Joining tree was constructed using the ARB software package (http://www.arb-home.de/), utilizing the Greengenes 16s rRNA gene database (greengenes.lbl.gov/cgi-bin/nph-citation.cgi).[111,112] A filter was created and applied to the aligned syntrophic sequences using the maximum frequency method with a 50% minimum cutoff.

Phylogenetic Relationships of Syntrophic Metabolizers

When the 16S rRNA gene sequences from bacteria capable of syntrophic metabolism are compared, it is evident that many of these microorganisms cluster with species in *Deltaproteobacteria* and the low G+C gram-positive bacteria (FIG. 2). Genera that contain syntrophic species within the Deltaproteobacteria include *Syntrophus, Syntrophobacter, Desulfoglaeba, Geobacter, Desulfovibrio,* and *Pelobacter*. Two other groupings of microbes that perform syntrophic metabolism fall into the low G+C gram-positive bacteria (FIG. 2). One group is composed of species within the genera *Desulfotomaculum, Pelotomaculum, Sporotomaculum,* and *Syntrophobotulus*. *Syntrophomonadaceae* comprises another group of microbes that perform syntrophic metabolism in the low G+C gram-positive bacteria and includes species in the genera *Syntrophomonas, Syntrophothermus,* and *Thermosyntropha*. Molecular phylogenetic analyses, ^{13}C lipid isotopic determinations, and microscopic mass spectrometric analysis identified two major groups of methanogen-related archaea (ANME-1 and ANME-2) that anaerobically oxidize methane.[42,43] The close physical association between methane-oxidizing archaea and sulfate-reducing bacteria suggests a syntrophic relationship.[44] Although the ANME microbes have not yet been isolated, it does appear that syntrophic metabolism occurs in both the archaeal and bacterial lines of descent.

Diversity and Ecology of Syntrophic Metabolizers

A wide variety of compounds, including saturated fatty acids, unsaturated fatty acids, alcohols, and hydrocarbons, are syntrophically degraded in methanogenic environments.[2,38,45–47] *Syntrophomonas wolfei* was the first bacterium described that syntrophically oxidizes fatty acids in coculture with a hydrogen/formate-using microorganism.[2,3] *S. wolfei* was isolated from anaerobic digester sludge[2,3] and oxidizes saturated fatty acids, ranging from C4 to C8 in length, and isoheptanoate in coculture with hydrogen-users.[2] *S. wolfei* was shown later to grow in pure culture with crotonate.[36] *Syntrophomonas* spp. are rod-shaped, slightly motile, mesophilic, and capable of utilizing a variety of fatty acids[48] (TABLE 2). These microorganisms group

TABLE 2. Characteristics of syntrophic bacteria specializing in fatty-acid metabolism

Organism	pH range[a]	Temperature range (°C)[a]	Spore formation	Substrates used in: Pure culture	Coculture	Reference
Syntrophomonas bryantii	6.5–7.5	28–34	Yes	C4:1[b]	C4–C11	49, 50, 53
Syntrophomonas wolfei subsp. *wolfei*	ND	(35–37)	No	C4:1–C6:1	C4–C8	2
Syntrophomonas wolfei subsp. *saponavida*	ND	ND	No	C4:1	C4–C18	54
Syntrophomonas sapovorans	6.3–8.1 (7.3)	25–45 (35)	No	None	C4–C18, C16:1, C18:1, C18:2	55
Syntrophomonas curvata	6.3–8.4 (7.5)	20–42 (35–37)	No	C4:1	C4–C18, C18:1	56
Syntrophomonas erecta subsp. *sporosyntropha*	5.5–8.4 (7.0)	20–48 (35–37)	Yes	C4:1	C4–C8	52
Syntrophomonas erecta subsp. *erecta*	(7.8)	*(37–40)	No	C4:1, C4 + C5:1, C4 + DMSO	C4–C8	57
Syntrophomonas zehnderi	ND	25–40 (37)	Yes	None	C4–C18, C16:1, C18:1, C18:2	41
Syntrophomonas cellicola	6.5–8.5 (7.0–7.5)	25–45 (37)	Yes	C4:1	C4–C8, C10	49
Thermosyntropha lipolytica	7.5–9.5 (8.1–8.9)	52–70 (60–66)	No	C4:1, yeast extract, tryptone, casamino acids, betaine, pyruvate, ribose, xylose	C4–C18, C18:1, C18:2; triglycerides	59
Syntrophothermus lipocalidus	6.5–7.0	45–60 (55)	No	C4:1	C4–C10; isobutyrate	58
Algorimarina butyrica	6.2–7.1	10–25 (15)	No	None	C4, isobutyrate	25

[a]Optimal condition is given in parentheses; ND = not determined.

[b]The number of carbons in the fatty acid is indicated; the number following the colon is the number of unsaturated bonds for unsaturated fatty acids. When a range of fatty acids is given, this means that the organism can use fatty acids within the indicated range of carbon numbers, but not all possibilities were tested.

phylogenetically with the low G+C gram-positive bacteria in the family Syntrophomonadaceae.[48] Despite grouping phylogenetically with the gram-positive bacteria, members of the genus *Syntrophomonas* have atypical cell walls ultrastructurally similar to gram-negative cell walls,[2,48] but lacking lipopolysaccharides. *Syntrophomonas* spp. are differentiated from each other based on substrate utilization pattern and spore formation[41,48,49] (TABLE 2). *Syntrophomonas* are capable of forming spores including *S. erecta* subsp. *sporosyntropha, S. cellicola, S. zehnderi*, and *S. bryantii*, which was recently reclassified and was formerly described as both *Clostridium bryantii* and *S. bryantii*.[41,49–53] All of the described species are able to grow in pure culture with crotonate except *S. zehnderi*.[41] *S. erecta* subsp. *sporosyntropha* has been shown to sporulate only in coculture with methanogens, not in coculture with a sulfate reducer or in pure culture.[52] *S. wolfei* subsp. *saponavida*,[54] *Syntrophomonas sapovorans*,[55] *S. zehnderi*,[41] and *Syntrophomonas curvata*[56] use C4–C18 fatty acids, and *S. bryantii* uses C4–C10 fatty acids in coculture (TABLE 2). *S. wolfei* subsp. *wolfei* has not been shown to form spores and use C4–C8 fatty acids.[2] One strain of *S. erecta* can grow in pure culture on a mixture of butyrate and pentanoate.[57]

Other members of the *Syntrophomonadaceae* include two thermophilic genera, *Syntrophothermus* and *Thermosyntropha* (TABLE 2). *Syntrophothermus lipocalidus* was isolated from granular sludge in a thermophilic up-flow anaerobic sludge blanket (UASB) reactor, grows optimally at a temperature of 55°C, and metabolizes saturated fatty acids ranging from C4 to C10 and isobutyrate in coculture with a thermophilic hydrogen-using methanogen.[58] *Thermosyntropha lipolytica* was isolated from alkaline hot springs in Kenya[59] and grows at pH values of 7.15 to 9.5 and temperatures of 52°C to 70°C.[59] This organism, unlike *Syntrophomonas* spp. and *S. lipocalidus*, can use yeast extract, tryptone, casamino acids, and betaine in pure culture.[59] *T. lipolytica* and *S. lipocalidus* both use crotonate.[58,59] *T. lipolytica* uses olive oil, triacylglycerols, and both saturated and unsaturated fatty acids ranging from C4 to C18 in syntrophic association with hydrogen-using microorganisms.[59] Both *S. lipocalidus* and *T.*

TABLE 3. Characteristics of propionate-degrading syntrophic bacteria

Organism	pH range[a]	Temperature range (°C)[a]	Substrates used in: Pure culture[b]	Substrates used in: Coculture[b]	Reference
Syntrophobacter wolinii	ND	ND	C3[b] + sulfate; fumarate	C3	45, 71
Syntrophobacter pfennigii	6.2–8.0 (7.0–7.3)	20–37 (37)	C3 + sulfate, sulfite, thiosulfate; lactate	C3, lactate, propanol	38
Syntrophobacter fumaroxidans	6.0–8.0 (7.0)	20–40 (37)	C3 + sulfate or fumarate; fumarate	C3	18
Syntrophobacter sulfatireducens	6.2–8.8 (7.0–7.6)	20–48 (37)	C3 + sulfate or thiosulfate; pyruvate,	C3	23
Smithella propionica	6.3–7.8 (6.5–7.5)	23–40 (33–35)	C4:1	C3, C4, malate, fumarate	70
Pelotomaculum schinkii	ND	ND	None	C3	39
Pelotomaculum thermopropionicum	6.0–8.2 (7.0)	37–70 (55)	Pyruvate, fumarate	C3, lactate, various alcohols	63
Desulfotomaculum thermobenzoicum subsp. *thermosyntrophicum*	(7.0)	42–62 (55)	Fumarate, pyruvate, C4:1	C3, C4, benzoate	60
Desulfotomaculum thermocisternum	(6.7)	41–75 (62)	Fumarate, pyruvate	C3, C4	62

[a]Optimal condition is given in parentheses. ND = not determined.
[b]The number of carbons in the fatty acid is indicated; the number following the colon is the number of unsaturated bonds for unsaturated fatty acids.

lipolytica stain gram-negative and neither form spores.[58,59]

Certain gram-positive, spore-forming, thermophilic, sulfate-reducing bacteria from the genus *Desulfotomaculum* and other closely related genera have also been shown to degrade a variety of compounds in syntrophic association with hydrogen-using microorganisms.[60] *Desulfotomaculum* spp. are found in a variety of environments, including freshwater sediments, marine sediments, and have also been observed in hydrocarbon-degrading enrichments.[60,61] *Desulfotomaculum thermocisternum* was the first described thermophile that is capable of oxidizing propionate in syntrophic association with a hydrogen-using methanogen (TABLE 3).[62] *Desulfotomaculum thermobenzoicum* subsp. *thermosyntrophicum* is also capable of degrading propionate syntrophically in thermal environments (TABLE 3).[60] This organism was isolated from a thermophilic anaerobic digester treating kraft-pulp wastewater, and can be distinguished from other thermophilic, syntrophic propionate oxidizers due to its ability to oxidize benzoate in the presence of sulfate.[60]

P. schinkii and *Pelotomaculum thermopropionicum* strain SI have recently been shown to degrade propionate in syntrophic association with hydrogen-using methanogens (TABLE 3).[39,63] *P. schinkii* is currently considered to be an obligate syntrophic organism.[39] *P. thermopropionicum* grows at 55°C and metabolizes fumarate and pyruvate in pure culture, and propionate, ethanol, lactate, 1-butanol, 1-pentanol, 1,3-propanediol, 1-propanol, and ethylene glycol in co-culture.[63] Other species of in the genus *Pelotomaculum* include *P. terephthalicicum* and *P. isophthalicicum* (TABLE 4). *P. isophthalicicum* has not been grown in pure culture, while *P. terephthalicicum* grows in pure culture with crotonate, 2,5-dihydroxybenzoate, and hydroquinone.[40] *P. terephthalicicum* and *P. isophthalicicum* metabolize a variety of phthalate isomers and other aromatic compounds in syntrophic association with hydrogen-using methanogens.[40] *P. schinkii*, *P. thermopropionicum*, *P. terephthalicicum*, and *P. isophthalicicum* have recently been shown to group with *Desulfotomaculum* subcluster Ih.[64] However, none of these organisms utilize sulfate as an electron acceptor.[39,40,63] Expression of *dsrAB*, which encode for the alpha and beta subunits of the dissimilatory sulfite reductase, by real-time polymerase chain reaction (PCR) was observed in only one of five propionate-degrading enrichments that contained propionate degraders that group with *Desulfotomaculum* subcluster Ih.[64] Pure cultures of *P. schinkii*, *P. thermopropionicum*, *P. terephthalicicum*, and *P. isophthalicicum*

TABLE 4. Characteristics of aromatic-degrading syntrophic bacteria

Organism	pH range[a]	Temperature range (°C)[a]	Substrates used in: Pure culture	Coculture	Reference
Syntrophus buswellii	6.5–7.5 (7.1–7.4)	ND	C4:1[b]; cinnamate; C4:1 + benzoate or 3-phenyl-propionate	C4:1, benzoate	67
Syntrophus gentianae	6.5–7.5 (7.1–7.4)	10–33 (28)	C4:1, hydroquinone, 2,5-diOH-benzoate	C4:1, benzoate, hydroquinone, 2,5-diOH-benzoate	69
Syntrophus aciditrophicus	6.5–7.5 (7.1–7.4)	25–42 (37)	C4:1, benzoate, cyclohex-1-ene carboxylate	Benzoate, fatty acids, unsaturated fatty acids	68
Sporotomaculum syntrophicum	6.0–7.5 (7.0–7.2)	20–45 (35–40)	C4:1; C4:1 + benzoate	Benzoate	65
Pelotomaculum terephthalicicum	6.5–7.5 (6.8–7.2)	25–45 (37)	C4:1, hydroquinone, 2,5-diOH-benzoate	Benzoate, phthalates, hydroxy-benzoates, 3-phenylpropionate	40
Pelotomaculum isophthalicicum	6.8–7.2 (7.0)	25–45 (37)	None	Benzoate, phthalates, 3-OH-benzoate	40

[a]Optimal condition is given in parentheses. ND = not determined.
[b]The number of carbons in the fatty acid is indicated; the number following the colon is the number of unsaturated bonds for unsaturated fatty acids.

also failed to yield a PCR product.[64] Therefore, the inability of the isolates and enrichments to couple propionate oxidation to sulfate reduction and the lack of *dsrAB* genes in all but one enrichment suggests that the *Desulfotomaculum* subcluster Ih consists of syntrophic metabolizers that may have lost their ability to reduce sulfate.[39,40,63,64]

Sporotomaculum syntrophicum groups with members of the *Desulfotomaculum* and metabolizes benzoate in syntrophic association with hydrogen-using methanogens (TABLE 4).[65] *S. syntrophicum* does not use sulfate as an electron acceptor.[65] *S. syntrophicum* grows in pure culture on crotonate.[65] One final example of a syntrophic metabolizer that groups with *Desulfotomaculum* is *Syntrophobotulus glycolicus*.[66] This microorganism oxidizes glycolate in syntrophic association with hydrogen-using methanogens.[66] *S. glycolicus* is most commonly observed in freshwater environments, and cannot couple the oxidation of glycolate to the reduction of sulfate.[66]

Some gram-negative bacteria affiliated with the Deltaproteobacteria are capable of syntrophic metabolism.[67–69] The first syntrophic propionate oxidizer described was *Syntrophobacter wolinii*,[45] which was isolated from primary anaerobic digestor sludge. Three other *Syntrophobacter* species have been described, *Syntrophobacter pfennigii* from an anaerobic sludge of a sewage plant,[38] *Syntrophobacter fumaroxidans* isolated from an anoxic sludge blanket reactor treating wastewater from a sugar refinery,[18] and two strains of *Syntrophobacter sulfatireducens* TB8106 and WZH410 isolated from the anoxic sludge of a reactor treating brewery wastewater or a reactor treating bean-curd wastewater, respectively.[23] *Syntrophobacter* spp. form a monophylogenetic group that is separate from other *Deltaproteobacteria*, but are most closely related to group 7 of the sulfate-reducing bacteria.[18,23,38,45,70,71] The four species are mesophilic, nonmotile, and non-spore-forming bacteria. In the absence of a methanogen, all four species are capable of axenic growth and oxidize propionate by using sulfate or fumarate as an electron acceptor (TABLE 3).[18,23,38,45,70,71] They can also grow in pure culture by fermenting fumarate, malate, or pyruvate. *Syntrophobacter* spp. have been observed in freshwater sediments, marine sediments, rice paddy sediments, acidic fens, and eutrophic bog and marsh sediments.[22,24,30,72] 13C-Propionate labeling studies showed that *Syntrophobacter* spp., *Smithella* spp., and *Pelotomaculum* spp. were active in syntrophic propionate oxidation in rice paddies.[30]

Smithella propionica is another propionate-degrading syntrophic microorganism in the *Deltaproteobacteria* (TABLE 3). It was isolated from an anaerobic filter inoculated with domestic sewage sludge enriched with propionate. Unlike *Syntrophobacter*, *S. propionica* is unable to use sulfate as an electron acceptor and needs the presence of a hydrogen user to degrade propionate.[70] *Syntrophobacter* species degrade propionate to acetate and CO_2 by using the methylmalonyl-CoA pathway.[73]

However, *S. propionica* ferments propionate to acetate with the production of traces of butyrate by a new pathway that involves the condensation of two molecules of propionate to form a six-carbon intermediate that is ultimately cleaved to form acetate and butyrate.[74] *S. propionica* grows with butyrate, malate, and fumarate in coculture with a methanogen and with crotonate in pure culture.

Syntrophus spp. are rod-shaped bacteria capable of degrading aromatic compounds in syntrophic association with hydrogen-using microorganisms (TABLE 4).[67–69] *Syntrophus* spp. are also affiliated with the *Deltaproteobacteria* and have frequently been isolated from sewage sludge.[67–69] *Syntrophus buswellii* metabolizes benzoate in coculture with hydrogen-using microorganisms and crotonate in pure culture.[67] *Syntrophus gentianae* syntrophically metabolizes benzoate, gentisate, and 3-phenylpropionate.[69] *Syntrophus aciditrophicus* also syntrophically metabolizes benzoate, but differs from other *Syntrophus* spp. in its ability to metabolize a variety of fatty acids in syntrophic association with hydrogen users.[68] *S. gentianae* grows in pure culture with crotonate, producing butyrate and acetate,[69] whereas *S. aciditrophicus* ferments crotonate to acetate and cyclohexane carboxylate.[75]

Molecular ecological studies suggest that *Syntrophus* spp. may play an important role in a number of environments.[16,22,28,29,76] Sequences related to *Syntrophus* spp. are commonly detected in clone libraries from hydrocarbon-contaminated sites.[16,20,29,76] Several studies suggest that *Syntrophus* spp. may be involved in the degradation of benzoate, which is an important intermediate in the anaerobic degradation of aromatic hydrocarbons in hydrocarbon-contaminated sites.[16,28,29,76,77] The concentration of short-chain fatty acids have been observed to increase as a result of the degradation of hydrocarbons, which could also explain why *Syntrophus* sequences are observed at these sites.[17] Benzoate has been shown to be an important intermediate in the degradation of 3-chlorobenzoate and 2-chlorophenol, and sequences related to *Syntrophus* spp. tend to appear as benzoate degradation begins.[27] *Syntrophus* spp. sequences have also been observed in dechlorinating enrichments that contained either a mixture of trichloroethene and methanol or a mixture of vinyl chloride and methanol.[78] The degradation of alkylbenzenes[79] and halogenated aromatic compounds[80] to methane involves a consortium of microorganisms and likely involves the activity of *Syntrophus* spp.

The genus *Pelobacter* also clusters within *Deltaproteobacteria*. *Pelobacter* spp. are predominant in sediments and sludge where syntrophic alcohol oxidation occur.[81] *Pelobacter venetianus*[82] and *Pelobacter acetylenicus*[83] have been shown to syntrophically metabolize ethanol.

Syntrophococcus sucromutans is a rumen bacterium that requires an exogenous electron acceptor, either formate, methoxylated aromatic compounds, or a hydrogen/formate-using microorganism, to oxidize carbohydrates.[84] *Tepidanaerobacter syntrophicus* was also isolated from thermophilic anaerobic digesters and degrades lactate and numerous alcohols in syntrophic association with hydrogen-using methanogens.[85] This organism groups with the firmicutes, and appears to be most closely related to *Thermosediminibacter* spp. and *Thermovenabulum ferriorganovorum*.[85] Syntrophic butyrate metabolism has been described in cocultures of *Algorimarina butyrica* and hydrogen-using methanogens.[25] These cocultures were enriched from psychrophilic bay sediments, and are phylogenetically related to sulfate-reducing bacteria from the genera *Desulfonema* and *Desulfosarcina*.[25] This same study also established syntrophic propionate enrichments from these cold sediments, but the authors were unable to isolate the propionate degrader.[25]

Recent molecular characterization of enrichments for syntrophic metabolizers from a eutrophic site within the Florida Everglades revealed the presence of bacterial and archaeal sequences that were either members of novel lineages or closely related to uncultured environmental clones.[22] Subsequent cultivation-based and molecular studies revealed the presence of a novel *Methanosaeta* sp. and fatty acid–oxidizing bacteria related to *Syntrophomonas* spp. and *Syntrophobacter* spp.[22]

Also, hydrocarbons can be degraded syntrophically. A bacterium has been described that syntrophically degrades toluene in coculture with a sulfate reducer.[86] In some environments, the anaerobic oxidation of methane involves a highly organized, multicellular structure of the methane-oxidizing and sulfate-reducing bacteria.[44] More recently, *Desulfoglaeba alkanexedens* has been isolated, which can syntrophically degrade alkanes in coculture with methanogens.[87] Hexadecane-degrading enrichments contain microorganisms phylogenetically related to *Syntrophus* spp.[88] Other studies have shown that hydrocarbon loss is coupled to methane production, which would indicate that syntrophic metabolism must be involved.[46,47]

Zinder and Koch described a thermophilic acetate-degrading coculture consisting of an acetate-degrading bacterium and a H_2-consuming methanogen.[89] The syntrophic acetate oxidizer, strain AOR, appeared to be a homoacetogen in pure culture, producing acetate from H_2 and CO_2 as well as syntrophically with a methanogen where acetate is oxidized to H_2 and CO_2, suggesting that the pathway is reversible.[90] Biochemical studies revealed that strain AOR uses the acetyl-CoA synthase pathway ("Wood–Ljungdahl" pathway), as do other homoacetogens.[90,91]

Other microbes that syntrophically oxidize acetate include *Geobacter sulfurreducens* with partners, such as *Wolinella succinogenes* or *Desulfovibrio desulfuricans*, and nitrate as the electron acceptor.[92] *Clostridium ultunense* strain BST was isolated in pure culture with substrates typically utilized by homoacetogenic bacteria, but can also syntrophically oxidize acetate.[93] Syntrophic acetate oxidation also has been observed in *Thermacetogenium phaeum*, which was isolated from thermophilic anaerobic digesters.[94] This microorganism is phylogenetically related to *Clostridium* and *Bacillus* spp.[94] A recent article described *Candidatus Contubernalis alkalaceticum*, which is capable of syntrophic acetate oxidation in coculture with *Desulfonatronum cooperativum*.[95] These microorganisms were isolated from a soda lake, and appear to group phylogenetically with uncultured low G+C gram-positive bacteria within the family *Syntrophomonadaceae*.[95] In addition to acetate, these organisms were also observed to syntrophically oxidize ethanol, propanol, isopropanol, serine, fructose, and isobutyrate.[95]

Evidence also shows that some homoacetogens are capable of syntrophic metabolism because their metabolism of methanol is affected by cocultivation with H_2-consuming anaerobes.[96,97] In pure culture, *Sporomusa acidovorans* ferments methanol and CO_2 to acetate,

$$4\,\text{Methanol} + 2\,HCO_3^- \rightarrow 3\,\text{Acetate}^- + H^+ + 4\,H_2O\,\Delta G^{\circ\prime} = -220\text{ kJ per mol} \quad (4)$$

In coculture with the H_2-consuming *Desulfovibrio desulfuricans* with nitrate, no acetate is formed, indicating that methanol is oxidized to CO_2 and H_2, a conversion that is only possible at a low hydrogen partial pressure.[97]

$$\text{Methanol} + 2\,H_2O \rightarrow HCO_3^- + H^+ + 3\,H_2 \quad \Delta G^{\circ\prime} = +24\text{ kJ per mol} \quad (5)$$

Another syntrophic homoacetogen example may be *Methanobacillus kuznezovii*, which was described as a methanogen, but is likely to be a syntrophic coculture of a homoacetogen and a methanogen, since *M. kuznezovii* produced acetate during methanol metabolism.[98]

Syntrophic methanol-degrading enrichments were obtained from thermophilic digestors using cobalt-deficient medium to suppress methanogenesis.[99] *Moorella mulderi* and a *Desulfotomaculum* species were isolated from the enrichment.[100,101] Although the sulfate reducer used methanol in pure culture, it appeared to use the hydrogen produced by *M. mulderi* when grown in coculture. A syntrophic methanol-degrading coculture of a *Moorella* species with a H_2-utilizing *Methanothermobacter* strain was obtained when sulfate was deleted from the cobalt-deficient medium.[102] The homoacetogen grew in pure culture with methanol only when cobalt was added to the medium.[102] The effect of cobalt on the growth of the *Moorella* species is not exactly clear. It is likely that, in the presence of cobalt, a corrinoid-containing methyltransferase is used for methanol degradation, while in the absence of cobalt, a methanol dehydrogenase is used to oxidize methanol to formaldehyde, which would not be further degraded by pure cultures. *Moorella thermoautotrophica* contains a methanol dehydrogenase with pyrroloquinoline quinone as the prosthetic group.[103]

Energetics

An intriguing aspect of the metabolism of syntrophically fermenting bacteria is the fact that they must catalyze reactions that are endergonic under standard reactions conditions (i.e., positive $\Delta G^{\circ\prime}$). The activity of the partner organism is required to maintain the concentrations of hydrogen, formate, and acetate low enough to permit the metabolism of the substrate by the syntrophic metabolizer to be sufficiently exergonic to support adenosine triphosphate (ATP) synthesis, anabolism, and growth. For this reason, syntrophically fermenting bacteria are excellent model organisms to study the minimum limits of microbial energy metabolism. For example, is there a minimum amount of free energy required to conserve energy in a biologically useful form to maintain viability?

Based on the energy released by the hydrolysis of ATP and the concentrations of adenylate molecules in growing bacteria, it is estimated that a free energy change of -60 to -70 kJ per mol is required for the biochemical synthesis of ATP under physiological conditions.[104] This amount of energy does not need to be supplied in one single step, as exemplified by the formation of ATP by substrate-level phosphorylation, but rather, can be accomplished in smaller increments, for example, by membrane-bound proton-translocating redox reactions or other exergonic reactions involving the transport of sodium ions[105] that culminate in ATP synthesis through a membrane-bound ATP synthase. A ratio of three protons translocated per ATP formed has been assumed,[4] although recent research on the structure and function of ATP synthases from different organisms indicates that this stoichiometry may vary between 3 and 5 H^+ per ATP formed or hydrolyzed.[106] Thus, the minimum increment of energy required for ATP synthesis may be as low as 12–15 kJ per mol, and values in the range of 15–20 kJ per mol reaction have been calculated for most syntrophic fermentations under *insitu* conditions.[4] In some cases, especially in syntrophic fermentation of propionate to acetate and hydrogen, the overall energetics appear to be lower, about −12 kJ per mol reaction, as calculated from *insitu* propionate, acetate, and hydrogen concentrations. This view may be incomplete, since formate exchange between the partners also appears to be required for syntrophic propionate fermentation.[35] In experiments with resting cells of the butyrate-fermenting *Syntrophus aciditrophicus* in buffer, an overall free energy change as low as −4 kJ per mol of butyrate was calculated,[10] although there was no evidence that ATP was synthesized under these conditions. The stoichiometry of ions translocated per mole substrate consumed by the syntrophic metabolizer in addition to the stoichiometry of ions consumed in support of ATP synthesis are critical issues that remain unresolved.

One specifically fascinating syntrophic system is syntrophic acetate oxidation via CO_2 and H_2 as intermediates, which was first demonstrated by Zinder and Koch[89] in a thermophilic reactor at 58°C. The overall reaction yields −35 kJ per mol for the entire two-step process under standard conditions (25°C); at 58°C, the energy yield increases to −42 kJ per mol, which is sufficient to support the growth of the two syntrophic partners. A coculture fermenting acetate syntrophically to methane and CO_2 at 37°C has been isolated.[93] The $\Delta G'$ for syntrophic acetate oxidation at 37°C is −36 kJ per mol of acetate. The hydrogen concentrations indicated that the free energy change for hydrogen production from acetate was very close to the value for hydrogen use by the methanogen, indicating that the coculture partners equally shared the available energy.[90,91] If this is the case, then the free energy available to each organism would be about −15 kJ per reaction. The growth yields also indicated that each partner equally shared the available energy. The doubling times of the coculture was in the range of 3–4 weeks, which indicates that a lower limit for efficient energy transformation was being reached under these conditions.

A new thermophilic acetate-fermenting coculture was isolated in Japan,[94] and enzyme measurements with this culture demonstrated that all enzymes of the Wood–Ljungdahl pathway were present under both acetate utilization and acetate formation conditions. Moreover, the coculture was able to immediately switch from syntrophic acetate oxidation to homoacetogenic acetate formation, indicating that the entire enzyme apparatus appears to operate in a reversible manner.[107] This is the first demonstrated case of a metabolism that is entirely reversible, thus demonstrating how close to the thermodynamic equilibrium such metabolism operates. Unresolved is how these bacteria manage to make ATP by each mode. There must be unique steps linked to ATP formation in one direction that are decoupled from ATP hydrolysis in the reverse direction, possibly some switch in electron flow in the membrane. So far, these steps have not yet been identified. The complete sequencing of the genome of the bacterium involved, *Thermacetogenium phaeum*, as well as other syntrophic acetate oxidizers, is underway and may provide insight into this exciting phenomenon soon.

The long-disputed process of anaerobic methane oxidation appears to be a syntrophic cooperation of two microbes.[44] One organism, related to methanogens, appears to operate its methanogenic-like pathway in reverse, thus oxidizing methane. The second organism is a sulfate-reducing bacterium. The calculated free energy change of this syntrophic association is very low:

$$CH_4 + SO_4^{2-} + H^+ \rightarrow CO_2 + HS^- + 2\,H_2O$$
$$\Delta G^{\circ\prime} = -18\ \text{kJ per mol} \qquad (6)$$

The free energy available for anaerobic methane oxidation appears to be insufficient to provide both partners with an adequate amount of energy. Only under *in situ* conditions, that is, about 100 atm CH_4, would the overall process provide enough energy to "feed" both partners, and it is only under these conditions that the process can be maintained in the laboratory.[13] The preceding experimental evidence indicates that the minimum amount of energy required for synthesis of ATP must occur in small increments within the range of −15 kJ per mol reaction. It may also explain why, despite numerous efforts, nobody succeeded in the past in cultivation of anaerobic methane oxidizers in the laboratory under conditions close to standard reaction conditions.

Genomes of Syntrophs

The complete genome sequences of *Syntrophus aciditrophicus*, *S. wolfei*, *Syntrophobacter fumaroxidans*, and *Pelotomaculum thermopropionicum* have been reported (TABLE 5)[108,109] (http://www.jgi.doe.gov, http://www.integratedgenomics.com). The genome

TABLE 5. Properties of the sequenced syntrophic bacterial genomes[a]

Organism	% G+C	Genome size (Mb)	% Coding region	No. of ORF	ORF with assigned function		ORF without assigned function		ORF without similarity[b]		rRNA operons	GenBank ID
					No.	%	No.	%	No.	%		
Syntrophus aciditrophicus	52	3.18	88	3,168	2,078	65.6	1,090	34.4	422	13.3	1	NC_008346
Syntrophobacter fumaroxidans	59	4.99	82	4,098	2,809	68.5	1,289	31.5	468	11.5	2	NC_008554
Syntrophomonas wolfei subsp. *wolfei*	45	2.85	87	2,574	1,507	58.5	1,067	41.5	404	13.5	3	NC_007759
Pelotomaculum thermopropionicum	52	3.03	85	2,920	1,767[c]	60.5	1,153[c]	39.5	ND		2	NC_009454

[a]ORF (open reading frames) statistics generated by Joint Genome Institute (www.img.jgi.gov). ND = not determined.
[b]Value generated by ERGO (www.integrated genomics.com).
[c]Number determined from NC_009454 GenBank file.

sizes are in the range of about 3 MB for *S. aciditrophicus*, *S. wolfei*, and *P. thermopropionicum*, while that of *S. fumaroxidans* is higher, about 5 MB. Approximately 50–60% of the open reading frames (ORFs) were assigned tentative annotations, with the remaining 40–50% having no assigned function. It is noteworthy that a significant fraction of the ORFs in each genome lack either annotated function or similarity to proteins in any other described organism. Also noteworthy is that a best reciprocal gene comparison among these strains reveals that fewer than one-third of the genes in one strain are significantly related to those in another.[109]

While the detailed analysis of the genomic content of three of these genomes has not been reported as of the writing of this article, the genetic inventory of the *S. aciditrophicus* genome was recently published.[109] *S. aciditrophicus* appears to be self-sufficient with respect to its anabolic pathways, but in contrast is highly specialized in catabolic ability, as genes for utilization of most carbon compounds by fermentation or respiration are absent. The genome of *S. aciditrophicus* is devoid of genes for electron transport proteins common to many anaerobic fermentative or respiratory bacteria, for example, the sulfate reducers, the nitrate reducers, or other organisms able to reduce other organic or inorganic electron acceptors. Furthermore, the genetic blueprint of *S. aciditrophicus* strain SB suggests unique and apparently undescribed mechanism(s) to metabolize its substrates (i.e., crotonate, benzoate, and cyclohexane carboxylate) to acetate and other products. A distinctive feature of syntrophic metabolism is the need for reverse electron transport. The presence of a unique ion-translocating electron transfer complex, menaquinone, and membrane-bound Fe-S proteins with associated heterodisulfide reductase domains in the genome of *S. aciditrophicus* suggest mechanisms to accomplish this task. Genomic analysis indicate that *S. aciditrophicus* has multiple mechanisms to create and use ion gradients such as ion-translocating ATP synthases, pyrophosphatases, decarboxylases, and hydrogenases, which would help modulate the energy status of the cells in response to varying thermodynamic conditions. The *S. aciditrophicus* genome contains genes for 17 sigma 54–interacting transcriptional regulators and 35 transcriptional regulators with a helix-turn-helix motif.[109] Other gram-negative microbes have a larger number of transcriptional regulators with a helix-turn-helix motif, suggesting that *S. aciditrophicus* appears to have adopted a regulatory strategy reliant on sigma-54 factor coupled signal transduction pathways. Interestingly, one of the operons involved in propionate metabolism in *P. thermopropionicum* also appears to involve sigma factor regulation.[108] The complete sequencing of genomes for other syntrophic metabolizers is under way, and the analysis of this genomic information should provide further insights into the complexity of this important microbial lifestyle.

Future Prospects

Syntrophic associations provide ideal model systems to study microbial interactions and their role in the maintenance of community structure and functional diversity within an ecosystem. Advances in cultivation techniques, molecular ecology, and genomics and functional genomics are rapidly merging and combined should allow a comprehensive approach to understand the syntrophic lifestyle at the edge of a minimal energy existence. With these techniques, it is quite likely that we will soon uncover the extent of the diversity of microorganisms capable of syntrophic metabolism with respect to the substrates that they degrade, the variety of their metabolic pathways, and their phylogenetic relatedness. We are beginning to reveal the molecular and biochemical details needed for the syntrophic lifestyle. In particular, the recent discovery of three genes, whose expression was altered during a shift from syntrophic metabolism to sulfate reduction, has provided clues about the origins of syntrophic interactions.[110] The combination of computational models with functional genomic information will allow us to interrogate the regulatory mechanisms involved in establishing and maintaining multispecies associations in order to quantify and predict the behavior of microorganisms and microbial communities in natural ecosystems.

Acknowledgments

This work was supported by grants from the National Science Foundation Award NSF EF-0333294 and from the U.S. Department of Energy, DE-FG02-96ER20214 (MM) and DE-FG03-86ER13498 (RG).

Conflict of Interest

The authors declare no conflicts of interest.

References

1. LOWE, S.E., M.K. JAIN & J.G. ZEIKUS. 1993. Biology, ecology, and biotechnological applications of anaerobic bacteria adapted to environmental stresses in temperature, pH, salinity, or substrates. Microbiol. Rev. **57:** 451–509.
2. MCINERNEY, M.J. *et al.* 1981. *Syntrophomonas wolfei* gen. nov. sp. nov., an anaerobic, syntrophic, fatty acid-oxidizing bacterium. Appl. Environ. Microbiol. **41:** 1029–1039.
3. MCINERNEY, M.J., M.P. BRYANT & N. PFENNIG. 1979. Anaerobic bacterium that degrades fatty acids in syntrophic association with methanogens. Arch. Microbiol. **122:** 129–135.
4. SCHINK, B. 1997. Energetics of syntrophic cooperation in methanogenic degradation. Microbiol. Mol. Biol. Rev. **61:** 262–280.
5. SCHINK, B. 2006. Syntrophic associations in methanogenic degradation. Prog. Mol. Subcell. Biol. **41:** 1–19.
6. DOLFING, J. & J.M. TIEDJE. 1988. Acetate inhibition of methanogenic, syntrophic benzoate degradation. Appl. Environ. Microbiol. **54:** 1871–1873.
7. MACKIE, R.I. & B.A. WHITE. 1990. Recent advances in rumen microbial ecology and metabolism: potential impact on nutrient output. J. Dairy Sci. **73:** 2971–2995.
8. BIEBL, H. & N. PFENNIG. 1978. Growth yields of green sulfur bacteria in mixed cultures with sulfur and sulfate reducing bacteria. Arch. Microbiol. **117:** 9–16.
9. ADAMS, C.J., M.C. REDMOND & D.L. VALENTINE. 2006. Pure-culture growth of fermentative bacteria, facilitated by H_2 removal: bioenergetics and H_2 production. Appl. Environ. Microbiol. **72:** 1079–1085.
10. JACKSON, B.E. & M.J. MCINERNEY. 2002. Anaerobic microbial metabolism can proceed close to thermodynamic limits. Nature **415:** 454–456.
11. KLEEREBEZEM, R. & A.J. STAMS. 2000. Kinetics of syntrophic cultures: a theoretical treatise on butyrate fermentation. Biotechnol. Bioeng. **67:** 529–543.
12. SCHOLTEN, J.C. & R. CONRAD. 2000. Energetics of syntrophic propionate oxidation in defined batch and chemostat cocultures. Appl. Environ. Microbiol. **66:** 2934–2942.
13. NAUHAUS, K. *et al.* 2007. In vitro cell growth of marine archaeal-bacterial consortia during anaerobic oxidation of methane with sulfate. Environ. Microbiol. **9:** 187–196.
14. BARKER, H.A. 1939. Studies upon the methane fermentation. IV. The isolation and culture of *Methanobacterium omelianskii*. Antonie Leeuwenhoek. **6:** 201–220.
15. BRYANT, M.P. *et al.* 1967. *Methanobacillus omelianskii*, a symbiotic association of two species of bacteria. Arch. Microbiol. **59:** 20–31.
16. ALLEN, J.P. *et al.* 2007. The microbial community structure in petroleum-contaminated sediments corresponds to geophysical signatures. Appl. Environ. Microbiol. **73:** 2860–2870.
17. COZZARELLI, I.M., R.P. EGANHOUSE & M.J. BAEDECKER. 1990. Transformation of monoaromatic hydrocarbons to organic acids in anoxic groundwater environment. Environ. Geol. **16:** 135–141.
18. HARMSEN, H.J. *et al.* 1998. *Syntrophobacter fumaroxidans* sp. nov., a syntrophic propionate-degrading sulfate-reducing bacterium. Int. J. Syst. Bacteriol. **48:** 1383–1387.
19. AHRING, B.K. & P. WESTERMANN. 1988. Product inhibition of butyrate metabolism by acetate and hydrogen in a thermophilic coculture. Appl. Environ. Microbiol. **54:** 2393–2397.
20. DWYER, D.F. *et al.* 1988. Bioenergetic conditions of butyrate metabolism by a syntrophic, anaerobic bacterium

in coculture with hydrogen-oxidizing methanogenic and sulfidogenic bacteria. Appl. Environ. Microbiol. **54:** 1354–1359.

21. D.P. Smith & P.L. McCarty. 1989. Energetic and rate effects on methanogenesis of ethanol and propionate in perturbed CSTRs. Biotechnol. Bioeng. **34:** 39–54.
22. Chauhan, A., A. Ogram & K.R. Reddy. 2004. Syntrophic-methanogenic associations along a nutrient gradient in the Florida Everglades. Appl. Environ. Microbiol. **70:** 3475–3484.
23. Chen, S., X. Liu & X. Dong. 2005. *Syntrophobacter sulfatireducens* sp. nov., a novel syntrophic, propionate-oxidizing bacterium isolated from UASB reactors. Int. J. Syst. Evol. Microbiol. **55:** 1319–1324.
24. Juottonen, H. *et al.* 2005. Methanogen communities and bacteria along an ecohydrological gradient in a northern raised bog complex. Environ. Microbiol. **7:** 1547–1557.
25. Kendall, M.M., Y. Liu & D.R. Boone. 2006. Butyrate- and propionate-degrading syntrophs from permanently cold marine sediments in Skan Bay, Alaska, and description of *Algorimarina butyrica* gen. nov., sp. nov. FEMS. Microbiol. Lett. **262:** 107–114.
26. Schwarz, J.I., W. Eckert & R. Conrad. 2007. Community structure of archaea and bacteria in a profundal lake sediment Lake Kinneret (Israel). Syst. Appl. Microbiol. **30:** 239–254.
27. Becker, J.G. *et al.* 2005. The role of syntrophic associations in sustaining anaerobic mineralization of chlorinated organic compounds. Environ. Health. Perspect. **113:** 310–316.
28. Dojka, M.A. *et al.* 1998. Microbial diversity in a hydrocarbon- and chlorinated-solvent-contaminated aquifer undergoing intrinsic bioremediation. Appl. Environ. Microbiol. **64:** 3869–3877.
29. Grabowski, A. *et al.* 2005. Microbial diversity in production waters of a low-temperature biodegraded oil reservoir. FEMS. Microbiol. Ecol. **54:** 427–443.
30. Lueders, T., B. Pommerenke & M.W. Friedrich. 2004. Stable-isotope probing of microorganisms thriving at thermodynamic limits: syntrophic propionate oxidation in flooded soil. Appl. Environ. Microbiol. **70:** 5778–5786.
31. Zindel, U. *et al.* 1988. *Eubacterium acidaminophilum* sp. nov., a versatile amino acid-degrading anaerobe producing or utilizing H_2 or formate. Arch. Microbiol. **150:** 254–266.
32. Dong, X., C.M. Plugge & A.J. Stams. 1994. Anaerobic degradation of propionate by a mesophilic acetogenic bacterium in coculture and triculture with different methanogens. Appl. Environ. Microbiol. **60:** 2834–2838.
33. Dong, X. & A.J. Stams. 1995. Evidence for H2 and formate formation during syntrophic butyrate and propionate degradation. Anaerobe **1:** 35–39.
34. Dong, X., G. Cheng & A.J. M. Stams. 1994. Butyrate oxidation by *Syntrophospora bryantii* in co-culture with different methanogens and in pure culture with pentenoate as electron acceptor. Appl. Microbiol. Biotechnol. **42:** 647–652.
35. de Bok, F.A., M.L. Luijten & A.J. Stams. 2002. Biochemical evidence for formate transfer in syntrophic propionate-oxidizing cocultures of *Syntrophobacter fumaroxidans* and *Methanospirillum hungatei*. Appl. Environ. Microbiol. **68:** 4247–4252.
36. Beaty, P.S. & M.J. McInerney. 1987. Growth of *Syntrophomonas wolfei* in pure culture on crotonate. Arch. Microbiol. **147:** 389–393.
37. Stams, A.J. *et al.* 1993. Growth of syntrophic propionate-oxidizing bacteria with fumarate in the absence of methanogenic bacteria. Appl. Environ. Microbiol. **59:** 1114–1119.
38. Wallrabenstein, C., E. Hauschild & B. Schink. 1995. *Syntrophobacter pfennigii* sp. nov., new syntrophically propionate-oxidizing anaerobe growing in pure culture with propionate and sulfate. Arch. Microbiol. **164:** 346–352.
39. de Bok, F.A. *et al.* 2005. The first true obligately syntrophic propionate-oxidizing bacterium, *Pelotomaculum schinkii* sp. nov., co-cultured with *Methanospirillum hungatei*, and emended description of the genus *Pelotomaculum*. Int. J. Syst. Evol. Microbiol. **55:** 1697–1703.
40. Qiu, Y.L. *et al.* 2006. *Pelotomaculum terephthalicicum* sp. nov. and *Pelotomaculum isophthalicicum* sp. nov.: two anaerobic bacteria that degrade phthalate isomers in syntrophic association with hydrogenotrophic methanogens. Arch. Microbiol. **185:** 172–182.
41. Sousa, D.Z. *et al.* 2007. *Syntrophomonas zehnderi* sp. nov., an anaerobe that degrades long-chain fatty acids in co-culture with *Methanobacterium formicicum*. Int. J. Syst. Evol. Microbiol. **57:** 609–615.
42. Hinrichs, K.U. *et al.* 1999. Methane-consuming archaebacteria in marine sediments. Nature **398:** 802–805.
43. Orphan, V.J. *et al.* 2001. Methane-consuming archaea revealed by directly coupled isotopic and phylogenetic analysis. Science **293:** 484–487.
44. Boetius, A. *et al.* 2000. A marine microbial consortium apparently mediating anaerobic oxidation of methane. Nature **407:** 623–626.
45. Boone, D.R. & M.P. Bryant. 1980. Propionate-degrading bacterium, *Syntrophobacter wolinii* sp. nov. gen. nov., from methanogenic ecosystems. Appl. Environ. Microbiol. **40:** 626–632.
46. Ficker, M. *et al.* 1999. Molecular characterization of a toluene-degrading methanogenic consortium. Appl. Environ. Microbiol. **65:** 5576–5585.
47. Townsend, G.T., R.C. Prince & J.M. Suflita. 2003. Anaerobic oxidation of crude oil hydrocarbons by the resident microorganisms of a contaminated anoxic aquifer. Environ. Sci. Technol. **37:** 5213–5218.
48. Sobieraj, M. & D.R. Boone. 2006. *Syntrophomonadaceae*. *In* The Prokaryotes: An Evolving Electronic Resource for the Microbiological Community, Vol. 4. M. Dworkin, *et al.*, Eds.: 1041–1046. New York: Springer-Verlag.
49. Wu, C., X. Liu & X. Dong. 2006. *Syntrophomonas cellicola* sp. nov., a spore-forming syntrophic bacterium isolated from a distilled-spirit-fermenting cellar, and assignment of *Syntrophospora bryantii* to *Syntrophomonas bryantii* comb. nov. Int. J. Syst. Evol. Microbiol. **56:** 2331–2335.
50. Stieb, M. & B. Schink. 1985. Anaerobic oxidation of fatty acids by *Clostridium bryantii* sp. nov., a sporeforming, obligately syntrophic bacterium. Arch. Microbiol. **140:** 387–390.

51. WU, C., X. DONG & X. LIU. 2007. *Syntrophomonas wolfei* subsp. *methylbutyratica* subsp. nov., and assignment of *Syntrophomonas wolfei* subsp. *saponavida* to *Syntrophomonas saponavida* sp. nov. comb. nov. Syst. Appl. Microbiol. doi: 10.1016/j.syapm.2006.12.001.
52. WU, C., X. LIU & X. DONG. 2006. *Syntrophomonas erecta* subsp. *sporosyntropha* subsp. nov., a spore-forming bacterium that degrades short chain fatty acids in co-culture with methanogens. Syst. Appl. Microbiol. **29:** 457–462.
53. ZHAO, H.X. *et al.* 1990. Assignment of *Clostridium bryantii* to *Syntrophospora bryantii* gen. nov., comb. nov. on the basis of a 16S rRNA sequence analysis of its crotonate-grown pure culture. Int. J. Syst. Bacteriol. **40:** 40–44.
54. LOROWITZ, W.H., H. ZHAO & M.P. BRYANT. 1989. *Syntrophomonas wolfei* subsp. *saponavida* subsp. nov., a long-chain fatty-acid-degrading, anaerobic, syntrophic bacterium. Int. J. Syst. Bacteriol. **39:** 122–126.
55. ROY, F. *et al.* 1986. *Synthrophomonas sapovorans* sp. nov., a new obligately proton reducing anaerobe oxidizing saturated and unsaturated long chain fatty acids. Arch. Microbiol. **145:** 142–147.
56. ZHANG, C., X. LIU & X. DONG. 2004. *Syntrophomonas curvata* sp. nov., an anaerobe that degrades fatty acids in co-culture with methanogens. Int. J. Syst. Evol. Microbiol. **54:** 969–973.
57. ZHANG, C., X. LIU & X. DONG. 2005. *Syntrophomonas erecta* sp. nov., a novel anaerobe that syntrophically degrades short-chain fatty acids. Int. J. Syst. Evol. Microbiol. **55:** 799–803.
58. SEKIGUCHI, Y. *et al.* 2000. *Syntrophothermus lipocalidus* gen. nov., sp. nov., a novel thermophilic, syntrophic, fatty-acid-oxidizing anaerobe which utilizes isobutyrate. Int. J. Syst. Evol. Microbiol. **50:** 771–779.
59. SVETLITSHNYI, V., F. RAINEY & J. WIEGEL. 1996. *Thermosyntropha lipolytica* gen. nov., sp. nov., a lipolytic, anaerobic, alkalitolerant, thermophilic bacterium utilizing short- and long-chain fatty acids in syntrophic coculture with a methanogenic archaeum. Int. J. Syst. Bacteriol. **46:** 1131–1137.
60. PLUGGE, C.M., M. BALK & A.J. STAMS. 2002. *Desulfotomaculum thermobenzoicum* subsp. *thermosyntrophicum* subsp. nov., a thermophilic, syntrophic, propionate-oxidizing, spore-forming bacterium. Int. J. Syst. Evol. Microbiol. **52:** 391–399.
61. RIOS-HERNANDEZ, L.A., L.M. GIEG & J.M. SUFLITA. 2003. Biodegradation of an alicyclic hydrocarbon by a sulfate-reducing enrichment from a gas condensate-contaminated aquifer. Appl. Environ. Microbiol. **69:** 434–443.
62. NILSEN, R.K., T. TORSVIK & T. LIEN. 1996. *Desulfotomaculum thermocisternum* sp. nov., a sulfate reducer isolated from a hot North Sea oil reservoir. Int. J. Syst. Bacteriol. **46:** 397–402.
63. IMACHI, H. *et al.* 2002. *Pelotomaculum thermopropionicum* gen. nov., sp. nov., an anaerobic, thermophilic, syntrophic propionate-oxidizing bacterium. Int. J. Syst. Evol. Microbiol. **52:** 1729–1735.
64. IMACHI, H. *et al.* 2006. Non-sulfate-reducing, syntrophic bacteria affiliated with *Desulfotomaculum* cluster I are widely distributed in methanogenic environments. Appl. Environ. Microbiol. **72:** 2080–2091.
65. QIU, Y.L. *et al.* 2003. *Sporotomaculum syntrophicum* sp. nov., a novel anaerobic, syntrophic benzoate-degrading bacterium isolated from methanogenic sludge treating wastewater from terephthalate manufacturing. Arch. Microbiol. **179:** 242–249.
66. FRIEDRICH, M. *et al.* 1996. Phylogenetic positions of *Desulfofustis glycolicus* gen. nov., sp. nov., and *Syntrophobotulus glycolicus* gen. nov., sp. nov., two new strict anaerobes growing with glycolic acid. Int. J. Syst. Bacteriol. **46:** 1065–1069.
67. FARDEAU, M.L. *et al.* 1993. Determination of the G+ C content of two *Syntrophus buswellii* strains by ultracentrifugation techniques. Curr. Microbiol. **26:** 185–189.
68. JACKSON, B.E. *et al.* 1999. *Syntrophus aciditrophicus* sp. nov., a new anaerobic bacterium that degrades fatty acids and benzoate in syntrophic association with hydrogen-using microorganisms. Arch. Microbiol. **171:** 107–114.
69. SZEWZYK, U. & B. SCHINK. 1989. Degradation of hydroquinone, gentisate, and benzoate by a fermenting bacterium in pure or defined mixed culture. Arch. Microbiol. **151:** 541–545.
70. LIU, Y. *et al.* 1999. Characterization of the anaerobic propionate-degrading syntrophs *Smithella propionica* gen. nov., sp. nov. and *Syntrophobacter wolinii*. Int. J. Syst. Bacteriol. **49:** 545–556.
71. WALLRABENSTEIN, C., E. HAUSCHILD & B. SCHINK. 1994. Pure culture and cytological properties of *Syntrophobacter wolinii*. FEMS Microbiol. Lett. **123:** 249–254.
72. LOY, A. *et al.* 2004. Microarray and functional gene analyses of sulfate-reducing prokaryotes in low-sulfate, acidic fens reveal cooccurrence of recognized genera and novel lineages. Appl. Environ. Microbiol. **70:** 6998–7009.
73. HOUWEN, F.P. *et al.* 1987. 13C-NMR study of propionate degradation by a methanogenic coculture. FEMS Microbiol. Lett. **41:** 269–274.
74. DE BOK, F.A.M. *et al.* 2001. Pathway of propionate oxidation by a syntrophic culture of *Smithella propionica* and *Methanospirillum hungatei*. Appl. Environ. Microbiol. **67:** 1800–1804.
75. MOUTTAKI, H., M.A. NANNY & M.J. MCINERNEY. 2007. Cyclohexane carboxylate and benzoate formation from crotonate in *Syntrophus aciditrophicus*. Appl. Environ. Microbiol. **73:** 930–938.
76. KASAI, Y. *et al.* 2005. Physiological and molecular characterization of a microbial community established in unsaturated, petroleum-contaminated soil. Environ. Microbiol. **7:** 806–818.
77. ELSHAHED, M.S. *et al.* 2001. Metabolism of benzoate, cyclohex-1-ene carboxylate, and cyclohexane carboxylate by *Syntrophus aciditrophicus* strain SB in syntrophic association with H2-using microorganisms. Appl. Environ. Microbiol. **67:** 1728–1738.
78. DUHAMEL, M. & E.A. EDWARDS. 2006. Microbial composition of chlorinated ethene-degrading cultures dominated by *Dehalococcoides*. FEMS Microbiol. Ecol. **58:** 538–549.
79. EDWARDS, E.A. & D. GRBIC-GALIC. 1994. Anaerobic degradation of toluene and *o*-xylene by a methanogenic consortium. Appl. Environ. Microbiol. **60:** 313–322.
80. HAGGBLOM, M.M., V.K. KNIGHT & L.J. KERKHOF. 2000. Anaerobic decomposition of halogenated aromatic compounds. Environ. Pollut. **107:** 199–207.

81. DUBOURGUIER, H.C. *et al.* 1986. Characterization of two strains of *Pelobacter carbinolicus* isolated from anaerobic digesters. Arch. Microbiol. **145:** 248–253.
82. SCHINK, B. & M. STIEB. 1983. Fermentative degradation of polyethylene glycol by a strictly anaerobic, gram-negative, nonsporeforming bacterium, *Pelobacter venetianus* sp. nov. Appl. Environ. Microbiol. **45:** 1905–1913.
83. SCHINK, B. 1985. Fermentation of acetylene by an obligate anaerobe, *Pelobacter acetylenicus* sp. nov. Arch. Microbiol. **142:** 295–301.
84. KRUMHOLZ, L.R. & M.P. BRYANT. 1986. *Syntrophococcus sucromutans* sp. nov. gen. nov. uses carbohydrates as electron donors and formate, methoxymonobenzenoids or *Methanobrevibacter* as electron acceptor systems. Arch. Microbiol. **143:** 313–318.
85. SEKIGUCHI, Y. *et al.* 2006. *Tepidanaerobacter syntrophicus* gen. nov., sp. nov., an anaerobic, moderately thermophilic, syntrophic alcohol- and lactate-degrading bacterium isolated from thermophilic digested sludges. Int. J. Syst. Evol. Microbiol. **56:** 1621–1629.
86. MECKENSTOCK, R.U. 1999. Fermentative toluene degradation in anaerobic defined syntrophic cocultures. FEMS Microbiol. Lett. **177:** 67–73.
87. DAVIDOVA, I.A. *et al.* 2006. *Desulfoglaeba alkanexedens* gen. nov., sp. nov., an *n*-alkane-degrading, sulfate-reducing bacterium. Int. J. Syst. Evol. Microbiol. **56:** 2737–2742.
88. ZENGLER, K. *et al.* 1999. Methane formation from long-chain alkanes by anaerobic microorganisms. Nature **401:** 266–269.
89. ZINDER, S.H. & M. KOCH. 1984. Non-aceticlastic methanogenesis from acetate: acetate oxidation by a thermophilic syntrophic coculture. Arch. Microbiol. **138:** 263–272.
90. LEE, M.J. & S.H. ZINDER. 1988. Isolation and characterization of a thermophilic bacterium which oxidizes acetate in syntrophic association with a methanogen and which grows acetogenically on H_2-CO_2. Appl. Environ. Microbiol. **54:** 124–129.
91. LEE, M.J. & S.H. ZINDER. 1988. Carbon monoxide pathway enzyme activities in a thermophilic anaerobic bacterium grown acetogenically and in a syntrophic acetate-oxidizing coculture. Arch. Microbiol. **150:** 513–518.
92. CORD-RUWISCH, R., D.R. LOVLEY & B. SCHINK. 1998. Growth of *Geobacter sulfurreducens* with acetate in syntrophic cooperation with hydrogen-oxidizing anaerobic partners. Appl. Environ. Microbiol. **64:** 2232–2236.
93. SCHNURER, A., B. SCHINK & B.H. SVENSSON. 1996. *Clostridium ultunense* sp. nov., a mesophilic bacterium oxidizing acetate in syntrophic association with a hydrogenotrophic methanogenic bacterium. Int. J. Syst. Bacteriol. **46:** 1145–1152.
94. HATTORI, S. *et al.* 2000. *Thermacetogenium phaeum* gen. nov., sp. nov., a strictly anaerobic, thermophilic, syntrophic acetate-oxidizing bacterium. Int. J. Syst. Evol. Microbiol. **50:** 1601–1609.
95. ZHILINA, T.N. *et al.* 2005. "*Candidatus contubernalis alkalaceticum*," an obligately syntrophic alkaliphilic bacterium capable of anaerobic acetate oxidation in a coculture with *Desulfonatronum cooperativum*. Microbiology **74:** 800–809.
96. HEIJTHUIJSEN, J. & T.A. HANSEN. 1986. Interspecies hydrogen transfer in co-cultures of methanol-utilizing acidogens and sulfate-reducing or methanogenic bacteria. FEMS Microbiol. Ecol. **38:** 57–64.
97. CORD-RUWISCH, R. & B. OLLIVIER. 1986. Interspecific hydrogen transfer during methanol degradation by *Sporomusa acidovorans* and hydrogenophilic anaerobes. Arch. Microbiol. **144:** 163–165.
98. PANTSKHAVA, E.S. & V.V. PCHYOLKINA. 1969. Methane fermentation of methanol by *Methanobacillus kuznecovii*. Appl. Environ. Microbiol. **5:** 416–420.
99. WEIJMA, J. & A.J. STAMS. 2001. Methanol conversion in high-rate anaerobic reactors. Water Sci. Technol. **44:** 7–14.
100. BALK, M. *et al.* 2003. Methanol utilization by a novel thermophilic homoacetogenic bacterium, *Moorella mulderi* sp. nov., isolated from a bioreactor. Arch. Microbiol. **179:** 315–320.
101. BALK, M. *et al.* 2007. Methanol utilizing *Desulfotomaculum* species utilizes hydrogen in a methanol-fed sulfate-reducing bioreactor. Appl. Microbiol. Biotechnol. **73:** 1203–1211.
102. JIANG, B. 2006. The effect of trace elements on the metabolism of methanogenic consortia. Ph.D. thesis, Wageningen University. Wageningen, the Netherlands.
103. WINTERS, D.K. & L.G. LJUNGDAHL. 1989. PQQ-Dependent methanol dehydrogenase from *Clostridium thermoautotrophicum*. *In* PQQ and Quinoproteins. J. Jongejan & J.A. Duine, Eds.: 35–39. Dordrecht, The Netherlands: Kluwer Academic Publishers.
104. THAUER, R.K., K. JUNGERMANN & K. DECKER. 1977. Energy conservation in chemotrophic anaerobic bacteria. Bacteriol. Rev. **41:** 100–180.
105. DIMROTH, P. & B. SCHINK. 1998. Energy conservation in the decarboxylation of dicarboxylic acids by fermenting bacteria. Arch. Microbiol. **170:** 69–77.
106. SCHINK, B. & A.J. M. STAMS. 2001. Syntrophism among prokaryotes. *In* The Prokaryotes: An Evolving Electronic Resource for the Microbiological Community. M. Dworkin, *et al.*, Eds. New York: Springer-Verlag.
107. HATTORI, S. *et al.* 2005. Operation of the CO dehydrogenase/acetyl coenzyme A pathway in both acetate oxidation and acetate formation by the syntrophically acetate-oxidizing bacterium *Thermacetogenium phaeum*. J. Bacteriol. **187:** 3471–3476.
108. KOSAKA, T. *et al.* 2006. Reconstruction and regulation of the central catabolic pathway in the thermophilic propionate-oxidizing syntroph *Pelotomaculum thermopropionicum*. J. Bacteriol. **188:** 202–210.
109. MCINERNEY, M.J. *et al.* 2007. The genome of *Syntrophus aciditrophicus*: life at the thermodynamic limit of microbial growth. Proc. Natl. Acad. Sci. USA **104:** 7600–7605.
110. SCHOLTEN, J.C. *et al.* 2007. Evolution of the syntrophic interaction between *Desulfovibrio vulgaris* and *Methanosarcina barkeri*: involvement of an ancient horizontal gene transfer. Biochem. Biophys. Res. Commun. **352:** 48–54.
111. DESANTIS, T.Z. *et al.* 2006. Greengenes, a Chimera-checked 16S rRNA gene database and workbench compatible with ARB. Appl. Environ. Microbiol. **72:** 5069–72.
112. LUDWIG, W. *et al.* 2004. ARB: a software environment for sequence data. Nucleic Acids Res. **32:** 1363–1371.

Comparative Genomics of Clostridia

Link between the Ecological Niche and Cell Surface Properties

HOLGER BRÜGGEMANN[a] AND GERHARD GOTTSCHALK[b]

[a]*Max Planck Institute for Infection Biology, Department of Molecular Biology, Chariteplatz 1, Berlin, Germany*

[b]*Göttingen Genomics Laboratory, Institute of Microbiology and Genetics, Georg August University, Grisebachstr., Göttingen, Germany*

Several clostridial genomes have been sequenced in the past few years, and new sequencing projects as well as postgenomics approaches to study this important anaerobic genus are under way. This wealth of sequence data extends our knowledge of the metabolic versatility of this genus and its potential biotechnological exploitation and also adds to our view of virulence traits of pathogenic clostridia. So far, little is known about the clostridial cell surface and its role in the interaction with the respective environment. This is of special interest in understanding the host interaction of intestinal clostridia, which are assumed to influence the integrity of the gastrointestinal tract. Here recently sequenced clostridial genomes are explored with respect to features of cell surface association. Several classes of proteins with cell wall–binding domains are found, including S-layer proteins and adhesins. It could be shown that most species are specifically equipped with surface-associated factors, resulting in distinctive host/matrix-interacting properties.

Key words: **genomics; clostridia; comparative genomics; surface; S-layer; cell wall; host interaction; energy metabolism**

Introduction

The genus *Clostridium* comprises more than 120 species and is the largest among the anaerobic spore-forming genera. Several clostridia have been or are still being used for biotechnological purposes, such as the solvent producer *Clostridium acetobutylicum*. Others have been subject to many investigations because they employ metabolic pathways and enzyme activities of interest, for example, *Clostridium pasteurianum*, nitrogenase, hydrogenase, and ferredoxins; *Clostridium thermocellum*, cellulases; *Clostridium tetanomorphum*, coenzyme B12-dependent reactions; *Clostridium kluyveri*, energy metabolism and reduction of enoates; *Clostridium formicoaceticum*, tungsten-containing formate dehydrogenase; and *Clostridium ljungdahlii*, growth with carbon monoxide.[1–3] In addition, the genus *Clostridium* contains at least 35 pathogenic species.[4] Several well-known diseases are caused by toxin-producing clostridia, such as gas gangrene and necrotic enteritis caused by *Clostridium perfringens*, diarrhea and pseudomembraneous colitis caused by *Clostridium difficile*, tetanus due to *Clostridium tetani*, and the food-borne botulism caused by *Clostridium botulinum*. The latter two organisms produce the most powerful neurotoxins known to mankind, the tetanus and the botulinum toxin, respectively.[5] Pathogenic bacteria, such as *Clostridium novyi*, are considered promising candidates in cancer therapy.[6] When injected intravenously, *C. novyi* spores produce substantial antitumor effects in experimental animals without excessive toxicity. The proliferation of *C. novyi* depends on anoxic conditions, since the vegetative form is extremely sensitive to oxygen. Thus, tumors with their large hypoxic regions provide an ideal growth environment for *C. novyi*.

In recent years, clostridia regained attention for various reasons, not only from microbiologists but also from researchers in other areas, such as biotechnology, genomics, cell biology, toxicology, pharmacology, and clinical research. Studies on clostridial toxins have been reinforced in the light of possible bioterrorist attacks. This led to a wealth of insights in structure–function

Address for correspondence: Holger Brüggemann, Max Planck Institute for Infection Biology, Department of Molecular Biology, Chariteplatz 1, D-10117 Berlin, Germany.
hbruegg@gwdg.de, ggottsc@gwdg.de

Ann. N.Y. Acad. Sci. 1125: 73–81 (2008).
doi: 10.1196/annals.1419.021

relationships of toxins, such as the rho-dependent glucosyltransferases TcdA and B of *C. difficile*,[7,8] or the binary toxins with ADP-ribosyltransferase activity of *C. perfringens* (ι-toxin) and *C. difficile*.[9,10] Cell biologists have benefited from this research, and use clostridial toxins as a tool to investigate elementary cellular processes, such as synaptic vesicle trafficking and retrograde transport pathways in motor neurons (tetanus and botulinum toxin)[11,12] or rho-dependent signaling pathways and their implication in cell morphology and endo- and exocytosis (TcdAB).[13,14] Botulinum toxin (sold commercially under the brand name Botox) is used for cosmetic purposes to smoothen facial lines, but is also used to treat such conditions as migraine headaches and cervical dystonia.[15] Biotechnologically exploited clostridia, such as the acetone- and butanol-producing *C. acetobulylicum*, have experienced a renaissance due to efforts to develop alternatives to petrol to reduce our reliance on finite crude-oil resources and to lower CO_2 emissions. In the course of large metagenomic projects to sequence the entire human microbiota it became apparent that clostridial species are major commensals of the human intestinal tract.[16,17] Their contribution to our well-being is still poorly understood. Other reasons why clostridial research has intensified are based on recent advances in developing new genetic tools to study this rather reluctant genus. A knock-out system has been developed, which has been applied successfully to some pathogenic clostridial species, including *C. botulinum* and *C. difficile*.[18] This technique and the approaches of modern systems biology depend on the availability of whole genome sequences. In this chapter, we present the progression and current status of clostridial genome projects. In the second part we present a genomic search for and comparative analysis of clostridial cell surface constituents. These structures have an important role in the interaction with the environment, for instance, in the cross talk between clostridial commensals and the human intestinal tract.

Genome Sequences of Clostridial Species

Prior to the genome-sequencing era, genetic information about clostridial species, if available, was largely restricted to a few loci, such as those encoding major toxin genes of pathogenic clostridia, or those encoding important fermentative pathways and unique or unusual enzymatic activities. Since 2000, the genomes of nine clostridial species have been completely sequenced, starting with the genome of the solvent-producing nonpathogenic *C. acteobutylicum* in 2001[19] (TABLE 1). Subsequently, the genomes of major clostridial pathogens were sequenced, including *C. perfringens*,[20] *C. tetani*,[21] *C. novyi*,[6] *C. botulinum*,[22] and *C. difficile*.[23] More recently, genome sequencing has been applied to deduce the "pan-genome" of a given species: recently, several different strains of *C. perfringens* and *C. botulinum* were completely sequenced (see http://www.ncbi.nlm.nih.gov or http://www.ebi.ac.uk/genomes/). This allows us to distinguish the core genome from strain-specific genomic variation (i.e., flexible gene pool) within a given species.[24] In addition, in the last few years, the genome sequences of several nonpathogenic clostridia, such as those of *Clostridium beijerinckii*, *C. thermocellum*, and *C. kluyveri*, have been obtained. These species have special metabolic properties, which are or might be exploited for biotechnological purposes: *C. beijerinckii* produces butanol, acetone, and/or isopropanol and has a broad substrate range, including pentoses, hexoses, and starch; *C. thermocellum* is a thermophilic bacterium capable of directly converting cellulosic substrates into ethanol; *C. kluyveri* grows anaerobically on ethanol and acetate as sole energy sources, with butyrate, caproate, and H_2 as fermentation products.[25] Draft assembly sequences currently exist for several other clostridial genomes, including pathogenic species, such as clinically relevant strains of *C. difficile*, and five additional strains each of *C. botulinum* and *C. perfringens*. Draft genome sequences of nonpathogenic clostridia include strains isolated from the human intestinal tract (see later in the chapter), and species, such as *Clostridium cellolyticum* and *Clostridium phytofermentans*, a cellulolytic bacterium that is capable of fermenting all major carbohydrate components of biomass, such as cellulose, pectin, starch, and xylan, while producing large amounts of ethanol and hydrogen (see http://www.jgi.doe.gov). A draft genome also exists for *C.* sp. OhILAs, a bacterium that can ferment glycerol, fructose, and lactate, as well as respire thiosulfate and arsenate, thereby tolerating high arsenate concentrations.

The availability of these sequence data allows us to get a deeper insight into the life of clostridial species, in order to understand and exploit their exceptional metabolism for biotechnological purposes. For those clostridial species that are in close contact with humans, research interests aim at the identification of traits that enable the microorganisms to persist, interact with, and/or harm the human body.

New Discoveries Concerning the Energy Metabolism of Clostridia

The substrate range of clostridial species is highly diverse. The most important substrates in nature are polymers, such as cellulose, starch, pectin, other polysaccharides, and proteins. Monosaccharides are broken down via the Embden–Meyerhof–Parnas pathway. Sugar acids such as gluconate are degraded via a modified Entner–Doudoroff pathway.[26,27] Pathogenic clostridia are usually proteolytic species. Fermented amino acids can serve as sole carbon, nitrogen, and energy sources. The central intermediate in the formation of fermentation products is pyruvate, which in most clostridial fermentations is converted to acetyl-coenzyme A (acetyl-CoA) by pyruvate:ferredoxin oxidoreductase. Electrons from reduced ferredoxin can be used for H_2 evolution catalyzed by hydrogenase. The fate of acetyl-CoA depends on the clostridial species. It can be converted into a mixture of ethanol, acetate, and/or butyrate, and adenosine triphosphate (ATP) is generated by substrate-level phosphorylation via acetate kinase and/or butyrate kinase. The ratio at which the fermentation products are formed depends on the amount of H_2 evolved.[28,29] In several butyrate- and acetate-producing clostridia, more H_2 is evolved than expected. The additional H_2 seems to originate from reduced nicotinamide adenine dinucleotide (NADH). It is assumed that a NADH:ferredoxin oxidoreductase is involved, which catalyzes the electron transfer from NADH to ferredoxin. Other clostridia, mainly ethanol/acetate-producing species, often evolve less H_2. Here, some of the reduced ferredoxin gained in the pyruvate:ferredoxin oxidoreductase reaction is apparently used to reduce oxidized NAD (NAD^+) to NADH, which is then employed for NADH-dependent reactions, such as those yielding ethanol from acetyl-CoA or butanol from butyryl-CoA. Membrane preparations of some clostridia such as *C. tetanomorphum* indeed show a high NADH dehydrogenase activity (Buckel *et al.*, unpublished results).

Until recently no protein complex acting as a NADH:ferredoxin oxidoreductase to catalyze the electron transfer between ferredoxin and NAD^+, and vice versa, could be identified in clostridia. The first indication that such a complex indeed exists was found in the genome sequence of *C. tetani*.[21] Here, a cluster of six genes was identified, whose proteins show close similarity to the *Rhodobacter*-specific nitrogen fixation (Rnf) system of *Rhodobacter capsulatus*, as well as to the NADH:quinone oxidoreductase (Nqr, not to confused with NDH-1 and NDH-2) present in many aerobic pathogens such as *Vibrio cholerae*, *Salmonella typhimurium*, *Yersinia pestis*, and *Haemophilus influenzae*.[30,31] In the latter organisms, the Nqr system functions as a sodium-motive NADH:quinone oxidoreductase. The six subunits (NqrA–F) harbor one 2Fe–2S cluster, one noncovalently bound flavin adenine dinucleotide (FAD), two covalently bound flavin mononucleotide (FMN) residues, and possibly also one ubiquinone-8 as prosthetic groups.[32] The Nqr activity is stimulated by sodium ions and is coupled by pumping Na^+, but not H^+. In *C. tetani*, the Nqr system may be the system that catalyzes electron flow from reduced ferredoxin via NADH to the NADH-consuming dehydrogenases of the butyrate pathway. Such a reaction would be coupled with the translocation of Na^+, which, in turn, is used for sodium-dependent substrate uptake.[33] In this respect, many Na^+-dependent amino acid symporters are encoded in the genome of *C. tetani*.

The sequences of additional clostridial genomes revealed that the Rnf system is widespread among clostridial species, but is lacking in *C. acetobutylicum* and *C. cellulyticum*. A function for Rnf/Nqr has been experimentally verified in *C. kluyveri*, whose genome contains the six genes for RnfA-F (Seedorf *et al.*, unpublished results). The organism uses ethanol as the main substrate and evolves H_2 up to a pressure of 10^5 Pa and higher.[25] The existence of an Rnf complex could help to explain how *C. kluyveri* manages to produce H_2 during fermentation of ethanol and acetate.

Thus, a number of clostridial species do not rely only on ATP synthesis by substrate-level phosphorylation. The transfer of electrons from reduced ferredoxin to NAD^+ catalyzed by the membrane-associated Rnf system will generate an ion-motive force, which can also be taken advantage of by the F_1F_O-ATPase for ATP synthesis.

Clostridia of the Human Intestine

The microbiota in the adult human intestine is enormous, comprising up to 100 trillion microorganisms.[34] Although the relationship of this flora with the host is often described as commensal, the nature of interaction and the potential mutual benefits are poorly understood. The majority of the microbial flora consists of uncultivated species and novel microorganisms. Thus, large-scale methods, such as the metagenomics approach, that is, the decipherment of the collective genome of all microbial species in an ecosystem (the microbiome), have recently been applied to the

TABLE 1. Genome features of sequenced clostridial genomes, with special emphasis on cell surface–associated coding sequences

General genome features	*C. kluyveri* DSM 555	*C. difficile* 630	*C. tetani* E88	*C. acetobutylicum* ATCC 824	*C. perfringens* 13	*C. novyi* NT	*C. botulinum* A (ATCC 3502)	*C. beijerinckii* NCIMB 8052	*C. thermocellum* ATCC 27405	*C. bolteae*[a] ATCC BAA-613	*C.* sp. L2–50[+] /
Chromosome size [bp]	3,964,618	4,290,252	2,799,250	3,940,880	3,031,430	2,547,720	3,886,916	6,000,632	3,843,301	6,547,592	2,972,901
G + C content [%]	32.0	29.1	28.6	30.0	28.6	28.9	28.2	29.0	38.0	49.1	41.3
Number of chromosomal CDS	38,38	3,776	2,368	3,740	2,660	2,325	3,650	5,094	3,191	nd	nd
Plasmids	pCKL555A	pCD630	pE88	pSOL1	pCP13	/	pBOT3502	/	/	nd	nd
Plasmid size [bp]	59,182	7,881	74,082	192,000	54,310	/	16,344	/	/	nd	nd
Coding bias [%]	76	82	82	78	83	83	83	83	80	nd	nd
Number of putative "alien" genes[b]	200	249	35	36	3	12	102	21	85	nd	nd
% putative "alien" genes	5.2	6.6	1.5	1.0	0.1	0.5	2.8	0.4	2.7	nd	nd
RNF complex	Yes	Yes	Yes	No	Yes	Yes	Yes	Yes	Yes	nd	nd
PF00037 – 4Fe-4S	49	49	22	24	18	24	40	62	30	Yes	Yes
Cell-wall binding[c]										39	12
PF04122 – Cw_binding_2	17	29	19	0	0	0	8	1	0	0	0
PF01473 – Cw_binding_1	0	3	0	0	6	0	0	66	0	39	3
LPxTG-motif proteins[d]	0	3	3	2	13	5	1	0	0	0	0
PF04203 – Sortase	0	1	1	1	4	2	1	0	0	1	0
PF00395 – SLH	0	0	1	0	0	0	0	1	26	2	0
PF01471 – PG_binding_1	3	2	5	11	1	4	5	5	7	2	1
Protein–protein/-matrix interaction[c]											
PF02368 – Big_2	1	0	5	18	1	0	1	6	3	0	14
PF07532 – Big_4	5	0	3	0	2	2	5	1	1	0	0
PF00560 – LRR_1	3	0	3	6	0	2	4	1	1	1	0
PF00754 – F5_F8_type_C (CBM_32)	1	1	1	1	17	0	1	1	1	0	0
PF00232 – GH1	0	9	0	7	1	0	2	11	2	4	0
PF00404 – Dockerin_1	0	1	0	10	4	0	0	0	74	0	1
Σ CBMs	nd	2	0	26	27	1	7	7	70	nd	nd
Σ GHs	nd	39	9	73	54	11	21	75	70	nd	nd

Continued

TABLE 1. Continued.

General genome features	*C. kluyveri* DSM 555	*C. difficile* 630	*C. tetani* E88	*C. acetobutylicum* ATCC 824	*C. perfringens* 13	*C. novyi* NT	*C. botulinum* A (ATCC 3502)	*C. beijerinckii* NCIMB 8052	*C. thermocellum* ATCC 27405	*C. bolteae*[a] ATCC BAA-613	*C.* sp. L2–50[+] /
Capsule/polysaccharide[c]											
PF00534 – GT1	8	7	6	24	19	9	13	13	17	11	4
PF00535 – GT2	9	9	4	35	9	1	13	16	8	15	12
PF04932 – Wzy_C	1	0	2	4	1	2	1	3	4	3	1
PF02706 – Wzz	0	0	0	2	2	1	1	1	3	2	1
PF02350 – epimerase_2	4	3	2	2	1	3	2	2	4	1	0
Σ GTs	nd	23	14	64	33	15	31	39	34	nd	nd

NOTE: To identify cell surface–associated proteins in clostridial proteomes, Hidden Markov Model (HMM) searches with known cell surface domains were performed. All HMMs except (2) were taken from http://pfam.janelia.org/ For the identification of all glycoside hydrolases (GHs), glycosyl transferases (GTs), and carbohydrate-binding module proteins (CBMs), see http://www.cazy.org

[a]Genome sequence incomplete—draft assemblies with >50 contigs.

[b]To identify alien genes and genomic islands within a genome, the program Colombo was used (http://www.tcs.informatik.uni-goettingen.de/colombo).

[c]Numbers represent all CDS in a given genome that contain the respective domain/motif (e-value cutoff < 0.1).

[d]LPxTG-HMM taken from http://bamics3.cmbi.kun.nl/cgi-bin/jos/sortase_substrates/index.py

intestine in order to disclose the species diversity and the genetic repertoire within these communities.[17,35] First sequencing analyses of this microbiome revealed that the human gut microbiota, while consisting of the highest cell densities recorded for any ecosystem, contain only very few prokaryotic divisions. In fact, most identified bacterial species belong to the phyla Firmicutes, in particular the genera *Clostridium* and *Eubacterium*, and Bacteroidetes (e.g., genus *Bacteroides*), which make up >99% of the identified phylotypes.[17] Subsequently, about 100 predominant members of the gut microbiota, mostly of the two phyla Firmicutes and Bacteroidetes, and a few Actinobacteria and Proteobacteria, the predominate archaeon, *Methanobrevibacter smithii*,[36] and a second archaeon, *Methanosphaera stadtmanae*,[37] have been or are currently being sequenced, and draft genome sequences are already available for the scientific community. Among the selected clostridial species are *Clostridium bartlettii, Clostridium bolteae, Clostridium leptum, Clostridium ramosum, Clostridium scindens, Clostridium spiroforme, Clostridium symbiosum*, and five additional clostridial strains (http://genome.wustl.edu/index.cgi). The genomes of these intestinal clostridia have a much higher guanine-cytosine content, between 40 and 50%, than environmental clostridia isolated from the soil (28–38%), which might already reflect an evolutionary adaptation to the gut microbiota.

Although first insights were gained, in particular concerning the contribution of the clostridia and other members of the microbiota to metabolic properties of the human host, such as the degradation of otherwise indigestible nutrients and the provision of vitamins and cofactors,[35] other issues remain puzzling, such as the interaction of the gut flora with the host immune system. As shown previously, intestinal bacteria are required for the development of gut-associated lymphoid tissues, which mediate a variety of host immune functions, such as mucosal immunity and oral tolerance.[38,39] Thus, the nature and extent of an innate or adaptive immune response triggered by members of the gut microbiota on contact with intestinal tissue is currently under intensive investigation. Another challenge ahead is the identification of immunogenic substances derived from intestinal bacteria. As shown in many previous studies, secreted factors and microbial surface components have a pivotal role as immunostimulatory agents,[40,41] representing possible antigens as well as ligands for immune receptors. Since only very little is known about host interactive properties of clostridia, we will have a close look at clostridial cell surface–embedded and –associated factors.

Clostridial Cell Surfaces

Properties of the cell surface and envelope play a crucial role in the interaction of clostridia with their environment with respect to protection (e.g., from oxygen), attachment, colonization and proliferation, substrate accessibility and utilization, and (the evasion of) immunological responses. Information on proteins involved in environment–bacterium interactions should be obtained from a comparative analysis of cell surface properties of clostridial species, deduced from sequenced genomes.

Cell-Wall-Bound Proteins

Most clostridia produce a surface layer, a paracrystalline array that surrounds the cell. Surface layers are usually composed of one or two surface-layer proteins (SLPs) or glycoproteins.[42] SLPs are expressed at very high levels, and are anchored in the peptidoglycan moiety of the cell wall. In some species the surface layer is a well-characterized virulence factor that undergoes antigenic variation. The surface layer of *C. difficile* has been shown to act as an adhesin to host tissue and to interfere with the immune response, perturbing the balance between inflammatory and regulatory cytokines.[43,44] The respective SLP of *C. difficile*, SlpA, contains three copies of the domain PF04122, which is predicted to be responsible for cell-wall binding and anchoring.[45,46] Interestingly, this domain also can be found in other cell surface proteins. Altogether, 28 coding sequences (CDS) can be identified in the genome of *C. difficile*, harboring such a domain in two or three copies (TABLE 1). PF04122 is also found in proteins encoded by other clostridial genomes; as examples, using a Hidden Markov Model (HMM)–based search, 19 such proteins were identified in *C. tetani* E88, 8 in *C. botulinum* A, and 17 in nonpathogenic *C. kluyveri*. Recently, the SLP of *C. tetani* has been identified,[47] which also contains two PF04122 domains near the N terminus, and a FIVAR domain at the C terminus. Other sequenced clostridial genomes, including the intestinal clostridia *C. bolteae* and *C.* sp L2–50, do not contain CDS harboring such domains (TABLE 1). They must employ different mechanisms to anchor surface proteins to the cell wall. For instance, another wall binding motif (PF01473) has been found, which is present in multiple copies in toxin A (TcdA) and toxin B (TcdB) of *C. difficile*.[48] This domain is especially overrepresented in proteins of *C. beijerinckii* (66 CDS) and *C. bolteae* (39 CDS), respectively, but absent from most other clostridia.

A characteristic of mainly pathogenic gram-positive bacterial species is the presence of surface-associated proteins with a C-terminal cell-wall anchor containing the LPxTG amino acid motif.[49,50] A HMM search revealed that such proteins can be found almost exclusively in pathogenic clostridia (with the exception of *C. acetobutylicum*). In *C. perfringens* such proteins are highly abundant, although strain-specific differences exist. For example, there are 7, 13, and 18 CDS in the sequenced genomes of strains SM101, strain 13 and ATCC 13124, respectively. In all clostridia harboring LPxTG-motif containing proteins, sortase-encoding CDS can be identified, which are responsible for attaching these proteins to the cell wall.[51] Conversely, the number of sortase paralogs in a given genome correlates with the number of LPxTG-motif containing sortase substrates.

Other putative cell-wall binding motifs can be found in clostridial proteomes, and some seem to be restricted to a certain species. For instance, *C. thermocellum* harbors 26 proteins with the S-layer homology domain PF00395,[52] but this is rarely found in other clostridial genomes. It is tempting to assume that such heterogeneity among clostridial surface proteins is a result of an adaptation process connected to the respective ecological niche.

Protein–Protein/-Matrix Interaction

What are the functions of clostridial cell-wall-bound proteins? As mentioned earlier, some of the proteins with the cell-wall binding repeat PF04122 constitute the S-layer, as shown experimentally for *C. difficile* and *C. tetani*.[45,47] Others are involved in host-tissue adhesion as it was shown for Cwp66 of *C. difficile*, which was isolated by immunoscreening.[46] The gene *cwp66* is clustered with *slpA* together with another gene encoding a PF04122-containing protein, Cwp84. The latter was shown to exhibit protease activity, and it was speculated that it is involved in the maturation of surface-associated adhesins of *C. difficile*.[53] For other cell-wall-bound proteins experimental data are lacking so far. However, a bioinformatics analysis revealed that such PF04122-containing proteins often harbor domains in their surface-exposed part that are related to protein protein interaction, suggesting a role in host/environment–bacterium interaction. For instance, domains such as leucine-rich repeats (PF00560), immunoglobulin-like domains (e.g., PF02368 and PF07532), tetratricopeptide repeats (PF00515), and characteristic repetitive amino acid motifs can be found.

Domain analysis of LPxTG-motif containing proteins suggest that at least some of these proteins have a role in host-tissue degradation, adhesion, or substrate uptake. An example of a possible adhesion is CTC00471 from *C. tetani*, which contains one collagen-binding (PF05737) and two Cna protein B-type domains (PF05738), all of which are present in the characterized collagen-binding surface protein of *Staphylococcus aureus*.[54] CTC01193 from *C. tetani* is probably acting as a siderophore, harboring seven iron transport-associated NEAT domains (PF05031). The genome of *C. perfringens* encodes LPxTG-motif containing hyaluronidases (nagK, nagI, nagJ),[20] which carry out random hydrolysis of 1–4-linkages between *N*-acetyl-β-D-glucosamine and D-glucuronate residues in hyaluronate. NagJ, possessing a glycoside hydrolase domain (GH84) and a carbohydrate binding module (CBM32, F5/8 domain), has been further characterized.[55] Binding studies indicated its attachment to carbohydrate-bearing surfaces via terminal LacNAc (β-D-galactosyl-1,4-β-D-*N*-acetylglucosamine) motifs common to the *O*-linked glycans of gastric mucin. Functionally related cell-wall anchor-containing proteins include CPE0289 and CPE0266 of *C. perfringens* 13 with predicted *endo*-β-*N*-acetylglucosaminidase and β-hexosamidase activities, respectively, and CPF_1087 and CPR_2607 of strains ATCC 13124 and SM101, coding for a predicted pullulanase and a laminarinase, respectively.[24] The arsenal of different cell-wall-bound glycoside hydrolases might explain why *C. perfringens* causes mucosal necrosis associated with severe enteritis. Indeed, the majority of these enzymes are predicted to have specificities appropriate for the degradation of complex glycans, suggesting that this bacterium is well equipped to attack the diverse sugar structures of the mucins. Among the sequenced clostridia, only *C. acetobutylicum*, *C. beijerincki* and *C. thermocellum*, all of which are specialized to degrade a variety of polysaccharides, contain more glycoside hydrolases than *C. perfringens* (TABLE 1). In addition, *C. perfringens* possesses more proteins with CBMs than all other clostridia except *C. thermocellum*.

Capsule

Another important factor for the colonization and persistence of bacteria, as well as for immunological aspects of host interaction, is the formation of a polysaccharide caspsule. The presence of a capsule-like structure in *C. difficile* has been reported.[56] Some strains possess a 30- to 80-nm-thick capsule visible in ultrathin sections, while others have a thinner polysaccharide layer (10–20 nm). In addition, most strains of *C. perfringens* possess serologically distinct and complex capsular structures, with several hundred serological types reported.[57] Comparison of the genomes of three different *C. perfringens* strains has revealed unique capsule-associated genetic loci in each strain, encoded in genomic islands.[24] The authors postulate a core set of clostridial capsule genes, comprising genes for capsular synthesis and export, and strain-specific capsule genes, for example, glycosyl transferases (GTs). Key components involved in capsule biosynthesis are present in almost all clostridia, such as *O*-antigen polymerase (PF04932), chain-length determinant protein (PF02706), and UDP-N-acetylglucosamine 2-epimerase (PF02350). Capsule biosynthesis gene clusters also contain various GTs, which catalyze the transfer of sugar moieties from activated donor molecules to specific acceptor molecules forming glycosidic bonds. An examination of clostridial proteomes has been carried out to identify all GTs (TABLE 1, and http://www.cazy.org). All clostrida possess multiple GTs, ranging from 14 in *C. tetani* to 64 in *C. acetobutylicum*. Although there are species and even strain-specific differences, all examined clostridia are equipped with genes of capsule/polysaccharide biosynthesis.

Conclusion and Outlook

Genome sequencing has given completely new in sights into all aspects of the anaerobic lifestyle of clostridia. At present little is known about the interaction between clostridia and their environment via cell surface-associated processes. Here we show that clostridia possess a range of surface-associated proteins, having structural as well as host-interactive roles, but only a few of them have been characterized so far. Species-specific traits exist, such as *C. perfringens*' cell surface containing proteins with host-tissue degradative functions. Preliminary data suggest that intestinal clostridia do not differ fundamentally from other clostridia regarding their surface properties. However, several genomes of intestinal clostridia are currently being sequenced, so that new information will be available soon for an in-depth analysis. A challenge ahead is to understand the interaction of clostridia with the host tissue for those species that are (or might be under certain conditions) in close contact with humans. Such interactions can be harmful (e.g., with *C. perfringens*), but also beneficial to the host (e.g., in the case of commensal clostridia of the intestine). Future research will also shed light on such questions as How does the intestinal immune system distinguish between a normal clostridial commensal and a potentially threatening pathogenic bacterium? and leads to closely related questions concerning gastrointestinal tract diseases.

Conflict of Interest

The authors declare no conflicts of interest.

References

1. DÜRRE, P. 2005. Handbook on Clostridia. Boca Raton, FL: CRC Press.
2. MINTON, N.P. & D.J. CLARKE. 1989. Clostridia. New York: Plenum Press.
3. GOTTSCHALK, G. 1979. Bacterial Metabolism. New York: Springer-Verlag.
4. STACKEBRANDT, E. & F.A. RAINEY. 1997. Phylogenetic relationships. *In* The Clostridia: Molecular Biology and Pathogenesis. J. Rood, B.A. McClane, J.G. Songer & R.W. Titball, Eds.: 3–19. San Diego, CA: Academic Press.
5. ARNON, S. 1997. Human tetanus and human botulism. *In* The Clostridia: Molecular Biology and Pathogenesis. J. Rood, B.A. McClane, J.G. Songer & R.W. Titball, Eds. San Diego, CA: Academic Press.
6. BETTEGOWDA, C. *et al.* 2006. The genome and transcriptomes of the anti-tumor agent Clostridium novyi-NT. Nat. Biotechnol. **24:** 1573–1580.
7. REINEKE, J. *et al.* 2007. Autocatalytic cleavage of Clostridium difficile toxin B. Nature **446:** 415–419.
8. JANK, T., T. GIESEMANN & K. AKTORIES. 2007. Rho-glucosylating Clostridium difficile toxins A and B: new insights into structure and function. Glycobiology **17:** 15R–22R.
9. SCHLEBERGER, C. *et al.* 2006. Structure and action of the binary C2 toxin from Clostridium botulinum. J. Mol. Biol. **364:** 705–715.
10. GIBERT, M. *et al.* 2007. Differential requirement for the translocation of clostridial binary toxins: iota toxin requires a membrane potential gradient. FEBS Lett. **581:** 1287–1296.
11. BOHNERT, S. & G. SCHIAVO. 2005. Tetanus toxin is transported in a novel neuronal compartment characterized by a specialized pH regulation. J. Biol. Chem. **280:** 42336–42344.
12. DEINHARDT, K. *et al.* 2006. Rab5 and Rab7 control endocytic sorting along the axonal retrograde transport pathway. Neuron **52:** 293–305.
13. AKTORIES, K. & I. JUST. 2005. Clostridial rho-inhibiting protein toxins. Curr. Top. Microbiol. Immunol. **291:** 113–145.
14. NOTTROTT, S. *et al.* 2007. Clostridium difficile toxin A induced apoptosis is p53-independent but depends on glucosylation of Rho GTPases. Apoptosis **12:** 1443–1453.
15. HACKETT, R. & P.C. KAM. 2007. Botulinum toxin: pharmacology and clinical developments: a literature review. Med. Chem. **3:** 333–345.
16. ECKBURG, P.B. *et al.* 2005. Diversity of the human intestinal microbial flora. Science **308:** 1635–1638.
17. GILL, S.R. *et al.* 2006. Metagenomic analysis of the human distal gut microbiome. Science **312:** 1355–1359.
18. HEAP, J.T. *et al.* 2007. The ClosTron: a universal gene knock-out system for the genus Clostridium. J. Microbiol. Methods **70:** 452–464.
19. NOLLING, J. *et al.* 2001. Genome sequence and comparative analysis of the solvent-producing bacterium Clostridium acetobutylicum. J. Bacteriol. **183:** 4823–4838.
20. SHIMIZU, T. *et al.* 2002. Complete genome sequence of Clostridium perfringens, an anaerobic flesh-eater. Proc. Natl. Acad. Sci. USA **99:** 996–1001.
21. BRUGGEMANN, H. *et al.* 2003. The genome sequence of Clostridium tetani, the causative agent of tetanus disease. Proc. Natl. Acad. Sci. USA **100:** 1316–1321.
22. SEBAIHIA, M. *et al.* 2007. Genome sequence of a proteolytic (Group I) Clostridium botulinum strain Hall A and comparative analysis of the clostridial genomes. Genome Res. **17:** 1082–1092.
23. SEBAIHIA, M. *et al.* 2006. The multidrug-resistant human pathogen Clostridium difficile has a highly mobile, mosaic genome. Nat. Genet. **38:** 779–786.
24. MYERS, G.S. *et al.* 2006. Skewed genomic variability in strains of the toxigenic bacterial pathogen, Clostridium perfringens. Genome Res. **16:** 1031–1040.
25. BARKER, H.A., M.D. KAMEN & B.T. BORNSTEIN. 1945. The synthesis of butyric and caproic acids from ethanol and acetic acid by Clostridium kluyveri. Proc. Natl. Acad. Sci. USA **31:** 373–381.
26. ANDREESEN, J.R. & G. GOTTSCHALK. 1969. The occurrence of a modified Entner-doudoroff pathway in Clostridium aceticum. Arch. Mikrobiol. **69:** 160–170.
27. ANDREESEN, J.R., H. BAHL & G. GOTTSCHALK. 1989. Introduction to the physiology and biochemistry of the genus Clostridium. *In* Clostridia. N.P. Minton, Ed. New York: Plenum Press.
28. THAUER, R.K. *et al.* 1968. The energy metabolism of Clostridium kluyveri. Eur. J. Biochem. **4:** 173–180.
29. SCHOBERTH, S. & G. GOTTSCHALK. 1969. Considerations on the energy metabolism of Clostridium kluyveri. Arch. Mikrobiol. **65:** 318–328.
30. SCHMEHL, M. *et al.* 1993. Identification of a new class of nitrogen fixation genes in Rhodobacter capsulatus: a putative membrane complex involved in electron transport to nitrogenase. Mol. Gen. Genet. **241:** 602–615.
31. HASE, C.C. *et al.* 2001. Sodium ion cycle in bacterial pathogens: evidence from cross-genome comparisons. Microbiol. Mol. Biol. Rev. **65:** 353–370, table of contents.
32. KERSCHER, S. *et al.* 2007. The three families of respiratory NADH dehydrogenases. *In* Results and Problems in Cell Differentiation. Berlin/Heidelberg: Springer. In press.
33. BRUGGEMANN, H. & G. GOTTSCHALK. 2004. Insights in metabolism and toxin production from the complete genome sequence of Clostridium tetani. Anaerobe **10:** 53–68.
34. O'HARA, A.M. & F. SHANAHAN. 2006. The gut flora as a forgotten organ. EMBO Rep. **7:** 688–693.
35. BACKHED, F. *et al.* 2005. Host-bacterial mutualism in the human intestine. Science **307:** 1915–1920.
36. SAMUEL, B.S. *et al.* 2007. Genomic and metabolic adaptations of Methanobrevibacter smithii to the human gut. Proc. Natl. Acad. Sci. USA **104:** 10643–10648.
37. FRICKE, W.F. *et al.* 2006. The genome sequence of Methanosphaera stadtmanae reveals why this human intestinal archaeon is restricted to methanol and H2 for methane formation and ATP synthesis. J. Bacteriol. **188:** 642–658.

38. Collier-Hyams, L.S. & A.S. Neish. 2005. Innate immune relationship between commensal flora and the mammalian intestinal epithelium. Cell. Mol. Life Sci. **62:** 1339–1348.
39. MacDonald, T.T. & G. Monteleone. 2005. Immunity, inflammation, and allergy in the gut. Science **307:** 1920–1925.
40. Weber, J.R., P. Moreillon & E.I. Tuomanen. 2003. Innate sensors for Gram-positive bacteria. Curr. Opin. Immunol. **15:** 408–415.
41. Mazmanian, S.K. *et al.* 2005. An immunomodulatory molecule of symbiotic bacteria directs maturation of the host immune system. Cell **122:** 107–118.
42. Sleytr, U.B. & T.J. Beveridge. 1999. Bacterial S-layers. Trends Microbiol. **7:** 253–260.
43. Calabi, E. *et al.* 2002. Binding of Clostridium difficile surface layer proteins to gastrointestinal tissues. Infect. Immun. **70:** 5770–5778.
44. Ausiello, C.M. *et al.* 2006. Surface layer proteins from Clostridium difficile induce inflammatory and regulatory cytokines in human monocytes and dendritic cells. Microbes Infect. **8:** 2640–2646.
45. Calabi, E. *et al.* 2001. Molecular characterization of the surface layer proteins from Clostridium difficile. Mol. Microbiol. **40:** 1187–1199.
46. Karjalainen, T. *et al.* 2001. Molecular and genomic analysis of genes encoding surface-anchored proteins from Clostridium difficile. Infect. Immun. **69:** 3442–3446.
47. Qazi, O. *et al.* 2007. Identification and characterization of the surface-layer protein of Clostridium tetani. FEMS Microbiol. Lett. **271:** 126–131.
48. Demarest, S.J. *et al.* 2005. Structural characterization of the cell wall binding domains of Clostridium difficile toxins A and B; evidence that Ca2+ plays a role in toxin A cell surface association. J. Mol. Biol. **346:** 1197–1206.
49. Schneewind, O., D. Mihaylova-Petkov & P. Model. 1993. Cell wall sorting signals in surface proteins of gram-positive bacteria. EMBO J. **12:** 4803–4811.
50. Dhar, G., K.F. Faull & O. Schneewind. 2000. Anchor structure of cell wall surface proteins in Listeria monocytogenes. Biochemistry **39:** 3725–3733.
51. Ton-That, H., L.A. Marraffini & O. Schneewind. 2004. Protein sorting to the cell wall envelope of Gram-positive bacteria. Biochim. Biophys. Acta **1694:** 269–278.
52. Zhao, G. *et al.* 2006. Binding of S-layer homology modules from Clostridium thermocellum SdbA to peptidoglycans. Appl. Microbiol. Biotechnol. **70:** 464–469.
53. Savariau-Lacomme, M.P. *et al.* 2003. Transcription and analysis of polymorphism in a cluster of genes encoding surface-associated proteins of Clostridium difficile. J. Bacteriol. **185:** 4461–4470.
54. Foster, T.J. & M. Hook. 1998. Surface protein adhesins of Staphylococcus aureus. Trends Microbiol. **6:** 484–488.
55. Ficko-Blean, E. & A.B. Boraston. 2006. The interaction of a carbohydrate-binding module from a Clostridium perfringens N-acetyl-beta-hexosaminidase with its carbohydrate receptor. J. Biol. Chem. **281:** 37748–37757.
56. Baldassarri, L. *et al.* 1991. Capsule-like structures in Clostridium difficile strains. Microbiologica **14:** 295–300.
57. Kalelkar, S. *et al.* 1997. Structure of the capsular polysaccharide of Clostridium perfringens Hobbs 5 as determined by NMR spectroscopy. Carbohydr. Res. **299:** 119–128.

Anaerobic Metabolism of Aromatic Compounds

GEORG FUCHS

Microbiology, Faculty of Biology, University of Freiburg, Freiburg, Germany

Aromatic compounds comprise a wide variety of natural and synthetic compounds that can serve as substrates for bacterial growth. So far, four types of aromatic metabolism are known. (1) The aerobic aromatic metabolism is characterized by the extensive use of molecular oxygen as cosubstrate for oxygenases that introduce hydroxyl groups and cleave the aromatic ring. (2) In the presence of oxygen, facultative aerobes use another so-called hybrid type of aerobic metabolism of benzoate, phenylacetate, and anthranilate (2-aminobenzoate). These pathways use coenzyme A thioesters of the substrates and do not require oxygen for ring cleavage; rather they use an oxygenase/reductase to dearomatize the ring. (3) In the absence of oxygen, facultative aerobes and phototrophs use a reductive aromatic metabolism. Reduction of the aromatic ring of benzoyl-coenzyme A is catalyzed by benzoyl-coenzyme A reductase. This Birch-like reduction is driven by the hydrolysis of 2 ATP molecules. (4) A completely different, still little characterized benzoyl-coenzyme A reductase operates in strict anaerobes, which cannot afford the costly ATP-dependent ring reduction.

Key words: **aromatic metabolism; benzoyl-CoA pathway; anaerobes; toluene; phenol; benzoate; oxygenases; benzoyl-CoA reductase**

Introduction

Aromatic compounds comprise a wide variety of natural and synthetic compounds that can serve as substrates for bacterial growth (FIG. 1). The interest in cellulose degradation has fostered scientific interest in the degradation of the aromatic copolymer lignin in lignocellulose. Whereas lignin is metabolized mainly by fungi, bacterial metabolism is thought to be limited to the metabolism of low-molecular-weight aromatic compounds. The diversity of aromatic metabolism is larger than previously thought. So far, besides the well-studied aerobic type, four types of aromatic metabolism are known. The distribution of these pathways depends on the availability of oxygen, on the availability of alternative electron acceptors for anaerobic respiration, and also on the rapid fluctuation of oxic and anoxic conditions.

The well-studied aerobic aromatic metabolism is characterized by the extensive use of molecular oxygen as cosubstrate for oxygenases, which introduce hydroxyl groups to facilitate the oxidative cleavage of the ring. Most importantly, the aromatic ring is cleaved by dioxygenases.

Under microaerobic conditions, facultative aerobes use another so-called hybrid type of aerobic metabolism of benzoate, phenylacetate, and anthranilate (2-aminobenzoate). This metabolism does not require oxygen for ring cleavage. All intermediates of these pathways are coenzyme A (CoA) thioesters. Dearomatization is catalyzed by an oxygenase/reductase acting on benzoyl-CoA, phenylacetyl-CoA, and 2-aminobenzoyl-CoA, respectively, followed by an oxygen-independent ring cleavage.

In anoxic water, groundwater, sediments and parts of soil, aromatic compounds are metabolized by facultative aerobes and phototrophs in the absence of oxygen in a purely reductive rather than oxidative process. Essential to anaerobic aromatic metabolism is the replacement of all oxygen-dependent steps by an alternative set of reactions and the formation of different central intermediates. These pathways involve a series of unprecedented enzymes. Notably, two-electron reduction of the aromatic ring of benzoyl-CoA is driven by the hydrolysis of two molecules of adenosine triphospate (ATP). The cyclic, nonaromatic product formed becomes hydrolytically opened and finally is oxidized to three molecules of acetyl-CoA.

Another reductive metabolism is found in strict anaerobes. Here again, benzoyl-CoA is a central

Address for correspondence: Georg Fuchs, Microbiology, Faculty of Biology, University of Freiburg, Schaenzelstr. 1, D-79104 Freiburg, Germany. Phone: +49 761 203 2649; fax: +49 761 203 2626. georg.fuchs@biologie.uni-freiburg.de

Dedicated to Lars Ljungdahl.

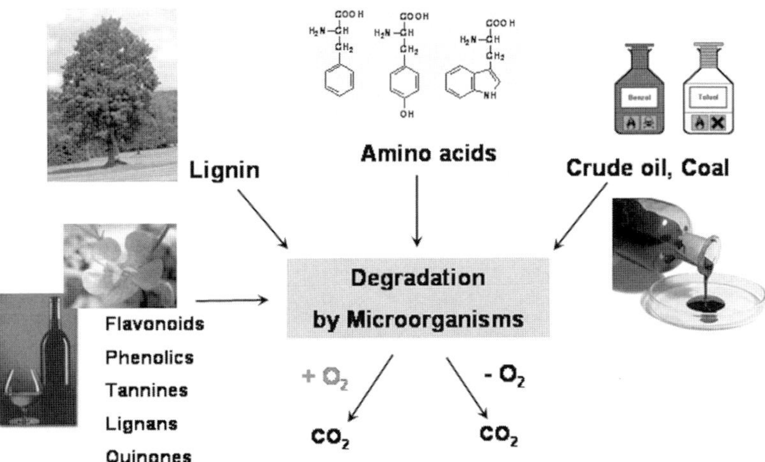

FIGURE 1. Sources of aromatic compounds in nature and man-made aromatic compounds derived from fossil material.

intermediate of aromatic metabolism. However, when one considers growth of strict anaerobes on benzoate, an energetic problem becomes evident: Facultative aerobes, such as denitrifyers and phototrophs, spend four ATP equivalents to activate benzoate as CoA thioester (two ATP equivalents) and to reductively dearomatize the ring (another two ATP equivalents). Their energy metabolism, anaerobic respiration, such as nitrate respiration, yields many more than four ATP equivalents per one benzoate metabolized; phototrophs conserve energy by photophosphorylation. In contrast, strict anaerobes gain fewer than four ATP equivalents out of one benzoate, which is metabolized via three molecules of acetyl-CoA plus one CO_2. Yet, they still require two ATP equivalents for benzoyl-CoA formation. However, they cannot spend another two ATP equivalents for the reductive dearomatization of benzoyl-CoA, because otherwise their energy metabolism would be energy-consuming rather than energy-providing. The outlines of this postulated new principle of benzoyl-CoA reduction are just emerging.

Most of the novel enzymatic reactions have counterparts in organic chemistry ("Chemistry-inspired Biology"). The chemical principles are modified according to biological constraints, for example, the limits of the redox potential of the cellular electron carriers, the use of water as solvent at relatively moderate temperature, or the low ambient substrate concentrations.

This short review gives an overview of the different strategies, with a focus on the anaerobic pathways. It does not cover the classic aerobic pathways and other important aspects, such as transport, regulation of enzymes, transcriptional control, genetic organization, distribution of the pathways, and ecological, evolutionary, and applied aspects.

Aerobic Aromatic Metabolism

The use of molecular oxygen in the cleavage of the aromatic ring and in hydroxylation reactions in general is widely distributed in nature, notably when inert or recalcitrant compounds and chemical bonds need to be attacked. Substrates, whose metabolism normally requires molecular oxygen, include aromatic, hydrocarbon, and ether compounds. Oxygenases and oxygen as cosubstrate in hydroxylation reactions have been known since the seminal work of Hayaishi (FIG. 2).[1] In the case of aromatic metabolism, the large variety of substrates is channeled via peripheral (upper) pathways into a few central intermediates. These peripheral pathways make extensive use of oxygen, which is required by monooxygenases and dioxygenases/reductases. The aromatic ring of the central intermediates contains two phenolic hydroxyl groups next to each other, or one hydroxyl group next to a carboxyl. Examples are catechol (1,2-dihydroxybenzene), protocatechuate (3,4-dihydroxybenzoic acid), and gentisate (2,5-dihydroxybenzoic acid). These free intermediates are substrates of ring cleaving dioxygenases of the central (lower) pathways. Ring cleavage may occur between the two hydroxyl groups (ortho cleavage) or next to one of the hydroxyl groups (meta clavage). This metabolism was established decades ago.[2–6]

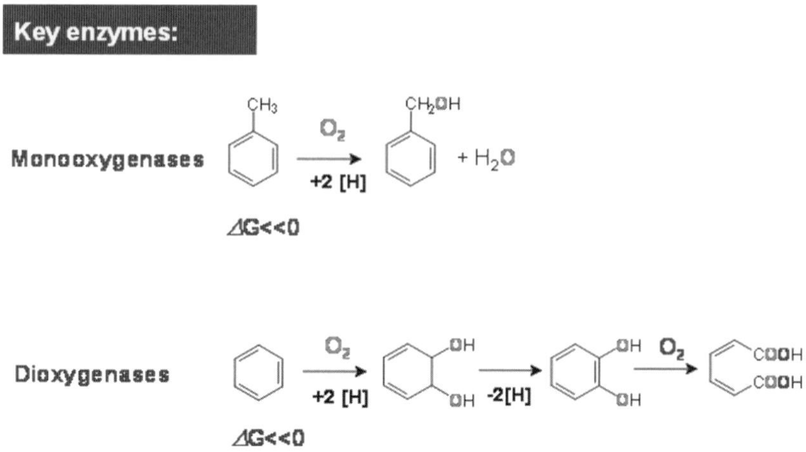

Key enzymes:

Monooxygenases

$\Delta G << 0$

Dioxygenases

$\Delta G << 0$

FIGURE 2. Aerobic monooxygenase and dioxygenase reactions in aromatic metabolism. Introduction of phenolic hydroxyl groups and ring cleavage both depend on molecular oxygen.

Aromatic Metabolism Under Microaerobic Conditions: The Hybrid Pathways

A few substrates have been recognized to be metabolized by facultative aerobes in a different way. These so-called hybrid pathways still make use of oxygen to introduce hydroxyl groups, as in the classic aerobic pathways. At the same time, the aromatic ring is reduced and CoA thioesters are used, as in the anaerobic metabolism (see later in the chapter). Ring cleavage also does not require oxygen. The new hybrid pathway of benzoate is shown in FIGURE 3.

Denitrifying facultative aerobes (and possibly others) convert phenylacetate,[7–14] benzoate,[15–18] and 2-aminobenzoate (anthranilate)[19–24] to their CoA thioesters first. Bacteria may even exclusively metabolize phenylacetate via this new principle; the more conventional pathway—via ring hydroxylation to homogentisate (2,5-dihydroxyphenylacetate) and ring cleavage of homogentisate—appears to be restricted to fungi. Benzoyl-CoA and phenylacetyl-CoA are then converted by dioxygenases/reductases to the corresponding nonaromatic *cis*-dihydrodiols. Normally, such intermediates would be rearomatized by oxidation, yielding dihydroxylated aromatic products. In this case, CoA thiosterification of the carboxyl group helps to promote the subsequent cleavage of the aromatic ring, whereby oxygen is not required. The postulated pathway of phenylacetyl-CoA degradation has not been solved yet, but the principle seems to be similar to the benzoate case (see FIG. 3). In the case of anthranilic acid, 2-aminobenzoyl-CoA is converted by a monooxygenase/reductase to a monohydroxylated nonaromatic intermediate. The subsequent ring cleavage reaction has not been studied yet.

These CoA thioester–dependent pathways may be advantageous under fluctuating oxic/anoxic conditions. The pathways allow flexibility and rapid adaptation to fluctuating oxygen levels, since both oxic and anoxic situations require substrate CoA thioesters. If the classic pathway of, for example, benzoate would operate, the shift from anoxic to oxic conditions would result in the accumulation of benzoyl-CoA, since the ring-reducing enzyme benzoyl-CoA reductase of the anoxic pathway is oxygen-sensitive and therefore would be inactive (see later in the chapter). All cellular CoA would be trapped in this dead-end product, since benzoate-CoA ligase is oxygen-insensitive and would still operate under oxic conditions. CoA depletion ultimately would be lethal. In addition, benzoyl-CoA oxygenase/reductase, the enzyme that dearomatizes benzoyl-CoA, has a high affinity for oxygen, and therefore can operate under microaerobic conditions. The energy spent in CoA thioester formation is not lost, but is later regained in the form of acetyl-CoA. Furthermore, the CoA ester intermediates may be less toxic than some intermediates of the classic pathways, notably those of some meta cleavage routes. Also, CoA thioester formation indirectly facilitates the transport of the aromatic acids.

Anaerobic Aromatic Metabolism by Phototrophs and by Facultative Aerobes with Anaerobic Respiration

Pioneering work on the anaerobic aromatic metabolism was done by Charles W. Evans (1911–1988)

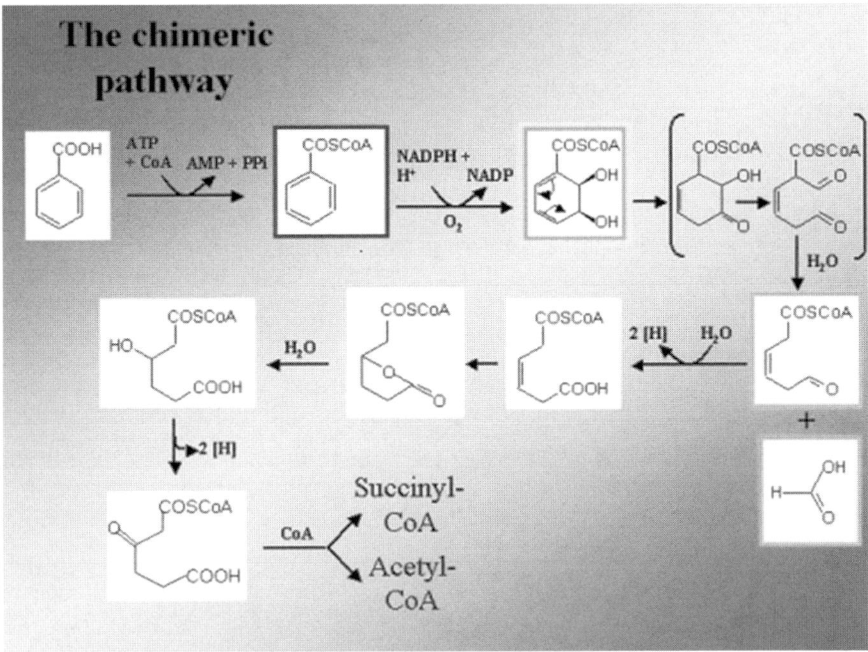

FIGURE 3. Hybrid (chimeric) aerobic pathway of benzoate metabolism carried out by some denitrifying bacteria. The reactions involve principles of aerobic and anaerobic aromatic metabolism.

and his group.[25,26] Several reviews have covered the state of the art.[27–42] Representative bacterial species include the phototrophic bacterium *Rhodopseudomonas palustris* and the denitrifying bacteria *Thauera aromatica, Azoarcus evansii, Azoarcus* strain EbN1 (renamed *Aromatoleum* sp.), and *Magnetospirillum* sp. These bacteria metabolize aromatic (and sometimes hydrocarbon) compounds anaerobically under phototrophic or denitrifying conditions. The energy yields (ATP) by metabolism in these bacteria is high, as compared to strict anaerobes such as iron-reducing, sulfate-reducing, or syntrophic-fermenting bacteria. Still, the aromatic molecule catabolism is performed anaerobically.

The aromatic substrates are mostly handled as CoA esters, and the peripheral pathways yield completely different central intermediates, such as benzoyl-CoA, resorcinol (1,3-dihydroxybenzene) or phloroglucinol (1,3,5-trihydroxybenzene), among others. In contrast to the aerobic pathways, where phenolic hydroxyl groups are introduced by oxygenases, phenolic hydroxyl groups are often reductively removed. The aromatic ring is reduced, affording alicyclic compounds. Whereas ring reduction in phloroglucinol can be accomplished by using pyridine nucleotides as reductant, the reduction of resorcinal already requires ferredoxin as reductant. The reduction of benzoyl-CoA, however, cannot be accomplished by ferredoxin alone; rather, ring reduction in addition requires the hydrolysis of one ATP per one electron transferred. The nonaromatic ring of such ring-reduction products is hydrolytically opened. Beta-oxidation yields acetyl-CoA as the final product.

To illustrate the new principles, the metabolism of phenol, toluene, and benzoate will be considered in more detail. These substrates are metabolized via the benzoyl-CoA pathway (for a description of the less sophisticated resorcinol and phloroglucinol pathways, see Refs. 33–35 and 38). Each of these pathways has its metabolic constraint (FIG. 4). Phenol carboxylation is unfavorable due to its unfavorable equilibrium constant. Toluene metabolism is difficult, since breaking C-H bonds by withdrawing a hydrogen atom from the hydrocarbon is energetically unfavorable. Reduction of the aromatic ring requires an enormous activation energy, because the transfer of the first electron requires an extremely low redox potential.

Anaerobic Phenol Metabolism

Anaerobic Phenol Metabolism in Facultative Aerobes

In *T. aromatica* and other facultative aerobes phenol is metabolized via transformation to benzoyl-CoA, whose aromatic ring then becomes reduced in the

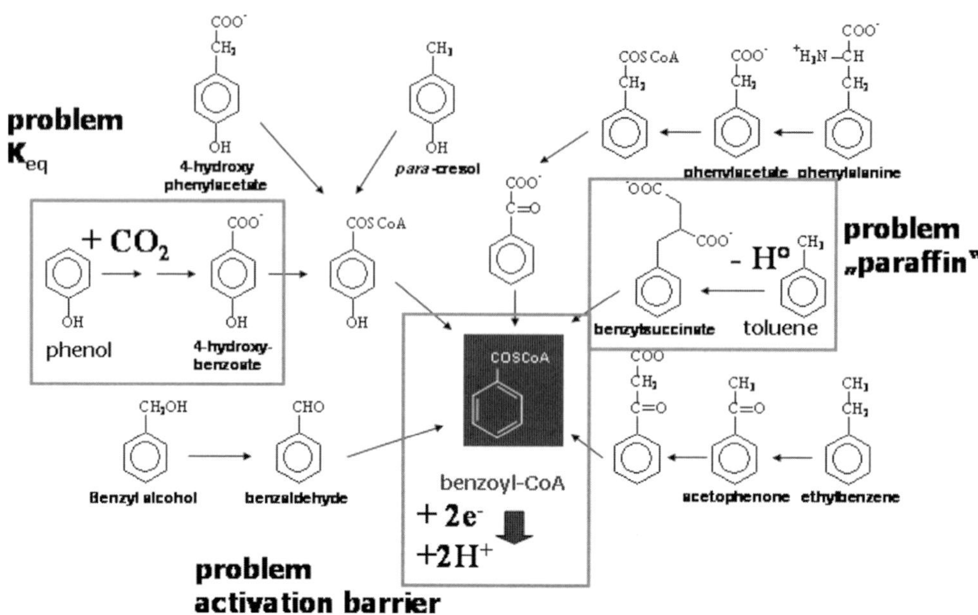

FIGURE 4. Anaerobic metabolism of aromatic compounds: three cases, three problems. Peripheral reactions in anaerobic aromatic metabolism. The conversion of phenol, toluene, and benzoyl-CoA are highlighted by *boxes*.

central benzoyl-CoA pathway. The peripheral phenol pathway leading to benzoyl-CoA obviously requires a carboxylation step, a CoA thioester activation of the carboxyl group, and a reductive removal of the phenolic hydroxyl group (FIG. 4). Such reactions have their counterparts in organic synthesis, for example, the Kolbe–Schmitt synthesis (phenol carboxylation) and the Birch reduction (reductive benzene ring dearomatization and the reductive dehydroxylation of phenols). The Kolbe–Schmitt synthesis (FIG. 5) requires high concentrations of phenol and CO_2. This process is well-studied and requires a phenolate anion as nucleophilic substrate, the electrophilic CO_2 as second substrate, as well as K^+ or Na^+ as cocatalyst.[43,44] Taking into account the concentrations of phenol and CO_2/bicarbonate under natural conditions, the carboxylation of phenol would yield 4-hydroxybenzoate at very low concentrations. If the assumed equilibrium constant is correct (FIG. 5), and given the assumed low concentrations of phenol and CO_2, theoretically fewer than one molecule 4-hydroxybenzoate per cell would result in the reaction equilibrium. This energetically unfavorable situation can be drastically improved by using ATP to phosphorylate the incoming phenol in an irreversible reaction (however, see later in the section for anaerobes that use an ATP-independent phenol carboxylation system). The product phenylphosphate, however, is a poor substrate (a poor nucleophile) for the electrophilic attack of CO_2 in the subsequent carboxylation step. Therefore, special care has to be taken to dephosphorylate phenylphosphate in the course of the carboxylation reaction without releasing phenol (which otherwise would result in a futile recycling of phenol coupled to ATP hydrolysis).

The first two steps in phenol metabolism in the denitrifying bacterium *T. aromatica* are phosphorylation of phenol to phenylphosphate by ATP, catalyzed by phenylphosphate synthase,[45–49] and subsequent carboxylation of phenylphosphate to 4-hydroxybenzoate with release of phosphate, catalyzed by phenylphosphate carboxylase.[49–51] At least 15 genes are phenol-induced and cotranscribed. They include the genes coding for seven purified proteins that are required for phenol carboxylation to 4-hydroxybenzoate.[51,52] The function of at least seven other genes of the phenol operon is unknown.

Phenylphosphate Synthase

The molecular and catalytic features of phenylphosphate synthase (FIG. 6) (E.C. 2.7.9.-) resemble those of phosphoenolpyruvate synthase (E.C. 2.7.9.2), albeit with interesting modifications. The reaction follows a

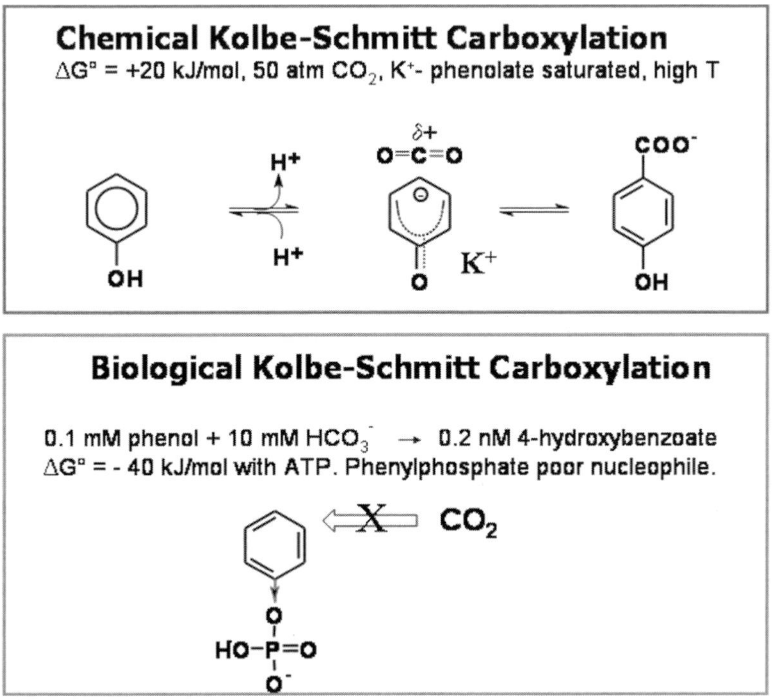
Chemical Kolbe-Schmitt Carboxylation
ΔG° = +20 kJ/mol, 50 atm CO2, K+- phenolate saturated, high T
H+
δ+
O=C=O
COO-
H+
K+
OH
O
OH
Biological Kolbe-Schmitt Carboxylation
0.1 mM phenol + 10 mM HCO3- → 0.2 nM 4-hydroxybenzoate
ΔG° = - 40 kJ/mol with ATP. Phenylphosphate poor nucleophile.
CO2
O
HO-P=O
O-

FIGURE 5. Comparison of the energy demands of chemical versus biological phenol carboxylation. Under natural conditions, low substrate and low temperature may limit carboxylation activity.

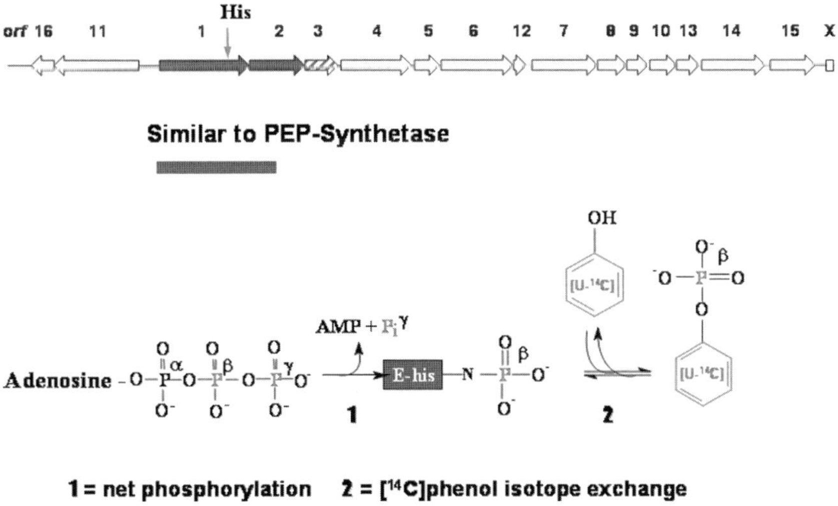
His
orf 16 11 1 2 3 4 5 6 12 7 8 9 10 13 14 15 X
Similar to PEP-Synthetase
OH
[U-14C]
AMP + Pi γ
Adenosine
E-his
1
2
1 = net phosphorylation 2 = [14C]phenol isotope exchange

FIGURE 6. Schematic presentation of phenylphosphate synthase (or synthetase) reaction and of the genes involved. For details, see the text.

Ping-Pong mechanism, and is described by Equation 1 (referred to as the *net phosphorylation reaction*). The whole reaction (Eq. 1) is understood as the sum of Equation 2 and Equation 3. In the course of net phenol phosphorylation, the enzyme becomes phosphorylated by ATP in an essentially irreversible step (Eq. 2). The phosphorylated enzyme E_1 subsequently transforms phenol to phenylphosphate in a reversible reaction (Eq. 3).

Consistent with this mechanism, the enzyme also catalyzes an exchange of free [^{14}C]phenol and the phenol moiety of phenylphosphate (Eq. 4, referred to as *phenol exchange reaction*). This suggests that enzyme E_1 becomes phosphorylated by phenylphosphate in the course of this phenol exchange reaction (Eqs. 5 + 6).

$$\text{Phenol} + \text{MgATP} + H_2O \rightarrow \text{phenylphosphate} + \text{MgAMP} + P_i \quad (1)$$

$$\text{MgATP} + E_1 + H_2O \rightarrow E_1\text{-phosphate} + \text{MgAMP} + P_i \quad (2)$$

$$E_1\text{-Phosphate} + \text{phenol} \rightarrow E_1 + \text{phenylphosphate} \quad (3)$$

$$\text{Phenylphosphate} + [^{14}C]\text{phenol} \rightarrow [^{14}C]\text{phenylphosphate} + \text{phenol} \quad (4)$$

$$\text{Phenylphosphate} + E_1 \rightarrow \text{phenol} + E_1\text{-phosphate} \quad (5)$$

$$E_1\text{-Phosphate} + [^{14}C]\text{phenol} \rightarrow E_1 + [^{14}C]\text{phenylphosphate} \quad (6)$$

Phenylphosphate synthase E_1 consists of three proteins whose genes are located adjacent to each other on the phenol operon.[48,53] Protein 1 (ORF1, 70 kDa) resembles the central part of phosphoenolpyruvate synthase, which contains a conserved histidine residue. It alone catalyzes the exchange of free [^{14}C]phenol and the phenol moiety of phenylphosphate (Eq. 4, which is the sum of Eqs. 5 and 6), but not the phosphorylation of phenol (Eq. 1). It interacts with the substrate phenol and transfers the phosphoryl group from the phosphorylated protein 1 to the substrate (Eq. 3). Phosphorylation of phenol requires protein 1, MgATP, and another protein 2 of 40 kDa, which resembles the N-terminal part of phosphoenolpyruvate synthase. Protein 2 (ORF2) catalyzes the phosphorylation of protein 1 (Eq. 2). The combination of proteins 1 + 2 affords the net phosphorylation reaction (Eq. 1). The phosphoryl group in phenylphosphate is derived from the β-phosphate group of ATP. It is suggested that protein 2 intermediately transfers a pyrophosphate group from ATP to the conserved histidine of protein 1, from which γ-phosphate is released. The overall reaction is stimulated severalfold by another protein 3 (ORF3, 24 kDa). The exact role of this protein is unknown; it may have a regulatory function since it has some similarity to adenosine monophosphate (AMP) binding proteins and contains a cystathionine beta-synthase (CBS) domain.

Phenylphosphate Carboxylase

Phenylphosphate synthase makes use of ATP to render the endergonic phenol carboxylation process unidirectional, even under the very low ambient concentrations of phenol (K_M 0.04 mM phenol) and of CO_2. At the same time, however, the electron-withdrawing phosphoryl group makes phenylphosphate a poor substrate for an electrophilic attack by CO_2. Hence, the subsequent phenylphosphate carboxylase (FIG. 7) (E.C. 4.1.1.-) E_2 is expected to exhibit special features.[49–51] The enzyme requires divalent metal ions (Mg^{2+} or Mn^{2+}) as well as K^+, and catalyzes the carboxylation of phenylphosphate to 4-hydroxybenzoate (Eq. 7) (referred to as *net carboxylation reaction*). The actual substrate is CO_2 rather than bicarbonate. Enzyme E_2 follows a Ping-Pong mechanism. The presumed E_2-phenolate intermediate is formed in an exergonic reaction from phenylphosphate (Eq. 8), followed by the reversible carboxylation reaction (Eq. 9). Consistent with this proposal, the enzyme also catalyzes an exchange of free $^{14}CO_2$ and the carboxyl group of 4-hydroxybenzoate (Eq. 10) (referred to as *CO_2 exchange reaction*), which is the sum of Equations 11 and 12. Free ^{14}C-phenol does not exchange with the phenol moiety of phenylphosphate.

$$\text{Phenylphosphate} + CO_2 \rightarrow 4-\text{hydroxybenzoate} + P_i \quad (7)$$

$$\text{Phenylphosphate} + E_2 \rightarrow E_2\text{-phenolate} + P_i \quad (8)$$

$$E_2\text{-Phenolate} + CO_2 \rightarrow E_2 + 4\text{-hydroxybenzoate} \quad (9)$$

$$4\text{-Hydroxybenzoate} + {}^{14}CO_2 \rightarrow [^{14}C]4\text{-hydroxybenzoate} + CO_2 \quad (10)$$

$$4\text{-Hydroxybenzoate} + E_2 \rightarrow CO_2 + E_2\text{-phenolate} \quad (11)$$

$$E_2 - \text{Phenolate} + {}^{14}CO_2 \rightarrow E_2 + [^{14}C]4\text{-hydroxybenzoate} \quad (12)$$

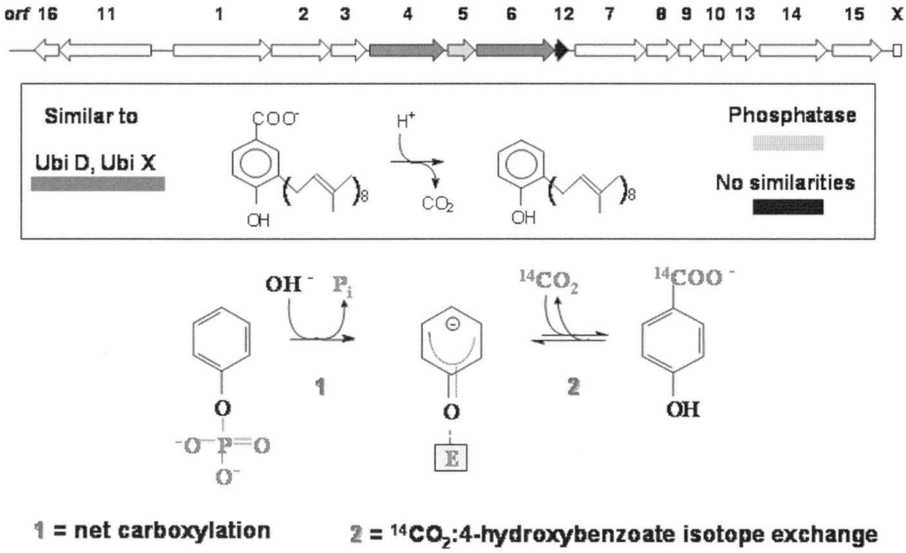

1 = net carboxylation **2 = $^{14}CO_2$:4-hydroxybenzoate isotope exchange**

FIGURE 7. Schematic presentation of phenylphosphate carboxylase reaction, of the genes involved, and of similar enzymes and reactions. For details, see the text.

Phenylphosphate carboxylase consists of four proteins whose genes are located adjacent to each other on the phenol gene cluster.[51,52] Three of the subunits ($\alpha\beta\gamma$, 54, 53, and 10 kDa) are sufficient to catalyze the CO_2 exchange reaction (Eq. 10, which is the sum of Eqs. 11 and 12), but not the net phenylphosphate carboxylation (Eq. 7). Phenylphosphate carboxylation is restored when the 18-kDa (δ) subunit is added. This 18-kDa phosphatase subunit alone also catalyzes a very slow hydrolysis of phenylphosphate. The 54- and 53-kDa subunits show similarity to UbiD, 3-octaprenyl-4-hydroxybenzoate carboxy lyase, which catalyzes the decarboxylation of a 4-hydroxybenzoate derivative in ubiquinone (*ubi*) biosynthesis.[51] The 18-kDa subunit belongs to a hydratase/phosphatase protein family. The 10 kDa is unique. The function of the remaining seven other genes of the phenol gene cluster, two genes related to *ubiD* and *ubiX*, respectively, is completely unknown.

The genomes of several bacteria contain genes related to phenol metabolism. Examples are *Magnetospirillum magnetotacticum*; *Magnetospirillum* sp. are among the dominant phenol-degrading denitrifyers.[54,55] The recent sequencing of the genome of *Azoarcus* strain EbN1,[56] which is closely related to *T. aromatica*, revealed a gene cluster very similar to the one found in *T. aromatica*. Phenol metabolism by iron-reducing or sulfate-reducing bacteria has not been studied in detail.

Anaerobic Phenol Metabolism in Strict Anaerobes

Enzymes related to the UbiD- and UbiX-like proteins are involved in phenol or hydroxybenzoate metabolism in strict anaerobes. The 54-kDa 4-hydroxybenzoate decarboxylase from *Clostridium hydroxybenzoicum*,[57,58] and subunits of various proven or putative vanillic acid (3-methoxy-4-hydroxybenzoic acid) decarboxylases from *E. coli*, *Bacillus subtilis*,[59] and *Streptomyces* sp.[60] show similarity with UbiD and UbiX. In some anaerobes these enzymes may function as decarboxylases, yielding phenolic compounds from the corresponding phenolic acids.[61] This reaction serves as a CO_2 source for acetogenic bacteria. However, clostridia appear to use the energetically unfavorable ATP-independent phenol carboxylation reaction even in phenol metabolism.[62–66] This follows from the time course of phenol consumption and product formation by whole cells and from the observation that ^{13}C labeled benzoate is formed from ^{13}C-labeled phenol; also, fluorinated phenolic compounds give rise to fluorinated benzoic acids. Obviously, these bacteria encounter higher phenol and CO_2 concentrations in their natural habitat. These substrates are formed there by other bacteria that decarboxylate hydroxybenzoic acids. Furthermore, effective consumption of 4-hydroxybenzoate in phenol-grown cells may indeed lower its concentration dramatically, thus enabling these anaerobes to live with phenol.

4-Hydroxybenzoate-CoA Ligase and 4-Hydroxybenzoyl-CoA Reductase (Dehydroxylating)

The product of phenol carboxylation, 4-hydroxybenzoate, is converted to its CoA thioester by a specific CoA ligase.[67,68] The dehydroxylation of 4-hydroxybenzoyl-CoA to benzoyl-CoA is essential for phenol metabolism, as the following enzyme in the pathway, benzoyl-CoA reductase, does not accept the *para*-hydroxylated compound as a substrate for mechanistic reasons. In contrast, benzoyl-CoA reductase can reduce the *ortho*- and *meta*-isomers of monohydroxylated benzoyl-CoA analogues.[69,70]

The reaction catalyzed by 4-hydroxybenzoyl-CoA reductase is shown in FIGURE 4. The enzyme from *T. aromatica* has a molecular mass of 270 kDa and consists of three subunits of 82 (a), 35 (b), and 17 kDa (c), suggesting an $(abc)_2$ composition.[71] The enzyme contains two [2Fe–2S] clusters, a [4Fe–4S] cluster, a flavin adenine dinucleotide (FAD), and a molybdopterin-cytosine dinucleotide cofactor per abc-trimer.[72] The genes coding for the three subunits of 4-hydroxybenzoyl-CoA reductase were identified in *T. aromatica*,[73] and in the phototrophic bacterium *R. palustris*.[74] The structure of 4-hydroxybenzoyl-CoA reductase confirmed that the enzyme belongs to the xanthine oxidase family of molybdenum enzymes.[75,76] The Mo-atom is coordinated by two sulfur atoms from the dithiolene group of the molybdopterin, by an oxo and a water ligand. A fifth ligand is most probably a sulfur atom, which, however, was artificially replaced by an oxo-ligand. Further structural and spectroscopic details of 4-hydroxybenzoyl-CoA reductase and comparisons with other members of the xanthine oxidase family have been discussed elsewhere.[77]

4-Hydroxybenzoyl-CoA reductase is the only member of this family whose function is to catalyze the reduction of substrate; the reversibility of this reaction could not be demonstrated. Buckel and Keese (1995) discussed the essential role of the thiol ester functionality for the reductive dehydroxylation reaction.[78] In analogy to the related process of benzene-ring reduction, they suggested that a ketyl radical anion is transiently formed. The electrochemical properties of the redox centers of 4-hydroxybenzoyl-CoA reductase also appear to be suited for a low-potential redox chemistry: a low potential reduced ferredoxin serves as electron donor; in addition, a unique [4Fe-4S] cluster and a Mo-cofactor with an unusually low redox potential are present.[79] In principle, the proposed catalytic cycle of 4-hydroxybenzoyl-CoA reductase runs counterclockwise to the one of xanthine oxidase members.[80]

Anaerobic Toluene Metabolism

In spite of their chemical inertness, hydrocarbons are degraded by microorganisms in the complete absence of oxygen. As all known aerobic hydrocarbon degradation pathways start with oxygen-dependent reactions, hydrocarbon catabolism in anaerobes must be initiated by novel biochemical reactions. Under anoxic conditions, a variety of reactions seems to be employed to overcome the activation barrier of different hydrocarbons. Examples include oxygen-independent hydroxylation, as employed in ethylbenzene metabolism, fumarate addition to methyl or methylene carbons in toluene or alkane degradation, and only recently discovered reactions, such as methylation of naphthalene or anaerobic methane oxidation via reverse methanogenesis.[80] All these reactions have in common a high-energy requirement of cleaving C-H sigma bonds in alkanes (FIG. 8).

Benzylsuccinate Synthase

The initial step in toluene degradation consists of the radical addition of fumarate to the methyl group of toluene, yielding (*R*)-benzylsuccinate, catalyzed by the glycine radical enzyme benzylsuccinate synthase (FIG. 9).[80–84] It consists of a large, glycyl-radical carrying subunit of 97 kDa and two very small subunits (8.5 and 6.5 kDa) of unknown function. The enzyme-bound radical is thought to abstract a hydrogen atom from the methyl group of toluene, generating a benzyl radical intermediate to which fumarate is added.[80] This process affords a benzylsuccinyl radical, which then abstracts the hydrogen atom from the enzyme to form benzylsuccinate; the enzyme radical is thereby regenerated. The formation of the active enzyme in the radical form requires activation by an activase enzyme, which uses *S*-adenosylmethionine and an electron donor as cosubstrates. A chaperon-like protein may be required for the assembly or activation of the system. These proteins are coded on a single operon. This unique metabolic capability has been discussed recently (see Refs. 80 and 85–87).

Oxidation of Benzylsuccinate to Benzoyl-CoA and Succinate

Benzylsuccinate is converted by a kind of beta-oxidation, which is initiated by CoA transfer from succinyl-CoA by a specific CoA transferase, forming 2-(*R*)-benzylsuccinyl-CoA. All enzymes of this peripheral pathway are encoded by a second operon. The reaction cycle is complete when succinate is oxidized to fumarate. Thus, the overall pathway brings about a six-electron oxidation of the methyl group of toluene to the carbonyl group of benzoyl-CoA. The reduced

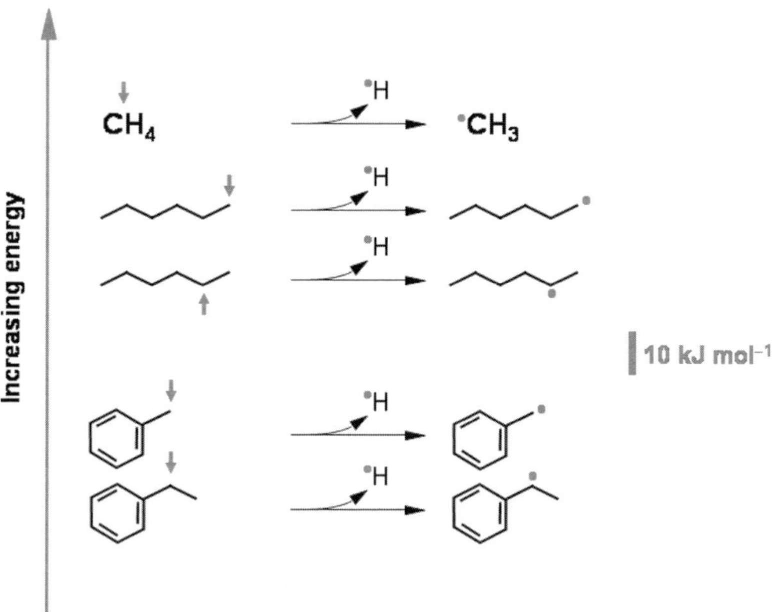

FIGURE 8. Energy requirement for C-H bond cleavage. Illustration of the comparative energy needs to homolytically cleave the C-H bond in various hydrocarbons by withdrawing a hydrogen atom via a radical mechanism. (After Friedrich Widdel, Bremen.)

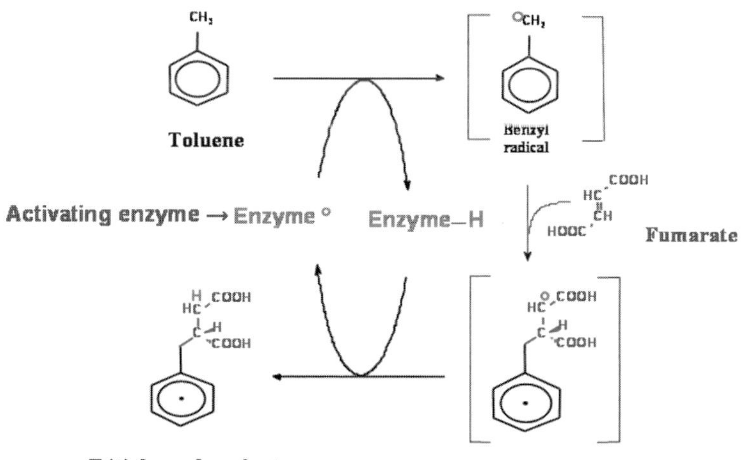

FIGURE 9. Proposed mechanism of benzylsuccinate synthase. Enzymatic attack of toluene by radical addition of fumarate. The same principle applies to many other hydrocarbons.

electron carriers are reoxidized in the course of the anaerobic respiration.

Anaerobic Benzoyl-CoA Reduction, and following Oxidation, to Acetyl-CoA

As can be seen from the scheme of the peripheral anaerobic metabolism (FIG. 4), quite different aromatic compounds, such as phenol, toluene, ethylbenzene, phenylacetate, some cresols, or benzyl alcohol, are all converted to benzoyl-CoA. Benzoyl-CoA reduction is the key step in the central metabolism of these and other, though not all, aromatic compounds.[25,26,88,89] As indicated earlier, resorcinol or phloroglucinol can be reduced more easily, and the metabolism of various

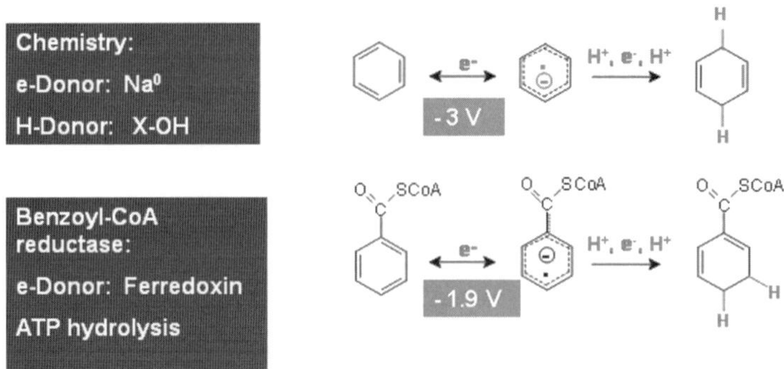

FIGURE 10. Birch reduction of the aromatic ring in benzene and enzymatic benzoyl-CoA reduction. Effect of the CoA thioester group of benzoyl-CoA on the redox potential of the first electron transfer reaction.

di- and trihydroxylated aromatic compounds proceeds via these intermediates.[34,35,38,39]

Benzoyl-CoA Reductase

The reduction of benzoyl-CoA is catalyzed by benzoyl-CoA reductase,[90,91] which has only been studied in some detail in the denitrifying bacterium *T. aromatica.* The genes for the benzoyl-CoA pathway were first detected in *R. palustris* and later in *T. aromatica.*[92,93] Benzoyl-CoA reduction seems to follow a Birch mechanism (FIG. 10), that is, a sequential transfer of single electrons and protons.[94] The product of the two-electron reduction of benzoyl-CoA is cyclohex-1,5-diene-1-carbonyl-CoA.[95] This reduction reaction is greatly facilitated by the use of the CoA thioester, because the thioester group lowers the redox potential difference of the first electron transfer step by almost one volt. This strong effect explains why the anaerobic pathways use CoA thioesters throughout, because at the end benzoyl-CoA rather than benzoate is needed for ring reduction. The formation of CoA thioesters is coupled to ATP hydrolysis to AMP and pyrophosphate, which normally is followed by hydrolysis of pyrophosphate. This drives the reaction forward and supports the transport of the substrate by keeping its intracellular pool concentration low. Whether the conjugated cyclic 1,5-diene is the general product of dearomatizing reductases from other organisms is not clear; in benzoyl-CoA reductase from *R. palustris,* a four-electron reduced monoene species (cyclohex-1-ene-1-carbonyl-CoA) is formed.[36]

The oxygen-sensitive benzoyl-CoA reductase from *T. aromatica* contains three [4Fe-4S] clusters as sole redox centers.[91,96,97] It has a modular composition: the 49- and 29-kDa subunits each contain one ATP-binding site, and a single [4Fe-4S] cluster is coordinated by two cysteine residues of each subunit. This arrangement is considered as the electron activation module. The other two subunits contain two [4Fe-4S] clusters; they are assigned to the aromatic ring-reduction module.

Benzoyl-CoA reductase couples the transfer of electrons from the electron donor, reduced ferredoxin, to the substrate benzoyl-CoA, to the hydrolysis of ATP[91,98]; one molecule of ATP is hydrolyzed per one electron transferred to the aromatic ring (i.e., two ATP are required to reduce benzoyl-CoA) (FIG. 11). In the catalytic cycle of ATP-dependent electron transfer, ATP hydrolysis yields a high-energy enzyme-phosphate linkage[99]; hydrolysis of the latter enables electron transfer to the substrate. During catalysis, two different ATP-dependent switches are involved in electron transfer to the substrate. The "nucleotide binding-switch" induces conformational changes in the vicinity of a special [4Fe-4S] cluster. The cluster serves as primary acceptor for electrons transferred from the external donor, which is reduced ferredoxin. In addition, binding of nucleotides switches the substrate-binding pocket into an open state, which enables benzoyl-CoA binding.[98] The "enzyme-phosphate hydrolysis" switch induces a low-spin/high-spin ($S = 7/2$) transition of a [4Fe-4S] cluster. This transition is assigned to a substantial lowering of the redox potential of the cluster.[97]

The proposed "Birch-like" reduction mechanism of benzoyl-CoA reductase was probed by kinetic and computational studies using a number of benzoyl-CoA analogues.[100] The redox potential for the one-electron reduction of benzoyl-CoA is $-1.9\,V$ (as determined

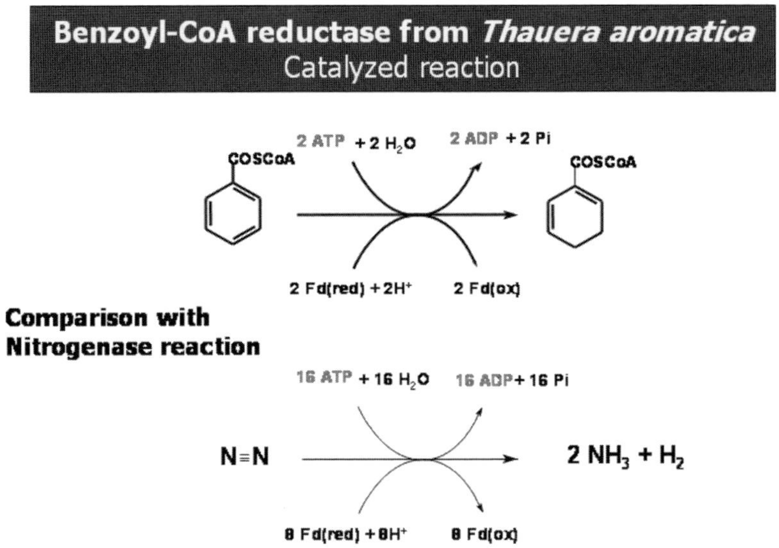

Benzoyl-CoA reductase from *Thauera aromatica*
Catalyzed reaction

COSCoA; 2 ATP + 2 H_2O → 2 ADP + 2 Pi; COSCoA; 2 Fd(red) + 2H^+ → 2 Fd(ox)

Comparison with Nitrogenase reaction

N≡N → 2 NH_3 + H_2; 16 ATP + 16 H_2O → 16 ADP + 16 Pi; 8 Fd(red) + 8H^+ → 8 Fd(ox)

FIGURE 11. Benzoyl-CoA reductase reaction and comparison to the nitrogenase reaction. Both reactions require ATP for electron transfer, though with a different stoichiometry of ATP per electron.

for the benzoic acid *S*-methylthiol ester),[95] which is far from the potential of the electron donor. However, a partial protonation of the carbonyl-oxygen of benzoyl-CoA could increase this redox potential considerably. In addition, the first electron transfer could be proton assisted, which would further facilitate the reaction. These steps may complement the action of ATP hydrolysis, which is thought to lower dramatically the redox potential of the reduced-electron transferring group. Single-turnover studies[97] and studies with ^{33}S- and ^{57}Fe-labeled enzymes suggested the presence of a sulfur-centered species, most probably a disulfide radical anion formed by two cysteine residues in close proximity of an active-site [4Fe–4S] cluster.

Following Oxidation of Cyclohex-1,5-diene-1-carbonyl-CoA to Acetyl-CoA

The oxidation of the product of benzoyl-CoA reductase, cyclohex-1,5-diene-1-carbonyl-CoA, follows a kind of beta-oxidation, including a hydrolytic opening of the alicyclic ring (FIG. 12). Finally, three molecules of acetyl-CoA and one molecule of CO_2 are formed. Denitrifying bacteria assimilate part of the acetyl-CoA into cell material, and most of the acetyl-CoA is oxidized completely to CO_2; they gain energy by electron-transport phosphorylation, using nitrate as an electron acceptor (anaerobic respiration). Phototrophs do not oxidize acetyl-CoA, but rather use it as a carbon source for biosynthesis; they obtain energy from photosynthesis.

Anaerobic Aromatic Metabolism in Strict Anaerobes

In strict anaerobes, benzoyl-CoA is again a central intermediate of aromatic metabolism. However, when one considers growth of strict anaerobes on benzoate, an energetic problem becomes evident: Facultative aerobes gain many more than four ATP equivalents from one benzoate metabolized than strict anaerobes and can therefore spend four ATP equivalents to activate benzoate as CoA thioester (two ATP equivalents) and to reductively dearomatize the ring (another two ATP equivalents). In contrast, strict anaerobes gain fewer than four ATP equivalents (probably three ATP equivalents) out of one benzoate, which is metabolized via three molecules of acetyl-CoA plus one CO_2. Yet, they still require two ATP equivalents for benzoyl-CoA formation. Clearly, they cannot spend another two ATP equivalents for the reductive dearomatization of benzoyl-CoA, because otherwise their energy metabolism would be energy consuming rather than energy providing (see FIG. 13). Consequently, they must use a less costly mechanism of ring reduction. In any case, benzoate needs to be activated to benzoyl-CoA by an ATP-dependent benzoate-CoA ligase (AMP plus PP_i forming). The sulfate-reducing bacterium *Desulfococcus multivorans*, when cultivated on benzoate and sulfate, requires selenium and molybdenum for growth, whereas growth on nonaromatic compounds does not require those trace elements (FIG. 14). In extracts of cells grown on benzoate in the presence

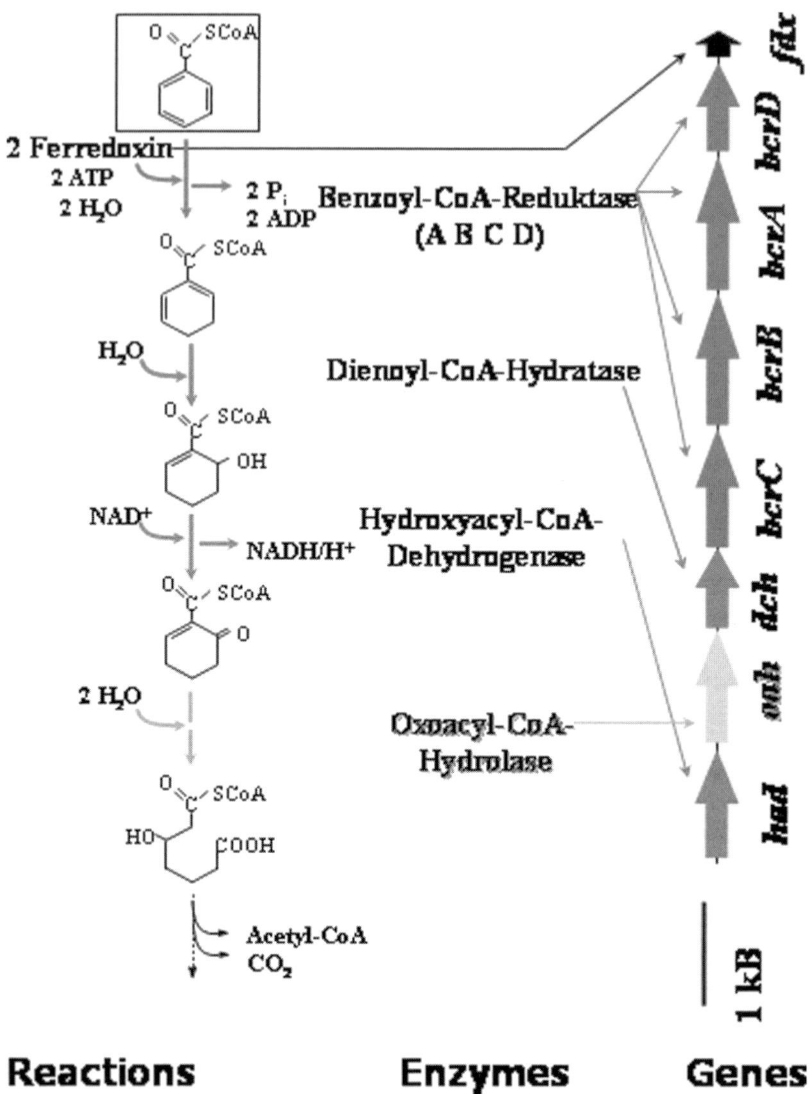

FIGURE 12. The anaerobic benzoyl-CoA pathway via two-electron reduction of benzoyl-CoA to cyclohex-1,5-diene-1-carbonyl-CoA, as studied in *T. aromatica*. The pathway in *R. palustris* differs in that the aromatic ring is reduced in a four-electron reduction, yielding cyclohex-1-ene-1-carbonyl-CoA, with corresponding modification of the subsequent steps.

of [^{75}Se]selenite, three radioactively labeled proteins with molecular masses of 100, 30, and 27 kDa were found. The 100- and 30-kDa selenoproteins were 5- to 10-fold induced in cells grown on benzoate compared to cells grown on lactate. These results suggest that the dearomatization process in *D. multivorans* is not catalyzed by the ATP-dependent Fe-S enzyme benzoyl-CoA reductase (as in facultative aerobes), but rather involves unknown molybdenum- and selenocysteine-containing proteins.[101]

Studies of the obligate anaerobic iron-reducing bacterium *Geobacter metallireducens* uncovered the genes coding for anaerobic benzoate metabolism. They are organized in two clusters comprising 44 genes. Induction of representative genes during growth on benzoate was confirmed by a reverse-transcription polymerase chain reaction. The results obtained suggest that benzoate is activated to benzoyl-CoA, which is then reductively dearomatized. However, in *G. metallireducens* (and most likely in other strict anaerobes) the process of reductive dearomatization of the benzene ring appears to be catalyzed by a set of completely different protein components comprising putative molybdenum- and selenocysteine-containing enzymes.[102]

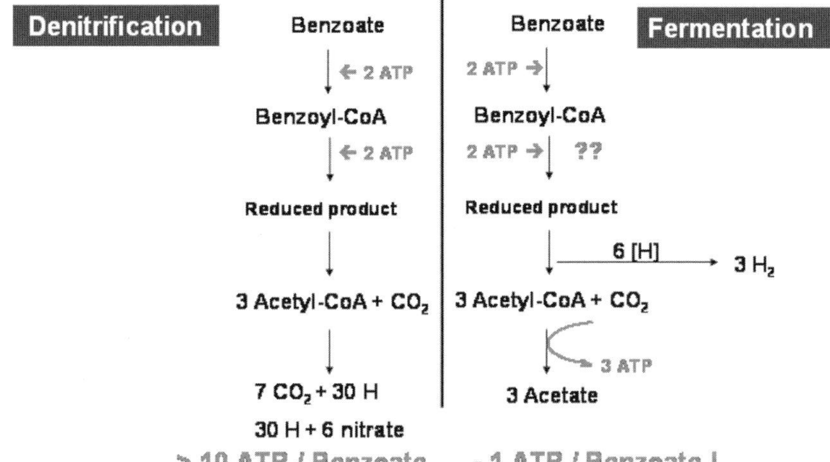

FIGURE 13. Energetic comparison of benzoate catabolic pathways in a denitrifying and a fermenting bacterium. Obviously, activation of benzoate and ATP-driven ring reduction consumes more energy than is gained in fermenting bacteria in the subsequent processes. This suggests an ATP-independent ring reduction in strict anaerobes.

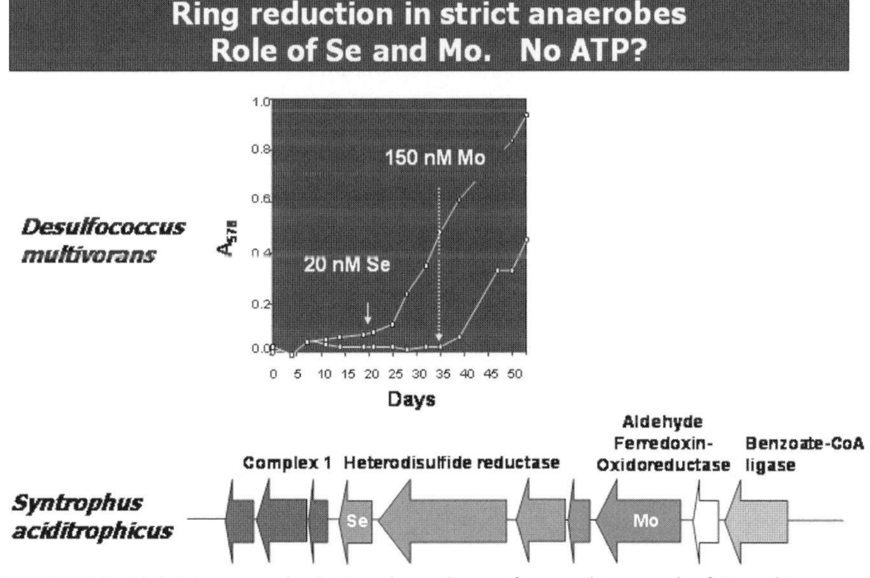

FIGURE 14. Molybdenum and selenium dependence of anaerobic growth of *D. multivorans* on benzoate and sulfate, and the gene cluster postulated to code for benzoate metabolism in *S. aciditrophicus*.

Surprisingly, the product of this new type of ring reduction is also cyclohex-1,5-diene-1-carbonyl-CoA, which must be formed by two-electron reduction of the aromatic ring. This follows from the fact that all genes coding for enzymes, which convert cyclohex-1,5-diene-1-carbonyl-CoA to three molecules of acetyl-CoA plus one molecule of CO_2, are present and the enzymes are active in those strict anaerobes; they are even highly similar to those found in denitrifying bacteria. This applies to *G. metallireducens* (iron reducing), *Syntrophus aciditrophicus* (fermenting), and probably also to *D. multivorans* (sulfate reducing).[103]

Perspective

Our view of aromatic metabolism has been widened by studying bacteria that do not live at high oxygen tensions. The metabolic diversity has been exemplified by considering the fate of benzoate under different conditions. The classic pathways using ring-cleaving dioxygenases represent just one principle to cope with the problem of breaking the stable aromatic ring. There are other options found by (facultative) aerobes to attack such molecules, even though they still use molecular oxygen to introduce hydroxyl groups at the ring. Interestingly, they use benzoyl-CoA as substrate for hydrolytic ring cleavage. This allows metabolic flexibility and rapid adaptation to fluctuating oxygen levels, since both oxic and anoxic types of metabolism use benzoyl-CoA as an intermediate. Under anoxic conditions, all oxygen-dependent steps need to be replaced by reductive steps. Depending on the energy yield of their metabolism, anaerobes use an ATP-driven or an ATP independent reduction of benzoyl-CoA. Our understanding of the different metabolic strategies is just at the beginning, as are the studies of the enzymes and reaction mechanisms. Their biotechnological potential has not been used. Many other aspects, such as regulation of the new pathways and their evolution, need to be addressed.

Acknowledgments

This work was funded by the Deutsche Forschungsgemeinschaft (DFG) and the Fonds der Chemischen Industrie.

Conflict of Interest

The author declares no conflicts of interest.

References

1. HAYAISHI, O. 1994. Tryptophan, oxygen, and sleep. Annu. Rev. Biochem. **63:** 1–24.
2. DAGLEY, S. 1978. Pathway for the utilization of organic growth substrates. *In* The Bacteria, Vol. 6. L.N. Ornston & J.R. Sokatch, Eds.: 305–388. New York: Academic Press.
3. DAGLEY, S. 1986. Biochemistry of aromatic hydrocarbon degradation in pseudomonads. *In* The Bacteria, Vol. 10. J.R. Sokatch & L.N. Ornston, Eds.: 527–555. New York: Academic Press.
4. HARAYAMA, S., M. KOK & E.L. NEIDLE. 1992. Functional and evolutionary relationships among diverse oxygenases. Annu. Rev. Microbiol. **446:** 565–601.
5. DAGLEY, S. 1971. Catabolism of aromatic compounds by microorganisms. Adv. Microbial. Physiol. **6:** 1–76.
6. HARWOOD, C.S. & R.E. PARALES. 1996. The β-ketoadipate pathway and the biology of self identity. Annu. Rev. Microbiol. **50:** 553–590.
7. OLIVERA, E.R., B. MINAMBERS, B. GARCIA, *et al.* 1998. Molecular characterization of the phenylacetic acid catabolic pathway in *Pseudomonas putida* U: the phenylacetyl-coenzyme A catabolon. Proc. Natl. Acad. Sci. USA. **95:** 6419–6424.
8. FERRANDEZ, A., B. MINAMBERS, B. GARCIA, *et al.* 1998. Catabolism of phenylacetic acid in *Escherichia coli*. J. Biol. Chem. **273:** 25974–25986.
9. DIAZ, E., A. FERRANDEZ, M.A. PRIETO & J.L. GARCIA. 2001. Biodegradation of aromatic compounds by *Escherichia coli*. Microbiol. Mol. Biol. Rev. **65:** 523–569.
10. MOHAMED, M.E., W. ISMAIL, J. HEIDER & G. FUCHS. 2002. Aerobic metabolism of phenylacetic acids in *Azoarcus evansii*. Arch. Microbiol. **178:** 180–192.
11. LUENGO, J.M., J.L. GARCÍA & E.R. OLIVERA. 2001. The phenylacetyl-CoA catabolon: a complex catabolic unit with broad biotechnological applications. Mol. Microbiol. **39:** 1434–1442.
12. ROST, R., S. HAAS, E. HAMMER, *et al.* 2002. Molecular analysis of aerobic phenylacetate degradation in *Azoarcus evansii*. Mol. Genet. Genomics **267:** 656–663.
13. ISMAIL, W., M.E. MOHAMED, B.L. WANNER, *et al.* 2003. Functional genomics by NMR spectroscopy: phenylacetate catabolism in *Escherichia coli*. Eur. J. Biochem. **270:** 3047–3054.
14. BARTOLOME-MARTIN, D., E.V. MARTINEZ-GARCIA, J. MASCARAQUE, *et al.* 2004. Characterization of a second functional gene cluster for the catabolism of phenylacetic acid in *Pseudomonas* sp. Strain Y2. Gene **341:** 167–179.
15. ZAAR, A., W. EISENREICH, A. BACHER & G. FUCHS. 2001. A novel pathway of aerobic benzoate catabolism in the bacteria *Azoarcus evansii* and *Bacillus stearothermophilus*. J. Biol. Chem. **276:** 24997–25004.
16. ZAAR, A., J. GESCHER, W. EISENREICH, *et al.* 2004. New enzymes in aerobic benzoate metabolism in *Azoarcus evansii*. Benzoyl-CoA oxygenase. Mol. Microbiol. **54:** 223–238.
17. GESCHER, J., W. EISENREICH, J. WOERTH, *et al.* 2005. Aerobic benzoyl-coenzyme A catabolic pathway in *Azoarcus evansii*: studies on the non-oxygenolytic ring cleavage enzyme. Mol. Microbiol. **56:** 1586–1600.
18. GESCHER, J., W. ISMAIL, E. OELGESCHLAEGER, *et al.* 2006. Aerobic benzoyl-coenzyme A (CoA) catabolic pathway in *Azoarcus evansii*: conversion of ring cleavage product by 3,4-dehydroadipyl-CoA semialdehyde dehydrogenase. J. Bacteriol. **188:** 2919–2927.
19. BUDER, R. & G. FUCHS. 1989. 2-Aminobenzoyl CoA monooxygenase/reductase, a novel type of flavoenzyme. Purification and some properties of the enzyme. Eur. J. Biochem. **185:** 629–635.
20. BUDER, R., K. ZIEGLER, G. FUCHS, *et al.* 1989. 2-Aminobenzoyl-CoA monooxygenase/reductase, a novel type of flavoenzyme. Studies on the stoichiometry and the course of the reaction. Eur. J. Biochem. **185:** 637–643.
21. LANGKAU, B., S. GHISLA, R. BUDER & G. FUCHS. 1990. 2-Aminobenzoyl-CoA monooxygenase/ reductase, a novel type of flavoenzyme. Identification of the reaction products. Eur. J. Biochem. **191:** 365–371.
22. LANGKAU, B., P. VOCK, V. MASSEY, *et al.* 1995. 2-Aminobenzoyl-CoA monooxygenase/reductase. Evidence for two distinct loci catalyzing substrate

monooxygenation and hydrogenation. Eur. J. Biochem. **230:** 676–685.

23. HARTMANN, S., C. HULTSCHIG, W. EISENREICH, *et al.* 1999. NIH-shift in flavin dependent monooxygenation. Mechanistic studies with 2-aminobenzoyl-CoA monooxygenase/reductase. Proc. Natl. Acad. Sci. USA **96:** 7831–7836.
24. SCHÜHLE, C., M. JAHN, S. GHISLA & G. FUCHS. 2001. Two similar gene clusters coding for the enzymes of a new type of aerobic 2-aminobenzoate (anthranilate) metabolism in the bacterium *Azoarcus evansii*. J. Bacteriol. **183:** 5268–5278.
25. EVANS, W.C. 1977. Biochemistry of the bacterial catabolism of aromatic compounds in anaerobic environments. Nature **270:** 17–22.
26. DUTTON, P.L. & W.C. EVANS. 1978. Metabolism of aromatic compounds by Rhodospirillaceae. *In* The Photosynthetic Bacteria. R.K. Clyton & W.R. Sistrom, Eds.: 719–726. New York: Plenum Press.
27. KAISER, J.P. & K.W. HANSELMANN. 1982. Aromatic chemicals through anaerobic microbial conversion of lignin monomers. Experientia **38:** 167–175.
28. SLEAT, R. & J.P. ROBINSON. 1984. The bacteriology of anaerobic degradation of aromatic compounds. J. Appl. Bacteriol. **57:** 381–394.
29. YOUNG, L.Y. 1984. Anaerobic degradation of aromatic compounds. *In* Microbial Degradation of Organic Compounds. D.T. Gibson, Ed.: 487–523. New York: Marcel Dekker.
30. GIBSON, D.T. & V. SUBRAMANIAN. 1984. Microbial degradation of aromatic compounds. *In* Microbial Degradation of Organic Compounds. D.T. Gibson, Ed.: 181–252. New York: Marcel Dekker.
31. BERRY, D.F., A.J. FRANCIS & J.M. BOLLAG. 1987. Microbial metabolism of homocyclic and heterocyclic aromatic compounds under anaerobic conditions. Microbiol. Rev. **51:** 43–59.
32. EVANS, W.C. & G. FUCHS. 1988. Anaerobic degradation of aromatic compounds. Annu. Rev. Microbiol. **42:** 289–317.
33. SCHINK, B. & A. TSCHECH. 1988. Fermentative degradation of aromatic compounds. *In* Microbial Metabolism and the Carbon Cycle. S.R. Hagedorn, R.S. Hanson & D.A. Kunz, Eds.: 213–226. Chur, Switzerland: Harwood Academic Publishers.
34. HEIDER, J. & G. FUCHS. 1997. Anaerobic metabolism of aromatic compounds. Eur. J. Biochem. **243:** 577–596.
35. HEIDER, J. & G. FUCHS. 1997. Microbial anaerobic aromatic metabolism. Anaerobe **3:** 1–22.
36. HARWOOD, C.S., G. BURCHHARDT, H. HERMANN & G. FUCHS. 1999. Anaerobic metabolism of aromatic compounds via the benzoyl-coenzyme A pathway. FEMS Microbiol. Rev. **22:** 439–458.
37. BOLL, M., G. FUCHS & J. HEIDER. 2002. Anaerobic oxidation of aromatic compounds and hydrocarbons. Curr. Opin. Biol. Chem. **6:** 604–611.
38. SCHINK, B., B. PHILIPP & J. MÜLLER. 2000. Anaerobic degradation of phenolic compounds. Naturwissenschaften **87:** 12–23.
39. GIBSON, J. & C.S. HARWOOD. 2002. Metabolic diversity in aromatic compound utilization by anaerobic microbes. Annu. Rev. Microbiol. **56:** 345–369.
40. BOLL, M., G. FUCHS & J. HEIDER. 2002. Anaerobic metabolism of aromatic compounds. Curr. Opin. Chem. Biol. **6:** 604–611.
41. BOLL, M. & G. FUCHS. 2005. Unusual reactions involved in the anaerobic metabolism of phenolic compounds. Biol. Chem. **386:** 989–997.
42. BOLL, M. 2005. Key enzymes in the anaerobic aromatic metabolism catalysing Birch-like reductions. Biochim. Biophys. Acta **1707:** 34–50.
43. HARUKI, E. 1982. Organic syntheses with carbon dioxide. *In* Organic and Bio-organic Chemistry of Carbon Dioxide. S. Inoue & N. Yamazaki, Eds.: 5–78. New York: Halsted Press.
44. KOSUGI, Y., Y. IMAOKA, F. GOTOH, *et al.* 2003. Carboxylations of alkali metal phenoxides with carbon dioxide. Org. Biomol. Chem. **1:** 817–821.
45. TSCHECH, A. & G. FUCHS. 1987. Anaerobic degradation of phenol by pure cultures of newly isolated denitrifying pseudomonads. Arch. Microbiol. **148:** 213–217.
46. TSCHECH, A. & G. FUCHS. 1989. Anaerobic degradation of phenol via carboxylation to 4-hydroxybenzoate: in vitro study of isotope exchange between $^{14}CO_2$ and 4-hydroxybenzoate. Arch. Microbiol. **152:** 594–599.
47. LACK, A. & G. FUCHS. 1994. Evidence that phenol phosphorylation to phenylphosphate is the first step in anaerobic phenol metabolism in a denitrifying *Pseudomonas* sp. Arch. Microbiol. **161:** 132–139.
48. SCHMELING, S., A. NARMANDAKH, O. SCHMITT, *et al.* 2004. Phenylphosphate synthase: a new phosphotransferase catalyzing the first step in anaerobic phenol metabolism in *Thauera aromatica*. J. Bacteriol. **186:** 8044–8057.
49. LACK, A. & G. FUCHS. 1992. Carboxylation of phenylphosphate by phenol carboxylase, an enzyme system of anaerobic phenol metabolism. J. Bacteriol. **174:** 3629–3636.
50. LACK, A., I. TOMMASI, M. ARESTA & G. FUCHS. 1991. Catalytic properties of phenol carboxylase. In vitro study of CO_2: 4-hydroxybenzoate isotope exchange reaction. Eur. J. Biochem. **197:** 473–479.
51. SCHÜHLE, K. & G. FUCHS. 2004. Phenylphosphate carboxylase: a new C-C lyase involved in anaerobic phenol metabolism in *Thauera aromatica*. J. Bacteriol. **186:** 4556–4567.
52. BREINIG, S. & G. FUCHS. 2000. Genes involved in the anaerobic metabolism of phenol in the bacterium *Thauera aromatica*. J. Bacteriol. **182:** 5849–5863.
53. NARMANDAKH, A., N. GAD'ON, F. DREPPER, *et al.* 2006. Phosphorylation of phenol by phenylphosphate synthase: role of histidine phosphate in catalysis. J. Bacteriol. **188:** 7815–7822.
54. SHINODA, Y., Y. SAKAI, M. UE, *et al.* 2000. Isolation and characterization of a new denitrifying spirillum capable of anaerobic degradation of phenol. Appl. Environ. Microbiol. **66:** 1286–1291.
55. SHINODA, Y., J. AKAGI, Y. UCHIHASHI, *et al.* 2005. Anaerobic degradation of aromatic compounds by *Magnetospirillum* strains: isolation and degradation genes. Biosci. Biotechnol. Biochem. **69:** 1483–1491.

56. Rabus, R., M. Kube, J. Heider, *et al.* 2005. The genome sequence of an anaerobic aromatic-degrading denitrifying bacterium, strain EbN1. Arch Microbiol. **183:** 27–36.
57. He, Z. & J. Wiegel. 1995. Purification and characterization of an oxygen-sensitive 4-hydroxybenzoate decarboxylase from *Clostridium hydroxybenzoicum.* Eur. J. Biochem. **229:** 77–82.
58. Huang, J., Z. He & J. Wiegel. 1999. Cloning, characterization, and expression of a novel gene encoding a 4-hydroxybenzoate decarboxylase from *Clostridium hydroxybenzoicum.* J. Bacteriol. **181:** 5119–5122.
59. Cavin, J.F., V. Dartois & C. Divies. 1998. Gene cloning, transcriptional analysis, purification, and characterization of phenolic acid decarboxylase from *Bacillus subtilis.* Appl. Environ. Microbiol. **64:** 1466–1471.
60. Chow, K.T., M.K. Pope & J. Davies. 1999. Characterization of a vanillic acid non-oxidative decarboxylation gene cluster from *Streptomyces* sp. D7. Microbiology **145:** 2393–2403.
61. Hsu, T., S.L. Daniel, M.F. Lux & H.L. Drake. 1990. Biotransformations of carboxylated aromatic compounds by the acetogen *Clostridium thermoaceticum*: generation of growth-supportive CO_2 equivalents under CO_2-limited conditions. J. Bacteriol. **172:** 212–217.
62. Zhang, X. & J. Wiegel. 1994. Reversible conversion of 4-hydroxybenzoate and phenol by *Clostridium hydroxybenzoicum.* Appl. Environ. Microbiol. **60:** 4182–4185.
63. Zhang, X., T.V. Morgan & J. Wiegel. 1990. Conversion of ^{13}C-1 phenol to ^{13}C-4 benzoate, an intermediate step in the anaerobic degradation of chlorophenols. FEMS Microbiol. Lett. **67:** 63–66.
64. Zhang, X. & J. Wiegel. 1990. Sequential anaerobic degradation of 2,4-dichlorophenol in freshwater sediments. Appl. Environ. Microbiol. **56:** 1119–1127.
65. Zhang, X. & J. Wiegel. 1992. The anaerobic degradation of 3-chloro-4-hydroxybenzoate in freshwater sediment proceeds via either chlorophenol or hydroxybenzoate to phenol and subsequently to benzoate. Appl. Environ. Microbiol. **58:** 3580–3585.
66. Sharak Genthner, B.R., G.T. Townsend & P.J. Chapman. 1990. Effect of fluorinated analogues of phenol and hydroxybenzoates on the anaerobic transformation of phenol to benzoate. Biodegradation **1:** 65–74.
67. Biegert, T., U. Altenschmidt, C. Eckerskorn & G. Fuchs. 1993. Enzymes of anaerobic metabolism of phenolic compounds. 4-Hydroxybenzoate-CoA ligase from a denitrifying *Pseudomonas* species. Eur. J. Biochem. **213:** 555–561.
68. Gibson, J., M. Dispensa, G.C. Fogg, *et al.* 1994. 4-Hydroxybenzoate-coenzyme A ligase from *Rhodopseudomonas palustris*: purification, gene sequence, and role in anaerobic degradation. J. Bacteriol. **176:** 634–41.
69. Laempe, D., M. Jahn, K. Breese, *et al.* 2001. Anaerobic metabolism of 3-hydroxybenzoate by the denitrifying bacterium *Thauera aromatica.* J. Bacteriol. **183:** 968–979.
70. Möbitz, H. & M. Boll. 2002. A Birch-like mechanism in enzymatic benzoyl-CoA reduction—A kinetic study of substrate analogues combined with an *ab initio* model. Biochemistry **41:** 1752–1758.
71. Brackmann, R. & G. Fuchs. 1993. Enzymes of anaerobic metabolism of phenolic compounds—4-Hydroxybenzoyl-CoA reductase (dehydroxylating) from a denitrifying *Pseudomonas* species. Eur. J. Biochem. **213:** 563–571.
72. Boll, M., G. Fuchs, C. Meier, *et al.* 2001. Redox centers of 4-hydroxybenzoyl-CoA reductase, a member of the xanthine oxidase family of molybdenum-containing enzymes. J. Biol. Chem. **276:** 47853–47862.
73. Breese, K. & G. Fuchs. 1998. 4-Hydroxybenzoyl-CoA reductase (dehydroxylating) from the denitrifying bacterium *Thauera aromatica.* Prosthetic groups, electron donor, and genes of a member of the molybdenum-flavin-iron-sulfur proteins. Eur. J. Biochem. **251:** 916–923.
74. Gibson, J., M. Dispensa & C.S. Harwood. 1997. 4-Hydroxybenzoyl-CoA reductase (dehydroxylating) is required for anaerobic degradation of 4-hydroxybenzoate by *Rhodopseudomonas palustris* and shares features with molybdenum-containing hydroxylases. J. Bacteriol. **179:** 634–642.
75. Unciuleac, M., E. Warkentin, C.C. Page, *et al.* 2004. Structure of a xanthine oxidase-related 4-hydroxybenzoyl-CoA reductase with an additional [4Fe-4S] cluster and an inverted electron flow. Structure **12:** 2249–2256.
76. Hille, R. 2005. Molybdenum-containing hydroxylases. Arch. Biochem. Biophys. **433:** 107–116.
77. Boll, M., B. Schink, A. Messerschmidt & P.M. Kroneck. 2005. Novel bacterial molybdenum and tungsten enzymes: three-dimensional structure, spectroscopy, and reaction mechanism. Biol. Chem. **386:** 999–1006.
78. Buckel, W. & R. Keese. 1995. One electron reactions of CoASH esters in anaerobic bacteria. Angew. Chem. Int. Ed. Engl. **34:** 1502–1506.
79. Boll, M., G. Fuchs, G. Tilley, *et al.* 2000. Unusual spectroscopic and electrochemical properties of the 2[4Fe-4S] ferredoxin of *Thauera aromatica.* Biochemistry **39:** 4929–4938.
80. Heider, J. 2007. Adding handles to unhandy substrates: anaerobic hydrocarbon activation mechanisms. Curr. Opin. Chem. Biol. **11:** 188–194.
81. Biegert, T., G. Fuchs & J. Heider. 1996. Evidence that anaerobic oxidation of toluene in the denitrifying bacterium *Thauera aromatica* is initiated by formation of benzylsuccinate from toluene and fumarate. Eur. J. Biochem. **238:** 661–668.
82. Beller, H.R. & A.M. Spormann. 1997. Anaerobic activation of toluene and *o*-xylene by addition of fumarate in denitrifying strain T. J. Bacteriol. **179:** 670–676.
83. Leuthner, B., C. Leutwein, H. Schulz, *et al.* 1998. Biochemical and genetic characterization of benzylsuccinate synthase from *Thauera aromatica*: a new glycyl radical enzyme catalyzing the first step in anaerobic toluene metabolism. Mol. Microbiol. **28:** 615–628.
84. Krieger, C.J., W. Roseboom, S.P. Albracht & A.M. Spormann. 2001. A stable organic free radical in anaerobic benzylsuccinate synthase of *Azoarcus* sp. strain T. J. Biol. Chem. **276:** 12924–12927.
85. Heider, J., A.M. Spormann, H.R. Beller & F. Widdel. 1999. Anaerobic bacterial metabolism of hydrocarbons. FEMS Microbiol. Rev. **22:** 459–473.

86. SPORMANN, A.M. & F. WIDDEL. 2000. Metabolism of alkylbenzenes, alkanes, and other hydrocarbons in anaerobic bacteria. Biodegradation **11:** 85–105.
87. WIDDEL, F. & R. RABUS. 2001. Anaerobic biodegradation of saturated and aromatic hydrocarbons. Curr. Opin. Biotechnol. **12:** 259–276.
88. KOCH, J. & G. FUCHS. 1992. Enzymatic reduction of benzoyl-CoA to alicyclic compounds, a key reaction in anaerobic aromatic metabolism. Eur. J. Biochem. **205:** 195–202.
89. KOCH, J., W. EISENREICH, A. BACHER & G. FUCHS. 1993. Products of enzymatic reduction of benzoyl-CoA, a key reaction in anaerobic aromatic metabolism. Eur. J. Biochem. **211:** 649–661.
90. BOLL, M. & G. FUCHS. 1995. Benzoyl-coenzyme A reductase (dearomatizing), a key enzyme of anaerobic aromatic metabolism. ATP dependence of the reaction, purification and some properties of the enzyme from *Thauera aromatica* strain K172. Eur. J. Biochem. **234:** 921–933.
91. BOLL, M., S.J.P. ALBRACHT & G. FUCHS. 1997. Benzoyl-CoA reductase (dearomatizing), a key enzyme of anaerobic aromatic metabolism. A study of adenosinephosphate activity, ATP stoichiometry of the reaction and EPR properties of the enzyme. Eur. J. Biochem. **244:** 840–851
92. EGLAND, P.G., D.A. PELLETIER, M. DISPENSA, *et al.* 1997. A cluster of bacterial genes for anaerobic benzene ring biodegradation. Proc. Natl. Acad. Sci. USA **94:** 6484–6489.
93. BREESE, K., M. BOLL, J. ALT-MÖRBE, *et al.* 1998. Genes coding for the benzoyl-CoA pathway of anaerobic aromatic metabolism in the bacterium *Thauera aromatica*. Eur. J. Biochem. **256:** 148–154.
94. BOLL, M., D. LAEMPE, W. EISENREICH, *et al.* 2000. Nonaromatic products from anoxic conversion of benzoyl-CoA with benzoyl-CoA reductase and cyclohexa-1,5-diene-1-carbonyl-CoA hydratase. J. Biol. Chem. **275:** 21889–21895.
95. BOLL, M., G. FUCHS, C. MEIER, *et al.* 2000. EPR and Mössbauer studies of benzoyl-CoA reductase. J. Biol. Chem. **275:** 31857–31868.
96. BOLL, M., G. FUCHS & D.J. LOWE. 2001. Single turnover EPR studies of benzoyl-CoA reductase. Biochemistry **40:** 7612–7620.
97. BIRCH, A.J., A.K. HINDE & L. RADOM. 1980. A theoretical approach to the Birch reduction. Structures and stabilities of the radical anions of substituted benzenes. J. Am. Chem. Soc. **102:** 3370–3376.
98. MÖBITZ, H., T. FRIEDRICH & M. BOLL. 2004. Substrate binding and reduction of benzoyl-CoA reductase: evidence for nucleotide-dependent conformational changes. Biochemistry **43:** 1376–85.
99. UNCIULEAC, M. & M. BOLL. 2001. Mechanism of ATP-driven electron transfer catalyzed by the benzene ring-reducing enzyme benzoyl-CoA reductase. Proc. Natl. Acad. Sci. USA **98:** 13619–13624.
100. MÖBITZ, H. & M. BOLL. 2002. A Birch-like mechanism in enzymatic benzoyl-CoA reduction: a kinetic study of substrate analogues combined with an ab initio model. Biochemistry **41:** 1752–1758.
101. PETERS, F., M. ROTHER & M. BOLL. 2004. Selenocysteine-containing proteins in anaerobic benzoate metabolism of *Desulfococcus multivorans*. J. Bacteriol. **186:** 2156–2163.
102. WISCHGOLL, S., D. HEINTZ, F. PETERS, *et al.* 2005. Gene clusters involved in anaerobic benzoate degradation of *Geobacter metallireducens*. Mol. Microbiol. **58:** 1238–1252.
103. PETERS, F., Y. SHINODA, M.J. MCINERNEY & M. BOLL. 2007. Cyclohexa-1,5-diene-1-carbonyl-coenzyme A (CoA) hydratases of *Geobacter metallireducens* and *Syntrophus aciditrophicus*: evidence for a common benzoyl-CoA degradation pathway in facultative and strict anaerobes. J. Bacteriol. **189:** 1055–1060.

Old Acetogens, New Light

HAROLD L. DRAKE,[a] ANITA S. GÖßNER,[a] AND STEVEN L. DANIEL[b]

[a]*Department of Ecological Microbiology, University of Bayreuth, Bayreuth, Germany*

[b]*Department of Biological Sciences, Eastern Illinois University, Charleston, Illinois, USA*

Acetogens utilize the acetyl-CoA Wood-Ljungdahl pathway as a terminal electron-accepting, energy-conserving, CO_2-fixing process. The decades of research to resolve the enzymology of this pathway (1) preceded studies demonstrating that acetogens not only harbor a novel CO_2-fixing pathway, but are also ecologically important, and (2) overshadowed the novel microbiological discoveries of acetogens and acetogenesis. The first acetogen to be isolated, *Clostridium aceticum*, was reported by Klaas Tammo Wieringa in 1936, but was subsequently lost. The second acetogen to be isolated, *Clostridium thermoaceticum*, was isolated by Francis Ephraim Fontaine and coworkers in 1942. *C. thermoaceticum* became the most extensively studied acetogen and was used to resolve the enzymology of the acetyl-CoA pathway in the laboratories of Harland Goff Wood and Lars Gerhard Ljungdahl. Although acetogenesis initially intrigued few scientists, this novel process fostered several scientific milestones, including the first ^{14}C-tracer studies in biology and the discovery that tungsten is a biologically active metal. The acetyl-CoA pathway is now recognized as a fundamental component of the global carbon cycle and essential to the metabolic potentials of many different prokaryotes. The acetyl-CoA pathway and variants thereof appear to be important to primary production in certain habitats and may have been the first autotrophic process on earth and important to the evolution of life. The purpose of this article is to (1) pay tribute to those who discovered acetogens and acetogenesis, and to those who resolved the acetyl-CoA pathway, and (2) highlight the ecology and physiology of acetogens within the framework of their scientific roots.

Key words: **acetogenesis; acetogenic bacteria; acetyl-CoA pathway; autotrophy; bioenergetics; *Clostridium aceticum*; electron transport; intercycle coupling; *Moorella thermoacetica*; nitrate dissimilation**

The Discoverers of Acetogens and Acetogenesis

Acetogenic bacteria, acetogens for short, are anaerobes that use the acetyl-CoA pathway for the reduction of CO_2 to the acetyl moiety of acetyl-coenzyme A (CoA), for the conservation of energy, and for the assimilation of CO_2 into cell carbon. Initially studied primarily for their novel CO_2-fixing properties, these prokaryotes are now known to be much more metabolically, ecologically, and phylogenetically diverse than once thought. Indeed, over 100 acetogenic species, representing 22 genera, have been isolated to date from a variety of habitats (e.g., soils, sediments, sludge, and the intestinal tracts of many animals, including humans and termites) (TABLE 1). Of the 22 different genera, *Acetobacterium* and *Clostridium* harbor the most known acetogenic species. Overall, acetogens as a group differ widely in their morphological, nutritional, and physiological properties. Furthermore, the acetyl-CoA pathway itself is widely distributed in nature, occurring in various forms in such microbial groups as methanogens and sulfate-reducing bacteria. These nonacetogens make use of metabolic pathways that are similar to the acetyl-CoA pathway for the assimilation of CO_2 into biomass or the oxidation of acetate.[1–3]

So, how did we get to this point relative to our understanding of acetogens and the acetyl-CoA pathway? Needless to say, the journey has been a long one, spanning some 60 years, with many stops and passengers along the way. The journey actually started in 1932 with the discovery of acetogenesis (i.e., the metabolic process by which two molecules of CO_2 are reduced to acetate) by F. Fischer and associates.[4] Their work demonstrated that microbial populations in sewage were competent in the formation of acetate from the H_2-dependent reduction of CO_2. Four years later, the Dutch microbiologist K. T. Wieringa (FIG. 1A)

Address for correspondence: Harold L. Drake, Department of Ecological Microbiology, University of Bayreuth, 95440 Bayreuth, Germany. hld@uni-bayreuth.de

TABLE 1. Acetogenic bacteria isolated to date[a]

Acetogen	Source of isolate	Gram type[b]	Cell morphology	Growth temperature[c]	Date of isolation; investigator
Acetitomaculum ruminis	Rumen fluid, steer	+	Rod	Mesophilic	1989; Greening and Leedle[183]
Acetoanaerobium noterae	Sediment	–	Rod	Mesophilic	1985; Sleat *et al.*[184]
"*Acetoanaerobium romashkovii*"	Oil field	+	Rod	Mesophilic	1992; Davydova-Charakhchyan *et al.*[185]
Acetobacterium bakii	Wastewater sediment	+	Rod	Psychrotrophic	1995; Kotsyurbenko *et al.*[186]
Acetobacterium carbinolicum	Freshwater sediment	+	Rod	Mesophilic	1984; Eichler and Schink[187]
"*Acetobacterium dehalogenans*"	Sewage digester sludge	+	Coccus	Mesophilic	1991; Traunecker *et al.*[166]
Acetobacterium fimetarium	Digested cattle manure	+	Rod	Psychrotrophic	1995; Kotsyurbenko *et al.*[186]
Acetobacterium malicum	Freshwater sediment	+	Rod	Mesophilic	1988; Tanaka and Pfennig[188]
Acetobacterium paludosum	Fen sediment	+	Rod	Psychrotrophic	1995; Kotsyurbenko *et al.*[186]
"*Acetobacterium psammolithicum*"	Subsurface sandstone	–	Rod	Mesophilic	1999; Krumholz *et al.*[95]
Acetobacterium tundrae	Tundra soil	+	Rod	Psychrotrophic	2000; Simankova *et al.*[92]
Acetobacterium wieringae	Sewage digester	+	Rod	Mesophilic	1982; Braun and Gottschalk[189]
Acetobacterium woodii	Marine sediment	+	Rod	Mesophilic	1977; Balch *et al.*[90]
Acetobacterium sp. AmMan1	Freshwater sediment	+	Rod	Mesophilic	1991; Dörner and Schink[190]
Acetobacterium sp. B10	Wastewater pond	+	Rod	Mesophilic	1989; Sembiring and Winter[191,192]
Acetobacterium sp. HA1	Sewage sludge	+	Rod	Mesophilic	1991; Schramm and Schink[193]
Acetobacterium sp. HP4	Lake sediment	+	Rod	Psychrotrophic	1989; Conrad *et al.*[194]
Acetobacterium sp. KoB58	Sewage sludge	+	Rod	Mesophilic	1988; Wagener and Schink[195]
Acetobacterium sp. LuPhet1	Sewage sludge	+	Rod	Mesophilic	1994; Frings and Schink[196]
Acetobacterium sp. LuTria3	Sewage sludge	+	Rod	Mesophilic	1994; Frings *et al.*[197]
Acetobacterium sp. MrTac1	Marine sediment	+	Rod	Mesophilic	1987; Emde and Schink[198]
Acetobacterium sp. OyTac1	Freshwater sediment	+	Rod	Mesophilic	1987; Emde and Schink[198]
Acetobacterium sp. RMMac1	Marine sediment	–	Rod	Mesophilic	1990; Schuppert and Schink[199]
Acetobacterium sp. 69	Sea sediment	+	Rod	Mesophilic	1992; Inoue *et al.*[200]
Acetobacterium sp.	Tundra wetland soil	+	Rod	Psychrotrophic	1996; Kotsyurbenko *et al.*[201]
Acetohalobium arabaticum	Saline lagoon	–	Rod	Mesophilic	1990; Zhilina and Zavarzin[202]
Acetonema longum	Wood-eating termite, gut	–	Rod	Mesophilic	1991; Kane and Breznak[203]
Bryantella formatexigens	Human feces	+	Rod	Mesophilic	2003; Wolin *et al.*[98]
"*Butyribacterium methylotrophicum*"	Sewage digester	+	Rod	Mesophilic	1980; Zeikus *et al.*[204]
Caloramator fervidus (?)	Hot spring	–	Rod	Thermophilic	1987; Patel *et al.*[205]
Clostridium aceticum	Soil	–	Rod	Mesophilic	1936; Wieringa,[5] Adamse,[9] Braun *et al.*[10]
"*Clostridium autoethanogenum*" (?)	Rabbit feces	+	Rod	Mesophilic	1994; Abrini *et al.*[158]
Clostridium carboxidivorans	Lagoon sediment	+	Rod	Mesophilic	2005; Liou *et al.*[89]
Clostridium coccoides	Mice feces, human feces	+	Coccoid rod	NR[d]	1976; Kaneuchi *et al.*,[206] Kamlage *et al.*[86]
Clostridium difficile AA1	Rumen, newborn lamb	+	Rod	Mesophilic	1998; Rieu-Lesme *et al.*[207]
Clostridium drakei	Coal mine pond sediment	+	Rod	Mesophilic	2000; Küsel *et al.*,[47] Liou *et al.*[89]
Clostridium formicaceticum	Sewage	–	Rod	Mesophilic	1967; El Ghazzawi,[17] Andreesen *et al.*[18]
Clostridium glycolicum 22	Sewage	+	Rod	Mesophilic	1977; Ohwaki and Hungate[88]

Continued.

TABLE 1. *Continued*

Acetogen	Source of isolate	Gram type[b]	Cell morphology	Growth temperature[c]	Date of isolation; investigator
Clostridium glycolicum RD-1	Sea-grass roots	+	Rod	Mesophilic	2001; Küsel *et al.*[87]
Clostridium ljungdahlii	Chicken waste	+	Rod	Mesophilic	1988; Barik *et al.*,[101] Tanner *et al.*[83]
Clostridium magnum	Freshwater sediment	−	Rod	Mesophilic	1984; Schink[208]
Clostridium mayombei	Soil-feeding termite, gut	+	Rod	Mesophilic	1991; Kane *et al.*[203]
Clostridium methoxybenzovorans	Olive oil mill wastewater	+	Rod	Mesophilic	1999; Mechichi *et al.*[209]
Clostridium scatologenes	Soil, coal mine pond sediment	+	Rod	Mesophilic	1927; Weinberg and Ginsbourg,[46] Küsel *et al.*[47]
Clostridium ultunense	Swine manure digester	+	Rod	Mesophilic	1996; Schnürer *et al.*[210]
Clostridium sp. CV-AA1	Sewage sludge	−	Rod	Mesophilic	1982; Adamse and Velzeboer[211]
Clostridium sp. M5a3	Human feces	+	Rod	NR	1996; Bernalier *et al.*,[127] Leclerc *et al.*[212,213]
Clostridium sp. F5a15	Human feces	+	Rod	NR	1996; Bernalier *et al.*,[127] Leclerc *et al.*[212,213]
Clostridium sp. Ag4f2	Human feces	+	Rod	NR	1996; Bernalier *et al.*[127]
Clostridium sp. TLN2	Human feces	+	Coccobacillus	NR	1996; Bernalier *et al.*[127]
Eubacterium aggregans	Olive oil mill Wastewater	+	Rod	Mesophilic	1998; Mechichi *et al.*[209]
Eubacterium limosum	Rumen fluid, sheep	+	Rod	Mesophilic	1981; Sharak Genthner *et al.*[214]
Holophaga foetida	Freshwater ditch mud	−	Rod	Mesophilic	1992; Bak *et al.*,[134] Liesack *et al.*[84]
Moorella glycerini	Hot spring sediment	+	Rod	Thermophilic	1997; Slobodkin *et al.*[118]
Moorella mulderi	Bioreactor	+	Rod	Thermophilic	2003; Balk *et al.*[215]
Moorella thermoacetica	Horse manure, soil	+/−	Rod	Thermophilic	1942; Fontaine *et al.*,[13] Gößner *et al.*[14–16]
Moorella thermoautotrophica	Hot spring	+/−	Rod	Thermophilic	1981; Wiegel *et al.*[216]
Moorella sp. F21	Soil	+	Rod	Thermophilic	2003; Karita *et al.*[133]
Moorella sp. HUC22-1	Mud	+	Rod	Thermophilic	2004; Sakai *et al.*[217]
Natroniella acetigena	Soda lake deposits	−	Rod	Mesophilic	1996; Zhilina *et al.*[97]
Natronincola histidinovorans	Soda lake deposits	+	Rod	Mesophilic	1998; Zhilina *et al.*[218]
Oxobacter pfennigii	Rumen fluid, steer	+	Rod	Mesophilic	1985; Krumholz and Bryant[219]
Ruminococcus hydrogenotrophicus	Human feces	+	Coccobacillus	Mesophilic	1996; Bernalier *et al.*[220,221]
Ruminococcus productus U1	Sewage digester	+	Coccus	Mesophilic	1984; Lorowitz and Bryant[222]
Ruminococcus productus Marburg	Sewage digester	+	Coccus	Mesophilic	1987; Geerligs *et al.*[223]
Ruminococcus schinkii	Rumen, 3-day-old lamb	+	Coccoid rod	Mesophilic	1996; Rieu-Lesme *et al.*[224]
Ruminococcus sp. TLF1	Human feces	+	Coccobacillus	NR	1996; Bernalier *et al.*[127]
Sporomusa acidovorans	Distillation waste water	−	Rod	Mesophilic	1985; Ollivier *et al.*[225]
Sporomusa aerivorans	Soil-eating termite, gut	−	Rod	Mesophilic	2003; Boga *et al.*[226]
Sporomusa malonica	Freshwater sediment	−	Rod	Mesophilic	1989; Dehning *et al.*[227]
Sporomusa ovata	Silage	−	Rod	Mesophilic	1984; Möller *et al.*[228]
Sporomusa paucivorans	Lake sediment	−	Rod	Mesophilic	1987; Hermann *et al.*[229]
Sporomusa silvacetica	Beech forest soil	+	Rod	Mesophilic	1997; Kuhner *et al.*[93]
Sporomusa sphaeroides	River mud	−	Rod	Mesophilic	1984; Möller *et al.*[228]
Sporomusa termitida	Wood-eating termite, gut	−	Rod	Mesophilic	1988; Breznak *et al.*[230]
Sporomusa sp. DR6	Rice field soil	+	Rod	NR	1999; Rosencrantz *et al.*[231]

TABLE 1. *Continued*

Acetogen	Source of isolate	Gram type[b]	Cell morphology	Growth temperature[c]	Date of isolation; investigator
Sporomusa sp. DR1/8	Rice field soil	+	Rod	NR	1999; Rosencrantz *et al.*[231]
Syntrophococcus sucromutans	Rumen fluid, steer	−	Coccus	Mesophilic	1986; Krumholz and Bryant[232]
Thermoacetogenium phaeum	Pulp wastewater reactor	+	Rod	Thermophilic	2000; Hattori *et al.*[233]
Thermoanaerobacter kivui	Lake sediment	−	Rod	Thermophilic	1981; Leigh *et al.*[234]
Tindallia californiensis	Alkaline lake sediment	+	Rod	Mesophilic	2003; Pikuta *et al.*[85]
Treponema azotonutricium	Termite, hindgut	NR	Spirochete	Mesophilic	1999; Leadbetter *et al.*,[129] Graber *et al.*[100]
Treponema primitia	Termite, hindgut	NR	Spirochete	Mesophilic	1999; Leadbetter *et al.*,[129] Graber *et al.*[100]
Unclassified					
AG (?)	Granular reactor sludge	+	Rod	Thermophilic	1996; Davidova and Stams[235]
AOR	Thermophilic digester	+	Rod	Thermophilic	1988; Lee and Zinder[106]
CS1Van	Human feces	+	Rod	Mesophilic	1993; Wolin and Miller[236]
CS3Glu	Human feces	+	Coccoid rod	Mesophilic	1993; Wolin and Miller[236]
CS7H	Human feces	+	Rod	Mesophilic	1993; Wolin and Miller[236]
D	Rumen fluid, deer	−	Rod	Mesophilic	1995; Rieu-Lesme *et al.*[237]
DMG58	River mud	−	Rod	Mesophilic	1984; Möller *et al.*[228]
EE121	Granular reactor sludge	+	Rod	NR	1990; Plugge *et al.*[238]
HA	Horse feces	−	Coccobacillus	NR	1995; Miller and Wolin[161]
I52	Human feces	−	Coccoid rod	Mesophilic	1994; Wolin and Miller[239]
S5a2	Human feces	+	Coccus	NR	1996; Bernalier *et al.*,[127] Leclerc *et al.*[212,213]
Ser8	Rumen, newborn lamb	NR	NR	NR	1995; Chaucheyras *et al.*[240]
SS1	406-m-deep sediment	+	Oval rod	Mesophilic	1993; Liu and Suflita[96]
TH-001	Sewage sludge	−	Rod	Mesophilic	1985; Frazer and Young[241]
VK64	Human feces	+	Coccus	NR	1996; Bernalier *et al.*[127]
X-8	Vegetable wastewater	−	Rod	Mesophilic	1982; Samain *et al.*[242]
ZT	Tundra soil	+	Rod	Psychrophilic	1992; Kotsyurbenko *et al.*,[243] Nozhevnikova *et al.*[244]
417/2	Oil field	−	Rod	Mesophilic	1992; Davydova-Charakhchyan *et al.*[185]
417/5	Oil field	−	Rod	Mesophilic	1992; Davydova-Charakhchyan *et al.*[185]
"New acetogenic bacterium"	Rumen, 15-hour-old lamb	+	Coccoid rod	Mesophilic	1996; Rieu-Lesme *et al.*[245]

[a]Bacteria listed appear to use the acetyl-CoA pathway for the synthesis of acetate and growth (modified from Drake,[246] Drake,[3] and Drake *et al.*[41,42]). If the acetogenic nature of an organism is uncertain, a question mark occurs after the name of the organism. Organisms not having validated names are enclosed in quotation marks.

[b]Gram type is based on electron microscopic analyses of the cell-wall structure, if reported. Otherwise, gram type is based on the gram stain reaction. (NOTE: Results of the gram stain reaction are not always in agreement with the electron microscopic analysis of the cell wall.) +/− indicates the gram type is variable.

[c]General temperature preference: psychrophilic (5–10°C), psychrotrophic (16–20°C), mesophilic (31–34°C), and thermophilic (58–62°C).

[d]NR, not reported.

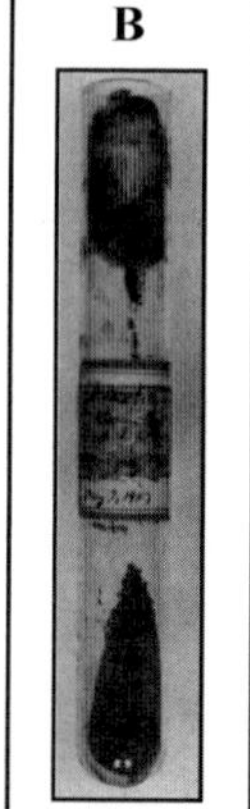

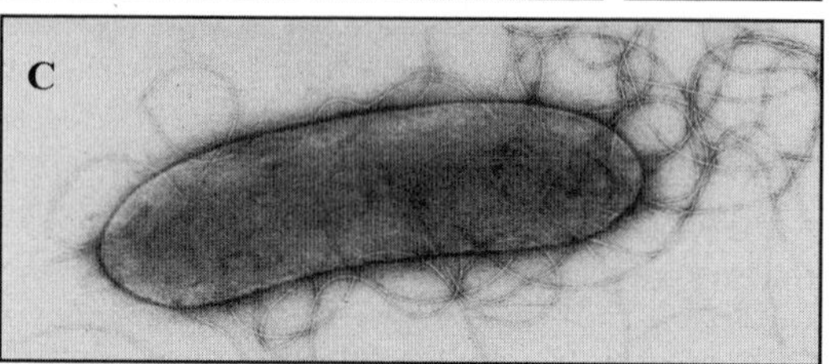

FIGURE 1. (**A**) Klaas Tammo Wieringa, who isolated the first acetogen, *Clostridium aceticum*, in 1936.[5] (**B**) A culture tube, dated 7 May 1947, containing spores of *C. aceticum* in dried soil. The tube was obtained from H. A. Barker and contained spores derived from Wieringa's culture of *C. aceticum*. These spores were used to revive the organism, as reported by Adamse in 1980[9] and Braun *et al.* in 1981.[10] (**C**) Electron micrograph of a peritrichiously flagellated cell of *C. aceticum*.[10] (Parts (B) and (C) are from Drake *et al.*[42] and are used with the kind permission of Springer.)

documented the isolation of the first acetogen, *Clostridium aceticum* (FIG. 1B and 1C; TABLE 1), from soil (i.e., ditch mud).[5] This spore-forming, mesophilic bacterium was shown to grow at the expense of H_2-CO_2 according to the following stoichiometry[5–7]:

$$4H_2 + 2CO_2 \rightarrow CH_3COOH + 2H_2O$$

In 1936, this reaction constituted a unique mechanism for the fixation of CO_2. Unfortunately, the culture of *C. aceticum* was later lost and no further work, except for one study to define its nutritional requirements,[8] was done with this acetogen until it was reisolated in the early 1980s (FIG. 1).[9–11] The life and career of K. T. Wieringa is encapsulated in the following paragraph and is largely based on information kindly provided by Wouter Middelhoven (Laboratory of Microbiology, Wageningen University, Wageningen, the Netherlands).

K. T. Wieringa was born into a well-to-do farmer's family on September 19, 1891 in Noordhorn, a village in the northeastern part of the Netherlands. He received his diploma at the University of Wageningen in 1916, spent some time at an agricultural research station in Groningen, and then returned to Wageningen in October 1918, when he was recruited by N. L. Söhngen [Professor of Microbiology and student of M. W. Beijerinck (Delft; Söhngen's dissertation is dated July 1906)][12] to join him at the Landbouwhogeschool (Agricultural University). Upon arrival in Wageningen, Wieringa participated in designing and constructing the Laboratory of Microbiology that opened in 1922. Wieringa received his doctoral degree from Wageningen in 1928 and taught both laboratory and lecture courses on a variety of subjects, including plant diseases, nitrogen fixation, and soil fertility, areas that were representative of his research interests. He remained active in the laboratory for many years subsequent to his retirement in 1956. Wieringa was a devoted member of the Mennonite church and passed away on July 28, 1980 in Wageningen.

Ironically, the discovery of an old culture tube containing spores of *C. aceticum* (FIG. 1B) led to the revival of Wieringa's culture of *C. aceticum* close to the time of his death.[9–11] The discovery of the first organism known to be capable of acetogenic autotrophy was most certainly Wieringa's greatest scientific accomplishment, work likely inspired by Söhngen's keen interest in methanogenesis and the microbial conversions of gases. Wieringa's collaboratrion with Söhngen is acknowledged by Wieringa in the introductory comments of his early works on *C. aceticium*[5–7] and is attested to by Wieringa's obituary on Söhngen, who passed away on December 24, 1934.[12] Indeed, it was H_2-CO_2 methanogenic enrichments that Söhngen and Wieringa had set up before Söhngen's death that yielded subtle hints toward H_2-dependent acetogenesis. A slight discrepancy in the amount of carbon recovered during the H_2-dependent formation of methane led Wieringa to conclude that transfers of H_2-CO_2 methanogenic enrichments must also contain a physiologically new type of organism:

> *Now it has been found that subcultures of this type sometimes may absorb large amounts of hydrogen without any formation of CH_4. Microscopical examination of such cultures shows the presence of club-shaped spore-bearing bacilli.*[6]

With a specially constructed gas-absorption apparatus that encouraged the growth of this novel, nonmethanogenic, H_2-consumming, spore-forming autotroph, Wieringa was able to enrich and isolate *C. aceticum*.[5] Despite the subsequent loss of *C. aceticum*

A

B

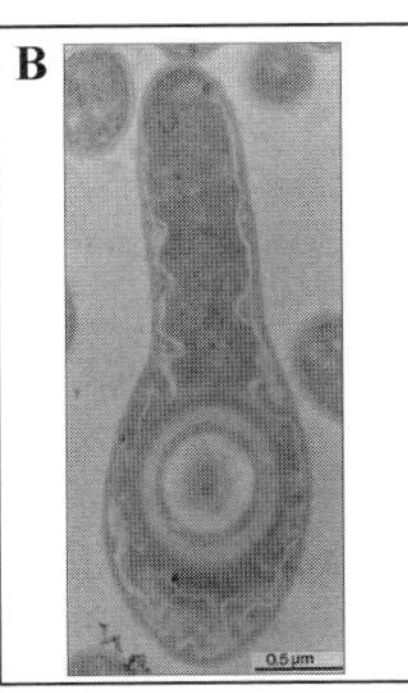

C

TABLE 7

Summary of glucose fermentations by C. thermoacetium

NUMBER	ORIGINAL GLUCOSE	GLUCOSE FERMENTED	ACETIC ACID	RATIO: ACETIC / GLUCOSE
	mM/100 ml.	*mM/100 ml.*	*mM/100 ml.*	
137-1	15.69	3.88	10.12	2.61
154-1	14.38	3.96	10.40	2.63
137-2	15.69	8.05	21.00	2.61
70-2a	12.22	8.19	21.00	2.57
56-304	13.60	9.37	23.50	2.51
70-4a	12.22	9.46	23.90	2.53
70-4b	12.22	10.20	26.40	2.59
189-123	12.24	10.29	26.20	2.54
200-155	12.15	11.26	28.40	2.53
200-154	12.15	11.54	30.05	2.60
Average of 21 fermentations....		8.64	22.06	2.55

FIGURE 2. (A) Francis Ephraim Fontaine, who isolated the second acetogen, *Clostridium thermoaceticum*, in 1942.[13] This picture was taken in the 1930s when Fontaine was a graduate student at the University of Wisconsin-Madison. **(B)** Electron micrograph of a sporulated cell of *C. thermoacoticum* ATCC 39073, which was reclassified in 1994 as *Moorella thermoacetica* (Collins *et al.*[48]).(From Drake[3] and used with kind permission of Springer.) **(C)** Original data from Fontaine *et al.*[13] These data document the near stoichiometric conversion of 1 mole of glucose to 3 moles of acetate by *C. thermoaceticum.* (Used with kind permission of the American Society of Microbiology.)

for many decades, it was within this historical context that the first acetogen was isolated.

In 1942, F. E. Fontaine (FIG. 2A) and co-workers isolated the second acetogen, *Clostridium thermoaceticum* (FIG. 2B), a spore-forming, thermophilic bacterium that catalyzed the near stoichiometric conversion of glucose to acetate (FIG. 2C)[13]:

$$C_6H_{12}O_6 \rightarrow 3CH_3COOH$$

C. thermoaceticum was isolated from horse manure. In retrospect, the mammalian gastrointestinal tract (as we know it) can hardly be considered a habitat that is suitable for colonization by thermophilic microorganisms. It is therefore more likely that *C. thermoaceticum*, as a transient organism voided from the horse's gut or as an organism introduced into the manure from the surrounding soil, was able to grow due to the elevated temperatures provided by the composting manure. Indeed, *C. thermoaceticum* has been isolated from Kansas prairie soils and Egyptian garden soils (FIG. 3),[14–16] and, thus, is widely distributed in soils that experience elevated (thermophilic) temperatures.

Given the abundance of acetogens in a wide variety of habitats (TABLE 1), it seems ironic that nearly three decades would pass after the isolation of *C. thermoaceticum* before J. R. Andreesen (FIG. 4A) and associates validated the isolation of *Clostridium formicoaceticum*, which was first reported by E. El Ghazzawi in 1967 and constituted the third published acetogen (FIG. 4B and 4C).[17,18] This spore-forming, mesophilic bacterium was isolated from sewage sludge and produced both formate and acetate during glucose-dependent fermentation.[17,18] In addition, C. *formicoaceticum* possesses other diverse metabolic potentials including the ability to (1) fix dinitrogen gas, (2) utilize the reductant derived from the oxidation of aromatic aldehyde groups for growth and acetate synthesis, and (3) dissimilate fumarate to acetate and succinate without engaging the acetyl-CoA pathway.[19–22] Surprisingly, with all of its metabolic capabilities relative to carbon and reductant flow, *C. formicoaceticum* is unable to grow at the expense of H_2-CO_2.[18,23]

With the loss of the *C. aceticum* culture and the absence of *C. formicoaceticum*, the isolation of *C. thermoaceticum* by Fontaine proved to be important and timely, given that this organism was the only acetogen available for laboratory study. This circumstance, in turn, ultimately set the stage for *C. thermoaceticum* to become the most historically important acetogen relative to the resolution of the acetyl-CoA pathway (TABLE 2).[3,24,25] Nonetheless, when *C. thermoaceticum* was isolated, it was unclear as to the nature of the metabolic process that could account for the formation of nearly 3 moles of acetate per mole of glucose (FIG. 2C). At that time, carbohydrate-based fermentations were usually known to yield such end-products as lactate, ethanol, or mixtures of one- and two-carbon products. The isolation and characterization of *C. thermoaceticum* constituted a portion of Fontaine's doctoral research in the Department of Biochemistry at the University of Wisconsin-Madison, and this matter was addressed by Fontaine on page 53 of his dissertation (dated May 19, 1941) entitled "The Fermentation of Cellulose and Glucose by Thermophilic Bacteria":

> *Since, in this fermentation, 2.5 mols of a two-carbon compound (acetic acid) are obtained from 1 mol of glucose it seems probable that either there is some primary cleavage of glucose other than the classical 3–3 split, or that a one-carbon compound is being reabsorbed. Of these two possibilities, the recent work on carbon dioxide uptake makes the latter seem more likely. Although the carbon*

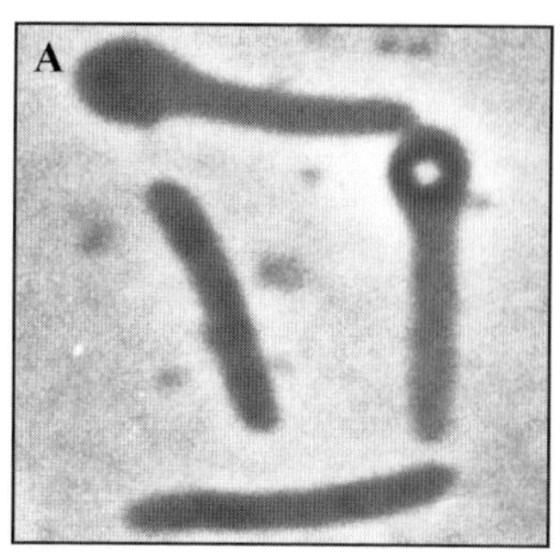

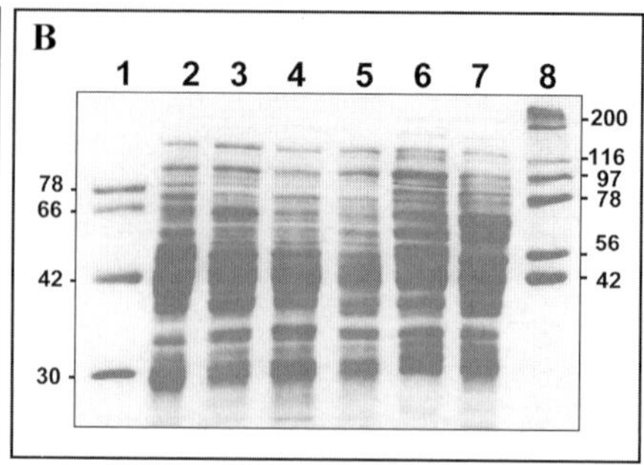

FIGURE 3. (**A**) *M. thermoacetica* PT1 (DSM 12993) obtained from Kansas prairie soil. (**B**) Sodium dodecyl sulfate-polyacrylamide gel electrophoretic (SDS-PAGE) analysis of *M. thermoacetica* isolates obtained from Kansas and Egyptian soils. Cells were cultivated on fructose; *Lanes 2–7* are protein profiles of the different isolates of *M. thermoacetica,* while *Lanes 1 and 8* are molecular-weight standards. All isolates have nearly identical metabolic capabilities to *M. thermoacetica* ATCC 39073 and grow chemolithoautotrophically at the expense of H_2-CO_2 or CO-CO_2 (Daniel *et al.*[43]). (Parts (A) and (B) are from Drake and Daniel[94] and are used with the kind permission of Elsevier.)

analyses show that there is no net gain or net loss of carbon dioxide, it is nevertheless possible that carbon dioxide is produced and then reabsorbed.[26]

The phrase "recent work on carbon dioxide uptake" was referring to various studies that assessed the uptake of carbon dioxide into organic carbon.[4,27–31] After completing his doctoral studies at Wisconsin, Fontaine went on to work at American Cyanamid's Lederle Laboratories division in Pearl River, New York. Fontaine was born in 1916 in Sheboygan, Wisconsin, and passed away in 1983 in Shohola, Pennsylvania.

The fixation of CO_2 proposed by Fontaine was demonstrated experimentally a few years later in 1945 with the landmark ^{14}C-studies of Barker and Kamen.[32,33] These studies were the first in biology to make use of carbon-14 and demonstrated that *C. thermoaceticum* incorporated $^{14}CO_2$ equally into both carbons of acetate.[32,33] That *C. thermoaceticum* was capable of synthesizing acetate from two molecules of CO_2 was later confirmed in 1952 by H. G. Wood using $^{13}CO_2$ and mass spectrometry.[34] On a collective basis, these early physiological and isotopic ('tracer') studies by Wieringa, Fontaine, Barker, Kamen, and Wood made it clear that acetogens possessed a new autotrophic mechanism for the fixation of CO_2. However, it would take nearly 40 more years of sustained research to determine the exact nature of the metabolic process by which acetogens convert 2 moles of CO_2 into 1 mole of acetate. *C. thermoaceticum* gave birth to many scientific milestones en route to resolving the acetyl-CoA pathway, including the discovery by Andreesen and Ljungdahl that tungsten is a biologically active metal.[35–37]

By the mid-to-late 1980s, the individual steps involved in the acetyl-CoA pathway were elucidated. This pathway is also referred to as the Wood–Ljungdahl pathway in recognition of the two biochemists H. G. Wood and L. G. Ljungdahl, who, together with their co-workers, resolved the chemical and enzymological features of the pathway using *C. thermoaceticum* as their model acetogen (FIG. 5).[3,24,25,38–42] Ironically, although we now know that the acetyl-CoA pathway represents a major autotrophic process that is central to carbon flow in various ecosystems, its biochemical details were resolved with *C. thermoaceticum*, an organism initially considered to be an obligate heterotroph. Interestingly, it was well after the enzymological details of the pathway were more or less established that autotrophic growth by *C. thermoaceticum* at the expense of H_2-CO_2 and CO-CO_2 was demonstrated.[43] Milestones leading up to the resolution of the acetyl-CoA pathway and autotrophic capabilities of *C. thermoaceticum* can be found in TABLE 2 and in numerous review articles.[3,24,25,38,41,42,44,45] Ironies are dispersed among the milestones. One rather large irony is that in 1927, M. Weinberg and B. Ginsbourg isolated a bacterium, *Clostridium scatologenes*, but did not discover a certain physiological novelty that it harbors.[46] In 2000, K. Küsel and co-workers[47] discovered that *C. scatologenes* is an acetogen, meaning that an acetogen had already been isolated 9 years prior to the isolation of *C. aceticum* by Wieringa.

In 1994, *C. thermoaceticum* was reclassified as *Moorella thermoacetica* when the taxonomy of the genus *Clostridium* was reorganized.[48] The organism will be referred to as *Moorella thermoacetica* throughout the remainder of this article.

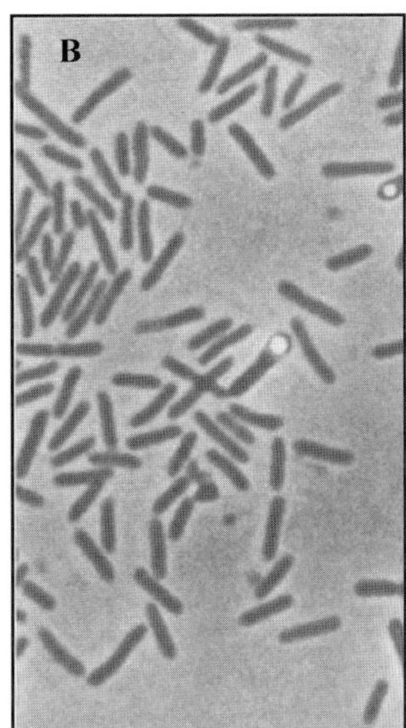

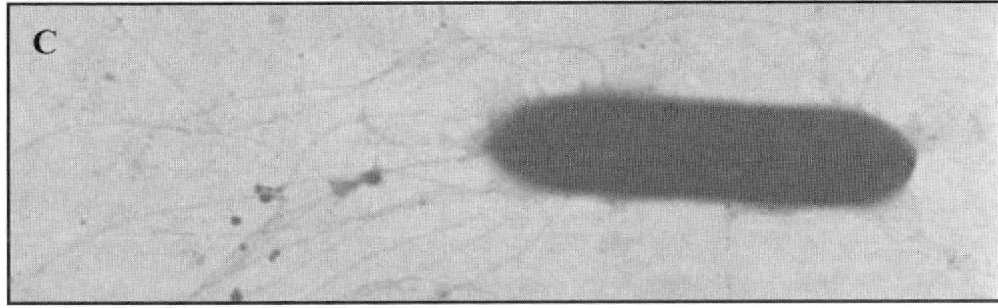

FIGURE 4. (**A**) Jan Remmer Andreesen, who in 1970 validated the isolation of the third acetogen, *Clostridium formicoaceticum*.[17,18] (**B**) *C. formicoaceticum* A1 obtained from an agar slant culture. (**C**) Electron micrograph showing the peritrichous flagellation of *C. formicoaceticum* A1. (Parts (B) and (C) are from Andreesen *et al.*[18] and are used with the kind permission of Springer.)

Acetyl-CoA Wood–Lungdahl Pathway: Evolution, Fixation of Carbon, Bioenergetics

The acetyl-CoA pathway is a reductive, linear, "one-carbon" process (FIG. 5), and is thus in marked contrast to cyclic CO_2-fixing processes (i.e., the Calvin cycle, the reductive tricarboxylic acid cycle, and the hydroxypropionate cycle) that are dependent upon recycled intermediates (i.e., ribulose bisphosphate, oxaloacetate, and acetyl-CoA, respectively) for the initial fixation of CO_2.[24,49] The relatively simple features of the linear acetyl-CoA pathway and the catalytic mechanisms by which carbon is chemically fixed in this pathway have been cited as reasons why the acetyl-CoA pathway might have been important in the evolution of life (i.e., the so-called primordial soup that existed on earth at the time the evolution of life was initiated might have fostered chemical processes indicative of those reflected in the key catalytic steps in the acetyl-CoA pathway by which organic molecules were initially formed).[50–53] These concepts were well conceived many years ago.[49] For example:

> *It is becoming apparent that the Acetyl-CoA Pathway plays a significant role in the carbon cycle. The direct combination of two CO_2 to form acetate may have been used by the earliest life forms rather than the more complicated cyclic mechanisms of autotrophism.*[49]

The usage of acetyl-CoA synthase for a variety of processes in nonacetogens, such as methanogens, sulphate reducers, hydrogenogens, and possibly anammox bacteria, illustrates how widely various features of the acetyl-CoA pathway are distributed among evolutionarily diverse functional groups of prokaryotes.[1,2,54–57]

As illustrated in FIGURE 5, much of Ljungdahl's work focused on understanding how CO_2 was reduced to a bound methyl unit on the methyl branch of the pathway, while much of Wood's work focused on resolving how CO_2 was reduced to a carbonyl unit on the carbonyl branch of the pathway. One should note that

TABLE 2. Milestones that led to resolving the acetyl-CoA Wood–Ljungdahl pathway and chemolithoautotrophic abilities of the model acetogen *Clostridium thermoaceticum*[a]

Year	Event
Events prior to the isolation of *Clostridium thermoaceticum*	
1927	Isolation of *Clostridium scatologenes* (Weinberg and Ginsbourg[46]); shown to be an acetogen in 2000 (Küsel *et al.*[47])
1932	H_2-dependent conversion of CO_2 to acetate in sewage sludge (Fischer *et al.*[4])
1936	Discovery of first acetogen, *Clostridium aceticum*; total synthesis of acetate from H_2-CO_2 (Wieringa[5–7]) (NOTE: Culture was lost.)
1942	Discovery of second acetogen, *Clostridium thermoaceticum*; conversion of glucose to 3 acetate (Fontaine *et al.*[13])
1944	Acetogenic conversion of pyruvate to acetate (Barker[247])
1945–1952	Synthesis of acetate from $^{14}CO_2$ (Barker and Kamen[33]) or $^{13}CO_2$ (Wood[34,132])
1955	Formate as a methyl-group precursor (Lentz and Wood[248])
1964	Methylcobalamin as methyl-group precursor (Poston *et al.*[249])
1965	Autotrophic synthesis of cell-carbon precursors from CO_2 (Ljungdahl and Wood[250])
1966–1969	Proposal of one-carbon pathway for the tetrahydrofolate/corrinoid-mediated synthesis of acetate from CO_2 (Ljungdahl *et al.*,[251] Ljungdahl and Wood[252])
1973–1976	Discovery that tungsten is a biologically active metal in formate dehydrogenase (Andreesen and Ljungdahl,[35] Ljungdahl and Andreesen,[37] Ljungdahl[36])
1973–1986	Resolution of the tetrahydrofolate pathway (reviewed in Ljungdahl[25])
1978–1980	Discovery of CO dehydrogenase as a nickel-containing enzyme (Diekert and Thauer,[60] Drake *et al.*[62])
1981	Resolution of enzmes required for synthesis of acetyl-CoA from pyruvate and methyltetrahydrofolate (Drake *et al.*[253])
1981–1982	Demonstration that CO replaces the carboxyl-group of pyruvate and undergoes an exchange reaction with acetyl-CoA (Drake *et al.*,[254] Hu *et al.*[61])
1982	Discovery of hydrogenase (Drake[137])
1983	Purification of CO dehydrogenase (Diekert and Ritter,[59] Ragsdale *et al.*[255])
1983	Use of H_2 and CO under organotrophic conditions (Kerby and Zeikus[256])
1984	Resolution of nutritional requirements (Lundie and Drake[257])
1984	Enzyme system for H_2-dependent synthesis of acetyl-CoA (Pezacka and Wood[258])
1984–1986	CO dehydrogenase is acetyl-CoA synthase (Pezacka and Wood,[258,259] Ragsdale and Wood[260]), and CO is the carbonyl precursor in the acetyl-CoA pathway under growth conditions (Diekert *et al.*,[58] Martin *et al.*[113])
1985–1991	Catalytic mechanism of acetyl-CoA synthase (reviewed in Ragsdale[64])
1986–1990	H_2- and CO-dependent electron-transport system coupled to the synthesis of ATP (Ivey and Ljungdahl,[261] Hugenholtz and Ljungdahl,[262,263] Das *et al.*[264])
1990	Chemolithoautotrophic growth on H_2-CO_2 and CO-CO_2 (Daniel *et al.*[43])
1991	Integrated model for catabolic, anabolic, and bioenergetic features of the acetyl-CoA Wood–Ljungdahl pathway (Wood and Ljungdahl[24])

SOURCE: Modified from Drake *et al.*[41,42]

[a] *Clostridium thermoaceticum* was reclassified to *Moorella thermoacetica* in 1994 (Collins *et al.*[48]).

the simplicity of the carbonyl branch on paper belies the complexity of the experiments needed to resolve it. The two branches of the acetyl-CoA pathway, and thus the joint efforts of Wood and Ljungdahl, merge at the synthesis of acetyl-CoA that is subsequently converted to either acetate or assimilated into biomass. Acetyl-CoA synthase not only catalyzes the reduction of CO_2 to CO and the synthesis of acetyl-CoA as shown in FIGURE 5, but can also oxidizes CO to CO_2; this latter reaction is the historical basis for referring to acetyl-CoA synthase as CO dehydrogenase.[58–62] It is now understood that the different subunits of acetyl-CoA synthase are capable of catalyzing different reactions. The enzymological features of the acetyl-CoA pathway that was resolved from *M. thermoacetica* can be found in several reviews and articles.[24,38–42,44,45,63–75]

Energy conservation occurs during the reductive synthesis of acetate by substrate-level phosphorylation and chemiosmotic processes. Four molecules of adenosine triphosphate (ATP) are produced by substrate-level phosphorylation (ATP_{SLP}) for each hexose (e.g., glucose) that is converted to three acetates (FIG. 6), a number that is higher than the amount of ATP_{SLP} formed by normal fermentations (e.g., homolactate

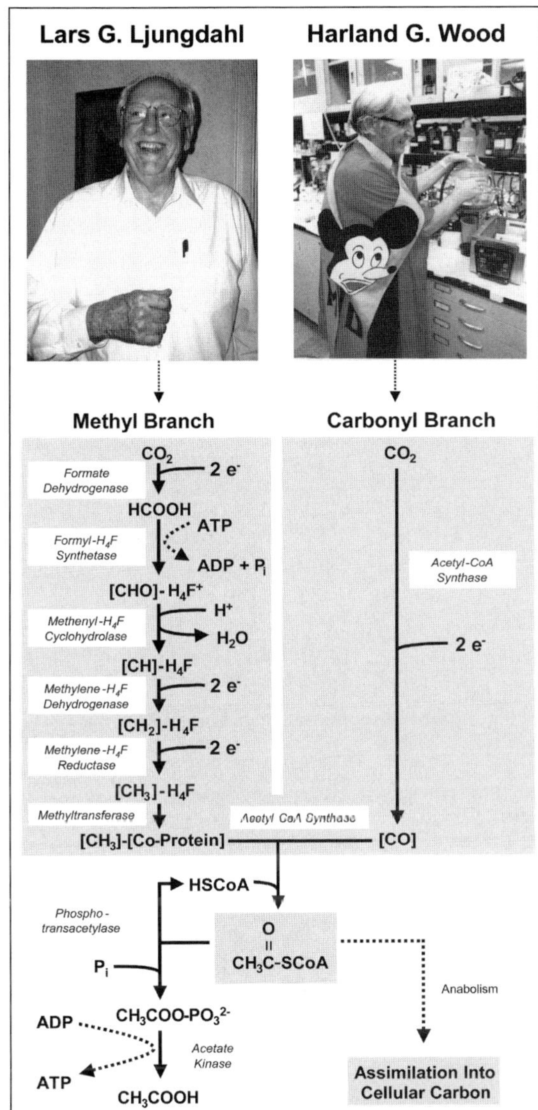

FIGURE 5. The acetyl-CoA Wood–Ljungdahl pathway. *Brackets* indicate that a particular C_1 unit is bound to a cofactor or structurally associated with an enzyme. ABBREVIATIONS: H_4F = tetrahydrofolate; HSCoA = coenzyme A; P_i = inorganic phosphate; e^- = electron; Co-Protein = corrinoid enzyme; ATP = adenosine 5′-triphosphate. (The pathway is from Müller *et al.*[77] and is used with the kind permission of Horizon Bioscience. The photographs of Harland Goff Wood and Lars Gerhard Ljungdahl are from Drake and Daniel[94] and are used with the kind permission of Elsevier.)

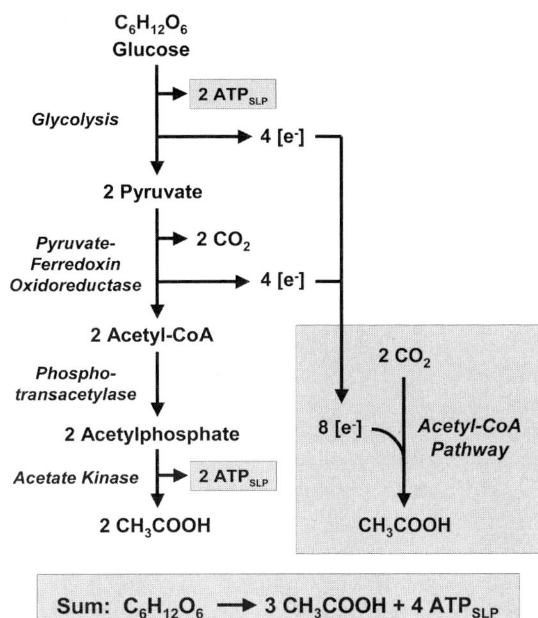

FIGURE 6. Homoacetogenic conversion of glucose to acetate. The two molecules of CO_2 that are reduced to acetate in the acetyl-CoA pathway can be derived from exogenous CO_2 rather than the CO_2 that is produced via the decarboxylation of pyruvate (see text). ABBREVIATIONS: ATP_{SLP} = ATP that is produced by substrate-level phosphorylation; [e^-], reducing equivalent. (From Muller *et al.*[77] and used with the kind permission of Horizon Bioscience.)

fermentation yields two ATP_{SLP} per glucose consumed). However, when the acetyl-CoA pathway operates under autotrophic conditions (e.g., during growth at the expense of H_2 and CO_2), there is no net gain in ATP_{SLP} (1 ATP_{SLP} is required for the activation of formate, and 1 ATP_{SLP} is produced when acetylphosphate is converted to acetate by acetate kinase) (FIG. 5). Thus, growth of acetogens under autotrophic conditions is strictly dependent upon chemiosmotic, energy-conserving processes that are coupled to the translocation of protons or sodium ions.[44,76,77] Certain acetogens (e.g., *M. thermoacetica*) utilize membranous electron transport systems that translocate protons out of the cell. Such acetogens have membranous, proton-translocating electron transport systems than contain cytochromes, menaquinones, and various oxidoreductases (e.g., hydrogenase); the formation of a proton motive force subsequently drives the cytoplasmic formation of ATP by proton-dependent ATPases.[39,44,75] Other acetogens (e.g., *Acetobacterium woodii*) lack membranous electron transport systems and translocate sodium ions concomitant to membranous transmethylation processes during acetate synthesis. The methyltransferase reaction at the terminal stage of the methyl branch of the acetyl-CoA pathway serves a duel function and translocates sodium ions out of the cell; the subsequent gradient of sodium ions drives the formation of ATP by sodium-dependent ATPase.[76,77]

Acetogens that utilize sodium pumping are dependent on sodium for growth at the expense of acetogenesis, while those acetogens that have proton-translocating electron transport systems do not require sodium for growth at the expense of acetogenesis.[78–80] Motility can likewise be dependent on sodium.[81] Some acetogens might also employ sodium-proton antiporters for generating electrochemical gradients.[82]

Organismal Diversity of Acetogenic Prokaryotes

Acetogens are defined as anaerobes that use the acetyl-CoA pathway for the (1) reductive synthesis of the acetyl moiety of acetyl-CoA from CO_2, (2) conservation of energy, and (3) assimilation of CO_2 into biomass.[3,41,42] Although the production of acetate is the classic hallmark of acetogens, the production of acetate is not a part of this definition, mainly because an acetogen might not form acetate (i.e., acetate formation is conditional and dependent upon both the acetogen and growth environment). Nonetheless, the ability of an organism to form acetate as the sole reduced end-product is compelling evidence that it utilizes the acetyl-CoA pathway per the definition given earlier and is an acetogen. As explained in detail elsewhere,[3,41,42] referring to microorganisms as acetogens when they form acetate by processes that do not involve the reductive synthesis of acetate from CO_2 yields unfortunate confusion in the literature.

The utilization of the acetyl-CoA pathway is the main unifying feature of acetogens. However, they display extreme genetic diversity, having genomic G + C contents that vary between 22 mol% (*Clostridium ljungdahlii*[83]) to 62 mol% (*Holophaga foetida*[84]). Acetogens have been assigned to 22 different bacterial genera[41,42]: *Acetitomaculum, Acetoanaerobium, Acetobacterium, Acetohalobium, Acetonema, Bryantella, "Butyribacterium," Caloramator, Clostridium, Eubacterium, Holophaga, Moorella, Natroniella, Natronincola, Oxobacter, Ruminococcus, Sporomusa, Syntrophococcus, Tindallia, Thermoacetogenium, Thermoanaerobacter, Treponema* (name in quotation marks has not been validated). In certain cases, the acetogenic nature of an organism that is characterized to be an acetogen is less than certain. For example, *Tindallia californiensis* appears to be an acetogen, in that it forms large amounts of acetate from various amino acids and pyrvuate, and cell extracts have hydrogenase and CO dehydrogenase activities.[85] However, (1) substrate/product stoichiometries have not been reported for this organism, (2) the occurrence of hydrogenase and CO dehydrogenase activities is not definitive evidence that an organism utilizes the acetyl-CoA pathway (e.g., cell extracts of *Clostridium pasteurianum* have both hydrogenase and CO dehydrogenase activities, but the organism is not an acetogen), and (3) the organism forms a mixture of products and is described as a fermentative organotroph. Thus, the acetogenic nature of this genus remains uncertain. Nonetheless, the potential occurrence of acetogenic strains in the genus *Tindallia* is noteworthy, as this genus is alkaliphilic.

Some acetogenic genera are monophyletic (i.e., all species of a genus that have been isolated to date are acetogens). *Moorella* and *Sporomusa* are examples of such genera. However, many acetogens are phylogenetically dispersed within genera that contain nonacetogenic species (FIG. 7). For example, *Clostridium* and *Ruminococcus* contain acetogens that are dispersed among closely related species that are not acetogens [e.g., the closest relative of the acetogen *C. formicoaceticum* is the nonacetogen *Clostridium felsineum* (99.3% 16S rRNA gene sequence similarity)]. Thus, the classification of new acetogens is sometimes problematic, in that the phylogenetic position of 16S rRNA gene sequences is inadequate for resolving the functional identity of a potential acetogen. In some cases, species that were not originally described as acetogens when first isolated are later discovered to be acetogens (e.g., *Clostridium coccoides*[86] and *C. scatologenes*[47]). In other cases, essentially identical organisms (i.e., organisms that have essentially identical 16S rRNA genes) may display opposite acetogenic capacities. For example, the type strain of *C. glycolicum* does not display acetogenic capabilities, but *C. glycolicum* strain RD-1[87] and *C. glycolicum* strain 22[88] are acetogens. Prolonged cultivation under certain conditions in the laboratory might cause certain acetogens to lose the capacity to engage the acetyl-CoA pathway and grow acetogenically.[41,42] DNA-DNA hydridation of genomic material can differentiate species-level differences between acetogens and nonacetogens where 16S rRNA-level differences are inadequate. For example, *C. scatologenes* SL1 was initially characterized as an acetogenic strain of the type strain of this species[47] and later shown to be a unique species by DNA-DNA hybridization and reclassified as *Clostridum drakei*.[89]

Acetogens have been isolated from very diverse habitats (TABLE 1), including sediments (e.g., *A. woodii* from blackish sediment of a marine estuary[90] and *H. foetida* from freshwater sediment[84]), soils (e.g., psychrotrophic *Acetobacterium tundrae* from tundra soil and Antarctic surficial material,[91,92] the mesophile *Sporomusa silvacetica* from forest soil,[93] and the thermophile *M. thermoacetica* from Egyptian and Kansas soils that experience thermophilic temperatures[14–16,94]), the subsurface (e.g., strain SS1 and "*Acetobacterium*

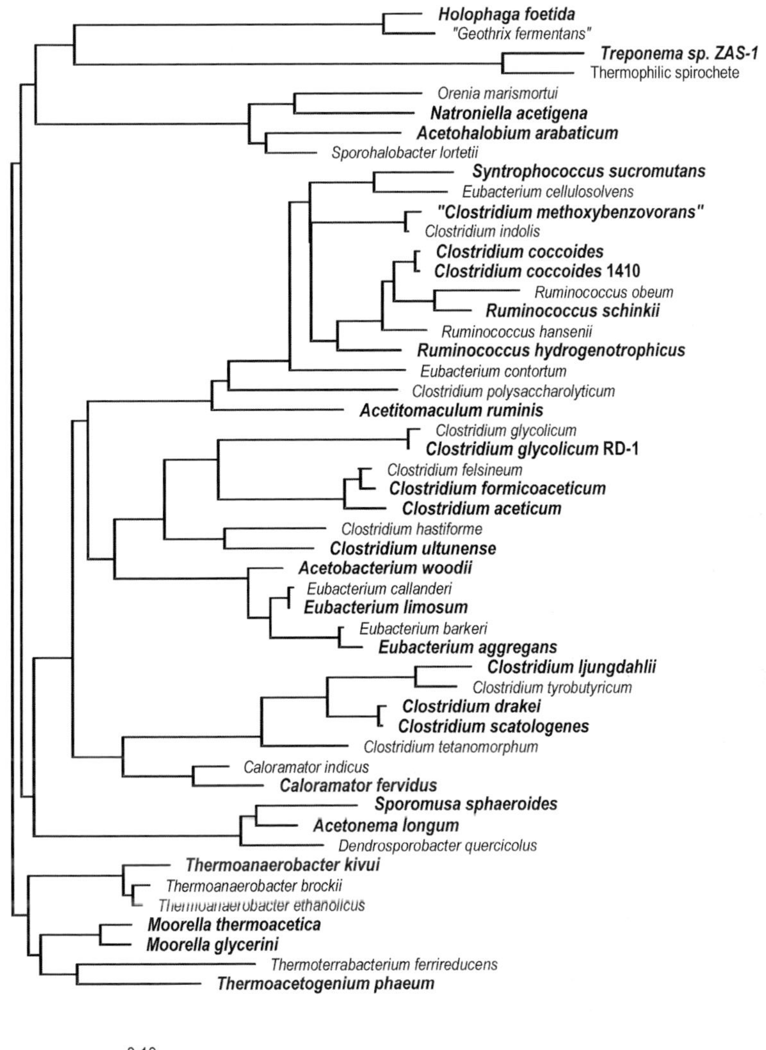

FIGURE 7. Parsimony tree of selected acetogenic bacteria (*bold font*) and their closest nonacetogenic relatives (*nonbold font*) based on full-length 16S rRNA sequences. *Bar* corresponds to 10 nucleotide substitutions per 100 sequence positions. (From Drake *et al.*[41,42] and used with the kind permission of Springer.)

psammolithicum" from subsurface sediment and sandstone, respectively[95,96]), acidic coal mine ponds (e.g., *C. drakei*[47,89]), salt-lake soda deposits (e.g., *Natroniella acetigena* from the soda deposits at Lake Magadi, Kenya[97]), fecal material (e.g., *C. coccoides* and *Bryantella formatexigens* from human feces,[86,98] *Treponema primitia* from the termite hindgut,[99,100] and *C. ljungdahlii* from chicken manure/waste[83,101]), and estuarine and salt-marsh plants (e.g., *C. glycolicum* RD-1 from the roots of the sea grass *Halodule wrightii*[87] and *Sporomusa rhizae* from the roots of the needlerush *Juncus roemerianus*[102]).

The diverse habitat range of acetogens demonstrates that organisms that can utilize the acetyl-CoA pathway are adapted to a broad range of *in situ* conditions. Many acetogens are spore-formers, a feature that likely aids in their *in situ* survival. Indeed, spores of *M. thermoacetica* have a decimal reduction time (i.e., the time required to decrease the population of viable spores by 90%) of nearly 2 hours at 121°C.[103] Some acetogens have connecting filaments (FIG. 8), structures that might aid cells in remaining close to one another for structural or communication purposes.

The usage of the acetyl-CoA pathway for autotrophic assimilation of carbon and acetate utilization

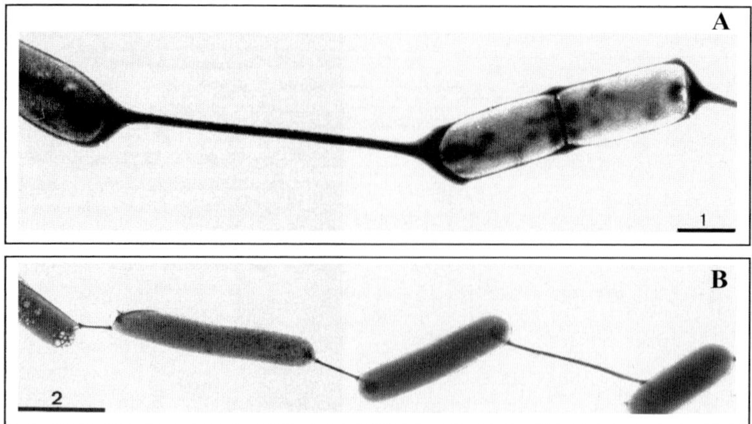

FIGURE 8. Electron micrographs of (**A**) the acetogen *Clostridium glycolicum* RD-1. (From Küsel *et al.*[87] and used with the kind permission of the American Society for Microbiology.) (**B**) The nitrogen-fixing bacterium *Clostridium akagii*. (From Drake *et al.*[41,42] and used with the kind permission of Springer.) Both organisms have the potential to form connecting filaments that tether cells to each other. The ultrastructure of connecting filaments is shown.[87,181,182] *Bars* are in micrometers.

by methanogens,[1,2,104,105] and the apparent reversibility of the pathway in some organisms (e.g., strain AOR[106] and *T. phaeum*[107]) makes it likely that Archaea exist that can grow via acetogenesis. Indeed, recent evidence demonstrates that the methanogenic archaeon *Methanosarcina acetivorans* C2A uses the acetyl-CoA pathway to convert carbon monoxide (CO) to both acetate and methane, that is, methane is not the sole reduced end-product of this methanogen under certain conditions (FIG. 9).[52,108] Similar observations have been made with the archaeon *Archaeoglobus fulgidus* VC16, that is, this archaea can grow via CO-dependent acetogenesis.[109] The capacity of anaerobic archaea to consume and oxidize CO has been known for decades,[110] and it is not without interest that this domain possesses organisms capable of engaging the acetyl-CoA pathway for the synthesis of acetate and the conservation of energy. Future studies must determine if these potentials occur under *in situ* environmental conditions, that is, are of ecological significance. The phylogenetic diversity of such organisms must likewise be determined.

Functional Diversity of Acetogenic Prokaryotes

The large phylogenetic diversity of acetogens, as well as the nonmonophyletic nature of many acetogenic genera, make it likely that acetogens possess broad functional diversity. Indeed, in contrast to the narrow substrate range of methanogens, acetogens utilize a wide variety of electron donors and electron acceptors (FIG. 10) and can engage alternative terminal electron-accepting processes when challenged with O_2. Thus, acetogens catalyze a variety of redox reactions by which they create fusion points in the carbon and other biological cycles (referred to below as "intercycle coupling").

Electron Acceptors and Intercycle Coupling

CO_2 and Acetogenesis

The growth of acetogens can be impaired when exogenous CO_2 is not available.[18,22,111,112] Given the fact that CO_2 is the terminal electron acceptor during acetogenesis (i.e., the reductive synthesis of acetate from CO_2), the importance of the availability of CO_2 might seem obvious. However, the stoichiometry of glucose-dependent acetogenesis (i.e., three acetate produced per glucose consumed) does not require CO_2 (i.e., CO_2 is not a substrate in the reaction). However, sugars are often not utilized optimally (indeed, might not be used at all) in the absence of supplemental CO_2. During glycolysis, oxidative reactions precede the decarboxylation of pyruvate (which yields CO_2). Thus, the recycling of electron carriers often necessitates an adequate supply of supplemental CO_2. Consistent with the flow of carbon predicted in FIGURE 6, growth of *M. thermoacetica* with uniformly labeled [^{14}C]glucose yields the following percent relative distribution of recovered ^{14}C: 31% [^{14}C]carbonates/CO_2, 62% [^{14}C]acetate, and 6% [^{14}C]biomass (Martin and Drake, unpublished

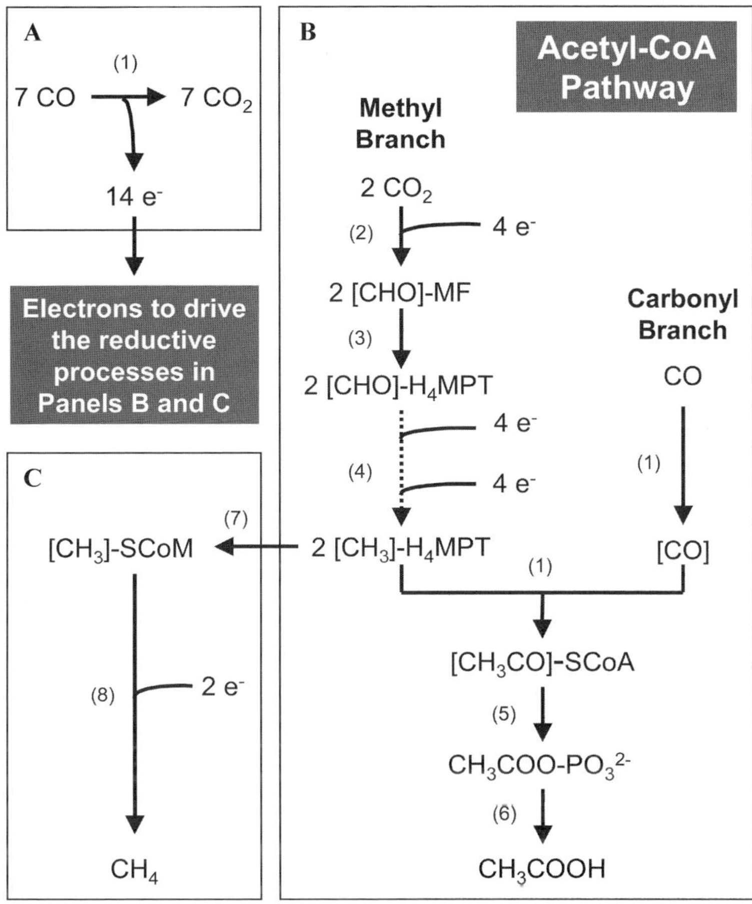

FIGURE 9. Metabolic processes by which carbon monoxide is utilized by *Methanosarcina acetivorans.* (**A**) Oxidation of carbon monoxide; (**B**) reductive synthesis of acetate via the acetyl-CoA pathway; (**C**) production of methane. *Parenthetical numbers* identify enzymes that catalyze the indicated reactions: 1, CO dehydrogenase/acetyl-CoA synthetase; 2, formyl-methanofuran dehydrogenase; 3, formyl-methanofuran:H_4MPT formyltransferase; 4, combined activities of methenyl-H_4MPT cyclohydrolase, methylene-H_4MPT dehydrogenase; methylene-H_4MPT reductase; 5, phosphotransacetylase; 6, acetate kinase; 7, methyl-H_4MPT:CoM methyltransferase; 8, combined activities of methyl-CoM reductase, membranous heterodisulfide reductase, and membranous $F_{420}H_2$ dehydrogenase complex (F_{420}, coenzyme F_{420}). ABBREVIATIONS: MF = methanofuran; H_4MPT = tetrahydromethanopterin; HSCoM = coenzyme M. [The figure is based on information in Lessner *et al.*[52] (further details on how these processes are coupled to the chemiosmotic conservation of energy can be found in this reference).]

data). Furthermore, large amounts of exogenous CO_2 are reduced to acetate during glucose-dependent acetogenesis.[113] The decarboxylation of carboxylated aromatic compounds by certain acetogens can augment the availability of CO_2 for acetogenesis.[112,114] Likewise, carbonic anhydrase might optimize the availability of intracellular CO_2 in acetogens.[115]

Alternative Electron Acceptors and Intercycle Coupling

Many acetogens can utilize one or more terminal electron-accepting processes in addition to acetogenesis (FIG. 10A). For example, nitrate is the preferred electron acceptor for *M. thermoacetica* and is dissimilated to nitrite and ammonium.[116,117] Closely related acetogens differ in their ability to utilize alternative electron acceptors. For example, the thermophilc acetogen *Moorella glycerini* is a close relative of *M. thermoacetica* but does not dissimilate nitrate.[118]

Nitrate dissimilation of *M. thermoacetica* is noteworthy, since the standard redox potential of the CO_2/acetate half-cell reaction is −290 mV (classically thought of as essential to the growth of this acetogen), while that of the nitrate/nitrite half-cell reaction

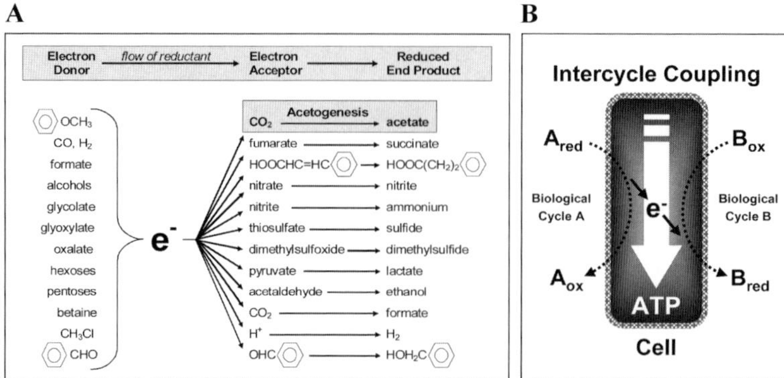

FIGURE 10. (**A**) Diverse redox couples that can be utilized by acetogens. For a more complete list of electron donors, see Drake *et al.*[41,42] ABBREVIATIONS: e^- = reducing equivalent. (**B**) Intercycle coupling (see text). (Modified from Drake *et al.*[122] and Drake and Küsel[40] and used with the kind permission of IOS Press and CRC Press.)

is 430 mV. Indeed, not only is nitrate dissimilation preferred to acetogenesis, growth of *M. thermoacetica* is significantly enhanced during nitrate dissimilation.[116,117] The dissimilation of nitrate significantly increases the growth efficiency of *M. thermoacetica*. H_2-dependent growth yields are enhanced eightfold when nitrate, rather than CO_2, is utilized as a terminal electron acceptor.[116] This enhancement of growth is consistent with the thermodynamics of these alternative terminal electron-accepting processes. The standard change in Gibbs free energy for H_2-dependent dissimilation of CO_2 to acetate via the acetyl-CoA pathway is −95 kJ per mol reaction, while that of the H_2-dependent dissimilation of nitrate to ammonium is −600 kJ per mol reaction.[119] *M. thermoacetica* cannot use ethanol or *n*-propanol as substrates for acetogenesis. However, both ethanol and *n*-propanol are readily utilized as electron donors when nitrate is available for dissimilation.[116] Nitrite can also be used by *M. thermoacetica* as an energy-conserving terminal electron acceptor.[120] These observations (1) demonstrate that *M. thermoacetica* is a facultative nitrate dissimilator rather than a so-called homoacetogen, (2) suggest that the acetogenic nature of this classic, model acetogen might have been overlooked if it had been isolated with a nitrate-rich cultivation medium, and (3) demonstrate that the acetyl-CoA pathway in *M. thermoacetica* is not constitutive. A membranous *b*-type cytochrome that is required on the methyl branch of the acetyl-CoA pathway is not present in the membrane when *M. thermoacetica* is grown in the presence of nitrate.[116] This cytochrome deficiency in the membrane appears to disable the acetyl-CoA pathway (FIG. 11). There are conflicting reports on the occurrence of active forms of the enzymes of the acetyl-CoA pathway in nitrate-dissimilating cells.[116,121] This matter has recently been addressed[94]:

> *The acetyl-CoA pathway is linked to anabolism, and the cell's inability to form acetyl-CoA from CO_2 (required for the assimilation of carbon when CO_2 is the sole source of carbon) during the dissimilation of nitrate is overcome with preformed methyl and carbonyl groups.*[116] *Thus, the ability of* M. thermoacetica *to synthesize acetyl-CoA via acetyl-CoA synthase is retained even when cells are dissimilating nitrate, as long as preformed methyl and carbonyl groups are available. This fact indicates that (a) the inability of the cell to assimilate CO_2 during nitrate dissimilation is not because acetyl-CoA synthase is repressed and (b) the control of electron flow is the primary reason why the catabolic function of the acetyl-CoA pathway is repressed when cells dissimilate nitrate.*[40,41,122]

The engagement of diverse redox couples enables acetogens to catalyze intercycle coupling, that is, to form junction points within and between biological cycles (FIG. 10B).[40] For example, the H_2-dependent reduction of CO_2 to acetate links the hydrogen and carbon cycles, and the oxidation of organic substrates via the dissimilation of nitrate fuses the carbon and nitrogen cycles. Intercycle coupling not only provides the cell with a means of conserving energy but forms a basis for linking biological cycles at the ecosystem level.

Not all of the redox couples illustrated in FIGURE 10A conserve energy. For example, the reduction of aldehyde groups of aromatic compounds appears to merely vent excess reductant under certain conditions and is not coupled to the conservation of energy. Nonetheless, their ability to facilitate a diverse number of redox reactions means that acetogens form a large variety of reduced end-products and not just acetate. As noted earlier, (1) exogenous CO_2 might determine how efficiently a particular electron donor is metabolized by an acetogen, and (2) an

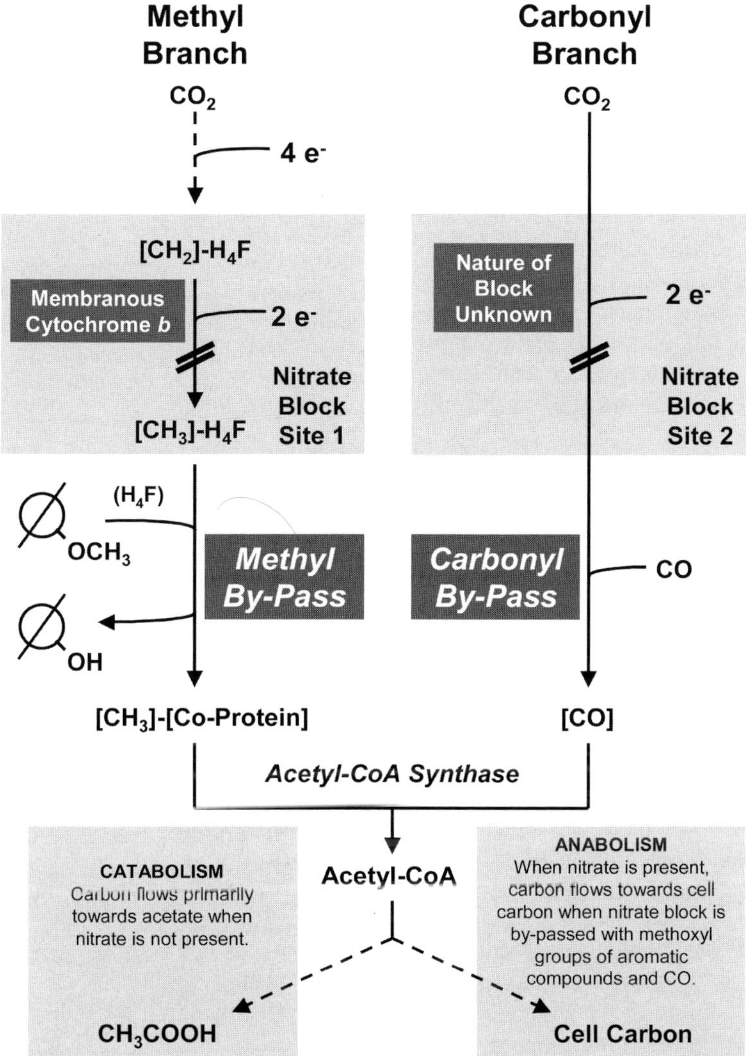

FIGURE 11. Scheme illustrating the proposed two major sites where the acetyl-CoA pathway is blocked when nitrate is dissimilated to ammonium by *M. thermoacetica*. The dissimilation of nitrite to ammonium appears to have the same affect.[120] ABBREVIATIONS: H_4F = tetrahydrofolate; CoA = coenzyme-A; Co-Protein = corrinoid enzyme. (Modified from Drake and Küsel[40] and Drake and Daniel[94] and used with the kind permission of Springer and Elsevier.)

acetogen may utilize alternative terminal electron acceptors in preference to CO_2. However, an acetogen might also use multiple terminal electron acceptors simultaneously. For example, the acetogen *Ruminococcus productus* simultaneously reduces phenylacrylates to phenylpropionates and CO_2 to acetate,[123] and also reduces CO_2 to acetate concomitant to lactate fermentation during growth at the expense of fructose.[124] *A. woodii* likewise has the potential to simultaneously utilize phenylacrylates (e.g., caffeate) and the acetyl-CoA pathway as terminal electron-accepting processes.[125] However, these two terminal processes might be selectively engaged by certain electron donors (e.g., *A. woodii* preferentially uses acetogenesis when reductant is derived from methanol).

Interspecies H_2 Transfer

Although energy might not be conserved by the reduction of protons when pure cultures of acetogens vent H_2 as a trace byproduct of acetogenesis, H_2-producing acetogens (e.g., *A. woodii*) can form close trophic associations with H_2-consuming partners.[126]

Such participation in interspecies H_2 transfer suggests that the production of H_2 by acetogens can be coupled to the conservation of energy under certain conditions. In contrast, acetogens from certain gastrointestinal tract systems (e.g., human and termite) can be the H_2-consuming partner in the interspecies transfer of H_2.[127–129]

Electron Donors

It is not possible that Fotaine and his co-workers could have known that the organism they would isolate, *M. thermoacetica*, would become not only the organism from which the acetyl-CoA pathway would be resolved (TABLE 2), but would also be the most metabolically robust acetogen characterized to the present date. Because the constraints of this presentation will not provide for an elaborate assessment of the diverse catabolic potentials of all of the acetogens known to date, this section will concentrate mostly on the ability of *M. thermoacetica* to activate reductant from diverse substates.

Most of the acetogens isolated to date can utilize hexoses or pentoses for growth and acetogenesis.[41,42,130,131] Tracer studies demonstrated that the glycolytic Embden–Meyerhof–Parnas (EMP) pathway is operative in *M. thermoacetica*.[132] However, there is limited information on the ability of acetogens to metabolize polymers (e.g., cellulose and lignin). Most acetogens isolated to date do not appear to be able to degrade high-molecular-weight polymers. However, a cellulose-degrading strain of *M. thermoacetica* was recently reported,[133] suggesting that the genus *Moorella* might contain cellulolytic strains. The acetogen *B. formatexigens* initially had the ability to use amorphous cellulose, but this ability was lost after prolonged growth under laboratory conditions.[98] It thus seems likely that future studies will resolve additional strains of acetogens that are able to use polymers of monosaccharides. Although most of the acetogens isolated to date are unable to degrade aromatic rings, the ability of the acetogen *H. foetida* to degrade aromatic rings[84,134,135] indicates that some acetogens have this metabolic potential. The degradation of lignin by an acetogen has not been reported.

In contrast to the chemical complexity of polymers, H_2 is chemically the simpliest source of reductant in nature, and H_2 is the only noncarbonaceous source of reductant that is currently known to be growth supportive for acetogens. Although H_2-dependent autotrophic acetogenesis is a distinguishing feature of acetogens, H_2-dependent growth (i.e., cell yields) is usually poor, and the ability of an acetogen to grow autotrophically at the expense of H_2 might therefore go undetected when the organism is first isolated (as was the case with *M. thermoacetica*[136,137]). Acetogens can contain multiple hydrogenases, and activity levels can vary with growth conditions.[136,137] Hydrogenase can be expressed during heterotrophic growth, but its physiological role under such conditions is unclear. During the heterotrophic dissimilation of nitrate by *M. thermoacetica*, the specific activity of hydrogenase is 14-fold lower than it is when growth is coupled to heterotrophic acetogenesis,[116] suggesting that hydrogenase is indeed important to the heterotrophic growth of this acetogen. Oxidoreductases that are detected by standard hydrogenase assays might be involved in intracellular reductant flow rather than the consumption or production of extracellular H_2, as has been proposed for the hydrogenase activity detected during the heterotrophic growth of *M. thermoacetica*.[94]

The one-carbon nature of the acetyl-CoA pathway provides for the efficient use of both one-carbon substrates and one-carbon side chains of aromatic compounds. For example, *M. thermoacetica* can use CO, formate, and methanol as sources of reductant, and is also able to utilize the methoxyl groups of a wide variety of aromatic compounds (e.g., 1,2,3-trimethoxybenzene, 4-hydroxy-3-methoxybenzyl alcohol, 2,3-dimethoxybenzoate, and 2,6-dimethoxyphenol).[94] One-carbon substrates can be either oxidized or be directly assimilated into the acetyl-CoA pathway. For example, as can be visualized in FIGURE 5, formate and CO can enter the methyl or carbonyl branches of the pathway, respectively, or be oxidized as sources of reductant. Methanol and methyl-level groups are disproportionated when utilized for acetogenesis.

CO-dependent acetogenesis yields the following stoichiometry:

$$4CO + 2H_2O \rightarrow CH_3COOH + 2CO_2$$

During this specialized process, three molecules of CO are oxidized to CO_2, yielding six electrons that are then used to reduce CO_2 on the methyl branch of the acetyl-CoA pathway. CO enters the carbonyl branch directly as a preformed carbonyl-level molecule. Thus, during CO-dependent acetogenesis, the methyl group of acetate is derived from CO_2, while the carboxyl group of acetate is derived from CO.[113] Traces of H_2 and methane can be produced during CO-dependent growth.[111]

Acetogens can oxidize a variety of alchohols and organic acids. For example, *M. thermoaetica* can utilize methanol, ethanol, *n*-propanol, *n*-butanol, formate, oxalate, glyoxylate, glycolate, pyruvate, and lactate.[94] As noted earlier, the potential to oxidize a particular substate may be dependent upon the electron acceptor utilized (e.g., short-chain alcohols are not growth supportive for *M. thermoacetica* when CO_2 is used as a terminal electron acceptor, but are growth supportive when nitrate is used as an terminal electron acceptor).

Although the use of glyoxylate, glycolate, and oxalate has not been widely demonstrated in acetogens or other anaerobes, these two-carbon compounds are readily used by *M. thermoacetica* according to the following reactions (kJ per mol calculated from Thauer *et al.*[119])[138,139]:

$$2\,^{-}OOC-CHO + 2\,H_2O \rightarrow CH_3COO^- + 2\,HCO_3{}^- + H^+(-86 \text{ kJ per mol glyoxylate})$$

$$4\,^{-}OOC-CH_2OH \rightarrow 3\,CH_3COO^- + 2\,HCO_3{}^- + H^+(-50 \text{ kJ per mol glycolate})$$

$$4\,^{-}OOC-COO^- + 5\,H_2O \rightarrow CH_3COO^- + 6\,HCO_3{}^- + OH^-(-41 \text{ kJ per mol oxalate})$$

The enzyme system by which *M. thermoacetica* catabolizes oxalate requires a utilizable electron acceptor.[140] Glyoxylate, glycolate, and oxalate appear to be metabolized by different mechanisms in *M. thermoacetica*. For example, glyoxylate and oxalate are utilized when nitrate is dissimilated, whereas glycolate is not utilized under this condition.[120,139,141] Reducing equivalents that are theoretically derived from these two-carbon substrates appear to be managed differently and yield dissimilar growth efficiencies.[94]

Response to O_2 and Oxidative Stress

Acetogens have been classically referred to as obligate, if not strict, anaerobes. Indeed, many enzymes central to acetogenesis are extremely sensitive to O_2 (i.e., oxidation), and the decades of work on resolving the acetyl-CoA pathway were very much impaired by this sensitivity. During those years, it was certainly logical to assume that acetogens are strict anaerobes. However, acetogens have been isolated from redox-unstable environments. Indeed, aerated soils and the rooting zones of estuarine and salt-marsh macrophytes that experience periods of O_2 enrichment harbor high numbers of culturable acetogens, including classic and novel acetogenic species.[14,93,102,142,143] Acetogens in redox-unstable habitats are likely challenged with O_2 and must cope with periods of oxidative stress. Thus, and in retrospect of the fact that the first acetogen, *C. aceticum*, was isolated from soil that was likely subject to periodic wetting and drying (i.e., aeration),[5,6] it is not surprising that recent studies have demonstrated that acetogens have several adaptation strategies by which they might deal with oxidative stress under *in situ* conditions (FIG. 12).

Reductive Removal of O_2

Acetogens contain numerous enzymes that can reductively remove O_2 and its toxic byproducts (e.g., superoxide and peroxide). These enzymes include peroxidase, reduced nicotinamide adenine dinucleotide (NADH) -oxidase, rubredoxin oxidoreductase (a superoxide reductase), rubrerythrin (a peroxidase), superoxide dismutase, catalase, and cytochrome *bd* oxidase.[144–147] These enzymes are effective in protecting acetogens from oxidative stress when the concentration of O_2 is relatively low.

Use of Alternative Electron Acceptors in Response to O_2

As noted earlier, acetogens can utilize a variety of terminal electron acceptors. Thus, some acetogens can shift reductant flow away from the acetyl-CoA pathway to alternative terminal electron-accepting processes that are less sensitive to O_2 and operate at higher redox potentials than does the acetyl-CoA pathway (as noted earlier, the standard redox potential of the CO_2/acetate half-cell reaction is -290 mV). For example, *C. glycolicum* RD-1 (isolated from sea-grass roots) is an aerotolerant acetogen that switches from acetogenesis to classic fermentation in response to O_2.[87] This acetogen tolerates up to 4% O_2 in the headspace of agitated cultures, during which sugars are metabolized via combined lactate–ethanol fermentation. The high standard redox potential of the nitrate/nitrite half-cell reaction (430 mV) suggests that nitrate dissimilation by *M. thermoacetica* would be less sensitive to O_2 than acetogenesis.

Trophic Interaction with O_2-consuming Partner

Acetogens can form symbiotic relationships with O_2-consuming microaerophiles and aerotolerant fermenters. Such relationships have been observed between the acetogen *M. thermoacetica* and the fermentative microaerophilic bacterium *Thermicanus aegyptius* (two thermophiles initially isolated as a co-culture from Egyptian soil[14]), and the acetogen *S. rhizae* and the aerotolerant fermenter *Clostridium intestinale* (two mesophiles initially isolated as a coculture from the

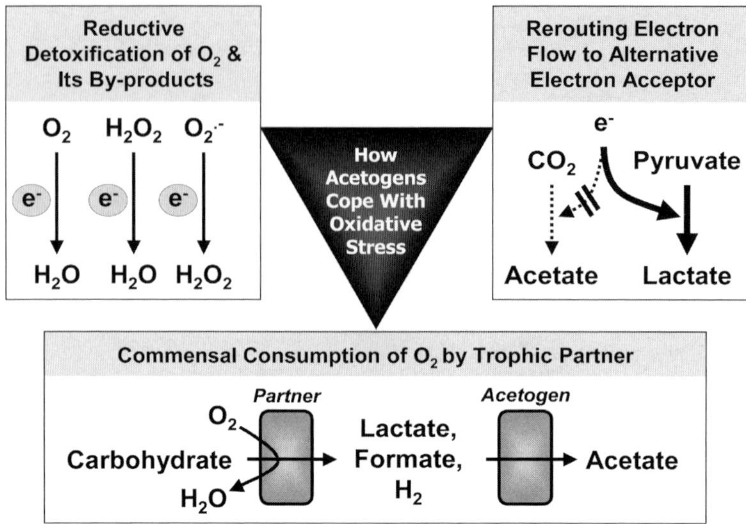

FIGURE 12. Mechanisms by which acetogens cope with oxidative stress. ABBREVIATIONS: X = products (e.g., H_2, formate, lactate) that are derived from the partial oxidation of carbohydrates [in some cases, short-chain polymers (e.g., stachyose) that are not substrates for the acetogen]; e^- = electron. (Modified from Müller *et al.*[77] and used with the kind permission of Horizon Bioscience.)

roots of the needlerush *J. roemerianus*[102]). In both cases, a fermentative nonacetogen that has the capacity to consume O_2 (thus protecting the acetogen from O_2) while simultaneously forming fermentation products (e.g., lactate, formate, and H_2) that can be used by the acetogen for acetogenesis. Although such partnerships between acetogens and O_2-consuming microorganisms have only been documented with laboratory cultures, such interactions might constitute a basis by which certain acetogens are protected from oxidative stress and form trophic linkages to other microorganisms under *in situ* conditions.

Harnessing the Functional Talents of Acetogens

Acetic acid is an important chemical. Its commercial production at the global level in 2001 approximated 10^{10} kg, and numerous studies have evaluated the potential use of acetogens to produce acetic acid or a salt thereof.[41,148–155] The acetogenic conversion of synthesis gas (i.e., H_2, CO, and CO_2) to acetic acid, ethanol, and butanol has also been investigated.[83,156–158] Unfortunately, acetogens are sensitive to acetate and acidic conditions. High concentrations of acetate and protons inhibit the growth of acetogens, mainly because a proton motive force and transmembrane electrical potential cannot be maintained under such conditions.[159] These limitations have hampered the commercialization of acetogens. Nonetheless, the existence of acetogens in acidic habitats (e.g., *C. drakei*[47,89]) suggests that new acetogens might be found that have higher tolerance to acidic conditions.

Despite the broad physiological activities of acetogens, their potential to degrade high-molecular-weight polymers (e.g., cellulose and lignin) appears to be limited. This limitation constitutes another problem for commercializing acetogenesis. However, normal strategies for isolating acetogens do not take such growth potentials into consideration. It is therefore noteworthy that two acetogens, *B. formatexigens*[98] and *M. thermoacetica* strain F21,[133] have recently been shown to degrade cellulose. Co-cultures of anaerobes might also offer promise for application. For example, cocultures of the cellulolytic thermophile *Clostridium thermocellum* and the thermophilic acetogen *Thermoanaerobacter kivui* can produce acetate from cellulose.[160] Cocultures of the cellulolytic mesophile *Ruminococcus albus* and the mesophilic acetogen HA have similar potentials; in coculture, the reducing equivalents derived from cellulose are utilized by HA via interspecies H_2 transfer.[161] *Clostridium lentocellum* strain SG6 forms high amounts of acetate from cellulose, and product stoichiometries suggest that this organism might utilize CO_2 as a terminal electron acceptor via the acetyl-CoA pathway.[162] However, it appears that significant amounts of acetyl-CoA are reduced to ethanol. The metabolism of such organisms might offer new strategies for conversion of cellulose to commercially useful chemicals.

The recovery of acetate from cultivation broths is another problem relative to the commercialization of acetogenesis. The concentration of acetate formed by acetogens is relatively low do to its inhibitory effects on growth. New strategies to efficiently recover acetate from cultivation broths might circumvent this problem.[163,164]

Acetogens and the enzymes that they produce might be useful in the bioremediation of certain anthropogenic compounds (e.g., trinitrotoluene)[165–167] or the production of fine chemicals (e.g., corrinoids) and enzymes (e.g., acetate kinase).[41,168–171] However, a commercial application of these potentials has not been reported.

Ecological Impact of Acetogens and the Acetyl-CoA Pathway

It is beyond the scope of this chapter to evaluate in detail the ecology of acetogens and acetogenesis. However, the following generalizations highlight both the ecological importance of acetogens and acetyl-CoA pathway, and also identify some of the challenges that future studies will be confronted by in this area.

In Situ *Information on Acetogens*

It has been estimated that 10^{12} kg of acetate are synthesized per year in sediments via acetogenesis.[24] Likewise, it has been estimated that 10^{12} kg of acetate are produced annually via acetogenesis in the hindgut of termites, a number that is fivefold greater than the annual amount of methane produced via the methanogenic reduction of CO_2.[172] Despite such estimations that make the ecological importance of acetogens seem obvious, assessing the *in situ* activity of acetogens is extremely problematic. Measuring the turnover of acetate is complicated, and, as noted earlier, acetogens catalyze a large number of redox reactions (i.e., the production of acetate is very likely not their only *in situ* activity). Plus, their interactions with other microbes is not restricted to carbon flow, but is also coupled to interspecies transfer of nutrients (e.g., folates).[173] Thus, although acetate is quantitatively an important trophic link in a wide variety of ecosystems, and although acetogens are important players in the carbon flow of many ecosystems, *in situ* information on acetogens is often conceptual.

Molecular Analysis

As noted earlier, acetogens are not monophyletic, in that many genera that contain acetogens also contain nonacetogens. Thus, broad-based analysis of acetogens as a distinct functional group by 16S rRNA-based approaches is problematic. Nonetheless, highly specific 16S rRNA-based probes and primers designed to target subsets of acetogenic taxa (e.g., a genus that only contains acetogens) have been developed (e.g., Küsel *et al.*[143]). Molecular approaches that are based on the analysis of functional genes central to the acetyl-CoA pathway (e.g., a gene for formyltetrahydrofolate synthetase) have also been developed for analyzing acetogenic bacteria.[174,175] Although these functional gene approaches have great promise and utility (e.g., Salmassi and Leadbetter,[176] Pester and Brune[177]), they, too, are compromised by problems of specificity.[178]

Phylogenetic and Global Distributions of the Acetyl-CoA Pathway

The *in situ* importance of the acetyl-CoA pathway is reflected in its wide distribution. As noted earlier, various forms of the acetyl-CoA pathway are used by both Bacteria and Archaea. However, an understanding of the distribution of the acetyl-CoA pathway is incomplete, as is an understanding of how the acetyl-CoA pathway is utilized in Prokaryotes. Recent evidence suggests that there are yet to be discovered processes that link the acetyl-CoA to the carbon flow of certain habitats. As noted earlier, the methanogen *M. acetivorans* uses the acetyl-CoA pathway to convert CO to both acetate and methane, that is, methane is not the sole reduced end-product of this methanogen (FIG. 9).[52,108] Likewise, the archaeon *A. fulgidus* VC16 can grow via CO-dependent acetogenesis.[109] Such fascinating observations point toward new ways in which the acetyl-CoA pathway might be metabolically linked to the carbon cycle via prokaryotic microbes.

Primary Production

Because the acetyl-CoA pathway and variants thereof facilitates the fixation of CO_2 in both domains of the Prokayotes, the pathway might be important to primary production in certain habitats. As reviewed elsewhere,[41,42,179,180] there is compelling evidence for that in the deep subsurface. Given the likelihood that chemical variations of the acetyl-CoA pathway were important to the evolution of life, that is, to early processes that were coupled to the fixation of carbon and the synthesis of organic molecules (see the second section of this chapter), future studies in this area would likely be very rewarding.

Conclusions

Acetogens and acetogenesis are microbiological discoveries. Those who made these early discoveries could not have known that their observations would foster thousands of studies on the biochemical, cellular, ecological, and evolutionary features of acetogens and the acetyl-CoA pathway. As one travels forward, the voyage made and those who were a part of it are well worth remembering. And on the occasion of your 80th birthday, we extend a special thanks to you, Lars.

Acknowledgments

The authors express their appreciation to (1) Alfons J. M. Stams and Wouter J. Middelhoven at Wageningen University for providing the picture of and historical information on K. T. Wieringa, (2) Gary Roberts and David Null at the University of Wisconsin-Madison for providing the picture of and historical information on F. E. Fontaine, and (3) Jan R. Andreesen at the University of Halle for providing the picture of himself (which can be seen on his current homepage).

Conflict of Interest

The authors declare no conflicts of interest.

References

1. FUCHS, G. 1994. Variations of the acetyl-CoA pathway in diversely related microorganisms that are not acetogens. *In* Acetogenesis. H.L. Drake, Ed.: 507–520. New York: Chapman & Hall.
2. FUCHS, G. 1986. CO_2 fixation in acetogenic bacteria: variations on a theme. FEMS Microbiol. Rev. **39:** 181–213.
3. DRAKE, H.L. 1994. Acetogenesis, acetogenic bacteria, and the acetyl-CoA "Wood/Ljungdahl" pathway: past and current perspectives. *In* Acetogenesis. H.L. Drake, Ed.: 3–60. New York: Chapman & Hall.
4. FISCHER, F., R. LIESKE & K. WINZER. 1932. Biologische Gasreaktionen. II. Über die Bildung von Essigsäure bei der biologischen Umsetzung von Kohlenoxyd und Kohlensäure mit Wasserstoff zu Methan. Biochem. Z. **245:** 2–12.
5. WIERINGA, K.T. 1936. Over het verdwijnen van waterstof en koolzuur onder anaerobe voorwaarden. Antonie Leeuwenhoek **3:** 263–273.
6. WIERINGA, K.T. 1939–1940. The formation of acetic acid from carbon dioxide and hydrogen by anaerobic spore-forming bacteria. Antonie Leeuwenhoek J. Microbiol. Serol. **6:** 251–262.
7. WIERINGA, K.T. 1941. Über die Bildung von Essigsäure aus Kohlensäure und Wasserstoff durch anaerobe Bazillen. Brennstoff-Chemie **22:** 161–164.
8. KARLSSON, J.L., B.E. VOLCANI & H.A. BARKER. 1948. The nutritional requirements of *Clostridium aceticum*. J. Bacteriol. **56:** 781–782.
9. ADAMSE, A.D. 1980. New isolation of *Clostridium aceticum* (Wieringa). Antonie Leeuwenhoek **46:** 523–531.
10. BRAUN, M., F. MAYER & G. GOTTSCHALK. 1981. *Clostridium aceticum* (Wieringa), a microorganism producing acetic acid from molecular hydrogen and carbon dioxide. Arch. Microbiol. **128:** 288–293.
11. GOTTSCHALK, G. & M. BRAUN. 1981. Revival of the name *Clostridium aceticum*. Int. J. Syst. Bacteriol. **31:** 476.
12. WIERINGA, K.T. 1935. In memorium Prof. Dr. Ir. N.L. Söhngen. Antonie Leeuwenhoek **2:** 1–7.
13. FONTAINE, F.E., W.H. PETERSON, E. MCCOY & M.J. JOHNSON. 1942. A new type of glucose fermentation by *Clostridium thermoaceticum* n. sp. J. Bacteriol. **43:** 701–715.
14. GÖßNER, A.S., R. DEVEREUX, N. OHNEMÜLLER, *et al.* 1999. *Thermicanus aegyptius* gen. nov., sp. nov., isolated from oxic soil, a fermentative microaerophile that grows commensally with the thermophilic acetogen *Moorella thermoacetica*. Appl. Environ. Microbiol. **65:** 5124–5133.
15. GÖSSNER, A. & H.L. DRAKE. 1997. Characterization of a new thermophilic acetogen (PT-1) isolated from aggregated Kansas prairie soil. Abstr. Ann. Meet. Am. Soc. Microbiol., p. 401, Abstr. N-122.
16. GOESSNER, A.S., K. KUESEL, R. DEVEREUX & H.L. DRAKE. 1998. Occurrence of thermophilic acetogens in Egyptian soils. Abstr. Ann. Meet. Am. Soc. Microbiol., p. 366, Abstr. N-1.
17. EL GHAZZAWI, E. 1967. Neuisolierung von *Clostridium aceticum* Wieringa und stoffwechselphysiologische Untersuchungen. Arch. Mikrobiol. **57:** 1–19.
18. ANDREESEN, J.R., G. GOTTSCHALK & H.G. SCHLEGEL. 1970. *Clostridium formicoaceticum* nov. spec. isolation, description and distinction from *C. aceticum* and *C. thermoaceticum*. Arch. Microbiol. **72:** 154–174.
19. BOGDAHN, M., J.R. ANDREESEN & D. KLEINER. 1983. Pathways and regulation of N_2, ammonium and glutamate assimilation by *Clostridium formicoaceticum*. Arch. Microbiol. **134:** 167–169.
20. DORN, M., J.R. ANDREESEN & G. GOTTSCHALK. 1978. Fermentation of fumarate and L-malate by *Clostridium formicoaceticum*. J. Bacteriol. **133:** 26–32.
21. GÖßNER, A., S.L. DANIEL & H.L. DRAKE. 1994. Acetogenesis coupled to the oxidation of aromatic aldehyde groups. Arch. Microbiol. **161:** 126–131.
22. MATTHIES, C., A. FREIBERGER & H.L. DRAKE. 1993. Fumarate dissimilation and differential reductant flow by *Clostridium formicoaceticum* and *Clostridium aceticum*. Arch. Microbiol. **160:** 273–278.
23. LUX, M.F. & H.L. DRAKE. 1992. Re-examination of the metabolic potentials of the acetogens *Clostridium aceticum* and *Clostridium formicoaceticum*: chemolithoautotrophic and aromatic-dependent growth. FEMS Microbiol. Lett. **95:** 49–56.
24. WOOD, H.G. & L.G. LJUNGDAHL. 1991. Autotrophic character of acetogenic bacteria. *In* Variations in Autotrophic Life. J.M. Shively & L.L. Barton, Eds.: 201–250. San Diego, CA: Academic Press.
25. LJUNGDAHL, L.G. 1986. The autotrophic pathway of acetate synthesis in acetogenic bacteria. Annu. Rev. Microbiol. **40:** 415–450.

26. FONTAINE, F.E. 1941. The fermentation of cellulose and glucose by thermophilic bacteria. Ph.D. dissertation. University of Wisconsin-Madison.
27. BARKER, H.A., S. RUBEN & J.V. BECK. 1940. Radioactive carbon as an indicator of carbon dioxide reduction. IV. The synthesis of acetic acid from carbon dioxide by *Clostridium acidi-urici*. Proc. Natl. Acad. Sci. USA **26:** 477–482.
28. PHELPS, A.S., M.J. JOHNSON & W.H. PETERSON. 1939. Carbon dioxide utilization during the dissimilation of glycerol by the propionic acid bacteria. Biochem. J. **33:** 726–728.
29. STEPHENSON, M. & L.H. STICKLAND. 1933. Hydrogenase. III. The bacterial formation of methane by the reduction of one-carbon compounds by molecular hydrogen. Biochem. J. **27:** 1517–1527.
30. WOOD, H.G. & C.H. WERKMAN. 1936. The utilization of carbon dioxide in the dissimilation of glycerol by the propionic acid bacteria. Biochem. J. **30:** 48–53.
31. WOOD, H.G. & C.H. WERKMAN. 1940. Heavy carbon as a tracer in bacterial fixation of carbon dioxide. J. Biol. Chem. **135:** 789–790.
32. KAMEN, M.D. 1963. The early history of carbon-14. J. Chem. Educ. **40:** 234–242.
33. BARKER, H.A., KAMEN, M.D. 1945. Carbon dioxide utilization in the synthesis of acetic acid by *Clostridium thermoaceticum*. Proc. Natl. Acad. Sci. USA **31:** 219–225.
34. WOOD, H.G. 1952. A study of carbon dioxide fixation by mass determination on the types of C^{13}-acetate. J. Biol. Chem. **194:** 905–931.
35. ANDREESEN, J.R. & L.G. LJUNGDAHL. 1973. Formate dehydrogenase of *Clostridium thermoaceticum*: incorporation of selenium-75, and the effects of selenite, molybdate, and tungstate on the enzyme. J. Bacteriol. **116:** 867–873.
36. LJUNGDAHL, L.G. 1976. Tungsten, a biologically active metal. Trends Biochem. Sci. **1:** 63–65.
37. LJUNGDAHL, L.G. & J.R. ANDREESEN. 1975. Tungsten, a component of active formate dehydrogenase from *Clostridium thermoaceticum*. FEBS Lett. **54:** 279–282.
38. RAGSDALE, S.W. 1997. The Eastern and Western branches of the Wood/Ljungdahl pathway: how the East and West were won. Biofactors **6:** 3–11.
39. DAS, A. & L.G. LJUNGDAHL. 2003. Electron-transport system in acetogens. *In* Biochemistry and Physiology of Anaerobic Bacteria. L.G. Ljungdahl, M.W. Adams, L.L. Barton, *et al.*, Eds.: 191–204. New York: Springer-Verlag.
40. DRAKE, H.L. & K. KÜSEL. 2005. Acetogenic clostridia. *In* Handbook on Clostridia. P. Dürre, Ed.: 719–746. Boca Raton, FL.: CRC Press.
41. DRAKE, H.L., K. KÜSEL & C. MATTHIES. 2004. Acetogenic prokaryotes. *In* The Prokaryotes An Evolving Electronic Resource for the Microbiological Community, 3rd ed. M. Dworkin, S. Falkow, E. Rosenberg, *et al.*, Eds.: Release 3.17, August 2004 http://springeronline.com. New York: Springer-Verlag.
42. DRAKE, H.L., K. KÜSEL & C. MATTHIES. 2006. Acetogenic prokaryotes. *In* The Prokaryotes, Vol. 2. M. Dworkin, S. Falkow, E. Rosenberg, *et al.*, Eds.: 354–420. New York: Springer-Verlag.
43. DANIEL, S.L., T. HSU, S.I. DEAN & H.L. DRAKE. 1990. Characterization of the H_2- and CO-dependent chemolithotrophic potentials of the acetogens *Clostridium thermoaceticum* and *Acetogenium kivui*. J. Bacteriol. **172:** 4464–4471.
44. LJUNGDAHL, L.G. 1994. The acetyl-CoA pathway and the chemiosmotic generation of ATP during acetogenesis. *In* Acetogenesis. H.L. Drake, Ed.: 63–87. New York: Chapman & Hall.
45. RAGSDALE, S.W. 2004. Life with carbon monoxide. Crit. Rev. Biochem. Mol. Biol. **39:** 165–195.
46. WEINBERG, M. & B. GINSBOURG. 1927. Données récéntes sur les microbes anaérobies et leur role en pathologie, pp. 1–291. Paris: Masson et Cie.
47. KÜSEL, K., T. DORSCH, G. ACKER, *et al.* 2000. *Clostridium scatologenes* strain SL1 isolated as an acetogenic bacterium from acidic sediments. Int. J. Syst. Evol. Microbiol. **50:** 537–546.
48. COLLINS, M.D., P.A. LAWSON, A. WILLEMS, *et al.* 1994. The phylogeny of the genus *Clostridium*: proposal of five new genera and eleven new species combinations. Int. J. Syst. Bacteriol. **44:** 812–826.
49. WOOD, H.G. 1991. Life with CO or CO_2 and H_2 as a source of carbon and energy. FASEB J. **5:** 156–63.
50. FERRY, J.G. & C.H. HOUSE. 2006. The stepwise evolution of early life driven by energy conservation. Mol. Biol. Evol. **23:** 1286–1292.
51. RUSSELL, M.J. & W. MARTIN. 2004. The rocky roots of the acetyl-CoA pathway. Trends Biochem. Sci. **29:** 358–363.
52. LESSNER, D.J., L. LI, Q. LI, *et al.* 2006. An unconventional pathway for reduction of CO_2 to methane in CO-grown *Methanosarcina acetivorans* revealed by proteomics. Proc. Natl. Acad. Sci. USA **103:** 17921–17926.
53. MIYAKAWA, S., H. YAMANASHI, K. KOBAYASHI, *et al.* 2002. Prebiotic synthesis from CO atmospheres: implications for the origins of life. Proc. Natl. Acad. Sci. USA **99:** 14628–14631.
54. SCHOUTEN, S., M. STROUS, M.M.M. KUYPERS, *et al.* 2004. Stable carbon isotopic fractionations associated with inorganic carbon fixation by anaerobic ammonium-oxidizing bacteria. Appl. Environ. Microbiol. **70:** 3785–3788.
55. SVETLITCHNYI, V., H. DOBBEK, T. MEINS, *et al.* 2004. A functional Ni-Ni-[4Fe-4S] cluster in the momomeric acetyl-CoA synthase from *Carboxydothermus hydrogenoformans*. Proc. Natl. Acad. Sci. USA **101:** 446–451.
56. SVETLITCHNYI, V., C. PESCHEL, G. ACKER & O. MEYER. 2001. Two membrane-associated NiFeS-carbon monoxide dehydrogenases from the anaerobic carbon-monoxide-utilizing eubacterium *Carboxydothermus hydrogenoformans*. J. Bacteriol. **183:** 5134–5144.
57. HENSTRA, A.M. & A.J.M. STAMS. 2004. Novel physiological features of *Carboxydothermus hydrogenoformans* and *Thermoterrabacterium ferrireducens*. Appl. Environ. Microbiol. **70:** 7236–7240.
58. DIEKERT, G., M. HANSCH & R. CONRAD. 1984. Acetate synthesis from $2CO_2$ in acetogenic bacteria: is carbon monoxide an intermediate? Arch. Microbiol. **138:** 224–228.
59. DIEKERT, G. & M. RITTER. 1983. Purification of the nickel protein carbon monoxide dehydrogenase of *Clostridium thermoaceticum*. FEBS Lett. **151:** 41–44.

60. DIEKERT, G.B. & R.K. THAUER. 1978. Carbon monoxide oxidation by *Clostridium thermoaceticum* and *Clostridium formicoaceticum*. J. Bacteriol. **136:** 597–606.
61. HU, S.I., H.L. DRAKE & H.G. WOOD. 1982. Synthesis of acetyl coenzyme A from carbon monoxide, methyltetrahydrofolate, and coenzyme A by enzymes from *Clostridium thermoaceticum*. J. Bacteriol. **149:** 440–448.
62. DRAKE, H.L., S.I. HU & H.G. WOOD. 1980. Purification of carbon monoxide dehydrogenase, a nickel enzyme from *Clostridium thermocaceticum*. J. Biol. Chem. **255:** 7174–80.
63. DAS, A. & L.G. LJUNGDAHL. 2000. Acetogenesis and acetogenic bacteria. *In* Encyclopedia of Microbiology, 2nd ed., Vol. 1. J. Lederberg, Ed.: 18–27. San Diego, CA: Academic Press.
64. RAGSDALE, S.W. 1991. Enzymology of the acetyl-CoA pathway of CO2 fixation. Crit. Rev. Biochem. Mol. Biol. **26:** 261–300.
65. RAGSDALE, S.W. 1994. CO dehydrogenase and the central role of this enzyme in the fixation of carbon dioxide by anaerobic bacteria. *In* Acetogenesis. H.L. Drake, Ed.: 88–126. New York: Chapman & Hall.
66. DIEKERT, G. & G. WOHLFARTH. 1994. Metabolism of homoacetogens. Antonie Leeuwenhoek **66:** 209–221.
67. RADFAR, R., R. SHIN, G.M. SHELDRICK, *et al.* 2000. The crystal structure of N(10)-formyltetrahydrofolate synthetase from *Moorella thermoacetica*. Biochemistry **39:** 3920–3926.
68. RAGSDALE, S.W. 2000. Nickel containing CO dehydrogenases and hydrogenases. Subcell. Biochem. **35:** 487–518.
69. LINDAHL, P.A. & B. CHANG. 2001. The evolution of acetyl-CoA synthase. Origins Life Evol. Biosphere. **31:** 403–434.
70. MAYNARD, E.L. & P.A. LINDAHL. 2001. Catalytic coupling of the active sites in Acetyl-CoA synthase, a bifunctional CO-channeling enzyme. Biochemistry **40:** 13262–13267.
71. LEAPHART, A.B., H.T. SPENCER & C.R. LOVELL. 2002. Site-directed mutagenesis of a potential catalytic and formyl phosphate binding site and substrate inhibition of N-10-formyltetrahydrofolate synthetase. Arch. Biochem. Biophys. **408:** 137–143.
72. LINDAHL, P.A. 2002. The Ni-containing carbon monoxide dehydrogenase family: light at the end of the tunnel? Biochemistry **41:** 2097–2105.
73. DARNAULT, C., A. VOLBEDA, E.J. KIM, *et al.* 2003. Ni-Zn-[Fe_4-S_4] and Ni-Ni-[Fe_4-S_4] clusters in closed and open a subunits of acetyl-CoA synthase/carbon monoxide dehydrogenase. Nat. Struct. Biol. **10:** 271–279.
74. RAGSDALE, S.W. 2003. Pyruvate ferredoxin oxidoreductase and its radical intermediate. Chem. Rev. **103:** 2333–2346.
75. DAS, A. & L.G. LJUNGDAHL. 1995. Bioenergetics of acetogenic thermophilic clostridia, the homoacetogens—An overview. Biochem. (Life Sci. Adv.) **14:** 33–38.
76. MÜLLER, V. 2003. Energy conservation in acetogenic bacteria. Appl. Environ. Microbiol. **69:** 6345–6353.
77. MÜLLER, V., F. INKAMP, A. RAUWOLF, K. KÜSEL & H.L. DRAKE. 2004. Molecular and cellular biology of acetogenic bacteria. *In* Strict and Facultative Anaerobes: Medical and Environmental Aspects. M. Nakano & P. Zuber, Eds.: 251–281. Norfolk: UK. Horizon Scientific Press.
78. GEERLIGS, G., P. SCHÖNHEIT & G. DIEKERT. 1989. Sodium dependent acetate formation from CO_2 in *Peptostreptococcus productus* (strain Marburg). FEMS Microbiol. Lett. **57:** 253–258.
79. HEISE, R., V. MÜLLER & G. GOTTSCHALK. 1989. Sodium dependence of acetate formation by the acetogenic bacterium *Acetobacterium woodii*. J. Bacteriol. **171:** 5473–5478.
80. YANG, H.C. & H.L. DRAKE. 1990. Differential effects of sodium on hydrogen- and glucose-dependent growth of the acetogenic bacterium *Acetogenium kivui*. Appl. Environ. Microbiol. **56:** 81–86.
81. MÜLLER, V. & S. BOWIEN. 1995. Differential effects of sodium ions on motility in the homoacetogenic bacteria *Acetobacterium woodii* and *Sporomusa sphaeroides*. Arch. Microbiol. **164:** 363–369.
82. TERRACCIANO, J.S., W.J.A. SCHREURS & E.R. KASHKET. 1987. Membrane H^+ conductance of *Clostridium thermoaceticum* and *Clostridium acetobutylicum*: evidence for electrogenic Na^+/H^+ antiport in *Clostridium thermoaceticum*. Appl. Environ. Microbiol. **53:** 782–786.
83. TANNER, R.S., L.M. MILLER & D. YANG. 1993. *Clostridium ljungdahlii* sp. nov., and acetogenic species in clostridial rRNA homology group I. Int. J. Sys. Bacteriol. **43:** 232–236.
84. LIESACK, W., F. BAK, J.-U. KREFT & E. STACKEBRANDT. 1994. *Holophaga foetida* gen. nov., sp. nov., a new, homoacetogenic bacterium degrading methoxylated aromatic compounds. Arch. Microbiol. **162:** 85–90.
85. PIKUTA, E.V., R.B. HOOVER, A.K. BEJ, *et al.* 2003. *Tindallia californiensis* sp. nov., a new anaerobic, haloalkaliphilic, spore-forming acetogen isolated from Mono Lake in California. Extremophiles **7:** 327–334.
86. KAMLAGE, B., B. GRUHL & M. BLAUT. 1997. Isolation and characterization of two new homoacetogenic hydrogen-utilizing bacteria from the human intestinal tract that are closely related to *Clostridium coccoides*. Appl. Environ. Microbiol. **63:** 1732–1738.
87. KÜSEL, K., A. KARNHOLZ, T. TRINKWALTER, *et al.* 2001. Physiological ecology of *Clostridium glycolicum* RD-1, an aerotolerant acetogen isolated from sea grass roots. Appl. Environ. Microbiol. **67:** 4734–4741.
88. OHWAKI, K. & R.E. HUNGATE. 1977. Hydrogen utilization by clostridia in sewage sludge. Appl. Environ. Microbiol. **33:** 1270–1274.
89. LIOU, J.S.-C., D.L. BALKWELL, G.R. DRAKE & R.S. TANNER. 2005. *Clostridium carboxidivorans* sp. nov., a solvent-producing clostridium isolated from an agricultural settling lagoon, and reclassification of the acetogen *Clostridium scatologenes* strain SL1 as *Clostridium drakei* sp. nov. Int. J. Syst. Evol. Microbiol. **55:** 2085–2091.
90. BALCH, W.E., S. SCHOBERTH, R.S. TANNER & R.S. WOLFE. 1977. *Acetobacterium*, a new genus of hydrogen-oxidizing, carbon dioxide-reducing, anaerobic bacteria. Int. J. Syst. Bacteriol. **27:** 355–361.
91. SATTLEY, W.M. & M.T. MADIGAN. 2007. Cold-active acetogenic bacteria from surficial sediments of perennially ice-covered Lake Fryxell, Antarctica. FEMS Microbiol. Lett.**9:** 1836–1841.

92. SIMANKOVA, M.V., O.R. KOTSYURBENKO, E. STACKEBRANDT, *et al.* 2000. *Acetobacterium tundrae* sp. nov., a new psychrophilic acetogenic bacterium from tundra soil. Arch. Microbiol. **174:** 440–447.
93. KUHNER, C.H., C. FRANK, A. GRIESSHAMMER, *et al.* 1997. *Sporomusa silvacetica* sp. nov., an acetogenic bacterium isolated from aggregated forest soil. Int. J. Syst. Bacteriol. **47:** 352–358.
94. DRAKE, H.L. & S.L. DANIEL. 2004. Physiology of the thermophilic acetogen *Moorella thermoacetica*. Res. Microbiol. **155:** 869–883.
95. KRUMHOLZ, L.R., S.H. HARRIS, S.T. TAY & S.M. SUFLITA. 1999. Characterization of two subsurface H2-utilizing bacteria, *Desulfomicrobium hypogeium* sp. nov. and *Acetobacterium psammolithicum* sp. nov., and their ecological roles. Appl. Environ. Microbiol. **65:** 2300–2306.
96. LIU, S. & J.M. SUFLITA. 1993. H_2/CO_2-dependent anaerobic O-demethylation activity in subsurface sediments and by an isolated bacterium. Appl. Environ. Microbiol. **59:** 1325–1331.
97. ZHILINA, T.N., G.A. ZAVARZIN, E.N. DETKOVA & F.A. RAINEY. 1996. *Natroniella acetigena* gen. nov. sp. nov., an extremely halophilic, homoacetogenic bacterium: a new member of *Haloanaerobiales*. Curr. Microbiol. **32:** 320–326.
98. WOLIN, M.J., T.L. MILLER, M.D. COLLINS & P.A. LAWSON. 2003. Formate-dependent growth and homoacetogenic fermentation by a bacterium from human feces: description of *Bryantella formatexigens* gen. nov., sp. nov. Appl. Environ. Microbiol. **69:** 6321–6326.
99. GRABER, J.R. & J.A. BREZNAK. 2004. Physiology and nutrition of *Treponema primitia*, an H_2-CO_2-acetogenic spirochete from termite hindguts. Appl. Environ. Microbiol. **70:** 1307 1314.
100. GRABER, J.R., J.R. LEADBETTER & J.A. BREZNAK. 2004. Description of *Treponema azotonutricium* sp. nov., and *Treponema primitia* sp. nov., the first spirochetes isolated from termite guts. Appl. Environ. Microbiol. **70:** 1315–1320.
101. BARIK, S., PRIETO, S.B. HARRISON, *et al.* 1988. Biological production of alcohols from coal through indirect liquefaction. Appl. Biochem. Biotech. **18:** 363–378.
102. GÖßNER, A.S., K. KÜSEL, G. ACKER, *et al.* 2006. Trophic interaction of the aerotolerant anaerobe *Clostridium intestinale* and the acetogen *Sporomusa rhizae* sp. nov. isolated from roots of the black needlerush *Juncus roemerianus*. Microbiology **152:** 1209–1219.
103. BYRER, D.E., F.A. RAINEY & J. WIEGEL. 2000. Novel strains of *Moorella thermoacetica* form unusually heat-resistant spores. Arch. Microbiol. **174:** 334–339.
104. FERRY, J.G. 1994. CO dehydrogenase of methanogens. *In* Acetogenesis. H.L. Drake, Ed.: 539–556. New York: Chapman & Hall.
105. WHITMAN, W.B. 1994. Autotrophic acetyl coenzyme A biosynthesis in methanogens. *In* Acetogenesis. H.L. Drake, Ed.: 521–538. New York: Chapman & Hall.
106. LEE, M.J. & S.H. ZINDER. 1988. Isolation and characterization of a thermophilic bacterium which oxidizes acetate in syntrophic association with a methanogen and which grows acetogenically on H_2-CO_2. Appl. Environ. Microbiol. **54:** 124–129.
107. HATTORI, S., A.S. GALUSHKO, Y. KAMAGATA & B. SCHINK. 2005. Operation of the CO dehydrogenase/acetyl coenzyme A pathway in both acetate oxidation and acetate formation by the syntrophically acetate-oxidizing bacterium *Thermoacetogenium phaeum*. J. Bacteriol. **187:** 3471–3476.
108. ROTHER, M. & W.W. METCALF. 2004. Anaerobic growth of *Methanosarcina acetivorans* C2A on carbon monoxide: an unusual way of life for a methanogenic archaeon. Proc. Natl. Acad. Sci. USA **101:** 16929–16934.
109. HENSTRA, A.M., C. DIJKEMA & A.J.M. STAMS. 2007. *Archaeoglobus fulgidus* couples CO oxidation to sulfate reduction and acetogenesis with transient formate accumulation. Environ. Microbiol. **9:** 1836–1841.
110. DANIELS, L., G. FUCHS, R.K. THAUER & J.G. ZEIKUS. 1977. Carbon monoxide oxidation by methanogenic bacteria. J. Bacteriol. **132:** 118–126.
111. SAVAGE, M.D., Z.G. WU, S.L. DANIEL, *et al.* 1987. Carbon monoxide-dependent chemolithotrophic growth of *Clostridium thermoautotrophicum*. Appl. Environ. Microbiol. **53:** 1902–1906.
112. HSU, T., S.L. DANIEL, M.F. LUX & H.L. DRAKE. 1990. Biotransformations of carboxylated aromatic compounds by the acetogen *Clostridium thermoaceticum*: generation of growth-supportive CO_2 equivalents under CO2-limited conditions. J. Bacteriol. **172:** 212–217.
113. MARTIN, D.R., A. MISRA & H.L. DRAKE. 1985. Dissimilation of carbon monoxide to acetic acid by glucose-limited cultures of *Clostridium thermoaceticum*. Appl. Environ. Microbiol. **49:** 1412–1417.
114. HSU, T.D., M.F. LUX & H.L. DRAKE. 1990. Expression of an aromatic-dependent decarboxylase which provides growth-essential CO2 equivalents for the acetogenic (Wood) pathway of *Clostridium thermoaceticum*. J. Bacteriol. **172:** 5901–5907.
115. BRAUS-STROMEYER, S.A., G. SCHNAPPAUF, G.H. BRAUS, *et al.* 1997. Carbonic anhydrase in *Acetobacterium woodii* and other acetogenic bacteria. J. Bacteriol. **179:** 7197–7200.
116. FRÖSTL, J.M., C. SEIFRITZ & H.L. DRAKE. 1996. Effect of nitrate on the autotrophic metabolism of the acetogens *Clostridium thermoautotrophicum* and *Clostridium thermoaceticum*. J. Bacteriol. **178:** 4597–4603.
117. SEIFRITZ, C., S.L. DANIEL, A. GÖßNER & H.L. DRAKE. 1993. Nitrate as a preferred electron sink for the acetogen *Clostridium thermoaceticum*. J. Bacteriol. **175:** 8008–8013.
118. SLOBODKIN, A., A.L. REYSENBACH, F. MAYER & J. WIEGEL. 1997. Isolation and characterization of the homoacetogenic thermophilic bacterium *Moorella glycerini* sp. nov. Int. J. Syst. Bacteriol. **47:** 969–974.
119. THAUER, R.K., K. JUNGERMANN & K. DECKER. 1977. Energy conservation in chemotrophic anaerobic bacteria. Bacteriol. Rev. **41:** 100–180.
120. SEIFRITZ, C., H.L. DRAKE & S.L. DANIEL. 2003. Nitrite as an energy-conserving electron sink for the acetogenic bacterium *Moorella thermoacetica*. Curr. Microbiol. **46:** 329–333.

121. Arendsen, A.F., M.Q. Soliman & S.W. Ragsdale. 1999. Nitrate-dependent regulation of acetate biosynthesis and nitrate respiration by *Clostridium thermoaceticum*. J. Bacteriol. **181:** 1489–1495.

122. Drake, H.L., S.L. Daniel, K. Küsel, *et al.* 1997. Acetogenic bacteria: What are the in situ consequences of their diverse metabolic versatilities? Biofactors **6:** 13–24.

123. Misoph, M., S.L. Daniel & H.L. Drake. 1996. Bidirectional usage of ferulate by the acetogen *Peptostreptococcus productus* U-1: CO_2 and aromatic acrylate groups as competing electron acceptors. Microbiology **142:** 1983–1988.

124. Misoph, M. & H.L. Drake. 1996. Effect of CO_2 on the fermentation capacities of the acetogen *Peptostreptococcus productus* U-1. J. Bacteriol. **178:** 3140–3145.

125. Dilling, S., F. Inkamp, S. Schmidt & V. Müller. 2007. Regulation of caffeate respiration in the acetogenic bacterium *Acetobacterium woodii*. Appl. Environ. Microbiol. **73:** 3630–3636.

126. Winter, J.U. & R.S. Wolfe. 1980. Methane formation from fructose by syntrophic associations of *Acetobacterium woodii* and different strains of methanogens. Arch. Microbiol. **124:** 73–39.

127. Bernalier, A., M. Lelait, V. Rochet, *et al.* 1996. Acetogenesis from H_2 and CO_2 by methane- and non-methane-producing human colonic bacterial communities. FEMS Microbiol. Immunol. **19:** 193–202.

128. Doré, J., P. Pochart, A. Bernalier, *et al.* 1995. Enumeration of H_2-utilizing methanogenic archaea, acetogenic and sulfate-reducing bacteria from human feces. FEMS Microbiol. Ecol. **17:** 279–284.

129. Leadbetter, J.R., T.M. Schmidt, J.R. Graber & J.A. Breznak. 1999. Acetogenesis from H_2 plus CO_2 by spirochetes from termite guts. Science **283:** 686–689.

130. Schink, B. 1994. Diversity, ecology, and isolation of acetogenic bacteria. *In* Acetogenesis. H.L. Drake, Ed.: 197–235. New York: Chapman & Hall.

131. Ljungdahl, L.G. 1983. Formation of acetate using homoacetate fermenting anaerobic bacteria. *In* Organic Chemicals from Biomass. D.L. Wise, Ed.: 219–248. Menlo Park, CA: Benjamin/Cummings Publishing Company.

132. Wood, H.G. 1952. Fermentation of 3,4-C^{14}- and 1-C^{14}-labeled glucose by *Clostridium thermoaceticum*. J. Biol. Chem. **199:** 579–583.

133. Karita, S., K. Nakayama, M. Goto, *et al.* 2003. A novel cellulolytic, anaerobic, and thermophilic bacterium, *Moorella* sp. strain F21. Biosci. Biotechnol. Biochem. **67:** 183–185.

134. Bak, F., K. Finster, F. Rothfuss. 1992. Formation of dimethylsulfide and methanethiol from methoxylated aromatic compounds and inorganic sulfide by newly isolated anaerobic bacteria. Arch. Microbiol. **157:** 529–534.

135. Kreft, J.U. & B. Schink. 1993. Demethylation and degradation of phenylmethylethers by the sulfide-methylating homoacetogenic bacterium strain TMBS 4. Arch. Microbiol. **159:** 308–315.

136. Kellum, R. & H.L. Drake. 1984. Effects of cultivation gas phase on hydrogenase of the acetogen *Clostridium thermoaceticum*. J. Bacteriol. **160:** 466–469.

137. Drake, H.L. 1982. Demonstration of hydrogenase in extracts of the homoacetate-fermenting bacterium *Clostridium thermoaceticum*. J. Bacteriol. **150:** 702–709.

138. Daniel, S.L. & H.L. Drake. 1993. Oxalate- and glyoxylate-dependent growth and acetogenesis by *Clostridium thermoaceticum*. Appl. Environ. Microbiol. **59:** 3062–3069.

139. Seifritz, C., J.M. Frostl, H.L. Drake & S.L. Daniel. 1999. Glycolate as a metabolic substrate for the acetogen *Moorella thermoacetica*. FEMS Microbiol. Lett. **170:** 399–405.

140. Daniel, S.L., C. Pilsl & H.L. Drake. 2004. Oxalate metabolism by the acetogenic bacterium *Moorella thermoacetica*. FEMS Microbiol. Lett. **231:** 39–43.

141. Seifritz, C., J.M. Fröstl, H.L. Drake & S.L. Daniel. 2002. Influence of nitrate on oxalate- and glyoxylate-dependent growth and acetogenesis by *Moorella thermoacetica*. Arch. Microbiol. **178:** 457–464.

142. Küsel, K., C. Wagner& H.L. Drake. 1999. Enumeration and metabolic product profiles of the anaerobic microflora in the mineral soil and litter of a beech forest. FEMS Microbiol. Ecology **29:** 91–103.

143. Küsel, K., H.C. Pinkart, H.L. Drake & R. Devereux. 1999. Acetogenic and sulfate-reducing bacteria inhabiting the rhizoplane and deep cortex cells of the sea grass *Halodule wrightii*. Appl. Environ. Microbiol. **65:** 5117–5123.

144. Das, A., E.D. Coulter, D.M. Kurtz & L.G. Ljungdahl. 2001. Five-gene cluster in *Clostridium thermoaceticum* consisting of two divergent operons encoding rubredoxin oxidoreductase-rubredoxin and rubrerythrin-type A flavoprotein-high-molecular-weight rubredoxin. J. Bacteriol. **183:** 1560–1567.

145. Karnholz, A., K. Küsel, A. Gößner, *et al.* 2002. Tolerance and metabolic response of acetogenic bacteria toward oxygen. Appl. Environ. Microbiol. **68:** 1005–1009.

146. Silaghi-Dumitrescu, R., E.D. Coulter, A. Das, *et al.* 2003. A flavodiiron protein and high molecular weight rubredoxin from *Moorella thermoacetica* with nitric oxide reductase activity. Biochemistry **42:** 2806–2815.

147. Das, A., R. Silaghi-Dumitrescu, L.G. Ljungdahl & D.M. Kurtz, Jr. 2005. Cytochrome *bd* oxidase, oxidative stress, and dioxygen tolerance of the strictly anaerobic bacterium *Moorella thermoacetica*. J. Bacteriol. **187:** 2020–2029.

148. Busche, R.M. 1991. Extractive fermentation of acetic acid: economic tradeoff between yield of *Clostridium* and concentration of *Acetobacter*. Appl. Biochem. Biotechnol. **28–29:** 605–621.

149. Cheryan, M., S. Parekh, M. Shah & K. Witjitra. 1997. Production of acetic acid by *Clostridium thermoaceticum*. Adv. Appl. Microbiol. **43:** 1–33.

150. Ljungdahl, L.G., J. Hugenholtz & J. Wiegel. 1989. Acetogenic and acid-producing clostridia. *In* Clostridia. N.P. Minton & D.J. Clarke, Eds.: 145–191. New York: Plenum Press.

151. LJUNGDAHL, L.G., L.H. CARREIRA, R.J. GARRISON, *et al.* 1985. Comparison of three thermophilic acetogenic bacteria for production of calcium magnesium acetate. Biotechnol. Bioeng. Symp. **15:** 207–223.
152. SCHWARTZ, R.D. & F.A. KELLER, JR. 1982. Isolation of a strain of *Clostridium thermoaceticum* capable of growth and acetic acid production at pH 4.5. Appl. Environ. Microbiol. **43:** 117–123.
153. SHAH, M.M. & M. CHERYAN. 1995. Improvement of productivity in acetic acid fermentation with *Clostridium thermoaceticum*. Appl. Biochem. Biotechnol. **51–52:** 413–422.
154. WIEGEL, J., L.H. CARREIRA, R.J. GARRISON, *et al.* 1990. Calcium magnesium acetate (CMA) manufacture from glucose by fermentation with thermophilic homoacetogenic bacteria. *In* Calcium Magnesium Acetate. D.L. Wise, Y. Levendis & M. Metghalchi, Eds.: 359–416. Amsterdam, The Netherlands: Elsevier.
155. WIEGEL, J. 1994. Acetate and the potential of homoacetogenic bacteria for industrial applications. *In* Acetogenesis. H.L. Drake, Ed.: 484–504. New York: Chapman & Hall.
156. GRETHLEIN, A.J. & M.K. JAIN. 1992. Bioprocessing of coal-derived synthesis gases by anaerobic bacteria. Trends Biotechnol. **10:** 418–423.
157. PHILLIPS, J.R., E.C. CLAUSEN & J.L. GADDY. 1994. Synthesis gas as substrate for the biological production of fuels and chemicals. Appl. Biochem. Biotechnol. **45/46:** 145–157.
158. ABRINI, J., H. NAVEAU & E.-J. NYNS. 1994. *Clostridium autoethanogenum*, sp. nov., an anaerobic bacterium that produces ethanol from carbon monoxide. Arch. Microbiol. **161:** 345–351.
159. BARONOFSKY, J.J., W.J.A. SCHREURS & E.R. KASHKET. 1984. Uncoupling by acetic acid limits growth of and acetogenesis by *Clostridium thermoaceticum*. Appl. Environ. Microbiol. **48:** 1134–1139.
160. LE RUYET, P., H.C. DUBOURGUIER & G. ALBAGNAC. 1984. Homoacetogenic fermentation of cellulose by a coculture of *Clostridium thermocellum* and *Acetogenium kivui*. Appl. Environ. Microbiol. **48:** 893–894.
161. MILLER, T.L. & M.J. WOLIN. 1995. Bioconversion of cellulose to acetate with pure cultures of *Ruminococcus albus* and a hydrogen-using acetogen. Appl. Environ. Microbiol. **61:** 3832–3835.
162. RAVINDER, T., M.V. SWAMY, G. SEENAYYA & G. REDDY. 2001. *Clostridium lentocellum* SG6—A potential organism for fermentation of cellulose to acetic acid. Bioresour. Technol. **80:** 171–177.
163. EGGEMAN, T. & D. VERSER. 2005. Recovery of organic acids from fermentation broths. Appl. Biochem. Biotechnol. **121–124:** 605–618.
164. EGGEMAN, T. & D. VERSER. 2006. The importance of utility systems in today's biorefineries and a vision for tomorrow. Appl. Biochem. Biotechnol. **129–132:** 361–381.
165. EGLI, C., T. TSCHAN, R. SCHOLTZ, *et al.* 1988. Transformation of tetrachloromethane to dichloromethane and carbon dioxide by *Acetobacterium woodii*. Appl. Environ. Microbiol. **54:** 2819–2824.
166. TRAUNECKER, J., A. PREUß & G. DIEKERT. 1991. Isolation and characterization of a methyl chloride utilizing, strictly anaerobic bacterium. Arch. Microbiol. **156:** 416–421.
167. DUHAMEL, M. & E.A. EDWARDS. 2007. Growth and yields of dechlorinators, acetogens, and methanogens during reductive dechlorination of chlorinated ethenes and dihaloelimination of 1,2-dicholoroethane. Environ. Sci. Technol. **41:** 2303–2310.
168. LEBLOAS, P., P. LOUBIERE & N.D. LINDLEY. 1994. Use of unicarbon substrate mixtures to modify carbon flux improves vitamin B12 production with the acetogenic methylotroph *Eubacterium limosum*. Biotechnol. Lett. **16:** 129–132.
169. SCHULZ, M., H. LEICHMANN, H. GÜNTHER & H. SIMON. 1995. Electromicrobial regeneration of pyridine nucleotides and other preparative redox transformations with *Clostridium thermoaceticum*. Appl. Microbiol. Biotechnol. **42:** 916–922.
170. KOESNANDAR, S. AGO, N. NISHIO & S. NAGAI. 1989. Production of extracellular 5-aminoleuvulinic acid by *Clostridium thermoaceticum* grown in minimal medium. Biotechnol. Lett. **11:** 567–572.
171. KOESNANDAR, N. NISHIO, A. YAMAMOTO & S. NAGAI. 1991. Enzymatic reduction of cystine into cysteine by cell-free extract of *Clostridium thermoaceticum*. J. Ferment. Bioeng. **72:** 11–14.
172. BREZNAK, J.A. & M.D. KANE. 1990. Microbial H_2/CO_2 acetogenesis in animal guts: nature and nutritional significance. FEMS Microbiol. Rev. **7:** 309–313.
173. GRABER, J.R. & J.A. BREZNAK. 2005. Folate cross-feeding supports symbiotic homoacetogenic spirochetes. Appl. Environ. Microbiol. **71:** 1883–1889.
174. LOVELL, C.R. 1994. Development of DNA probes for the detection and identification of acetogenic bacteria. *In* Acetogenesis. H.L. Drake, Ed.: 236–253. New York: Chapman & Hall.
175. LOVELL, C.R. & A.B. LEAPHART. 2005. Community-level analysis: key genes of CO_2-reductive acetogenesis. Methods Enzymol. **397:** 454–469.
176. SALMASSI, T.M. & J.R. LEADBETTER. 2003. Analysis of genes of tetrahydrofolate-dependent metabolism from cultivated spirochaetes and the gut community of the termite *Zootermopsis angusticollis*. Microbiology **149:** 2529–2537.
177. PESTER, M. & A. BRUNE. 2006. Expression profiles of fhs (FTHFS) genes support the hypothesis that spirochaetes dominate reductive acetogenesis in the hindgut of lower termites. Appl. Environ. Microbiol. **8:** 1261–1270.
178. LEAPHART, A.B., M.J. FRIEZ & C.R. LOVELL. 2003. Formyltetrahydrofolate synthetase sequences from salt marsh plant roots reveal a diversity of acetogenic bacteria and other bacterial functional groups. Appl. Environ. Microbiol. **69:** 693–696.
179. KOTELNIKOVA, S. 2002. Microbial production and oxidation of methane in deep subsurface. Earth Sci. Rev. **58:** 367–395.
180. KRUMHOLZ, L.R. 2000. Microbial communities in the deep subsurface. Hydrogeol. J. **8:** 4–10.
181. KUHNER, C.H., C. MATTHIES, G. ACKER, *et al.* 2000. *Clostridium akagii* sp. nov. and *Clostridium acidisoli* sp. nov.:

acid-tolerant, N2-fixing clostridia isolated from acidic forest soil and litter. Int. J. Syst. Evol. Microbiol. **50:** 873–881.

182. MATTHIES, C., C.H. KUHNER, G. ACKER & H.L. DRAKE. 2001. *Clostridium uliginosum* sp nov., a novel acid-tolerant anaerobic bacterium with connecting filaments. Int. J. Syst. Evol. Microbiol. **51:** 1119–1125.
183. GREENING, R.C. & J.A.Z. LEEDLE. 1989. Enrichment and isolation of *Acetitomaculum ruminis*, gen. nov., sp. nov.: acetogenic bacteria from the bovine rumen. Arch. Microbiol. **151:** 399–406.
184. SLEAT, R., R.A. MAH & R. ROBINSON. 1985. *Acetoanaerobium noterae* gen. nov., sp. nov.: an anaerobic bacterium that forms acetate from H_2 and CO_2. Int. J. Syst. Bacteriol. **35:** 10–15.
185. DAVYDOVA-CHARAKHCHYAN, I.A., A.N. MILEEVA, L.L. MITYUSHINA & S.S. BELYAEV. 1992. Acetogenic bacteria from oil fields of Tataria and western Siberia. Mikrobiologiya **61:** 306–315.
186. KOTSYURBENKO, O.R., M.V. SIMANKOVA, A.N. NOZHEVNIKOVA, *et al.* 1995. New species of psychrophilic acetogens: *Acetobacterium bakii* sp. nov., *A. paludosum* sp. nov., *A. fimetarium* sp. nov. Arch. Microbiol. **163:** 29–34.
187. EICHLER, B. & B. SCHINK. 1984. Oxidation of primary aliphatic alcohols by *Acetobacterium carbinolicum* sp. nov., a homoacetogenic anaerobe. Arch. Microbiol. **140:** 147–152.
188. TANAKA, K. & N. PFENNIG. 1988. Fermentation of 2-methoxyethanol by *Acetobacterium malicum* sp. nov. and *Pelobacter venetianus*. Arch. Microbiol. **149:** 181–187.
189. BRAUN, M. & G. GOTTSCHALK. 1982. *Acetobacterium wieringae* sp. nov., a new species producing acetic acid from molecular hydrogen and carbon dioxide. Zentralbl.Bakteriol.Hyg., I. Abt.Originale C 3368–376.
190. DÖRNER, C. & B. SCHINK. 1991. Fermentation of mandelate to benzoate and acetate by a homoacetogenic bacterium. Arch. Microbiol. **156:** 302–306.
191. SEMBIRING, T. & J. WINTER. 1989. Anaerobic degradation of o-phenylphenol by mixed and pure cultures. Appl. Microbiol. Biotechnol. **31:** 89–92.
192. SEMBIRING, T. & J. WINTER. 1990. Demethylation of aromatic compounds by strain B10 and complete degradation of 3-methoxybenzoate in co-culture with *Desulfosarcina* strains. Appl. Microbiol. Biotechnol. **33:** 233–238.
193. SCHRAMM, E. & B. SCHINK. 1991. Ether-cleaving enzyme and diol dehydratase involved in anaerobic polyethylene glycol degradation by a new *Acetobacterium* sp. Biodegradation **2:** 71–79.
194. CONRAD, R., F. BAK, H.J. SEITZ, B. THEBRATH, *et al.* 1989. Hydrogen turnover by psychrotrophic homoacetogenic and mesophilic methanogenic bacteria in anoxic paddy soil and lake sediment. FEMS Microbiol. Ecol. **62:** 285–294.
195. WAGENER, S. & B. SCHINK. 1988. Fermentative degradation of nonionic surfactants and polyethylene glycol by enrichment cultures and by pure cultures of homoacetogenic and propionate-forming bacteria. Appl. Environ. Microbiol. **54:** 561–565.
196. FRINGS, J. & B. SCHINK. 1994. Fermentation of phenoxyethanol to phenol and acetate by a homoacetogenic bacterium. Arch. Microbiol. **162:** 199–204.
197. FRINGS, J., C. WONDRAK & B. SCHINK. 1994. Fermentative degradation of triethanolamine by a homoacetogenic bacterium. Arch. Microbiol. **162:** 103–107.
198. EMDE, R. & B. SCHINK. 1987. Fermentation of triacetin and glycerol by *Acetobacterium* sp. No energy is conserved by acetate excretion. Arch. Microbiol. **149:** 142–148.
199. SCHUPPERT, B. & B. SCHINK. 1990. Fermentation of methoxyacetate to glycolate and acetate by newly isolated strains of Acetobacterium sp. Arch. Microbiol. **153:** 200–204.
200. INOUE, K., S. KAGEYAMA, K. MIKI, *et al.* 1992. Vitamin B12 production by *Acetobacterium* sp. and its tetrachloromethane-resistant mutants. J. Ferment. Bioeng. **73:** 76–78.
201. KOTSYURBENKO, O.R., A.N. NOZHEVNIKOVA, T.I. SOLOVIOVA & G.A. ZAVARIN. 1996. Methanogenesis at low temperatures by microflora of tundra wetland soil. Antonie Leeuwenhoek **69:** 75–86.
202. ZHILINA, T.N. & G.A. ZAVARZIN. 1990. Extremely halophilic, methylotrophic, anaerobic bacteria. FEMS Microbiol. Rev. **87:** 315–322.
203. KANE, M.D. & J.A. BREZNAK. 1991. *Acetonema longum* gen. nov. sp. nov., an H_2/CO_2 acetogenic bacterium from the termite, *Pterotermes occidentis*. Arch. Microbiol. **156:** 91–98.
204. ZEIKUS, J.G., L.H. LYND, T.E. THOMPSON, *et al.* 1980. Isolation and characterization of a new, methylotrophic, acidogenic anaerobe, the Marburg strain. Curr. Microbiol. **3:** 381–386.
205. PATEL, B.K.C., C. MONK, H. LITTLEWORTH, *et al.* 1987. *Clostridium fervidus* sp. nov., a new chemoorganotrophic acetogenic thermophile. Int. J. Syst. Bacteriol. **37:** 123–126.
206. KANEUCHI, C., Y. BENNO & T. MITSUOKA. 1976. *Clostridium coccoides*, a new species from the feces of mice. Int. J. Syst. Bacteriol. **26:** 482–486.
207. RIEU-LESME, F., C. DAUGA, G. FONTY & J. DORÉ. 1998. Isolation from the rumen of a new acetogenic bacterium phylogenetically closely related to *Clostridium difficile*. Anaerobe **4:** 89–94.
208. SCHINK, B. 1984. *Clostridium magnum* sp. nov., a non-autotrophic homoacetogenic bacterium. Arch. Microbiol. **137:** 250–255.
209. MECHICHI, T., M. LABAT, T.H.S. WOO, *et al.* 1998. *Eubacterium aggregans* sp nov., a new homoacetogenic bacterium from olive mill wastewater treatment digestor. Anaerobe **4:** 283–291.
210. SCHNÜRER, A., B. SCHINK & B.H. SVENSSON. 1996. *Clostridium ultunense* sp. nov., a mesophilic bacterium oxidizing acetate in syntrophic association with a hydrogenotrophic methanogenic bacterium. Int. J. Syst. Bacteriol. **46:** 1145–1152.
211. ADAMSE, A.D. & C.T.M. VELZEBOER. 1982. Features of a *Clostridium*, strain CV-AA1, an obligatory anaerobic bacterium producing acetic acid from methanol. Antonie Leeuwenhoek **48:** 305–313.

212. LECLERC, M., A. BERNALIER, G. DONADILLE & M. LELAIT. 1997. H_2/CO_2 metabolism in acetogenic bacteria isolted from the human colon. Anaerobe **3:** 307–315.
213. LECLERC, M., A. BERNALIER, M. LELAIT & J.-P. GRIVET 1997. ^{13}C-NMR study of glucose and pyruvate catabolism in four acetogenic species isolated from the human colon. FEMS Microbiol. Lett. **146:** 199–204.
214. SHARAK GENTHNER, B.R., C.L. DAVIES & M.P. BRYANT. 1981. Features of rumen and sewage sludge strains of *Eubacterium limosum*, a methanol- and H_2-CO_2-utilizing species. Appl. Environ. Microbiol. **42:** 12–19.
215. BALK, M., J. WEIJMA, M.W. FRIEDRICH & A.J. STAMS. 2003. Methanol utilization by a novel thermophilic homoacetogenic bacterium, *Moorella mulderi* sp. nov., isolated from a bioreactor. Arch. Microbiol. **179:** 315–320.
216. WIEGEL, J., M. BRAUN & G. GOTTSCHALK. 1981. *Clostridium thermoautotrophicum* species novum, a thermophile producing acetate from molecular hydrogen and carbon dioxide. Curr. Microbiol. **5:** 255–260.
217. SAKAI, S., Y. NAKASHIMADA, H. YOSHIMOTO, *et al.* 2004. Ethanol production from H2 and CO2 by a newly isolated thermophilic bacterium, *Moorella* sp. HUC22-1. Biotechnol. Lett. **26:** 1607–1612.
218. ZHILINA, T.N., E.N. DETKOVA, F.A. RAINEY, *et al.* 1998. *Natronoincola histidinovorans* gen. nov., sp. nov., a new alkaliphilic acetogenic anaerobe. Curr. Microbiol. **37:** 177–185.
219. KRUMHOLZ, L.R. & M.P. BRYANT. 1985. *Clostridium pfennigii* sp. nov. uses methoxyl groups of monobenzenoids and produces butyrate. Int. J. Syst. Bacteriol. **35:** 454–456.
220. BERNALIER, A., V. ROCHET, M. LECLERC, *et al.* 1996. Diversity of H_2/CO_2-utilizing acetogenic bacteria from feces of non-methane-producing humans. Curr. Microbiol. **33:** 94–99.
221. BERNALIER, A., A. WILLEMS, M. LECLERC, *et al.* 1996. *Ruminococcus hydrogenotrophicus* sp. nov., a new H2/CO2-utilizing acetogenic bacterium isolated from human feces. Arch. Microbiol. **166:** 176–183.
222. LOROWITZ, W.H. & M.P. BRYANT. 1984. *Peptostreptococcus productus* strain that grows rapidly with CO as the energy source. Appl. Environ. Microbiol. **47:** 961–964.
223. GEERLIGS, G., H.C. ALDRICH, W. HARDER & G. DIEKERT. 1987. Isolation and characterization of a carbon monoxide utilizing strain of the acetogen *Peptostreptococcus productus*. Arch. Microbiol. **148:** 305–313.
224. RIEU-LESME, F., B. MORVAN, M.D. COLLINS, *et al.* 1996. A new H_2/CO_2-using acetogenic bacterium from the rumen: description of *Ruminococcus schinkii* sp. nov. FEMS Microbiol. Lett. **140:** 281–286.
225. OLLIVIER, B., R. CORDRUWISCH, A. LOMBARDO & J.-L. GARCIA. 1985. Isolation and characterization of *Sporomusa acidovorans* sp. nov., a methylotrophic homoacetogenic bacterium. Arch. Microbiol. **142:** 307–310.
226. BOGA, H.I., W. LUDWIG & A. BRUNE. 2003. *Sporomusa aerivorans* sp. nov., an oxygen-reducing homoacetogenic bacterium from the gut of a soil-feeding termite. Int. J. Syst. Evol. Microbiol. **53:** 1397–1404.
227. DEHNING, I., M. STIEB & B. SCHINK. 1989. *Sporomusa malonica* sp. nov., a homoacetogenic bacterium growing by decarboxylation of malonate or succinate. Arch. Microbiol. **151:** 421–426.
228. MÖLLER, B., R. OßMER, B.H. HOWARD, *et al.* 1984. *Sporomusa*, a new genus of gram-negative anaerobic bacteria including *Sporomusa sphaeroides* spec. nov. and *Sporomusa ovata* spec. nov. Arch. Microbiol. **139:** 388–396.
229. HERMANN, M., M.-R. POPOFF & M. SEBALD. 1987. *Sporomusa paucivorans* sp. nov., a methylotrophic bacterium that forms acetic acid from hydrogen and carbon dioxide. Int. J. Syst. Bacteriol. **37:** 93–101.
230. BREZNAK, J.A., J.M. SWITZER & H.J. SEITZ. 1988. *Sporomusa termitida* sp. nov., an H2/CO2-utilizing acetogen isolated from termites. Arch. Microbiol. **150:** 282–288.
231. ROSENCRANTZ, D., F.A. RAINEY & P.H. JANSSEN. 1999. Culturable populations of *Sporomusa* spp. and *Desulfovibrio* spp. in the anoxic bulk soil of flooded rice microcosms. Appl. Environ. Microbiol. **65:** 3526–3533.
232. KRUMHOLZ, L.R. & M.P. BRYANT. 1986. *Syntrophococcus sucromutans* sp. nov. gen. nov. uses carbohydrates as electron donors and formate, methoxymonobenzenoids or *Methanobrevibacter* as electron acceptor systems. Arch. Microbiol. **143:** 313–318.
233. HATTORI, S., Y. KAMAGATA, S. HANADA & H. SHOUN. 2000. *Thermoacetogenium phaeum* gen. nov., sp. nov., a strictly anaerobic, thermophilic, syntrophic acetate-oxidizing bacterium. Int. J. Syst. Evol. Microbiol. **50:** 1601–1609.
234. LEIGH, J.A., F. MAYER & R.S. WOLFE. 1981. *Acetogenium kivui*, a new thermophilic hydrogen-oxidizing, acetogenic bacterium. Arch. Microbiol. **129:** 275–280.
235. DAVIDOVA, I.A. & A.J.M. STAMS. 1996. Sulfate reduction with methanol by a thermophilic consortium obtained from a methanogenic reactor. Appl. Microbiol. Biotechnol. **46:** 297–302.
236. WOLIN, M.J. & T.L. MILLER. 1993. Bacterial strains from human feces that reduce CO_2 to acetic acid. Appl. Environ. Microbiol. **59:** 3551–3556.
237. RIEU-LESME, F., G. FONTY & J. DORÉ. 1995. Isolation and characterization of a new hydrogen-utilizing bacterium from the rumen. FEMS Microbiol. Lett. **125:** 77–82.
238. PLUGGE, C.M., J.T.C. GROTENHUIS & A.J.M. STAMS. 1990. Isolation and characterization of an ethanol-degrading anaerobe from methanogenic granular sludge. *In* Microbiology and Biochemistry of Strict Anaerobes Involved in Interspecies Hydrogen Transfer. J.-P. Belaich, M. Bruschiand & J.-L. Garcia, Eds.: 439–442. FEMS Symposium No. 54. New York: Plenum Press.
239. WOLIN, M.J. & T.L. MILLER. 1994. Acetogenesis from CO2 in the human colonic ecosystem. *In* Acetogenesis. H.L. Drake, Ed.: 365–385. New York: Chapman & Hall.
240. CHAUCHEYRAS, F., G. FONTY, G. BERTIN & P. GOUET. 1995. In vitro H_2 utilization by a ruminal acetogenic bacterium cultivated alone or in association with an archaea methanogen is stimulated by a probiotic strain of *Saccharomyces cerevisiae*. Appl. Environ. Microbiol. **61:** 3466–3467.
241. FRAZER, A.C. & L.Y. YOUNG. 1985. A gram-negative anaerobic bacterium that utilizes O-methyl substituents of aromatic acids. Appl. Environ. Microbiol. **49:** 1345–1347.

242. SAMAIN, E., G. ALBANGNAC, H.C. DUBOURGUIER & J.-P. TOUZEL. 1982. Characterization of a new propionic acid bacterium that ferments ethanol and displays a growth factor-dependent association with a gram-negative homoacetogen. FEMS Microbiol. Lett. **15:** 69–74.

243. KOTSYURBENKO, O.R., M.V. SIMANKOVA, N.P. BOLOTINA, *et al.* 1992. Psychrotrophic homoacetogenic bacteria from several environments. 7th Int. Symp. C1 Compounds, Abstr. C136.

244. NOZHEVNIKOVA, A., O.R. KOTSYURBENKO & M.V. SIMANKOVA. 1994. Acetogenesis at low temperature. *In* Acetogenesis. H.L. Drake, Ed.: 416–431. New York: Chapman & Hall.

245. RIEU-LESME, F., C. DAUGA, B. MORVAN, *et al.* 1996. Acetogenic coccoid spore-forming bacteria isolated from the rumen. Res. Microbiol. **147:** 753–764.

246. DRAKE, H.L. 1992. Acetogenesis and acetogenic bacteria. *In* Encyclopedia of Microbiology, Vol. 1. J. Lederberg, Ed.: 1–15. San Diego, CA: Academic Press.

247. BARKER, H.A. 1944. On the role of carbon dioxide in the metabolism of *Clostridium thermoaceticum.* Proc. Natl. Acad. Sci. USA **30:** 88–90.

248. LENTZ, K. & H.G. WOOD. 1955. Synthesis of acetate from formate and carbon dioxide by Clostridium thermoaceticum. J. Biol. Chem. **215:** 645–654.

249. POSTON, J.M., E.R. STADTMAN & K. KURATOMI. 1964. Methyl-vitamin B_{12} as source of methyl groups for synthesis of acetate by cell-free extracts of *Clostridium thermoaceticum.* Ann. N.Y. Acad. Sci. **112:** 804–806.

250. LJUNGDAHL, L. & H.G. WOOD. 1965. Incorporation of C^{14} from carbon dioxide into sugar phosphates, carboxylic acids, and amino acids by *Clostridium thermoaceticum.* J. Bacteriol. **89:** 1055–1064.

251. LJUNGDAHL, L., E. IRION & H.G. WOOD. 1966. Role of corrinoids in the total synthesis of acetate from CO_2 by *Clostridium thermoaceticum.* Fed. Proc. **25:** 1642–1648.

252. LJUNGDAHL, L.G. & H.G. WOOD. 1969. Total synthesis of acetate from CO_2 by heterotrophic bacteria. Annu. Rev. Microbiol. **23:** 515–538.

253. DRAKE, H.L., S.I. HU & H.G. WOOD. 1981. Purification of five components from *Clostridium thermoaceticum* which catalyze synthesis of acetate from pyruvate and methyltetrahydrofolate. Properties of phosphotransacetylase. J. Biol. Chem. **256:** 11137–11144.

254. DRAKE, H.L., S.-I. HU & H.G. WOOD. 1981. The synthesis of acetate from carbon monoxide plus methyltetrahydrofolate and the involvement of the nickel enzyme, CO dehydrogenase. Abstr. Ann. Meet. Am. Soc. Microbiol., p. 144, Abstr. K-42.

255. RAGSDALE, S.W., J.E. CLARK, L.G. LJUNGDAHL, *et al.* 1983. Properties of purified carbon monoxide dehydrogenase from *Clostridium thermoaceticum,* a nickel, iron-sulfur protein. J. Biol. Chem. **258:** 2364–2369.

256. KERBY, R. & J.G. ZEIKUS. 1983. Growth of *Clostridium thermoaceticum* on H_2/CO_2 or CO as energy source. Curr. Microbiol. **8:** 27–30.

257. LUNDIE, L.L., JR. & H.L. DRAKE. 1984. Development of a minimally defined medium for the acetogen *Clostridium thermoaceticum.* J. Bacteriol. **159:** 700–7003.

258. PEZACKA, E. & H.G. WOOD. 1984. The synthesis of acetyl-CoA by *Clostridium thermoaceticum* from carbon dioxide, hydrogen, coenzyme A and methyltetrahydrofolate. Arch. Microbiol. **137:** 63–69.

259. PEZACKA, E. & H.G. WOOD. 1984. Role of carbon monoxide dehydrogenase in the autotrophic pathway used by acetogenic bacteria. Proc. Natl. Acad. Sci. USA **81:** 6261–6265.

260. RAGSDALE, S.W. & H.G. WOOD. 1985. Acetate biosynthesis by acetogenic bacteria. Evidence that carbon monoxide dehydrogenase is the condensing enzyme that catalyzes the final steps of the synthesis. J. Biol. Chem. **260:** 3970–3977.

261. IVEY, D.M. & L.G. LJUNGDAHL. 1986. Purification and characterization of the F1-ATPase from *Clostridium thermoaceticum.* J. Bacteriol. **165:** 252–257.

262. HUGENHOLTZ, J. & L.G. LJUNGDAHL. 1989. Electron transport and electrochemical proton gradient in membrane vesicles of *Clostridium thermoautotrophicum.* J. Bacteriol. **171:** 2873–2875.

263. HUGENHOLTZ, J. & L.G. LJUNGDAHL. 1990. Metabolism and energy generation in homoacetogenic clostridia. FEMS Microbiol. Rev. **7:** 383–389.

264. DAS, A., J. HUGENHOLTZ, H. VAN HALBEEK & L.G. LJUNGDAHL. 1989. Structure and function of a menaquinone involved in electron transport in membranes of *Clostridium thermoautotrophicum* and *Clostridium thermoaceticum.* J. Bacteriol. **171:** 5823–5829.

Enzymology of the Wood–Ljungdahl Pathway of Acetogenesis

STEPHEN W. RAGSDALE

Department of Biological Chemistry, University of Michigan, Ann Arbor, Michigan, USA

The biochemistry of acetogenesis is reviewed. The microbes that catalyze the reactions that are central to acetogenesis are described and the focus is on the enzymology of the process. These microbes play a key role in the global carbon cycle, producing over 10 trillion kilograms of acetic acid annually. Acetogens have the ability to anaerobically convert carbon dioxide and CO into acetyl-CoA by the Wood–Ljungdahl pathway, which is linked to energy conservation. They also can convert the six carbons of glucose stoichiometrically into 3 mol of acetate using this pathway. Acetogens and other anaerobic microbes (e.g., sulfate reducers and methanogens) use the Wood–Ljungdahl pathway for cell carbon synthesis. Important enzymes in this pathway that are covered in this review are pyruvate ferredoxin oxidoreductase, CO dehydrogenase/acetyl-CoA synthase, a corrinoid iron-sulfur protein, a methyltransferase, and the enzymes involved in the conversion of carbon dioxide to methyl-tetrahydrofolate.

Key words: **acetogenic bacteria; carbon dioxide fixation; carbon monoxide; cobalamin**

Introduction

Some anaerobic microbes, including acetogenic bacteria and methanogenic archaea, convert CO_2 to cellular carbon by the Wood–Ljungdahl pathway (FIG. 1).[1,2] The global importance of acetogens is covered fully in Harold Drake's chapter.[3] Acetogenic bacteria use this pathway as their major means of generating energy for growth. *Moorella thermoacetica*, isolated in 1942,[4] is the model acetogen and is the organism on which most studies of the Wood–Ljungdahl have been performed; its genome was recently sequenced. Methanogens growing on H_2/CO_2 use the pathway for generating cell carbon; however, those that can grow on acetate, essentially run the pathway in reverse and generate energy by oxidizing acetate to 2 mol of CO_2.[5] Acetoclastic methanogens also convert acetate into acetyl-CoA for cell carbon synthesis through the combined actions of acetate kinase[6,7] and phosphotransacetylase.[8]

The Wood–Ljungdahl pathway contains an *Eastern* (in red) and a *Western* (in blue) branch (FIG. 1), as originally described.[9] The Eastern branch is essentially the folate-dependent one-carbon metabolic pathway that is present from bacteria to humans and recapitulated with methanopterin in methanogens. The Western branch is unique to acetogens, methanogens, and sulfate reducers, and exhibits novel mechanistic features. Acetogenic bacteria (e.g., acetogens or homoacetogens) synthesize acetic acid as their sole or primary metabolic end-product. Globally, acetogens produce over 10^{13} kg (100 billion U.S. tons) of acetic acid annually,[10] which dwarfs the total output of the world's chemical industry.

FIGURE 1 depicts growth of acetogens on glucose. However, these organisms can use a variety of substrates, including the biodegradation products of most natural polymers, such as cellulose, lignin (sugars, alcohols, aromatic compounds), and inorganic gases (CO, H_2, CO_2). When acetogens grow on H_2/CO_2, carbon enters the Wood–Ljungdahl pathway at the CO_2 reduction step, with H_2 serving as the electron donor. Acetogens are important in the biology of the soil, of extreme environments, and of organisms that house them in their digestive tract, such as humans, termites, and ruminants.[11–13]

Pyruvate Ferredoxin Oxidoreductase

Pyruvate ferredoxin oxidoreductase (PFOR) was reviewed relatively recently.[14] Catalyzing the oxidative decarboxylation of pyruvate to form acetyl-CoA and CO_2, PFOR links heterotrophic metabolism to the Wood–Ljungdahl pathway (FIG. 1). PFOR is also key to cell carbon synthesis since, besides its catabolic function, PFOR catalyzes pyruvate formation by reductive carboxylation of acetyl-CoA.[15,16] Pyruvate can then

Address for correspondence: Stephen W. Ragsdale, University of Michigan Medical School, 1150 W. Medical Center Dr., 5301 MSRB III, Ann Arbor, MI 48109.
sragsdal@umich.edu

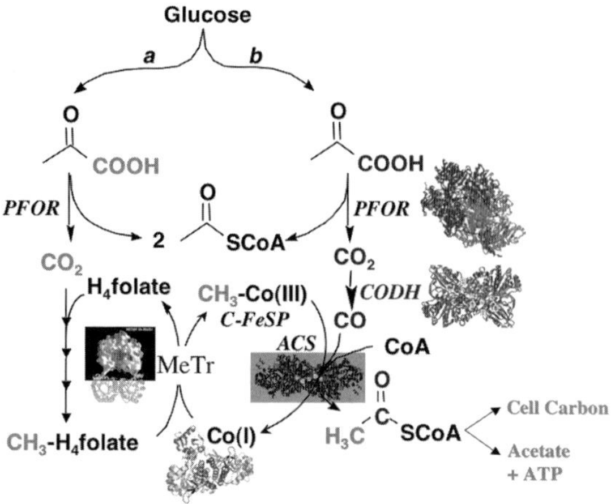

FIGURE 1. The Wood–Ljungdahl pathway with the Eastern and Western branches depicted in red and blue, respectively. (In color in *Annals* on line.)

enter the incomplete tricarboxylic acid cycle[17,18] to generate intermediates for cell carbon synthesis. PFOR is found in archaea, bacteria, and even anaerobic protozoa like Giardia.[19]

$$H_3C\text{-}C(=O)\text{-}CO_2^{\ominus} + \text{CoASH} \underset{}{\overset{E_0 = -540\text{ mV}}{\rightleftharpoons}} H_3C\text{-}C(=O)\text{-SCoA} + CO_2 + H^+ + 2e^- \quad (1)$$

The structure of the *Desulfovibrio africanus* PFOR reveals seven domains.[20] Thiamine pyrophosphate (TPP) and a proximal [4Fe-4S] cluster (Cluster A) are deeply buried within the protein, while the other two clusters (B and C) lead to the surface (FIG. 2). The TPP binding site consists of two domains that bear strong structural similarity to those in other TPP enzymes[21] and contains conserved residues that interact with Mg^{++}, pyrophosphate, and the thiazolium ring. Rapid kinetic studies indicate that the initial steps in oxidative decarboxylation of pyruvate by PFOR are similar to those of other TPP enzymes.[22] Deprotonation of TPP yields the active ylide, which forms an adduct with pyruvate. Then, CO_2 is released, forming the 2α-hydroxyethylidene TPP adduct (HE-TPP).[23] In acetogens, the CO_2 feeds into the Wood–Ljungdahl pathway[24,25] (FIG. 1) and, based on isotope-exchange studies,[26] may be channeled directly to CO dehydrogenase/acetyl-CoA synthase (CODH/ACS).

After forming the HE-TPP intermediate, the negative charge on C-2 of HE-TPP promotes one-electron reduction of a proximal FeS cluster, forming an HE-TPP radical intermediate (FIG. 2). Based on the crystal structure, this intermediate was proposed to be a novel sigma-type acetyl radical[27]; however, recent studies show that it is a pi radical with spin density delocalized over the aromatic thiazolium ring, as shown in the figure.[28,29]

In the next step of the PFOR mechanism, the HE-TPP radical transfers an electron to the Fe-S electron-transfer chain, presumably through the oxidized A-cluster (FIG. 2). The rate of this electron-transfer reaction is CoA dependent.[30] In the absence of CoA, the half-life of the HE-TPP radical intermediate is ~2 min; however, in the presence of CoA, the rate of radical decay increases 100,000-fold.[22] By studying various CoA analogues, it was shown that the *thiol* group of CoA *alone* lowers the transition state barrier for electron transfer by 40.5 kJ/mol. The final step in the PFOR mechanism is electron transfer through Clusters A, B, and C (FIG. 2) to ferredoxin,[20] the terminal electron acceptor for PFOR. This electron-transfer reaction occurs extremely rapidly, with a second-order rate constant of $2–7 \times 10^7\ M^{-1}\ s^{-1}$.[22,31]

CODH/ACS

CODH/ACS was recently reviewed,[32] so this part of the Wood–Ljungdahl pathway will be treated rather briefly. As shown in FIGURE 1, when acetogens are grown heterotrophically, the CO_2 and electrons generated by the PFOR reaction are utilized by CODH/ACS and formate dehydrogenase to generate CO and formate, respectively. When they are grown on CO, CODH generates CO_2, which is then converted to formate in the Eastern branch of the pathway and CO is incorporated directly as the carbonyl group of acetyl-CoA. Both CO and CO_2 are unreactive without a catalyst, but the enzyme-catalyzed reactions are fast, with turnover numbers as high as $40{,}000\ s^{-1}$ reported for CO oxidation by the Ni-CODH from *Carboxydothermus hydrogenoformans* at its physiological growth temperature.[33] Even the least active CODHs catalyze CO oxidation at rates of $\sim 50\ s^{-1}$.[34] There are two major classes of CODHs: the aerobic Mo-Cu-Se CODH from carboxydobacteria and the Ni-CODHs. Found in aerobic bacteria that oxidize CO with O_2,[35,36] the Mo-CODH contains FAD,[37] Fe/S centers, Cu, and 2 Mo atoms bound by molybdopterin cytosine dinucleotide, and its structure has been solved.[38] This enzyme will not be discussed further in this chapter, since only the Ni-CODH/ACS is involved in the Wood–Ljungdahl pathway. Ni-CODHs are divided into two classes: the monofunctional nickel CODH, which catalyzes the reaction shown in Equation 2, and the bifunctional CODH/ACS, which couples Equation 2 (CO formation) with Equation 3 (acetyl-CoA synthesis).

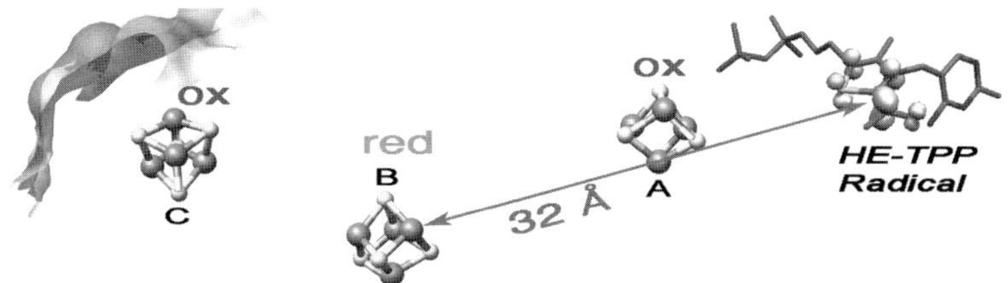

FIGURE 2. PFOR state including the HE-TPP radical, with the structure based on spectroscopic results, showing highly delocalized spin distribution (From Astashkin *et al.*[29] Used with permission.), and the distance from HE-TPP radical to coupled cluster. Location of the clusters is based on the structure (PDB 1KEK).

The monofunctional CODH functions physiologically in the direction of CO oxidation, allowing microbes to take up and oxidize CO at the low levels found in the environment, while the CODH in the bifunctional protein converts CO_2 into acetyl-CoA.

$$CO_2 + 2\,H^+ + 2\,e^- \rightleftharpoons CO + H_2O \qquad \Delta E^{\circ\prime} = -520\,mV \tag{2}$$

$$CO + CH_3{-}CFeSP + CoA \rightarrow acetyl - CoA + CFeSP \tag{3}$$

The crystal structures of the monofunctional NI-CODH and the CODH component of the bifunctional enzymes are very similar.[39–42] These mushroom-shaped, homodimeric enzymes contain five metal clusters per dimer: two C-clusters, two B-clusters, and a bridging D-cluster. The C-cluster is the catalytic site for CO oxidation and is buried 18 † below the surface. This cluster can be described as a [3Fe-4S] cluster bridged to a binuclear NiFe cluster. The *Rhodospirillum rubrum* CODH structure with a bridging Cys between Ni and a special iron atom called ferrous component II (FCII), while the *C. hydrogenoformans* protein appears to have a sulfide bridge between Ni and FCII.[43,44] Furthermore, there is evidence for a catalytically important persulfide at the C-cluster.[45] Issues related to the different structures are discussed in a recent review.[46] Cluster B is a typical $[4Fe\text{-}4S]^{2+/1+}$ cluster, while Cluster D is a $[4Fe\text{-}4S]^{2+/1+}$ cluster that bridges the two identical subunits, similar to the $[4Fe\text{-}4S]^{2+/1+}$ cluster in the iron protein of nitrogenase.

The CODH mechanism involves a Ping-Pong reaction: CODH is reduced by CO in the "ping" step and the reduced enzyme transfers electrons to an external redox mediator like ferredoxin in the "pong" step. The reduced electron acceptors then couple to other reduced nicolinamide adenine dinucleotide phosphate (NAD(P)H) or ferredoxin-dependent cellular processes that require energy. Details of the CODH reaction have been reviewed.[39] Recent studies of CODH linked to a pyrolytic graphite electrode show complete reversibility of CO oxidation and CO_2 reduction; in fact, at low pH values, the rate of CO_2 reduction exceeds that of CO oxidation.[47]

The association of ACS with CODH forms a bifunctional CODH/ACS machine that is encoded by the *acsA*/*acsB* genes, respectively, and plays the key role in the Wood–Ljungdahl pathway.[32,48,49] The CODH component catalyzes the conversion of CO_2 to CO (Eq. 2) to generate CO as a metabolic intermediate.[25] The CO_2 comes from the growth medium or from decarboxylation of pyruvate.[25,30] Then, ACS catalyzes the condensation of CO,[50] CoA, and the methyl group of a methylated corrinoid iron–sulfur protein (CFeSP) to generate acetyl-CoA (Eq. 3),[25] a precursor of cellular material and a source of energy.

CODH/ACS contains a 140-† channel that delivers CO generated at the C-cluster to the A-cluster.[40,51,52] The only metallocenter in ACS is the A-cluster, which consists of a [4Fe-4S] cluster bridged to a Ni site (Ni_p) that is thiolate bridged to another Ni ion in a thiolato- and carboxamido-type N_2S_2 coordination environment.[40,42,53] Thus, one can describe the A-cluster as a binuclear NiNi center bridged by a cysteine residue (Cys509) to a [4Fe-4S] cluster, an arrangement similar to the Fe-Fe hydrogenases in which a [4Fe-4S] cluster and a binuclear Fe site are bridged by a Cys residue.[87,88]

Two mechanisms for acetyl-CoA synthesis have been proposed that differ mainly in the electronic structure of the intermediates: one proposes a paramagnetic Ni(I)-CO species as a central intermediate,[2] and the other proposes a Ni(0) intermediate.[54,55] A generic mechanism that emphasizes the organometallic nature of this reaction sequence is described in FIGURE 3.

Step 1, as shown in the figure, involves the migration of CO, derived from CO_2 reduction, through the intersubunit channel to bind to the Ni_p site in the A-cluster. The binding of CO to ACS forms an organometallic complex, called the NiFeC species that has been characterized by a number of spectroscopic approaches.[2]

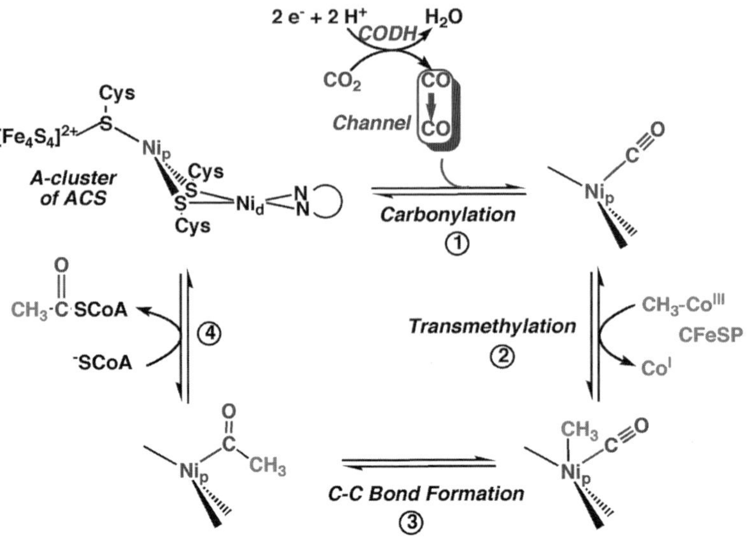

FIGURE 3. ACS mechanism emphasizing the organometallic nature of the reaction sequence and the channel to deliver CO from the CODH active site to the A-cluster.

The electronic structure of the NiFeC species is described as a $[4Fe\text{-}4S]^{2+}$ cluster linked to a Ni^{1+} center at the Ni_p site, while the other Ni apparently remains redox-inert in the Ni^{2+} state.[56] Lindahl's group has argued that the NiFeC species is not a true catalytic intermediate in acetyl-CoA synthesis, that it may represent an inhibited state, and that the Ni(0) state is the catalytically relevant one.[54,55] See the recent review for a discussion of these issues.[32] Step 2 in the ACS mechanism involves methylation of the A-cluster,[57] which involves the conversion of one organometallic species (methyl-Co) to another (methyl-Ni). There is evidence that the methyl group binds to the Ni_p site of the A-cluster.[58–60] Carbon-carbon bond formation, Step 3 in the catalytic cycle, occurs by condensation of the methyl and carbonyl groups to form an acetyl-metal species. In the last step, CoA binds to ACS, triggering thiolysis of the acetyl-metal bond to form the C-S bond of acetyl-CoA, completing the reactions of the Western branch of the Wood–Ljungdahl pathway. FIGURE 3 indicates, for simplicity, that the ACS reaction sequence is ordered with CO binding before the methyl group. Conversely, Lindahl has argued for a strictly ordered binding mechanism, with methyl binding first, then CO, and finally CoA, as described in a recent review.[55] In the author's opinion, there is insufficient evidence to exclude the 1991 proposal that the carbonylation and methylation steps occur randomly[61] (FIG. 4).

Tetrahydrofolate-Dependent Enzymes

Most of the work on the folate enzymes involved in the Eastern branch of the pathway has been performed in the Ljungdahl laboratory. The methyl group of acetyl-CoA is formed by the six-electron reduction of CO_2 in the reactions of the Eastern branch of the acetyl-CoA pathway (FIG. 1).[2,62] First, formate dehydrogenase converts CO_2 to formate,[63] which is condensed with H_4folate to form 10-formyl-H_4folate.[64,65] The latter is then converted by a cyclohydrolase to 5,10-methenyl-H_4folate. Next, a dehydrogenase reduces methenyl- to 5,10-methylene-H_4folate,[66] which is reduced to (6*S*)-5-CH_3-H_4folate by a reductase.[67,68]

Methyltransferase (MeTr, AcsE) and Corrinoid Iron Sulfur Protein (CFeSP, AcsCD)

The methyl group of CH_3-H_4folate is transferred to the cobalt site in the cobalamin cofactor bound to the CFeSP[69,70] to form an organometallic methyl-Co(III) intermediate in the Wood–Ljungdahl pathway (FIG. 1). This reaction is catalyzed by MeTr, encoded by the *acsE* gene.[24] MeTr belongs to the B_{12}-dependent methyltransferase family that includes methionine synthase and related enzymes from methanogens.[71] We have cloned, sequenced, and actively overexpressed MeTr[72,73] and the CFeSP[74] in *E coli*, making them amenable for site-directed mutagenesis studies. In collaboration with Cathy Drennan (MIT, Cambridge, Massachusetts), we also have determined the structure of MeTr and a site-directed variant in its uncomplexed[75] and CH_3-H_4folate–bound[76] states.

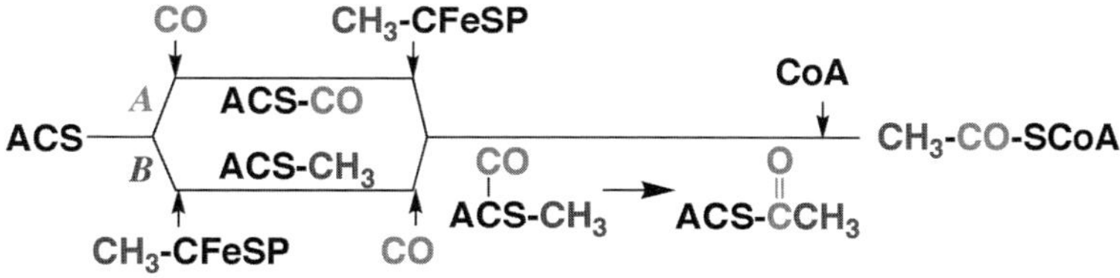

FIGURE 4. Random mechanism of acetyl-CoA synthesis.

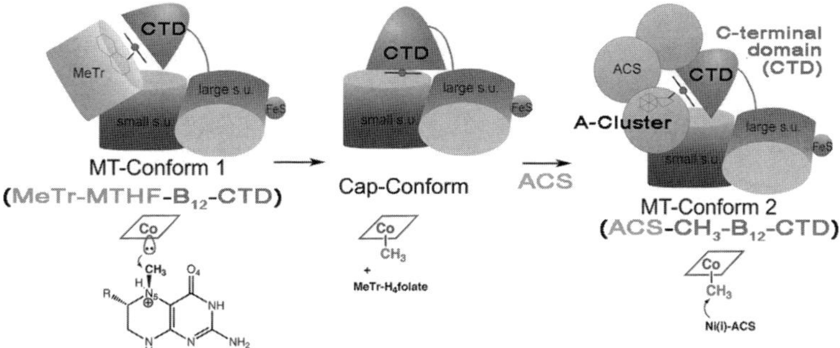

FIGURE 5. Proposed conformers and interactions of the CFeSP. (Modified from Svetlitchnaia *et al.*[79])

It was concluded that CH_3-H_4folate binds tightly ($K_d < 10\,\mu M$[77]) to MeTr within a negatively charged crevice of the triose phosphate isomerase (TIM) barrel.[78] The structure of the CFeSP was also recently solved.[79]

A key step in the MeTr mechanism is activation of the methyl group of CH_3-H_4folate, since the reaction involves displacement at a tertiary amine and because the CH_3-N bond is much stronger than the product CH_3-Co bond. Of the activation mechanisms that have been considered, protonation at N5 of the pterin seems to be most plausible.[71] Generation of a positive charge on N5 would lower the activation barrier for nucleophilic displacement of the methyl group by the Co(I) nucleophile. There is significant experimental support for protonation at N5 of the pterin, including proton uptake studies,[77,80] pH dependencies of the steady state, and transient reaction kinetics of MeTr[81] and methionine synthase,[82] and studies of variants that are compromised in acid–base catalysis.[76] A question that has not been resolved is whether the proton transfer takes place upon formation of the binary complex, as indicated by studies with MeTr from *M. thermoacetica*,[49,77] or the ternary complex (with the methyl acceptor), as concluded from studies on *E. coli* methionine synthase.[80] Recent studies indicate that this protonation step relies on an H-bonding network, instead of a single acid–base catalyst and that an Asn residue is a key component of that network.[76]

As shown in FIGURE 1, the CFeSP interfaces between CH_3-H_4folate/MeTr and CODH/ACS. This 88-kDa heterodimeric protein contains a $[4Fe\text{-}4S]^{2+/1+}$ cluster and a cobalt cobamide.[70,74] The Fe-S cluster plays a role in reductive activation of the cobalt to the active Co(I) state.[83,84] Svetlitchnaia *et al.*[79] proposed that the C-terminal domain (FIG. 5) of the large subunit is a mobile element that interacts alternatively with the A-cluster domain of ACS and with MeTr. Three major conformers or complexes are described: (1) a methylation complex, in which the Co(I)-CFeSP binds MeTr and accepts the methyl group of CH_3-H_4folate; (2) the methylated CFeSP; (3) a complex between ACS and the methylated CFeSP. A fourth conformation, which is not shown here, would be a reductive activation conformer, in which the corrinoid is in the inactive Co(II) state. This molecular juggling proposed for the CFeSP shown in FIGURE 5 has precedent in the related mechanism involving the various domains of methionine synthase, as shown by the elegant structure–function studies of Matthews and Ludwig.[82,85,86]

Summary

Studies of the enzymes involved in the Wood–Ljungdahl pathway have elucidated new roles of metal ions in biology (including the formation of bioorganometallic intermediates, discovery of new heterometallic clusters, and nucleophilic metal ions), uncovered novel substrate-derived radical intermediates, and revealed channeling of gaseous substrates. These new outcomes and mechanisms will likely be applicable to other currently less well-studied metal-dependent enzyme systems. Now that detailed structures of PFOR, CODH/ACS, the CFeSP, and MeTr are available to provide a structural framework for these novel and important chemical reactions, mechanistic hypotheses can be posed and tested at a deeper level using a variety of biochemical and biophysical methods.

Acknowledgments

Work on the enzymology of acetogenesis has been funded by the NIH (GM39451). I am grateful for the contributions of Javier Seravalli and Elizabeth Pierce in my laboratory to the recent work on acetogenesis described here.

Conflict of Interest

The author declares no conflicts of interest.

References

1. Ragsdale, S.W. 2006. Metalloenzymes in the reduction of one-carbon compounds. *In* Biological Inorganic Chemistry: Structure and Reactivity. Bertini, I., *et al.*, Eds.: 452–467. University Science Books. Mill Valley, CA.
2. Ragsdale, S.W. 2004. Life with carbon monoxide. CRC Crit. Rev. Biochem. Mol. Biol. **39:** 165–195.
3. Drake, H.L. 2008. Old acetogens, new light. Ann. N.Y. Acad Sci. Incredible Anaerobes: From Physiology to Genomics to Fuels. In Press.
4. Fontaine, F.E. *et al.* 1942. A new type of glucose fermentation by *Clostridium thermoaceticum*. J. Bacteriol. **43:** 701–715.
5. Ferry, J.G. 1992. Biochemistry of methanogenesis. Crit. Rev. Biochem. Mol. Biol. **27:** 473–502.
6. Ingram-Smith, C. *et al.* 2005. Characterization of the acetate binding pocket in the Methanosarcina thermophila acetate kinase. J. Bacteriol. **187:** 2386–2394.
7. Gorrell, A. *et al.* 2005. Structural and kinetic analyses of arginine residues in the active site of the acetate kinase from Methanosarcina thermophila. J. Biol. Chem. **280:** 10731–10742.
8. Iyer, P.P. *et al.* 2004. Crystal structure of phosphotransacetylase from the methanogenic archaeon Methanosarcina thermophila. Structure **12:** 559–567.
9. Ragsdale, S.W. 1997. The Eastern and Western branches of the Wood/Ljungdahl pathway: how the East and West were won. BioFactors **9:** 1–9.
10. Drake, H.L. *et al.* 1994. Acetogenesis, acetogenic bacteria, and the acetyl-CoA pathway: past and current perspectives. *In* Acetogenesis. Drake, H.L., Ed.: 3–60. Chapman and Hall. New York.
11. Lajoie, S.F. *et al.* 1988. Acetate production from hydrogen and [^{13}C]carbon dioxide by the microflora of human feces. Appl. Environ. Microbiol. **54:** 2723–2727.
12. Breznak, J.A. *et al.* 1986. Acetate synthesis from H_2 plus CO_2 by termite gut microbes. Appl. Environ. Microbiol. **52:** 623–630.
13. Breznak, J.A. *et al.* 1990. Microbial H_2/CO_2 acetogenesis in animal guts: nature and nutritional significance. FEMS Microbiol, Rev. **87:** 309–314.
14. Ragsdale, S.W. 2003. Pyruvate:ferredoxin oxidoreductase and its radical intermediate. Chem. Rev. **103:** 2333–2346.
15. Furdui, C. *et al.* 2000. The role of pyruvate: ferredoxin oxidoreductase in pyruvate synthesis during autotrophic growth by the Wood-Ljungdahl pathway. J. Biol. Chem. **275:** 28494–28499.
16. Bock, A.K. *et al.* 1996. Catalytic properties, molecular composition and sequence alignments of pyruvate: ferredoxin oxidoreductase from the methanogenic archaeon Methanosarcina barkeri (strain Fusaro). Eur. J. Biochem. **237:** 35–44.
17. Simpson, P.G. *et al.* 1993. Anabolic pathways in methanogens. *In* Methanogenesis: Ecology Physiology, Biochemistry & Genetics. Ferry, J.G., Ed.: 445–472. Chapman & Hall. London.
18. Yoon, K.S. *et al.* 1999. Rubredoxin from the green sulfur bacterium Chlorobium tepidum functions as an electron acceptor for pyruvate ferredoxin oxidoreductase. J. Biol. Chem. **274:** 29772–29778.
19. Horner, D.S. *et al.* 1999. A single eubacterial origin of eukaryotic pyruvate: ferredoxin oxidoreductase genes: Implications for the evolution of anaerobic eukaryotes. Mol. Biol. Evol. **16:** 1280–1291.
20. Chabriere, E. *et al.* 1999. Crystal structures of the key anaerobic enzyme pyruvate:ferredoxin oxidoreductase, free and in complex with pyruvate. Nat. Struct. Biol. **6:** 182–190.
21. Muller, Y.A. *et al.* 1993. A thiamin diphosphate binding fold revealed by comparison of the crystal structures of transketolase, pyruvate oxidase and pyruvate decarboxylase. Structure **1:** 95–103.
22. Furdui, C. *et al.* 2002. The roles of coenzyme A in the pyruvate:ferredoxin oxidoreductase reaction mechanism: rate enhancement of election transfer from a radical intermediate to an iron-sulfur cluster. Biochemistry **41:** 9921–9937.
23. Breslow, R. 1957. Rapid deuterium exchange in thiazolium salts. J. Am. Chem. Soc. **79:** 1762–1763.
24. Drake, H.L. *et al.* 1981. Purification of five components from *Clostridium thermoacticum* which catalyze synthesis of acetate from pyruvate and methyltetrahydrofolate. Properties of phosphotransacetylase. J. Biol. Chem. **256:** 11137–11144.
25. Menon, S. *et al.* 1996. Evidence that carbon monoxide is an obligatory intermediate in anaerobic acetyl-CoA synthesis. Biochemistry **35:** 12119–12125.
26. Schulman, M. *et al.* 1973. Total synthesis of acetate from CO_2. VII. Evidence with *Clostridium thermoaceticum* that the carboxyl of acetate is derived from the carboxyl of pyruvate by transcarboxylation and not by fixation of CO_2. J. Biol. Chem. **248:** 6255–6261.
27. Chabriere, E. *et al.* 2001. Crystal structure of the free radical intermediate of pyruvate:ferredoxin oxidoreductase. Science **294:** 2559–2563.
28. Mansoorabadi, S.O. *et al.* 2006. EPR spectroscopic and computational characterization of the hydroxyethylidene-thiamine pyrophosphate radical intermediate of pyruvate: ferredoxin oxidoreductase. Biochemistry **45:** 7122–7131.

29. ASTASHKIN, A.V. *et al.* 2006. Pulsed electron paramagnetic resonance experiments identify the paramagnetic intermediates in the pyruvate ferredoxin oxidoreductase catalytic cycle. J. Am. Chem. Soc. **128:** 3888–3889.
30. MENON, S. *et al.* 1997. Mechanism of the *Clostridium thermoaceticum* pyruvate:ferredoxin oxidoreductase: evidence for the common catalytic intermediacy of the hydroxyethylthiamine pyropyrosphate radical. Biochemistry **36:** 8484–8494.
31. PIEULLE, L. *et al.* 1999. Structural and kinetic studies of the pyruvate-ferredoxin oxidoreductase/ferredoxin complex from Desulfovibrio africanus. Eur. J. Biochem. **264:** 500–508.
32. RAGSDALE, S.W. 2007. Nickel and the carbon cycle. J. Inorg. Biochem. **101:** 1657–1666.
33. SVETLITCHNYI, V. *et al.* 2001. Two membrane-associated NiFeS-carbon monoxide dehydrogenases from the anaerobic carbon-monoxide-utilizing eubacterium Carboxydothermus hydrogenoformans. J. Bacteriol. **183:** 5134–5144.
34. MEYER, O. *et al.* 2000. The role of Se, Mo and Fe in the structure and function of carbon monoxide dehydrogenase. Biol. Chem. **381:** 865–876.
35. MEYER, O. *et al.* 1993. Biochemistry of the aerobic utilization of carbon monoxide. *In* Microbial Growth on C_1 Compounds. Murrell, J.C. & D.P. Kelly, Eds.: 433–459. Intercept, Ltd. Andover, MA.
36. GNIDA, M. *et al.* 2003. A novel binuclear [CuSMo] cluster at the active site of carbon monoxide dehydrogenase: characterization by X-ray absorption spectroscopy. Biochemistry **42:** 222–230.
37. BATES, D.M. *et al.* 2000. Substitution of leucine 28 with histidine in the Escherichia coli transcription factor FNR results in increased stability of the [4Fe-4S](2+) cluster to oxygen. J. Biol. Chem. **275:** 6234–6240.
38. DOBBEK, H. *et al.* 1999. Crystal structure and mechanism of CO dehydrogenase, a molybdo iron-sulfur flavoprotein containing S-selanylcysteine. Proc. Natl. Acad. Sci. USA **96:** 8884–8889.
39. DRENNAN, C.L. *et al.* 2001. Life on carbon monoxide: X-ray structure of Rhodospirillum rubrum Ni-Fe-S carbon monoxide dehydrogenase. Proc. Natl. Acad. Sci. USA **98:** 11973–11978.
40. DOUKOV, T.I. *et al.* 2002. A Ni-Fe-Cu center in a bifunctional carbon monoxide dehydrogenase/acetyl-CoA synthase. Science **298:** 567–572.
41. DOBBEK, H. *et al.* 2001. Crystal structure of a carbon monoxide dehydrogenase reveals a [Ni-4Fe-5S] cluster. Science **293:** 1281–1285.
42. DARNAULT, C. *et al.* 2003. Ni-Zn-[Fe(4)-S(4)] and Ni-Ni-[Fe(4)-S(4)] clusters in closed and open alpha subunits of acetyl-CoA synthase/carbon monoxide dehydrogenase. Nat. Struct. Biol. **10:** 271–279.
43. DOBBEK, H. *et al.* 2004. Carbon monoxide induced decomposition of the active site [Ni-4Fe-5S] cluster of CO dehydrogenase. J. Am. Chem. Soc. **126:** 5382–5387.
44. SUN, J. *et al.* 2007. Sulfur ligand substitution at the nickel(II) sites of cubane-type and cubanoid NiFe3S4 clusters relevant to the C-clusters of carbon monoxide dehydrogenase. Inorg. Chem. **46:** 2691–2699.
45. KIM, E.J. *et al.* 2004. Evidence for a proton transfer network and a required persulfide-bond-forming cysteine residue in ni-containing carbon monoxide dehydrogenases. Biochemistry **43:** 5728–5734.
46. DRENNAN, C.L. *et al.* 2004. The metalloclusters of carbon monoxide dehydrogenase/acetyl-CoA synthase: a story in pictures. J. Biol. Inorg. Chem. **9:** 511–515.
47. PARKIN, A. *et al.* 2007. Rapid electrocatalytic CO2/CO interconversions by Carboxydothermus hydrogenoformans CO dehydrogenase I on an electrode. J. Am. Chem. Soc. **129:** 10328–10329.
48. FONTECILLA-CAMPS, J.-C. *et al.* 1999. Nickel-iron-sulfur active sites: hydrogenase and CO dehydrogenase. *In* Advances in Inorganic Chemistry, Vol. 47. Sykes, A.G. & R. Cammack, Eds.: 283–333. Academic Press, Inc. San Diego.
49. SERAVALLI, J. *et al.* 1999. Mechanism of transfer of the methyl group from (6S)-methyltetrahydrofolate to the corrinoid/iron-sulfur protein catalyzed by the methyltransferase from Clostridium thermoaceticum: a key step in the Wood-Ljungdahl pathway of acetyl-CoA synthesis. Biochemistry **38:** 5728–5735.
50. RAGSDALE, S.W. *et al.* 1982. EPR evidence for nickel substrate interaction in carbon monoxide dehydrogenase from *Clostridium thermoaceticum*. Biochem. Biophys. Res. Commun. **108:** 658–663.
51. SERAVALLI, J. *et al.* 2000. Channeling of carbon monoxide during anaerobic carbon dioxide fixation. Biochemistry **39:** 1274–1277.
52. MAYNARD, E.L. *et al.* 1999. Evidence of a molecular tunnel connecting the active sites for CO_2 reduction and acetyl-CoA synthesis in acetyl-CoA synthase from *Clostridium thermoaceticum*. J. Am. Chem. Soc. **121:** 9221–9222.
53. SVETLITCHNYI, V. *et al.* 2004. A functional Ni-Ni-[4Fe-4S] cluster in the monomeric acetyl-CoA synthase from Carboxydothermus hydrogenoformans. Proc. Natl. Acad. Sci. USA **101:** 446–451.
54. BRAMLETT, M.R. *et al.* 2006. Mossbauer and EPR study of recombinant acetyl-CoA synthase from Moorella thermoacetica. Biochemistry **45:** 8674–8685.
55. LINDAHL, P.A. 2004. Acetyl-coenzyme A synthase: the case for a Ni_p^0-Based mechanism of catalysis. J. Biol. Inorg. Chem. **9:** 516–524.
56. BRUNOLD, T.C. 2004. Spectroscopic and computational insights into the geometric and electronic properties of the A cluster of acetyl-coenzyme A synthase. J. Biol. Inorg. Chem. **9:** 533–541.
57. SERAVALLI, J. *et al.* 2002. Rapid kinetic studies of acetyl-CoA synthesis: evidence supporting the catalytic intermediacy of a paramagnetic NiFeC species in the autotrophic Wood-Ljungdahl pathway. Biochemistry **41:** 1807–1819.
58. SHIN, W. *et al.* 1993. Heterogeneous nickel environments in carbon monoxide dehydrogenase from *Clostridium thermoaceticum*. J. Am. Chem. Soc. **115:** 5522–5526.
59. BARONDEAU, D.P. *et al.* 1997. Methylation of carbon monoxide dehydrogenase from Clostridium thermoaceticum and mechanism of acetyl coenzyme A synthesis. J. Am. Chem. Soc. **119:** 3959–3970.

60. Seravalli, J. *et al.* 2004. Evidence that Ni-Ni acetyl-CoA synthase is active and that the Cu-Ni enzyme is not. Biochemistry **43:** 3944–3955.
61. Lu, W.P. *et al.* 1991. Reductive activation of the coenzyme A/acetyl-CoA isotopic exchange reaction catalyzed by carbon monoxide dehydrogenase from *Clostridium thermoaceticum* and its inhibition by nitrous oxide and carbon monoxide. J. Biol. Chem. **266:** 3554–3564.
62. Ragsdale, S.W. 1991. Enzymology of the acetyl-CoA pathway of CO_2 fixation. CRC Crit. Rev. Biochem. Mol. Biol. **26:** 261–300.
63. Ljungdahl, L.G. *et al.* 1978. Formate dehydrogenase, a selenium-tungsten enzyme from *Clostridium thermoaceticum.* Methods Enzymol. **53:** 360–372.
64. Lovell, C.R. *et al.* 1988. Cloning and expression in *Escherichia coli* of the *Clostridium thermoaceticum* gene encoding thermostable formyltetrahydrofolate synthetase. Arch. Microbiol. **149:** 280–285.
65. Lovell, C.R. *et al.* 1990. Primary structure of the thermostable formyltetrahydrofolate synthetase from *Clostridium thermoaceticum.* Biochemistry **29:** 5687–5694.
66. Moore, M.R. *et al.* 1974. Purification and characterization of nicotinamide adenine dinucleotide-dependent methylenetetrahydrofolate dehydrogenase from *Clostridium formicoaceticum.* J. Biol. Chem. **249:** 5250–5253.
67. Clark, J.E. *et al.* 1984. Purification and properties of 5,10-methylenetetrahydrofolate reductase, an iron-sulfur flavoprotein from Clostridium formicoaceticum. J. Biol. Chem. **259:** 10845–10889.
68. Park, E.Y. *et al.* 1991. 5,10-methylenetetrahydrofolate reductases: iron-sulfur-zinc flavoproteins of two acetogenic clostridia. *In* Chemistry and Biochemistry of Flavoenzymes, Vol. 1. Miller, F., Ed.: 389–400. CRC Press. Boca Raton, FL.
69. Hu, S.-I. *et al.* 1984. Acetate synthesis from carbon monoxide by *Clostridium thermoaceticum.* Purification of the corrinoid protein. J. Biol. Chem. **259:** 8892–8897.
70. Ragsdale, S.W. *et al.* 1987. Mössbauer, EPR, and optical studies of the corrinoid/Fe-S protein involved in the synthesis of acetyl-CoA by *Clostridium thermoaceticum.* J. Biol. Chem. **262:** 14289–14297.
71. Banerjee, R. *et al.* 2003. The many faces of vitamin B_{12}: catalysis by cobalamin-dependent enzymes. Ann. Rev. Biochem. **72:** 209–247.
72. Roberts, D.L. *et al.* 1989. Cloning and expression of the gene cluster encoding key proteins involved in acetyl-CoA synthesis in *Clostridium thermoaceticum*: CO dehydrogenase, the corrinoid/Fe-S protein, and methyltransferase. Proc. Natl. Acad. Sci. USA **86:** 32–36.
73. Roberts, D.L. *et al.* 1994. The reductive acetyl-CoA pathway: sequence and heterologous expression of active CH_3-H_4folate:corrinoid/iron sulfur protein methyltransferase from Clostridium themoaceticum. J. Bacteriol. **176:** 6127–6130.
74. Lu, W.-P. *et al.* 1993. Sequence and expression of the gene encoding the corrinoid/iron-sulfur protein from *Clostridium thermoaceticum* and reconstitution of the recombinant protein to full activity. J. Biol. Chem. **268:** 5605–5614.
75. Doukov, T. *et al.* 1995. Preliminary X-ray diffraction analysis of the methyltetrahydrofolate:corrinoid/iron sulfur protein methyltransferase from Clostridium themoaceticum. Acta Crystallographa. **D51: Part 6:** 1092–1093.
76. Doukov, T.I. *et al.* 2007. Structural and kinetic evidence for an extended hydrogen bonding network in catalysis of methyl group transfer: role of an active site asparagine residue in activation of methyl transfer by methyltransferases. J. Biol. Chem. **282:** 6609–6618.
77. Seravalli, J. *et al.* 1999. Binding of (6R,S)-methyltetrahydrofolate to methyltransferase from Clostridium thermoaceticum: role of protonation of methyltetrahydrofolate in the mechanism of methyl transfer. Biochemistry **38:** 5736–5745.
78. Doukov, T. *et al.* 2000. Crystal structure of a methyltetrahydrofolate and corrinoid dependent methyltransferase. Structure **8:** 817–830.
79. Svetlitchnaia, T. *et al.* 2006. Structural insights into methyltransfer reactions of a corrinoid iron-sulfur protein involved in acetyl-CoA synthesis. Proc. Natl. Acad. Sci. USA **103:** 14331–14336.
80. Smith, A.E. *et al.* 2000. Protonation state of methyltetrahydrofolate in a binary complex with cobalamin-dependent methionine synthase. Biochemistry **39:** 13880–13890.
81. Zhao, S. *et al.* 1995. Mechanistic studies of the methyltransferase from Clostridium thermoaceticum: origin of the pH dependence of the methyl group transfer from methyltetrahydrofolate to the corrinoid/iron-sulfur protein. Biochemistry **34:** 15075–15083.
82. Matthews, R.G. 2001. Cobalamin-dependent methyltransferases. Acc. Chem. Res. **34:** 681–689.
83. Menon, S. *et al.* 1999. The role of an iron-sulfur cluster in an enzymatic methylation reaction: methylation of CO dehydrogenase/acetyl-CoA synthase by the methylated corrinoid iron-sulfur protein. J. Biol. Chem. **274:** 11513–11518.
84. Menon, S. *et al.* 1998. Role of the [4Fe-4S] cluster in reductive activation of the cobalt center of the corrinoid iron-sulfur protein from *Clostridium thermoaceticum* during acetyl-CoA synthesis. Biochemistry **37:** 5689–5698.
85. Evans, J.C. *et al.* 2004. Structures of the N-terminal modules imply large domain motions during catalysis by methionine synthase. Proc. Natl. Acad. Sci. USA **101:** 3729–3736.
86. Taurog, R.E. *et al.* 2006. Synergistic, random sequential binding of substrates in cobalamin-independent methionine synthase. Biochemistry **45:** 5083–5091.
87. Peters, J.W. *et al.* 1998. X-ray crystal structure of the Fe-only hydrogenase (CpI) from Clostridium pasteurianum to 1.8 angstrom resolution. Science **282:** 1853–1858.
88. Nicolet, Y. *et al.* 2000. A novel FeS cluster in Fe-only hydrogenases. Trends Biochem. Sci. **25:** 138–143.

Discovery of a Ferredoxin:NAD$^+$-Oxidoreductase (Rnf) in Acetobacterium woodii

A Novel Potential Coupling Site in Acetogens

VOLKER MÜLLER, FRANK IMKAMP,[a] EVA BIEGEL, SILKE SCHMIDT, AND SABRINA DILLING[b]

Molecular Microbiology & Bioenergetics, Institute of Molecular Biosciences, Johann Wolfgang Goethe University, Frankfurt am Main, Germany

Acetogens use the Wood–Ljungdahl pathway for reduction of carbon dioxide to acetate. This pathway not only allows reoxidation of reducing equivalents during heterotrophic growth but also supports chemolithoautotrophic growth on $H_2 + CO_2$. The latter argues for this pathway being a source for net energy conservation, but the mechanism involved remains unknown. In addition to CO_2, acetogens can use alternative electron acceptors, such as nitrate or caffeate. Caffeate respiration in the model acetogen *Acetobacterium woodii* is coupled to energy conservation via a chemiosmotic mechanism, with Na^+ as coupling ion. The pathway and its bioenergetics were solved in some detail very recently. This review focuses on the regulation of caffeate respiration, describes the enyzmes involved, summarizes the evidence for a potential Na^+-translocating ferredoxin:NAD$^+$-oxidoreductase (Rnf complex) as a new coupling site, and hypothesizes on the role of this Rnf complex in the Wood–Ljungdahl pathway.

Key words: **acetogens; *Acetobacterium woodii*; energy conservation; Rnf; caffeate respiration; carbonate respiration**

Energy Conservation in the Wood–Ljungdahl Pathway in Acetogens

Organisms able to reduce CO_2 to acetate via the acetyl-CoA pathway, also defined as Wood–Ljungdahl pathway (FIG. 1), are termed *acetogens*.[1,2] This metabolic capability differentiates acetogens from organisms that synthesize acetate by other metabolic pathways. Acetogens can grow on a variety of different substrates, such as, for example, hexoses, C_2, and C_1 compounds. Hexoses are converted exclusively to acetate according to:

$$1C_6H_{12}O_6 \rightarrow 3CH_3COOH \quad (1)$$

and therefore this fermentation is also referred to as homoacetogenesis. Hexose fermentation proceeds via the Embden–Meyerhof–Parnas pathway to pyruvate, which is then oxidized by pyruvate:ferredoxin-oxidoreductase to acetyl-CoA, reduced ferredoxin, and CO_2. The acetyl-CoA is then converted to acetate via acetyl phosphate. The oxidative branch of the pathway (Eq. 2) is coupled to the synthesis of 4 mol of ATP by substrate-level phosphorylation (SLP):

$$C_6H_{12}O_6 + 4ADP + 4P_i \rightarrow 2CH_3COOH + 2CO_2 + 4ATP + 8[H] \quad (2)$$

The reducing equivalents gained during glycolysis and pyruvate:ferredoxin-oxidoreductase are reoxidized by reducing the two mol of CO_2 to another mol of acetate via the Wood–Ljungdahl pathway:

$$2CO_2 + 8[H] + nADP + nP_i \rightarrow CH_3COOH + nATP \quad (3)$$

The Wood–Ljungdahl pathway also enables growth on $H_2 + CO_2$, according to:

$$2CO_2 + 4H_2 + nADP + nP_i \rightarrow CH_3COOH + 2H_2O + nATP \quad (4)$$

and therefore it must be coupled to net adenosine triphosphate (ATP) synthesis. The overall free-energy change of the reaction ($\Delta G^{\circ\prime} = -95$ kJ/mol) could allow the synthesis of 1–2 mol of ATP. One mol of ATP is produced by SLP in the acetate kinase reaction, but one

Address for correspondence: Volker Müller, Max-von-Laue-Str. 9, 60439 Frankfurt, Germany. Voice: 49-69-798-29507; fax: 49-69-798-29306.
vmueller@bio.uni-frankfurt.de

[a]Present adress: Institute of Molecular Biology and Biophysics, ETH Zürich, Switzerland.

[b]Present adress: Institute of Microbiology, ETH Zürich, Switzerland.

Ann. N.Y. Acad. Sci. 1125: 137–146 (2008). © 2008 New York Academy of Sciences.
doi: 10.1196/annals.1419.011

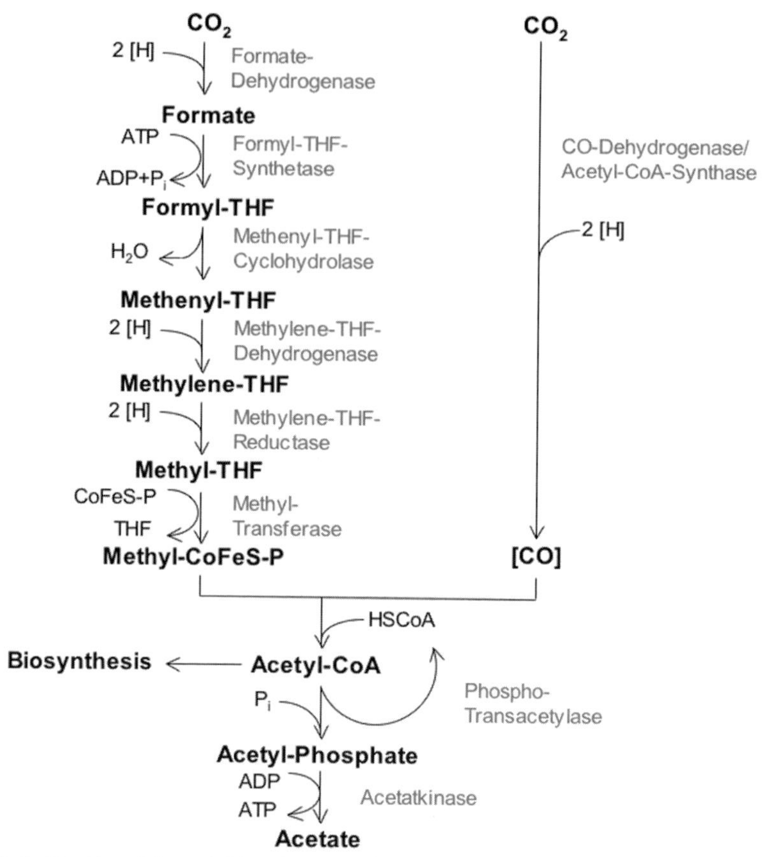

FIGURE 1. The Wood–Ljungdahl pathway. Reductants are derived from oxidation of hydrogen or organic substrates. ABBREVIATIONS: [H] = reducing equivalent; THF = tetrahydrofolate; CoFeS-P = corrinoid iron sulfur protein; [CO]= enzyme-bound CO.

mol of ATP is consumed in the formyl-THF synthetase reaction (FIG. 1). Therefore, the net ATP gain by SLP is zero and ion gradient-driven phosphorylation must occur as well (because the organisms grow chemolithoautotrophically). In recent years experimental evidence was presented that ion gradient-driven phosphorylation indeed occurs in acetogens during the operation of the Wood–Ljungdahl pathway, but the sites of energy conservation and the mechanisms employed are still unknown, but must be different. Briefly, from a bioenergetic point of view, acetogens can be divided into two groups, the Na^+-dependent ones with *Acetobacterium woodii*[3] and the H^+-dependent ones with *Moorella thermoacetica* (formerly *Clostridium thermoaceticum*) as model organisms.[4] The latter group contains cytochromes and a membrane-bound, H^+-motive electron-transport chain, as well as a $\Delta\mu H^+$-coupled F_1F_O ATP synthase. The Na^+-dependent acetogens lack cytochromes, but have membrane-bound corrinoids. Furthermore, experiments with *A. woodii* revealed the coupling of the Wood–Ljungdahl pathway to primary and electrogenic translocation of Na^+ across the cytoplasmic membrane. The ion gradient established is used by a Na^+-translocating F_1F_O ATP synthase. For a discussion of the energetics of the Wood–Ljungdahl pathway, the reader is referred to recent reviews (see, e.g., Refs. 5 and 6).

Caffeate, an Alternative Electron Acceptor for *Acetobacterium woodii*

In addition to CO_2 alternative electron acceptors,—such as aromatic acrylate groups,—fumarate, dimethyl sulfoxide, nitrate, or nitrite, can be used by some acetogens. Under these conditions, the Wood–Ljungdahl pathway may be turned off completely and only 2 mol of acetate are formed. Furthermore, acetate synthesis may be blocked completely.[1] A typical substrate combination that does not yield acetate is the oxidation of a methyl group coupled to the reduction of a phenylacrylate.[7]

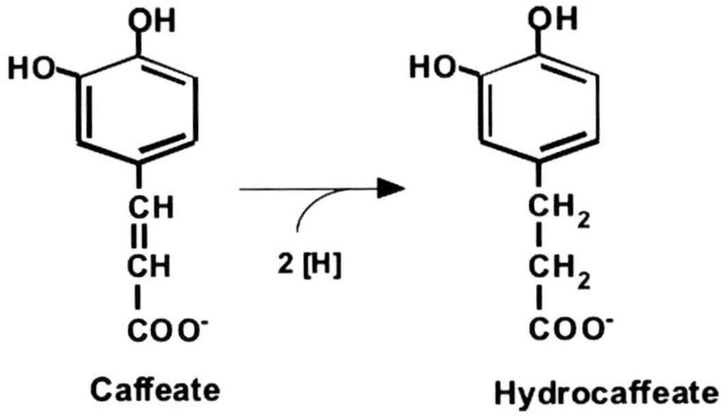

FIGURE 2. Caffeate reduction to hydrocaffeate by *A. woodii*. The reducing equivalents may derive from different donors, such as, for example, hydrogen, sugars or C_1 compounds.

A. woodii is known to reduce the carbon-carbon double bond of phenylacrylates, such as caffeate, as shown in FIGURE 2.[8,9] The electrons for caffeate reduction can be derived from various donors, such as, for example, fructose, methanol, or hydrogen.[10] It is important to note that *A. woodii* cannot use caffeate either as carbon or as energy source, but only reduces the double bond, yielding hydrocaffeate. Hydrocaffeate also is neither a carbon nor an energy source for *A. woodii*.

Caffeate is widespread in soils and derives from the degradation of lignin. The utilization of caffeate is not constitutive but induced by caffeate[8] and requires *de novo* protein synthesis.[11] Induction was observed in the presence of both CO_2 and caffeate as electron acceptors and fructose as electron donor, and once induced, caffeate and CO_2 were used simultaneously as electron acceptors. Induction of caffeate reduction was also observed during acetogenesis from $H_2 + CO_2$, but not during oxidation of methyl groups derived from methanol or betaine, where acetogenesis was the preferred energy conserving pathway. The differential flow of reductants toward either caffeate or CO_2 was also observed with suspensions of resting cells in which caffeate reduction was induced prior to harvesting the cells. These cell suspensions utilized caffeate and CO_2 simultaneously with fructose or hydrogen as electron donors, but CO_2 was preferred over caffeate during methyl-group oxidation. Resting cells of caffeate-induced cells could reduce caffeate but also *p*-coumarate or ferulate, with hydrogen as electron donor. *p*-Coumarate or ferulate also served as inducers for caffeate reduction. Interestingly, caffeate-induced cells reduced ferulate in the absence of an external reductant, indicating that caffeate also induces the enzymes required for oxidation of the methyl group of ferulate.[11]

Cell yield measurements with cells grown on fructose or methyl-group containing substrates gave evidence that caffeate is not only used as an electron sink, but, in addition, caffeate reduction is coupled to energy conservation.[8] Clear evidence for ATP synthesis coupled to caffeate reduction was obtained with resting cells of *A. woodii* in which hydrogen-dependent caffeate reduction was accompanied with the synthesis of ATP.[12] Recently, it was shown that ATP synthesis occurred by a chemiosmotic mechanism.[13] Most interestingly, like CO_2 reduction, hydrogen-dependent caffeate reduction as well as ATP synthesis coupled to caffeate reduction were strictly Na^+ dependent and the latter was dependent on a transmembrane Na^+ gradient. These studies were fully compatible with the following sequence of events: caffeate reduction → generation of a transmembrane Na^+ gradient → generation of ATP by the Na^+ F_1F_O ATP synthase.[13]

Dissection of the Electron-transfer Pathway Involved in Caffeate Respiration: A Ferredoxin:NAD$^+$-Oxidoreductase (Rnf Complex) as Potential Coupling Site in *Acetobacterium woodii*

From the finding that caffeate reduction with electrons derived from hydrogen is coupled to the generation of a transmembrane Na^+ gradient it is evident that one of the reactions must catalyze electrogenic Na^+ transport across the membrane. Cytochromes or quinones were not detected in caffeate-grown cultures.[8] To unravel the enzymes involved in hydrogen-dependent caffeate reduction, subcellular systems were used. A cell-free extract of *A. woodii* catalyzed H_2-dependent caffeate reduction. This reaction is strictly ATP dependent, but can also be activated by acetyl coenzyme A, indicating the activation of caffeate to caffeoyl-CoA prior to its reduction.[14] This is an interesting observation, since it would argue for uptake of caffeate into the cell and subsequent intracellular reduction of caffeoyl-CoA to hydrocaffeoyl-CoA.

To identify proteins involved in caffeate reduction, cells were grown with fructose either in the absence or presence of caffeate, followed by subsequent comparison of their cellular protein content by 2D gel electrophoresis. Two proteins were identified by ESI-MS/MS and the genes were cloned. They are very similar to subunits α (EtfA) and β (EtfB) of electron-transfer flavoproteins (Etf) present in various anaerobic bacteria. Etf are widespread in nature and are

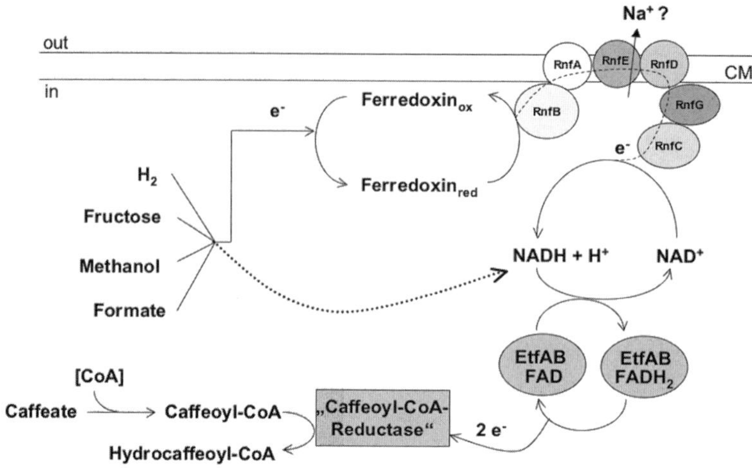

FIGURE 3. Caffeate respiration as carried out by *A. woodii*. Note that the "caffeoyl-CoA reductase" is speculative. Hydrogenase or pyruvate:ferredoxin oxidoreductase metabolism generates reduced ferredoxin that may be reoxidized by Rnf. NADH generated during, for example, glycolysis may be used directly as electron donor for caffeoyl-CoA reduction. For further explanations, see text. ABBREVIATION: CM = cytoplasmic membrane.

found in archaea, bacteria, and eukarya.[15] They are heterodimers of a large α-subunit (EtfA) and a small β-subunit (EtfB) that contain a noncovalently bound flavin adenine dinucleotide (FAD). The role of Etf's is to mediate electron flow between donor and acceptor pairs. Etf's are involved in fatty-acid oxidation, where they funnel the electrons into the membrane-bound electron transport chain at the level of ubiquinone,[16,17] but also in reduced nicotinamide adenine dinucleotide (NADH)-dependent reduction of electron acceptors in cytosolic redox reactions, as, for example, butyric acid fermentation.[18–20] Western blotting analyses revealed that EtfAB of *A. woodii* was present in the cytoplasm.[14] This finding prompted us to test for NADH-dependent caffeate reduction in the cytoplasm, and indeed, the cytoplasmic fraction reduced caffeate with NADH as reductant with 60% of the rate of cell-free extract. Furthermore, membranes neither contained nor stimulated the activity. These findings suggest a cytoplasmic electron flow from NADH via EtfAB to caffeoyl-CoA.

An interesting analogy to caffeate reduction as catalyzed by *A. woodii* is found in some clostridia. In *Clostridium propionicum* the acryloyl-CoA reductase catalyzes the reduction of acryloyl-CoA to propionyl-CoA with electrons derived from NADH via Etf.[20] The reductase and Etf constitute an enzyme complex. An analogous reaction is catalyzed by butyryl-CoA dehydrogenases involved in butyrate fermentation, which also receives electrons from NADH via Etf.[19,21,22]

Remaining candidates for a membrane localization and thus for the Na^+-translocating enzyme include the enzymes involved in hydrogen-dependent NAD^+ reduction. *A. woodii* was shown some time ago to have a ferredoxin-dependent, iron-only, soluble hydrogenase.[23] We could confirm the cytosolic localization of the hydrogenase in *A. woodii* DSM 1030, arguing against a role of hydrogenase in primary energy coupling. Therefore, we searched for an enzyme that could couple oxidation of reduced ferredoxin (generated by hydrogenase) with the reduction NAD^+. Indeed, membranes catalyzed NADH oxidation with hexacyanoferrate or benzylviologen as acceptor, but also ferredoxin-dependent NAD^+ reduction. The ferredoxin:NAD^+-oxidoreductase in membranes of *A. woodii* was the only membrane-bound enzyme found, and thus became a very likely candidate for the Na^+-translocating enzyme system. A membrane-bound enzyme that could catalyze electron transfer from ferredoxin to NAD^+ coupled to electrogenic ion transport across the membrane may be the Rnf complex found recently in various bacteria and archaea. Indeed, evidence for a Rnf complex in *A. woodii* was provided by molecular tools. Using primers directed against conserved regions of *rnfC*, we amplified a DNA fragment that encodes a peptide with similarity to RnfC. Further cloning strategies revealed additional *rnf* genes adjacent to *rnfC* in *A. woodii*.[14] A summary of the pathway of electron flow from various donors to caffeate is shown in FIGURE 3.

The Biochemistry and Bioenergetics of Rnf Complexes

The Rnf complex was first discovered in the purple nonsulfur bacterium *Rhodobacter capsulatus*.[24] Through analysis of the defined insertion and deletion mutants of *R. capsulatus*, six genes were identified (*rnfABCDEF*) that were required for nitrogen fixation (Rnf = *Rhodobacter* nitrogen *f*ixation). *rnfABCDE* were found to be organized in a cluster, whereas *rnfF* was not part of it. Corrections were made to the DNA sequence of *R. capsulatus* and two additional *rnf* genes were identified, *rnfG* and *rnfH*. Therefore, the *rnf* cluster consists of seven genes.[30] Bioinformatic analyses revealed that the encoded proteins are similar to subunits of Na^+-translocating NADH:quinone-oxidoreductases (Nqr). Nqr was first found in the marine bacterium *Vibrio alginolyticus*,[25] but are widespread in nature and have attracted considerable interest.[25–28] Like Rnf, Nqr complexes are short-chain (6 subunits) NADH dehydrogenases. They mediate electrogenic Na^+ translocation, which is coupled to electron flow from NADH to ubiquinone. The enzyme purified from *V. alginolyticus* contained one non-covalently bound FAD, an iron–sulfur cluster and ubiquinone. After reconstitution into liposomes, it catalyzed electrogenic Na^+ transport, but the subunits involved and the mechanism used remain obscure.[29]

For *R. capsulatus* it was hypothesized that the *rnf* genes encode a membrane bound electron transport chain that is required for electron transport to nitrogenase. This hypothesis was corroborated by the finding that strains of *R. capsulatus* overexpressing the *rnf* genes showed enhanced nitrogenase activity,[30] indicating that the supply of reductants through the Rnf complex might be rate limiting for nitrogenase activity *in vivo*. When grown in light, the reducing power of the primary acceptor of the photosystem, an ubiquinone ($E^{\circ\prime} = -150\,mV$)[31] is too low to meet the requirement of the nitrogenase system. Therefore, it was speculated that the Rnf complex of *Rhodobacter* catalyzes a reverse electron transport of electrons derived from NADH to ferredoxin driven by the electrochemical proton potential across the membrane.

In *R. capsulatus* the Rnf complex is encoded by the seven genes *rnfABCDEGH*.[30] RnfC of *R. capsulatus* is a soluble protein with a predicted molecular mass of 55 kDa. It contains cysteine motifs typical for 2 [4Fe-4S]-type ferredoxin-like domains, and indications for an NADH- and FMN-binding site were found.[32] RnfA, RnfD, and RnfE are predicted to be transmembrane proteins. A hydrophobicity plot of the amino acid sequence of RnfA (20.4 kDa) suggested that it contains 5–6 membrane-spanning domains, whereas RnfD (38 kDa) contains 9–12 and RnfE (25.8 kDa) is predicted to contain 7–8 transmembrane helices. Immunoblot as well as activity measurements with translational fusions confirmed the membrane localization of RnfA.[32] It is suggested that RnfE might represent the coupling site for sodium ion translocation.[32] Similar to NqrB, NqrD, and NqrE, the hydrophobic subunits that may form the ion channel of the Na^+-translocating Nqr in *V. alginolyticus*, RnfA, RnfD and RnfE might form a subcomplex as well.[26] RnfD is similar to NqrB, and RnfE and RnfA are related to NqrD and NqrE, respectively. RnfB is mainly hydrophilic, but contains a short hydrophobic stretch at the N terminus. The predicted molecular mass of RnfB is 19 kDa. In addition, recombinant RnfB isolated from *Escherichia coli* was found to contain one [2Fe-2S] cluster based on absorption spectra, electron paramagnetic resonance (EPR) spectroscopy data, and iron content.[30] By immunoblot analysis it was shown that RnfB and RnfC were tightly bound to the membrane despite the prediction that they are mostly hydrophilic.[30] RnfG of *R. capsulatus* has an apparent molecular mass of 23 kDa. The amino acid sequence of RnfG is mainly hydrophilic despite the N-terminal region that contains a short, hydrophobic region. This suggests that the protein might be anchored in the membrane. A potential flavin mononucleotide (FMN)–binding site was detected in RnfG through database searches for conserved domains. In addition RnfG was found to be similar to NqrC[30]. The amino acid sequences of RnfH with a predicted molecular mass of 9.5 kDa shows mainly hydrophilic domains in the hydrophobicity plot, indicating that RnfH might be a soluble protein.

The overall similar subunit organization as well as the similarity of single subunits of Rnf to Nqr has led to the idea that Rnf complexes mediate electrogenic Na^+ transport driven by electron transport from reduced ferredoxin to NAD^+. However, although this is an attractive idea, it must be emphasized that there has been, up to now, no experimental proof that the Rnf complex of any organism indeed translocates ions across the cytoplasmic membrane. The first Rnf complex, from *Clostridium tetanomorphum*, was purified recently.[33] The enzyme purified under anaerobic conditions contained six subunits, NfoABCDEG (Nfo is homologous to Rnf and stands for NADH:ferredoxin-oxidoreductase). Along with noncovalently and covalently bound flavins iron was detected in the enyzme. Sequence analyses suggest a polyferredoxin with four

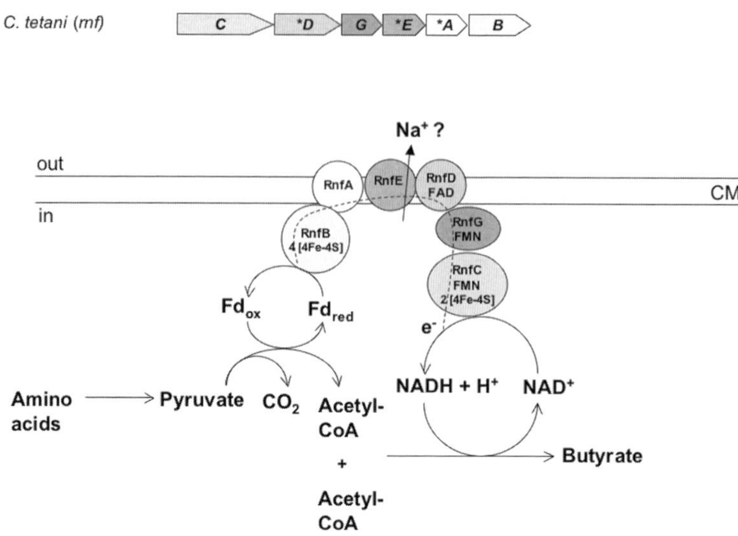

FIGURE 4. The Rnf complex of *C. tetani*.[34] The Rnf complex of *C. tetani* comprises six subunits RnfCDGEAB. Its physiological function is suggested to be the reoxidation of reduced ferredoxin derived from pyruvate oxidation coupled to NAD^+ reduction and Na^+ export. The NADH thus generated is used as reductant in butyrate fermentation.[34] Similar subunits have similar color coding. The cofactors/redox centers where assigned based on the data for *C. tetanomorphum* and *R. capsulatus*.[30,32,33]

ferredoxin-like [4Fe-4S] centers in NfoB (RnfB).[33] As suggested by Boiangiu *et al.*[33] and Brüggemann *et al.*[34] the polyferredoxin in RnfB accepts the electrons from ferredoxin and transfers them to the membrane domain, through which they are transported to RnfC, which contains the NADH binding site (FIG. 4). Again, it should be mentioned that there is no proof that the Rnf complex translocates ions. The experimental validation of the predicted function is a challenging task for future studies.

The Genetic Organization of Rnf Complexes

Rnf complexes are widespread in nature and are found to be encoded by genomes of a variety of organisms, such as the Gram-negative *E. coli*, *Pseudomonas aeruginosa*, *Azotobacter vinelandii*, and *Haemophilus influenzae*, but also by Gram-positive bacteria, such as *Clostridium tetani* and *Clostridium perfringens*, just to mention a few (FIG. 5). Interestingly, the genomes of many pathogenic bacteria harbor *rnf* genes, but a possible role of these complexes in pathogenicity is not documented.[34] The genome of *A. vinelandii* contains two *rnf*-like gene clusters.[35] The organism is a free-living, aerobic bacterium that can fix nitrogen. The *rnf1* operon consists of seven genes (*ABCDGEH*) and the *rnf2* operon of six genes (*ABCDGE*). The corresponding proteins are very similar to each other.[35] Expression of *rnf1* is coregulated with the *nif* genes, whereas expression of *rnf2* is constitutive. In anaerobes, such as *C. tetani* and others, the order of genes is slightly different (FIG. 5). Furthermore, *Methanosarcina acetivorans* contains additional genes. The physical linkage of the *rnf* genes implies a coregulation, but whether or not they form an operon has not been addressed. However, iron limitation in *R. capsulatus* leads to an increase in cellular levels of RnfB.[30]

The Variation of Subunit Composition and Physiological Function of Rnf Complexes

An interesting variation of a Rnf complex is found in the archaeon *M. acetivorans*, where the *rnf* gene cluster has eight genes.[36] It is speculated that in cells grown on acetate the *rnf* complex mediates oxidation of reduced ferredoxin (generated during acetyl-CoA oxidation) with reduction of methanophenanzine that then reduces the heterodisulfide. Since the electron acceptor is methanophenazine and not NAD^+, one would expect the electron output module of this complex to be different from others. Indeed, in contrast to any other Rnf complex described so far, an additional subunit encoding a cytochrome *c* is present. This cytochrome *c* is suggested to be involved in methanophenanzine

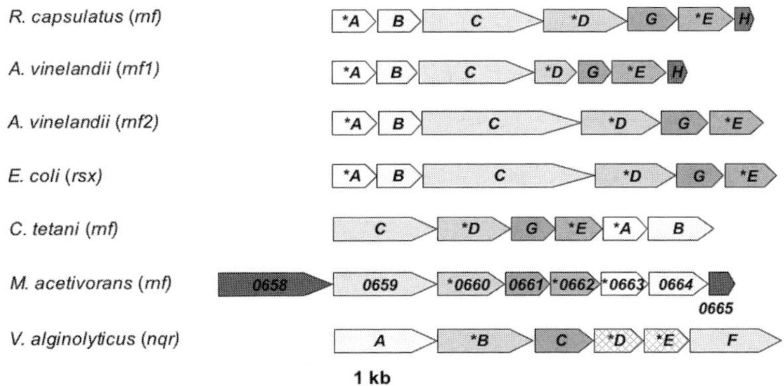

FIGURE 5. The genetic organization of Rnf and Nqr complexes. Sequences were derived from databases. Similar subunits have similar color coding. The membrane integral subunits are marked by an *asterisk*.

reduction. The *rnf* operon of *C. tetani* also contains six genes (*rnfCDGEAB*). The function of the Rnf complex is the production of NADH required for the reduction of acetoacetyl-CoA to butyrate. Again, it is suggested that electron transfer from Fd_{red} to NAD^+ generates an Na^+ potential.[34] Like the Rnf complex of *R. capsulatus*, an Rnf complex is speculated to be involved in ferredoxin reduction in the synthrophic bacterium *Synthrophus aciditrophicus*.[37] When growing on benzoate or fatty acids, this organism has to synthesize pyruvate from acetyl-CoA and CO_2. The ferredoxin required might be generated via reversed electron flow from NADH driven by the transmembrane electrochemical ion (H^+, Na^+) gradient.

Energetics of Rnf-Mediated Reactions

The difference in redox potential between ferredoxin ($E^{\circ\prime} =$ ca. -420 mV) and NADH ($E^{\circ\prime} = -320$ mV) is rather small and may not exceed -100 mV, equivalent to approx. -20 kJ/mol. Assuming an electrochemical ion potential around -200 mV across the cytoplasmic membrane, the free-energy change of this reaction would allow for the translocation of only one ion across the membrane. This clearly is at or near the thermodynamic limit to permit energy conservation. A lower electrochemical ion potential would increase the number of ions that can be pumped. Anyway, if one assumes the energetically most unfavorable value of one ion exported per mol of ferredoxin oxidized, three mol of reduced ferredoxin must be oxidized to give the three ions required by the ATP synthase to synthesize one mol of ATP.

Again, it should be stressed that there is no biochemical proof for ion export coupled to electron transport through the Rnf complex. However, our finding that caffeate respiration is coupled to ATP synthesis by a chemiosmotic mechanism, with Na^+ as coupling ion and the finding that a ferredoxin:NAD^+-oxidoreductase was the only membrane-bound enzyme detected in the pathway of hydrogen-dependent caffeate reduction makes it likely that the Rnf complex of *A. woodii* is indeed a sodium ion pump. Purification and characterization of the enzyme is currently under way.

The Rnf Complex, A Potential Coupling Site Also in the Wood–Ljungdahl Pathway?

As outlined earlier, the acetyl-CoA pathway is coupled to the generation of a transmembrane ion gradient that in turn drives the synthesis of ATP, but the reactions involved are still unknown. The sodium ion-dependent organisms with *A. woodii* as the model organism do not have cytochromes or quinones but membrane-bound corrinoids, like methanogenic archaea.[5,38,39] The latter have a membrane-bound, corrinoid-containing methyltransferase that couples the transfer of the methyl group from methyltetrahydromethanopterin to coenzyme M with the electrogenic translocation of Na^+ into the medium, thus creating an electrochemical sodium ion potential across the membrane.[40] The methyl transfer moiety of the enyzme is not membrane integral but attached to the complex on the cytosolic side. All the enzymatic activities of the Wood–Ljungdahl pathway were found in the cytoplasmic fraction, and it was always speculated that the treatments required for the isolation of the cytoplasmic membranes sheared off the enzymes or their catalytic moieties of the membrane. Although it was tried for some time to obtain experimental hints for a membrane-bound methyltransferase in Na^+-dependent acetogens, the experiments did not reveal such an enzyme.[5]

Despite the still not answered question whether one of the enzymes catalyzing the carbon flow in the Wood–Ljungdahl pathway in Na^+-dependent acetogens is localized in the cytoplasmic membrane, the Rnf complex could also represent a (the) coupling site in carbonate respiration via the Wood–Ljungdahl pathway. The hydrogenase (during chemolithoautotrophic growth) and the pyruvate:ferredoxin-oxidoreductase (during heterotrophic growth) both generate reduced ferredoxin. Electrons have to flow from reduced ferredoxin to NAD^+ during the operation of the Wood–Ljungdahl pathway, since the three reductive steps leading from CO_2 to the methylated corrinoid-iron sulfur protein use NADH or reduced nicotinamide adenine dinucleotide phosphate (NADPH) as reductant. The involvement of the Rnf complex in the Wood–Ljungdahl pathway is an interesting idea that clearly needs to be further addressed by appropriate experiments. Unfortunately, a genetic system is not available for *A. woodii*, so knockouts cannot be generated. However, the Rnf activity was found in cells grown on fructose in the absence of caffeate. This would argue for a constitutive expression and would be consistent with a function of Rnf in the electron flow during operation of the Wood–Ljungdahl pathway.

It should be mentioned that the reduction of CO_2 to CO mediated by the acetyl-CoA synthase/CO dehydrogenase complex is an endergonic reaction with NADH as reductant. The redox potential of $NAD^+/NADH + H^+$ of $-320\,mV$[41] is too electropositive to be used as a reductant for CO_2 ($E^{\circ\prime}\, CO_2/CO = -518\,mV$).[41] However, both the hydrogenase and the pyruvate:ferredoxin-oxidoreductase use ferredoxin as electron acceptor, which can be used directly as reductant for CO_2 reduction during chemolithoautotrophic and chemoorganoheterotrophic growth. In addition, the reduced ferredoxin required for carboxylation of acetyl-CoA to pyruvate ($E^{\circ\prime} = -498\,mV$),[41] an essential reaction in the anabolism during chemolithoautotrophic growth, can be generated directly from hydrogen by the iron-only soluble hydrogenase of *A. woodii*.

Conclusions

Although the elucidation of the mechanisms of energy conservation in acetogens is still in its infancy, it turned out that *A. woodii* is one of the rare cases in which the entire energetics is based on a Na^+ current across the cytoplasmic membrane. After decades of searching for the Na^+-motive coupling site in acetogens, there is now some light on the horizon. Although the experimental evidence is not yet conclusive, the Rnf complex may represent the long-sought-after sodium pump in acetogens. Clearly, a better understanding of the Na^+ bioenergetics in acetogens is in sight.

An interesting observation is the apparent difference in subunit composition of Rnf complexes in nature. Depending on the electron acceptor and donor system used by a given organism, different modules might be used in Rnf complexes from different sources. This would give Rnf's considerable versatility among diverse organisms. For example, the Rnf complex of *M. acetivorans* is suggested to reduce methanophenazine, which requires a different module compared to an enzyme that reduces NAD^+. That may be the reason why this Rnf complex has a cytochrome *c* in addition.

Genome sequencing revealed genes encoding Rnf complexes in a number of anaerobes. Since the interconversion of ferredoxin and NADH is of pivotal importance in the physiology of anaerobes, Rnf complexes are considered to have central functions in their metabolism. Anaerobes often have the problem that the energy span between electron donor and acceptor pairs is low, often close to the thermodynamic limit, and highly efficient machines and considerable consumption of substrate are required to use this small energetic difference to generate sufficient transmembrane ion potential. Such machines are, for example, the Ech hydrogenase[42] and the Rnf-type NADH dehydrogenase. The former enzyme has been characterized to some detail, but up to now, ion pumping coupled to catalytic activity was not demonstrated. In comparison, there is only little information on the biochemistry of Rnf complexes in general, and in particular, almost nothing is known about Rnf complexes and their role in anaerobes. All that is discussed in the literature stems from "molecular hallucinations" that derive from DNA sequences and predicted protein functions deliniated from comparisons to similar oxidoreductases, that is, Nqr-type NADH dehydrogenases. However, the time is now ripe to address the structure and function of these unique NADH dehydrogenases and to unravel the nature, if any, of the ion translocated during electron transport.

Acknowledgments

We are indebted to the Deutsche Forschungsgemeinschaft for continous and generous support. We are very grateful to Prof. W. Buckel, Marburg, Germany, for stimulating discussions and the introduction into the art of measuring ferredoxin:NAD$^+$-oxidoreductase, as well as the gift of ferredoxin.

Conflict of Interest

The authors declare no conflicts of interest.

References

1. DRAKE, H.L., K. KÜSEL & C. MATTHIES. 2006. Acetogenic prokaryotes. *In* The Prokaryotes. M. Dworkin *et al.*, Eds.: 354–420. New York: Springer.
2. DIEKERT, G. & G. WOHLFARTH. 1994. Metabolism of homoacetogens. Antonie Leeuwenhoek Int. J. Gen. Microbiol. **66:** 209–221.
3. MÜLLER, V. & G. GOTTSCHALK. 1994. The sodium ion cycle in acetogenic and methanogenic bacteria: generation and utilization of a primary electrochemical sodium ion gradient. *In* Acetogenesis. H.L. Drake, Ed.: 127–156. New York: Chapman & Hall.
4. LJUNGDAHL, L.G. 1994. The acetyl-CoA pathway and the chemiosmotic generation of ATP during acetogenesis. *In* Acetogenesis. H.L. Drake, Ed.: 63–87. New York: Chapman & Hall.
5. MÜLLER, V. 2003. Energy conservation in acetogenic bacteria. Appl. Environ. Microbiol. **69:** 6345–6353.
6. MÜLLER, V. *et al.* 2004. Molecular and cellular biology of acetogenic bacteria. *In* Strict and Facultative Anaerobes. Medical and Enviromental Aspects. M.M. Nakano & P. Zuber, Eds.: 251–281. Norfolk, VA: Horizon Biosciences.
7. DRAKE, H.L. *et al.* 1994. Acetogenesis: reality in the laboratory, uncertainty elsewhere. *In* Acetogenesis. H.L. Drake, Ed.: 273–302. New York: Chapman & Hall.
8. TSCHECH, A. & N. PFENNIG. 1984. Growth yield increase linked to caffeate reduction in *Acetobacterium woodii*. Arch. Microbiol. **137:** 163–167.
9. BACHE, R. & N. PFENNIG. 1981. Selective isolation of *Acetobacterium woodii* on methoxylated aromatic acids and determination of growth yields. Arch. Microbiol. **130:** 255–261.
10. DAVIES, E.T. & G.M. STEPHENS. 1998. Effect of growth substrate and electron donor on hydrogenation of carbon-carbon double bounds by *Acetobacterium woodii*. Enzyme Microb. Technol. **23:** 129–132.
11. DILLING, S. *et al.* 2007. Regulation of caffeate respiration in the acetogenic bacterium *Acetobacterium woodii*. Appl. Environ. Microbiol. **73:** 3630–3636.
12. HANSEN, B. *et al.* 1988. ATP formation coupled to caffeate reduction by H_2 in *Acetobacterium woodii* NZva16. Arch. Microbiol. **150:** 447–451.
13. IMKAMP, F. & V. MÜLLER. 2002. Chemiosmotic energy conservation with Na^+ as the coupling ion during hydrogen-dependent caffeate reduction by *Acetobacterium woodii*. J. Bacteriol. **184:** 1947–1951.
14. IMKAMP, F. *et al.* 2007. Dissection of the caffeate respiratory chain in the acetogen *Acetobacterium woodii*: identification of a Rnf-type NADH dehydrogenase as potential coupling site. J. Bacteriol. **189:** 8145–8153.
15. TSAI, M.H. & M.H. SAIER, JR. 1995. Phylogenetic characterization of the ubiquitous electron transfer flavoprotein families ETF-α and ETF-β. Res. Microbiol. **146:** 397–404.
16. FRERMAN, F.E. 1987. Reaction of electron-transfer flavoprotein ubiquinone oxidoreductase with the mitochondrial respiratory chain. Biochim. Biophys. Acta **893:** 161–169.
17. RAMSAY, R.R., D.J. STEENKAMP & M. HUSAIN. 1987. Reactions of electron-transfer flavoprotein and electron-transfer flavoprotein: ubiquinone oxidoreductase. Biochem. J. **241:** 883–892.
18. KOMUNIECKI, R. *et al.* 1989. Electron-transfer flavoprotein from anaerobic *Ascaris suum* mitochondria and its role in NADH-dependent 2-methyl branched-chain enoyl-CoA reduction. Biochim. Biophys. Acta **975:** 127–131.
19. O'NEILL, H., S.G. MAYHEW & G. BUTLER. 1998. Cloning and analysis of the genes for a novel electron-transferring flavoprotein from *Megasphaera elsdenii*. Expression and characterization of the recombinant protein. J. Biol. Chem. **273:** 21015–21024.
20. HETZEL, M. *et al.* 2003. Acryloyl-CoA reductase from *Clostridium propionicum*. An enzyme complex of propionyl-CoA dehydrogenase and electron-transferring flavoprotein. Eur. J. Biochem. **270:** 902–910.
21. BOYNTON, Z.L., G.N. BENNET & F.B. RUDOLPH. 1996. Cloning, sequencing, and expression of clustered genes encoding β-hydroxybutyryl-coenzyme A (CoA) dehydrogenase, crotonase, and butyryl-CoA dehydrogenase from *Clostridium acetobutylicum* ATCC 824. J. Bacteriol. **178:** 3015–3024.
22. ASANUMA, N. *et al.* 2005. Characterization and transcription of the genes involved in butyrate production in *Butyrivibrio fibrisolvens* type I and II strains. Curr. Microbiol. **51:** 91–94.
23. RAGSDALE, S.W. & L.G. LJUNGDAHL. 1984. Hydrogenase from *Acetobacterium woodii*. Arch. Microbiol. **139:** 361–365.
24. SCHMEHL, M. *et al.* 1993. Identification of a new class of nitrogen fixation genes in *Rhodobacter capsulatus*: a putative membrane complex involved in electron transport to nitrogenase. Mol. Gen. Genet. **241:** 602–615.
25. TOKUDA, H. & T. UNEMOTO. 1981. A respiration-dependent primary sodium extrusion system functioning at alkaline pH in the marine bacterium *Vibrio alginolyticus*. Biochem. Biophys. Res. Commun. **102:** 265–271.
26. RICH, P.R., B. MEUNIER & F.B. WARD. 1995. Predicted structure and possible ionmotive mechanism of the sodium-linked NADH-ubiquinone oxidoreductase of *Vibrio alginolyticus*. FEBS Lett. **375:** 5–10.
27. STEUBER, J. 2001. Na^+ translocation by bacterial NADH:quinone oxidoreductases: an extension to the complex-I family of primary redox pumps. Biochim. Biophys. Acta **1505:** 45–56.
28. BOGACHEV, A.V. & M.I. VERKHOVSKY. 2005. Na^+-translocating NADH:quinone oxidoreductase: progress achieved and prospects of investigations. Biochemistry (Mosc.) **70:** 143–149.
29. PFENNINGER-LI, X.D. *et al.* 1996. NADH:ubiquinone oxidoreductase of *Vibrio alginolyticus*: purification, properties, and reconstitution of the Na^+ pump. Biochemistry **35:** 6233–6242.
30. JOUANNEAU, Y. *et al.* 1998. Overexpression in *Escherichia coli* of the *rnf* genes from *Rhodobacter capsulatus*—

Characterization of two membrane-bound iron-sulfur proteins. Eur. J. Biochem. **251:** 54–64.

31. PRINCE, R.C. & P.L. DUTTON. 1978. Protonation and the reducing potential of the primary electron receptor. *In* The Photosynthetic Bacteria. R.K. Clayton & W.R. Sistrom, Eds.: 439–453. New York: Plenum Press.
32. KUMAGAI, H. *et al.* 1997. Membrane localization, topology, and mutual stabilization of the *rnfABC* gene products in *Rhodobacter capsulatus* and implications for a new family of energy-coupling NADH oxidoreductases. Biochemistry **36:** 5509–5521.
33. BOIANGIU, C.D. *et al.* 2005. Sodium ion pumps and hydrogen production in glutamate fermenting anaerobic bacteria. J. Mol. Microbiol. Biotechnol. **10:** 105–119.
34. BRÜGGEMANN, H. *et al.* 2003. The genome sequence of *Clostridium tetani*, the causative agent of tetanus disease. Proc. Natl. Acad. Sci. USA **100:** 1316–1321.
35. CURATTI, L. *et al.* 2005. Genes required for rapid expression of nitrogenase activity in *Azotobacter vinelandii*. Proc. Natl. Acad. Sci. USA **102:** 6291–6296.
36. LI, Q. *et al.* 2006. Electron transport in the pathway of acetate conversion to methane in the marine archaeon *Methanosarcina acetivorans*. J. Bacteriol. **188:** 702–710.
37. MCINERNEY, M.J. *et al.* 2007. The genome of *Syntrophus aciditrophicus*: life at the thermodynamic limit of microbial growth. Proc. Natl. Acad. Sci. USA **104:** 7600–7605.
38. MÜLLER, V., M. BLAUT & G. GOTTSCHALK. 1993. Bioenergetics of methanogenesis. *In* Methanogenesis. J.G. Ferry, Ed.: 360–406. New York: Chapman & Hall.
39. DANGEL, W. *et al.* 1987. Occurrence of corrinoid-containing membrane proteins in anaerobic bacteria. Arch. Microbiol. **148:** 52–56.
40. GOTTSCHALK, G. & R.K. THAUER. 2001. The Na^+-translocating methyltransferase complex from methanogenic archaea. Biochim. Biophys. Acta **1505:** 28–36.
41. THAUER, R.K., K. JUNGERMANN & K. DECKER. 1977. Energy conservation in chemotrophic anaerobic bacteria. Bacteriol. Rev. **41:** 100–180.
42. HEDDERICH, R. 2004. Energy-converting [NiFe] hydrogenases from archaea and extremophiles: ancestors of complex I. J. Bioenergy. Biomembr. **36:** 65–75.

Methanogenesis in Marine Sediments

JAMES G. FERRY AND DANIEL J. LESSNER

Department of Biochemistry and Molecular Biology, The Pennsylvania State University, University Park, Pennsylvania, USA

The anaerobic conversion of complex organic matter to CH_4 is an essential link in the global carbon cycle. In freshwater anaerobic environments, the organic matter is decomposed to CH_4 and CO_2 by a microbial food chain that terminates with methanogens that produce methane primarily by reduction of the methyl group of acetate and also reduction of CO_2. The process also occurs in marine environments, particularly those receiving large loads of organic matter, such as coastal sediments. The great majority of research on methanogens has focused on marine and freshwater CO_2-reducing species, and freshwater acetate-utilizing species. Recent molecular, biochemical, bioinformatic, proteomic, and microarray analyses of the marine isolate *Methanosarcina acetivorans* has revealed that the pathway for acetate conversion to methane differs significantly from that in freshwater methanogens. Similar experimental approaches have also revealed striking contrasts with freshwater species for the pathway of CO-dependent CO_2 reduction to methane by *M. acetivorans*. The differences in both pathways reflect an adaptation by *M. acetivorans* to the marine environment.

Key words: **marine; methane; *Methanosarcina*; anaerobic; acetate; proteomics; carbon monoxide**

Ecology and Microbiology of the Anaerobic Decomposition of Organic Matter

The conversion of complex organic matter to CH_4 is an essential link in the global carbon cycle (FIG. 1). Most organic matter is cycled back to CO_2 for photosynthesis by oxygen-requiring aerobes; however, a portion of the organic matter enters a diversity of anaerobic (oxygen-free) environments, where it is decomposed by a consortia of microbes.[1,2] In freshwater anoxic environments the organic matter is decomposed to CH_4 and CO_2 by a microbial food chain made up of at least three metabolic groups. The fermentative group decomposes the complex organic matter to H_2, formate, CO, acetate, and higher volatile fatty acids. The acetogenic group further decomposes the higher volatile fatty acids to acetate, H_2, and formate that are substrates for growth and methanogenesis by the methanogen group. About two-thirds of all CH_4 derives from reduction of the methyl group of acetate and the remaining one-third from the reduction of CO_2 with electrons derived from the oxidation of either H_2 or formate. Examples of anoxic methanogenic environments include the rumen and lower intestinal tract of animals, anaerobic digesters of sewage treatment plants, landfills, and the sediments of freshwater wetlands, rice paddies, ponds, streams, and lakes. The CH_4 escapes from sediments into the overlying water column where aerobic (O_2-requiring) microbes oxidize it to CO_2, thereby completing the carbon cycle. However, recent reports suggest that a portion of the CH_4 is oxidized to CO_2 anaerobically by microbes using nitrate as the electron acceptor.[3] The CH_4 that escapes both aerobic and anaerobic CH_4 oxidizers enters the atmosphere, contributing to the greenhouse effect.

In many anaerobic marine environments, sulfate-reducing microbes out compete methanogens by virtue of a greater affinity for H_2 and acetate; however, methanogenesis predominates where sulfate is depleted by high organic matter loading.[1] Continental shelf environments that receive high organic loading produce between 0.7 and 14×10^9 kilograms of CH_4 annually, which approximates the estimated range for freshwater lakes. One example with high methanogenic activity is Scripps Canyon adjacent to the Scripps Institution of Oceanography in La Jolla, California.[1] The canyon sediment contains copious amounts of decaying organic matter contributed in large part to the macroalga *Macrocystis pyrifera*. The CH_4 profile in the overlying water column ranges from 4.7 nM at the surface to 870 nM at 23 m, indicating that the CH_4 originates from the sediment. The

Address for correspondence: James G. Ferry, Department of Biochemistry and Molecular Biology, The Pennsylvania State University, University Park, PA 16802. Voice: 814-863-5721.
jgf3@psu.edu

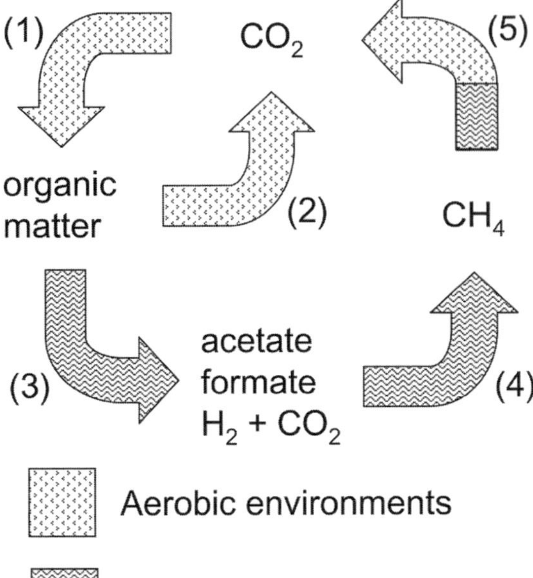

FIGURE 1. The global carbon cycle. Steps: (1) Fixation of CO_2 into organic matter; (2) aerobic (oxygen-dependent) decomposition of organic matter to CO_2; (3) deposition of organic matter into anaerobic (oxygen-free) environments and decomposition to metabolic end-products by fermentative and obligate hydrogen-producing anaerobes; (4) conversion of the end-products to CH_4 by the methanoarchaea and escape of the CH_4 to aerobic environments; (5) oxidation of CH_4 to CO_2 by aerobic and anaerobic methanotrophs.

surface concentration is four-fold greater than the predicted solubility equilibrium with air, indicating a net flux of CH_4 into the atmosphere. Open ocean waters also show a supersaturation of CH_4 in surface waters and an increase of dissolved CH_4 with depth. The estimated annual flux into the atmosphere from open oceans is 4–6.7 × 10^9 kilograms, based on a supersaturation of 1.3 and concentration of 4.7 × 10^{-5} mL CH_4 liter^{-1} seawater at the surface.[1] Estimates of the annual CH_4 release form marine sources are only a fraction of the total biogenic production on earth of nearly one billion metric tonnes; however, the total production in marine environments is an underestimate since it is oxidized to CO_2 by aerobic and anaerobic methanotrophs.[4–6]

Approximately one-third of the described species of methanogens are of marine origin, occurring in 4 of the 5 orders within the kingdom Euryarchaeota of the Archaea domain.[1,7] Marine isolates include psychrophiles, thermophiles, acidophiles, extreme halophiles, autotrophs, and heterotrophs. All described species in the order Methanococcales are of marine origin, autotrophic, and only produce CH_4 by reducing CO_2. The order Methanomicrobiales contains both nonmarine and marine species that produce CH_4 by oxidizing either formate or H_2 and reducing CO_2, although some species also use secondary alcohols as electron donors. The order Methanopyrales includes only one marine species, the extreme thermophile *Methanopyrus kandleri*, that only grows and produces CH_4 by oxidation of H_2 and reduction of CO_2. Members of the order Methanosarcinales contain both nonmarine and marine species and are the most catabolically diverse, reducing CO_2 with H_2 or CO and converting the methyl groups of methylotrophic substrates (methanol, methylamines and dimethylsulfide) to CH_4. These compounds are not generally utilized by sulfate-reducing species and, therefore, are "non-competitive" substrates for marine methanogens. Members of this order also convert acetate to CH_4 and CO_2. *Methanosarcina* species are able to utilize all these substrates for growth and methanogenesis, whereas other genera within the order Methanosarcinales are unable to reduce CO_2 or utilize acetate for growth and methanogenesis. Although a large volume of literature exists on the microbiology, molecular biology, and biochemistry of CO_2-reducing and methylotrophic marine methanogens, little has been published concerning the conversion of acetate or CO to methane by marine isolates. Indeed, only two species isolated from marine environments are able to utilize acetate, *Methanosarcina acetivorans* and *Methanosarcina siciliae*.[8,9] The following sections present recent advances in understanding the unusual pathways for conversion of acetate and CO to methane by *M. acetivorans*.

M. acetivorans strain C2A (DSM 2834 and ATCC 35395) was isolated from CH_4-evolving sediment in the Sumner branch of Scripps Canyon where the acetotrophic CH_4-producing population ranges from 530 to 4300/mL in the first 20 cm of sediment.[1,9] *M. acetivorans* cells from the exponential phase of growth are irregularly shaped cocci (FIG. 2) that aggregate in the early stationary phase and become differentiated into communal cysts that release individual cocci when ruptured or transferred to fresh medium (not shown). Cells possess a single fimbria-like structure (FIG. 2), although cells are nonmotile. Thin sections show a monolayered cell wall approximately 10 nm thick comprising protein subunits (FIG. 2). Internal membrane-like vesicles are present in late exponential phase cultures. *M. siciliae* strain C2J was also isolated from the Sumner branch of Scripps Canyon.[8] The morphology and physiology of *M. siciliae* closely resembles *M. acetivorans*. *M. acetivorans* contains a plasmid (pC2A) of approximately

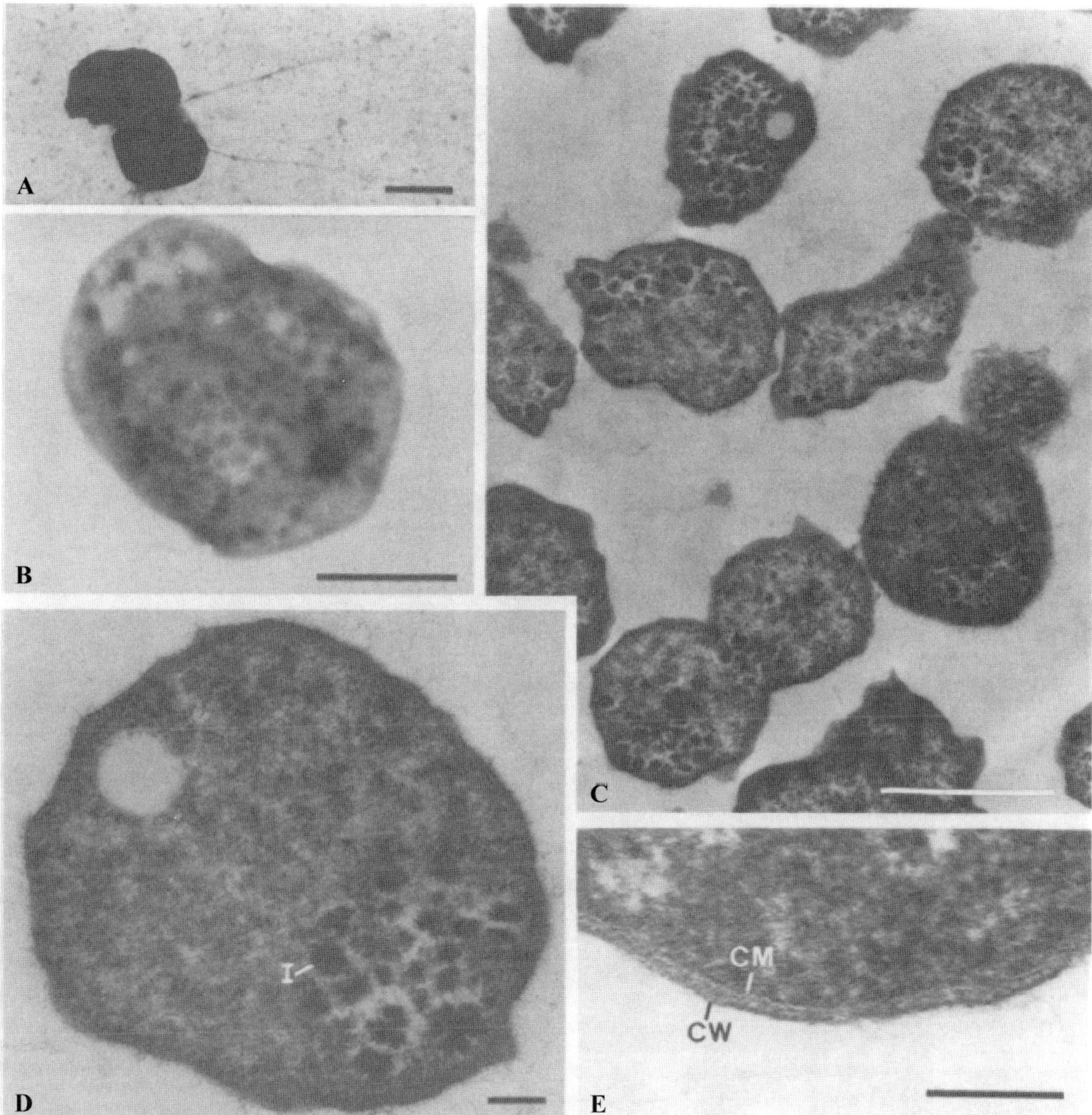

FIGURE 2. Electron micrographs of *M. acetivorans* grown on acetate. (**A**) Negative stain of whole cells showing fimbria-like structures (bar = 1 μm). (**B**) Negative stain of a whole cell showing irregular shape and inclusions (bar = 0.5 μm). (**C**) Thin sections showing inclusions (I) (bar = 1 μm). (**D**) Thin sections showing inclusions (bar = 0.2 μm). (**E**) Thin section illustrating the cell wall (CW) and cytoplasmic membrane (CM) (bar, 100 nm). (From Sowers *et al.*[9] Reprinted by permission.)

5.1 kilobase pairs, which is present in a low copy number of six plasmids per genome.[10] This plasmid is the basis for a genetic exchange system that has accelerated *M. acetivorans* to the forefront of genetic analyses of the Methanosarcinales.[11] Indeed, an unprecedented host of genetic tools has been developed for *M. acetivorans* which include plasmid shuttle vectors, high efficiency transformation, random *in vivo* transposon mutagenesis, directed mutagenesis of specific genes, multiple selectable markers, and plasmid-mediated reporter fusions.[12–16] The sequenced genome at 5,751,492 base pairs, containing 4524 open reading frames (ORFs), is one of the largest known archaeal genomes.[17] These properties, coupled with the extraordinary metabolic diversity and morphological transformations, make *M. acetivorans* an outstanding model organism for investigating marine methanogenesis and, in general, archaeal biology.

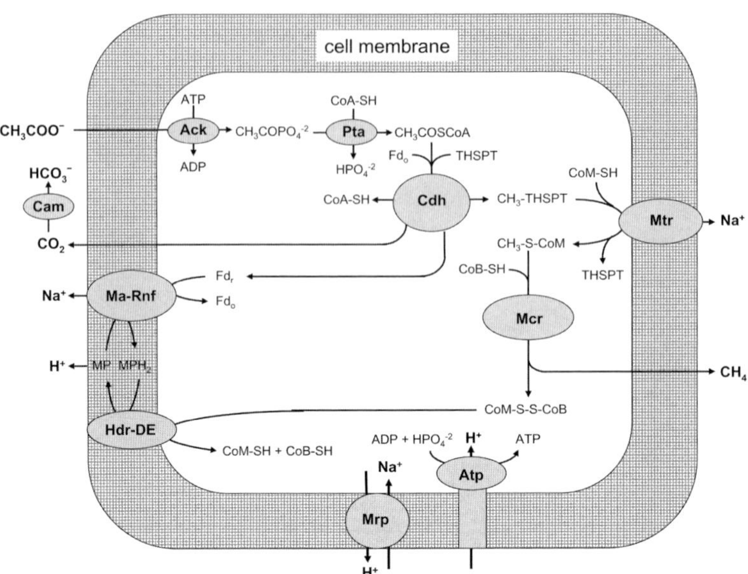

FIGURE 3. Pathway proposed for the conversion of acetate to CH_4 by *M. acetivorans*. *Abbreviations:* Ack, acetate kinase; Pta, phosphotransacetylase; CoA-SH, coenzyme A; THSPT, tetrahydrosarcinapterin; Fd_r, reduced ferredoxin; Fd_o, oxidized ferredoxin; Cdh, CO dehydrogenase/acetyl–CoA synthase; CoM-SH, coenzyme M; Mtr, methyl-THSPT:CoM-SH methyltransferase; CoB-SH, coenzyme B; Cam, carbonic anhydrase; Ma-Rnf, *M. acetivorans* Rnf; MP, methanophenazine; Hdr-DE, heterodisulfide reductase; Mrp, multiple resistance/pH regulation Na^+/H^+ antiporter; Atp, H^+-translocating ATP synthase.

Conversion of Acetate to CH_4

The pathway for conversion of acetate to CH_4 in marine sediments has only been investigated with *M. acetivorans* (FIG. 3), primarily by analysis of the proteome and transcriptome of acetate-grown cells versus methanol-grown cells and modeled with reference to the well-characterized pathway in the freshwater isolates *Methanosarcina thermophila* and *Methanosarcina barkeri* (FIG. 4).[18–20] The results reveal similarities and significant differences between *M. acetivorans* and freshwater *Methanosarcina* species that are described in this section.

The Pathway in Freshwater Methanosarcina

The pathway for conversion of acetate to CH_4 in freshwater *Methanosarcina* species is shown in FIGURE 4. Acetate is transported into the cell by an unknown mechanism, where it is converted to acetyl-CoA catalyzed by acetate kinase (reaction 1) and phosphotransacetylase (reaction 2).

$$CH_3COO^- + ATP \rightarrow CH_3CO_2PO_3{}^{-2} + ADP \quad (1)$$

$$CH_3CO_2PO_3{}^{-2} + HS\text{-}CoA \rightarrow CH_3COSCoA + Pi \quad (2)$$

Genes encoding both enzymes are up-regulated in acetate-grown *M. thermophila* and the freshwater isolate *Methanosarcina mazei* compared to methanol-grown cells results, supporting the proposed functions.[21,22] Biochemical characterization and analyses of the crystal structures for both enzymes from *M. thermophila* has revealed the catalytic mechanisms.[23–28] Kinetic and biochemical analyses of site-specific amino acid variants of acetate kinase has identified residues essential for catalysis and established a direct in-line mechanism for transfer of the phosphate from ATP to acetate. Similar experimental approaches have revealed a concerted mechanism for phosphotransacetylase that proceeds through base-catalyzed generation of $^-$S-CoA, followed by nucleophilic attack of the thiolate anion on the carbonyl carbon of acetyl phosphate.

The five-subunit CO dehydrogenase/acetyl-CoA synthase (Cdh) complex catalyzes reaction 3, which is central to the pathway in the freshwater *Methanosarcina*.

$$CH_3COSCoA + THSPT + H_2O + Fd^{ox} \rightarrow CH_3\text{-}THSPT + Fd^{red} + CO_2 + HS\text{-}CoA \quad (3)$$

The complex cleaves the C-C and C-S bonds of acetyl-CoA, transfers the methyl group to the cofactor tetrahydrosarcinapterin (THSPT), oxidizes the

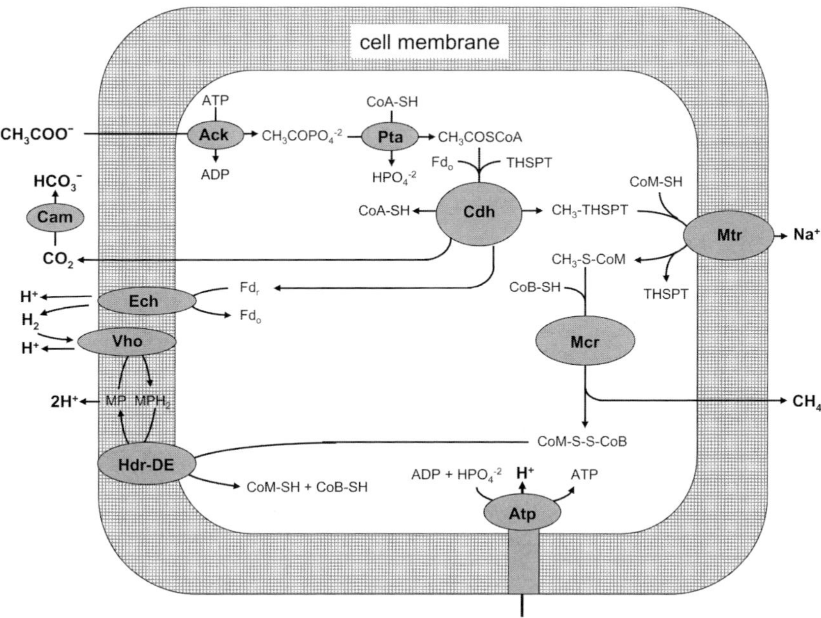

FIGURE 4. Pathway proposed for the conversion of acetate to CH_4 by freshwater *Methanosarcina* species. *Abbreviations:* Ack, acetate kinase; Pta, phosphotransacetylase; CoA-SH, coenzyme A; THSPT, tetrahydrosarcinapterin; Fd_r, reduced ferredoxin; Fd_o, oxidized ferredoxin; Cdh, CO dehydrogenase/acetyl-CoA synthase; CoM-SH, coenzyme M; Mtr, methyl-THSPT:CoM-SH methyltransferase; CoB-SH, coenzyme B; Cam, carbonic anhydrase; Ech, Ech hydrogenase; Vho, Vho hydrogenase; Hdr-DE, heterodisulfide reductase; Atp, H^+-translocating ATP synthase.

carbonyl group to CO_2, and reduces ferredoxin (Fd). Each of these activities has been demonstrated for the complexes isolated from acetate-grown *M. thermophila* and *M. barkeri*.[29–35] The complex from *M. thermophila* is more abundant in acetate-grown cells versus methanol-grown cells, and genes encoding subunits of the complex are up-regulated in acetate- versus methanol-grown *M. mazei* results, which further supports a role in conversion of acetate to CH_4.[22,29] The CO dehydrogenase/acetyl-CoA synthase complexes from *M. thermophila* and *M. barkeri* have been biochemically characterized. The five-subunit (αβδγε) complex from *M. thermophila* is separable into three components.[36] The nickel/iron-sulfur component comprises subunits CdhA (α) and CdhB (ε), whereas the corrinoid/iron-sulfur component contains subunits CdhD (δ) and CdhE (γ). The third component is subunit CdhC (β), where C-C bond cleavage is proposed to occur on metal cluster "A" composed of a 4Fe4S center bridged to a binuclear nickel-nickel site (FIG. 5).[37–40] The methyl group is thought to be transferred to the corrinoid/iron-sulfur component that methylates THSPT.[41,42] Oxidation of the carbonyl group and reduction of ferredoxin is catalyzed by the nickel/iron-sulfur component.[36]

A carbonic anhydrase, Cam, is postulated to convert CO_2 to bicarbonate outside the cell (reaction 4) to facilitate removal from the cytoplasm by converting CO_2 to the less membrane-impermeable bicarbonate.[43,44]

$$CO_2 + H_2O \rightarrow HCO_3{}^- + H^+ \tag{4}$$

A role for Cam in the conversion of acetate to CH_4 is supported by a greater abundance in acetate- versus methanol-grown *M. thermophila* and up-regulation of the gene in acetate- versus methanol-grown *M. mazei*.[22,44] Cam is the archetype of an independently evolved class (γ) of carbonic anhydrase, for which the crystal structure reveals a left-handed β-helical fold.[43,45] Unlike all other carbonic anhydrases, Cam contains iron at the active site.[46] Despite the distinct fold and unique metal requirement, the mechanism of Cam resembles that of the well-studied human CAII, belonging to the α class, in which a hydrogen bond network coordinates the metal-bound hydroxyl for nucleophilic attack on CO_2.[47–49]

The methyl group of CH_3-THSPT is transferred to coenzyme M (HS-CoM) catalyzed by the eight-subunit methyl-tetrahydrosarcinapterin:coenzyme M methyltransferase (Mtr) (reaction 5).

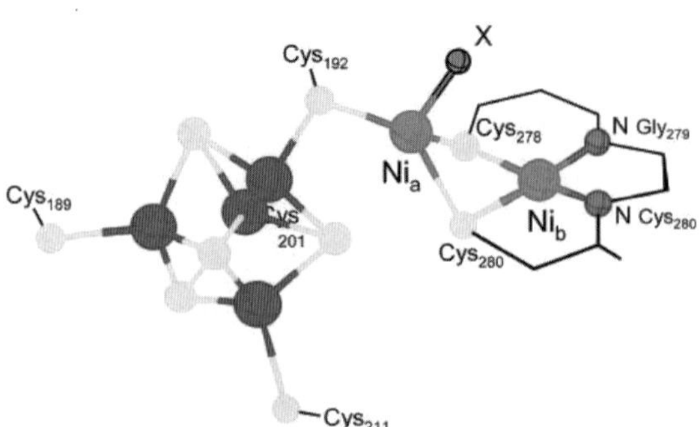

FIGURE 5. Model of the active site in the carbon monoxide/acetyl-CoA synthase complex from *M. thermophila*. (From Gu *et al.*[38] Reproduced by permission.)

$$CH_3\text{-THSPT} + \text{HS-CoM} \rightarrow CH_3\text{-S-CoM} + \text{THSPT} \quad (5)$$

Deletion of the *mtr* operon eliminates growth of *M. barkeri* on acetate, supporting the proposed role for Mtr.[50] The membrane-bound enzyme, characterized from acetate-grown *M. mazei* and *M. barkeri*, translocates sodium ions from the cytoplasm to outside the cell membrane, generating a sodium gradient; however, a specific role for the gradient has not been documented.[51–55] Methyl transfer and sodium translocation is dependent on a corrinoid cofactor. The final step in conversion of the methyl group of acetate to CH_4 is the reductive demethylation of CH_3-S-CoM to CH_4 (reaction 6).

$$CH_3\text{-S-CoM} + \text{HS-CoB} \rightarrow CH_4 + \text{CoM-S-S-CoB} \quad (6)$$

The reaction is catalyzed by methyl-coenzyme M methylreductase (Mcr). Evidence for a role in acetate conversion to CH_4 is reported for *M. barkeri* and *M. thermophila*.[56–59] The homodimeric enzyme from CO_2-reducing methanogens has been extensively characterized. The crystal structure reveals two independent active sites containing the nickel-containing cofactor F_{430}, which accepts the methyl group from CH_3-S-CoM for reduction to CH_4.[60–64] The reductive demethylation of CH_3-S-CoM requires coenzyme B (HS-CoB) as the electron donor, which upon oxidation, forms the heterodisulfide (CoM-S-S-CoB) with CoM. The enzyme in acetate-grown *M. thermophila* requires HS-CoB and contains F_{430}.[59,65]

The heterodisulfide CoM-S-S-CoB produced in reaction 6 is reduced to the corresponding sulfhydryl forms of the cofactors by Hdr (reaction 7).

$$\text{CoM-S-S-CoB} + 2e^- + 2H^+ \rightarrow \text{HS-CoM} + \text{HS-CoB} \quad (7)$$

Heterodisulfide reductase activity has been characterized in acetate-grown *M. thermophila*.[66] The enzyme has been purified and characterized from *M. thermophila* and *M. barkeri*.[67–69] The enzyme from *M. thermophila* has been characterized in biochemical detail and shown to accept electrons from 2-hydroxyphenazine, a water-soluble analog of methanophenazine (MP).[68,69] Methanophenazine is a membrane-bound electron transport cofactor unique to methanogens that functions similarly to quinones in the translocation of protons from the cytoplasm to outside the membrane coupled to electron transfer.[70]

The heterodisulfide CoM-S-S-CoB is the terminal electron acceptor of a membrane-bound electron transport chain coupled to formation of an electrochemical proton gradient that drives adenosine triphosphate (ATP) synthesis. In *M. barkeri* it has been proposed that reduced ferredoxin (Fd_r) donates electrons to Ech hydrogenase (Ech), which produces H_2 and translocates protons from the cytoplasm to outside the membrane.[71,72] This proposal is supported by gene knockout experiments showing that Ech is essential for growth with acetate.[72] Further, a H_2:heterodisulfide oxidoreductase complex purified from acetate-grown *M. barkeri* was shown to contain hydrogenase activity consistent with a role for H_2 as an intermediate in electron transport.[67] It is hypothesized that the hydrogenase (Vho) reoxidizes the H_2 produced by Ech, and that methanophenazine mediates electron transfer to the heterodisulfide reductase.[72] The heterodisulfide reductase purified from acetate-grown cells of *M. thermophila* is also reported to have associated hydrogenase activity consistent with a role for H_2 as an intermediate in electron transport for this species.[68] Isf (iron–sulfur flavoprotein) has been proposed to transfer electrons from Fd_r to the membrane-bound electron transport chain of *M. thermophila*.[73] The crystal structure of Isf reveals an unusual compact cysteine motif ligating a 4Fe4S cluster that accepts electrons from Fd_r and donates to the flavin (FMN) cofactor of Isf.[74] The expression of genes encoding Ech are up-regulated in acetate- versus methanol-grown *M. mazei*, suggesting that Ech functions as proposed for *M. barkeri*.[22] The mechanism for transfer of electrons from Ech to CoMS-SCoB in *M. mazei* is unknown, although the gene encoding Isf is up-regulated in response to growth on acetate, consistent with a role for Isf in this species.[22] Isf from *M.*

thermophila was shown to reduce O_2 and H_2O_2 to water, consistent with a role in the oxidative stress response.[75] Further, Isf from *M. thermophila* is encoded in a gene cluster containing homologues of genes encoding several enzymes involved in the oxidative stress response, a result suggesting that Isf may function in oxidative stress as opposed to electron transport.[75,76] It has been proposed that up-regulation of Isf in response to growth on acetate may be a general stress response.[20] Thus, although a role for Ech is suggested, the pathway for transport of electrons from Ech to CoMS-SCoB and the generation of ion gradients is largely unknown for freshwater *Methanosarcina* species.

The Pathway in Methanosarcina acetivorans

The pathway for acetate conversion to CH_4 in an acetotrophic marine isolate has only been investigated in *M. acetivorans* (FIG. 3). Global proteomic analyses revealing the relative abundance of proteins in acetate-grown cells versus methanol-grown cells indicate that the steps and enzymes for conversion of the methyl group of acetate to CH_4 are similar to freshwater *Methanosarcina* species (FIG. 4).[70] Peptides from Ack, Pta, and subunits of Cdh, enzymes that do not function in the well-characterized pathway of methanol conversion to CH_4 of freshwater *Methanosarcina* species, were shown to be at least 10-fold more abundant in acetate-grown cells.[7] Peptides from Cam were nine-fold more abundant in acetate-grown cells, consistent with a role for this carbonic anhydrase proposed for freshwater *Methanosarcina* species. Peptides from subunits of Mtr and Mcr, essential to the pathway of methanol conversion to CH_4 in freshwater *Methanosarcina*, were shown to be at approximately the same levels in acetate-grown *M. acetivorans*. Furthermore, the amino acid sequence deduced from the encoding genes have greater than 73% identity between the putative enzymes from *M. acetivorans* and the enzymes characterized from freshwater *Methanosarcina* species, suggesting that fundamental properties of the marine and freshwater enzymes are similar.

Contrarily, bioinformatic, biochemical, and proteomic analyses of *M. acetivorans* suggest that enzymes and proteins involved in electron transport are fundamentally different between *M. acetivorans* and freshwater *Methanosarcina*.[77] The combined results suggest that H_2 is not an intermediate in electron transport leading from Fd^{red} to the heterodisulfide in *M. acetivorans*. The genome does not encode a functional Ech and acetate-grown cells have very low H_2-dependent methyl-CoM methylreductase activity.[17,56] Proteomic analyses show that subunits of an electron-transfer complex (Rnf) are at least 10-fold more abundant in acetate- versus methanol-grown *M. acetivorans*, consistent with an electron transport role for this complex in the conversion of acetate to CH_4.[77] Moreover, deletion of the genes encoding the Rnf complex confirm that it is essential for growth on acetate (W. Metcalf, personal communication). Rnf was first discovered in *Rhodobacter capsulatus*, where the six-subunit complex oxidizes reduced nicotinamide adenine dinucleotide (NADH) and reduces the ferredoxin that supplies electrons to nitrogenase.[78,79] Rnf homologs are also proposed to couple electron transport to generation of a Na^+ gradient that drives energy-requiring processes.[80,81] Thus, it is postulated that the Rnf complex in *M. acetivorans* oxidizes Fd_r with electrons ultimately transferred to the heterodisulfide reductase, and that that Rnf pumps sodium ions generating a gradient that is high outside the membrane (FIG. 3).[77] The six subunits of Rnf are encoded in a transcriptional unit containing the *rnfCDGEAB* genes and two additional flanking ORFs.[77] One of the flanking ORFs encode a cytochrome *c* for which spectroscopic analysis of membrane fractions show a greater abundance in acetate-grown cells versus methanol-grown cells. Thus, it is postulated that the cytochrome mediates electron transfer between the six-subunit Rnf complex and the heterodisulfide reductase. A role is also postulated for methanophenazine, mediating electron transfer between the cytochrome and the heterodisulfide reductase (FIG. 3). Proteomic analyses also show that the proton-translocating A_1A_0 ATP synthase is present in acetate-grown *M. acetivorans*.[77] Proteomic analyses further show that subunits of a Na^+/H^+ antiporter (Mrp, multiple resistance/pH) are at least 30-fold more abundant in acetate- versus methanol-grown *M. acetivorans*, consistent with a role during growth on acetate. Thus, a potential role for Mrp is to exchange the Na^+ gradient for a proton gradient (high outside the membrane) that drives ATP synthesis by the A_1A_0 ATP synthase (FIG. 3). The finding that *M. acetivorans* evolved an electron transport pathway distinct from freshwater *Methanosarcina* species not involving H_2 is consistent with the marine environment where sulfate-reducing species out compete methanogens for H_2.

A combined proteomic and microarray analysis of acetate- versus methanol-grown *M. acetivorans* further supports the pathway shown in FIGURE 3 and broadens the understanding of growth on acetate.[20] Duplicate Cdh complexes are synthesized in acetate-grown cells, suggesting both have roles in cleavage of acetyl-CoA. A greater abundance of proteins are detected in acetate-grown cells that respond to stress, including enzymes specific for polyphosphate accumulation and oxidative stress, suggesting that growth on

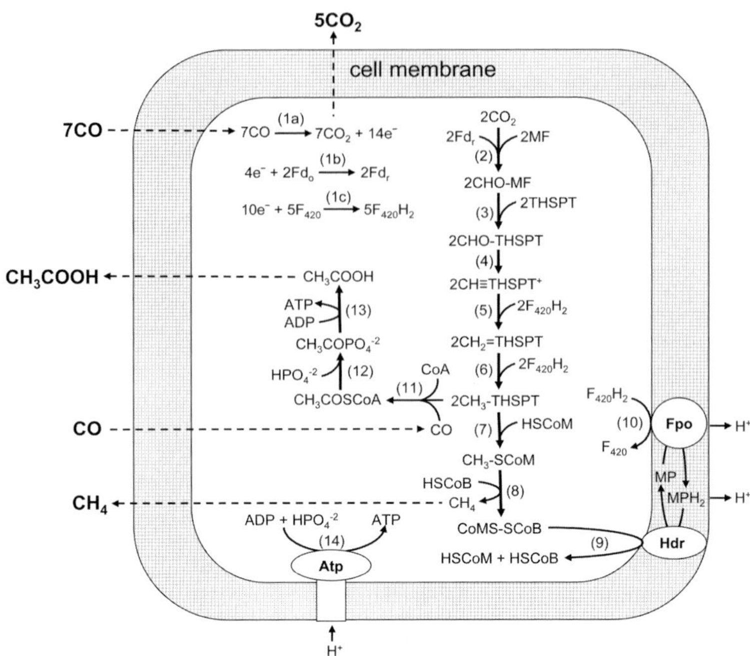

FIGURE 6. Pathway proposed for the conversion of CO to acetate and CH_4 by *M. acetivorans*. *ABBREVIATIONS*: Fd_o, oxidized ferredoxin; Fd_r, reduced ferredoxin; F_{420}, coenzyme F_{420}; MF, methanofuran; THSPT, tetrahydrosarcinapterin; HSCoM, coenzyme M, HSCoB, coenzyme B; Fpo, $F_{420}H_2$ dehydrogenase complex; MP, methanophenazine; Hdr, heterodisulfide reductase.

acetate induces a general stress response. The analyses also indicate the up-regulation of genes in acetate-grown cells that encode several regulatory proteins with identity to the PhoU, MarR, GlnK, and TetR families that are common to the Bacteria domain. The roles for these regulatory proteins have yet to be investigated.

Conversion of Carbon Monoxide to CH_4 by *Methanosarcina acetivorans*

M. acetivorans was isolated from marine sediments containing kelp, the gas bladders of which contain CO that is a growth substrate for *M. acetivorans*.[9,82,83] Acetate and formate are produced in addition to CH_4, suggesting that the metabolism of CO and mechanisms for energy conservation by *M. acetivorans* are unique among CH_4-producing Archaea.[82,83] Further, H_2/CO_2 does not support growth of *M. acetivorans* and H_2 is not detected during growth with CO, although in freshwater methanogens CO is first converted to H_2, which is subsequently oxidized for reduction of CO_2 to CH_4.[9,82,83] A genomewide proteomic approach to identify proteins differentially abundant in CO- versus methanol- or acetate-grown *M. acetivorans* suggests that the pathway for CH_4 formation involves novel enzymes and a mechanism for energy-conservation not previously reported for CO_2-reducing pathways in methanogens (FIG. 6).[82] In the pathway, oxidation of CO to CO_2 supplies electrons (FIG. 6, steps 1a–1c) for reduction of CO_2 to CH_3-THSPT (FIG. 6, steps 2–5) catalyzed by enzymes utilized by all other CO_2-reducing methanogens.[7] However, the proteomic results suggest that transfer of the methyl group from CH_3-THSPT to CoM-SH in CO-grown *M. acetivorans* involves novel methyltransferases other than the membrane-bound Mtr complex (FIG. 6, step 7). The results further suggest that a coenzyme $F_{420}H_2$:heterodisulfide oxidoreductase system (FIG. 6, steps 9 and 10), not previously described for all other CO_2-reducing methanogens, generates a proton gradient for ATP synthesis (FIG. 6, step 14). Finally, the results indicate that acetate is synthesized from CH_3-THSPT and CO (FIG. 6, steps 11–13) by a reversal of initial steps in the pathway for conversion of acetate to CH_4 that yields ATP by substrate-level phosphorylation. As opposed to freshwater CO-utilizing species, the H_2-independent pathway in *M. acetivorans* is again consistent with the marine environment where H_2-utilizing sulfate-reducing anaerobes out compete methanogens for H_2.

Conclusions

Bioinformatic, proteomic, microarray, and biochemical approaches investigating the marine species *M. acetivorans* have revealed differences in the metabolism of acetate and CO to methane for freshwater and marine methanogens. The differences encountered in *M. acetivorans* appear to be an adaptation to the marine environment to avoid competition with H_2-utilizing sulfate-reducers. The experimental approaches have also gained a deeper understanding of the physiology of *M. acetivorans* and opened several avenues for investigating electron transport, methyl transfer, the stress response, and regulation of gene expression.

Acknowledgments

The work performed in the laboratory of the authors was supported by the Department of Energy and the National Science Foundation.

Conflict of Interest

The authors declare no conflicts of interest.

References

1. Sowers, K.R. & J.G. Ferry. 2002. Methanogenesis in the marine environment. *In* The Encyclopedia of Environmental Microbiology. G. Bitton, Ed.: 1913–1923. New York: John Wiley & Sons.
2. Barber, R.D. & J.G. Ferry. 2001. Methanogenesis. *In* Encyclopedia of Life Sciences: On-line www.els.net London: Nature Publishing Group.
3. Raghoebarsing, A.A. *et al.* 2006. A microbial consortium couples anaerobic methane oxidation to denitrification. Nature **440:** 918–921.
4. Schlesinger, W.H. 2000. The global carbon cycle. *In* Biogeochemistry: 308–321. San Diego, CA: Academic Press.
5. Orphan, V.J. *et al.* 2001. Methane-consuming archaea revealed by directly coupled isotopic and phylogenetic analysis. Science **293:** 484–487.
6. Boetius, A. *et al.* 2000. A marine microbial consortium apparently mediating anaerobic oxidation of methane. Nature **407:** 623–626.
7. Ferry, J.G. & K.A. Kastead. 2007. Methanogenesis. *In* Archaea: Molecular Cell Biology. R. Cavicchioli, Ed.: 288–314. Washington, D.C.: ASM Press.
8. Elberson, M.A. & K.R. Sowers. 1997. Isolation of an aceticlastic strain of *Methanosarcina siciliae* from marine canyon sediments and emendation of the species description for *Methanosarcina siciliae*. Int. J. Syst. Bacteriol. **47:** 1258–1261.
9. Sowers, K.R., S.F. Baron & J.G. Ferry. 1984. *Methanosarcina acetivorans* sp. nov., an acetotrophic methane-producing bacterium isolated from marine sediments. Appl. Environ. Microbiol. **47:** 971–978.
10. Sowers, K.R. & R.P. Gunsalus. 1988. Plasmid DNA from the acetotrophic methanogen *Methanosarcina acetivorans*. J. Bacteriol. **170:** 4979–4982.
11. Metcalf, W.W. *et al.* 1997. A genetic system for Archaea of the genus *Methanosarcina*. Liposome-mediated transformation and construction of shuttle vectors. Proc. Natl. Acad. Sci. USA **94:** 2626–2631.
12. Boccazzi, P., J.K. Zhang & W.W. Metcalf. 2000. Generation of dominant selectable markers for resistance to pseudomonic acid by cloning and mutagenesis of the *ileS* gene from the archaeon *Methanosarcina barkeri* fusaro. J. Bacteriol. **182:** 2611–2618.
13. Zhang, J.K. *et al.* 2000. *In vivo* transposon mutagenesis of the methanogenic archaeon *Methanosarcina acetivorans* C2A using a modified version of the insect mariner-family transposable element Himar1. Proc. Natl. Acad. Sci. **97:** 9665–9670.
14. Zhang, J.K. *et al.* 2002. Directed mutagenesis and plasmid-based complementation in the methanogenic archaeon *Methanosarcina acetivorans* C2A demonstrated by genetic analysis of proline biosynthesis. J. Bacteriol. **184:** 1449–1454.
15. Pritchett, M.A., J.K. Zhang & W.W. Metcalf. 2004. Development of a markerless genetic exchange method for *Methanosarcina acetivorans* C2A and its use in construction of new genetic tools for methanogenic archaea. Appl. Environ. Microbiol. **70:** 1425–1433.
16. Apolinario, E.E., K.M. Jackson & K.R. Sowers. 2005. Development of a plasmid-mediated reporter system for *in vivo* monitoring of gene expression in the archaeon *Methanosarcina acetivorans*. Appl. Environ. Microbiol. **71:** 4914–4918.
17. Galagan, J.E. *et al.* 2002. The genome of *M. acetivorans* reveals extensive metabolic and physiological diversity. Genome Res. **12:** 532–542.
18. Li, Q. *et al.* 2005. The proteome of *Methanosarcina acetivorans*. Part I. An expanded view of the biology of the cell. J. Proteome Res. **4:** 112–128.
19. Li, Q. *et al.* 2005. The proteome of *Methanosarcina acetivorans*. Part II. Comparison of protein levels in acetate- and methanol-grown cells. J. Proteome Res. **4:** 129–136.
20. Li, L. *et al.* 2007. Quantitative proteomic and microarray analysis of the archaeon *Methanosarcina acetivorans* grown with acetate versus methanol. J. Proteome Res. **6:** 759–771.
21. Singh-Wissmann, K. & J.G. Ferry. 1995. Transcriptional regulation of the phosphotransacetylase-encoding and acetate kinase-encoding genes (*pta* and *ack*) from *Methanosarcina thermophila*. J. Bacteriol. **177:** 1699–1702.
22. Hovey, R. *et al.* 2005. DNA microarray analysis of *Methanosarcina mazei* Go1 reveals adaptation to different methanogenic substrates. Mol. Genet. Genomics **273:** 225–239.
23. Ingram-Smith, C. *et al.* 2005. Identification of the acetate binding site in the *Methanosarcina thermophila* acetate kinase. J. Bacteriol. **187:** 2386–2394.

24. GORRELL, A., S.H. LAWRENCE & J.G. FERRY. 2005. Structural and kinetic analyses of arginine residues in the active-site of the acetate kinase from *Methanosarcina thermophila*. J. Biol. Chem. **280:** 10731–10742.
25. BUSS, K.A. *et al.* 2001. Urkinase: structure of acetate kinase, a member of the ASKHA superfamily of phosphotransferases. J. Bacteriol. **183:** 680–686.
26. LAWRENCE, S.H. & J.G. FERRY. 2006. Steady-state kinetic analysis of phosphotransacetylase from *Methanosarcina thermophila*. J. Bacteriol. **188:** 1155–1158.
27. LAWRENCE, S.H. *et al.* 2006. Structural and functional studies suggest a catalytic mechanism for the phosphotransacetylase from *Methanosarcina thermophila*. J. Bacteriol. **188:** 1143–1154.
28. IYER, P.P. *et al.* 2004. Crystal structure of phosphotransacetylase from the methanogenic archaeon *Methanosarcina thermophila*. Structure **12:** 559–567.
29. TERLESKY, K.C., M.J.K. NELSON & J.G. FERRY. 1986. Isolation of an enzyme complex with carbon monoxide dehydrogenase activity containing a corrinoid and nickel from acetate-grown *Methanosarcina thermophila*. J. Bacteriol. **168:** 1053–1058.
30. TERLESKY, K.C. & J.G. FERRY. 1988. Ferredoxin requirement for electron transport from the carbon monoxide dehydrogenase complex to a membrane-bound hydrogenase in acetate-grown *Methanosarcina thermophila*. J. Biol. Chem. **263:** 4075–4079.
31. ABBANAT, D.R. & J.G. FERRY. 1990. Synthesis of acetyl-CoA by the carbon monoxide dehydrogenase complex from acetate-grown *Methanosarcina thermophila*. J. Bacteriol. **172:** 7145–7150.
32. RAYBUCK, S.A. *et al.* 1991. Demonstration of carbon-carbon bond cleavage of acetyl coenzyme A by using isotopic exchange catalyzed by the CO dehydrogenase complex from acetate-grown *Methanosarcina thermophila*. J. Bacteriol. **173:** 929–932.
33. GRAHAME, D.A. 1991. Catalysis of acetyl-CoA cleavage and tetrahydrosarcinapterin methylation by a carbon monoxide dehydrogenase-corrinoid enzyme complex. J. Biol. Chem. **266:** 22227–22233.
34. GRAHAME, D.A. & E. DEMOLL. 1996. Partial reactions catalyzed by protein components of the acetyl-CoA decarbonylase synthase enzyme complex from *Methanosarcina barkeri*. J. Biol. Chem. **271:** 8352–8358.
35. BHASKAR, B., E. DEMOLL & D.A. GRAHAME. 1998. Redox-dependent acetyl transfer partial reaction of the acetyl-CoA decarbonylase/synthase complex: kinetics and mechanism. Biochemistry **37:** 14491–14499.
36. ABBANAT, D.R. & J.G. FERRY. 1991. Resolution of component proteins in an enzyme complex from *Methanosarcina thermophila* catalyzing the synthesis or cleavage of acetyl-CoA. Proc. Natl. Acad. Sci. USA **88:** 3272–3276.
37. GENCIC, S. & D.A. GRAHAME. 2003. Nickel in subunit β of the acetyl-CoA decarbonylase/synthase multienzyme complex in methanogens. J. Biol. Chem. **278:** 6101–6110.
38. GU, W.W. *et al.* 2003. The A-cluster in subunit beta of the acetyl-CoA decarbonylase/synthase complex from *Methanosarcina thermophila*: Ni and Fe K-Edge XANES and EXAFS analyses. J. Am. Chem. Soc. **125:** 15343–15351.
39. FUNK, T. *et al.* 2004. Chemically distinct Ni sites in the A-cluster in subunit beta of the acetyl-CoA decarbonylase/synthase complex from *Methanosarcina thermophila*: Ni L-edge absorption and x-ray magnetic circular dichroism analyses. J. Am. Chem. Soc. **126:** 88–95.
40. MURAKAMI, E. & S.W. RAGSDALE. 2000. Evidence for intersubunit communication during acetyl-CoA cleavage by the multienzyme CO dehydrogenase/acetyl-CoA synthase complex from *Methanosarcina thermophila*. Evidence that the beta subunit catalyzes C-C and C-S bond cleavage. J. Biol. Chem. **275:** 4699–4707.
41. JABLONSKI, P.E. *et al.* 1993. Characterization of the metal centers of the corrinoid/iron- sulfur component of the CO dehydrogenase enzyme complex from *Methanosarcina thermophila* by EPR spectroscopy and spectroelectrochemistry. J. Biol. Chem. **268:** 325–329.
42. MAUPIN-FURLOW, J. & J.G. FERRY. 1996. Characterization of the *cdhD* and *cdhE* genes encoding subunits of the corrinoid iron-sulfur enzyme of the CO dehydrogenase complex from *Methanosarcina thermophila*. J. Bacteriol. **178:** 340–346.
43. ALBER, B.E. & J.G. FERRY. 1994. A carbonic anhydrase from the archaeon *Methanosarcina thermophila*. Proc. Nat. Acad. Sci. USA **91:** 6909–6913.
44. ALBER, B.E. & J.G. FERRY. 1996. Characterization of heterologously produced carbonic anhydrase from *Methanosarcina thermophila*. J. Bacteriol. **178:** 3270–3274.
45. KISKER, C. *et al.* 1996. A left-handed beta-helix revealed by the crystal structure of a carbonic anhydrase from the archaeon *Methanosarcina thermophila*. EMBO J. **15:** 2323–2330.
46. TRIPP, B.C. *et al.* 2004. A role for iron in an ancient carbonic anhydrase. J. Biol. Chem. **279:** 6683–6687.
47. ZIMMERMAN, S.A. & J.G. FERRY. 2006. Proposal for a hydrogen bond network in the active site of the prototypic gamma-class carbonic anhydrase. Biochemistry **45:** 5149–5157.
48. TRIPP, B.C. & J.G. FERRY. 2000. A structure-function study of a proton transport pathway in a novel gamma-class carbonic anhydrase from *Methanosarcina thermophila*. Biochemistry **39:** 9232–9240.
49. IVERSON, T.M. *et al.* 2000. A closer look at the active site of gamma-carbonic anhydrases: high resolution crystallographic studies of the carbonic anhydrase from *Methanosarcina thermophila*. Biochemistry **39:** 9222–9231.
50. WELANDER, P.V. & W.W. METCALF. 2005. Loss of the mtr operon in *Methanosarcina* blocks growth on methanol, but not methanogenesis, and reveals an unknown methanogenic pathway. Proc. Natl. Acad. Sci. USA **102:** 10664–10669.
51. FISCHER, R. *et al.* 1992. N5-Methyltetrahydromethanopterin: coenzyme-M methyltransferase in methanogenic archaebacteria is a membrane protein. Arch. Microbiol. **158:** 208–217.
52. BECHER, B., V. MULLER & G. GOTTSCHALK. 1992. N^5 Methyl-tetrahydromethanopterin: coenzyme M methyltransferase of *Methanosarcina* strain Go1 is an Na+-translocating membrane protein. J. Bacteriol. **174:** 7656–7660.
53. LIENARD, T. *et al.* 1996. Sodium ion translocation by N^5-methyltetrahydromethanopterin: coenzyme M methyltransferase from *Methanosarcina mazei* Go1 recon-

stituted in ether lipid liposomes. Eur. J. Biochem. **239:** 857–864.

54. GOTTSCHALK, G. & R.K. THAUER. 2001. The Na+ translocating methyltransferase complex from methanogenic archaea. Biochim. Biophys. Acta **1505:** 28–36.
55. LU, W.P. *et al.* 1995. Electron paramagnetic resonance spectroscopic and electrochemical characterization of the partially purified N^5-methyltetrahydromethanopterin: coenzyme M methyltransferase from *Methanosarcina mazei* Go1. J. Bacteriol. **177:** 2245–2250.
56. NELSON, M.J.K. & J.G. FERRY. 1984. Carbon monoxide-dependent methyl coenzyme M methylreductase in acetotrophic *Methanosarcina* spp. J. Bacteriol. **160:** 526–532.
57. BARESI, L. & R.S. WOLFE. 1981. Levels of coenzyme F420, coenzyme M, hydrogenase, and methylcoenzyme M methylreductase in acetate-grown *Methanosarcina*. Appl. Environ. Microbiol. **41:** 388–391.
58. KRZYCKI, J.A., L.J. LEHMAN & J.G. ZEIKUS. 1985. Acetate catabolism by *Methanosarcina barkeri*: evidence for involvement of carbon monoxide dehydrogenase, methyl coenzyme M, and methylreductase. J. Bacteriol. **163:** 1000–1006.
59. JABLONSKI, P.E. & J.G. FERRY. 1991. Purification and properties of methyl coenzyme M methylreductase from acetate grown *Methanosarcina thermophila*. J. Bacteriol. **173:** 2481–2487.
60. ERMLER, U. *et al.* 1997. Crystal structure of methyl-coenzyme M reductase: the key enzyme of biological methane formation. Science **278:** 1457–1462.
61. GRABARSE, W. *et al.* 2001. On the mechanism of biological methane formation: structural evidence for conformational changes in methyl-coenzyme M reductase upon substrate binding. J. Mol. Biol. **309:** 315–330.
62. FINAZZO, C. *et al.* 2003. Coenzyme B induced coordination of coenzyme M via its thiol group to Ni(I) of F430 in active methyl-coenzyme M reductase. J. Am. Chem. Soc. **125:** 4988–4989.
63. GOENRICH, M. *et al.* 2004. Probing the reactivity of Ni in the active site of methyl-coenzyme M reductase with substrate analogues. J. Biol. Inorg. Chem. **9:** 691–705.
64. GOENRICH, M. *et al.* 2005. Temperature dependence of methyl-coenzyme M reductase activity and of the formation of the methyl-coenzyme M reductase red2 state induced by coenzyme B. J. Biol. Inorg. Chem. **10:** 333–342.
65. CLEMENTS, A.P., R.H. WHITE & J.G. FERRY. 1993. Structural characterization and physiological function of component-B from *Methanosarcina thermophila*. Arch. Microbiol. **159:** 296–300.
66. PEER, C.W. *et al.* 1994. Characterization of a CO: heterodisulfide oxidoreductase system from acetate-grown *Methanosarcina thermophila*. J. Bacteriol. **176:** 6974–6979.
67. HEIDEN, S. *et al.* 1993. Purification of a cytochrome-*b* containing H_2-heterodisulfide oxidoreductase complex from membranes of *Methanosarcina barkeri*. Eur. J. Biochem. **213:** 529–535.
68. SIMIANU, M. *et al.* 1998. Purification and properties of the heme- and iron-sulfur-containing heterodisulfide reductase from *Methanosarcina thermophila*. Biochemistry **37:** 10027–10039.
69. MURAKAMI, E., U. DEPPENMEIER & S.W. RAGSDALE. 2001. Characterization of the intramolecular electron transfer pathway from 2-hydroxyphenazine to the heterodisulfide reductase from *Methanosarcina thermophila*. J. Biol. Chem. **276:** 2432–2439.
70. ABKEN, H.-J. *et al.* 1998. Isolation and characterization of methanophenazine and the function of phenazines in membrane-bound electron transport of *Methanosarcina mazei* Go1. J. Bacteriol. **180:** 2027–2032.
71. HEDDERICH, R. 2004. Energy-converting [NiFe] hydrogenases from Archaea and extremophiles: ancestors of complex I. J. Bioenerg. Biomembr. **36:** 65–75.
72. MEUER, J. *et al.* 2002. Genetic analysis of the archaeon *Methanosarcina barkeri* Fusaro reveals a central role for Ech hydrogenase and ferredoxin in methanogenesis and carbon fixation. Proc. Natl. Acad. Sci. USA **99:** 5632–5637.
73. LATIMER, M.T., M.H. PAINTER & J.G. FERRY. 1996. Characterization of an iron-sulfur flavoprotein from *Methanosarcina thermophila*. J. Biol. Chem. **271:** 24023–24028.
74. ANDRADE, S.L.A. *et al.* 2005. Structures of the iron-sulfur flavoproteins from *Methanosarcina thermophila* and *Archaeoglobus fulgidus*. J. Bacteriol. **187:** 3848–3854.
75. CRUZ, F.C. & J.G. FERRY. 2006. Interaction of iron-sulfur flavoprotein with oxygen and hydrogen peroxide. Biochim. Biophys. Acta **1760:** 858–864.
76. DING, Y.H. & J.G. FERRY. 2004. Flavin mononucleotide-binding flavoprotein family in the domain Archaea. J. Bacteriol. **186:** 90–97.
77. LI, Q. *et al.* 2006. Electron transport in the pathway of acetate conversion to methane in the marine archaeon *Methanosarcina acetivorans*. J. Bacteriol. **188:** 702–710.
78. SCHMEHL, M. *et al.* 1993. Identification of a new class of nitrogen fixation genes in *Rhodobacter capsulatus*: a putative membrane complex involved in electron transport to nitrogenase. Mol. Gen. Genet. **241:** 602–615.
79. SAEKI, K. & H. KUMAGAI. 1998. The *rnf* gene products in *Rhodobacter capsulatus* play an essential role in nitrogen fixation during anaerobic DMSO-dependent growth in the dark. Arch. Microbiol. **169:** 464–467.
80. BOIANGIU, C.D. *et al.* 2005. Sodium ion pumps and hydrogen production in glutamate fermenting anaerobic bacteria. J. Mol. Microbiol. Biotechnol. **10:** 105–119.
81. BRUGGEMANN, H. *et al.* 2003. The genome sequence of *Clostridium tetani*, the causative agent of tetanus disease. Proc. Natl. Acad. Sci. USA **100:** 1316–1321.
82. LESSNER, D.J. *et al.* 2006. An unconventional pathway for reduction of CO_2 to methane in CO-grown *Methanosarcina acetivorans* revealed by proteomics. Proc. Natl. Acad. Sci. USA **103:** 17921–17926.
83. ROTHER, M. & W.W. METCALF. 2004. Anaerobic growth of *Methanosarcina acetivorans* C2A on carbon monoxide: an unusual way of life for a methanogenic archaeon. Proc. Natl. Acad. Sci. USA **101:** 16929–16934.

Methane as Fuel for Anaerobic Microorganisms

RUDOLF K. THAUER AND SEIGO SHIMA

Max Planck Institute for Terrestrial Microbiology, Marburg, Germany

Methane has long been known to be used as a carbon and energy source by some aerobic alpha- and delta-proteobacteria. In these organisms the metabolism of methane starts with its oxidation with O_2 to methanol, a reaction catalyzed by a monooxygenase and therefore restricted to the aerobic world. Methane has recently been shown to also fuel the growth of anaerobic microorganisms. The oxidation of methane with sulfate and with nitrate have been reported, but the mechanisms of anaerobic methane oxidation still remains elusive. Sulfate-dependent methane oxidation is catalyzed by methanotrophic archaea, which are related to the *Methanosarcinales* and which grow in close association with sulfate-reducing delta-proteobacteria. There is evidence that anaerobic methane oxidation with sulfate proceeds at least in part via reversed methanogenesis involving the nickel enzyme methyl-coenzyme M reductase for methane activation, which under standard conditions is an endergonic reaction, and thus inherently slow. Methane oxidation coupled to denitrification is mediated by bacteria belonging to a novel phylum and does not involve methyl-coenzyme M reductase. The first step in methane oxidation is most likely the exergonic formation of 2-methylsuccinate from fumarate and methane catalyzed by a glycine-radical enzyme.

Key words: **anaerobic oxidation of methane; methyl-coenzyme M reductase; nickel cofactor F_{430}; glycyl-radical enzymes; methylsuccinate synthase; methanotrophic archaea; methanotrophic bacteria**

Introduction

The anaerobic oxidation of methane (AOM) by microorganisms has long been thought to be impossible, not because the dehydrogenation of methane to methanol ($E^{\circ\prime} = +0.166$ V) with, for example, nitrate ($E^{\circ\prime} = +0.43$ V) as electron acceptor is thermodynamically not possible (TABLE 1), but because the C-H bond in methane is not polarized, and therefore the abstraction of a hydride in water at neutral pH is mechanistically not possible. This is only possible in superacids or with strongly electrophilic metal-containing species in the absence of water.[1,2] Therefore, the general view is that in biological systems the first step in methane oxidation must be its oxidation to a methyl radical.

The dissociation energy of the C-H bond in methane is +439 kJ/mol (TABLE 2), and thus larger than that in other organic molecules. Only the C-H bond in benzene (−473 kJ/mol) is stronger. There is only one radical of biological relevance, the O-H radical (the dissociation energy of the O-H bond of water is 497 kJ/mol), that reacts with methane under standard conditions in an exergonic reaction.

$$CH_4 + OH\,(\text{radical}) = CH_3\,(\text{radical}) + H_2O \qquad \Delta G^{\circ\prime} = -58\,\text{kJ/mol} \tag{1}$$

In the dark the OH radical cannot be generated from water since the redox potential $E^{\circ\prime} = +2.33$ V of the OH/H_2O couple is more positive than that of all other electron acceptors that are stable in water. It can, however, be formed by reduction of O_2. Three electrons are required for the reduction, one of which can be provided by the methyl radical.

$$O_2 + 3e + 3H^+ = OH(\text{radical}) + H_2O \qquad E^{\circ\prime} = +0.29\text{ V} \tag{2}$$

$$CH_4 + O_2 + 2e + 2H^+ = CH_3OH + H_2O \qquad E^{\circ\prime} = +1.45\text{ V} \tag{3}$$

Reaction 3 (reaction 1 + reaction 2) describes correctly how methane is oxidized to methanol in aerobic methanotrophic bacteria, while reactions 1 and 2 are simplifications of the catalytic mechanism. They disregard that reaction 3 is catalyzed by metalloenzymes and that the primary attack of the methane

Address for correspondence: Prof. Dr. R. Thauer, Max Planck Institute for Terrestrial Microbiology, Karl-von-Frisch-Strasse, D-35043 Marburg, Germany. Voice: +49 6421 178101; fax: +49 6421 178109.
thauer@mpi-marburg.mpg.de

Ann. N.Y. Acad. Sci. 1125: 158–170 (2008). © 2008 New York Academy of Sciences.
doi: 10.1196/annals.1419.000

TABLE 1. Redox potentials E°′ of electron acceptors that could be used by microorganisms for the anaerobic oxidation of methane

Redox couple	*n*	E°′ (V)
CO_2/CH_4	**8**	**−0.24**
S°/H_2S*	2	−0.27 (−0.12)[a]
SO_4^{2-}/HS^-	8	−0.22 (−200)[b]
SO_3H^-/HS^-	6	−0.12
APS/SO_3H^-	2	−0.06
Glycine/acetate$^-$ + NH_4^+	2	−0.01
$CH_3SH/CH_4 + H_2S$	**2**	**+0.03**
Fumarate/succinate	2	+0.03
AsO_4^{3-}/AsO_2^-	2	+0.13
Trimethylamine N-oxide/trimethylamine	2	+0.13
Dimethylsulfoxide/dimethylsulphide	2	+0.16
CH_3OH/CH_4	**2**	**+0.17**
$Fe(OH)_3 + HCO_3^-/FeCO_3^c$	1	+0.2
NO_2^-/NH_3	6	+0.33
NO_2^-/NO	1	+0.34
NO_3^-/NH_3	8	+0.36
Mn^{4+}/Mn^{2+}	2	+0.41
NO_3^-/NO_2^-	2	+0.43
$2NO_3^-/N_2$	10	+0.76
$2NO_2^-/N_2$	6	+0.95
$2NO/N_2O$	2	+1.2
N_2O/N_2	2	+1.36
CH_3(radical)/CH_4	1	+2.1

E°′at pH 7.0 are given for H_2, CO_2, CO, CH_4, and O_2 in the gaseous state at 10^5 Pa, for S° in the solid state, and for all other compounds in aqueous solution at 1 M concentration. The values in brackets are E′ values calculated for physiological substrate and product concentrations. The E°′ values were calculated from the ΔG°′ values: $\Delta G^{\circ\prime} = -nF\Delta E$, where *n* is the number of electrons and F = 96 487 J/mol/V. Except were indicated, ΔG°′ values were taken from Thauer *et al.*[91]

[a]Calculated for a $[HS^-] = 0.1$ mM.

[b]Calculated for [sulphate] = 30 mM and $[HS^-] = 0.1$ mM.

[c] From Ehrenreich and Widdel.[92]

is by a high-valent metal-oxo species rather than by a free OH radical.[3–5] But they can best explain why both O_2 and the reducing equivalents are required for the oxidation of methane to methanol with O_2.

In methanotrophic aerobic bacteria the methanol formed from methane is further oxidized to CO_2. The three two-electron steps from methanol via formaldehyde and formate to CO_2 are not dependent on O_2 and can proceed with electron acceptors readily available in both aerobic and anaerobic organisms.[6]

$$CH_3OH + H_2O = CO_2 + 6e + 6H^+ \quad E^{\circ\prime} = -0.38\ V \tag{4}$$

TABLE 2. Gas-phase bond-dissociation energies and radicals that are relevant to the enzymatic systems discussed in this review

Bond	Dissociation energy (kJ/mol)	Radicals or related radicals of biological importance formed by bond dissociation[a]
H-OH	497	OH radical (+2.33 V)[b]
H-C_6H_5	473	
H-CH_3	**439**	
H-H	436	
H-n-C_4H_9	425	
H-n-C_3H_7	423	
H-C_2H_5	423	
H-CH_2COCH_3	411	
H-s-C_4H_9	411	
H-s-C_3H_7	409	
H-CH_2COOH	418[c]	
H-CH_2COCH_3	411	
H-CH_2CHO	395	
H-CH_2CH_2OH	390?	5′-Deoxyadenosine radical (+1.4 V)
H-SCH_3	365	Thiyl radical (+1.33 V)
H-$CH_2C_6H_5$	376	
H-OOH	369	O_2^- radical (+0.89 V)[d]
H-OC_6H_5	361	Tyrosyl radical (+0.94 V)
H-C of glycine	350[e]	Glycyl radical (+1.22 V)
CH_3-CH_3	376	
CH_2=CH_2	352[f]	

Bond dissociation energies, except where otherwise indicated, were taken from the *Handbook of Chemistry and Physics*, 84th Edition 2003–2004, CRC Press.

[a]The numbers in brackets are redox potentials E°′ of the radical/undissociated compound couple in aqueous solution at pH 7. In the case of the glycyl radical, the pH was 10.5. The redox potentials were taken from Stubbe and van der Donk.[93] The redox potential of 5′-deoxyadenosyl radical formation is an estimate (see Wang and Frey[94]).

[b]The OH radical can also be generated from H_2O_2 by a one-electron reduction: $H_2O_2 + e^- + H^+ = OH + H_2O$ ($E^{\circ\prime} = +0.38$).

[c]From Sandala *et al.* [95] The C-H bond-dissociation energy was calculated in this paper to be 408 kJ/mol. Since the H-C bond-dissociation energy for methane was calculated via the same method to be only 429 kJ/mol rather than 439 kJ/mol, the value was increased by 10 kJ/mol.

[d]O_2^- radical can also be generated from O_2 by a one-electron reduction: $O_2 + e^- = O_2^-$ ($E^{\circ\prime} = -0.33$ V).

[e]From Armstrong *et al.* [96]; 334 kJ/mol in Sandala *et al.* [95]

[f]Bond-dissociation energy for CH_2=CH_2 to CH_2-CH_2. Calculated from the energy of 728 kJ/mol for the dissociation of CH_2=CH_2 to 2 CH_2 minus the energy of 376 kJ/mol for the dissociation of CH_3-CH_3 to 2 CH_3.

Because of these theoretical considerations, the fact that methane is oxidized to CO_2 in marine sediments in the complete absence of O_2 has long been ignored. The anaerobic oxidation in the sediments is coupled to dissimilatory sulfate reduction.

$$CH_4 + SO_4^{2-} + H^+ = CO_2 + HS^- + 2H_2O \quad \Delta G^{\circ\prime} = -21 \text{ kJ/mol} \quad (5)$$

It has been shown that AOM with sulfate is a quantitatively very important process in the global carbon cycle (for an extensive review, see Ref. 7).

It was recently found that AOM is not restricted to sulfate as electron acceptor. A year ago microbes were enriched, which coupled methane oxidation with denitrification.[8]

$$5CH_4 + 8NO_3^- + 8H^+ = 5CO_2 + 4N_2 + 14H_2O \quad \Delta G^{\circ\prime} = -765 \text{ kJ/mol } CH_4 \quad (6)$$

$$3CH_4 + 8NO_2^- + 8H^+ = 3CO_2 + 4N_2 + 10H_2O \quad \Delta G^{\circ\prime} = -928 \text{ kJ/mol } CH_4 \quad (7)$$

In this article we describe what is currently known about how anaerobic microbes oxidize methane and which microbes are involved. The available evidence indicates that AOM with sulfate proceeds via a different mechanism than AOM with nitrate, which is why the two processes are discussed separately. Mostly, the literature of the last two years will be considered. For earlier references the reader is referred to two reviews on the subject by Shima and Thauer[9] and by Thauer and Shima.[10]

Anaerobic Oxidation of Methane with Sulfate

Rate, Yields, and Apparent K_m

The microorganisms involved in AOM with sulfate could, until now, only be grown in complex cultures consisting of archaea and bacteria.[11,12] Growth is extremely slow and has a low yield. Doubling times of several months and growth yields of 0.6 g (dry weight) per mol methane oxidized appear typical. Only 1% of the consumed methane is channeled into the synthesis of consortia biomass. The growth rate is almost linearly dependent on the methane concentration in the range tested (up to 1.4×10^6 Pa), indicating that the apparent K_m for methane is above 10 mM. The specific rate of methane oxidation is 10 nmol methane per min and mg cells (dry weight) at a CH_4 partial pressure of 1.4×10^6 Pa.[12] *In situ* measurements suggest that in marine sediments and in the Black Sea chimneys fueled by methane hydrates[13] the specific methane oxidation rates are of the same order.[14–17]

The Black Sea microbial mats catalyzing AOM in the laboratory at a specific rate of 1 nmol per min (mU) per mg protein (at 1×10^5 Pa CH_4) mediated the formation of methane from methanogenic substrates, such as H_2/CO_2, formate, methanol, methylamines, and/or acetate at specific rates less than 0.01 mU/mg protein, indicating that the mats are dedicated to AOM rather than to methanogenesis.[18] There is, however, evidence that the mats simultaneously oxidize and produce methane.[19] The contemporaneous activity of AOM and methanogenesis also has been reported at the Gulf of Mexico cold seeps.[20]

Involvement of Archaea and Sulfate-Reducing Bacteria

Available evidence indicates that the archaea involved in AOM with sulfate are phylogenetically most closely related to methanogenic archaea of the order *Methanosarcinales* and that the archaea are the ones that catalyze methane oxidation. Three lineages have been identified that are referred to as ANME-1, ANME-2, and ANME-3.[21–23] Whether these archaea also catalyze the reduction of sulfate is still a much-debated question. In their natural environment they are always associated with sulfate-reducing bacteria (SRB) belonging taxonomically to the delta group of proteobacteria.[24–28] Whether the archaea depend on the SRB for growth remains to be shown.

Syntrophy?

H_2, formate, and acetate have been considered to be involved in interspecies electron transfer from the methanotrophic archaea to the SRB. None of these compounds have been found to inhibit AOM or to stimulate sulfate reduction in mats catalyzing AOM with sulfate, which for thermodynamic reasons they should if they were intermediates, as exemplified for H_2 in reactions 8 and 9.

$$CH_4 + 2H_2O = CO_2 + 4H_2 \quad \Delta G^{\circ\prime} = +131 \text{ kJ/mol} \quad (8)$$

$$4H_2 + SO_4^{2-} + 2H^+ = H_2S + 4H_2O \quad \Delta G^{\circ\prime} = -152 \text{ kJ/mol} \quad (9)$$

In this respect it is of interest that methanogenic archaea have been shown to require a free energy change under physiological conditions (ΔG) of at least −10 kJ/mol and SRB one of at least −19 kJ/mol to support their metabolism *in situ*.[29,30] The free energy change of −21 kJ/mol associated with AOM with sulfate (reaction 5) is therefore probably not sufficiently large to fuel the energy metabolism of two organisms in tandem.

Syntrophy of methanotrophic archaea and SRB could also proceed by extracellular electron transfer involving "nanowires,"[31,32] which would, however, require each archaeon to be in physical contact with an SRB. In the microbial mats investigated so far, especially in those in which ANME-1 cells dominate, this is not always the case. In the water column ANME-2 also appear as single cells.[33] The present view, therefore, is that at least in some of the archaea, methane oxidation and sulfate reduction occur in the same cells, although this is not backed up by metagenomic data. It has not yet been possible to demonstrate the presence of gene homologues for the enzymes required for dissimilatory sulfate reduction in archaea of any of the ANME clusters.

All attempts to grow the SRB in the absence of the methanotrophic archaea, with which they are associated in the microbial mats, have failed so far. Combinations of sulfate with H_2, formate, acetate, lactate, glucose, and fructose were tried. Dissimilatory sulfate reduction with these electron donors is strongly exergonic (e.g., reaction 9), and therefore should sustain growth. The hypothesis has therefore been put forward that the SRB in the mats might be fueled by the oxidation of the glycocalix surrounding the methanotrophic archaea and gluing the cells in the mats together. If this hypothesis is correct, growth of the SRB would be dependent on the presence of complex sugar mixtures, which are growth requirements difficult to mimic in the laboratory. One might argue that sulfate reducers generally do not ferment sugars, one of the few exceptions being *Desulfovibrio fructosovorans*. A look in the sequenced genomes indicates, however, that many of the sulfate reducers contain all the genes required to metabolize sugars.

Involvement of Methyl-Coenzyme M Reductase in Anaerobic Oxidation of Methane

Methanotrophic archaea have been shown to contain gene homologues for methyl-coenzyme M reductase (MCR) and for other enzymes involved in methanogenic archaea in methane formation from CO_2, suggesting that these enzymes are involved in AOM[34,35] (for review, see Ref. 6). This is also indicated by the observation that AOM is inhibited by bromoethane sulfonate (BES),[22] which is considered to be a specific inhibitor of MCR.[36]

Recently, however, it has been shown that BES also effectively inhibits growth of *Xantobacter autotrophicus*, metabolizing propylene by specific inactivation of one of the enzymes involved in propylene degradation.[37] Thus inhibition of the growth of an organism by BES is not necessarily an indication of the presence of MCR. Contrarily, noninhibition of the growth by BES cannot be taken as evidence for the absence of MCR, since MCR is a cytoplasmic enzyme, and therefore BES has to be transported into the cells in order to exert its inhibitory effect. Uptake generally is by a coenzyme M transporter,[38] which some methanogenic archaea lack. These methanogens are resistant to inhibition by BES.

The most convincing evidence for the involvement of MCR in AOM comes from the biochemical analysis of microbial mats from the Black Sea, which catalyze AOM and which comprise more than 50% archaea from the ANME-1 cluster.[18] These mats were found to contain high concentrations (10% of the protein extracted from the mats) of two enzymes, Ni-protein I and II. The nickel proteins were identified as MCR by their subunit composition, primary structure, and nickel cofactor F_{430}. Metagenome analyses (see later in the chapter) revealed that the genes coding for the two nickel-proteins belong to the genomes of the ANME organisms.[18]

The *mcrA* gene is a taxonomic marker for methanogenic archaea. Their phylogenetic relationship is reflected in the primary structures of the MCR from these organisms.[39] Detection of the *mcrA* gene in microbial communities catalyzing AOM is also currently used as an indication of the involvement of MCR.[34,35,40–42] But without knowing whether the *mcrA* gene is expressed, one has to be careful not to overinterpret these findings.[43]

Methyl-Coenzyme M Reductase from Methanogenic Archaea

MCR from methanogenic archaea is composed of three different subunits in an $\alpha_2\beta_2\gamma_2$ arrangement and contains 2 mol of a nickel porphinoid, designated coenzyme F_{430} (FIG. 1) as prosthetic group, which has to be in the Ni(I) oxidation state for the enzyme to be active (for review, see Ref. 44). The crystal structure of MCR with F_{430} in the Ni(II) oxidation state was resolved to 1.16 Å. It revealed that the subunits are intertwined such that they form two structurally interlinked active sites made up of the subunits α, $\alpha'\beta$, and γ and α', α, β', and γ', respectively. Each active site harbors one F_{430} molecule buried deep within the protein and accessible from the outside only via a 50-Å-long channel made up mainly of hydrophobic amino acid residues. Near to the active site are five modified amino acids: a thioglycine, an *N*-methyl-histidine, an *S*-methyl cysteine, a 5-(*S*)-methyl arginine, and a 2-(*S*)-methyl

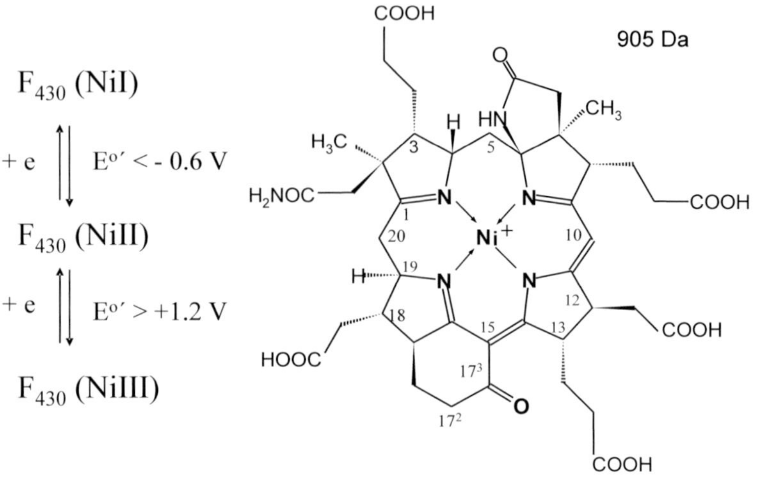

F_{430} (NiI)

+e ⇅ $E^{o\prime} < -0.6$ V

F_{430} (NiII)

+e ⇅ $E^{o\prime} > +1.2$ V

F_{430} (NiIII)

905 Da

FIGURE 1. Structure of F_{430}, the prosthetic group of methyl-coenzyme M reductase from methanogenic archaea. View from the β-face. In methyl-coenzyme M reductase from methanotrophic archaea of the ANME-1 cluster, 17^2-methylthio-F_{430} with a molecular mass of 951 Da rather than F_{430} with a mass of 905 Da is present.

glutamine.[45] Labeling studies have shown that the methyl groups are biosynthetically derived from the methyl group of methionine and not from the methyl group of methyl-coenzyme M.[46] The five modified amino acids are highly conserved. However, in MCR from *Methanosarcina* the respective glutamine is not methylated.[47] Generally, the genes coding for the three MCR subunits are coded in a transcription unit *mcrBDCGA*. The function of McrC and McrD, which do not copurify with MCR, is not known. In some methanogens the *mcrC* and/or *mcD* genes are not located together with the *mcrAGB* genes.

Methyl-Coenzyme M Reductase from ANME-1 and ANME-2

The archaea present in the microbial mats catalyzing AOM in the Black Sea contain at least two different MCR, designated Ni-protein I and Ni-protein II, that can be separated by anion exchange chromatography.[18] Ni-protein I contains a modified F_{430} with a molecular mass of 951 Da, whereas Ni-protein II contains the normal F_{430} with a molecular mass of 905 Da.

The structure of the modified F_{430} with a molecular mass of 951 Da has recently been elucidated to be 17^2-methylthio-F_{430} (S. Mayr, S. Shima, M. Krüger, R. Thauer, and B. Jaun, unpublished results). Ni-protein I is present in a concentration of 7% of the extracted soluble proteins, and Ni-protein II in a concentration of up to 3%. Via the N-terminal amino acid sequences of the three subunits, the encoding genes were identified in a metagenome library of the mats. The codon usage and tetranucleotide signature of the three genes in the cluster *mcrBGA* revealed that the three genes coding for Ni-protein I are located on the genome of the dominant ANME-1 archaeon present in the mats[18] and those coding for Ni-protein II in the genome of an ANME-2 archaeon (A. Meyerdierks, S. Shima, J. Kahnt, and M Krüger, unpublished results). Immunogold labeling of microbial mats with a specific antibody revealed that MCR was located in both ANME-1 and ANME-2 archaea. The data also show that the MCR-like enzymes are not only encoded in the genomes of ANME 1 and ANME-2 archaea but are, in fact, expressed at a high level.[48]

Ni-protein I from ANME-1 archaea differs from Ni-protein II from ANME-2 archaea not only in that the former harbors a modified F_{430} and the latter the normal F_{430}. In the α-subunit of Ni-protein I there is a valine, where there is a glutamine in Ni-protein II and in MCR from methanogenic archaea, which can be methylated at C2. Two amino acids downstream of the valine, there is a cysteine-rich sequence CCX_4CX_5C not present in Ni-protein II from ANME-2 and in MCR from methanogenic archaea. There is probably another interesting difference. In the sequence of Ni-protein I there appears to be a normal glycine, while in that of MCR from methanogens there is principally a thioglycine. Whether the thioglycine is also absent in Ni-protein II is not yet known.

Methyl-Coenzyme M Reductase Catalyzed Reactions

MCR from methanogenic archaea catalyzes the reduction of methyl-coenzyme M (CH_3-S-CoM)

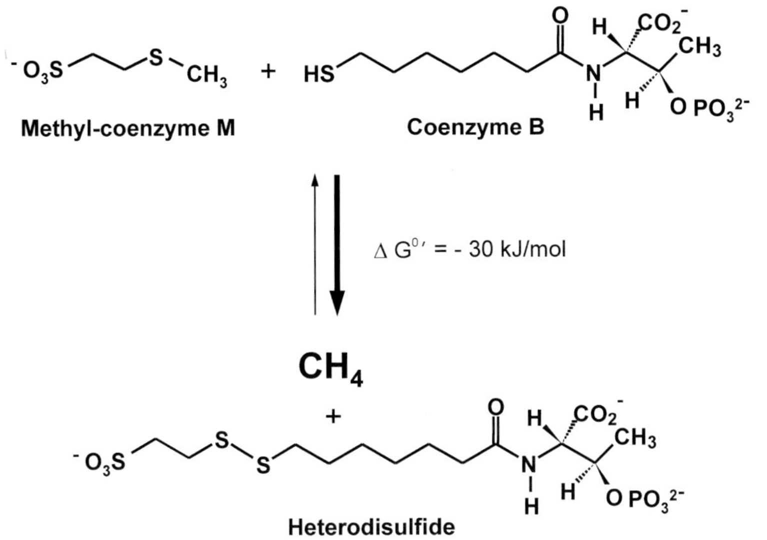

Methyl-coenzyme M + Coenzyme B

$\Delta G^{0\prime}$ = - 30 kJ/mol

CH_4 + Heterodisulfide

FIGURE 2. Reaction catalyzed by methyl-coenzyme M reductase from methanogenic archaea. $\Delta G^{\circ} = -30$ kJ/mol for the MCR-catalyzed reaction is obtained from the difference in the free energy changes associated with several reactions.[54,89] One of these reactions is the reduction of CoM-S-S-CoB with H_2. ΔG° for this reaction was revised recently due to the finding that the redox potential of the CoM-S-S-CoB/ HS-CoB + HS-CoM couple is -143 ± 10 mV rather than -200 mV.[90] As a result, ΔG° for methyl-coenzyme M reduction with coenzyme B decreased from -45 kJ/mol to -33 kJ/mol. This value also has some uncertainty, since it is in part based on $\Delta G^{\circ} = -28$ kJ/mol associated with methyl-coenzyme M formation from methanol and coenzyme M, which was calculated from differences in bond energies, which neglects differences in solvation energies.[89] ΔG° for methyl-coenzyme M reduction is best given as being -30 ± 10 kJ/mol.

with coenzyme B (HS-CoB) to methane and CoM-S-S-CoB ($\Delta G^{\circ\prime} = -30$ kJ/mol). This is the methane-forming reaction in all methanogenic archaea. The MCR catalyzed reaction is exergonic (FIG. 2). Under standard conditions (nongaseous substrates and products with a concentration of 1 M, CH_4 at 10^5 Pa pressure, and a pH of 7.0) the free energy change ($\Delta G^{\circ\prime}$) associated with the reaction is estimated to be -30 kJ/mol methane. The free energy change under physiological conditions ($\Delta G'$) is obtained from $\Delta G' = \Delta G^{\circ\prime} + RT\ln$ [Products]/[Substrates] $= -30 + 5.7 \log$ [Products]/[Substrates]. The equation predicts that the back reaction becomes exergonic when the product-to-substrate concentration ratio is approximately 10^5. Such a ratio is physiologically not unrealistic. At 10^5 Pa methane, the ratio is 10^5 when the intracellular concentration of CoM-S-S-CoB is, for example, 1 mM (10^{-3} M) and that of CH_3-S-CoM and HS-CoB is 0.1 mM (10^{-4} M) each. Consistent with this, methanogenic archaea have been shown to be capable of very slow methane oxidation.[49,50] In some of the earlier reports,[51,52] however, it has to be considered that the ^{14}C methane used to follow AOM was generated from $^{14}CO_2$ by methanogens and was therefore most likely contaminated by ^{14}CO.[53]

Rate of the Forward and Backward Reactions

MCR from *Methanothermobacter marburgensis* catalyzes methane formation from methyl-coenzyme M with a maximal specific activity of approximately 100 U/mg protein.[54] Exponentially growing *M. marburgensis* can produce methane at a specific rate of up to 5 U/mg protein. Consistently, such grown methanogenic archaea contain MCR at concentrations of 5–10% of the soluble cell proteins. Due to experimental reasons (equilibrium already reached after a few turnovers), the rate of the back reaction catalyzed by the enzyme has not yet been determined experimentally. The specific rate can be estimated, however, employing the Haldane equation, which correlates the equilibrium constant (K_{eq}) of a reaction with the catalytic efficiency (k_{cat}/K_m) of an enzyme to catalyze the forward and the back reaction: K_{eq} = catalytic efficiency (forward reaction)/catalytic efficiency (backwards reaction). The K_{eq} for the MCR catalyzed reaction is calculated from $\Delta G^{\circ\prime} = -RT\ln K_{eq} = -30$ kJ/mol to be near 10^5. Assuming the K_m values of MCR for its substrates and products to be all, for example, 0.1 mM, the Haldane equation predicts that MCR with a V_{max} for methyl-coenzyme M reduction of 100 U/mg catalyzes methane oxidation at a maximal specific rate of 1 mU/mg ($10^{-5} \times 100$ U/mg). As indicated in the legend to FIGURE 2, $\Delta G^{o\prime} = -30$ kJ/mol of the MCR catalyzed reaction is only known with an uncertainty of ± 10 kJ/mol. Therefore, and because the K_m values are most probably not all the same, the maximal specific rate of methane oxidation could be as high as 100 mU/mg and as low as 0.01 mU/mg.

Conformational Change Induced by Coenzyme B

The active-site structure indicates that methyl-coenzyme M reduction to methane takes place in a hydrophobic pocket from which water is completely excluded. When entering the active site via the hydrophobic channel, methyl-coenzyme M is stripped of water, and after the reaction the heterodisulfide is expelled into the water phase. The reaction most probably starts with a conformational change within the active site that is induced upon binding of coenzyme B. This is indicated by the finding that upon addition of coenzyme B to active MCR in the presence of coenzyme M, the enzyme is converted from the MCRred1c state, exhibiting a Ni(I)-derived axial electron paramagnetic resonance (EPR) signal, into the MCRred2 state, exhibiting a Ni(I)-derived highly rhombic EPR signal.[55,56] In single-turnover experiments methane formation from methyl-coenzyme M was found to be dependent on coenzyme B.[57]

Coupling of the Two Methyl-Coenzyme M Reductase Active Sites

There is evidence that the two active sites are structurally and functionally interlinked. The two α subunits in the enzyme are intertwined such that a conformational change in the one active site (made up of the subunits α, α′, β, and γ) can be transmitted to the other active site (made up of the subunits α′, α, β′, and γ′) and vice versa.[45] An indication for the coupling of the two active sites is the finding that at most 50% of the enzymes are converted from the MCRred1c state into the MCRred2 state upon addition of coenzyme B.[55] MCR thus shows "half-of-the-sites" reactivity. Based on these findings, it has been proposed that the enzyme operates according to a dual-stroke engine mechanism. This would allow the coupling of the endergonic steps of the catalytic cycle in one active site to the exergonic steps in the other site. The coupling is predicted to lower the activation energy for both the forward and the backward reactions[55] (FIG. 3).

Catalytic Mechanisms Discussed for Methyl-Coenzyme M Reductase

Three mechanisms for MCR are currently considered.[44,54] In **mechanism 1** the methyl group of methyl-coenzyme M reacts with the Ni(I) in a nucleophilic substitution reaction, yielding methyl-Ni(III) and coenzyme M, which in turn react to methyl-Ni(II) and the thiyl radical of coenzyme M. Methyl-Ni(II) then reacts with a proton in an electrophilic substitution reaction to methane and Ni(II), and the coenzyme M thiyl radical reacts with coenzyme B, yielding a disulfide anion radical, which is a strong reductant and which reduces Ni(II) back to Ni(I), thus closing the catalytic cycle. This mechanism is mainly supported by the findings that MCR-catalyzed methyl-coenzyme M reduction proceeds with the inversion of stereoconfiguration,[58] that enzyme bound Ni(I)F_{430} reacts with 3-bromopropane sulfonate to alkyl Ni(III),[59] and that free Ni(I)F_{430} in aprotic solvents reacts with methylbromide to form methyl-Ni(II)F_{430}, which subsequently can be protonolyzed to methane and Ni(II)F_{430}.[60] Also, the finding that MCR catalyzes the reduction of ethyl-coenzyme M with less than 1% of the catalytic efficiency of methyl-coenzyme M reduction is consistent with a nucleophilic substitution as the first step in the catalytic cycle.[55]

Looking at the back reaction, the oxidation of methane, this mechanism has a problem. Methane oxidation would start by a nucleophilic attack of methane on Ni(II) of F_{430}. This is most unlikely since Ni(II) of F_{430} is not electrophilic enough to be able to attack methane with a pK_a of above 40. The low electrophilicity of F_{430}(Ni II) is reflected by the low redox potential $E^{o\prime} = <-0.6$ V of the Ni(II)F_{430}/Ni(I)F_{430} couple[60,61] (FIG. 1).

Density function calculations have revealed that mechanism 1 is energetically not favorable,[62,63] although the calculations have not taken into account that the two active sites could be energetically coupled. Based on their calculations, Ghosh[64] and Siegbahn[62,63] have proposed **mechanism 2**, in which, as the first step in the catalytic cycle, the methyl thioether bond in methyl-coenzyme M is reductively cleaved, yielding a Ni(II) thiolate and a methyl radical, which in turn reacts with HS-CoB, yielding methane and a CoB-S· thiyl radical. The latter reacts with coenzyme M thiolate to form the disulfide anion radical, which, as in mechanism 1, is used to re-reduce the Ni(II)F_{430} in MCR to the Ni(I) oxidation state. Mechanism 2 is backed up by the experimental finding that in active MCR coenzyme M reversibly coordinates with its thiol sulfur to Ni(I) of F_{430} when coenzyme B is present.[65,66]

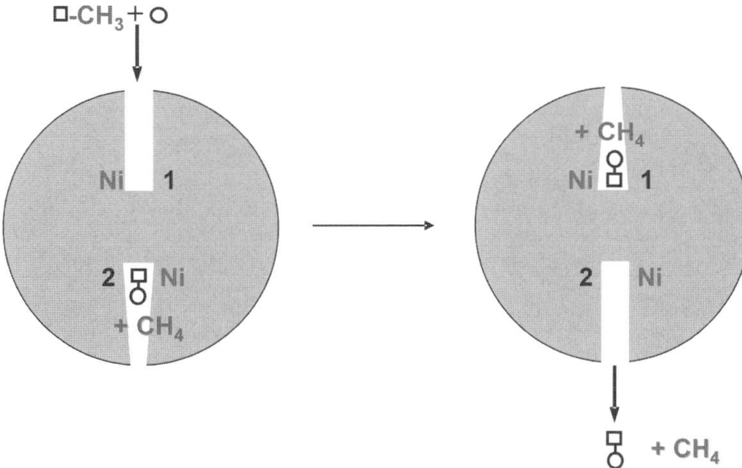

FIGURE 3. Dual-stroke engine mechanism proposed for methyl-coenzyme M reductase. The mechanism allows the coupling of endergonic steps of the catalytic cycle in the one active site to exergonic steps in the other active site. The coupling is predicted to lower the activation energy both in the forward and the back reaction.

But this mechanism also exhibits a flaw when looking at the back reaction. Methane oxidation would start with the reaction of methane with the CoB-S$^{\cdot}$ thiyl radical. The bond-dissociation energy of the C-H bond in methane is 439 kJ/mol, as compared to that of the S-H bond of only 365 kJ/mol (TABLE 2), which makes the reaction of methane with a thiyl radical yielding a methyl radical and a thiol thermodynamically very unfavorable. The activation energy for the back reaction would be unusually high (on the order of 80 kJ/mol), but possibly could be lowered by coupling the two active sites in MCR, as previously described.

Mechanism 3 is a modification of mechanism 1. Methyl-Ni(III)F_{430} generated from Ni(I)F_{430} and methyl-coenzyme M reacts with a proton, yielding methane and Ni(III)F_{430}. Free Ni(III)F_{430} is a very strong electrophile ($E^{\circ\prime} = > +1$ V),[60] and this is probably also true for enzyme bound Ni(III)F_{430}, although the EPR spectrum and the UV/visible spectrum of free Ni(III)F_{430} and those of Ni(III)F_{430} in MCRox are very different, indicating major differences in the coordination sphere.[67,68] The back reaction, the oxidation of methane, would therefore start, with the electrophilic metalation of methane by reaction of methane either end-on or side-on with the high-valent Ni(III) complex in MCR, as described for the activation of C-H bonds by other high-valent metal complexes.[1]

Why Methyl-Coenzyme M Reductase and Not Methylsuccinate Synthase?

As is outlined below, AOM with nitrate does not involve MCR. Available evidence indicates that the first step is the formation of 2-methylsuccinate from methane and fumarate catalyzed by a glycyl radical enzyme. From bond-dissociation energy differences it can be predicted that the formation of 2-methylsuccinate from methane and fumarate is an exergonic reaction with $\Delta G^{\circ\prime}$ between -10 and -20 kJ/mol[69] (TABLE 2). The radical mechanism is, however, such that this step cannot be coupled with energy conservation (see later in this paper). Therefore, most if not all of the free energy generated during methane oxidation with sulfate ($\Delta G^{\circ\prime} = -21$ kJ/mol) would be dissipated as heat in the first step, leaving not enough energy to drive the phosphorylation of ADP in other steps. However, this argument holds true only for AOM with sulfate as an electron acceptor. With nitrate or nitrite (reactions 6 and 7) the free energy change associated with AOM would be more than sufficient to allow the formation of 2-methylsuccinate, and thus dissipation of 15 ± 5 kJ/mol as heat in the first step.

Based on the same arguments, AOM with sulfate is predicted to start with a reaction operating under physiological conditions near thermodynamic equilibrium, and thus without the loss of energy. The MCR-catalyzed reaction perfectly fulfills this requirement.

Anaerobic Oxidation of Methane with Nitrate

Only one convincing publication describing AOM with electron acceptors other than sulfate has appeared. This is the paper by Raghoebarsing *et al.*,[8] in which they report that AOM can be coupled to NO_3^- or NO_2^- reduction to N_2 (reactions 6 and 7). The mixed culture mediating these reactions consisted to 10% of archaea related to the methanotrophic archaea (ANME) of the ANME clusters and to 90% of bacteria belonging to a novel phylum. From their results the authors concluded that the archaea in the mixed culture could be the organisms oxidizing CH_4, and if so that then MCR would probably be the enzyme catalyzing the first step in methane oxidation.

Noninvolvement of Methyl-Coenzyme M Reductase

There was, however, one finding that already argued strongly against the conclusion that MCR could be involved in AOM with nitrate. The culture growing at a doubling rate for over 100 days oxidized methane with a specific rate of 2 nmol/min and mg protein and an apparent K_m for methane of $<1\ \mu M$.[8] The catalytic efficiency ($kcat/K_m$) of AOM with nitrate was thus more than 1000 times higher than that of AOM with sulfate (apparent K_m for methane >1 mM), which, for reasons discussed earlier (Haldane equation), is not consistent with MCR being involved in AOM with nitrate.[70]

But there was also a second finding, which is difficult to explain, assuming methane in the mixed culture is oxidized by the 10% archaea. In labeling experiments with $^{13}CH_4$ only the bacterial lipids and not the lipids of the archaea became enriched in ^{13}C indicating that in the time span of the experiment (3 and 6 days) the archaeal population did not proliferate.[8]

It was therefore not unexpected when it was found that during further cultivation of the mixed culture the percentage of the archaea continuously decrease from 10% to well below 1%, as determined microscopically by fluorescence *in situ* hybridization (FISH) of the archaea. Parallelly, the F_{430} content of the cells (archaea plus bacteria) decreased below the detection limit, excluding that MCR is involved in AOM with nitrate in the culture of concern.[71]

Mechanism of Anaerobic Oxidation of Methane with Nitrate

What could be the mechanism of AOM with nitrate? Could methane directly react with one of the intermediates of denitrification, which are known to be strong oxidants? Probably not! The NO/N_2O couple has a redox potential $E^{\circ\prime}$ of +1.17 V and the N_2O/N_2 couple has one of +1.36 V. But both redox potentials are still more negative than that of the CH_3/CH_4 couple, which is more positive than +2 V (TABLES 1 and 2). But a direct reaction of CH_4 with a nitrogen oxide cannot be completely excluded. In nonaqueous solutions, methane can react with NO^+ and NO_2^+ by electrophilic substitution and insertion, respectively.[72,73]

An indication comes from the observation that the denitrifying cultures also catalyzed the oxidation of ethane, propane, and butane.[71] This finding suggests that methane oxidation could proceed via the mechanism of anaerobic oxidation of propane and of longer chain alkanes in sulfate-reducing and -denitrifying bacteria.

In recent years sulfate-reducing and -denitrifying bacteria have been characterized that can grow on alkanes with more than one carbon (for reviews, see Refs. 74–76). None of these organisms can metabolize methane and only a few can oxidize ethane. The first step in the anaerobic oxidation of alkanes has been shown to be the exergonic formation of 2-alkylsuccinate from alkane and fumarate. The functionalized alkane is then degraded via β-oxidation regenerating fumarate.[77–81]

2-Alkylsuccinate formation is catalyzed by a glycyl radical enzyme. In the proposed catalytic mechanism the glycyl radical (for the EPR signal, see Ref. 82) extracts an H atom from the alkane via a thiyl radical, preferentially from the C2 position, yielding an alkyl radical, which reacts with fumarate, yielding a 2-alkylsuccinyl radical. This radical reacts with the cysteine residue to 2-alkylsuccinate, regenerating the thiyl radical, and thus the glycyl radical of the enzyme.[83–85]

The dissociation energy of the C-H bonds of C2 of propane and of other alkanes is near +410 kJ/mol and thus 60 kJ/mol higher than that of the C-H bond in the glycine residue of the alkylsuccinate synthase (TABLE 2). The difference of 60 kJ/mol evidently can be overcome in the transition state of the enzyme; otherwise, the reaction would not be catalyzed. The question is, could such a mechanism also work for AOM?

In the reaction of methane with fumarate to 2-methylsuccinate, a methyl radical (439 kJ/mol) would have to be formed at the expense of a glycyl radical (350 kJ/mol). The difference in dissociation energies is almost 90 kJ/mol, which appears to be too large to be overcome in the transition state of the enzyme. It has to be considered, however, that glycyl radical enzymes are, just like MCR, functional dimers and show half-of-the-site reactivity. By coupling the endergonic steps of the catalytic cycle in one active site with exergonic steps in the catalytic cycle of the second active site, the relatively large difference in dissociation energies might be overcome.

Metabolic Cycles Regenerating Fumarate

The formation of 2-methyl succinate from fumarate and methane is metabolically attractive since fumarate can theoretically be regenerated from 2-methylsuccinate via at least two metabolic cycles in which CH_4 is oxidized to CO_2 and in which only known biochemical reactions and enzymes are involved (for more information, see the References in, e.g., Erb *et al.*[86]).

(i) Fumarate + CH_4 → 2-methylsuccinate → mesaconate + 2H → citramalate → citramalyl-CoA → pyruvate + acetyl-CoA → → → oxaloacetate → citrate → isocitrate → 2-oxoglutarate + CO_2 + 2H → succinyl-CoA + CO_2 + 2H → succinate → fumarate + 2H.

(ii) Fumarate + CH_4 → 2-methylsuccinate → 2-methylsuccinyl-CoA → 2-ethylmalonyl-CoA → butyryl-CoA + CO_2 → crotonyl-CoA + 2H → 3-hydroxybutyryl-CoA → 4-hydroxybutyryl-CoA → 4-hydroxybutyrate → succinate semialdehyde + 2H → succinyl-CoA + 2H → succinate → fumarate + 2H.

The following reactions could provide the cells with precursors for biosynthesis: fumarate + CH_4 → 2-methylsuccinate → mesaconate + 2H → citramalate → citramalyl-CoA → pyruvate + acetyl-CoA → → → oxaloacetate. Oxaloacetate would be formed from pyruvate via carboxylation and pyruvate formed from acetyl-CoA via reductive carboxylation.

Why Not Methyl-Coenzyme M Reductase Involved in Anaerobic Oxidation of Methane with Nitrate?

MCR from methanogenic archaea is only active when its prosthetic group F_{430} is in the Ni(I) oxidation state. The redox potential ($E^{\circ\prime}$) of the Ni(II)F_{430}/Ni(I)F_{430} couple has been determined to be below −0.6 V.[60,61] Due to the negative redox potential, which is more than 0.2 V below that of the hydrogen electrode at pH 7.0, and due to the fact that in the enzyme F_{430} is not completely electrically insulated from electron acceptors present in the solvent, Ni(I)F_{430} in MCR slowly oxidizes to Ni(II)F_{430}, even under strictly anoxic conditions. The rate of MCR inactivation increases with increasing redox potetial in its environment.[44] Considering that re-reduction of MCR in the cells requires ATP,[54] the negative redox potential of F_{430} could preclude the operation of MCR in cells, in which the redox potential of the terminal electron acceptors is more positive than 0 V as in the case of the nitrate/nitrite couple ($E^{\circ\prime}$ = +0.43 V), the NO_2^-/NO couple ($E^{\circ\prime}$ = +0.34 V), the NO/N_2O couple ($E^{\circ\prime}$ = +1.17 V), and the N_2O/N_2 couple ($E^{\circ\prime}$ = +1.36 V). This is probably one reason why AOM coupled to denitrification does not involve MCR. On the contrary, the redox potential of the adenylylsulfate (APS)/sulfite couple ($E^{\circ\prime}$ = −0.06 V) and that of the sulfite/H_2S couple ($E^{\circ\prime}$ = −0.12 V) involved in dissimilatory sulfate reduction are evidently sufficiently negative to allow the function of MCR in the presence of these electron acceptors.

Following these arguments one can predict that, if found, AOM with sulfur could involve MCR since the redox potential of the S°/H_2S couple is −0.12 V at physiological HS^- concentrations (TABLE 1). And, AOM with Fe(III) (+0.2 V) or Mn(IV) (+0.41 V), if once found, is predicted not to involve MCR.

Alkane Carboxylation and the Biological Formation of Ethane and Propane

Recently evidence has been published that in some SRB the anaerobic oxidation of alkanes could be initiated by the subterminal carboxylation of the alkane at the C3 position.[87] The carboxylation of alkanes is, under standard conditions, an endergonic reaction ($\Delta G^{\circ\prime}$ = +28 kJ/mol). For it to proceed under physiological conditions, the fatty acid concentration would have to be in the μM range. Carboxylation of methane to acetate is thermodynamically even less favorable ($\Delta G^{\circ\prime}$ = +36 kJ/mol). The findings by So *et al.*[87] should therefore be taken only as a reminder that there could be more than two ways to anaerobically oxidize methane.

In this respect the finding that ethane and propane are biologically formed in the deep marine subsurface is of interest. Reduction of acetate to ethane is one feasible mechanism. Propane is enriched in ^{13}C relative to ethane. The amount is consistent with derivation of the third C from CO_2. At typical sedimentary conditions, the reactions yield free energy sufficient for growth.[88]

Conclusion

Until now AOM by microorganisms has only been reported with sulfate or nitrate/nitrite as electron acceptors. AOM with sulfate involves archaea and most probably MCR, whereas AOM with nitrate involves bacteria and possibly a glycyl radical enzyme. However, it has to be considered that the organisms mediating AOM have not yet been grown in pure culture and that therefore most conclusions are only based on circumstantial evidence. From thermodynamics of AOM with different electron acceptors one can predict that AOM should also be possible with S°, arsenate, Fe(III), and Mn(IV). Organisms mediating these reactions will have to looked for in appropriate niches

Acknowledgments

This work was supported by the Max Planck Society and by a grant from the Fonds der Chemischen Industrie. We thank Martin Krüger and Fritz Widdel from the Max Planck Institute for Marine Microbiology in Bremen for a fruitful collaboration.

Conflict of Interest

The authors declare no conflicts of interest.

References

1. SHILOV, A.E. & G.B. SHUL'PIN. 1997. Activation of C-H bonds by metal complexes. Chem. Rev. **97:** 2879–2932.
2. OLAH, G.A. 2001. 100 years of carbocations and their significance in chemistry. J. Org. Chem. **66:** 5943–5957.
3. LIEBERMAN, R.L. & A.C. ROSENZWEIG. 2004. Biological methane oxidation: regulation, biochemistry, and active site structure of particulate methane monooxygenase. Crit. Rev. Biochem. Mol. Biol. **39:** 147–164.
4. SHAIK, S. *et al.* 2004. The "Rebound controversy": an overview and theoretical modeling of the rebound step in C-H hydroxylation by cytochrome P450. Eur. J. Inorg. Chem. 207–226.
5. DALTON, H. 2005. The Leeuwenhoek Lecture 2000—The natural and unnatural history of methane-oxidizing bacteria. Philos. Trans. R. Soc. Lond.B Biol. Sci. **360:** 1207–1222.
6. CHISTOSERDOVA, L., J.A. VORHOLT & M.E. LIDSTROM. 2005. A genomic view of methane oxidation by aerobic bacteria and anaerobic archaea. Genome Biol. **6:** 208.1–208.6.
7. REEBURGH, W.S. 2007. Oceanic methane biogeochemistry. Chem. Rev. **107:** 486–513.
8. RAGHOEBARSING, A.A. *et al.* 2006. A microbial consortium couples anaerobic methane oxidation to denitrification. Nature **440:** 918–921.
9. SHIMA, S. & R.K. THAUER. 2005. Methyl-coenzyme M reductase (MCR) and the anaerobic oxidation of methane (AOM) in methanotrophic archaea. Curr. Opin. Microbiol. **8:** 643–648.
10. THAUER, R.K. & S. SHIMA. 2007. Methyl-coenzyme M reductase in methanogens and methanotrophs. *In* Archaea, Evolution, Physiology and Molecuar Biology. R. Garrett & H.-P. Klenk, Eds.: 275–283. Malden, MA: Blackwell Publishing.
11. GIRGUIS, P.R., A.E. COZEN & E.F. DELONG. 2005. Growth and population dynamics of anaerobic methane-oxidizing archaea and sulfate-reducing bacteria in a continuous-flow bioreactor. Appl. Environ. Microbiol. **71:** 3725–3733.
12. NAUHAUS, K. *et al.* 2007. *In vitro* cell growth of marine archaeal-bacterial consortia during anaerobic oxidation of methane with sulfate. Environ. Microbiol. **9:** 187–196.
13. TISHCHENKO, P. *et al.* 2005. Calculation of the stability and solubility of methane hydrate in seawater. Chem. Geol. **219:** 37–52.
14. KRUEGER, M. *et al.* 2005. Microbial methane turnover in different marine habitats. Palaeogeogr. Palaeoclimatol. **227:** 6–17.
15. TREUDE, T. *et al.* 2005. Subsurface microbial methanotrophic mats in the Black Sea. Appl. Environ. Microb. **71:** 6375–6378.
16. TREUDE, T. *et al.* 2005b. Anaerobic oxidation of methane and sulfate reduction along the Chilean continental margin. Geochim. Cosmochim. Acta **69:** 2767–2779.
17. NIEMANN, H. *et al.* 2006. Microbial methane turnover at mud volcanoes of the Gulf of Cadiz. Geochim. Cosmochim. Acta **70:** 336–5355.
18. KRÜGER, M. *et al.* 2003. A conspicuous nickel protein in microbial mats that oxidize methane anaerobically. Nature **426:** 878–881.
19. SEIFERT, R. *et al.* 2006. Methane dynamics in a microbial community of the Black Sea traced by stable carbon isotopes *in vitro*. Org. Geochem. **37:** 1411–1419.
20. ORCUTT, B. *et al.* 2005. Molecular biogeochemistry of sulfate reduction, methanogenesis and the anaerobic oxidation of methane at Gulf of Mexico cold seeps. Geochim. Cosmochim. Acta **69:** 4267–4281.
21. KNITTEL, K. *et al.* 2005. Diversity and distribution of methanotrophic archaea at cold seeps. Appl. Environ. Microbiol. **71:** 467–479.
22. NAUHAUS, K. *et al.* 2005. Environmental regulation of the anaerobic oxidation of methane: a comparison of ANME-I and ANME-II communities. Environ. Microbiol. **7:** 98–106.
23. STADNITSKAIA, A. *et al.* 2005. Biomarker and 16S rDNA evidence for anaerobic oxidation of methane and related carbonate precipitation in deep-sea mud volcanoes of the Sorokin Trough, Black Sea. Mar. Geol. **217:** 67–96.
24. BOETIUS, A. *et al.* 2000. A marine microbial consortium apparently mediating anaerobic oxidation of methane. Nature **407:** 623–626.
25. ORPHAN, V.J. *et al.* 2001. Comparative analysis of methane-oxidizing archaea and sulfate-reducing bacteria in anoxic marine sediments. Appl. Environ. Microbiol. **67:** 1922–1934.
26. ELVERT, M. *et al.* 2003. Characterization of specific membrane fatty acids as chemotaxonomic markers for sulfate-reducing bacteria involved in anaerobic oxidation of methane. Geomicrobiol. J. **20:** 403–419.
27. KNITTEL, K. *et al.* 2003. Activity, distribution, and diversity of sulfate reducers and other bacteria in sediments above gas hydrate (Cascadia margin, Oregon). Geomicrobiol. J. **20:** 269–294.
28. BLUMENBERG, M. *et al.* 2005. In vitro study of lipid biosynthesis in an anaerobically methane-oxidizing microbial mat. Appl. Environ. Microbiol. **71:** 4345–4351.

29. HOEHLER, T. *et al.* 2001. Apparent minimum free energy requirements for methanogenic Archaea and sulfate-reducing bacteria in an anoxic marine sediment. FEMS Microbiol. Ecol. **38:** 33–41.
30. DALE, A.W., P. REGNIER & P. VAN CAPPELLEN. 2006. Bioenergetic controls on anaerobic oxidation of methane (AOM) in coastal marine sediments: a theoretical analysis. Am. J. Sci. **306:** 246–294.
31. REGUERA, G. *et al.* 2005. Extracellular electron transfer via microbial nanowires. Nature **435:** 1098–1101.
32. STAMS, A.J.M. *et al.* 2006. Exocellular electron transfer in anaerobic microbial communities. Environ. Microbiol. **8:** 371–382.
33. DURISCH-KAISER, E. *et al.* 2005. Evidence of intense archaeal and bacterial methanotrophic activity in the black sea water column. Appl. Environ. Microb. **71:** 8099–8106.
34. HALLAM, S.J. *et al.* 2003. Identification of methyl coenzyme M reductase A (mcrA) genes associated with methane-oxidizing archaea. Appl. Environ. Microbiol. **69:** 5483–5491.
35. HALLAM, S.J. *et al.* 2004. Reverse methanogenesis: testing the hypothesis with environmental genomics. Science **305:** 1457–1462.
36. GOENRICH, M. *et al.* 2004. Probing the reactivity of Ni in the active site of methyl-coenzyme M reductase with substrate analogues. J. Biol. Inorg. Chem. **9:** 691–705.
37. BOYD, J.M., A. ELLSWORTH & S.A. ENSIGN. 2006. Characterization of 2-bromoethanesulfonate as a selective inhibitor of the coenzyme M-dependent pathway and enzymes of bacterial aliphatic epoxide metabolism. J. Bacteriol. **188:** 8062–8069.
38. SANTORO, N. & J. KONISKY. 1987. Characterization of bromoethanesulfonate-resistant mutants of *Methanococcus voltae*—Evidence of a coenzyme-M transport-system. J. Bacteriol. **169:** 660–665.
39. FRIEDRICH, M.W. 2005. Methyl-coenzyme M reductase genes: unique functional markers for methanogenic and anaerobic methane-oxidizing Archaea. Methods Enzymol. **397:** 428–442.
40. INAGAKI, F. *et al.* 2004. Characterization of C1 metabolizing prokaryotic communities in methane seep habitats at the Kuroshima Knoll, southern Ryukyu Arc, by analyzing pmoA, mmoX, mxaF, mcrA, and 16S rRNA genes. Appl. Environ. Microbiol. **70:** 7445–7455.
41. NERCESSIAN, O. *et al.* 2005. Diversity of functional genes of methanogens, methanotrophs and sulfate reducers in deep-sea hydrothermal environments. Environ. Microbiol. **7:** 118–132.
42. NUNOURA, T. *et al.* 2006. Quantification of mcrA by quantitative fluorescent PCR in sediments from methane seep of the Nankai Trough. FEMS Microbiol. Ecol. **57:** 149–157.
43. OREMLAND, R.S. *et al.* 2005. Whither or wither geomicrobiology in the era of 'community metagenomics.' Nat. Rev. Microbiol. **3:** 572–578.
44. JAUN, B. & R.K. THAUER. 2007. Methyl-coenzmye M reductase and its nickel corphin coenzyme F430 in methanogenic archaea. *In* Nickel and Its Surprising Impact in Nature, Vol. 2 of Metal Ions in Life Sciences. A. Sigel, H. Sigel & R.K.O. Sigel, Eds.: 323–356. Chichester, UK: John Wiley & Sons.
45. ERMLER, U. *et al.* 1997. Crystal structure of methyl-coenzyme M reductase: the key enzyme of biological methane formation. Science **278:** 1457–1462.
46. SELMER, T. *et al.* 2000. The biosynthesis of methylated amino acids in the active site region of methyl-coenzyme M reductase. J. Biol. Chem. **275:** 3755–3760.
47. GRABARSE, W. *et al.* 2000. Comparison of three methyl-coenzyme M reductases from phylogenetically distant organisms: unusual amino acid modification, conservation and adaptation. J. Mol. Biol. **303:** 329–344.
48. HELLER, C. *et al.* 2007. Immunological localization of coenzyme M reductase in anaerobic methane-oxidizing archaea. BIOspektrum. Special Issue 2007, 60:KE 014.
49. SHILOV, A.E. *et al.* 1999. Methanogenesis is reversible: the formation of acetate in methane carboxilation by bacteria of methanigenic biocenose. Dokl. Akad. Nauk. **367:** 557–559.
50. MORAN, J.J. *et al.* 2004. Trace methane oxidation studied in several Euryarchaeota under diverse conditions. Archaea **1:** 303–309.
51. ZEHNDER, A.J.B. & T. BROCK. 1979. Methane formation and methane oxidation by methanogenic bacteria. J. Bacteriol. **137:** 420–432.
52. ZEHNDER, A.J.B. & T.D. BROCK. 1980. Anaerobic methane oxidation—Occurrence and ecology. Appl. Environ. Microbiol. **39:** 194–204.
53. CONRAD, R. & R.K. THAUER. 1983. Carbon monoxide production by *Methanobacterium thermoautotrophicum*. FEMS Microbiol. Lett. **20:** 229–232.
54. THAUER, R.K. 1998. Biochemistry of methanogenesis: a tribute to Marjory Stephenson. Microbiology **144:** 2377–2406.
55. GOENRICH, M. *et al.* 2005. Temperature dependence of methyl-coenzyme M reductase activity and of the formation of the methyl-coenzyme M reductase red2 state induced by coenzyme B. J. Biol. Inorg. Chem. **10:** 333–342.
56. KERN, D.I. et al. 2007. Two sub-strates of the red2 state of methyl-coenzyme M reductase revealed by high-field EPR spectrospcopy. J. Biol. Inorg. Chem. **12**(8): 1097–1105.
57. HORNG, Y.C., D.F. BECKER & S.W. RAGSDALE. 2001. Mechanistic studies of methane biogenesis by methyl-coenzyme M reductase: evidence that coenzyme B participates in cleaving the C-S bond of methyl-coenzyme M. Biochemistry **40:** 12875–12885.
58. AHN, Y., J.A. KRZYCKI & H.G. FLOSS. 1991. Steric course of the reduction of ethyl coenzyme M to ethane catalyzed by methyl coenzyme M reductase from *Methanosarcina barkeri*. J. Am. Chem. Soc. **113:** 4700–4701.
59. HINDERBERGER, D. *et al.* 2006. A nickel-alkyl bond in an inactivated state of the enzyme catalyzing methane formation. Angew. Chem. Int. Ed. Engl. **45:** 3602–3607.
60. JAUN, B. 1993. Methane formation by methanogenic bacteria: redox chemistry of coenzyme F430. *In* Metal Ions in Biological Systems, Vol. 29, Properties of Metal Alkyl Derivatives. H. Sigel & A. Sigel, Eds.: 287–337. New York: Marcel Dekker.
61. PISKORSKI, R. & B. JAUN. 2003. Direct determination of the number of electrons needed to reduce coenzyme F430 pentamethyl ester to the Ni(I) species exhibiting the electron paramagnetic resonance and ultraviolet-visible

spectra characteristic for the MCR(red1) state of methyl-coenzyme M reductase. J. Am. Chem. Soc. **125:** 13120–13125.

62. PELMENSCHIKOV, V. *et al.* 2002. A mechanism from quantum chemical studies for methane formation in methanogenesis. J. Am. Chem. Soc. **124:** 4039–4049.
63. PELMENSCHIKOV, V. & P.E. SIEGBAHN. 2003. Catalysis by methyl-coenzyme M reductase: a theoretical study for heterodisulfide product formation. J. Biol. Inorg. Chem. **8:** 653–662.
64. GHOSH, A., T. WONDIMAGEGN & H. RYENG. 2001. Deconstructing F430: quantum chemical perspectives of biological methanogenesis. Curr. Opin. Chem. Biol. **5:** 744–750.
65. FINAZZO, C. *et al.* 2003. Coenzyme B induced coordination of coenzyme M via its thiol group to Ni(I) of F430 in active methyl-coenzyme M reductase. J. Am. Chem. Soc. **125:** 4988–4989.
66. FINAZZO, C. *et al.* 2003. Characterization of the MCRred2 form of methyl-coenzyme M reductase: a pulse EPR and ENDOR study. J. Biol. Inorg. Chem. **8:** 586–593.
67. CRAFT, J.L. *et al.* 2004. Nickel oxidation states of F430 cofactor in methyl-coenzyme M reductase. J. Am. Chem. Soc. **126:** 4068–4069.
68. DUIN, E.C. *et al.* 2004. Spectroscopic investigation of the nickel-containing porphinoid cofactor F430. Comparison of the free cofactor in the +1, +2 and +3 oxidation states with the cofactor bound to methyl-coenzyme M reductase in the silent, red and ox forms. J. Biol. Inorg. Chem. **9:** 563–574.
69. LI, L. & E.N.G. MARSH. 2006. Mechanism of benzylsuccinate synthase probed by substrate and isotope exchange. J. Am. Chem. Soc. **128:** 16056–16057.
70. THAUER, R.K. & S. SHIMA. 2006. Biogeochemistry: methane and microbes. Nature **440:** 878–879.
71. ETTWIG, K.F. *et al.* 2007. Freshwater microbial consortium couples anaerobic methane oxidation to denitrification. BIOspektrum. Special Issue 2007, 58:KE 007.
72. OLAH, G.A. *et al.* 1995. Electrophilic substitution of methane revisited. J. Am. Chem. Soc. **117:** 1336–1343.
73. OLAH, G.A., P. RAMAIAH & G.K.S. PRAKASH. 1997. Electrophilic nitration of alkanes with nitronium hexafluorophosphate. Proc. Natl. Acad. Sci. USA **94:** 11783–11785.
74. SPORMANN, A.M. & F. WIDDEL. 2000. Metabolism of alkylbenzenes, alkanes, and other hydrocarbons in anaerobic bacteria. Biodegradation **11:** 85–105.
75. BONIN, P. *et al.* 2004. The anaerobic hydrocarbon biodegrading bacteria: an overview. Ophelia **58:** 243–254.
76. SHENNAN, J.L. 2006. Utilisation of C-2-C-4 gaseous hydrocarbons and isoprene by microorganisms. J. Chem. Technol. Biotechnol. **81:** 237–256.
77. RABUS, R., T. HANSEN & F. WIDDEL. 2001. Dissimilatory sulfate- and sulfur-reducing prokaryotes. *In* The Prokaryotes: An Evolving Electronic Resource for the Microbiological Community. S. Dworkin *et al.*, Eds.: New York:Springer-Verlag.release 3.3, http://link.springer-ny.com/link/service/books/10125.
78. WILKES, H. *et al.* 2003. Formation of n-alkane- and cycloalkane-derived organic acids during anaerobic growth of a denitrifying bacterium with crude oil. Org. Geochem. **34:** 1313–1323.
79. CRAVO-LAUREAU, C. *et al.* 2005. Anaerobic n-alkane metabolism by a sulfate-reducing bacterium, *Desulfatibacillum aliphaticivorans* strain CV2803. Appl. Environ. Microb. **71:** 3458–3467.
80. DAVIDOVA, I.A. *et al.* 2005. Stable isotopic studies of n-alkane metabolism by a sulfate-reducing bacterial enrichment culture. Appl. Environ. Microbiol. **71:** 8174–8182.
81. DAVIDOVA, I.A. *et al.* 2006. *Desulfoglaeba alkanexedens* gen. nov., sp nov., an n-alkane-degrading, sulfate-reducing bacterium. Int. J. Syst. Evol. Micr. **56:** 2737–2742.
82. KACPRZAK, S., R. REVIAKINE & M. KAUPP. 2007. Understanding the electon paramagnetic resonance parameters of protein-bound glycyl radicals. J. Phys. Chem. B **111:** 820–831.
83. SELMER, T., A.J. PIERIK & J. HEIDER. 2005. New glycyl radical enzymes catalysing key. Metabolic steps in anaerobic bacteria. Biol. Chem. **386:** 981–988.
84. LI, L. & E.N.G. MARSH. 2006. Deuterium isotope effects in the unusual addition of toluene to fumarate catalyzed by benzylsuccinate synthase. Biochemistry **45:** 13932 13938.
85. BUCKEL, W. & B.T. GOLDING. 2006. Radical enzymes in anaerobes. Annu. Rev. Microbiol. **60:** 27–49.
86. ERB, T.J. *et al.* 2007. Synthesis of C5-dicarboxylic acids from C2-units involving crotonyl-CoA carboxylase/reductase. Proc. Natl. Acad. Sci. USA **104**: 10631–10636.
87. SO, C.M., C.D. PHELPS & L.Y. YOUNG. 2003. Anaerobic transformation of alkanes to fatty acids by a sulfate-reducing bacterium, strain Hxd3. Appl. Environ. Microbiol. **69:** 3892–3900.
88. HINRICHS, K.U. *et al.* 2006. Biological formation of ethane and propane in the deep marine subsurface. Proc. Natl. Acad. Sci. USA **103:** 14684–14689.
89. KELTJENS, J.T. & C. VAN DER DRIFT. 1986. Electron-transfer reactions in methanogens. FEMS Microbiol. Rev. **39:** 259–303.
90. TIETZE, M. *et al.* 2003. Redox potentials of methanophenazine and CoB-S-S-CoM, factors involved in electron transport in Methanogenic archaea. Chembiochem. **4:** 333–335.
91. THAUER, R.K., K. JUNGERMANN & K. DECKER. 1977. Energy conservation in chemotrophic anaerobic bacteria. Bact. Rev. **41:** 100–180.
92. EHRENREICH, A. & F. WIDDEL. 1994. Anaerobic oxidation of ferrous iron by purple bacteria, a new-type of phototrophic metabolism. Appl. Environ. Microb. **60:** 4517–4526.
93. STUBBE, J.A. & W.A. VAN DER DONK. 1998. Protein radicals in enzyme catalysis. Chem. Rev. **98:** 705–762.
94. WANG, S.C. & P.A. FREY. 2007. S-Adenosylmethionine as an oxidant: the radical SAM superfamily. Trends Biochem. Sci. **32:** 101–110.
95. SANDALA, G.M. *et al.* 2007. Toward an improved understanding of the glutamate mutase system. J. Am. Chem. Soc. **129:** 1623–1633.
96. ARMSTRONG, D.A., D. YU & A. RAUK. 1996. Oxidative damage to the glycyl alpha-carbon site in proteins: an ab initio study of the C-H bond dissociation energy and the reduction potential of the C-centered radical. Can. J. Chem. **74:** 1192–1196.

Metabolic, Phylogenetic, and Ecological Diversity of the Methanogenic Archaea

YUCHEN LIU AND WILLIAM B. WHITMAN

Department of Microbiology, University of Georgia, Athens, Georgia, USA

Although of limited metabolic diversity, methanogenic archaea or methanogens possess great phylogenetic and ecological diversity. Only three types of methanogenic pathways are known: CO_2-reduction, methyl-group reduction, and the aceticlastic reaction. Cultured methanogens are grouped into five orders based upon their phylogeny and phenotypic properties. In addition, uncultured methanogens that may represent new orders are present in many environments. The ecology of methanogens highlights their complex interactions with other anaerobes and the physical and chemical factors controlling their function.

Key words: **methane; methanogen; methanogenesis; hydrogenotrophic methanogenesis; methylotrophic methanogenesis; aceticlastic methanogenesis**

Introduction

Biological methane production or methanogenesis is an important process in the global carbon cycle, processing about 1.6% of the carbon fixed every year by plants and algae.[1] Biologically generated methane can either serve as a substrate for aerobic or anaerobic methane oxidation or be emitted to the atmosphere. Methane emitted to the atmosphere is a major greenhouse gas, whose atmospheric concentration has increased by threefold in the last 200 years.[1] The current global methane emission is 500–600 Tg CH_4 y^{-1}.[2] About 74% of the emitted methane is derived from biological methanogenesis.[3] The major sources of methane emissions are listed in TABLE 1.

Methanogens are the microorganisms that produce methane as the end-product of their anaerobic respiration. All methanogens are strictly anaerobic archaea belonging to the *Euryarchaeota*. They are a large and diverse group, all of which are obligate methane-producers that obtain all or most of their energy from methanogenesis. The methanogenesis pathway is complex, requiring a number of unique coenzymes and membrane-bound enzyme complexes, the details of which have been recently reviewed.[1]

Methanogens have been cultivated from a wide variety of anaerobic environments. In addition to temperate habitats, they are also common in environments of extreme temperatures, salinity, and pH. The common methanogenic habitats include marine sediments, freshwater sediments, flooded soils, human and animal gastrointestinal tracts, termites, anaerobic digestors, landfill, geothermal systems, and heartwood of trees.

Methanogenic Substrates

Although the methanogens are very diverse, they can only utilize a restricted number of substrates. The substrates are limited to three major types: CO_2, methyl-group containing compounds, and acetate (TABLE 2). Most organic substances, for instance, carbohydrates and long-chain fatty acids and alcohols, are not substrates for methanogenesis. Instead, these compounds must first be processed by anaerobic bacteria or eukaryotes to produce the substrates actually used by the methanogens. Thus, in most methanogenic environments, most of the energy available for growth is utilized by these nonmethanogenic organisms. It is an interesting physiological mystery as to why methanogens do not directly utilize complex organic matter, bypassing their dependency on other organisms.

The first type of substrate is CO_2. Most methanogens are hydrogenotrophs that can reduce CO_2 to methane with H_2 as the primary electron donor. Many hydrogenotrophic methanogens can also use formate as the major electron donor. In this case, four molecules of formate are oxidized to CO_2 by formate dehydrogenase (Fdh) before one molecule of CO_2 is reduced to methane. In hydrogenotrophic methanogenesis, CO_2 is reduced successively to methane through the formyl, methylene,

Address for correspondence: Dr. William B. Whitman, Department of Microbiology, University of Georgia, 541 Biological Sciences Building, Athens, GA 30605.
whitman@uga.edu

Ann. N.Y. Acad. Sci. 1125: 171–189 (2008). © 2008 New York Academy of Sciences.
doi: 10.1196/annals.1419.019

TABLE 1. Sources of global methane emissions from identified sources

Sources	Methane emission (Tg of CH_4 per year)	Percentage (%)[a]
Natural sources		
Wetlands	92–237	15–40
Termites	20	3
Ocean	10–15	2–3
Methane hydrates	5–10	1–2
Subtotal	127–282	21–47
Anthropogenic sources		
Ruminants	80–115	13–19
Energy generation[b]	75–110	13–18
Rice agriculture	25–100	7–17
Landfills	35–73	6–12
Biomass burning	23–55	4–9
Waste treatment	14–25	2–4
Subtotal	267–478	45–80
Total sources	500–600	

Source: Modified from Lowe[2] and Prather and Ehhalt.[140]

[a]Estimates of the relative contribution of methane emission from a source to the total global emissions of 600 Tg of CH_4 per year.

[b]Methane deposits released by coal mining, petroleum drilling, and petrochemical production.

and methyl levels (FIG. 1A). The C-1 moiety is carried by special coenzymes, methanofuran (MFR), tetrahydromethanopterin (H_4MPT), and coenzyme M (CoM). Initially, CO_2 binds to MFR and is reduced to the formyl level. In this first reduction step, ferredoxin (Fd), which is reduced with H_2, is the direct electron donor. The formyl group is then transferred to H_4MPT, forming formyl-H_4MPT. The formyl group is then dehydrated to methenyl group, which is subsequently reduced to methylene-H_4MPT and then to methyl-H_4MPT. Reduced F_{420} ($F_{420}H_2$) is the direct electron donor in these two reduction steps. The methyl group is then transferred to CoM, forming methyl-CoM. The last reduction step reduces methyl-CoM to methane by methyl coenzyme M reductase (Mcr), which is the key enzyme in methanogenesis. Coenzyme B (CoB) is the direct electron donor in this reduction, and the oxidized CoB forms a heterodisulfide with CoM (CoM-S-S-CoB). Finally, the heterodisulfide is reduced to regenerate the thiols. Thermodynamically, two reactions—the methyl transfer from H_4MPT to CoM and the reduction of the heterodisulfide—are exergonic and involved in energy conservation. The methyl transfer reaction is catalyzed by methyl-H_4MPT:HS-CoM methyltransferase (Mtr), which is a membrane-bound complex. The reduction of the heterodisulfide is catalyzed by heterodisulfide reductase (Hdr), which is a membrane-bound complex in *Methanosarcina* and coupled to $F_{420}H_2$ dehydrogenase (Fpo) when H_2 is present. The reduction of CO_2 to formyl-MFR is endergonic and driven by ion gradient via the membrane-bound energy conserving hydrogenase (Ech).

Some hydrogenotrophic methanogens can also use secondary alcohols, such as 2-propanol, 2-butanol, and cyclopentanol, as electron donors. A small number can use ethanol.[4–7] The secondary alcohols are oxidized to ketones via coenzyme F_{420}-dependent secondary alcohol dehydrogenases (Adf).[8] Ethanol is oxidized to acetate via a nicotinamide adenine dinucleotide phosphate (NADP)–dependent alcohol dehydrogenase.[9] Although the growth on alcohols is poor compared to that on H_2, it is an important exception to the generalization that methanogens cannot directly metabolize most organic compounds. Even in this case, where the substrate is obviously assimilated, the oxidation is incomplete, and methane is derived from CO_2 reduction.

Two species have been shown to utilize CO as a reductant for methanogenesis from CO_2.[10,11] In *Methanothermobacter thermoautotrophicus* and *Methanosarcina barkeri*, four molecules of CO are oxidized to CO_2 using CO dehydrogenase (CODH) before one molecule of CO_2 is reduced to methane.[11] H_2 is an intermediate in this reaction and serves as the direct electron donor for the reduction of CO_2. Growth with CO is poor, and the doubling time is more than 200 h for *M. thermoautotrophicus* and 65 h for *M. barkeri*. In contrast, *Methanosarcina acetivorans* grows on CO by an entirely different pathway to be discussed later in this chapter.

The second type of substrate is methyl-group containing compounds, including methanol, methylated amines (monomethylamine, dimethylamine, trimethylamine, and tetramethylammonium), and methylated sulfides (methanethiol and dimethylsulfide). Methanogens that are able to use methylated compounds, or methylotrophic methanogens, are limited to the order Methanosarcinales, except for *Methanosphaera* species, which belong to the order Methanobacteriales. During methanogenesis, the methyl-groups from methylated compounds are transferred to a cognate corrinoid protein and then to CoM (FIG. 1B).[12–14] Methyl-CoM subsequently enters the methanogenesis pathway and is reduced to methane. The activation and transfer of the methyl group requires substrate-specific methyltransferases. Interestingly, all known methylamine methyltransfereases contain a UAG (amber codon)-encoded L-pyrrolysine designated the twenty-second genetically encoded amino acid. This implies a connection between the

TABLE 2. Free energies and typical organisms of methanogenesis reactions

Reaction	$\Delta G^{\circ\prime\, a}$ (kJ/mol CH_4)	Organisms
I. CO_2-type		
$4\ H_2 + CO_2 \rightarrow CH_4 + 2\ H_2O$	−135	Most methanogens
$4\ HCOOH \rightarrow CH_4 + 3\ CO_2 + 2\ H_2O$	−130	Many hydrogenotrophic methanogens
CO_2 + 4 isopropanol → CH_4 + 4 acetone + 2 H_2O	−37	Some hydrogenotrophic methanogens
$4\ CO + 2H_2O \rightarrow CH_4 + 3\ CO_2$	−196	*Methanothermobacter* and *Methanosarcina*
II. Methylated C1 compounds		
$4\ CH_3OH \rightarrow 3\ CH_4 + CO_2 + 2\ H_2O$	−105	*Methanosarcina* and other methylotrophic methanogens
$CH_3OH + H_2 \rightarrow CH_4 + H_2O$	−113	*Methanomicrococcus blatticola* and *Methanosphaera*
$2\ (CH_3)_2\text{-}S + 2\ H_2O \rightarrow 3\ CH_4 + CO_2 + 2\ H_2S$	−49	Some methylotrophic methanogens
$4\ CH_3\text{-}NH_2 + 2\ H_2O \rightarrow 3\ CH_4 + CO_2 + 4\ NH_3$	−75	Some methylotrophic methanogens
$2\ (CH_3)_2\text{-}NH + 2\ H_2O \rightarrow 3\ CH_4 + CO_2 + 2\ NH_3$	−73	Some methylotrophic methanogens
$4\ (CH_3)_3\text{-}N + 6\ H_2O \rightarrow 9\ CH_4 + 3\ CO_2 + 4\ NH_3$	−74	Some methylotrophic methanogens
$4\ CH_3NH_3Cl + 2\ H_2O \rightarrow 3\ CH_4 + CO_2 + 4\ NH_4Cl$	−74	Some methylotrophic methanogens
III. Acetate		
$CH_3COOH \rightarrow CH_4 + CO_2$	−33	*Methanosarcina* and *Methanosaeta*

SOURCE: Modified from Hedderich and Whitman[1] and Zinder.[43]

[a]The standard changes in free energies were calculated from the free energy of formation of the most abundant ionic species at pH 7. For instance, CO_2 is $HCO_3^- + H^+$ and HCOOH is $HCOO^- + H^+$.

abilities of amber codon translation and methylamine utilization.[15–17] In most methylotrophic methanogens, the electrons required for the reduction of the methyl groups to methane are obtained from the oxidation of additional methyl-groups to CO_2, which proceeds stepwise as the reverse of hydrogenotrophic methanogenesis. In methylotrophic methanogenesis, three methyl groups are reduced to methane for every molecule of CO_2 formed. This process is termed a disproportionation, since the oxidation of a portion of the substrate is used to reduce the remainder. Different from this mode of growth, the methylotrophic growth of *Methanomicrococcus blatticola* and *Methanosphaera* species is H_2-dependent.[18–23] They are obligate methylotrophic and hydrogenotrophic methanogens that are specialized to reduce methyl groups with H_2. The metabolism of *Methanosphaera* is restricted to methanol, while *M. blatticola* can use both methanol and methylamine.

The third type of substrate is acetate. Acetate is a major intermediate in the anaerobic food chain, and as much as two-thirds of the biologically generated methane is derived from acetate. Surprisingly, only two genera are known to use acetate for methanogenesis: *Methanosarcina* and *Methanosaeta*. They carry out an aceticlastic reaction that splits acetate, oxidizing the carboxyl-group to CO_2 and reducing the methyl group to CH_4 (FIG. 1C). *Methanosarcina* is a relative generalist that prefers methanol and methylamine to acetate, and many species also utilize H_2. *Methanosaeta* is a specialist that uses only acetate. *Methanosaeta* is a superior acetate utilizer in that it can use acetate at concentrations as low as 5–20 μM, while *Methanosarcina* requires a minimum concentration of about 1 mM.[24] The difference of acetate affinity is probably due to differences in the first step of acetate metabolism. *Methanosarcina* uses the low-affinity acetate kinase (AK)-phosphotransacetylase (PTA) system to activate acetate to acetyl-CoA, while *Methanosaeta* uses the high-affinity adenosine monophosphate (AMP)–forming acetyl-CoA synthetase.[24–27] Moreover, based on their genome sequences, these two genera probably have different modes of electron transfer and energy conservation, even though the main steps in the methanogenesis pathway are likely to be similar.[27]

The metabolism of *M. acetivorans* grown on CO is unconventional. It is distinct from the CO metabolism in *M. barkeri* and *M. thermoautotrophicus*. First, although the oxidation of CO to CO_2 via CODH is commonly required for CH_4 production, H_2 is not generated as an intermediate in *M. acetivorans*.[28] *M. acetivorans* is also unable to grow with H_2/CO_2 and lacks hydrogenase activities. Second, proteomic analyses suggest that CO-dependent methanogenesis in *M. acetivorans* requires an unusual mechanism for energy conservation. Although the steps of CO_2 reduction to methyl groups are similar as those in hydrogenotrophic methanogens, a coenzyme $F_{420}H_2$:heterodisulfide oxidoreductase system is responsible for energy conservation.[29,30] This system is used in *Methanosarcina* during methanogenesis with

A
Formate
$F_{420}H_2$
CO_2
H_2
Fd_{red}
MFR
Fd_{ox}
CHO-MFR
H_4MPT
MFR
CHO-H_4MPT
H_2O
CH≡H_4MPT
H_2
$F_{420}H_2$
F_{420}
CH_2=H_4MPT
H_2
$F_{420}H_2$
F_{420}
CH_3-H_4MPT
CoM-SH
H_4MPT
CH_3-SCoM
CoB-SH
[2H]
CoM-S-S-CoB
CH_4
B
CO_2
H_2
Fd_{red}
MFR
Fd_{ox}
CHO-MFR
H_4MPT
MFR
CHO-H_4MPT
CH≡H_4MPT
$F_{420}H_2$
F_{420}
CH_2=H_4MPT
$F_{420}H_2$
F_{420}
CH_3-H_4MPT
H_4MPT
CH_3-OH
CH_3-SCoM
CoM-SH
CoB-SH
[2H]
CoM-S-S-CoB
CH_4
C
CH_3COOH
ATP
SH-CoA
CH_3-CO-S-CoA
[CH_3]
H_4MPT
[CO-S-CoA]
Fd_{ox}
CH_3-H_4MPT
Fd_{red}
H_2
CoM-SH
H_4MPT
CO_2+HS-CoA
CH_3-SCoM
CoB-SH
[2H]
CoM-S-S-CoB
CH_4

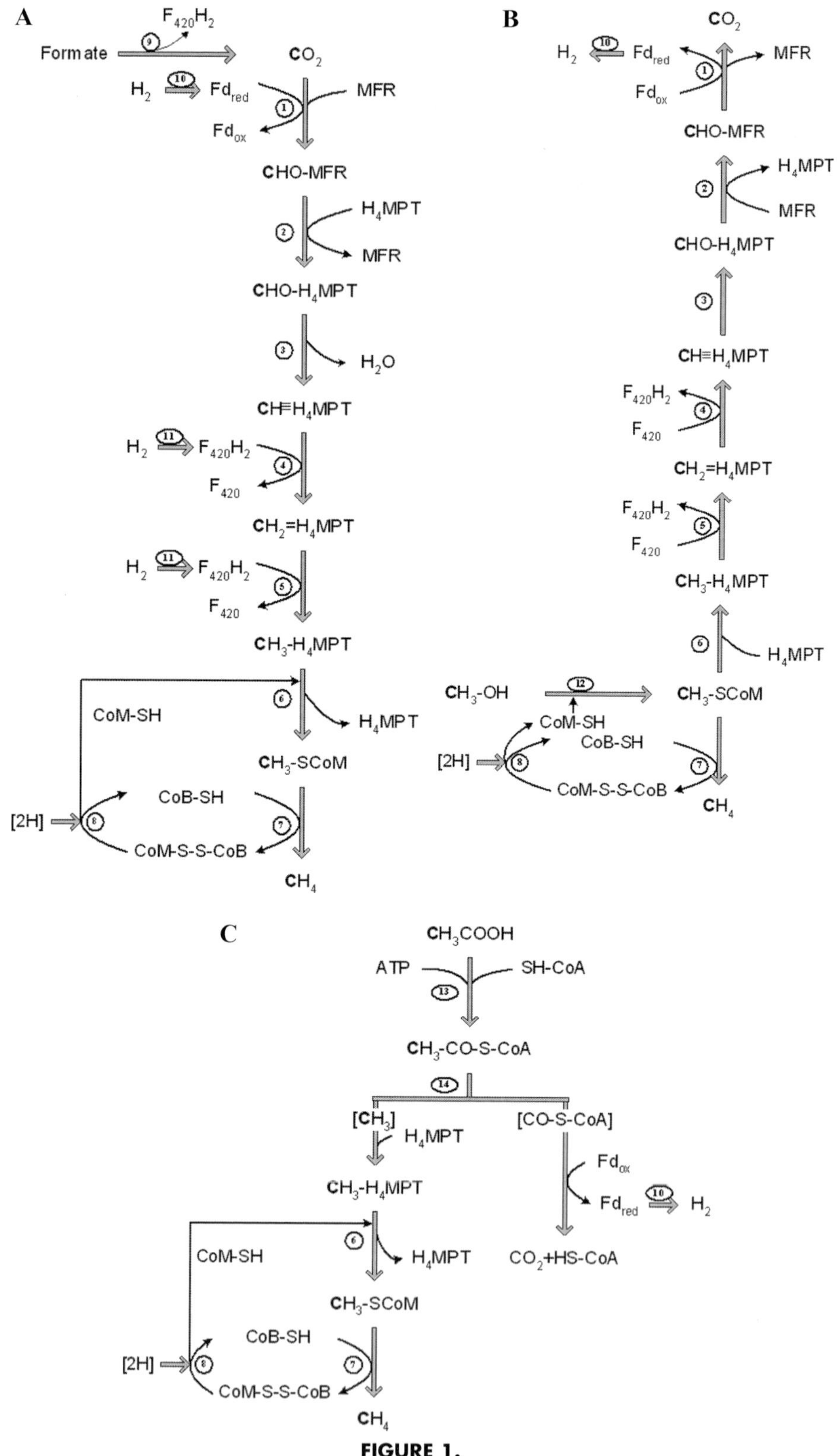

FIGURE 1.

methylated compounds in the absence of H_2, but it has not been identified in CO_2-dependent methanogenesis. Third, novel methyltransferases, which are used in place of methyl-H_4MPT:CoM methyltransferase (Mtr), are involved in the transfer of methyl group from H_4MPT to CoM.[29] Fourth, substantial amounts of acetate and formate are produced in addition to methane during the growth on CO in *M. acetivorans*. Mutational analysis suggests that acetate is formed via the AK-PTA system, which is the reverse of the initial steps of methanogenesis from acetate in *Methanosarcina*. Acetate production may generate adenosine triphosphate (ATP) via substrate level phophorylation. Thus, the growth on CO in *M. acetivorans* may conserve more energy than that in *M. barkeri* and *M. thermoautotrophicus*, which can be an explanation of the faster growth of *M. acetivorans* on CO (doubling time about 24 h). The mechanism of formate production is unclear yet. Since the formate dehydrogenase activity is absent in *M. acetivorans*, novel reactions are presumably involved.

Systematics of Methanogenic Archaea

Despite their restricted substrate range, methanogens are phylogenetically diverse. They are classified into five well-established orders: Methanobacteriales, Methanococcales, Methanomicrobiales, Methanosarcinales, and Methanopyrales. Organisms from different orders have less than 82% 16S rRNA sequence similarity. Methanogens belonging to different orders also possess different cell-envelope structure, lipid composition, substrate range, and other biological properties. The methanogen orders are further divided into 10 families and 31 genera. Organisms with less than 88–93% and less than 93–95% 16S rRNA sequence similarity are separated into different families and genera, respectively. Detailed descriptions of the systematics and characteristics of the methanogen taxa are reviewed in *Bergey's Manual of Systematic Bacteriology*[31] and *The Prokaryotes*.[3,32–35] Some characteristics of the methanogenic taxa are listed in TABLE 3.

Recent culture-independent studies have revealed the presence of novel phylogenetic groups of methanogens. They are not closely related to known organisms and probably represent new orders. For instance, the rice cluster I (RC-I), which is abundant in rice-field soil, forms a separate lineage within the phylogenetic radiation of Methanosarcinales and Methanomicrobiales. The 16S rRNA sequences of RC-I have similarities of less than 82% with known organisms.[36–39] Investigations of rumen methanogens have found a novel monophyletic group of at least two families.[40] The 16S rRNA sequences of this group have similarities closest to, but less than 80%, with those of Methanosarcinales. These studies illustrate our incomplete understanding of methanogen diversity.

Methanobacteriales

Members of the order Methanobacteriales generally produce methane using CO_2 as the elector acceptor and H_2 as the electron donor. Some species can use formate, CO, or secondary alcohols as electron donors. The genus *Methanosphaera* can only reduce methanol with H_2. In most genera, the cells are short to long rods with a length of 0.6–25 μm. They often form filaments up to 40 μm in length. The cell wall contains pseudomurein. The cellular lipids contain caldarchacol, archacol, and, in some species, hydroxyarchaeol as core lipids. The polar lipids can contain glucose, *N*-acetylglucosamine, *myo*-inositol, ethanolamine, and serine, depending on the species. Most species are nonmotile. However, members of the genus *Methanothermus* are motile via peritrichous flagella. They are widely distributed in anaerobic habitats, such as marine and freshwater sediments, soil, animal gastrointestinal tracts, anaerobic sewage digestors, and geothermal habitats.

The order of Methanobacteriales is divided into two families, Methanobacteriaceae and Methanothermaceae. The family Methanobacteriaceae contains three mesophilic genera, *Methanobacterium*, *Methanobrevibacter*, and *Methanosphaera*, and one extremely thermophilic

FIGURE 1. Pathways of methanogenesis. (**A**) Methanogenesis from H_2/CO_2 or formate. (**B**) Methanogenesis from methanol. (**C**) Methanogenesis from acetate. ABBREVIATIONS: Fd_{red} = reduced form of ferredoxin; Fd_{ox} = oxidized form of ferredoxin; $F_{420}H_2$ = reduced form coenzyme F_{420}; MFR - methanofuran; H_4MPT = tetrahydromethanopterin; CoM-SH = coenzyme M; CoB-SH = coenzyme B; CoM-S-S-CoB = heterodisulfide of CoM and CoB; CoA-SH = coenzyme A. Enzymes: 1. formyl-MFR dehydrogenase (Fmd); 2. formyl-MFR:H_4MPT formyltransferase (Ftr); 3. methenyl-H_4MPT cyclohydrolase (Mch); 4. methylene-H_4MPT dehydrogenase (Hmd); 5. methylene-H_4MPT reductase (Mer); 6. methyl-H_4MPT:HS-CoM methyltransferase (Mtr); 7. methyl-CoM reductase (Mcr); 8. heterodisulfide reductase (Hdr); 9. formate dehydrogenase (Fdh); 10. energy-conserving hydrogenase (Ech); 11. F_{420}-reducing hydrogenases; 12. methyltransferase; 13. acetate kinase (AK)-phosphotransacetylase (PTA) system in *Methanosarcina*; AMP-forming acetyl-CoA synthetase in *Methanosaeta*; 14. CO dehydrogenase/acetyl-CoA synthase (CODH/ACS).

TABLE 3. Properties of major taxonomic groups of methanogens

Order	Family	Genus	Methanogenesis Substrates[a]	Temperature Optimum (°C)	Typical Habitats
Methanobacteriales	Methanobacteriaceae	*Methanobacterium*	H_2, (formate)	37–45	Anaerobic digestors, freshwater sediments, marshy soils, rumen
		Methanobrevibacter	H_2, formate	37–40	Animal gastrointestinal tracts, anaerobic digestors, rice paddies, decaying woody tissues
		Methanosphaera[b]	H_2 + methanol	37	Animal gastrointestinal tracts
		Methanothermobacter	H_2, (formate)	55–65	Anaerobic digestors
	Methanothermaceae	*Methanothermus*	H_2	80–88	Hot springs
Methanococcales	Methanococcaceae	*Methanococcus*	H_2, formate	35–40	Marine sediments
		Methanothermococcus	H_2, formate	60–65	Marine geothermal sediments
	Methanocaldococcaceae	*Methanocaldococcus*	H_2	80–85	Marine geothermal sediments
		Methanotorris	H_2	88	Marine geothermal sediments
Methanomicrobiales	Methanomicrobiaceae	*Methanomicrobium*	H_2, formate	40	Anaerobic digestors, groundwater, rumen
		Methanoculleus	H_2, formate	20–55	Anaerobic digestors, marine sediments, freshwater sediments, rice paddies, oil fields, hot springs
		Methanofollis	H_2, formate	37–40	Anaerobic digestors
		Methanogenium	H_2, formate	15–57	Marine sediments, freshwater sediments, rice paddies, animal gastrointestinal tracts
		Methanolacinia	H_2	40	Marine sediments
		Methanoplanus	H_2, formate	32–40	Oil fields
	Methanospirillaceae	*Methanospirillum*	H_2, formate	30–37	Anaerobic digestors, marine sediments

continued.

TABLE 3. ***Continued.***

Order	Family	Genus	Methanogenesis Substrates[a]	Temperature Optimum (°C)	Typical Habitats
	Methanocorpusculaceae	*Methanocorpusculum*	H_2, formate	30–40	Anaerobic digestors, freshwater sediments
		Methanocalculus[b]	H_2, formate	30–40	Oil fields
Methanosarcinales	Methanosarcinaceae	*Methanosarcina*	(H_2), $MeNH_2$, Acetate	35–60	Anaerobic digestors, marine sediments, freshwater sediments, rumen
		Methanococcoides	$MeNH_2$	23–35	Marine sediments
		Methanohalobium	$MeNH_2$	40–55	Hypersaline sediments
		Methanohalophilus	$MeNH_2$	35–40	Hypersaline sediments
		Methanolobus	$MeNH_2$	37	Hypersaline sediments
		Methanomethylovorans	$MeNH_2$	20–50	Freshwater sediments, anaerobic digestors
		Methanimicrococcus	$H_2 + MeNH_2$	39	Animal gastrointestinal tracts
		Methanosalsum	$MeNH_2$	35–45	Hypersaline sediments
	Methanosaetaceae	*Methanosaeta*	Acetate	35–60	Anaerobic digestors, freshwater sediments
Methanopyrales	Methanopyraceae	*Methanopyrus*	H_2	98	Marine geothermal sediments

[a]Major substrates utilized for methanogenesis. $MeNH_2$ is methylamine. Parentheses means utilized by some, but not all species or strains.
[b]Placement in higher taxon is tentative.

genus *Methanothermobacter*. The family Methanothermaceae is represented by one hyperthermophilic genus, *Methanothermus*, which has only been isolated from thermal springs.

Methanococcales

Members of the order Methanococcales produce methane using CO_2 as the electron acceptor and H_2 or formate as the electron donor. The cells are irregular cocci with a diameter of 1–3 µm. The cell wall is composed of S-layer proteins. Glycoproteins and cell-wall carbohydrates are generally absent or of low abundance. The cellular lipids contain archaeol, caldarchaeol, hydroxyarchaeol, and macrocyclic archaeal, depending upon the species. The polar lipids can contain glucose, *N*-acetylglucosamine, serine, and ethanolamine. They are motile by means of flagella. The temperature range for growth varies from mesophilic to hyperthermophilic. They have all been isolated from marine habitats and require sea salts for optimal growth.

The order of Methanococcales has been divided into two families distinguished by their growth temperatures. The family Methanocaldococcaceae includes two hyperthermophilic genera, *Methanocaldococcus* and *Methanotorris*. The family Methanococcaceae includes the mesophilic genus *Methanococcus* and the extremely thermophilic genus *Methanothermococcus*.

Methanomicrobiales

Members of the order Methanomicrobiales use CO_2 as the electron acceptor and H_2 as electron donor. Most species can use formate, and many species also use secondary alcohols as alternative electron donors. Their morphology is diverse, including cocci, rods, and sheathed rods. Most cells have protein cell walls, and some cells are surrounded by a sheath containing glycoproteins. The cellular lipids contain archaeol and caldarchaeol as core lipids. Glucose, galactose, aminopentanetetrol, and glycerol are common polar lipids. Motility varies between species. They are widely distributed in anaerobic habitats, including marine and freshwater sediments, anaerobic sewage digestors, and animal gastrointestinal tracts.

The order of Methanomicrobiales is divided into three families, Methanomicrobiaceae, Methanospirillaceae, and Methanocorpusculaceae. The Methanospirillaceae is distinguished from the other two families by the curved rod shape of the cells. The Methanomicrobiaceae and Methanocorpusculaceae are difficult to distinguish by their physiological characteristics.

Methanosarcinales

Members of the order Methanosarcinales have the widest substrate range among methanogens. Most can produce methane by disproportionating the methyl-group containing compounds or by splitting acetate. Some species can reduce CO_2 with H_2, but formate is not used as an electron donor. Their cellular morphologies are diverse, including cocci, pseudosarcinae, and sheathed rods. Most cells have protein cell walls, and some cells are surrounded by a sheath or acidic heteropolysaccharides. The cellular lipids contain archaeol, hydroxyarchaeol, and caldarchaeol. Polar lipids can contain glucose, galactose, mannose, *myo*-inositol, ethanolamine, serine, and glycerol, depending upon the species. All cells are nonmotile. They are widely distributed in marine and freshwater sediments, anaerobic sewage digestors, and animal gastrointestinal tracts.

The order of Methanosarcinales is divided into two families, Methanosarcinaceae and Methanosaetaceae. All members of the family Methanosarcinaceae are able to grow with methyl group–containing compounds. The family Methanosaetaceae is represented by only one genus, *Methanosaeta*, and all cells produce methane by splitting acetate.

Methanopyrales

The order of Methanopyrales is represented by only one species, *Methanopyrus kandleri*. Cells reduce CO_2 with H_2 for methanogenesis. They are rod-shaped. Cell walls contain pseudomurein, and lipids contain archaeol. The cells are motile via flagella arranged as polar tufts. *M. kandleri* is hyperthermophilic with a growth temperature range of 84–110°C. It inhabits marine hydrothermal system.

Sequenced Genomes of Methanogens

Genome sequencing facilitates an understanding of the physiology of methanogens. Hitherto completed methanogen genomes include 18 cultured strains (TABLE 4). The genome sizes vary from 1.6 Mbp to 5.8 Mbp. The *Methanosarcina* species have the largest genomes, presumably due to their relatively versatile lifestyle. A large portion of genes in methanogen genomes have unknown functions, which will have to be characterized experimentally.

Metagenomic approaches lend insight into the physiology of uncultured methanogens. The genome of an uncultured RC-1 representative was reconstructed (TABLE 4) and revealed genes encoding enzymes for oxygen detoxification, carbohydrate metabolism, and assimilatory sulfate reduction, which are rare

TABLE 4. Currently completely sequenced methanogen genomes

Species	Strain	Length (bp)	GC content (%)	Protein coding genes	RNA genes[a]	References
Methanothermobacter thermautotrophicus	Delta H	1751377	49.54	1845	46	141
Methanobrevibacter smithii	ATCC 35061	1853160	31.03	1795	42	111
Methanosphaera stadtmanae	DSM 3091	1767403	27.63	1534	52	22
Methanococcus maripaludis	S2	1661137	33.10	1728	48	142
Methanococcus maripaludis	C5	1789046	32.98	1845	46	
Methanococcus maripaludis	C7	1766675	33.21	1819	37	
Methanococcus vannielii	SB	1693095	31.19	1694	38	
Methanococcus aeolicus	Nankai-3	1568349	30.02	1510	40	
Methanocaldococcus jannaschii	DSM 2661	1739916	31.29	1789	43	143
Methanoculleus marisnigri	JR1	2478101	62.06	2506	51	
Methanospirillum hungatei	JF-1	3544738	45.15	3239	65	
Methanocorpusculum labreanum	Z	1804962	50.01	1765	61	
Methanosaeta thermophila	PT	1879471	53.55	1730	50	
Methanosarcina barkeri	Fusaro	4873766	39.23	3758	71	144
Methanosarcina mazei	Go1	4096345	41.48	3398	67	145
Methanosarcina acetivorans	C2A	5751492	42.68	4721	56	146
Methanococcoides burtonii	DSM 6242	2575032	40.76	2431	61	147
Methanopyrus kandleri	AV19	1694969	61.16	1727	38	148
Uncultured methanogenic archaeon RC-1$_{MRE50}$		3179916	54.60	3103	63	38

[a]Include genes for the rRNAs and tRNAs.

in methanogens.[38] The metagenome derived from human gut suggested lower divergence between *Methanobrevibacter smithii* strains than *Bifidobacterium longum strains*, but differences in metabolic capacity and surface properties between different *M. smithii* strains were still substantial.[41,42]

Ecology of Methanogens

Methanogens are abundant in habitats where electron acceptors such as O_2, NO_3^-, Fe^{3+}, and SO_4^{2-} are limiting. When electron acceptors other than CO_2 are present, methanogens are outcompeted by the bacteria that utilize them, such as sulfate-reducing bacteria, denitrifying bacteria, and iron-reducing bacteria. This phenomenon probably occurs because these compounds are better electron acceptors, and their reductions are thermodynamically more favorable than CO_2 reduction to methane. However, because CO_2 is generated during fermentations, it is seldom limiting in anaerobic environments. Besides methanogens, homoacetogens are another group of anaerobes that can reduce CO_2 for energy production. During acetate production or acetogenesis, CO_2 is reduced with H_2 or other substances, for instance, sugars, alcohols, methylated compounds, CO, and organic acids. However, acetogenesis with H_2 is thermodynamically less favorable than methanogenesis. Therefore, homoacetogens do not compete well with methanogens in many habitats. However, homoacetogens outcompete methanogens in some environments, such as the hindgut of certain termites and cockroaches. Possible explanations may be their metabolic versatility as well as lower sensitivity to O_2.

In methanogenic habitats, complex organic matter is degraded to methane by the cooperation of different groups of anaerobes. A general scheme of the anaerobic food chain is shown in FIGURE 2. The organic polymers are initially degraded by specialized bacteria to simple sugars, lactate, volatile fatty acids, and alcohols. These products are further fermented by syntrophs and related bacteria to acetate, formate, H_2, and CO_2, which are substrates for methanogenesis. Methanogens catalyze the terminal step in the anaerobic food chain by converting methanogenic substrates to methane. The reactions catalyzed by the syntrophic bacteria, for instance, the conversion of volatile fatty acids and alcohols to acetate, CO_2, and H_2, is only favorable at H_2 partial pressures below 10^2 Pa.[43] When methanogens are present, H_2 is rapidly metabolized and maintained at concentrations below 10 Pa.[1] Therefore, these syntrophic bacteria depend on the association with methanogens or another hydrogenotrophic organisms for energy production. The interaction between H_2-producing organisms (e.g. propionate-oxidizing bacteria) and H_2-consuming organisms (e.g. hydrogenotrophic methanogens) is named interspecies hydrogen transfer.

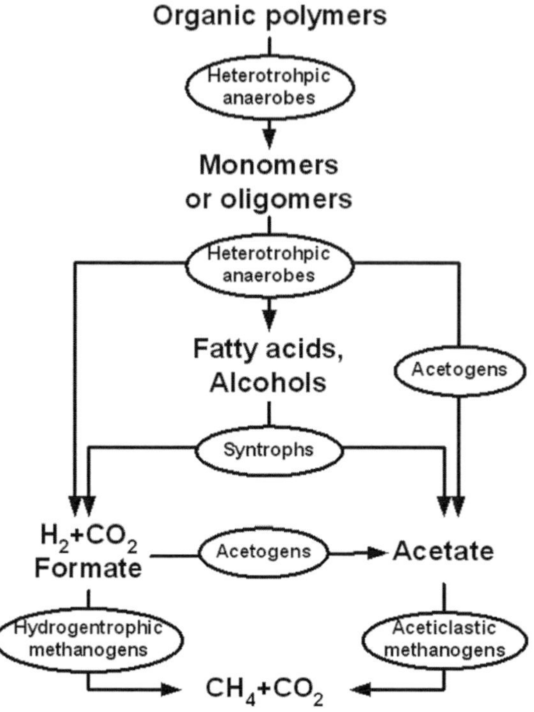

FIGURE 2. Anaerobic food chain for the conversion of organic matter to methane. Major microbial groups catalyzing the reactions are in ellipses.

Marine Sediments

Methanogenesis in marine habitats is a significant process that produces between 75 and 320 Tg CH_4 year^{-1}, but nearly all the methane is anaerobically oxidized to CO_2 instead of escaping to the atmosphere.[44] Sulfate, which is commonly present at 20–30 mM in seawater, is an important factor that controls the distribution of marine methanogens.[45] Sulfate-reducing bacteria outcompete methanogens in sulfate-rich marine sediments for H_2 and acetate. For instance, in the sediments of Cape Lookout Bight (North Carolina), the H_2 in the sulfate-rich sediments is mainly utilized by sulfate-reducing bacteria and kept at a level of 0.1–0.3 Pa, which is below the minimum values utilized by hydrogenotrophic methanogens.[46,47] Therefore in upper layers of sediments, or sulfate-reducing zones, methanogenesis is limited and accounts for less than 0.1% of total carbon turnover.[45] In sediments with high input of organic matter, sulfate can be depleted with depth, and methanogenesis can become the predominant terminal process in the anaerobic food chain.

Based on studies of carbon- and hydrogen-stable isotopes, CO_2-reduction by H_2 is a predominant source of methanogenesis in deep marine sediments.[48–50] In methanogenic zones, which are usually beneath the sulfate-reducing zones, the dissolved bicarbonate pool is replenished from oxidation of carbon compounds in upper sediments and can obtain concentrations greater than 100 mM.[49] CO_2-reducing methanogens in marine sediments include members within the orders Methanococcales, Methanomicrobiales, and Methanobacteriales.[51–53] They gain energy strictly by CO_2 reduction coupled with H_2 or formate oxidation. Moreover, hydrogenotrophic methanogens belonging to Methanomicrobiales can be detected from all depths of the sediments under certain circumstances[53]. Possibly, these methanogens gain energy from syntrophic growth with H_2 producers.[54]

Methylotrophic methanogens are major contributors to the limited methane production in sulfate-rich sediments. Identified organisms include members of the genera *Methanococcoides*, *Methanosarcina*, and *Methanolobus*.[52,53,55] Methylated compounds are generated in marine sediments from osmolytes of marine bacteria, algae, phytoplankton, and some plants. For instance, dimethylsulfide and trimethylamine are derived from dimethylsulfoniopropionate and betaine glycine, respectively. These compounds are not utilized efficiently by the sulfate-reducing bacteria and are termed noncompetitive substrates.[56] Because of the presence of a considerable amount of these substrates, obligately methylotrophic methanogens, such as *Methanococcoides*, can be cultivated from all depths in these sediments.[53]

Acetate is a minor substrate for methanogenesis in most marine sediments.[48,51] Only a few aceticlastic methanogens belonging to *Methanosarcina* have been isolated.[57–59] The minimum concentration of acetate utilized by *Methanosarcina* is typically 1 mM.[43] However, pore water acetate concentrations in marine sediments are usually below 20 μM.[50,52] Therefore, the aceticlastic methanogens that have been isolated probably also use methylated compounds for methanogenesis under environmental conditions. As measured by isotopes, the rate of acetate oxidation to CO_2 exceeds that of CH_4 production by the aceticlastic reaction in sulfate-reducing sediments.[50] Therefore, in these sediments, acetate is mainly metabolized by acetate oxidizers, for instance, sulfate-reducing bacteria.

Freshwater Sediments

Freshwater environments have lower sulfate concentrations (100–200 μM) than marine sediments.[45] Therefore methanogenesis in freshwater sediments proceeds uninhibited in anoxic zones and replaces sulfate reduction as the most important terminal process in the anaerobic degradation of organic matter. Due to the absence of competition by sulfate-reducing bacteria, the acetate pool is available for methanogenesis and is the dominant substrate. In most investigated freshwater environments, aceticlastic and hydrogenotrophic methanogenesis are responsible for about 70% and 30% of the CH_4 production, respectively.[49,60] The relative contribution of acetate and H_2/CO_2 in methanogenesis is close to the theoretical expectation. The fermentation of hexose yields 4 H_2, 2 acetate, and 2 CO_2. In addition, 4 H_2 are required to reduce CO_2 to methane. Thus, the expected ratio of methane from H_2 and acetate is 1:2. Methanogenesis from methylated compounds is minor, possibly reflecting the absence of these substrates in freshwater sediments.[45] The methanogenic communities are usually dominated by the aceticlastic family Methanosaetaceae and the hydrogenotrophic families Methanomicrobiaceae and Methanobacteriaceae. Methanosarcinaceae may also be present, and they may utilize either H_2/CO_2 or acetate.[61–65]

A few factors have been investigated that affect the relative contribution of aceticlastic and hydrogenotrophic methanogenesis as well as the relative abundance of different types of methanogens in freshwater sediments. First, hydrogenotrophic methanogenesis decreases at low pH.[66] For instance, in Lake Knaack sediments (pH 6.8), only 4% of the CH_4 production is derived from H_2/CO_2.[66] In Lake Grosse Fuchskuhle (pH < 5), only aceticlastic methanogens belonging to Methanosarcinaceae can be detected.[67] Low pH provides a selective advantage to homoacetogens, which reduce CO_2 to acetate instead of methane. This limits hydrogenotrophic methanogenesis. Second, the relative contribution of hydrogenotrophic methanogenesis decreases with temperature, even though the absolute rate of aceticlastic methanogenesis also decreases.[62,68–70] A similar effect of temperature is also observed in rice paddies.[71] This effect can be explained by better adaptation of homoacetogens to low temperatures.[68,69] Moreover, H_2 production by syntrophic bacteria decreases, as H_2-producing processes are less favorable at low temperature. Therefore, hydrogenotrophic methanogenesis is inhibited at low temperature due to an insufficient supply of substrates. Third, the relative contribution of hydrogenotrophic methanogenesis changes with depth in some cases. In Lake Dagow sediments, the relative contribution of hydrogenotrophic methanogenesis increases from 22% to 38% from 0 to 18 cm.[61] Coincidently, the relative abundance of Methanomicrobiales increases slightly with depth, while the abundance of Methanosaetaceae decreases. In contrast, in Lake Rotsee sediments, hydrogenotrophic methanogenesis is only observed in the upper 2 cm of the sediments.[72] In this case, hydrogenotrophic methanogens possibly live as symbionts of ciliates. Fourth, the abundance of other H_2 and acetate-consumers greatly influences the methanogenic communities. In Lake Constance sediments, the presence of H_2-consuming sulfate-reducing bacteria and absence of acetate-consuming sulfate-reducing bacteria, together with low temperature (4°C), results in 100% CH_4 production from aceticlastic methanogenesis.[68,73] In Lake Kinneret sediments, CH_4 is produced exclusively by hydrogenotrophic methanogens (Methanomicrobiaceae and Methanobacteriaceae) through syntrophic association with acetate-oxidizers.[74]

Rice-field Soils

Rice fields are major anthropogenic sources of methane emission. In rice paddies, upon flooding, oxygen, nitrate, ferric iron, and sulfate are rapidly exhausted due to high input of plant carbon, developing the conditions favoring methanogenesis. Methanogenesis is a significant process in rice fields, and about 3–6% of the photosynthetically fixed CO_2 is converted to methane.[75] Methanogens inhabit both the soils and the root surfaces. Methane is emitted mostly through the vascular system of the rice plant, which also supplies O_2 to the roots and adjacent rhizosphere soil.[76] Since rice-field soils are transiently oxic and exposed to drying, soil methanogens are proposed to be more resistant to O_2 and desiccation than methanogens in aquatic habitats.[77]

Methane in rice fields is derived from both H_2/CO_2 and acetate. Dominant methanogens are affiliated with Methanomicrobiaceae, Methanobacteriaceae, Methanosarcinaceae, Methanosaetaceae, and RC-I.[78–80] A few factors have been suggested that affect methanogenesis and/or methanogenic populations in rice fields. First, the relative populations of methanogens remain constant upon flooding and seasonal drying.[76,80–82] However, hydrogenotrophic methanogenesis dominates immediately after flooding, while aceticlastic methanogenesis increases later, reaching a maximum at about 70–80 days after flooding.[80,83] This shift may be due to activation of hydrogenotrophic methanogens immediately upon flooding and not changes of the methanogenic populations. Second, the methanogen community structures on rice roots differ from those in the bulk soils. On rice roots, RC-I is usually the dominant organism, and CH_4 is mainly derived from H_2/CO_2 (60–80%).[78,79] In the bulk soil, aceticlastic methanogens are dominant, and CH_4 is mainly derived from acetate cleavage (50–83%).[60,76,79,84,85] One possible reason for this difference is that the hydrogenotrophic methanogens on rice roots could be more resistant to O_2, which has a higher concentration on the roots than in the bulk soils. Third, the population of hydrogenotrophic methanogens increases with temperature ($\geq 30°C$).[86,87] Fourth, phosphate (≥ 20 mM) specifically inhibits aceticlastic methanogens.[78,79,88]

RC-I methanogens form a distinct phylogenetic lineage within the radiation of Methanosarcinales and Methanomicrobiales, based upon analyses of both 16S rRNA and *mcrA* genes (review in Ref. 37). These hydrogenotrophs are ubiquitous and abundant in rice paddies, representing 20–50% of the total methanogens. Incubations of rice roots or rice-field soils with RC-I methanogens show that they are preferentially active at low H_2 concentrations, moderately high temperatures (45–50°C), and with additional supplies of sugars or amino acids. Other speculations to explain their prevalence in rice fields include their superior adaptations to temporary drought and oxic conditions. Indeed, the genome sequence of RC-I suggests the presence of enzymes that detoxify highly reactive oxygen species.[38] Recently, isolation of RC-I in pure culture has been reported.[89] This novel isolate, strain SANAE, can utilize H_2/CO_2 or formate as methanogenesis substrates. Acetate cannot be used as an energy source. The doubling time with H_2/CO_2 is about 4.2 days.

Human and Animal Gastrointestinal Tracts

In the human colon, organic substrates that have escaped digestion in the upper intestinal tract are fermented by anaerobic microbial communities into short-chain fatty acids and CO_2. H_2 is produced for disposing of reducing equivalents and kept at very low partial pressures to meet the thermodynamic requirements for anaerobic fermentation processes. H_2 is mainly removed through reutilization by hydrogenotrophic microorganisms. For instance, in some individuals about 1 L of H_2 is excreted in breath and flatus per day, while 16 L of H_2 is used to make 4 L of CH_4 per day.[90] The predominant H_2-consuming populations in the human colon are methanogens, acetogens, and sulfate-reducing bacteria.

About one-third of healthy human adults are strong methane-producers and possess breathe methane levels >1 ppm above the atmospheric methane level.[90,91] Most of the produced CH_4 gas is excreted as flatus and some is absorbed in the blood and excreted in breath. Methanogens number 10^8–10^{10} cells/g feces in these individuals. Individuals with low levels of methane in their breathe contain $<10^2$ to 5×10^6 methanogen cells/g feces.[90,92] Acetogens predominate in these individuals, and the numbers of acetogens have a negative correlation with that of methanogens.[93] Although acetogens offer nutritional benefits to the hosts by producing acetate, methanogens can successfully outcompete acetogens in the colon of some individuals, possibly due to their lower threshold of H_2 utilization.[94] Sulfate-reducing bacteria outcompete methanogens in many other habitats, but they do not exclude methanogens in human feces.[95–98] Neither the cell numbers of sulfate-reducing bacteria nor the concentrations of sulfate and sulfide in feces are significantly different between methane-producing and nonproducing individuals.[96,97] Moreover, methane-producing feces consumes H_2 more rapidly than nonproducing feces, suggesting that methanogens can outcompete other H_2-consumers in the human colon.[95,97,99] What determines methanogen abundance in feces is still a mystery. Methanogens are not detected in feces of children under 27 months of age; while in methane-producing adults, the levels of methanogens are stable over time.[91,100] The development of methanogens is not directly related to the introduction of particular foods.[100] No significant autosomal genetic effects have yet been identified.[101] Shared environment is one of the determinants of methane production.[101] Methanogenesis is negatively correlated with the frequency of bowel movements.[91] Likewise, methanogen abundance is negatively correlated to the fecal concentrations of butyrate, but not to acetate or total short-chain fatty acids.[102] The lack of significant correlation between methanogen abundance and the concentration of acetate is unexpected, since low methanogen abundance enhances acetate production by hydrogenotrophic acetogens. Low acetate concentration in the colon is probably due to efficient absorption by the host and consumption by other anaerobes, such as butyrate-producing bacteria.[103–105]

Methane produced in the human colon is mainly derived from H_2/CO_2. Aceticlastic methanogenesis is generally limited, because the short retention time of the intestine does not allow the slow growth of methanogens on acetate.[90] Only two methanogens species have been isolated from human feces: *Methanobrevibacter smithii* and *Methanosphaera stadtmanae*.[106,107] Both of them belong to the order Methanobacteriales. *M. smithii* is the predominant methanogen species and constitutes up to 10% of all anaerobes in the human colon.[90,108] It produces methane from H_2/CO_2 or formate, but grows poorly with formate.[106] *M. stadtmanae* is present in lower numbers than *M. smithii*. It can only produce methane by reducing methanol with H_2 and depends on acetate as a carbon source. Methanol in human gut is derived from pectin degradation by *Bacteroides* and other

TABLE 5. Methanogens cultured from human and animal gastrointestinal tracts

Hosts	Representative methanogen species
Human	*Methanobrevibacter smithii*; *Methanosphaera stadtmanae*
Bovine	*Methanobrevibacter ruminantium*; *Methanobrevibacter thaueri*; *Methanobrevibacter smithii*; *Methanosarcina barkeri*; *Methanobacterium formicicum*; *Methanobrevibacter millerae*; *Methanobacterium mobilis*
Ovine	*Methanosarcina barkeri*; *Methanobacterium formicicum*; *Methanobrevibacter olleyae*; *Methanobrevibacter wolinii*
Termite	*Methanobrevibacter cuticularis; Methanobrevibacter curvatus; Methanobrevibacter filiformis*; *Methanobacterium bryantii*
Cockroach	*Methanomicrococcus blatticola*

anaerobes.[109,110] The restricted metabolism of *M. stadtmanae* can be explained by the absence of genes for CO dehydrogenase/acetyl-CoA synthase (CODH/ACS) and molybdopterin biosynthesis proteins, which are common among other methanogens.[22] The genome of both *M. smithii* and *M. stadtmanae* encode enzymes for the synthesis of surface glycans resembling the components of the mucosal layer of the hosts.[111] Their genomes also encode adhesion-like proteins.[111] These features are shared among bacteria in gastrointestinal tracts and may contribute to their adaptation to the human colon.

The rumen of herbivores is the primary location for microbial fermentation of plant material that cannot be digested by the host enzymes. A wide variety of microorganisms, such as bacteria, fungi, and protozoa, ferment plant biomass into H_2, short-chain fatty acids, CO_2, and CH_4. Methane production from the rumen contributes to up to 20% of the global methane emissions to the atmosphere. In addition, 6–10% of the energy value of the food ingested by host animal is lost as methane.[112] Therefore, manipulation of methane production in rumen serves as a way to limit methane emission from anthropogenic sources as well as improve animal production. Daily methane production is different among ruminant species. For instance, an adult cow produces about 200 L of CH_4 per day, while sheep under generous grazing conditions produce 35–50 L of CH_4 per day.[90,113]

In the rumen, hydrogenotrophic methanogens predominate. Methanogenesis limits hydrogenotrophic acetogenesis by lowering the H_2 concentration to below the minimal level required for acetogenesis.[114] Methanogens in the bovine rumen typically number 10^8–10^{10} cells/g of rumen content, and the population densities of methanogens are influenced by the type of diet, especially the fiber content.[115,116] *Methanobrevibacter* species, including *M. ruminantium*, *M. thaueri*, and *M. smithii*, are the predominant methanogens isolated and cultured from the bovine and ovine rumen[117] (TABLE 5). Recent 16S rRNA gene-based studies reveal that, although *Methanobrevibacter* are usually dominant, a wide diversity of ruminal methanogens coexist in the rumen.[117–119] *Methanosphaera* species similar to *M. stadtmanae* are likely to be common.[112,117] Methanogens belonging to the Methanosarcinales and Methanococcales are present in the rumen of some individuals, but they are not detectable in most individuals.[118] Substantial numbers of the 16S rRNA sequences obtained from the rumen are not related to known methanogens, and probably represent a novel order of archaea.[40,119,120] Some of these uncultured microorganisms can be established on methanogen-selective medium with H_2/CO_2, but do not compete well with *Methanobrevibacter*.[40] These studies indicate that uncultured methanogens may have significant population densities in the rumen.

The insect hindgut is another important habitat of methanogens. About 3% of the total global methane emissions or 11% the methane emissions from natural sources are from the termite hindgut. Methanogenesis and acetogenesis occur simultaneously in the termite hindgut, and the relative activities are related to host diets. In wood-feeding termites, acetogenesis outpaces methanogenesis; by contrast, in soil-feeding and fungus-feeding termites, methanogenesis outpaces acetogenesis.[121,122] It is interesting that acetogens can outcompete methanogens in termite guts, though the mechanism is still under study and may vary between different termite species. A few proposed mechanisms include: (1) Acetogens are metabolically versatile and able to use organic substrates in addition to H_2. Their mixtrophic lifestyles give them competitive advantages over methanogens. However, it is necessary to explain why these advantages would not be applied in other habitats. (2) The H_2 levels in termite guts are not limiting for acetogenesis. For instance, in *Reticulitermes flavipes*, a wood-feeding lower termite, the H_2 partial pressure is about 5 kPa, which is two orders of magnitude higher than the minimum level of H_2 for acetogenesis.[123] (3) Acetogens and methanogens may be spatially separated in different compartments of the termite gut to avoid direct competition for H_2. Termite guts are highly structured and characterized by steep gradients of O_2, H_2, and pH. Some regions may form microniches that favor acetogenesis.[122,124]

Methanogens identified from termite hindguts include members of the Methanobacteriaceae and Methanosarcinaceae species.[125,126] The relationship between methanogen community compositions and termite diet is not clear yet. Methanobacteriaceae are predominant in most termite species, but in four studied termite species Methanosarcinaceae are the dominant methanogen group.[125] The dominance of Methanosarcinaceae is uncommon in gastrointestinal tracts of animals. Possibly, their ability to use a variety of substrates, including H_2/CO_2, acetate, and methylated compounds, may provide an advantage in the termite hindgut.

Anaerobic Digestors

Anaerobic digestors are widely used for the degradation of organic wastes such as animal manure and municipal wastewater to methane. Compared with aerobic treatments, the anaerobic treatments have the advantage of generating large quantities of renewable fuel in the form of biogas. Moreover, they produce much less sludge, a costly by-product of the aerobic process.

The anaerobic conversion of wastes to methane requires cooperation of at least three groups of microorganisms.[127] The first step, hydrolysis, involves enzyme-mediated conversion of organic polymers, such as polysaccharides, lipids, proteins, and fats, into soluble organic monomers. This step is carried out by anaerobic bacteria such as *Bacterioides*, *Clostridium*, and *Streptococcus*. The second step, acidogenesis, involves anaerobic fermentation of monomers into acetate, H_2, CO_2, as well as short chain fatty acids (propionate and butyrate), which are subsequently converted to acetate and H_2. The third step, methanogenesis, involves conversion of acetate, H_2, and CO_2 into CH_4 by methanogens. Methanogenesis is considered the rate-limiting step, and high activity of the methanogens is important for maintaining efficient anaerobic digestion and avoiding the accumulation of H_2 and short chain fatty acids. Moreover, this step is most vulnerable to parameters such as temperature, pH, and inhibitory chemicals. Therefore, enhancement of methanogenesis is a major route for improving the performance of anaerobic digestors.

Acetate is a major product of the fermentation and accounts for two-thirds of the methane production in anaerobic digestors.[43] Usually, only one aceticlastic methanogen group, *Methanosaeta* or *Methanosarcina*, dominates each digestor, depending upon the type of waste and digestor.[128] *Methanosaeta* have slower growth rates and higher affinity for acetate, while *Methanosarcina* have faster growth rates and lower affinity for acetate. The relative abundance of these two groups is not only regulated by acetate concentrations, as in other environments, but also by feeding rates.[129,130] *Methanosaeta* perform better at high feeding-rate digestors, such as an upward-flow anaerobic sludge blanket (UASB), presumably due to their efficient adhesion and granulation.[131,132] In contrast, *Methanosarcina* are more sensitive to turbulence and shear, and they frequently dominate in fixed- and stirred-tank digestors.[133]

The H_2 partial pressures in anaerobic digestors range from 2 Pa to 1200 Pa.[129] Low H_2 levels indicate efficient hydrogenotrophic methanogenesis and are usually associated with stable performance. High H_2 levels (>10 Pa) lead to inhibition of the anaerobic fermentation, and accumulation of electron sinks as lactate, ethanol, propionate, and butyrate. Therefore, efficient interspecies hydrogen transfer is quite important for good performance of anaerobic digestors. Formation of granular sludge facilitates interspecies hydrogen transfer by reducing the distance between H_2-producing bacteria and H_2-consuming methanogens.[134] A wide variety of hydrogenotrophic methanogens belonging to Methanomicrobiales and Methanobacteriales are detected in and cultured from anaerobic digestors.[43,135]

Temperature and pH are two main parameters that influence methanogenesis in anaerobic digestors. Slightly thermophilic temperatures (50–60°C) are desired in certain anaerobic digestors to increase reaction rates and decrease retention times. The most common hydrogenotrophic methanogens in thermophilic digestors include *M. thermoautotrophicus* and *Methanoculleus thermophilicum*.[43,136] Aceticlastic methanogens generally cannot grow at temperatures higher than 65°C, thus limiting the temperature range of digestors. The most common aceticlastic methanogens in thermophilic digestors include *Methanosarcina thermophila* and thermophilic *Methanosaeta*.[43,137] The pH is another important parameter. Increases of loading rates lead to increases in the concentration of fatty acids and decreases in pH. Most methanogens grow optimally under neutral to slightly alkaline conditions (pH 6.8–8.5). Thus, acid tolerant methanogens are desired to improve the stability of anaerobic digestion. *Methanobrevibacter acididurans*, a methanogen isolated from a sour anaerobic digestor, grows in the pH range of 5.0–7.5.[138] Additions of *M. acididurans* to acidic digestors show better methanogenesis and decreases in accumulation of fatty acids.[139]

Conclusions

Methanogens are very diverse in terms of phylogeny and ecology. Their ability to use methanogenesis as an anaerobic respiration allows them to occupy a physiologically unique niche that is unavailable to the Bacteria. They play an important role in the anaerobic food chain, driving anaerobic fermentation through removal of excess H_2 and formate. Their physiological diversity makes them truly cosmopolitan in anaerobic environments. The relative abundance of different types of methanogens is regulated by the availability of substrates and other parameters, for instance, temperature, pH, and salinity. The diversity of methanogens is underestimated as indicated by recent culture-independent studies. Further physiological and ecological investigations are required to fully reveal the diversity and importance of methanogens.

Conflict of Interest

The authors declare no conflicts of interest.

References

1. Hedderich, R. & W. Whitman. 2006. Physiology and biochemistry of the methane-producing Archaea. *In* The Prokaryotes, Vol. 2, 3rd ed. M. Dworkin, *et al.*, Eds.: 1050–1079. New York: Springer Verlag.
2. Lowe, D.C. 2006. Global change: a green source of surprise. Nature **439:** 148–149.
3. Whitman, W., T. Bowen & D. Boone. 2006. The methanogenic bacteria. *In* The Prokaryotes, Vol. 3, 3rd ed. M. Dworkin, *et al.*, Eds.: 165–207. New York: Springer Verlag.
4. Widdel, F. 1986. Growth of methanogenic bacteria in pure culture with 2-propanol and other alcohols as hydrogen donors. Appl. Environ. Microbiol. **51:** 1056–1062.
5. Frimmer, U. & F. Widdel. 1989. Oxidation of ethanol by methanogenic bacteria. Arch. Microbiol. **152:** 479–483.
6. Widdel, F., P.E. Rouvière & R.S. Wolfe. 1988. Classification of secondary alcohol-utilizing methanogens including a new thermophilic isolate. Arch. Microbiol. **150:** 477–481.
7. Bleicher, K., G. Zellner & J. Winter. 1989. Growth of methanogens on cyclopentanol/CO2 and specificity of alcohol dehydrogenase. FEMS Microbiol. Lett. **59:** 307–312.
8. Widdel, F. & R.S. Wolfe. 1989. Expression of secondary alcohol dehydrogenase in methanogenic bacteria and purification of the F420-specific enzyme from *Methanogenium thermophilum* strain TCI. Arch. Microbiol. **152:** 322–328.
9. Berk, H. & R.K. Thauer. 1997. Function of coenzyme F420-dependent NADP reductase in methanogenic archaea containing an NADP-dependent alcohol dehydrogenase. Arch. Microbiol. **168:** 396–402.
10. O'Brien, J.M. *et al.* 1984. Association of hydrogen metabolism with unitrophic or mixotrophic growth of *Methanosarcina barkeri* on carbon monoxide. J. Bacteriol. **158:** 373–375.
11. Daniels, L. *et al.* 1977. Carbon monoxide oxidation by methanogenic bacteria. J. Bacteriol. **132:** 118–126.
12. Burke, S.A. & J.A. Krzycki. 1997. Reconstitution of monomethylamine:coenzyme M methyl transfer with a corrinoid protein and two methyltransferases purified from *Methanosarcina barkeri*. J. Biol. Chem. **272:** 16570–16577.
13. Ferguson, D.J., Jr. *et al.* 2000. Reconstitution of dimethylamine:coenzyme M methyl transfer with a discrete corrinoid protein and two methyltransferases purified from *Methanosarcina barkeri*. J. Biol. Chem. **275:** 29053–29060.
14. Sauer, K., U. Harms & R.K. Thauer. 1997. Methanol:-coenzyme M methyltransferase from *Methanosarcina barkeri*. Purification, properties and encoding genes of the corrinoid protein MT1. Eur. J. Biochem. **243:** 670–677.
15. Hao, B. *et al.* 2002. A new UAG-encoded residue in the structure of a methanogen methyltransferase. Science **296:** 1462–1466.
16. Krzycki, J.A. 2005. The direct genetic encoding of pyrrolysine. Curr. Opin. Microbiol. **8:** 706–712.
17. Mahapatra, A. *et al.* 2006. Characterization of a *Methanosarcina acetivorans* mutant unable to translate UAG as pyrrolysine. Mol. Microbiol. **59:** 56–66.
18. Sprenger, W.W. *et al.* 2000. *Methanomicrococcus blatticola* gen. nov., sp. nov., a methanol- and methylamine-reducing methanogen from the hindgut of the cockroach *Periplaneta americana*. Int. J. Syst. Evol. Microbiol. **50:** 1989–1999.
19. Sprenger, W.W., J.H.P. Hackstein & J.T. Keltjens. 2005. The energy metabolism of *Methanomicrococcus blatticola*: physiological and biochemical aspects. Antonie Leeuwenhoek. **87:** 289–299.
20. Sprenger, W.W., J.H.P. Hackstein & J.T. Keltjens. 2007. The competitive success of *Methanomicrococcus blatticola*, a dominant methylotrophic methanogen in the cockroach hindgut, is supported by high substrate affinities and favorable thermodynamics. FEMS Microbiol. Lett. **60:** 266–275.
21. Biavati, B., M. Vasta & J.G. Ferry. 1988. Isolation and characterization of "*Methanosphaera cuniculi*" sp. nov. Appl. Environ. Microbiol. **54:** 768–771.
22. Fricke, W.F. *et al.* 2006. The genome sequence of *Methanosphaera stadtmanae* reveals why this human intestinal archaeon is restricted to methanol and H2 for methane formation and ATP synthesis. J. Bacteriol. **188:** 642–658.
23. Miller, T.L. & M.J. Wolin. 1985. *Methanosphaera stadtmaniae* gen. nov., sp. nov.: a species that forms methane by reducing methanol with hydrogen. Arch. Microbiol. **141:** 116–122.
24. Jetten, M.S.M., A.J.M. Stams & A.J.B. Zehnder. 1992. Methanogenesis from acetate: a comparison of the acetate metabolism in *Methanothrix soehngenii* and *Methanosarcina* spp. FEMS Microbiol. Lett. **88:** 181–197.
25. Singh-Wissmann, K. & J.G. Ferry. 1995. Transcriptional regulation of the phosphotransacetylase-encoding and acetate kinase-encoding genes (*pta* and *ack*) from

Methanosarcina thermophila. J. Bacteriol. **177:** 1699–1702.

26. TEH, Y.L. & S. ZINDER. 1992. Acetyl-coenzyme A synthetase in the thermophilic, acetate-utilizing methanogen *Methanothrix* sp. strain CALS-1. FEMS Microbiol. Lett. **98:** 1–7.
27. SMITH, K.S. & C. INGRAM-SMITH. 2007. *Methanosaeta*, the forgotten methanogen? Trends Microbiol. **15:** 150–155.
28. ROTHER, M. & W.W. METCALF. 2004. Anaerobic growth of *Methanosarcina acetivorans* C2A on carbon monoxide: an unusual way of life for a methanogenic archaeon. Proc. Natl. Acad. Sci. USA **101:** 16929–16934.
29. LESSNER, D.J. *et al.* 2006. An unconventional pathway for reduction of CO2 to methane in CO-grown *Methanosarcina acetivorans* revealed by proteomics. Proc. Natl. Acad. Sci. USA **103:** 17921–17926. Epub 2006 Nov 13.
30. ROTHER, M., E. OELGESCHLÄGER & W.W. METCALF. 2007. Genetic and proteomic analyses of CO utilization by *Methanosarcina acetivorans*. Arch. Microbiol. **188**(5): 463–472.
31. WHITMAN, W.B. *et al.* 2001. Taxonomy of methanogeic archaea. *In* Bergey's Manual of Systematic Bacteriology, D.R. BOONE, R.W. CASTENHOLTZ & G.M. GARRITY, Vol. 1. Eds.: 211–213. New York: Springer.
32. GARCIA, J.-L., B. OLLIVIER & W. WHITMAN. 2006. The order *Methanomicrobiales*. *In* The Prokaryotes, Vol. 3, 3rd ed. M. Dworkin, *et al.*, Eds.: 208–230. New York: Springer Verlag.
33. BONIN, A. & D. BOONE. 2006. The order *Methanobacteriales*. *In* The Prokaryotes, Vol. 3, 3rd ed. M. Dworkin, *et al.*, Eds.: 231–243. New York: Springer Verlag.
34. KENDALL, M. & D. BOONE. 2006. The order *Methanosarcinales*. *In* The Prokaryotes, Vol. 3, 3rd ed. M. Dworkin, *et al.*, Eds.: 244–256. New York: Springer Verlag.
35. WHITMAN, W. & C. JEANTHON. 2006. *Methanococcales*. *In* The Prokaryotes, Vol. 3, 3rd ed. M. Dworkin, *et al.*, Eds.: 257–273. New York: Springer Verlag.
36. LUEDERS, T. *et al.* 2001. Molecular analyses of methyl-coenzyme M reductase alpha-subunit (mcrA) genes in rice field soil and enrichment cultures reveal the methanogenic phenotype of a novel archaeal lineage. Environ. Microbiol. **3:** 194–204.
37. CONRAD, R., C. ERKEL & W. LIESACK. 2006. Rice Cluster I methanogens, an important group of Archaea producing greenhouse gas in soil. Curr. Opin. Microbiol. **17:** 262–267.
38. ERKEL, C. *et al.* 2006. Genome of Rice Cluster I Archaea—The key methane producers in the rice rhizosphere. Science **313:** 370–372.
39. GRO KOPF, R., S. STUBNER & W. LIESACK. 1998. Novel euryarchaeotal lineages detected on rice roots and in the anoxic bulk soil of flooded rice microcosms. Appl. Environ. Microbiol. **64:** 4983–4989.
40. NICHOLSON, M., P. EVANS & K. JOBLIN. 2007. Analysis of methanogen diversity in the rumen using temporal temperature gradient gel electrophoresis: identification of uncultured methanogens. Microb. Ecol. **54**(1): 141–150.
41. GILL, S.R. *et al.* 2006. Metagenomic analysis of the human distal gut microbiome. Science **312:** 1355–1359.
42. WALKER, A. 2007. Say hello to our little friends. Nat. Rev. Micro. **5:** 572–573.
43. ZINDER, S.H. 1993. Physiological ecology of methanogens. *In* Methanogenesis: Ecology, Physiology, Biochemistry and Genetics. J.G. FERRY, Ed.: 128–206. New York: Chapman & Hall.
44. VALENTINE, D. 2002. Biogeochemistry and microbial ecology of methane oxidation in anoxic environments: a review. Antonie Leeuwenhoek **81:** 271–282.
45. CAPONE, D.G. & R.P. KIENE. 1988. Comparison of microbial dynamics in marine and freshwater sediments: contrasts in anaerobic carbon catabolism. Limnol. Oceanogr. **33:** 725–749.
46. HOEHLER, T.M. *et al.* 1998. Thermodynamic control on hydrogen concentrations in anoxic sediments. Geochim. Cosmochim. Acta. **62:** 1745–1756.
47. HOEHLER, T.M. *et al.* 2001. Apparent minimum free energy requirements for methanogenic Archaea and sulfate-reducing bacteria in an anoxic marine sediment. FEMS Microbiol. Ecol. **38:** 33–41.
48. WHITICAR, M.J., E. FABER & M. SCHOELL. 1986. Biogenic methane formation in marine and freshwater environments: CO2 reduction vs. acetate fermentation—Isotope evidence. Geochim. Cosmochim. Acta. **50:** 693–709.
49. WHITICAR, M.J. 1999. Carbon and hydrogen isotope systematics of bacterial formation and oxidation of methane. Chem. Geol. **161:** 291–314.
50. PARKES, R.J. *et al.* 2007. Biogeochemistry and biodiversity of methane cycling in subsurface marine sediments (Skagerrak, Denmark). Environ. Microbiol. **9:** 1146–1161.
51. NEWBERRY, C.J. *et al.* 2004. Diversity of prokaryotes and methanogenesis in deep subsurface sediments from the Nankai Trough, Ocean Drilling Program Leg 190. Environ. Microbiol. **6:** 274–287.
52. KENDALL, M.M. & D.R. BOONE. 2006. Cultivation of methanogens from shallow marine sediments at Hydrate Ridge, Oregon. Archaea. **2:** 31–38.
53. KENDALL, M.M. *et al.* 2007. Diversity of Archaea in marine sediments from Skan Bay, Alaska, including cultivated methanogens, and description of *Methanogenium boonei* sp. nov. Appl. Environ. Microbiol. **73:** 407–414.
54. KENDALL, M.M., Y. LIU & D.R. BOONE. 2006. Butyrate- and propionate-degrading syntrophs from permanently cold marine sediments in Skan Bay, Alaska, and description of *Algorimarina butyrica* gen. nov., sp. nov. FEMS Microbiol. Lett. **262:** 107–114.
55. LYIMO, T.J. *et al.* 2000. *Methanosarcina semesiae* sp. nov., a dimethylsulfide-utilizing methanogen from mangrove sediment. Int. J. Syst. Evol. Microbiol. **50:** 171–178.
56. OREMLAND, R.S. & S. POLCIN. 1982. Methanogenesis and sulfate reduction: competitive and noncompetitive substrates in estuarine sediments. Appl. Environ. Microbiol. **44:** 1270–1276.
57. ELBERSON, M.A. & K.R. SOWERS. 1997. Isolation of an aceticlastic strain of *Methanosarcina siciliae* from marine canyon sediments and emendation of the species

description for *Methanosarcina siciliae*. Int. J. Syst. Bacteriol. **47:** 1258–1261.

58. SOWERS, K.R., S.F. BARON & J.G. FERRY. 1984. *Methanosarcina acetivorans* sp. nov., an acetotrophic methane-producing bacterium isolated from marine sediments. Appl. Environ. Microbiol. **47:** 971–978.
59. VON KLEIN, D. *et al*. 2002. *Methanosarcina baltica*, sp. nov., a novel methanogen isolated from the Gotland Deep of the Baltic Sea. Extremophiles **6:** 103–110.
60. CONRAD, R. 1999. Contribution of hydrogen to methane production and control of hydrogen concentrations in methanogenic soils and sediments. FEMS Microbiol. Ecol. **28:** 193–202.
61. CHAN, O.C. *et al.* 2005. Vertical distribution of structure and function of the methanogenic archaeal community in Lake Dagow sediment. Environ. Microbiol. **7:** 1139–1149.
62. GLISSMAN, K. *et al.* 2004. Methanogenic pathway and archaeal community structure in the sediment of Eutrophic Lake Dagow: effect of temperature. Microb. Ecol. **48:** 389–399.
63. CHAN, O.C. *et al.* 2002. Methanogenic archaeal community in the sediment of an artificially partitioned acidic bog lake. FEMS Microbiol. Ecol. **42:** 119–129.
64. BRIEE, C., D. MOREIRA & P. LOPEZ-GARCIA. 2007. Archaeal and bacterial community composition of sediment and plankton from a suboxic freshwater pond. Res. Microbiol. **158:** 213–227.
65. MACGREGOR, B.J. *et al.* 1997. Crenarchaeota in Lake Michigan sediment. Appl. Environ. Microbiol. **63:** 1178–1181.
66. PHELPS, T.J. & J.G. ZEIKUS. 1984. Influence of pH on terminal carbon metabolism in anoxic sediments from a mildly acidic lake. Appl. Environ. Microbiol. **48:** 1088–1095.
67. CASPER, P. *et al.* 2003. Methane in an acidic bog lake: the influence of peat in the catchment on the biogeochemistry of methane. Aquat. Sci. **65:** 36–46.
68. SCHULZ, S. & R. CONRAD. 1996. Influence of temperature on pathways to methane production in the permanently cold profundal sediment of Lake Constance. FEMS Microbiol. Ecol. **20:** 1–14.
69. CONRAD, R. *et al.* 1989. Hydrogen turnover by psychrotrophic homoacetogenic and mesophilic methanogenic bacteria in anoxic paddy soil and lake sediment. FEMS Microbiol. Lett. **62:** 285–293.
70. SCHULZ, S., H. MATSUYAMA & R. CONRAD. 1997. Temperature dependence of methane production from different precursors in a profundal sediment (Lake Constance). FEMS Microbiol. Ecol. **22:** 207–213.
71. CHIN, K.-J., T. LUKOW & R. CONRAD. 1999. Effect of temperature on structure and function of the methanogenic archaeal community in an anoxic rice field soil. Appl. Environ. Microbiol. **65:** 2341–2349.
72. ZEPP FALZ, K. *et al.* 1999. Vertical distribution of methanogens in the anoxic sediment of Rotsee (Switzerland). Appl. Environ. Microbiol. **65:** 2402–2408.
73. BAK, F. & N. PFENNIG. 1991. Sulfate-reducing bacteria in littoral sediment of Lake Constance. FEMS Microbiol. Lett. **85:** 43–52.
74. NUSSLEIN, B. *et al.* 2001. Evidence for anaerobic syntrophic acetate oxidation during methane production in the profundal sediment of subtropical Lake Kinneret (Israel). Environ. Microbiol. **3:** 460–470.
75. DANNENBERG, S. & R. CONRAD. 1999. Effect of rice plants on methane production and rhizospheric metabolism in paddy soil. Biogeochemistry **45:** 53–71.
76. SCHUTZ, H., W. SEILER & R. CONRAD. 1989. Processes involved in formation and emission of methane in rice paddies. Biogeochemistry **7:** 33–53.
77. FETZER, S., F. BAK & R. CONRAD. 1993. Sensirivity of methanogenic bacteria from paddy soil to oxygen and desiccation. FEMS Microbiol. Ecol. **12:** 107–115.
78. CHIN, K.J. *et al.* 2004. Archaeal community structure and pathway of methane formation on rice roots. Microb. Ecol. **47:** 59–67.
79. LU, Y. *et al.* 2005. Detecting active methanogenic populations on rice roots using stable isotope probing. Environ. Microbiol. **7:** 326–336.
80. KRUGER, M. *et al.* 2005. Activity, structure and dynamics of the methanogenic archaeal community in a flooded Italian rice field. FEMS Microbiol. Ecol. **51:** 323–331.
81. PETER MAYER, H. & R. CONRAD. 1990. Factors influencing the population of methanogenic bacteria and the initiation of methane production upon flooding of paddy soil. FEMS Microbiol. Lett. **73:** 103–111.
82. LUEDERS, T. & M. FRIEDRICH. 2000. Archaeal population dynamics during sequential reduction processes in rice field soil. Appl. Environ. Microbiol. **66:** 2732–2742.
83. ROY, R., H.D. KLUBER & R. CONRAD. 1997. Early initiation of methane production in anoxic rice soil despite the presence of oxidants. FEMS Microbiol. Ecol. **24:** 311–320.
84. ROTHFUSS, F. & R. CONRAD. 1992. Vertical profiles of CH4 concentrations, dissolved substrates and processes involved in CH4 production in a flooded Italian rice field. Biogeochemistry **18:** 137–152.
85. JOULIAN, C. *et al.* 1998. Phenotypic and phylogenetic characterization of dominant culturable methanogens isolated from ricefield soils. FEMS Microbiol. Ecol. **25:** 135–145.
86. FEY, A., K.J. CHIN & R. CONRAD. 2001. Thermophilic methanogens in rice field soil. Environ. Microbiol. **3:** 295–303.
87. CHIN, K.-J. & R. CONRAD. 1995. Intermediary metabolism in methanogenic paddy soil and the influence of temperature. FEMS Microbiol. Ecol. **18:** 85–102.
88. CONRAD, R., M. KLOSE & P. CLAUS. 2000. Phosphate inhibits acetotrophic methanogenesis on rice roots. Appl. Environ. Microbiol. **66:** 828–831.
89. SAKAI, S. *et al.* 2007. Isolation of key methanogens for global methane emission from rice paddy fields: a novel isolate affiliated with the clone cluster Rice Cluster I. Appl. Environ. Microbiol. **73:** 4326–4331.
90. MILLER, T. & M. WOLIN. 1986. Methanogens in human and animal intestinal tracts. Syst. Appl. Microbiol. 223–229.
91. LEVITT, M.D. *et al.* 2006. Stability of human methanogenic flora over 35 years and a review of insights obtained

from breath methane measurements. Clin. Gastroenterol. Hepatol. **4:** 123–129.

92. EL OUFIR, L. *et al.* 1996. Relations between transit time, fermentation products, and hydrogen consuming flora in healthy humans. Gut **38:** 870–877.
93. BERNALIER, A. *et al.* 1996. Acetogenesis from H_2 and CO_2 by methane- and non-methane-producing human colonic bacterial communities. FEMS Microbiol. Ecol. **19:** 193–202.
94. LECLERC, M. *et al.* 1997. H_2/CO_2 metabolism in acetogenic bacteria isolated from the human colon. Anaerobe **3:** 307–315.
95. STROCCHI, A. *et al.* 1994. Methanogens outcompete sulphate reducing bacteria for H2 in the human colon. Gut **35:** 1098–1101.
96. POCHART, P. *et al.* 1992. Interrelations between populations of methanogenic archaea and sulfate-reducing bacteria in the human colon. FEMS Microbiol. Lett. **77:** 225–228.
97. STROCCHI, A. *et al.* 1991. Competition for hydrogen by human faecal bacteria: evidence for the predominance of methane producing bacteria. Gut **32:** 1498–1501.
98. STEWART, J.A., V.S. CHADWICK & A. MURRAY. 2006. Carriage, quantification, and predominance of methanogens and sulfate-reducing bacteria in faecal samples. Lett. Appl. Microbiol. **43:** 58–63.
99. STROCCHI, A. & M.D. LEVITT. 1992. Factors affecting hydrogen production and consumption by human fecal flora. The critical roles of hydrogen tension and methanogenesis. J. Clin. Invest. **89:** 1304–1311.
100. RUTILI, A. *et al.* 2006. Intestinal methanogenic bacteria in children of different ages. New Microbiol. **19:** 227–234.
101. FLORIN, T.H.J. *et al.* 2000. Shared and unique environmental factors determine the ecology of methanogens in humans and rats. Am. J. Gastroenterol. **95:** 2872–2879.
102. ABELL, G.C.J., M.A. CONLON & A.L. MCORIST. 2006. Methanogenic archaea in adult human faecal samples are inversely related to butyrate concentration. Microb. Ecol. Health Dis. **18:** 154–160
103. MACFARLANE, G.T. & G.R. GIBSON. 1994. Metabolic activities of the normal colonic flora. *In* Human Health. The Contribution of Microorganisms. G. SAW, Ed. London: Springer Verlag.
104. DUNCAN, S.H. *et al.* 2002. Acetate utilization and butyryl coenzyme A (CoA):acetate-CoA transferase in butyrate-producing bacteria from the human large intestine. Appl. Environ. Microbiol. **68:** 5186–5190.
105. HOLD, G.L. *et al.* 2003. Oligonucleotide probes that detect quantitatively significant groups of butyrate-producing bacteria in human feces. Appl. Environ. Microbiol. **69:** 4320–4324.
106. MILLER, T.L. *et al.* 1982. Isolation of *Methanobrevibacter smithii* from human feces. Appl. Environ. Microbiol. **43:** 227–232.
107. MILLER, T.L. & M.J. WOLIN. 1985. *Methanosphaera stadtmaniae* gen. nov., sp. nov.—A species that forms methane by reducing methanol with hydrogen. Arch. Microbiol. **141:** 116–122.
108. ECKBURG, P.B. *et al.* 2005. Diversity of the human intestinal microbial flora. Science **308:** 1635–1638.
109. DONGOWSKI, G., A. LORENZ & H. ANGER. 2000. Degradation of pectins with different degrees of esterification by *Bacteroides thetaiotaomicron* isolated from human gut flora. Appl. Environ. Microbiol. **66:** 1321–1327.
110. JENSEN, N.S. & E. CANALE-PAROLA. 1986. *Bacteroides pectinophilus* sp. nov. and *Bacteroides galacturonicus* sp. nov.: two pectinolytic bacteria from the human intestinal tract. Appl. Environ. Microbiol. **52:** 880–887.
111. SAMUEL, B.S. *et al.* 2007. Genomic and metabolic adaptations of *Methanobrevibacter smithii* to the human gut. Proc. Natl. Acad. Sci. USA **104:** 10643–10648.
112. WHITFORD, M., R. TEATHER & R. FORSTER. 2001. Phylogenetic analysis of methanogens from the bovine rumen. BMC Microbiol. **1:** 5.
113. PINARES-PATINO, C.S. *et al.* 2003 Persistence of differences between sheep in methane emission under generous grazing conditions. J. Agric. Sci. **140:** 227–233.
114. LE VAN, T.D. *et al.* 1998. Assessment of reductive acetogenesis with indigenous ruminal bacterium populations and *Acetitomaculum ruminis*. Appl. Environ. Microbiol. **64:** 3429–3436.
115. JOHNSON, K.A. & D.E. JOHNSON. 1995. Methane emissions from cattle. J. Anim. Sci. **73:** 2483–2492.
116. JOBLIN, K. 2005. Methanogenic archaea. *In* Methods in Gut Microbial Ecology for Ruminants: 47–53. New York: Springer Verlag.
117. SKILLMAN, L.C. *et al.* 2006. 16S rDNA directed PCR primers and detection of methanogens in the bovine rumen. Lett. Appl. Microbiol. **42:** 222–228.
118. LIN, C., L. RASKIN & D.A. STAHL. 1997. Microbial community structure in gastrointestinal tracts of domestic animals: comparative analyses using rRNA-targeted oligonucleotide probes. FEMS Microbiol. Ecol. **22:** 281–294.
119. WRIGHT, A.-D.G. *et al.* 2004. Molecular diversity of rumen methanogens from sheep in Western Australia. Appl. Environ. Microbiol. **70:** 1263–1270.
120. TAJIMA, K. *et al.* 2001. Phylogenetic analysis of archaeal 16S rRNA libraries from the rumen suggests the existence of a novel group of archaea not associated with known methanogens. FEMS Microbiol. Lett. **200:** 67–72.
121. BRAUMAN, A. *et al.* 1992. Genesis of acetate and methane by gut bacteria of nutritionally diverse termites. Science **257:** 1384–1387.
122. THOLEN, A. & A. BRUNE. 1999. Localization and in situ activities of homoacetogenic bacteria in the highly compartmentalized hindgut of soil-feeding higher termites (*Cubitermes* spp.). Appl. Environ. Microbiol. **65:** 4497–4505.
123. EBERT, A. & A. BRUNE. 1997. Hydrogen concentration profiles at the oxic-anoxic interface: a microsensor study of the hindgut of the wood-feeding lower termite *Reticulitermes flavipes* (Kollar). Appl. Environ. Microbiol. **63:** 4039–4046.
124. BRUNE, A. 1998. Termite guts: the world's smallest bioreactors. Trends Biotechnol. **16:** 16–21.
125. BRAUMAN, A. *et al.* 2001. Molecular phylogenetic profiling of prokaryotic communities in guts of termites with different feeding habits. FEMS Microbiol. Ecol. **35:** 27–36.

126. DONOVAN, S.E. *et al.* 2004. Comparison of Euryarchaea strains in the guts and food-soil of the soil-feeding termite *Cubitermes fungifaber* across different soil types. Appl. Environ. Microbiol. **70:** 3884–3892.
127. YADVIKA *et al.* 2004. Enhancement of biogas production from solid substrates using different techniques—A review. Bioresour. Technol. **95:** 1–10.
128. LECLERC, M., J.-P. DELGENES & J.-J. GODON. 2004. Diversity of the archaeal community in 44 anaerobic digesters as determined by single strand conformation polymorphism analysis and 16S rDNA sequencing. Environ. Microbiol. **6:** 809–819.
129. AIYUK, S. *et al.* 2006. Anaerobic and complementary treatment of domestic sewage in regions with hot climates—A review. Bioresour. Technol. **97:** 2225–2241.
130. CONKLIN, A., H.D. STENSEL & J. FERGUSON. 2006. Growth kinetics and competition between *Methanosarcina* and *Methanosaeta* in mesophilic anaerobic digestion. Water Environ. Res. **78:** 486–496.
131. GROTENHUIS, J.T. *et al.* 1991. Bacteriological composition and structure of granular sludge adapted to different substrates. Appl. Environ. Microbiol. **57:** 1942–1949.
132. SEKIGUCHI, Y. *et al.* 1999. Fluorescence in situ hybridization using 16S rRNA-targeted oligonucleotides reveals localization of methanogens and selected uncultured bacteria in mesophilic and thermophilic sludge granules. Appl. Environ. Microbiol. **65:** 1280–1288.
133. KOBAYASHI, H.A. *et al.* 1988. Direct characterization of methanogens in 2 high-rate anaerobic biological reactors. Appl. Environ. Microbiol. **54:** 693–698.
134. GROTENHUIS, J.T.C. *et al.* 1990. Effect of interspecies hydrogen transfer on the bacteriological structure of methanogenic granular sludge. *In* Proc. of the Int. Symp. on Physiology of Immobilized Cell, Wageningen, the Netherlands, pp. 95–98.
135. MCHUGH, S. *et al.* 2003. Methanogenic population structure in a variety of anaerobic bioreactors. FEMS Microbiol. Lett. **219:** 297–304.
136. HORI, T. *et al.* 2006. Dynamic transition of a methanogenic population in response to the concentration of volatile fatty acids in a thermophilic anaerobic digester. Appl. Environ. Microbiol. **72:** 1623–1630.
137. ZINDER, S.H., T. ANGUISH & S.C. CARDWELL. 1984. Effects of temperature on methanogenesis in a thermophilic (58 °C) anaerobic digestor. Appl. Environ. Microbiol. **47:** 808–813.
138. SAVANT, D.V. *et al.* 2002. *Methanobrevibacter acididurans* sp. nov., a novel methanogen from a sour anaerobic digester. Int. J. Syst. Evol. Microbiol. **52:** 1081–1087.
139. SAVANT, D.V. & D.R. RANADE. 2004. Application of *Methanobrevibacter acididurans* in anaerobic digestion. Water Sci. Technol. **50:** 109–114.
140. PRATHER, M. & D. EHHALT. 2001. Atmospheric chemistry and greenhouse gases. *In* Climate Change 2001: The Sicentific Basis. J.T. HOUGHTON *et al.*, Eds.: 239–287. Cambridge: Cambridge University Press.
141. SMITH, D.R. *et al.* 1997. Complete genome sequence of *Methanobacterium thermoautotrophicum deltaH*: functional analysis and comparative genomics. J. Bacteriol. **179:** 7135–7155.
142. HENDRICKSON, E.L. *et al.* 2004. Complete genome sequence of the genetically tractable hydrogenotrophic methanogen *Methanococcus maripaludis*. J. Bacteriol. **186:** 6956–6969.
143. BULT, C.J. *et al.* 1996. Complete genome sequence of the methanogenic archaeon, *Methanococcus jannaschii*. Science **273:** 1058–1073.
144. MAEDER, D.L. *et al.* 2006. The *Methanosarcina barkeri* genome: comparative analysis with *Methanosarcina acetivorans* and *Methanosarcina mazei* reveals extensive rearrangement within methanosarcinal genomes. J. Bacteriol. **188:** 7922–7931.
145. DEPPENMEIER, U. *et al.* 2002. The genome of *Methanosarcina mazei*: evidence for lateral gene transfer between bacteria and archaea. J. Mol. Microbiol. Biotechnol. **4:** 453–461.
146. GALAGAN, J.E. *et al.* 2002. The genome of *Methanosarcina acetivorans* reveals extensive metabolic and physiological diversity. Genome Res. **12:** 532–542.
147. SAUNDERS, N.F.W. *et al.* 2003. Mechanisms of thermal adaptation revealed from the genomes of the Antarctic archaea *Methanogenium frigidum* and *Methanococcoides burtonii*. Genome Res. **13:** 1580–1588.
148. SLESAREV, A.I. *et al.* 2002. The complete genome of hyperthermophile *Methanopyrus kandleri* AV19 and monophyly of archaeal methanogens. Proc. Natl. Acad. Sci. USA **99:** 4644–4649.

Promiscuous Anaerobes

New and Unconventional Metabolism in Methanogenic Archaea

LAURA L. GROCHOWSKI AND ROBERT H. WHITE

Department of Biochemistry, Virginia Polytechnic Institute and State University, Blacksburg, Virginia, USA

The development of an oxygenated atmosphere on earth resulted in the polarization of life into two major groups, those that could live in the presence of oxygen and those that could not—the aerobes and the anaerobes. The evolution of aerobes from the earliest anaerobic prokaryotes resulted in a variety of metabolic adaptations. Many of these adaptations center on the need to sustain oxygen-sensitive reactions and cofactors to function in the new oxygen-containing atmosphere. Still other metabolic pathways that were not sensitive to oxygen also diverged. This is likely due to the physical separation of the organisms, based on their ability to live in the presence of oxygen, which allowed for the independent evolution of the pathways. Through the study of metabolic pathways in anaerobes and comparison to the more established pathways from aerobes, insight into metabolic evolution can be gained. This, in turn, can allow for extra- polation to those metabolic pathways occurring in the Last Universal Common Ancestor (LUCA). Some of the unique and uncanonical metabolic pathways that have been identified in the archaea with emphasis on the biochemistry of an obligate anaerobic methanogen, *Methanocaldococcus jannaschii* are reviewed.

Key words: **archaea; central metabolism; purine biosynthesis; pyrimidine biosynthesis; amino acid biosynthesis; carbohydrate metabolism; isoprenoid biosynthesis; archaeal lipids**

Introduction

Life arose and first evolved on earth under anaerobic conditions,[1] where the effect of toxic oxygen gas was not able to influence the nature of reactions leading to the first living system(s). It may be thus expected that the most likely place to gain insights into the earliest metabolic reactions would be through studies with anaerobes. We would also expect to find a much larger number of radical mediated reactions occurring in anaerobes since these reactions are readily quenched by the presence of O_2.[2] Many but not all of these anaerobic reactions would be expected to be lost in present day aerobes either because they could not be protected from O_2 or because an alternate reaction that was not influenced by oxygen evolved. We see many examples in present-day biochemistry where important anaerobic reactions that likely evolved in the absence of oxygen have been modified in some way to allow them to function in the present-day aerobic environment. Central among these reactions are the all-important Fe-S cluster–containing enzymes,[3] many of which are oxygen sensitive, but can be repaired after oxidative damage.[4–7] A very important Fe-S enzyme is nitrogenase[8] that still is only able to function in an anaerobic environment because the enzyme is irreversibly inhibited by molecular oxygen.[9] When nitrogenase must function in an aerobic environment it does so inside a specialized structure that excludes O_2, such as the root nodules of Rhizobia[10] or heterocysts in the cyanobacteria.[11] Nitrogenase as well as hydrogenase and carbon monoxide dehydrogenase/acetyl-coenzyme A synthase are all examples of oxygen-sensitive enzymes containing a modified $[Fe_4S_4]$ cluster that each likely evolved from simpler clusters.[12,13] This finding clearly indicates that they likely also evolved under anaerobic conditions.[14]

Some facultative anaerobes, such as *Escherichia coli*, have exploited the O_2 sensitivity of Fe-S clusters and utilize them as oxygen sensors. One example of this is the fumarate and nitrate reduction regulator (FNR), an oxygen-responsive transcription regulator that

Address for correspondence: Robert H. White, Department of Biochemistry (0308), Virginia Polytechnic Institute and State University, Blacksburg, VA 24061. Voice: +1-(540) 231-6605; fax: +1-(540) 231-9070.
rhwhite@vt.edu

Ann. N.Y. Acad. Sci. 1125: 190–214 (2008). © 2008 New York Academy of Sciences.
doi: 10.1196/annals.1419.001

controls the expression of anaerobic genes in facultative anaerobes, such as *E. coli*.[15] In the absence of oxygen the iron-sulfur cluster of FNR is in the $[4Fe\text{-}4S]^{2+}$ state, which allows for dimer formation and DNA binding. Upon exposure to oxygen, the cluster is oxidized to the $[4Fe\text{-}4S]^{3+}$ state, which rapidly decomposes, resulting in protein conformational changes. The oxidized protein cannot form the active dimer and can no longer bind to DNA.[16]

Many other enzyme systems function with a Fe^{2+} as the catalytic metal that is not involved with Fe-S centers, and that must be continuously reduced after air or enzymatic oxidation to Fe^{3+}. A prime example of such a system is found in the enzyme NADH-methemoglobin reductase.[17] Still other enzymes have been found to be "opportunistic" and are capable of utilizing either Fe^{2+} or Zn^{2+}, a redox inactive metal. The literature is full of examples of enzymes that at first were considered to be Zn^{2+} enzymes, but on closer examination were found to be Fe^{2+}-dependent and to work even better with Fe^{2+}. Among these is *E. coli* cytosine deaminase,[18] an enzyme that was only shown to require Fe^{2+} after its initial characterization.[19] *E. coli* peptide deformylase was initially considered to be a Zn^{2+} enzyme, but later Fe^{2+} was found to be the physiologically relevant metal.[20] To be added to this list are thioesterases containing Fe^{2+},[21] methionyl aminopeptidase,[22] and atrazine chlorohydrolases.[23] We have recently characterized a new type of guanosine triphosphate (GTP) cyclohydrolase from *Methanocaldococcus jannaschii*, designated MptA to indicate that it catalyzes the first step in methanopterin biosynthesis.[24] MptA was found to be a unique GTP cyclohydrolase in that it forms 7,8-dihydro-D-neopterin 2′,3′-cyclic phosphate as its reaction product, and that it is a Fe^{2+}-dependent enzyme. Other known GTP cyclohydrolases have been shown to utilize Zn^{2+}. We propose that most or all of these enzymes, which evolved under anaerobic conditions in the presence of high amounts of Fe^{2+}, used Fe^{2+} as their catalytic metal. This is expected considering the much larger amount of Fe^{2+} in the anaerobic ocean as compared to Zn^{2+}. This exposure of cells to an abundant supply of stable soluble Fe^{2+}, would have continued for the first billion years of life's evolution on the anaerobic earth. As oxygen was introduced into the atmosphere, some of these proteins adapted to Zn^{2+} dependency, which has about the same acid dissociation constant of the aquo complex as Fe^{2+}. Both Zn^{2+} or Fe^{2+}, but not Fe^{3+}, would allow for the generation of the metal-bound hydroxide that functions as the nucleophile in these enzymes. The Zn^{2+} enzyme has the added advantage that it will not undergo oxidation to an inactive species, as occurs with the Fe^{2+} enzyme. We think that this switch to using Zn^{2+} occurred with the aerobic GTP cyclohydrolases.

While some enzymes modified their metal requirements, others exhibited more dramatic changes in response to an oxygen-rich environment. Ribonucleotide reductase is an excellent example of an enzyme evolving from one completely inhibited by oxygen to one that actually uses oxygen. The originally anaerobically evolved enzyme, most closely related to the current Class III enzymes present in anaerobes,[25] require FeS centers and *S*-adenosylmethionine to generate the radical required for catalysis, and is completely inactivated by oxygen. After the introduction of oxygen into the atmosphere the enzyme evolved into the Class II group of enzymes that use the more complex B_{12} coenzyme for radical generation and are not affected by oxygen. Finally, the Class I enzymes evolved to use oxygen for the required radical formation. Although extensive changes in the protein sequence of these enzymes have occurred during evolution, they all still retain the same active site structure and reaction mechanism.[26]

An excellent example of changes in the enzymes in a metabolic pathway with the introduction of oxygen can be found in the aerobic and anaerobic biosynthesis of ubiquinone in *E. coli*. As can be seen in FIGURE 1, the intermediates in both pathways are the same and begin with 4-hydroxybenzoic acid. As shown, three of the steps require the hydroxylation of different positions of the aromatic ring. In the aerobic pathway these reactions proceed using specific monooxygenases encoded by *ubiB, ubiH,* and *ubiF* genes, respectively, and incorporate a single oxygen from $^{18}O_2$ with each hydroxylation.[27] Under anaerobic conditions the reactions proceed in the absence of oxygen,[28] but at present the genes/enzymes involved are unknown. Thus this could be an example of anaerobic evolution of the pathway followed by modification to use more efficient oxygen-dependent hydroxylations once oxygen was introduced in the atmosphere.

Most biochemical enzymes/reactions are not affected by the presence of oxygen, and as a result one cannot develop criteria for whether or not the reaction catalyzed evolved before or after the introduction of oxygen in the biosphere. Some reactions/enzymes are in fact oxygen sensitive and their evolution only under anaerobic conditions seems reasonable. Considering the effect of oxygen on specific reactions gives insight as to the conditions under which they evolved.

One group of strict anaerobes are the methanogens, which are defined by their unique biochemistry and ability to make methane.[29] Methanogens belong to the domain Archaea. The major groups in this domain include (a) the methanogenic Archaea, (b) the

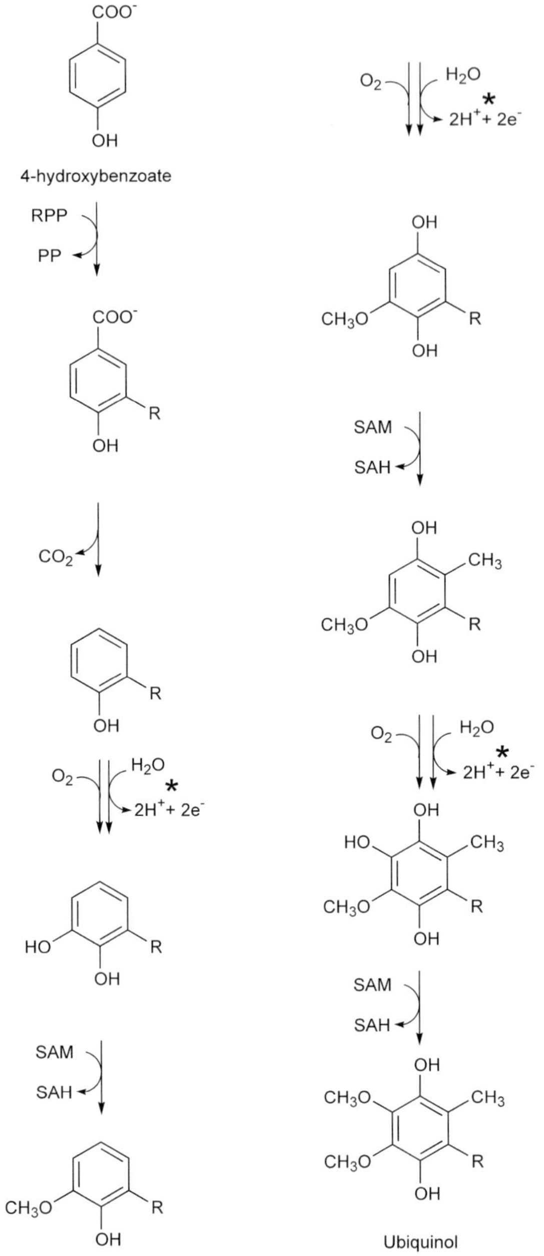

FIGURE 1. The anaerobic and aerobic biosynthetic pathways to ubiquinone in *E. coli*. Both pathways use the same intermediates, but each of the required three hydroxylation steps uses a different monooxygenase in the aerobic pathways and unknown enzymes for the anaerobic pathway. These three steps are indicated with an *asterisk*.

sulfate-reducing Archaea, (c) the extremely halophilic Archaea, (d) the Archaea lacking cell walls, and (e) the extremely thermophilic S°-metabolizing Archaea. Of all these groups only the methanogens are obligate anaerobes.

In addition to the unique biochemistry of methanogenesis, uncannonical biochemistry in archaeal central metabolic pathways has also been observed, including differences in sugar, nucleotide, isoprene, and amino acid biosynthesis. For this reason, we have concentrated our efforts on understanding the unique metabolism of these organisms. In this chapter we will review exciting new findings on metabolism found from the methanogens, concentrating specifically on *M. jannaschii*. Other exciting areas as they relate to these findings are also covered.

Nucleoside–Nucleotide Biosynthesis

Purine Biosynthesis

Purines play a central role in the metabolism of life on earth. They represent indispensable structural units of nucleic acids and are essential parts of many coenzymes including nicotinamide adenine dinucleotide (NAD), flavin adenine dinucleotide (FAD), coenzyme A, adenosine triphosphate (ATP), GTP, and molybdopterin guanine dinucleotide. They also serve as biosynthetic precursors to the chemically active parts of the coenzymes FAD, flavin mononucleotide (FMN), folate, methanopterin, F_{420}, and molybdopterin guanine dinucleotide. Unexpectedly, these final coenzymes can serve as precursors to other coenzymes, such as occurs in the conversion of NAD to the thiamin thiazole[30] and FMN to the 5,6-dimethylbenzimidazole of vitamin B_{12},[31] and to the stereochemically modified FAD occurring in alcohol oxidase.[32] NAD also serves as a donor for the posttranslational adenosine diphosphate–ribosylation (ADP-ribosylation) of eukaryotic proteins.[33] An intermediate in purine biosynthesis, aminoimidazole ribonucleotide (AIR) also serves as the precursor for the generation of the pyrimidine of thiamin.[34]

Work on establishing the biosynthesis of purines began in the late 1930s with studies on the biosynthesis of uric acid in pigeons. The chemistry of the pathway was largely established by 1950,[35] the enzymes by 1960, and the genes encoding the enzymes catalyzing the 10 separate reactions by 1980.[36] Among these enzymes was PurH, responsible for catalyzing the last two steps in the pathway: the conversion of 5-aminoimidazole-4-carboxamide ribonucleotide (AICAR) to 5-formaminoimidazole-4-carboxamide ribonucleotide (FAICAR) and its subsequent cyclization to inosine 5′-monophosphate (IMP) (FIG. 2). In bacteria and eukaryotes these reactions are catalyzed by a single bifunctional enzyme, aminoimidazole carboxamide ribonucleotide transformylase/inosine monophosphate cyclohydrolase (ATIC), encoded by

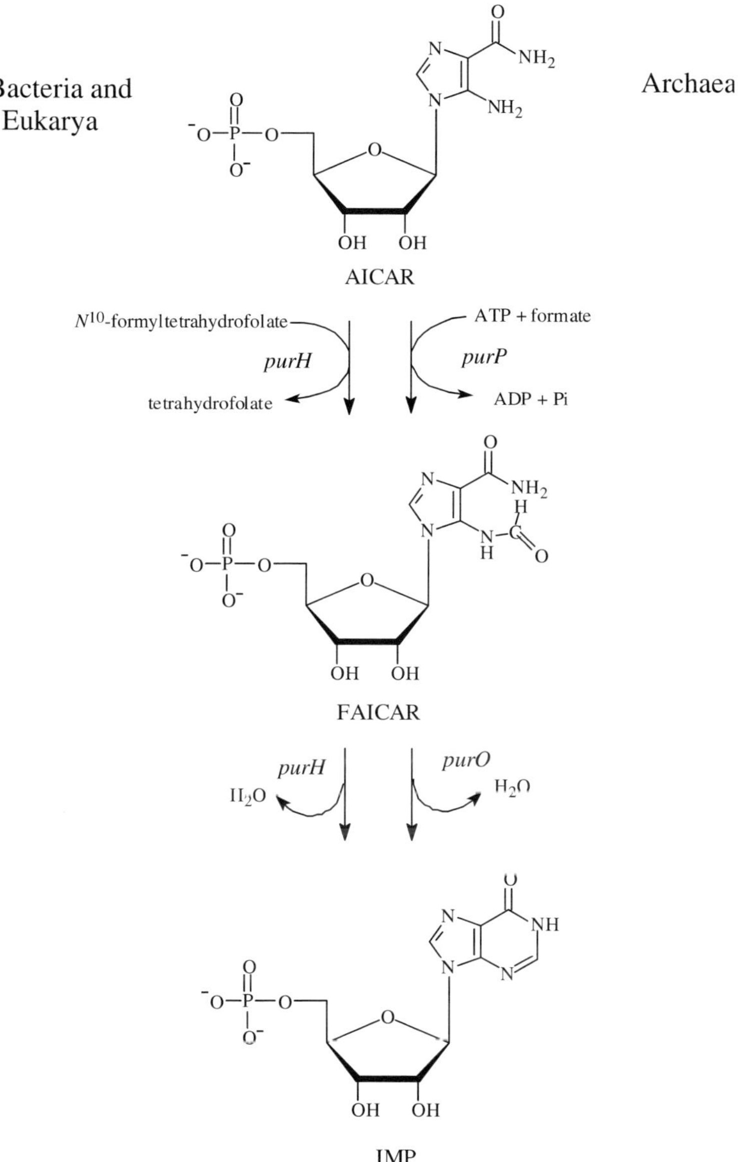

FIGURE 2. The currently known different enzymology of the last two steps in purine biosynthesis.

the *purH* gene. The first reaction is catalyzed by AICAR transformylase, which resides in the carboxyl-terminus of the enzyme and transfers the formyl group of N^{10}-formyltetrahydrofolate to AICAR with the formation of FAICAR. The product of this reaction is then cyclized by the IMP cyclohydrolase that resides in the amino-terminus of the enzyme.[37] The two-domain structure of ATIC has been confirmed by the X-ray crystal structure of the avian enzyme.[38]

The certainty and universality of the purine biosynthetic pathway and its enzymes, PurH in particular, appeared firmly established until biochemical and genomic data indicated several things were different with the PurH-catalyzed reactions in the Archaea. These data include the observation that N^{10}-formyltetrahydrofolate, required for the aminoimidazole carboxamide ribonucleotide transformylase reaction to FAICAR, was not required for this reaction in archaea[39]; genomic DNA sequence comparisons that indicated the absence of *purH* genes in archaeal genomes[40–42]; and the identification of *purO* (MJ0626), which encodes a new IMP cyclohydrolase that cyclizes FAICAR to IMP in *M. jannaschii*.[43] Attempts to identify the preceding AICAR transformylase gene

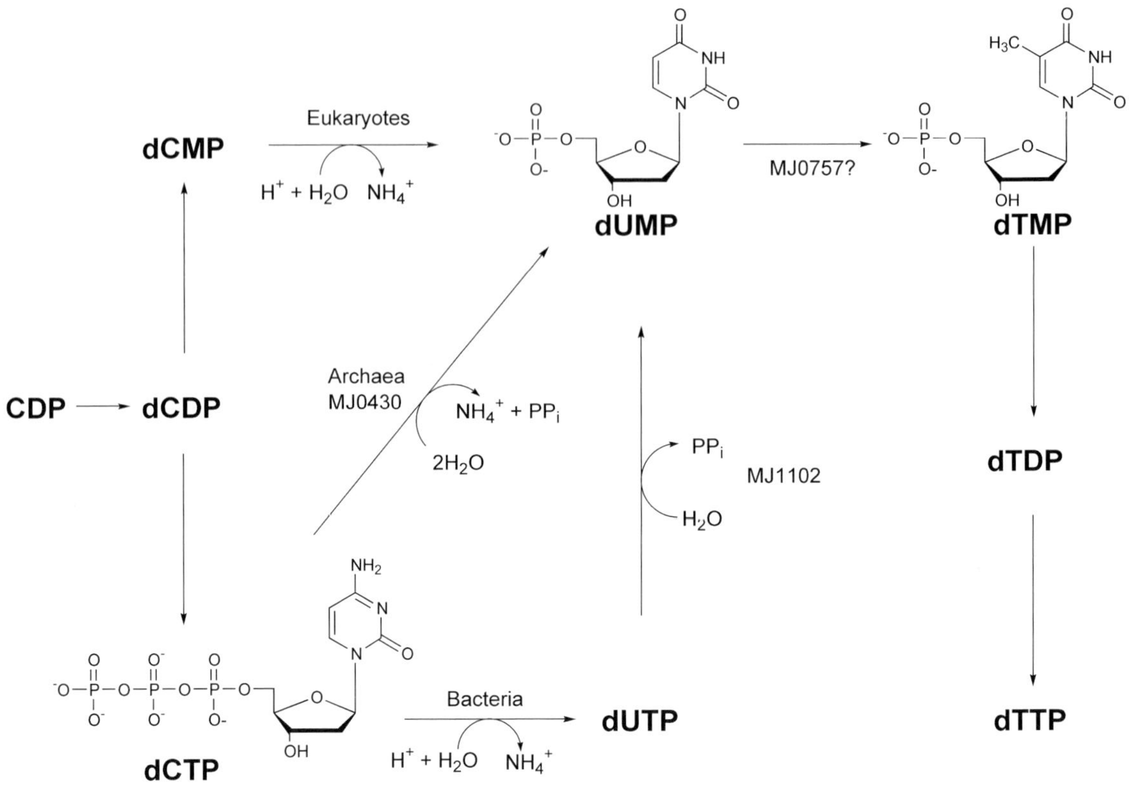

FIGURE 3. The altered steps in pyrimidine biosynthesis found in *M. jannaschii*.

by searching for genes close to *purO* in other genomes did not prove fruitful, in part due to the low abundance of *purO* in the currently sequenced genomes. Attempts at the isolation and purification of the enzyme using the identified reaction that used ATP and formate as cosubstrates also proved futile. The enzyme was finally identified by testing for the ability of annotated ATP-grasp enzymes present in the *M. jannaschii* genome for their ability to catalyze the reaction.[44] The enzyme, designated PurP, is derived from the MJ0136 gene and is a new member of the ATP-grasp enzyme family. PurP catalyzes the ATP and formate-dependent formylation of AICAR to FAICAR in the absence of folates and proceeds through a formyl-P intermediate.[44]

Having identified the function of both the PurO and PurP, a study was conducted on the distribution of these genes in other archaea and it was found that some of the archaea contained no *purO* gene, but did contain two *purP* genes. The second *purP* from *Pyrococcus furiosus* (PF1517) was designated *purP2*, cloned, recombinantly expressed, and no PurO or PurP activity has been observed (unpublished data). Subsequent structural work on the MthPurO from *Methanobacterium thermoautotrophicum* has shown that it has a four-layered αββα core structure, showing an N-terminal nucleophile (NTN) hydrolase fold.[45] The X-ray crystal structures of PurP and PurP2 complexed with various substrates and products are now completed (unpublished data). Both enzymes are trimers, although hexamers for both are found in the crystal structure. The PurP and PurP2 protomers are very similar and most of the active site residues are conserved. The ATP binding sites are highly homologous and only four substitutions in the first active site shell differentiate the two activities. The rare occurrence in which homologous enzymes catalyze consecutive steps of a biosynthetic pathway suggests the possibility of an ancestral enzyme capable of catalyzing both reactions.

Pyrimidine Biosynthesis

Most bacteria produce the dUMP precursor for thymine nucleotide biosynthesis using two enzymes: a dCTP deaminase that catalyzes the formation of dUTP followed by a dUTP diphosphatase that catalyzes the hydrolytic release of pyrophosphate (FIG. 3). Although these two hydrolytic enzymes appear to catalyze very different reactions, they are encoded by homologous genes. The *M. jannaschii* genome has two members of this gene family. One gene, at locus MJ1102, encodes a dUTP diphosphatase, which can scavenge deoxyuridine nucleotides that inhibit archaeal DNA polymerases. The second gene, at locus MJ0430, encodes a novel dCTP deaminase that releases dUMP, ammonia, and pyrophosphate.[46] Therefore this enzyme can singly catalyze both steps in dUMP biosynthesis, precluding the formation of free, mutagenic dUTP. Besides differing from the previously characterized *Salmonella typhimurium* dCTP deaminase in its reaction products, this archaeal enzyme has a higher affinity for dCTP and its steady-state turnover is faster than the bacterial enzyme. Kinetic studies suggested that the archaeal enzyme specifically recognizes dCTP; dCTP deamination and dUTP diphosphatase activities occur independently at the same active site; and both activities depend on Mg^{2+}. The bifunctional activity of this *M. jannaschii* enzyme illustrates the evolution of a suprafamily of related enzymes that catalyze mechanistically distinct reactions. The structural basis for substrate recognition and catalysis has been established.[47]

A currently confusing aspect of the pyrimidine biosynthesis pathway in the archaea is the true nature of the thymidylate synthase catalyzed reaction occurring in the conversion of dUMP to dTMP. In many archaeal genomes no clear gene is annotated for this enzyme. Through applying a novel procedure, ORF (an acronym for ostensible recognition of folds), MJ0757 was predicted[48] and subsequently shown to catalyze this reaction with tetrahydrofolate as the coenzyme.[49] The enzyme is distinct from the canonical enzyme as well as the recently discovered flavin-dependent thymidylate synthase.[50] Due to the generally recognized absence of folates in the methanogens, it was surprising to find that this enzyme required folate for the conversion of dUMP to dTMP. It was proposed that a nonmethylated methanopterin-like molecule was the true cofactor.

Amino Acids

M. jannaschii is an autotrophic anaerobe and as such is capable of biosynthesizing all its amino acids and vitamins required for growth from CO_2, H_2, and other trace inorganic compounds containing N, S, P, Fe, and Mo. Analysis of the *M. jannaschii* genome revealed that orthologs of some of the enzymes involved in the biosynthesis of amino acids were not present. This suggested that this archaea utilized nonorthologous replacements and/or novel biosynthetic pathways for the biosynthesis of some of the amino acids. Subsequent research into archaeal amino acid biosynthesis has revealed examples of both scenarios as described later in this paper.

Methionine and Cysteine Biosynthesis

Analysis of archaeal genomes has generally failed to show the presence of genes encoding any of the canonical enzymes known to be involved in the metabolism of cysteine and homocysteine (FIG. 4) in both anaerobes and aerobes.[51] Some exceptions to this rule do exist. For example, an *O*-acetylserine sulfhydrylase, CysK, catalyzing Reaction 14 in FIGURE 4, was isolated and characterized from *Methanosarcina thermophila*[52] and the genes for both the *O*-acetylserine sulfhydrylase and the transacetylase (reaction 13, FIG. 4) were also identified in this organism.[53] Additional work to identify the *O*-acetylserine sulfhydrylase in *Methanosarcina barkeri* strain Fusaro by complementing an *E. coli cysK* mutation confirmed the presence of this gene in this methanogen. These workers concluded that these genes for cysteine biosynthesis were likely acquired bacterial genes.[54] An extremely thermostable *O*-acetylserine sulfhydrylase was also isolated and characterized from *Aeropyrum pernix* K1.[55,56]

The pathways for the biosynthesis of cysteine and homocysteine in cell extracts of *M. jannaschii* were examined using gas chromatography–mass spectrometry stable isotope dilution methods to identify and quantitate the amounts of cysteine and homocysteine produced from different precursors.[57] It was proposed that the first step in the pathway, and the one likely responsible for incorporation of sulfur into both cysteine and methionine, was the reaction between *O*-phosphohomoserine and a currently unidentified sulfur source present in cell extracts, to produce L-homocysteine. This sulfur source was shown not to be sulfide or polysulfides. The resulting L-homocysteine then reacts with *O*-phosphoserine to form L-cystathionine, which is subsequently cleaved to L-cysteine. These observations would explain the formation of both cysteine and homocysteine, the precursor to methionine. The proposed pathway was based in part on the known transsulfuration pathway, and in total would define the following sequence of reactions: *O*-phosphohomoserine → homocysteine → cystathionine → cysteine + α-ketobutyrate. Subsequent work, however, failed to demonstrate the formation of α-ketobutyrate during the cleavage of the cystathionine, calling into question the operation of the pathway to produce cysteine.

After this work was published, additional work showed that *M. jannaschii* generates the cysteine-$tRNA^{Cys}$ used for protein biosynthesis directly from *O*-phosphoserine-$tRNA^{Cys}$, as shown in FIGURE 5.[58] The *O*-phosphoserine-$tRNA^{Cys}$ is formed by the charging of the $tRNA^{Cys}$ with *O*-phosphoserine. The required L-*O*-phosphoserine in turn is generated by

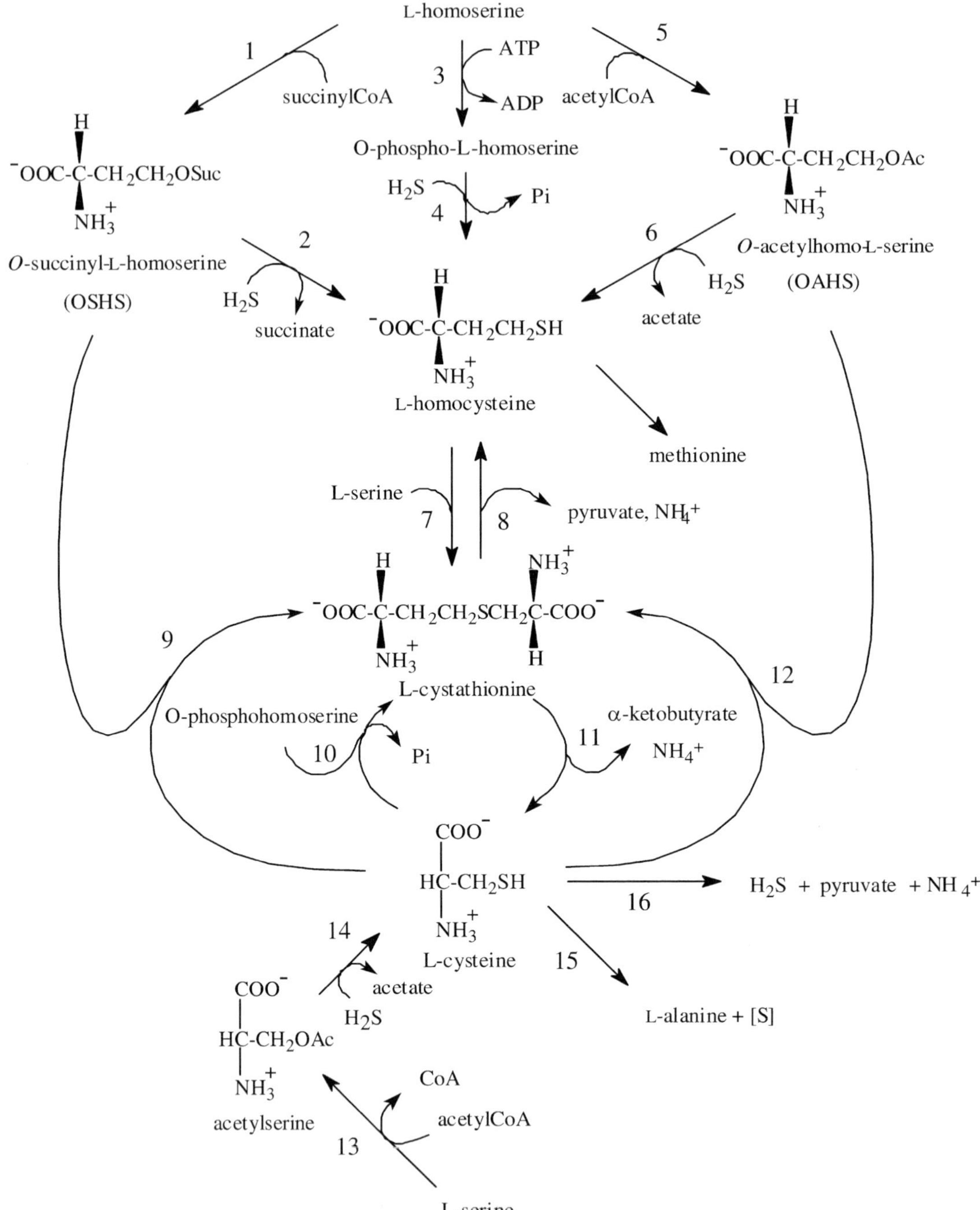

FIGURE 4. Established pathways for the metabolism of cysteine and homocysteine in bacteria and eukarya. The reactions indicated are catalyzed using the following enzymes: (1) succinyl-CoA: L-homoserine *O*-succinyltransferase, [EC 2.3.1.46]; (2) *O*-succinyl-L-homoserine succinate-lyase (adding cysteine), [EC 4.2.99.9]; (3) L-homoserine kinase; (4) *O*-phosphohomoserine (thiol)-lyase, [EC 4.2.99.9]; (5) acetyl-CoA:L-homoserine *O*-acetyltransferase, [EC 2.3.1.31]; (6) *O*-acetylhomoserine (thiol)-lyase, [E.C. 4.2.99.10]; (7) L-serine hydro-lyase (adding homocysteine), [EC 4.2.1.22]; (8) cystathionine L-homocysteine-lyase (deaminating) [EC 4.4.1.8]; (9) *O*-succinylhomoserine (thiol)-lyase, [EC 4.2.99.9]; (10) *O*-phosphohomoserine (thiol)-lyase, [EC 4.2.99.9]; (11) L-cystathionine cysteine-lyase (deaminating) [EC 4.4.1.15]; (12). *O*-acetylhomoserine (thiol)-lyase [EC 4.2.99.9]; (13) acetyl-CoA:L-serine *O*-acetyltransferase, [EC 2.3.1.30]; (14). *O*-acetylserine (thiol)-lyase-A or B, [EC 4.2.99.8]; (15) cysteine desulfurase; (16) cysteine/cystine *C-S* lyase.

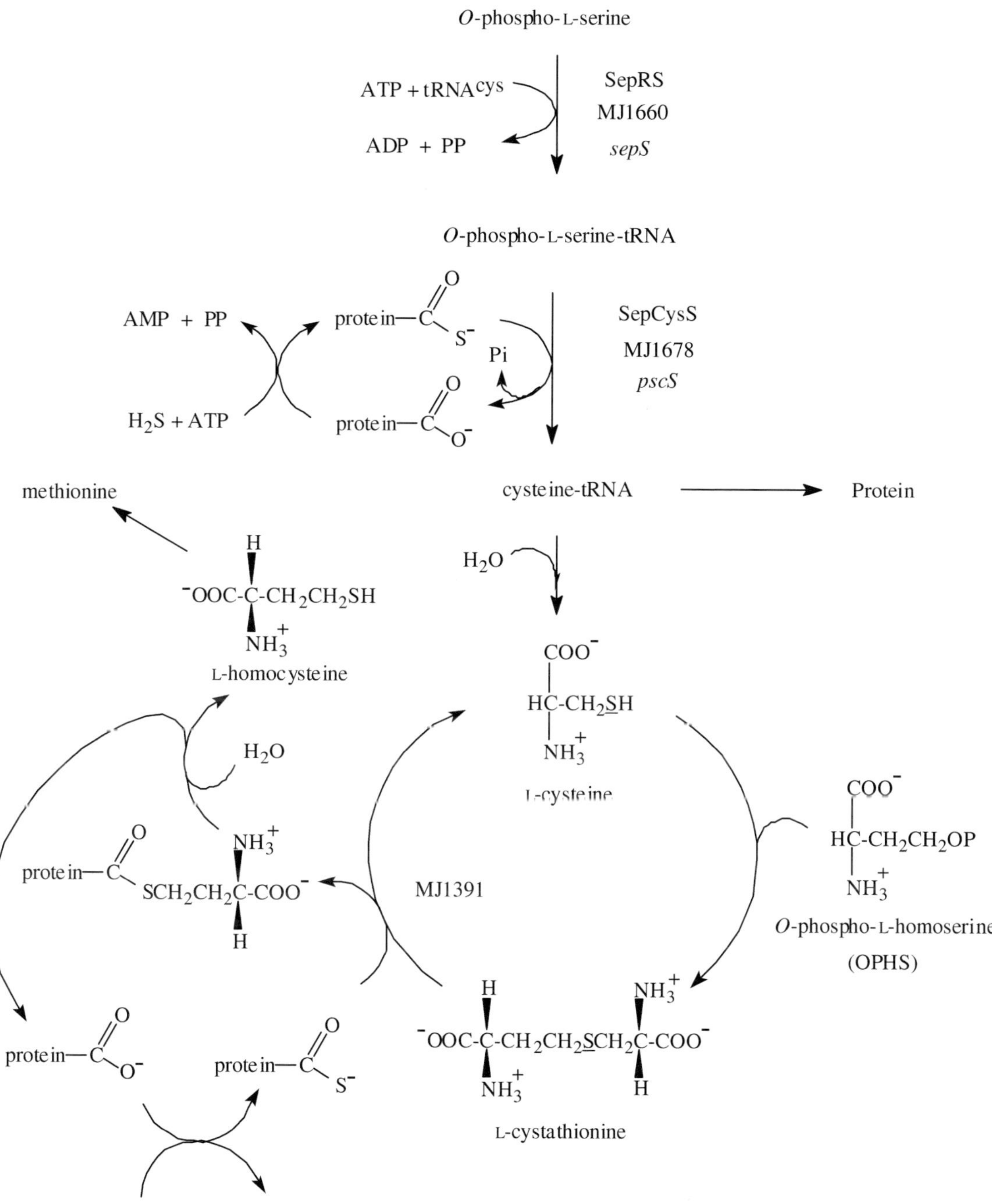

FIGURE 5. Proposed pathway for the formation of cysteine and methionine in *M. jannaschii*. The true sulfur source for the reactions catalyzed by the MJ1391 and MJ1678 gene products has not yet been established but is proposed to be a protein containing a thiocarboxylate.

the transamination of 3-phosphohydroxypyruvate.[59] This proposal, however, left no route for the production of cysteine that would be required for other metabolic processes in the cell, such as the synthesis of the sulfur-containing coenzymes thiamine, coenzyme M, and coenzyme B. The resolution of these issues came from the study of a new pyridoxal phosphate (PLP)-dependent enzyme that catalyzed the cleavage of cystathionine to cysteine and an *S*-substituted homocysteine. The enzyme was isolated from cell extracts of *M. jannaschii* by following production of cysteine from L-cystathionine, and, based on its sequence, is derived from the MJ1391 gene. Based on sequence comparisons and predicted tertiary structure, this enzyme is a member of the Asp-AT Ib subfamily of the α-family of PLP–dependent enzymes. In the presence of thioacetic acid the product is *S*-acetylhomocysteine that can subsequently be hydrolysed to acetate and homocysteine. We propose that thioacetic acid is substituting for true reactant, which is a sulfide carrier protein containing the sulfur as a carboxy-terminal thiocarboxylate, as shown in FIGURE 5. Thus cleavage of cystathionine produced cysteine that functions as a carrier coenzyme for the aminobutyl group of the homoserine in the production of homocysteine. The pathway is consistent with the original data for the conversion of L-*O*-phosphohomoserine to homocysteine with no sulfide incorporation. The cycle indicated in FIGURE 5 requires only a small amount of cysteine for its operation, since the cysteine is continuously regenerated. Since it has been recently demonstrated that cysteine is not the source of the Fe-S centers in this organism (unpublished data) and may in fact not be the precursor to the other sulfur-containing compounds in the cells, then the hydrolysis of only a small amount of the cysteine-$tRNA^{Cys}$ would be required. The pathway proposed is like that in the established transsulfuration pathway for the conversion of cysteine to homocysteine, but here the sulfur of the cysteine is not transferred from the cysteine to the homocysteine. Other enzymes confirmed by our lab in the methionine biosynthetic pathway include the conversion of aspartate semialdehyde to L-homoserine by homoserine dehydrogenase; the *M. jannaschii* MJ1602 gene product; the phosphorylation of L-homoserine by homoserine kinase; the MJ1104 gene product; and the methylation of L-homocysteine by a cobalamin-independent methionine synthase, the MJ1473 gene product using methyl-tetrahydromethanopterin as the methyl donor.

From the work on cysteine metabolism in *M. jannaschii*, a new enzyme catalyzing the decomposition of cysteine was also discovered and characterized.[60] The enzyme catalyzed decomposition of cysteine to hydrogen sulfide, ammonia, and pyruvate. The protein is the product of the *M. jannaschii* MJ1025 gene, is devoid of PLP, and contains a [4Fe-4S] center when active. The enzyme is rapidly inactivated upon exposure to air, and reactivated by reaction with Fe^{2+} and dithiothreitol in the absence of air. The exact role of this enzyme in cysteine metabolism in anaerobes is not clear at this time, but the enzyme is the first example of an enzyme using a [4Fe-4S] center to catalyze the decomposition of cysteine.

The true sulfur donor for these reactions has not yet been experimentally verified, but based on the lack of incorporation of ^{34}S-sulfide into cysteine and homocysteine in cell extracts, it is clear that the true sulfide donor is neither sulfide nor a persulfide. The lack of involvement of persulfides in the reaction is based on the assumption that they would readily exchange with sulfide. Considering the present data and what is currently known about the biochemistry of thioacids as sulfur donors, this leaves a sulfide-carrier protein containing the sulfur at a carboxy-terminal of thiocarboxylate as the most likely source of the sulfur. This group may not be expected to exchange with sulfide as readily as would a persulfide. It would also not be formed in the absence of ATP. This functional group has been shown to be involved in cysteine biosynthesis in *Mycobacterium tuberculosis*,[61] in the biosynthesis of the thiazole ring of thiamine,[62,63] and in biosynthesis of molybdopterin.[64] Thus the cystathionine bound to the PLP would react with the thiol of the thiocarboxylate of the protein, and the resulting thiol ester-bound homocysteine would undergo hydrolysis either with or without the N-S shift. The reaction would be just like that observed for cysteine biosynthesis in *M. tuberculosis* where a PLP intermediate undergoes a reaction with the thiol acid of the protein, followed by hydrolysis, but here the sulfur is adding to the γ-carbon of the cystathionine. If the reaction should occur at either end of the cystathionine, the net product would be the same, that is, cysteine and homocysteine.

Aromatic Amino Acids and Para Aminobenzoic Acid (pAB)

No orthologs are present in the genomes of many archaea for the first two established steps in the biosynthesis of the aromatic amino acids leading to 3-dehydroquinate (DHQ) (FIG. 6).[65] The absence of these genes prompted the examination of the nature of the reactions involved in the archaeal pathway leading to DHQ in *M. jannaschii*. It was found that 6-deoxy-5-ketofructose 1-phosphate (DKFP) and L-aspartate semialdehyde (ASA) were the precursors to DHQ[66] (FIG. 7). This is unlike the established pathway shown in FIGURE 6 and explains why the genes are different. The sugar, which is ultimately derived from glucose 6-P, supplies a "hydroxyacetone" fragment, which, via a transaldolase reaction, undergoes an aldol condensation with ASA to form 2-amino-3,7-dideoxy-D-*threo*-hept-6-ulosonic acid (ADH). Despite the fact that both hydroxyacetone and hydroxyacetone-P were measured in the cell extracts and confirmed to arise from

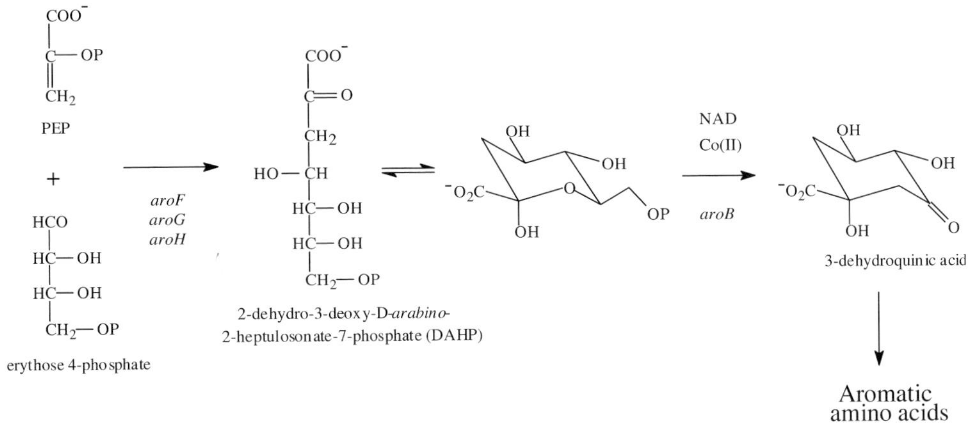

FIGURE 6. Established pathway for aromatic amino acid biosynthesis.

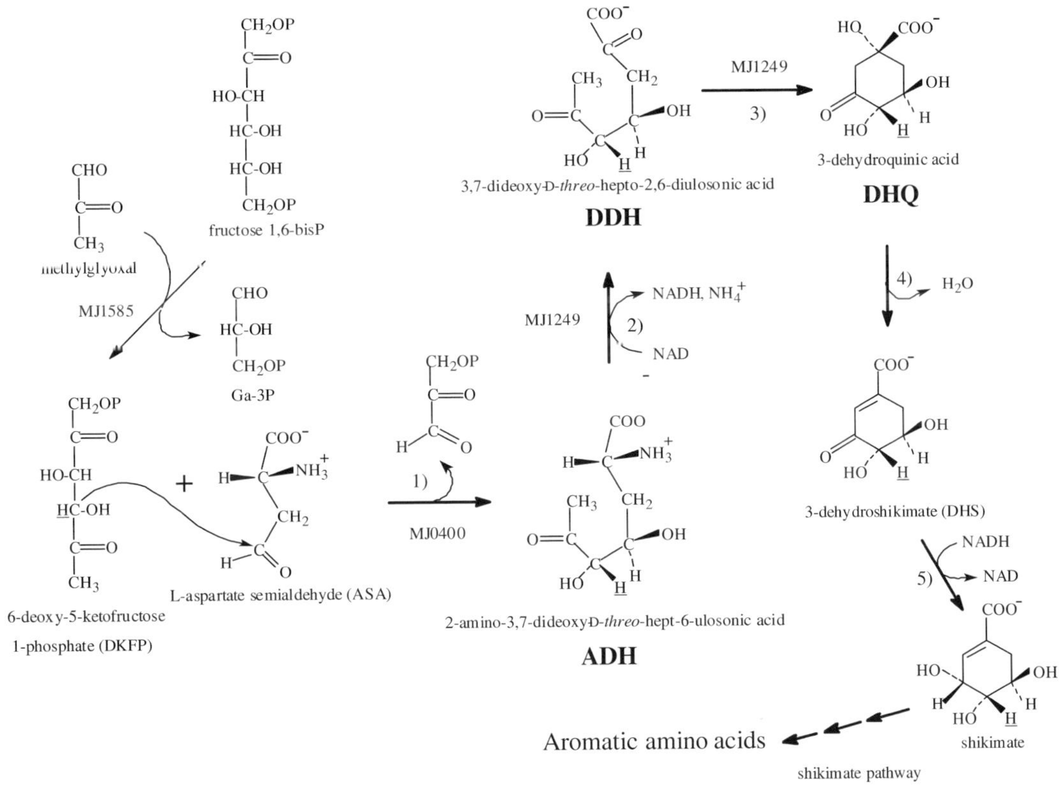

FIGURE 7. Altered pathway for the biosynthesis of DHQ occurring in some of the archaea. The steps after DHQ are the same as the canonical shikimate aromatic amino acid biosynthesis pathway.

glucose 6-P, neither compound was found to serve as a precursor to DHQ. The ADH amino sugar then undergoes a NAD-dependent oxidative deamination to produce 3,7-dideoxy-D-*threo*-hept-2,6-diulosonic acid (DDH), which cyclizes to DHQ. The protein product of the *M. jannaschii* MJ0400 gene catalyzes the transaldolase reaction, and the protein product of the MJ1249 gene catalyzes the oxidative deamination and the cyclization reactions. The DHQ is readily converted into dehydroshikimate and shikimate in *M. jannaschii* cell extracts, consistent with the remaining steps and genes in the pathway being the

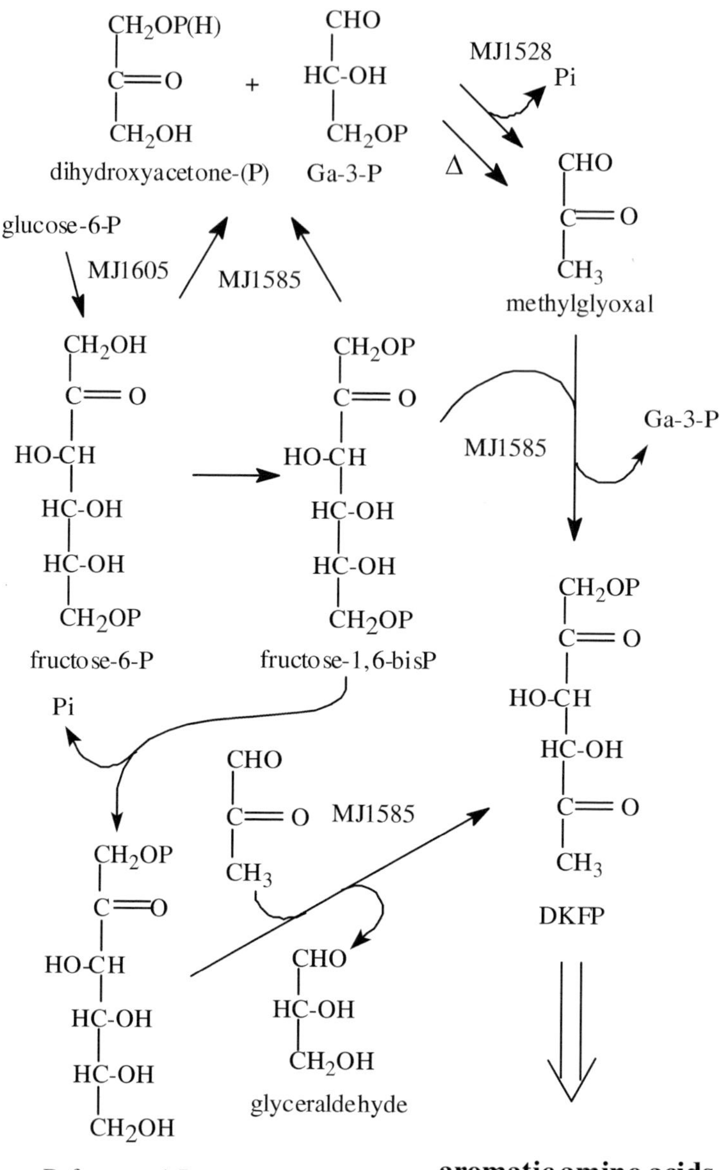

FIGURE 8. Involvement of methylglyoxal in the biosynthesis of DKFP, the precursor to the aromatic amino acids in the archaea.

same as in the established shikimate pathway. Evidence also showed that DHQ is the precursor to 4-aminobenzoic acid, for methanopterin biosynthesis, via a pathway different from the established pathway.[67]

It was demonstrated that DKFP arises from a transaldolase reaction between methylglyoxal and fructose-1,6-bisP or fructose-1-P catalyzed by the MJ1585 gene product (FIG. 8).[68] Methylglyoxal was shown in this work to be produced from glyceraldehyde-3-P by eliminating phosphate by both enzymatic and chemical routes. Methylglyoxal, after a H_2F_{420}-dependent reduction, produces L-lactaldehyde that is oxidized by NAD to L-lactate for F_{420} biosynthesis.[69] This work represents the first biochemical example of toxic methylglyoxal functioning as a central metabolite.[68]

β-*Glutamate*

Although the early steps for biosynthesis of the aromatic amino acids is different in the anaerobic archaea, these modified reactions do not have to

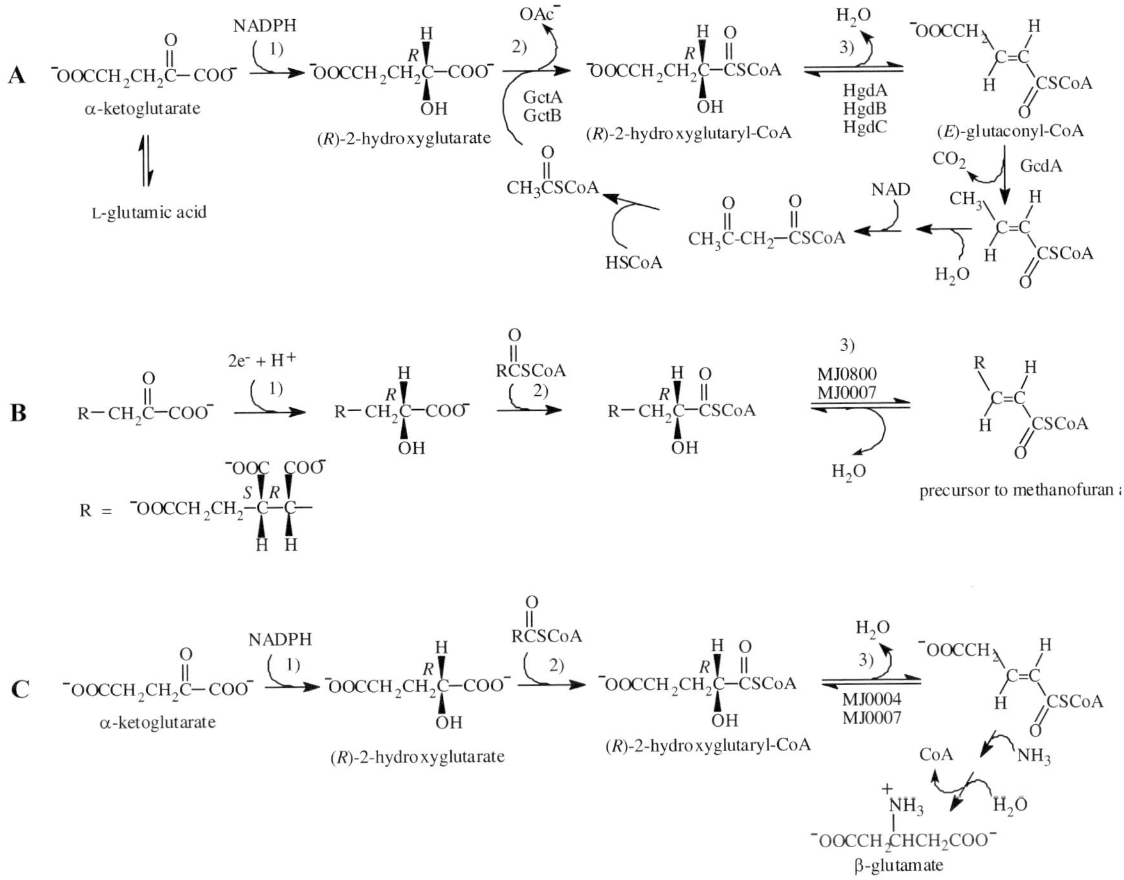

FIGURE 9. Biochemical examples of anaerobic reactions involved in the elimination of water from CoA thioesters of α-hydroxyacids.

function under anaerobic conditions. The biosynthesis of β-glutamate, on the other hand, has been found to proceed using oxygen-sensitive enzymes. β-Glutamate, an unusual nonchiral amino acid, is found in several marine methanogens,[70] as well as in anoxic sea sediments.[71–73] The compound is believed to serve as an osmolyte for methanogens,[74] and its occurrence in these and other microorganisms[72,75–77] likely accounts for its presence in marine sediments. Roberts' ^{13}C-NMR biosynthetic experiments[78–80] demonstrate that the carbon chain of the molecule derives from one acetate and three CO_2; the acetate methyl carbon incorporates at C-3 and the acetate carboxylate carbon is equally distributed between C-2 and C-4. The labeling patterns observed in these experiments are consistent with the carbon chain being derived from α-ketoglutarate, which, in turn, is derived from acetate via the following sequence: acetate, acetyl-CoA, pyruvate, oxalacetate, malate, fumarate, succinate, succinyl-CoA, and, finally, α-ketoglutarate. This sequence of transformations, involving two reductive carboxylations, is an established pathway for acetate metabolism in archaea.[81–84]

As part of our work to define the steps in the biosynthesis of *meso*-1,3,4,6-hexanetetracarboxylic acid (HTCA),[82] a moiety of methanofuran[85] and also an osmolyte in some archaea,[86] we recognized similarities among the steps involved in the hydroxyglutarate pathway for glutamate fermentation in *Acidaminococcus fermentans*, *Peptostreptococcus asaccharolyticus*, and *Clostridium synbiosum*[87–89] (FIG. 9, pathway (A); the latter steps of HTCA biosynthesis[90,91] (FIG. 9, pathway (B); and the steps in a possible pathway for the formation of β-glutamate (FIG. 9, pathway (C). The similarities among these pathways include the following: the NAD(P)H-dependent reduction of an α-ketoacid to an (*R*)-2-hydroxyacid, the formation of the (*R*)-2-hydroxyacid's CoA thioester, and the subsequent dehydration of the CoA thioester. Pathways (A) and (C) also have in common the addition of water and ammonia, respectively, to the α,β-unsaturated acyl-CoA ester. The first four intermediates in these two pathways are the same: α-ketoglutarate → (*R*)-2-hydroxyglutarate → (*R*)-2-hydroxyglutaryl-CoA → (*E*)-glutaconyl-CoA. In the case of *A. fermentans*, the pathway for glutamate fermentation begins with an NAD-dependent dehydrogenase,[92] which is a member of the D-2-hydroxyacid dehydrogenase protein family[93] related to D-lactate dehydrogenases.[94] The pathway continues with a CoA-transferase followed by a (*R*)-2-hydroxyglutaryl-CoA dehydratase[89] that catalyzes the reversible *syn*-elimination of water from (*R*)-2-hydroxyglutaryl-CoA, forming (*E*)-glutaconyl-CoA.[95]

A search of the genome of *M. jannaschii*, an organism that contains large amounts of β-glutamate,[70] revealed three homologs to known genes involved in the glutamate fermentation pathway. These are MJ0007, homologous to both the α (HgdA) and β (HgdB) subunits of the 2-hydroxyglutaryl-CoA dehydratase (EC 4.2.1.-), and MJ0004 and MJ0800, both homologous to the activator protein (HgdC).[88] A combination of the MJ0007-derived protein and either the MJ0004- or MJ0800-derived protein could catalyze the third step in pathway (C) (FIG. 9), as is observed in the anaerobic pathway for glutamate metabolism. The combination of the MJ0007-derived protein with the other nonfunctioning activator proteins could then catalyze the third step in pathway (B) (FIG. 9), in HTCA biosynthesis.

We have shown that the protein products of the MJ0004 and MJ0007 genes catalyze the dehydration of (*R*)-2-hydroxyglutaryl-CoA and have demonstrated the presence of an NADP-dependent dehydrogenase in *M. jannaschii* that reduces α-ketoglutarate to (*R*)-2-hydroxyglutaric acid. We have also demonstrated that the MJ0590-derived enzyme can produce (*R*)-2-hydroxyglutaryl-CoA from (*R*)-2-hydroxyglutarate, ATP, and CoA, and that the enzymatic addition of NH_3 to glutaconyl-CoA forms β-glutamate (Fig. 10).

Based on these observations we proposed the pathway shown in FIGURE 10 for anaerobic β-glutamate production. The complete pathway for β-glutamate synthesis in *M. jannaschii* begins with the reduction of α-ketoglutarate to (*R*)-2-hydroxyglutarate, followed by formation of the CoA thioester and elimination of water, and ending with the addition of ammonia and hydrolysis of the CoA thioester. While not all of the genes for the enzymes involved have been determined, some of the properties of the functioning enzymes could be identified, providing an avenue to later determine the genes. This pathway is quite different from that in *Clostridium difficile*, where a newly identified 2,3-glutamate mutase is operating.[96]

The enzyme produced by the MJ0590 gene has recently been proposed to function as an acyl-CoA synthetase,[97] and we have shown that it can, in fact, catalyze the formation of CoA thioesters of glutaconate, hydroxyglutarate, or acetate when incubated with these acids, ATP, and coenzyme A. The formation of the CoA thioester of 2-hydroxyglutaric acid could also be carried out by several different routes that have been identified in archaea. These routes could involve the sequential action of enzymes related to acetate kinase and phosphotransacetylase[98]; an enzyme related to acetyl-CoA synthetase (ADP-forming) [EC6.2.1.1.3],[99] an enzyme related to the coenzyme A transferases,[100] or, most likely, an enzyme related to acetyl-CoA synthase.[101,102] The most recently identified CoA transferase is not found in the genome of *M. jannaschii.*[103] Based on our results we conclude that the MJ0590-derived enzyme catalyzes the formation of (*R*)-2-hydroxyglutarate-CoA.

An unusual aspect of this anaerobic metabolism is the use of two stereoisomers to serve as precursors to two different compounds, here methanopterin and β-glutamate. The NADH reduction of α-ketoglutarate, catalyzed by the MJ1425 gene product, produces (*S*)-hydroxyglutaric acid[104] used in the biosynthesis of methanopterin[105,106] and the other (*R*)-isomer is used to generate β-glutamate. At present we do not know the enzyme that catalyzes the NADPH-dependent reduction of α-ketoglutarate to (*R*)-hydroxyglutaric acid, leading to β-glutamate, but it produces much more product than the MJ1425-derived enzyme because of the much larger amount of β-glutamate than methanopterin in the cells.

Polyamines

The established route for spermidine polyamine biosynthesis is shown in FIGURE 11. The genome *M. jannaschii* contains homologs of many of the genes required for polyamine biosynthesis. Yet genomes from neither this organism nor any other anaerobic euryarchaeon have orthologs of the pyridoxal-5-phosphate-dependent ornithine or arginine decarboxylase genes required to produce putrescine, the precursor to spermidine. Instead these organisms have been shown to function with a new class of arginine decarboxylase (PvlArgDC) formed by the self-cleavage of a proenzyme into a 5-kDa subunit and a 12-kDa subunit that contains a reactive pyruvoyl group.[107] The enzyme is extremely thermostable and has no significant sequence similarity to previously characterized proteins. The conserved active site residues are similar to those of the pyruvoyl-dependent histidine decarboxylase enzyme, and its subunits form a similar

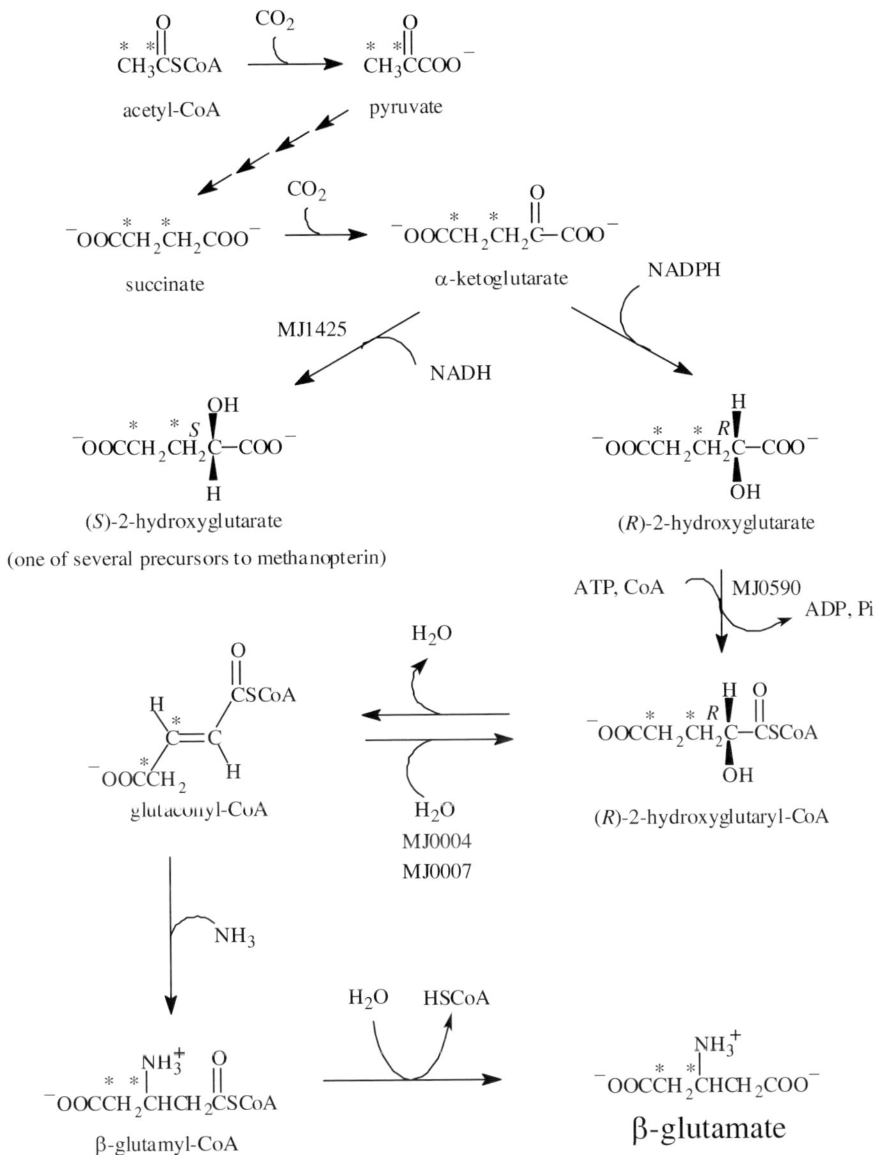

FIGURE 10. Proposed pathway for the formation of β–glutamate. (The *asterisks* indicate the positions of incorporation of labeled acetate.)

$(\alpha\beta)_3$ complex. Insights into the mechanism of the self-cleavage of this enzyme were obtained from the crystal structures of the self-cleaved and S53A proenzyme.[108]

As previously discussed, in the case of aromatic amino acid biosynthesis we have found many pathways in the anaerobic methanogens where the first committed step in the pathways are catalyzed by noncanonical enzymes. In the archaeal polyamine pathway the first step in one of the branches is catalyzed by the noncanonical PvlArgDC, and the first step in the second branch is catalyzed by a novel *S*-adenosylmethionine synthetase (MAT), representing a newly discovered class of MAT proteins.[109] Not only is the archaeal polyamine pathway different as a result of these changes, it is also unique since it contains two steps that use pyruvoyl-dependent enzymes. The other pyruvoyl enzyme is the *S*-adenosylmethionine decarboxylase encoded by the MJ0315 gene.[110] Having such a concentration of pyruvoyl-dependent enzymes in this pathway may be an indication that the pyruvoyl-dependent enzymes evolved in the anaerobes before the more complex PLP-dependent enzymes that

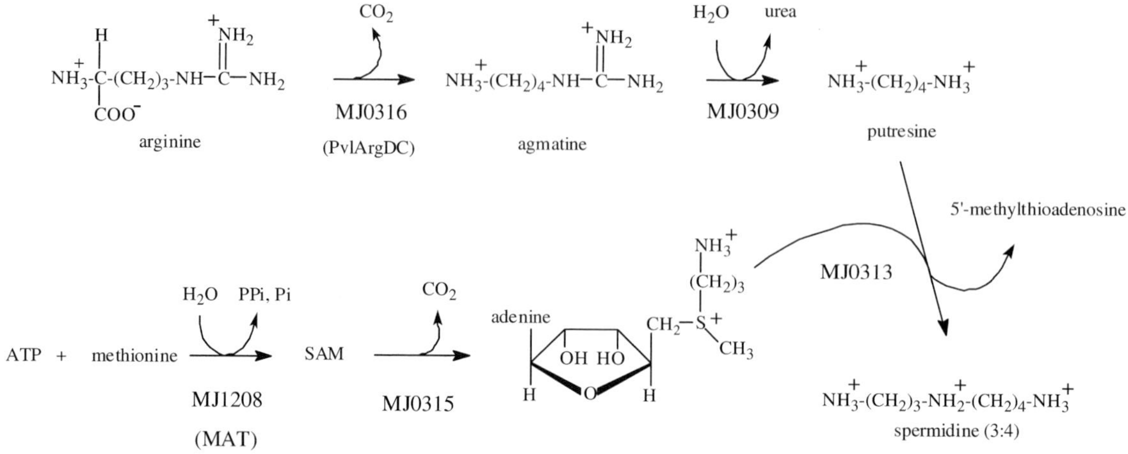

FIGURE 11. The pathway for the biosynthesis of the polyamine spermidine in *M. jannaschii*.

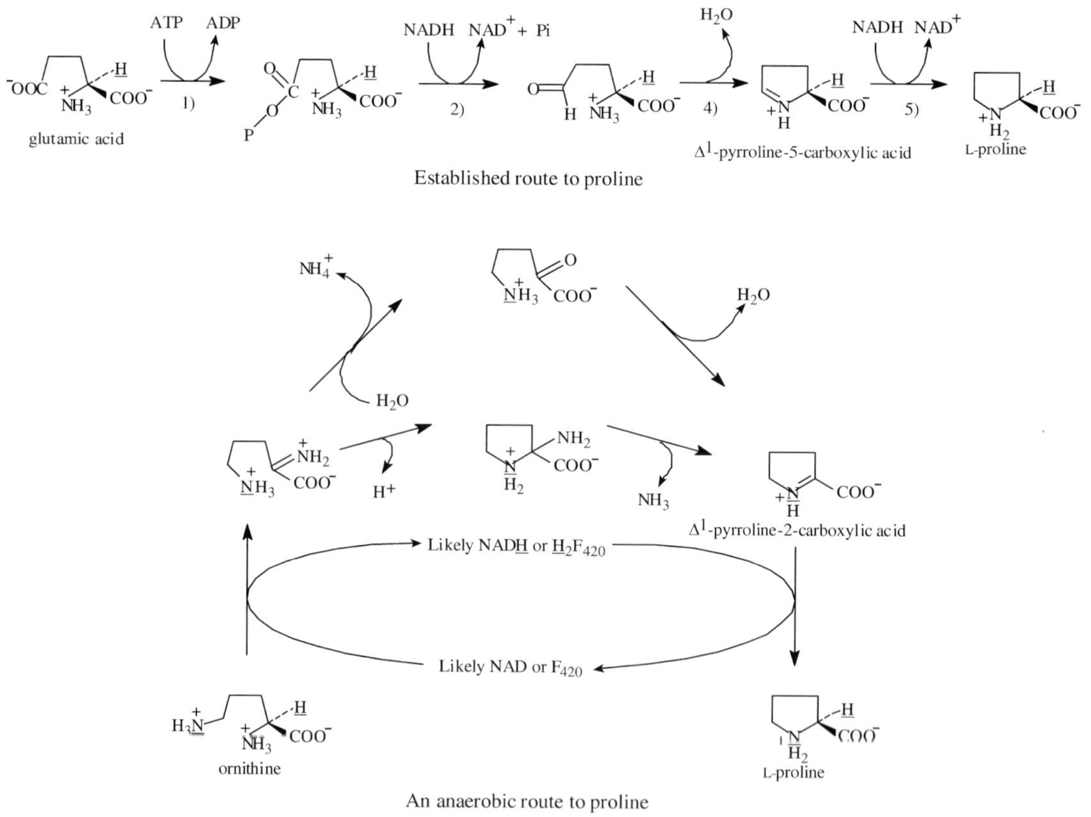

FIGURE 12. The established and new anaerobic route to proline.

require an oxygen-dependent step in the formation of PLP.

Proline

The genome of *M. jannaschii*, as well as many other archaea, does not contain genes encoding enzymes required to produce proline via the established route from glutamic acid (top pathway in FIG. 12). The absence of the established pathway for proline biosynthesis was confirmed by the absence of the incorporation of L-[3, 3′,4,4′]-glutamic acid into proline in cell extracts of *M. jannaschii* supplemented with ATP and NADH.[111] The absence of this pathway was further supported by the inability to demonstrate the last step in the established pathway, that is, the NADH- or NADPH-dependent enzymatic reduction of Δ^1-pyrroline-5-carboxylic acid to proline in the cell extracts (FIG. 12, reaction 5). On the other hand, L-ornithine was found to be readily converted into proline when incubated with cell extracts of *M. jannaschii.* When the incubation was conducted with L-[3, 3′,4, 4′, 5, 5′-$^2H_6$]-ornithine or [5-^{15}N]-ornithine all six deuteriums or the C-5 nitrogen were retained in the generated L-proline (FIG. 12). Incubation of the cell extract containing 40% D_2O with L-ornithine produced proline containing no deuterium. These results demonstrate that the C-2 hydrogen of ornithine, the removal of which would be required for conversion of ornithine to proline, was reincorporated back at the C-2 position of the proline without mixing with the water protons. The data are consistent with the biosynthesis of the L-proline proceeding by a mechanism analogous to that observed by ornithine cyclodeaminase.[112] The nature of the carrier coenzyme involved in the C-2 retained hydride was not established, but could be either NAD, as occurs in the ornithine cyclodeaminase, or F_{420}. This route for the formation of ornithine was also confirmed in *Methanobacterium thermoautotrophicum* strain ΔH and strain Marlburg, but was not observed in *Methanosarcina thermophila* strain TM-1, showing that not all anaerobic archaea generate proline in this manner. Despite much work, the gene(s) encoding the required enzyme(s) in *M. jannaschii* has(have) yet to be identified.

Sugar Biosynthesis

Gluconeogenesis

During the analysis of the first archaeal genome sequenced, that from *M. jannaschii,* the apparent absence of many genes homologous to those involved in central metabolic pathways was revealed. Subsequent studies into the carbohydrate metabolism of archaea has demonstrated that, in many cases, archaea use alternate or modified pathways for their carbohydrate metabolism, thus explaining the absence of many gene homologs.

Studies have shown that many archaea use-modified versions of either the Embden–Meyerhof (EM) or the Entner–Doudoroff (ED) pathways for glucose catabolism.[113,114] Although the individual chemical reactions in archaeal gluconeogenesis are the same as the canonical pathway, a wide spectrum of enzymes from those resembling their bacterial and eukaryal counterparts to those that are uniquely archaeal are found. Although homologs for classic glyceraldehyde-3-phosphate dehydrogenases are found in the archaea, much more diversity is found for enzymes for the other steps. An example of this is phosphoglucose isomerase, where some archaea, such as *M. jannaschii,* contain homologs of the bacterial enzyme, while other archaea utilize a novel metal-dependent phosphoglucose isomerase or a bifunctional phosphoglucose/phosphomannose isomerase.[113] Archaeal phosphoglycerate mutase was found to be a distant relative to the superfamily of proteins containing the 2,3-diphosphoglycerate-independent mutase; however, only the metal binding domain was expected to be conserved.[115] An example of an enzyme that was found to be unique to archaea is the fructose bisphosphate aldolase that has no sequence similarity to the bacterial or eukaryal enzymes.[116]

Ribose Biosynthesis and the Ribulose Monophospate Pathway

Another example where the biochemical characterization of an archaeal metabolic pathway revealed unpredicted results is in the biosynthesis of ribose. The identification of an archaeal aromatic amino acid biosynthetic pathway that does not utilize erythrose-4-phosphate (E-4-P) raised the question of the role of this sugar in archaeal metabolism. In canonical metabolism, E-4-P serves as an intermediate in aromatic amino acid biosynthesis (FIG. 6) and ribose biosynthesis through the pentose phosphate pathway. The genomes of only a few archaea, however, have been found to have homologs of all the nonoxidative pentose phosphate enzymes.[117] Taken together, the alternate aromatic amino acid biosynthetic pathway and the apparent absence of pentose phosphate genes in archaea suggest that E-4-P is not involved in archaeal metabolism. Consistent with this hypothesis, independent studies on two anaerobic, hyperthermophilic archaea have provided evidence that in many archaea ribose biosynthesis is not dependent on the nonoxidative pentose phosphate pathway.[118,119] Further support for this hypothesis came from the observation that key pentose phosphate intermediates, E-4-P, xylose-5-phosphate, and sedoheptulose-7-phosphate, were not detected in *M. jannaschii,*[118] even though this methanogen is one of only a few archaea to possess homologs of all nonoxidative pentose phosphate genes.[117]

An alternate route for ribose biosynthesis in the archaea is the ribulose-5-phosphate (RuMP) pathway shown in FIGURE 13. The RuMP pathway was originally described in methylotrophic bacteria and has been shown to be involved in formaldehyde

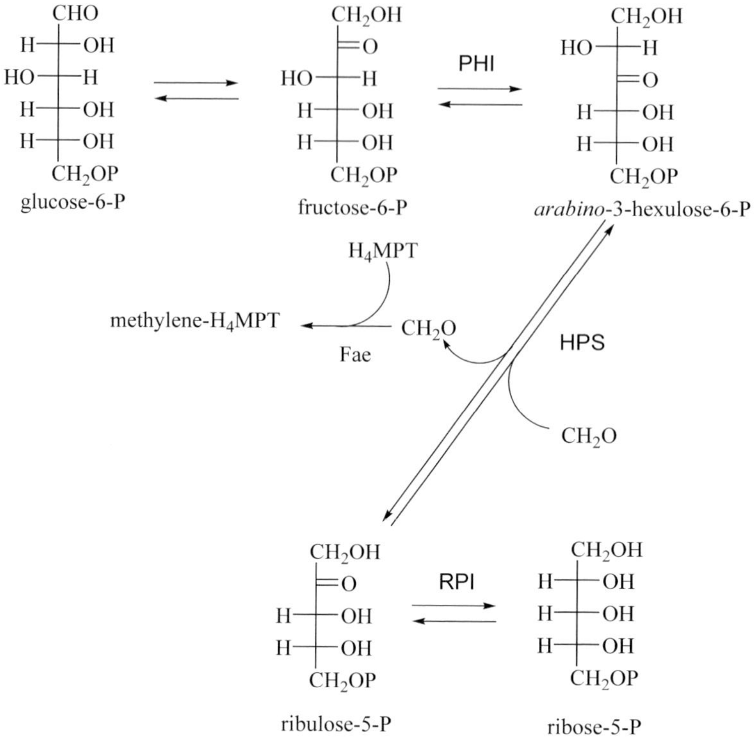

FIGURE 13. The ribulose-5-phosphate pathway.

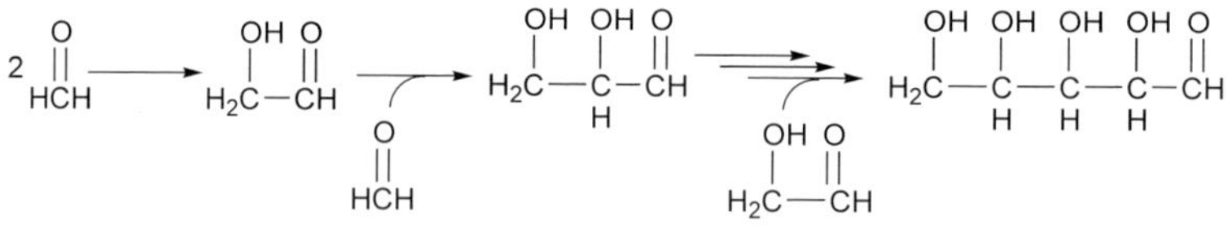

FIGURE 14. The prebiotic formose reaction.

detoxification.[120] As seen in FIGURE 13, the RuMP pathway involves the conversion of fructose-6-phosphate to ribose-5-phosphate and involves three enzymes: 3-hexulose-6-phosphate isomerase, 3-oxohexulose-6-phosphate synthase (HPS), and ribose-5-phosphate isomerase. While most archaea lack gene homologs for complete nonoxidative pentose phosphate pathways, they do contain homologs for the ribulose monophosphate enzymes. The archaeal pathway was proposed to operate in the direction to generate ribose-5-P from hexose-5-P with the released formaldehyde combining with tetrahydromethanopterin (H_4MPT) to form methylene-H_4MPT.[117] The role of the RuMP pathway in archaeal ribose-5-P metabolism has been verified experimentally.[118,119]

In several archaea, HPS is fused to formaldehyde activating enzyme (Fae).[121] Fae likely transfers the liberated formaldehyde to H_4MPT to form methylene-H_4MPT. The reverse reaction, which involves the condensation of formaldehyde and ribulose-5-P to form a ketohexose-6-P, is the first example of the involvement of the prebiotic formose reaction in biochemistry. The formose reaction was proposed by Decker *et al.*[122] and involves the sequential condensation of formaldehyde to form sugars (FIG. 14).

Isoprene Biosynthesis

Archaeal lipids differ from their bacterial and eukaryotic counterparts in several important regards. The archaeal lipids are composed of a *sn*-glycerol-1-phosphate backbone connected to saturated isoprenoid hydrocarbon chains via ether linkages. This is in contrast to bacterial and eukaryl lipids that consist of *sn*-glycerol-3-phosphate backbones connected to typically unbranched saturated and *cis* unsaturated

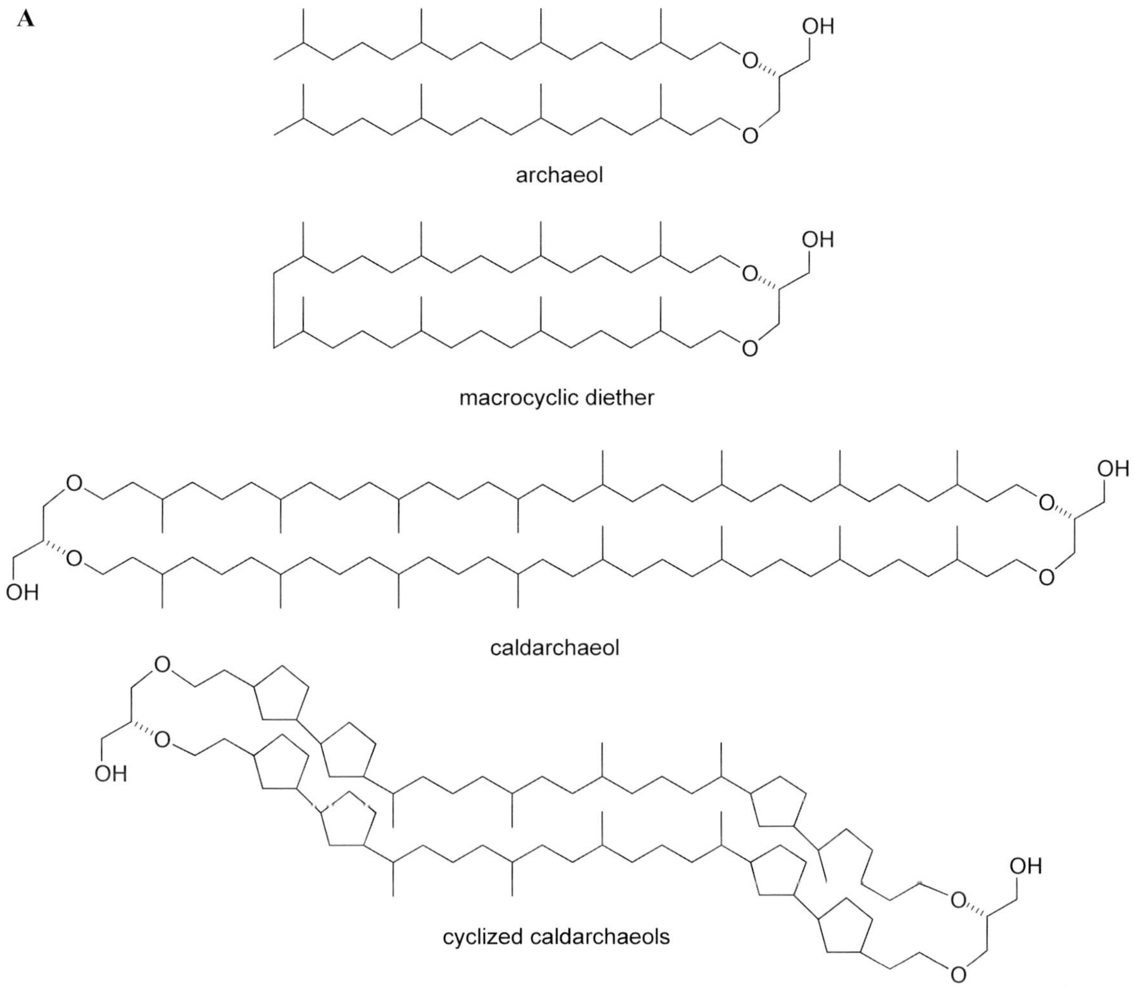

FIGURE 15. (A) Structures of the di- and tetraether lipids of archaeal origin. **(B)** Structures of the di- and tetraether lipids of anaerobic bacterial origin.[145,146,149,150]

fatty-acid chains through ester linkages. Cyclization of the isoprenoid chains and the formation of one or more cyclopentyl rings have also been observed in archaeal lipids.[123,124] In addition, many archaea contain bipolar tetraether lipids, such as caldarchaeol, in which two C40 lipids are joined at the tails (FIG. 15A).

Due to the integral importance of isoprenes in archaeal lipids, their biosynthesis is a subject of much interest. *In vivo*, *in vitro*, and bioinformatic evidence supports the use of the mevalonate pathway for archaeal biosynthesis with no evidence for the deoxyxyulose pathway.[125] Archaeal orthologs of several enzymes of the mevalonate pathway, identified through genomic analysis, have been cloned and characterized, including 3-hydroxy-3-methylglutaryl-coenzyme A reductase, and mevalonate kinase.[126–128] Genes for three of the remaining steps in the mevalonate pathway (phosphomevalonate kinase, diphosphomevalonate decarboxylase, and isopentenyl diphosphate (IPP) isomerase) were not initially identified, however. Subsequent research resulted in the identification of two of the remaining enzymes.

The first of the three "missing" enzymes in the archaeal mevalonate pathway was diphosphomevalonate kinase. Although diphosphomevalonate has not been identified as an intermediate in the pathway, its role was assumed based on analogy to the established eukaryotic/eubacterial pathway. Genomic analysis by Smit and Mushegian[129] allowed for the identification of a putative kinase that is consistently clustered with mevalonate pathway genes and was proposed to be the missing archaeal phosphomevalonate kinase. Subsequent cloning and characterization of the identified kinase from *M. jannaschii* showed that the enzyme had

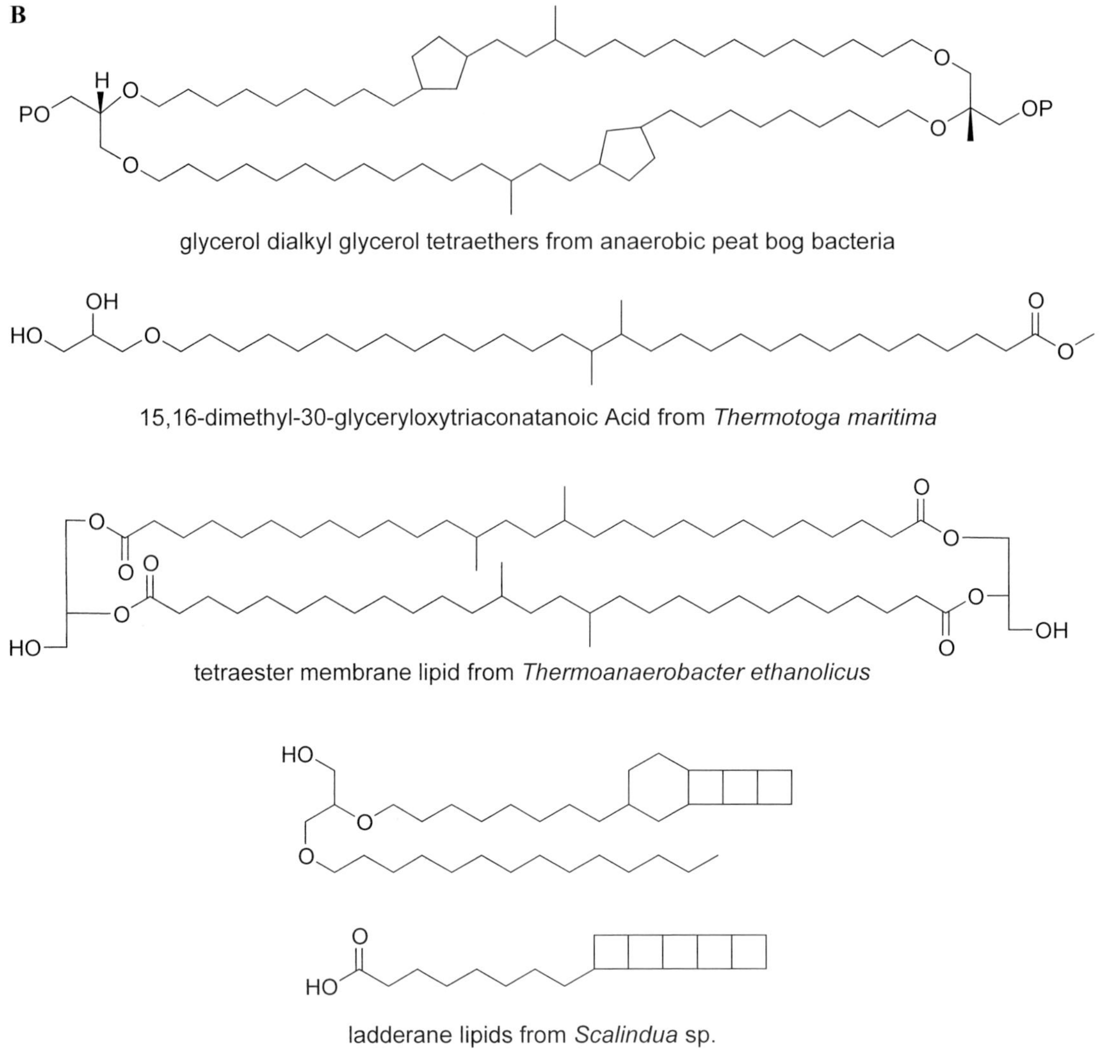

B

glycerol dialkyl glycerol tetraethers from anaerobic peat bog bacteria

15,16-dimethyl-30-glyceryloxytriaconatanoic Acid from *Thermotoga maritima*

tetraester membrane lipid from *Thermoanaerobacter ethanolicus*

ladderane lipids from *Scalindua* sp.

FIGURE 15. *Continued*

isopentenyl phosphate kinase (IP) activity rather than phosphomevalonate kinase activity.[130] The characterization of this enzyme led to the proposal of a modified mevalonate pathway that may be functioning in the archaea (FIG. 16). The modified pathway would differ from the established pathway only in the final steps, where mevalonate phosphate rather than mevalonate diphosphate serves as the substrate for an unknown decarboxylase. The product of the decarboxylation reaction would be IP. Isopentenyl monophosphate kinase would then phosphorylate IP to form IPP. The proposed mevalonate phosphate decarboxylase has not yet been identified and is required to validate the modified pathway.

Following the formation of IPP, the last step in the mevalonate pathway is the isomerization of IPP to dimethylallyl diphosphate (DMAPP). DMAPP is required for polyprenyl diphosphate synthesis, where it is condensed with multiple IPP units to form geranyl, farnesyl, geranylgeranyl, or farnesylgeranyl diphosphate. Initial sequencing and annotation of archaeal genomes did not identify any orthologs of IPP isomerase. The characterization of a new type of IPP isomerase (type II IPP isomerase) from *Streptomyces* sp. Strain CL190,[131] however, allowed for the discovery of the archaeal homolog. Type II IPP isomerases have been identified in cyanobacteria and a few gram-positive bacteria as well as the archaea.[132–134] Archaeal type II IPP isomerases from *Methanothermobacter thermautotrophicus* and *Sulfolobus shibatae* have been described.[134,135] Type II IPP isomerases require Mg^{2+} as well as H_2FMN for activity. The cofactor has been observed to oxidize in the

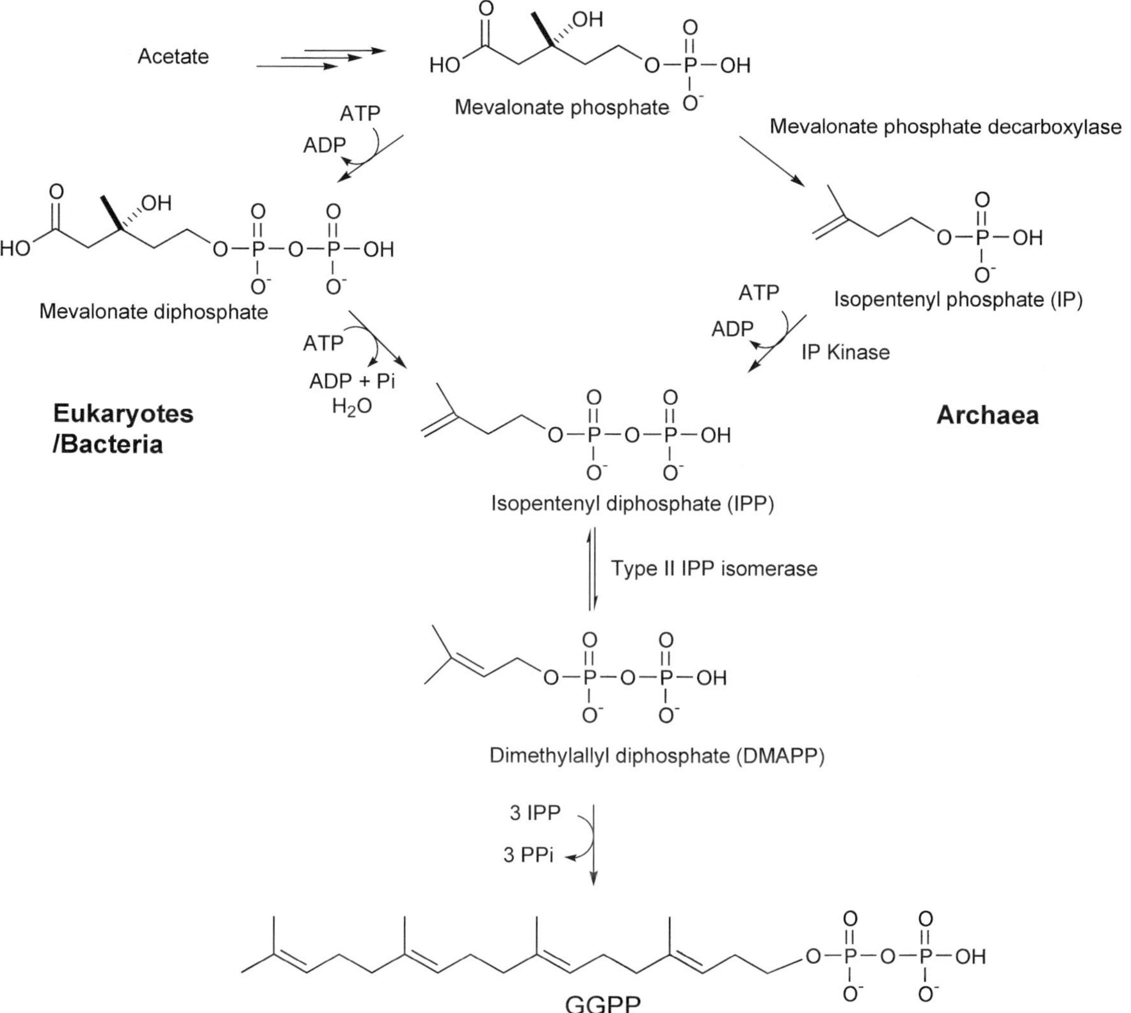

FIGURE 16. The mevalonate pathway in eukaryotes and bacteria (*left side of the figure*) and the proposed archaeal modifications (*right side of the figure*).

presence of oxygen, and the enzyme requires the addition of a reducing agent such as NADPH or dithionite to restore activity. The catalytic mechanism of type II IPP isomerase has not been elucidated; however, both protonation/deprotonation and radical mechanisms have been proposed.[136–138]

IPP and DMAPP are condensed to form prenyl chains by the activity of several prenyltransferases. The resulting geranylgeranyl diphosphate then serves as a precursor to the archaeal diether lipids. It is believed that saturation of the isoprenoid chains by geranylgeranyl reductase occurs after attachment of the glycerol moiety.[139] One of the most intriguing features of archael lipid membranes is the presence of cyclized and tetraether polar lipids (FIG. 15A). Cyclization and tetraether formation have been shown to reduce the fluidity and increase the stability of the lipid bilayer in the archaea. These adaptations are believed to be employed to adapt to the extreme environmental conditions inhabited by the archaea. A specific correlation between increased tetraether composition and tolerance toward increased temperature or acidic growth conditions has been observed.[140,141] Numerous studies have demonstrated the increased rigidity, stability, and lower hydrogen mobility of archaeal lipids.[142–144] These results are consistent with the importance of the cellular membrane in maintaining transmembrane proton and ionic gradients.

The formation of the cyclopentyl and tetraether lipids is of particular interest because they involve the formation of carbon-carbon bonds between unactivated methyl groups. This reaction is currently without precedent in known biochemical reactions. Much of the biochemistry of tetraether formation has yet to be established; however, it has been speculated that it proceeds through a radical mechanism from a saturated archaeol-containing lipid.[125]

The presence of tetraether lipids was thought to be an archaeal-specific trait; however, Weijers *et al.* recently reported the identification of a bacterial type 1,2-di-*O*-alkyl-*sn*-glycerol lipid from anaerobic peat-bog bacteria[145] (FIG. 15B). The bacterial lipids were also found to contain cyclopentyl moieties, another archaeal feature. This was not the first report of either the occurrence of ether-linked lipids or lipids containing cyclic moieties in bacteria. One of the most interesting groups of such lipids is the ladderanes (FIG. 15B) that have been isolated from anaerobic ammonia-oxidizing (anammox) bacteria.[146,147] The alkyl chains of ladderanes contain 3–5 linear concatenated cyclobutyl rings and both ester and ether linkages have been observed. Ladderanes are the primary lipid found in the membrane of a unique bacterial organelle, the anammoxosome, where anaerobic ammonia oxidation takes place. Like the archaeal tetraether lipids, the structural properties of ladderanes result in the formation of an extremely dense membrane. The anammoxosome membrane provides a diffusion barrier to both contain valuable anammox intermediates and protect the cell from DNA damage by the toxic intermediate hydrazine. The ladderane membrane also allows for the maintenance of a proton gradient that can be used to drive ATP synthesis.[148]

Archaeal isoprene biosynthesis is a mosaic of features found in bacterial and/or eukaryotic pathways, as well as several features that appear to be unique to archaea. Of particular interest to this chapter are the two steps in the pathway, IPP isomerization and tetraether formation, that may utilize radical mechanisms. In light of the proposed biosynthetic pathway involving radicals, it may be significant that tetraether, cyclopentyl, and ladderane lipids have only been found in anaerobes. The reactivity of dioxygen to radical intermediates would be expected to hinder the biosynthesis of the cyclic, tetraether, and cyclopentyl lipids in aerobes.[2]

Conclusions

The elucidation of metabolic pathways occurring in the methanogenic archaea *M. jannaschii* has yielded a wealth of information on new anaerobic metabolism. It is becoming increasingly apparent that many anaerobes utilize a mixture of canonical, modified and unique biosynthetic pathways. As the work outlined here demonstrates, the methanogenic archaea are truly remarkable organisms with many novel, unexpected, and unpredictable aspects to their metabolism. We eagerly anticipate more new and exciting discoveries from these organisms in the future.

Conflict of Interest

The authors declare no conflicts of interest.

References

1. IMLAY, J.A. 2006. Iron-sulphur clusters and the problem with oxygen. Mol. Microbiol. **59:** 1073–1082.
2. BUCKEL, W. & B.T. GOLDING. 2006. Radical enzymes in anaerobes. Annu. Rev. Microbiol. **60:** 27–49.
3. JOHNSON, D.C. *et al.* 2005. Structure, function, and formation of biological iron-sulfur clusters. Annu. Rev. Biochem. **74:** 247–281.
4. CAMBA, R. & F.A. ARMSTRONG. 2000. Investigations of the oxidative disassembly of Fe-S clusters in *Clostridium pasteurianum* 8Fe ferredoxin using pulsed-protein-film voltammetry. Biochemistry **39:** 10587–10598.
5. ROGERS, P.A. *et al.* 2003. Reversible inactivation of *E. coli* endonuclease III via modification of its [4Fe-4S] cluster by nitric oxide. DNA Repair (Amst). **2:** 809–817.
6. DJAMAN, O., F.W. OUTTEN & J.A. IMLAY. 2004. Repair of oxidized iron-sulfur clusters in *Escherichia coli*. J. Biol. Chem. **279:** 44590–44599.
7. OUTTEN, F.W., O. DJAMAN & G. STORZ. 2004. A suf operon requirement for Fe-S cluster assembly during iron starvation in *Escherichia coli*. Mol. Microbiol. **52:** 861–872.
8. RUBIO, L.M. & P.W. LUDDEN. 2005. Maturation of nitrogenase: a biochemical puzzle. J. Bacteriol. **187:** 405–414.
9. POSTGATE, J.R. 1998. Nitrogn Fixation. Cambridge: Cambridge University Press.
10. DOWNIE, J.A. 2005. Legume haemoglobins: symbiotic nitrogen fixation needs bloody nodules. Curr. Biol. **15:** R196–R198.
11. GOLDEN, J.W. & H.S. YOON. 2003. Heterocyst development in *Anabaena*. Curr. Opin. Microbiol. **6:** 557–563.
12. REES, D.C. & J.B. HOWARD. 2003. The interface between the biological and inorganic worlds: iron-sulfur metalloclusters. Science **300:** 929–931.
13. DRENNAN, C.L. & J.W. PETERS. 2003. Surprising cofactors in metalloenzymes. Curr. Opin. Struct. Biol. **13:** 220–226.
14. BRODA, E. & G.A. PESCHEK. 1983. Nitrogen fixation as evidence for the reducing nature of the early biosphere. Biosystems **16:** 1–8.
15. KILEY, P.J. & H. BEINERT. 1999. Oxygen sensing by the global regulator, FNR: the role of the iron-sulfur cluster. FEMS Microbiol. Rev. **22:** 341–352.
16. CRACK, J.C. *et al.* 2006. Detection of sulfide release from the oxygen-sensing [4Fe-4S] cluster of FNR. J. Biol. Chem. **281:** 18909–18913.
17. MANSOURI, A. & A.A. LURIE. 1993. Concise review: methemoglobinemia. Am. J. Hematol. **42:** 7–12.

18. IRETON, G.C. *et al.* 2002. The structure of *Escherichia coli* cytosine deaminase. J. Mol. Biol. **315:** 687–697.
19. PORTER, D.J. & E.A. AUSTIN. 1993. Cytosine deaminase. The roles of divalent metal ions in catalysis. J. Biol. Chem. **268:** 24005–24011.
20. RAJAGOPALAN, P.T., X.C. YU & D. PEI. 1997. Peptide deformylase: a new type of mononuclear iron protein. J. Am. Chem. Soc. **119:** 12418.
21. ZHU, J. *et al.* 2003. S-Ribosylhomocysteinase (LuxS) is a mononuclear iron protein. Biochemistry. **42:** 4717–4726.
22. D'SOUZA V.M. & R.C. HOLZ. 1999. The methionyl aminopeptidase from *Escherichia coli* can function as an iron(II) enzyme. Biochemistry **38:** 11079–11085.
23. SEFFERNICK, J.L. *et al.* 2002. Atrazine chlorohydrolase from Pseudomonas sp. strain ADP is a metalloenzyme. Biochemistry **41:** 14430–14437.
24. GROCHOWSKI, L.L. *et al.* 2007. Characterization of an Fe^{2+}-dependent archaeal specific GTP cyclohydrolase, MptA, from *Methanocaldococcus jannaschii*. Biochemistry **46:** 6658–6667.
25. FONTECAVE, M., E. MULLIEZ & D.T. LOGAN. 2002. Deoxyribonucleotide synthesis in anaerobic microorganisms: the class III ribonucleotide reductase. Prog. Nucleic Acid Res. Mol. Biol. **72:** 95–127.
26. NORDLUND, P. & P. REICHARD. 2006. Ribonucleotide reductases. Annu. Rev. Biochem. **75:** 681–706.
27. ALEXANDER, K. & I.G. YOUNG. 1978. Three hydroxylations incorporating molecular oxygen in the aerobic biosynthesis of ubiquinone in *Escherichia coli*. Biochemistry **17:** 4745–4750.
28. ALEXANDER, K. & I.G. YOUNG. 1978. Alternative hydroxylases for the aerobic and anaerobic biosynthesis of ubiquinone in *Escherichia coli*. Biochemistry **17:** 4750–1755.
29. DEPPENMEIER, U. 2002. The unique biochemistry of methanogenesis. Prog. Nucleic Acid Res. Mol. Biol. **71:** 223–283.
30. CHATTERJEE, A. *et al.* 2007. Biosynthesis of thiamin thiazole in eukaryotes: conversion of NAD to an advanced intermediate. J. Am. Chem. Soc. **129:** 2914–2922.
31. TAGA, M.E. *et al.* 2007. BluB cannibalizes flavin to form the lower ligand of vitamin B_{12}. Nature **446:** 449–453.
32. KELLOGG, R. *et al.* 1992. Structural analysis of a stereochemical modification of flavin adenine dinucleotide in alcohol oxidase from methylotrophic yeasts. Tetrahedron **48:** 4147–4162.
33. WALSH, C. 2006. Posttranslational Modification of Proteins Expanding Nature's Inventory. Englewood, CO: Roberts and Company Publishers.
34. LAWHORN, B.G., R.A. MEHL & T.P. BEGLEY. 2004. Biosynthesis of the thiamin pyrimidine: the reconstitution of a remarkable rearrangement reaction. Org. Biomol. Chem. **2:** 2538–2546.
35. BUCHANAN, J.M. & S.C. HARTMAN. 1959. Enzymatic reactions in the synthesis of purines. Adv. Enzymol. **21:** 199–261.
36. ZALKIN, H. & J.E. DIXON. 1992. De novo purine nucleotide biosynthesis. Prog. Nucleic Acid Res. Mol. Biol. **42:** 259–287.
37. RAYL, E.A., B.A. MOROSON & G.P. BEARDSLEY. 1996. The human *purH* gene product, 5-aminoimidazole-4-carboxamide ribonucleotide formyltransferase/IMP cyclohydrolase. Cloning, sequencing, expression, purification, kinetic analysis, and domain mapping. J. Biol. Chem. **271:** 2225–2233.
38. GREASLEY, S.E. *et al.* 2001. Crystal structure of a bifunctional transformylase and cyclohydrolase enzyme in purine biosynthesis. Nat. Struct. Biol. **8:** 402–406.
39. WHITE, R.H. 1997. Purine biosynthesis in the domain Archaea without folates or modified folates. J. Bacteriol. **179:** 3374–3377.
40. SELKOV, E. *et al.* 1997. A reconstruction of the metabolism of *Methanococcus jannaschii* from sequence data. Gene **197:** GC11–26.
41. KOONIN, E.V., R.L. TATUSOV & M.Y. GALPERIN. 1998. Beyond complete genomes: from sequence to structure and function. Curr. Opin. Struct. Biol. **8:** 355–363.
42. KOIKE, H., T. KAWASHIMA & M. SUZUKI. 1999. Enzymes identified using genomic DNA sequences suggests some atypical characteristics of *de novo* biosynthesis of purines in archaea. Proc. Jpn. Acad. **75, Ser.B:** 263–268.
43. GRAUPNER, M., H. XU & R.H. WHITE. 2002. New Class of IMP Cyclohydrolases in *Methanococcus jannaschii*. J. Bacteriol. **184:** 1471–1473.
44. OWNBY, K., H. XU & R. WHITE. 2005. A *Methanocaldococcus jannaschii* archeal signature gene encodes for a 5-formaminoimidazole-4-carboxamide-1-β-D-ribofuranosyl 5′-monophosphate synthetase: a new enzyme in purine biosynthesis. J. Biol. Chem. **280:** 10881–10887.
45. KANGA, Y.-N. *et al.* 2007. A novel function for the NTN hydrolase fold demonstrated by the structure of an archeal inosine monophosphate cyclohydrolase. Biochemistry **46:** 5050–5062.
46. LI, H. *et al.* 2003. The *Methanococcus jannaschii* dCTP deaminase is a bifunctional deaminase and diphosphatase. J. Biol. Chem. **278:** 11100–11106.
47. HUFFMAN, J.L. *et al.* 2003. Structural basis for recognition and catalysis by the bifunctional dCTP deaminase and dUTPase from *Methanococcus jannaschii*. J. Mol. Biol. **331:** 885–896.
48. AURORA, R. & G.D. ROSE. 1998. Seeking an ancient enzyme in *Methanococcus jannaschii* using ORF, a program based on predicted secondary structure comparisons. Proc. Natl. Acad. Sci. USA **95:** 2818–2823.
49. XU, H. *et al.* 1999. Identifying two ancient enzymes in Archaea using predicted secondary structure alignment. Nat. Struct. Biol. **6:** 750–754.
50. LEDUC, D. *et al.* 2004. Two distinct pathways for thymidylate (dTMP) synthesis in (hyper)thermophilic bacteria and Archaea. Biochem. Soc. Trans. **32:** 231–235.
51. HIGUCHI, S., T. KAWASHIMA & M. SUZUKI. 1999. Comparison of pathways for amino acid biosynthesis in archaebacteria using their genomic DNA sequences. Proc. Jpn. Acad. **75, Ser. B:** 241–245.
52. BORUP, B. & J.G. FERRY. 2000. *O*-Acetylserine sulfhydrylase from *Methanosarcina thermophila*. J. Bacteriol. **182:** 45–50.

53. Borup, B. & J.G. Ferry. 2000. Cysteine biosynthesis in the Archaea: *Methanosarcina thermophila* utilizes *O*-acetylserine sulfhydrylase. FEMS Microbiol. Lett. **189:** 205–210.

54. Kitabatake, M. *et al.* 2000. Cysteine biosynthesis pathway in the archaeon *Methanosarcina barkeri* encoded by acquired bacterial genes? J. Bacteriol. **182:** 143–145.

55. Mino, K. & K. Ishikawa. 2003. Characterization of a novel thermostable *O*-acetylserine sulfhydrylase from *Aeropyrum pernix* K1. J. Bacteriol. **185:** 2277–2284.

56. Mino, K. *et al.* 2003. Crystallization and preliminary X-ray diffraction analysis of *O*-acetylserine sulfhydrylase from *Aeropyrum pernix* K1. Acta Crystallogr. D Biol. Crystallogr. **59:** 338–340.

57. White, R.H. 2003. The biosynthesis of cysteine and homocysteine in *Methanococcus jannaschii*. Biochim. Biophys. Acta **1624:** 46–53.

58. Sauerwald, A. *et al.* 2005. RNA-dependent cysteine biosynthesis in archaea. Science **307:** 1969–1972.

59. Helgadottir, S. *et al.* 2007. Biosynthesis of phosphoserine in the *Methanococcales*. J. Bacteriol. **189:** 575–582.

60. Tchong, S.-I., H. Xu & R. White. 2004. Cysteine desulfidase: an [4Fe-4S] enzyme isolated from *Methanocaldococcus jannaschii* that catalyzes the breakdown of cysteine into pyruvate, ammonia and sulfide. Biochemistry **44:** 1659–1670.

61. Burns, K.E. *et al.* 2005. Reconstitution of a new cysteine biosynthetic pathway in *Mycobacterium tuberculosis*. J. Am. Chem. Soc. **127:** 11602–11603.

62. Wang, C. *et al.* 2001. Solution structure of ThiS and implications for the evolutionary roots of ubiquitin. Nat. Struct. Biol. **8:** 47–51.

63. Park, J.H. *et al.* 2003. Biosynthesis of the thiazole moiety of thiamin pyrophosphate (vitamin B1). Biochemistry **42:** 12430–12438.

64. Rudolph, M.J. *et al.* 2001. Crystal structure of molybdopterin synthase and its evolutionary relationship to ubiquitin activation. Nat. Struct. Biol. **8:** 42–46.

65. Daugherty, M. *et al.* 2001. Archaeal shikimate kinase, a new member of the GHMP-kinase family. J. Bacteriol. **183:** 292–300.

66. White, R.H. 2004. L-Aspartate semialdehyde and a 6-deoxy-5-ketohexose 1-phosphate are the precursors to the aromatic amino acids in *Methanocaldococcus jannaschii*. Biochemistry **43:** 7618–7627.

67. Porat, I. *et al.* 2006. Biochemical and genetic characterization of an early step in a novel pathway for the biosynthesis of aromatic amino acids and *p*-aminobenzoic acid in the archaeon *Methanococcus maripaludis*. Mol. Microbiol. **62:** 1117–1131.

68. White, R.H. & H. Xu. 2006. Methylglyoxal is an intermediate in the biosynthesis of 6-deoxy-5-ketofructose-1-phosphate: a precursor for aromatic amino acid biosynthesis in *Methanocaldococcus jannaschii*. Biochemistry **45:** 12366–12379.

69. Grochowski, L.L., H. Xu & R.H. White. 2006. Identification of lactaldehyde dehydrogenase in *Methanocaldococcus jannaschii* and its involvement in the production of lactate for F_{420} biosynthesis. J. Bacteriol. **188:** 2836–2844.

70. Robertson, D.E. *et al.* 1990. Occurrence of β-glutamate, a novel osmolyte, in marine methanogenic bacteria. Appl. Environ. Microbiol. **56:** 1504–1508.

71. Burdige, D. & C. Martens. 1984. Amino acid cycling in an organic-rich marine sediment. EOS Trans. Am. Geophys. Union. **65:** 960.

72. Henrichs, S. & R. Chel. 1985. Occurrence of β-aminoglutaric acid in marine bacteria. Appl. Environ. Microbiol. **50:** 543–545.

73. Henrichs, S. & J. Farrington. 1987. Early diagenesis of amino acids and organic matter in two coastal marine sediments. Geochim. Cosmochim. Acta **51:** 1–15.

74. Robertson, D.E. & M.F. Roberts. 1991. Organic osmolytes in methanogenic archaebacteria. Biofactors **3:** 1–9.

75. Roberts, M.F. 2000. Osmoadaptation and osmoregulation in archaea. Front. Biosci. **5:** D796–812.

76. Martin, D.D., R.A. Ciulla & M.F. Roberts. 1999. Osmoadaptation in archaea. Appl. Environ. Microbiol. **65:** 1815–1825.

77. Roberts, M.F. 2004. Osmoadaptation and osmoregulation in archaea: update 2004. Front. Biosci. **9:** 1999–2019.

78. Robertson, D.E., S. Lesage & M.F. Roberts. 1989. β-Aminoglutaric acid is a major soluble component of *Methanococcus thermolithotrophicus*. Biochim. Biophys. Acta **992:** 320–326.

79. Roberts, M.F., M.C. Lai & R.P. Gunsalus. 1992. Biosynthetic pathways of the osmolytes N^{ε}-acetyl-β-lysine, β-glutamine, and betaine in *Methanohalophilus* strain FDF1 suggested by nuclear magnetic resonance analyses. J. Bacteriol. **174:** 6688–6693.

80. Robertson, D.E., D. Noll & M.F. Roberts. 1992. Free amino acid dynamics in marine methanogens. β-Amino acids as compatible solutes. J. Biol. Chem. **267:** 14893–14901.

81. Fuchs, G. & E. Stupperich. 1978. Evidence for an incomplete reductive carboxylic acid cycle in *Methanobacterium thermoautotrophicum*. Arch. Microbiol. **118:** 121–125.

82. White, R.H. 1987. Biosynthesis of the 1,3,4,6-hexanetetracarboxylic acid subunit of methanofuran. Biochemistry **26:** 3163–3167.

83. White, R.H. 1989. A novel biosynthesis of medium chain length α-ketodicarboxylic acids in methanogenic archaebacteria. Arch. Biochem. Biophys. **270:** 691–697.

84. Danson, M.J. 1993. Central metabolism of the archaea. *In* The Biochemistry of the Archaea (Archaebacteria). M. Kates, D.J. Kushner & A.T. Matheson, Eds.: 1–24. Amsterdam, The Netherlands: Elsevier.

85. Leigh, J.A., K.L. Rinehart, Jr. & R.S. Wolfe. 1985. Methanofuran (carbon dioxide reduction factor), a formyl carrier in methane production from carbon dioxide in *Methanobacterium*. Biochemistry **24:** 995–999.

86. Gorkovenko, A., M.F. Roberts & R.H. White. 1994. Identification, biosynthesis, and function of 1,3,4,6-hexanetetracarboxylic acid in *Methanobacterium thermoautotrophicum* strain ΔH. Appl. Environ. Microbiol. **60:** 1249–1253.

87. Buckel, W. & H.A. Barker. 1974. Two pathways of glutamate fermentation by anaerobic bacteria. J. Bacteriol. **117:** 1248–1260.

88. HANS, M. *et al.* 1999. 2-Hydroxyglutaryl-CoA dehydratase from *Clostridium symbiosum*. Eur. J. Biochem. **265:** 404–414.
89. KIM, J. *et al.* 2004. Dehydration of (*R*)-2-hydroxyacyl-CoA to enoyl-CoA in the fermentation of α-amino acids by anaerobic bacteria. FEMS Microbiol. Rev. **28:** 455–468.
90. WHITE, R.H. 2001. Biosynthesis of the methanogenic cofactors. *In* Vitamins and Hormones, Vol. 61. T. Begley, Ed.: 299–337. New York: Academic Press.
91. GRAHAM, D.E. & R.H. WHITE. 2002. Elucidation of methanogenic coenzyme biosynthesis: from spectroscopy to genomics. Nat. Prod. Rep. **19:** 133–147.
92. MARTINS, B.M. *et al.* 2004. Structural basis for stereospecific NAD^+-dependent (*R*)-2-hydroxyglutarate dehydrogenase catalysis in from *Acidaminococcus fermentans*. FEBS J. **272:** 269–281.
93. GRANT, G.A. 1989. A new family of 2-hydroxyacid dehydrogenases. Biochem. Biophys. Res. Commun. **165:** 1371–1374.
94. BUNCH, P.K. *et al.* 1997. The ldhA gene encoding the fermentative lactate dehydrogenase of *Escherichia coli*. Microbiology **143**(Pt 1): 187–195.
95. LOCHER, K.P. *et al.* 2001. Crystal structure of the *Acidaminococcus fermentans* 2-hydroxyglutaryl-CoA dehydratase component A. J. Mol. Biol. **307:** 297–308.
96. RUZICKA, F.J. & P.A. FREY. 2007. Glutamate 2,3-aminomutase: a new member of the radical SAM superfamily of enzymes. Biochim. Biophys. Acta **1774:** 286–296.
97. SANCHEZ, L.B., M.Y. GALPERIN & M. MULLER. 2000. Acetyl-CoA synthetase from the amitochondriate eukaryote *Giardia lamblia* belongs to the newly recognized superfamily of acyl-CoA synthetases (nucleoside diphosphate-forming). J. Biol. Chem. **275:** 5794–5803.
98. LATIMER, M.T. & J.G. FERRY. 1993. Cloning, sequence analysis, and hyperexpression of the genes encoding phosphotransacetylase and acetate kinase from *Methanosarcina thermophila*. J. Bacteriol. **175:** 6822–6829.
99. GLASEMACHER, J. *et al.* 1997. Purification and properties of acetyl-CoA synthetase (ADP-forming), an archaeal enzyme of acetate formation and ATP synthesis, from the hyperthermophile *Pyrococcus furiosus*. Eur. J. Biochem. **244:** 561–567.
100. JACOB, U. *et al.* 1997. Glutaconate CoA-transferase from *Acidaminococcus fermentans*: the crystal structure reveals homology with other CoA-transferases. Structure **5:** 415–426.
101. EGGEN, R.I. *et al.* 1991. Cloning, sequence analysis, and functional expression of the acetyl coenzyme A synthetase gene from *Methanothrix soehngenii* in *Escherichia coli*. J. Bacteriol. **173:** 6383–6389.
102. JETTEN, M. *et al.* 1989. Isolation and characterization of acetyl-coenzyme A synthetase from *Methanothrix soehngenii*. J. Bacteriol. **171:** 5430–5435.
103. HEIDER, J. 2001. A new family of CoA-transferases. FEBS Lett. **509:** 345–349.
104. GRAUPNER, M., H. XU & R.H. WHITE. 2000. Identification of an archaeal 2-hydroxy acid dehydrogenase catalyzing reactions involved in coenzyme biosynthesis in methanoarchaea. J. Bacteriol. **182:** 3688–3692.
105. WHITE, R.H. 1990. Biosynthesis of methanopterin. Biochemistry **29:** 5397–5404.
106. WHITE, R.H. 1996. Biosynthesis of methanopterin. Biochemistry **35:** 3447–3456.
107. GRAHAM, D.E., H. XU & R.H. WHITE. 2002. *Methanococcus jannaschii* uses a pyruvoyl-dependent arginine decarboxylase in polyamine biosynthesis. J. Biol. Chem. **277:** 23500–23507.
108. TOLBERT, W.D. *et al.* 2003. Pyruvoyl-dependent arginine decarboxylase from *Methanococcus jannaschii*. Crystal structures of the self-cleaved and S53A proenzyme forms. Structure (Camb.). **11:** 285–294.
109. LU, Z.J. & G.D. MARKHAM. 2002. Enzymatic properties of S-adenosylmethionine synthetase from the archaeon *Methanococcus jannaschii*. J. Biol. Chem. **277:** 16624–16631.
110. KIM, A.D. *et al.* 2000. S-Adenosylmethionine decarboxylase from the archaeon *Methanococcus jannaschii*: identification of a novel family of pyruvoyl enzymes. J. Bacteriol. **182:** 6667–6672.
111. GRAUPNER, M. & R.H. WHITE. 2001. *Methanococcus jannaschii* generates L-proline by cyclization of L-ornithine. J. Bacteriol. **183:** 5203–5205.
112. MUTH, W.L. & R.N. COSTILOW. 1974. Ornithine cyclase (deaminating). III. Mechanism of the conversion of ornithine to proline. J. Biol. Chem. **249:** 7463–7467.
113. SIEBERS, B. & P. SCHONHEIT. 2005. Unusual pathways and enzymes of central carbohydrate metabolism in Archaea. Curr. Opin. Microbiol. **8:** 695–705.
114. VERHEES, C.H. *et al.* 2003. The unique features of glycolytic pathways in Archaea. Biochem. J. **375:** 231–246.
115. GRAHAM, D.E., H. XU & R.H. WHITE. 2002. A divergent archaeal member of the alkaline phosphatase binuclear metalloenzyme superfamily has phosphoglycerate mutase activity. FEBS Lett. **517:** 190–194.
116. SIEBERS, B. *et al.* 2001. Archaeal fructose-1,6-bisphosphate aldolases constitute a new family of archaeal type class I aldolase. J. Biol. Chem. **276:** 28710–28728.
117. SODERBERG, T. & R.C. ALVER. 2004. Transaldolase of *Methanocaldococcus jannaschii*. Archaea **1:** 255–262.
118. GROCHOWSKI, L.L., H. XU & R.H. WHITE. 2005. Ribose-5-phosphate biosynthesis in *Methanocaldococcus jannaschii* occurs in the absence of a pentose-phosphate pathway. J. Bacteriol. **187:** 7382–7389.
119. ORITA, I. *et al.* 2006. The ribulose monophosphate pathway substitutes for the missing pentose phosphate pathway in the archaeon *Thermococcus kodakaraensis*. J. Bacteriol. **188:** 4698–4704.
120. VORHOLT, J.A. *et al.* 2000. Novel formaldehyde-activating enzyme in *Methylobacterium extorquens* AM1 required for growth on methanol. J. Bacteriol. **182:** 6645–6650.
121. ORITA, I. *et al.* 2005. The archaeon *Pyrococcus horikoshii* possesses a bifunctional enzyme for formaldehyde fixation via the ribulose monophosphate pathway. J. Bacteriol. **187:** 3636–3642.
122. DECKER, P., H. SCHWEER & R. POLHMANN. 1982. Identification of formose sugars, presumable prebiotic metabolites, using capillary gas chromatography/gas chromatography-mass spectrometry of n-butoximine trifluoroacetates on OV-225. J. Chromatogr. **244:** 281–291.

123. De Rosa, M. *et al.* 1980. Structure of calditol, a new branched-chain nonitol, and of the derived tetraether lipids in thermoacidopile archaebacteria of the *Caldariella* group. Phytochemistry **19:** 249–254.
124. De Rosa, M. *et al.* 1977. Chemical structure of the ether lipids of thermophilic acidophilic bacteria of the *Caldariella* goup. Phytochemistry **16:** 1961–1965.
125. Koga, Y. & H. Morii. 2007. Biosynthesis of ether-type polar lipids in archaea and evolutionary considerations. Microbiol. Mol. Biol. Rev. **71:** 97–120.
126. Bischoff, K.M. & V.W. Rodwell. 1996. 3-Hydroxy-3-methylglutaryl-coenzyme A reductase from *Haloferax volcanii*: purification, characterization, and expression in *Escherichia coli*. J. Bacteriol. **178:** 19–23.
127. Huang, K.X., A.I. Scott & G.N. Bennett. 1999. Overexpression, purification, and characterization of the thermostable mevalonate kinase from *Methanococcus jannaschii*. Protein Exp. Purif. **17:** 33–40.
128. Bochar, D.A. *et al.* 1997. 3-Hydroxy-3-methylglutaryl coenzyme A reductase of *Sulfolobus solfataricus*: DNA sequence, phylogeny, expression in *Escherichia coli* of the hmgA gene, and purification and kinetic characterization of the gene product. J. Bacteriol. **179:** 3632–3638.
129. Smit, A. & A. Mushegian. 2000. Biosynthesis of isoprenoids via mevalonate in archaea: the lost pathway. Genome Res. **10:** 1468–1484.
130. Grochowski, L.L., H. Xu & R. White. 2006. *Methanocaldococcus jannaschii* uses a modified mevalonate pathway for the biosynthesis of isopentenyl diphosphate. J. Bacteriol. **188:** 3192–3198.
131. Kaneda, K. *et al.* 2001. An unusual isopentenyl diphosphate isomerase found in the mevalonate pathway gene cluster from *Streptomyces* sp. strain CL190. Proc. Natl. Acad. Sci. USA **98:** 932–937.
132. Siddiqui, M.A. *et al.* 2005. Enzymatic and structural characterization of type II isopentenyl diphosphate isomerase from hyperthermophilic archaeon *Thermococcus kodakaraensis*. Biochem. Biophys. Res. Commun. **331:** 1127–1136.
133. Laupitz, R. *et al.* 2004. Biochemical characterization of *Bacillus subtilis* type II isopentenyl diphosphate isomerase, and phylogenetic distribution of isoprenoid biosynthesis pathways. Eur. J. Biochem. **271:** 2658–2669.
134. Barkley, S.J., R.M. Cornish & C.D. Poulter. 2004. Identification of an archaeal type II isopentenyl diphosphate isomerase in *Methanothermobacter thermautotrophicus*. J. Bacteriol. **186:** 1811–1817.
135. Yamashita, S. *et al.* 2004. Type 2 isopentenyl diphosphate isomerase from a thermoacidophilic archaeon *Sulfolobus shibatae*. Eur. J. Biochem. **271:** 1087–1093.
136. Hoshino, T. *et al.* 2006. Inhibition of type 2 isopentenyl diphosphate isomerase from *Methanocaldococcus jannaschii* by a mechanism-based inhibitor of type 1 isopentenyl diphosphate isomerase. Bioorg. Med. Chem. **14:** 6555–6559.
137. Hemmi, H. *et al.* 2004. Catalytic mechanism of type 2 isopentenyl diphosphate:dimethylallyl diphosphate isomerase: verification of a redox role of the flavin cofactor in a reaction with no net redox change. Biochem. Biophys. Res. Commun. **322:** 905–910.
138. Rothman, S.C., T.R. Helm & C.D. Poulter. 2007. Kinetic and spectroscopic characterization of type II isopentenyl diphosphate isomerase from *Thermus thermophilus*: evidence for formation of substrate-induced flavin species. Biochemistry **46:** 5437–5445.
139. Murakami, M. *et al.* 2007. Geranylgeranyl reductase involved in the biosynthesis of archaeal membrane lipids in the hyperthermophilic archaeon *Archaeoglobus fulgidus*. FEBS J. **274:** 805–814.
140. Sprott, G.D., M. Meloche & J.C. Richards. 1991. Proportions of diether, macrocyclic diether, and tetraether lipids in *Methanococcus jannaschii* grown at different temperatures. J. Bacteriol. **173:** 3907–3910.
141. Macalady, J.L. *et al.* 2004. Tetraether-linked membrane monolayers in *Ferroplasma* spp: a key to survival in acid. Extremophiles **8:** 411–419.
142. Shinoda, K. *et al.* 2004. Comparative molecular dynamics study of ether- and ester-linked phospholipid bilayers. J. Chem. Phys. **121:** 9648–9654.
143. Bartucci, R. *et al.* 2005. Bipolar tetraether lipids: chain flexibility and membrane polarity gradients from spin-label electron spin resonance. Biochemistry **44:** 15017–15023.
144. van de Vossenberg, J.L., A.J. Driessen & W.N. Konings. 1998. The essence of being extremophilic: the role of the unique archaeal membrane lipids. Extremophiles **2:** 163–170.
145. Weijers, J.W. *et al.* 2006. Membrane lipids of mesophilic anaerobic bacteria thriving in peats have typical archaeal traits. Environ. Microbiol. **8:** 648–657.
146. Sinninghe Damste, J.S. *et al.* 2004. A mixed ladderane/n-alkyl glycerol diether membrane lipid in an anaerobic ammonium-oxidizing bacterium. Chem. Commun. **22:** 2590–2591.
147. Sinninghe Damste, J.S. *et al.* 2002. Linearly concatenated cyclobutane lipids form a dense bacterial membrane. Nature **419:** 708–712.
148. van Niftrik, L.A. *et al.* 2004. The anammoxosome: an intracytoplasmic compartment in anammox bacteria. FEMS Microbiol. Lett. **233:** 7–13.
149. De Rosa, M. *et al.* 1988. A new 15,16-dimethyl-30-glyceryloxytriacontanoic acid from lipids of *Termotoga maritima*. J. Chem. Soc., Chem. Commun. 1300–1301.
150. Jung, S., J.G. Zeikus & R.I. Hollingsworth. 1994. A new family of very long chain α, ω-dicarboxylic acids is a major structural fatty acyl component of the membrane lipids of *Thermoanaerobacter ethanolicus* 39E. J. Lipid Res. **35:** 1057–1065.

Tungsten, the Surprisingly Positively Acting Heavy Metal Element for Prokaryotes

JAN R. ANDREESEN AND KATHRIN MAKDESSI

Institute of Biology/Microbiology, Martin-Luther-University Halle-Wittenberg, Halle, Germany

The history and changing function of tungsten as the heaviest element in biological systems is given. It starts from an inhibitory element/anion, especially for the iron molybdenum-cofactor (FeMoCo)–containing enzyme nitrogenase involved in dinitrogen fixation, as well as for the many "metal binding pterin" (MPT)-, also known as tricyclic pyranopterin– containing classic molybdoenzymes, such as the sulfite oxidase and the xanthine dehydrogenase family of enzymes. They are generally involved in the transformation of a variety of carbon-, nitrogen- and sulfur-containing compounds. But tungstate can serve as a potential positively acting element for some enzymes of the dimethyl sulfoxide (DMSO) reductase family, especially for CO_2-reducing formate dehydrogenases (FDHs), formylmethanofuran dehydrogenases and acetylene hydratase (catalyzing only an addition of water, but no redox reaction). Tungsten even becomes an essential element for nearly all enzymes of the aldehyde oxidoreductase (AOR) family. Due to the close chemical and physical similarities between molybdate and tungstate, the latter was thought to be only unselectively cotransported or cometabolized with other tetrahedral anions, such as molybdate and also sulfate. However, it has now become clear that it can also be very selectively transported compared to molybdate into some prokaryotic cells by two very selective ABC-type of transporters that contain a binding protein TupA or WtpA. Both proteins exhibit an extremely high affinity for tungstate (K_D < 1 nM) and can even discriminate between tungstate and molybdate. By that process, tungsten finally becomes selectively incorporated into the few enzymes noted above.

Key words: **tungstate; molybdate; tungstoproteins; molybdoenzymes; molbindin**

Tungsten (W) (Swedish: "tung + sten" = heavy stone or W = German "Wolfram" or wolflike nature of wolframite) is in the sixth group within the transition elements in the periodic system like chromium and molybdenum. No element of the third row of transition elements (in the sixth period) is known to exert besides a negative or neutral even a physiological positive role, to always become bound via a unique cofactor to some enzymes. Normally, the biologically important ions of the transition elements, such as vanadium, manganese, iron, nickel, copper, and zinc, are found only in the fourth period in the first row, whereas chromium seems to play only a minor and uncertain role within the sixth group, in contrast to molybdenum. Molybdenum (Mo) is recognized as an essential constituent of a steadily increasing number of enzymes for more than 50 different proteins.[1] Thus, tungsten is the heaviest element with a biological function.[2,3] Molybdenum (Mo) and tungsten (W) are nowadays mostly present in the form of their tetrahedral oxyanions, molybdate and tungstate. The close chemical and physical similarities of tungstate and molybdate are based on equal atomic and ionic radii, similar electronegativity, and coordination characteristics[4–8] which lead to a lack of discrimination by biological systems and, thus, to the toxicity of tungstate. These characteristics were often used in former times to reveal the nature of a given protein as a molybdoenzyme by differentiating their effect on certain enzymes after growth in the presence of a high concentration of molybdate or tungstate. Both oxyanions display, at low pH values, the peculiarity of forming very complex iso- or heteropolyanions,[9,10] which seem to possess no major biological function.

Generally, the content of molybdenum is somewhat higher in natural abundance than that of tungsten, being the 54th and 55th elements when ranking them according to their general abundance on earth. Natural concentration values can vary in both directions up to an order of magnitude. Thus, special habitats, such as hot black smokers or sulfide-enriched waters, contain higher concentrations of tungsten. Whereas sulfides of

Address for correspondence: Jan R. Andreesen, Kurt Mothes Strasse 3, 06120 Halle, Germany.
jan.andreesen@mikrobiologie.uni-halle.de

Ann. N.Y. Acad. Sci. 1125: 215–229 (2008). © 2008 New York Academy of Sciences.
doi: 10.1196/annals.1419.003

Mo are generally very insoluble, sulfides of W are more soluble, especially in their form as thiotungstates.[7,11] The molybdenum isotopes give evidence of the lack of oxygen in the depths of mid-proterozoic oceans (at about 1.8 to 1 billion years ago), although oxygen was already present in the atmosphere due to oxygenic photosynthesis.[12] The final oxidation of these reduced sulfidic Mo-forms to molybdate led to an enormous increase in the availability of molybdenum, thus promoting the formation of Mo-containing enzymes. In light of this hypothesis, W-containing enzymes are considered to be more ancient than Mo-containing enzymes. Today, tungsten is commonly known as the essential component of conventional light bulbs and of special high-quality steel.

The first report about a positive biological function of tungsten besides molybdenum was given on nitrogen fixation by Bortels in Germany in 1930.[13] However, this report was modified by him, for molybdenum is the best element for biological nitrogen fixation, for example, by some *Azotobacter* species, whereas vanadium (V) is (still) second, and, at that time, tungsten acted only positively under certain conditions, such as suboptimal Mo and V provision.[14] Later, in the early 1950s, tungstate became known as a selective inhibitor for molybdoenzymes, starting with fungal nitrate reductase,[15] bacterial formate dehydrogenase (FDH),[16] chicken xanthine oxidase and aldehyde oxidase,[17] and rat sulfite oxidase.[18]

The label of tungsten as an antagonist of molyboenzymes proved to be so strong that the first review on experiments showing positive actions of tungstate on certain formate dehydrogenases[19] was not well accepted by the scientific community. However, this changed around the 1990s, after the isolation of some hyperthermophilic archaea growing at very high temperatures (i.e., 80°C or even up to 113°C). Many of these organisms require the addition of tungstate for growth due to its involvement in the formation of essential enzymes, such as aldehyde oxidoreductase and formylmethanofuran dehydrogenase.[20,21] Since that time, it has become accepted that tungsten can also be a positively acting element for certain enzymes, even being essential for some prokaryotes.[22,23] A book dealing with many different aspects of these two elements has recently been compiled to compare the different aspects and roles of molybdenum and tungsten in biological systems.[24]

In 1996, the first comprehensive reviews appeared on tungsten's involvement in biological systems, about tungstoenzymes and its relation to mononuclear molybdenum enzymes, respectively.[25–29] So far, molybdenum or tungsten is never directly bound by a protein, but always via a cofactor, called FeMoCo (in the only known case of nitrogenase) or MPT. The molybdopterin-containing cofactor MPT is now called "*m*etal binding *pt*erin" to emphasize the importance of this common pterin-cofactor in binding tungsten as well as of molybdenum (FIG. 1). These enzymes attracted many research groups due to their higher stability as compared to nitrogenase.[1,3,25–30] However, interest in the various Fe-Mo-cofactor–containing nitrogenases is still high due to the large economic impact of dinitrogen fixation to ammonia and the biochemical puzzle of its maturation,[31] despite its generally observed extreme lability as an isolated enzyme. Few bacteria, such as *Azotobacter vinelandii*, are actually able to form three genetically distinct and active nitrogenases, depending on the availability of molybdenum (*nif*), vanadium (*vnf*), or iron-only (*anf*).[32] The tungsten-substituted form is inactive in reducing dinitrogen.[33] The transport of molybdate into the cells is the topic of some reviews.[34–37] Most recently a review appeared that deals with a variety of aspects of molybdate and tungstate metabolism, starting from their uptake, and includes the proposed final steps of molybdenum insertion into the adenylated MPT to form the final active cofactor.[38]

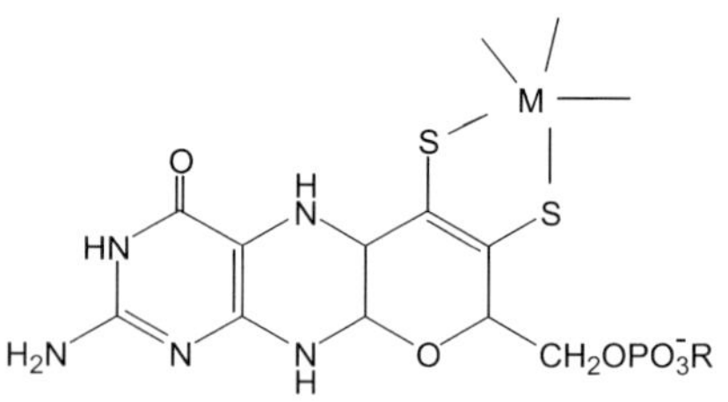

M = Mo or W; R = H or CMP or GMP

FIGURE 1. Structure of the molybdopterin cofactor (MoCo).

Vanadium is chemically related to molybdenum and tungsten by a diagonal relationship ("Schrägbeziehung") within the periodic system of the elements. Thus, the so-called "molybdenum-free" nitrate reductase of the gammaproteobacterium *Thioalkalivibrio nitratireducens* contains vanadium (besides heme c) instead of molybdenum and directly forms ammonium.[39] The reported tungstate requirement for nitrate reduction by *Pyrobaculum aerophilum*[40] is not due to an unusual nitrate reductase, but to the special growth conditions of that archaeon, preferably using peptides as the carbon source that require four different W-containing aldehyde oxidoreductases (AORs) and an unusual 7Fe ferredoxin of very low redox potential.[41]

Discovery of the First Tungstoenzyme: Formate Dehydrogenase of *Moorella thermoacetica* (*Clostridium thermoaceticum*)

The first experiments showing a definite positive action of tungstate were carried out about 1971 in the laboratory of Lars G. Ljungdahl at the University of Georgia in Athens, Georgia. Instead of obtaining a strong inhibitory effect of tungstate, as Jane Pinsent had found in 1954 for the FDH of *Escherichia coli*,[16] we could reproducibly show that tungsten, instead of molybdenum, is a positively acting element besides selenium for the nicotinamide adenine dinucleotide phosphate (NADP)–dependent FDH of *Moorella thermoacetica*, even by growing the organism in the very complex medium used in those days.[42,43] These results stood in sharp contrast to studies using *E. coli* that showed a strict requirement for molybdate and selenite to express an active FDH if the organism is grown in a defined medium, whereas addition of tungstate is a strong antagonist for FDH activity.[16,44] Until that time, tungstate was supposed to act only as an inhibitor for molybdoenzymes, such as nitrate reductase, xanthine dehydrogenase, aldehyde oxidase, and sulfite oxidase.[4,45] Thus, only a few people became convinced that tungstate even acts positively on a biological system. Of course, tungsten is, even more than molybdenum, in an eccentric position within the periodic table of the chemical elements, being close to the ions of the heaviest toxic heavy metals (mercury, thallium, lead, and bismuth) and all radioactive elements.

Starting with the work on the FDH of *M. thermoacetica*, the specific activity of the NADP-dependent enzyme was considered to be too low to fulfill its crucial function as the first enzyme involved in the "Wood–Ljungdahl pathway" of CO_2 reduction via formate to acetate.[46] Its NADP-dependent activity could just be increased by purification to 0.018 U/mg of protein [one unit (U) is defined as 1 μmol/min],[47] but the starting activity was already increased to 0.086 U/mg of protein using a high concentration of molybdate (0.5 mM) and selenite (1 μM) in the growth medium.[48] Purification starts from 0.455 U/mg of protein if tungstate (36 μM) is present during growth.[49] Using the β-emitter ^{185}W tungstate as a label and anaerobic conditions throughout the purification procedure, the FDH activity and radioactivity finally coelute in one single band at a fixed ratio. The further addition of different concentrations of molybdate (0.1 μM to 1 mM) at a constant concentration of ^{185}W tungstate (10 μM) does not shift the ratio: counts per minute (cpm) of ^{185}W to FDH activity, indicating a selective incorporation of tungsten into the active FDH protein.[50] Using more rigorous anaerobic conditions, the final specific activity increases up to 37 U/mg[51] and later even to 170 U/mg of protein,[52] again showing a strict coelution of FDH activity and ^{185}W radioactivity. This final preparation exhibits four protein bands as well as one major and one minor FDH activity band, whereas ^{185}W coelutes only with the major band. The chemical analysis indicates that the amount of W is about twice that of Mo (0.3 and 0.16 g-atom per mol, respectively), whereas the selenium content is higher (0.6 g-atom). Iron and labile sulfur are detected in high amounts (70 and 36 g-atom per 300 kDa, respectively). The final proof that the FDH of *M. thermoacetica* is indeed a tungsten-selenocysteine-containing iron-sulfur protein was again accomplished in the laboratory of Lars Ljungdahl using rigorous anaerobic conditions in a glove box and fast-acting purification methods.[53] The final enzyme preparation now exhibits 1050 U/mg of protein in the NADP-dependent reaction (thus being even five orders of magnitude higher than first reported by Li *et al.*[47]). SDS PAGE reveals two bands of 96 and 76 kDa, corresponding to the alpha subunit of 893 and the beta subunit of 707 amino acids according to gene annotations, respectively. Whereas the larger subunit contains the selenocysteine codon (TGA) and Fe/S-clusters, the β-subunit represents the NADP binding site, according to sequence annotations done later. The genome reveals a second FDH of 722 amino acids and a gamma subunit close to the alpha subunit. The gamma subunit encodes 226 amino acids and is annotated as the cytochrome b subunit of FDH. *M. thermoacetica* actually contains cytochrome b,[54] which is absent in nitrate-grown cells because nitrate represses key enzymes of the Wood–Ljungdahl pathway.[55] The annotated *fdhD* gene of *M. thermoacetica* encodes 260 amino acids, and the derived amino acid sequence is quite similar to FdhD of *Eubacterium acidaminophilum*, probably representing an FDH-specific accessory protein/chaperone.[56] Both proteins, although genetically associated with an FDH, do not copurify with the respective two subunits of the NADP-dependent enzyme. No indications of the presence of cytochrome b or flavins in highly purified preparations of FDH are obtained.[53] The specific activity of the FDH in different preparations is clearly correlated to the increased content of tungsten and selenium, but it becomes lower if the molybdenum content is increased. Thus, a completely purified active enzyme should contain (per 340 kDa heterotetramer) 2 g-atom of tungsten and 2 g-atom of selenium, 36 g-atom of Fe, and as much as 50 mol of inorganic sulfur, but no molybdenum. The form A of the MPT can be detected by fluorescence spectroscopy. Thus, tungsten will probably be present in the bis-form of the MPT dinucleotide form, as generally observed for an FDH of the DMSO family. No reduced NADP-flavin (NADPH) oxidoreductase activity is detected in this highly purified preparation, whereas the enzyme is able to reduce bicarbonate or CO_2 to formate by NADPH, a thermodynamically very unfavorable reaction, also shown to occur in crude lysates.[57,58]

The lower redox potential of W^{4+} compared to Mo^{4+}[4,59] may be one reason for the preference of using tungsten instead of molybdenum in this type of FDH to catalyze a bidirectional reaction, for the reduction of CO_2 to formate should become facilitated by the readiness of tungsten to be oxidized to the +6 form and by the large number of Fe/S centers, backing up the delivery of electrons for reducing CO_2.

Identification of Tungsten in Other Similar Enzymes: Bidirectionally Acting Formate Dehydrogenases in Other Acetogenic Mesophilic Bacteria

Clostridium formicoaceticum was isolated in an attempt to retrieve the autotrophic acetogenic anaerobe *Clostridium aceticum*, which was thought to be lost.[60,61] However, *C. formicoaceticum* is generally a heterotrophic acetogen, forming or consuming formate, depending on the growth conditions, and requiring just biotin as a vitamin. Therefore, this organism was the first choice to examine if tungstate, not molybdate, along with selenite is stimulatory for the viologen-dependent FDH activity after growth in their presence. The results clearly indicate the superiority of tungstate compared to molybdate, as it forms a more active enzyme at lower tungstate concentrations in spite of using a complex or a defined medium.[62,63] The optimal concentration of tungstate to obtain the highest FDH activity in defined media is at least lower by orders of magnitude. Using ^{185}W tungstate as a marker indicates about a 100-fold concentration factor by cells of *C. formicoaceticum*. ^{185}W is incorporated into at least two protein fractions, eluting at about 250 kDa (FDH) and 88 kDa (probably the AOR, also called carboxylate reductase[64]), whereas the small 5.5-kDa peak is only occasionally observed.[63] Molybdate is without any significant influence, so tungsten is the active ingredient. The influence of the composition of the medium seems to reflect the different high- or low-affinity uptake systems induced or repressed after growth on defined or complex medium components, respectively.[36,37] Therefore, different concentrations of these metal ions are actually required to obtain, for example, optimal FDH activity. *Acetobacterium woodii* is a known autotrophic acetogen that grows readily on hydrogen and carbon dioxide.[65] Its FDH is likewise affected by these metal ions in such a way that it expresses an active enzyme.[66]

The studies on the FDHs of the two acetogenic bacteria, *M. thermoacetica* and *C. formicoaceticum*, were the basis for the first review on tungsten as a biologically active metal,[19] later extended after the additional positive role of selenium became obvious for other enzymes beside the FDH,[45] such as for the classic molybdoenzyme xanthine dehydrogenase (XDH) of certain anaerobic acetogenic purinolytic bacteria, such as *C. acidiurici* and *Clostridium cylindrosporum*, as well as for the XDH and the similar nicotinic acid dehydrogenase (NDH) of the acetogen *Eubacterium* (*Clostridium*) *barkeri*. In contrast to the FDH, where selenium is generally part of the selenocysteine moiety, selenium in XDH and NDH is not present as selenocysteine, but is more loosely bound and cyanolysable as part of the MPT-Mo complex.[67–74] Purinolytic anaerobes are often strictly dependent on purines as growth substrates that are interconverted into xanthine by XDH, a classic molybdoenzyme.[75–78] Therefore, these bacteria should be used to demonstrate a classic conflict: requiring selenite for both FDH and XDH, but molybdate for XDH and tungstate for FDH activity. The first experiments indicated a differentiation between the two purinolytic species: *C. acidiurici* requires tungstate and selenite for FDH, whereas *C. cylindrosporum* seemed to prefer molybdate and selenite.[72] The newly isolated species *Clostridium purinolyticum* is also able to decompose adenine by first deaminating it to hypoxanthine before adenine becomes hydroxylated to insoluble derivatives.[79,80] However, the differentiation and assignment of 14 strains to one of these three species was only possible by DNA/DNA hybridization, but not by phenotypic means or by a specific requirement of FDH for tungstate or molybdate along with selenite.[81] In further studies, it became evident that both classic purinolytic clostridia are able to take up ^{185}W-tungstate very efficiently, even in the presence of a high molar surplus of molybdate. Surprisingly, *C. acidiurici* seems to be somewhat less well adapted compared to *C. cylindrosporum* to take up tungstate when a high molar excess of molybdate is also provided. Thus, the observed difference in a molybdate or tungstate requirement might be due to different efficient transport systems, which causes *C. acidiurici* to prefer the presence of tungstate, for it might not discriminate as sharply as *C. cylindrosporum* does.[82] However, in both organisms the FDH activity comigrates with the ^{185}W radioactivity, whereas ^{185}W always forms a sharp trough at the position where the XDH activity elutes, indicating a clear discrimination between tungsten and molybdenum, at least in the final stage of incorporation of W or Mo into the MPT-containing enzymes, FDH or XDH. The second specifically ^{185}W-labeled protein peak should correspond to a protein like AOR.

Another newly isolated, obligately purine-fermenting organism, *Eubacterium angustum*, does not require the addition of selenite, molybdate, or tungstate, although it is able to grow very fast in a completely synthetic medium with uric acid, xanthine (t_d 45 min on both substrates), or guanine as the sole substrate, requiring only the addition of thiamine.[83] However, both XDH and FDH activities increased two- to fivefold by the addition of these ions, depending on the growth substrate used. This behavior indicates the presence of a highly effective transport system for these ions, although in this study strict precautions were met to avoid a contamination, for example, by an earlier use of the glass containers.

Peptostreptococcus barnesae is a new anaerobic isolate from chicken feces that requires selenite, molybdate, and tungstate for growth on uric acid, xanthine, 6,8-dihydroxypurine, guanine, hypoxanthine, and to some degree on adenine and glycine.[84] Bile acids and some yeast extract improve growth, probably simulating natural conditions. Many strains of the former genus *Peptostreptococcus* available in the mid 1980s are able to utilize mucins, arginine, serine, glycine, and some uric acid.[85] Especially, glycine metabolism requires selenite to produce the more effective glycine reductase; however, the serine/glycine interconversion circumvents a selenite requirement if its availability is too low.[86]

All available strains of the above mentioned purinolytic clostridia are in principle able to utilize glycine, the general intermediate of the degradation of the purine xanthine[78] if selenite and a low concentration of a preferably non-readily utilizable purine compound are provided.[80,87] The latter fact is quite important and can be rationalized because these organisms are generally specialized on the degradation of purines. Therefore, they are hardly able to synthesize purines *de novo*. Thus, these bacteria have to recognize the purine in such a case not as a substrate, but as a "vitamin" for nucleotide synthesis like thiamine.

Due to greatly increased prizes for purines and their low solubility, a new isolate, *E. acidaminophilum* strain al-2, obtained from Fritz Widdel, became the model organism for studying selenium and tungsten metabolism.[88] The important reactions involving tungsten are given in FIGURE 2. The organism belongs to

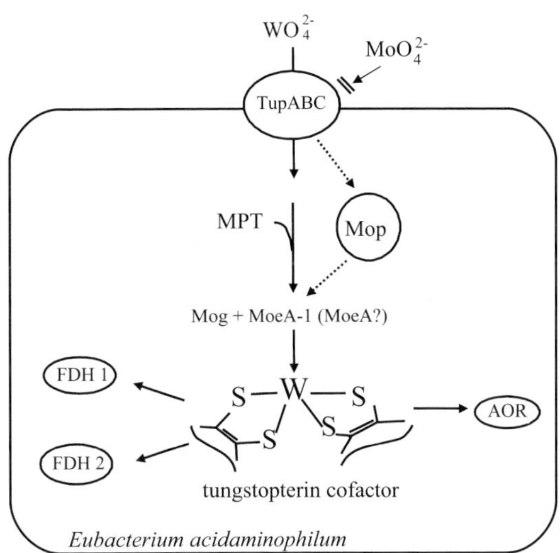

FIGURE 2. Scheme of a supposed flux of tungstate into the metal-binding pterin (MPT) of tungstoproteins from *E. acidaminophilum*.

cluster XI of the clostridia, and was originally isolated to just oxidize different amino acids and transfer the electrons via an interspecies hydrogen or formate transfer to a second partner, an acetogenic, methanogenic, or sulfate-reducing organism, thus representing a Stickland reaction divided up between two prokaryotic partners. However, this separation did not work out in the case of glycine as substrate if selenite is also present, for *E. acidaminophilum* now grows without an additional partner, using glycine as both electron donor and electron acceptor.[88–90] Although selenite has to be added in large amounts in order to observe optimal growth (1 μM; to provide selenocysteine, especially for the proteins GrdA and GrdB of glycine reductase, the peroxiredoxin PrxU, and the electron-transferring protein PrpU[90]), no requirement for tungstate, molybdate, or vanadate could be detected after numerous trials. An active FDH as well as a CO_2 reduction to formate can be observed in crude extracts.[88] Oxygen-sensitive, viologen-dependent FDH activity was purified to a certain extent using formate/betaine-grown cells, to obtain an amino acid sequence from the N terminus of the 95-kDa subunit and the knowledge that tungsten plus a pterin cofactor and an Fe/S-cluster, rather than molybdenum, are part of that protein fraction.[91,92] Although the enzyme activity of the FDH was mostly inactivated, the benzyl viologen-dependent activity was still 890 U/mg of protein, the methyl viologen activity was 540 U/mg, and the CO_2/bicarbonate reductase activity was about 24% of the latter, also indicating a high potential for the catalysis of this generally

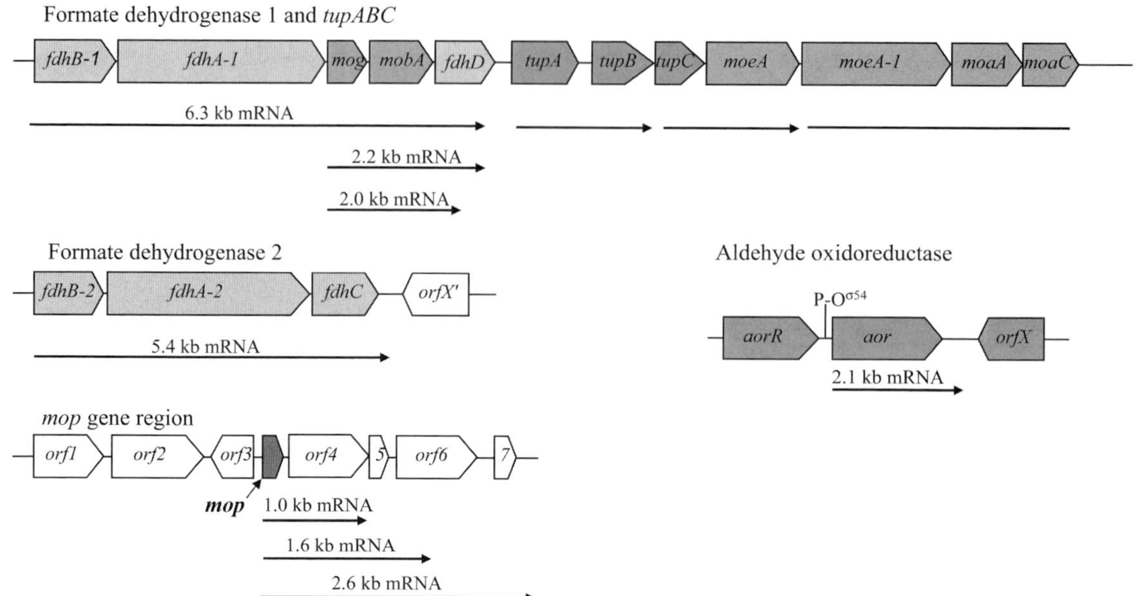

FIGURE 3. Genetic organization of tungsten/molybdenum related genes of *E. acidaminophilum*. Sequenced clusters of known and hypothetical genes for tungstoenzymes and proteins involved in cofactor synthesis, tungstate transport, and homeostasis are shown. Putative mRNA transcripts were identified using RT-PCR, Northern blotting, and primer extension experiments, and are indicated by *arrows*.

unfavored reaction. Attempts to purify the FDH by labeling it with ^{185}W indicate that the FDH fraction of *E. acidaminophilum,* eluting at a molecular mass of 157,000, contains only 0.4% of the incorporated radioactivity, whereas a second protein of about 74 kDa (the AOR, according to later studies) became strongly labeled. The presence of a third labeled protein of about 40 kDa (the Mop protein) depends on the concentration of tungstate and/or the presence of molybdate, in contrast to the two first mentioned proteins. Thus, only these first two proteins are genuine tungstoproteins. The incorporation of ^{185}W tungsten into the cells and these two protein fractions are already saturated at about 10^{-8} M tungstate, added to the growth medium and independent of the absence or presence of even a high molar surplus of molybdate. This indicates the presence of a highly selective transport system for tungstate with an unusually low affinity constant,[91,93] which explains the observed inability to obtain the influence of tungstate on growth.[88]

The highly selective tungstate uptake (Tup) system became known after sequencing the DNA in the area of the FDH 1 (FIG. 3) was obtained. By sequencing the genome in both directions, genes upstream of *fdhA1* encode the smaller FDH subunit FdhB1 and an iron-only hydrogenase I, which constitutes four genes (*hymABCD*), of which the derived HymD might be an integral membrane protein containing six transmembrane helices. The latter is missing in the second related cluster of genes encoding a second Fe-only hydrogenase Hys at a different location.[56] The organization of the redox-active components of this first iron-only hydrogenase Hym is quite similar to the four subunits of the NADH-dependent hydrogenase of *Thermoanaerobacter tengcongensis*.[94] Downstream of the genes for FDH 1 are *mogA* and *mobA*, both of which are involved in MPT biosynthesis, as well as *fdhD*, an accessory protein related to *fdhD* of *M. thermoacetica*. Further downstream are the genes of the highly specific tungstate *tupABC* transporter, an interesting topic at the end of this article, followed by two different *moeA* genes, *moaA* and *moaC*, which also are involved in the MPT-cofactor synthesis.[93] A second gene cluster (*fdhB2,A2,C*) encoding a different FDH 2 is present in the genome of *E. acidaminophilum*, exhibiting 61 and 68% identity to the respective subunits of FDH 1. This FDH 2 can be purified from serine/formate-grown cells, and the two subunits copurify with two subunits of the iron-only hydrogenase HymA and HymB, which is present in the cluster of FDH 1.[56] The benzyl viologen-dependent activity of FDH 2 is about 800 U/mg of protein, and the CO_2/bicarbonate reduction rate is about 27% of the latter FDH activity, which is quite similar to that of FDH 1. Tungsten, selenium, and iron, but no molybdenum, are detected by neutron activation analysis. Thus, both enzymes seem to be quite similar, except that a potential formate transporter, FdhC, is cotranscribed with FDH 2. As a pecularity, both FDHs of *E. acidaminophilum* contain in a large catalytically active 98-kDa subunit five potential Fe/S-clusters at the N-terminal region, as in *M. thermoacetica* and *Ralstonia eutropha* FdsA, plus the selenocysteine-decoding motif and the two MPT-binding motifs. Surprisingly, in both cases the smaller FDH subunits correspond to the N-terminal region observed for the FDH of the large subunits, also containing five instead of four Fe/S-clusters. By this special arrangement of electron-transferring clusters, a capacitor function is conceivable to store many electrons, allowing a more efficient reduction of the W^{+VI} to the unfavored W^{+IV} state, thus facilitating a CO_2 reduction to formate.

Quite similar results are obtained for the two W-containing FDH of the deltaproteobacterium *Syntrophobacter fumaroxidans*. Both enzymes catalyze also a strong CO_2-reductase activity and contain tungsten and selenium as well as Fe/S-clusters.[95] The crystal structure of the tungsten-containing FDH from *Desulfovibrio gigas* shows a structural homology with a large subunit of the molybdenum-containing FDH-H of *E. coli* in holding one 4Fe-4S center and selenocysteine, and only functions in the oxidation of formate.[96] The alphaproteobacterium *Methylobacterium extorquens* AM1 has a tungsten-containing NAD-dependent FDH that also exhibits tungsten in the catalytically active alpha subunit, but no selenocysteine. Nevertheless, it displays its highest sequence identities to the corresponding subunit of *E. acidaminophilum* and *M. thermoacetica*.[97] The betaproteobacterium *R. eutropha* contains many FDHs.[98] Besides the well-characterized NAD-dependent soluble FDH,[99] the organism also contains a membrane-bound, tungsten-containing FDH.[100] A comparison of structural, functional, and spectroscopic properties of FDH has been compiled by Moura *et al.*[101]

Aldehyde Oxidoreductase Is the Second Tungstoenzyme and Subordinate to Obtain Tungsten in *E. acidaminophilum*

As emphasized before, the AOR constitutes a family of typical tungstoenzymes in contrast to the FDHs belonging to the DMSO family. However, the superiority of providing tungsten for the two FDH enzymes of *E. acidaminophilum* is indicated by the complete lack of AOR activity after the prolonged depletion of the organism with tungstate, whereas the viologen-dependent FDH activity is still present at about 18% of its maximal value.[93] The addition of 10^{-5} M molybdate induces an AOR activity corresponding to that obtained after growth in the presence of 10^{-9} M tungstate. By this fact, the natural contamination of tungstate in molybdate (p.a.) can be estimated to be about 1 in 10,000. The activity of AOR is the highest in extracts of serine-formate–grown cells, as observed for the FDH, too. The AOR is purified as a 67-kDa protein containing 6 g-atom of iron, 1.1 g-atom of tungsten, but no molybdenum, being a monomer[102] in contrast to the dimeric nature of most other AORs, for example, of *Pyrococcus furiosus* that contains five different AORs.[25,27,38,103–105] The physiological function of the more general AOR, which exhibits a broad substrate spectrum in contrast to the other four known AORs, is often uncertain. However, it might be related to the variety of amino acids utilized by *E. acidaminophilum*.[88] It has been reported that a 2-oxoacid ferredoxin oxidoreductase might act—under conditions where there is a shortage of an electron acceptor such as ferredoxin—as a CoA-dependent decarboxylase,[106] thereby producing a corresponding aldehyde plus CO_2. However, Fe/S proteins generally act as natural electron acceptors of AORs. Thus, the situation might not become improved by this side reaction. The AOR of *E. acidaminophilum* utilizes a wide variety of aldehydes and this pattern does not change during purification or after growth on different substrates, indicating the presence of just one AOR in the genome.[102] Formaldehyde and even formate are also oxidized. A reverse reaction, however, the reduction of a nonactivated free carboxylic acid, is not observed, whereas this unusual property is reported for the AORs of *C. formicoaceticum* and *M. thermoacetica*. The former organism contains two AORs, one containing molybdenum, the other tungsten. Both enzymes differ in their sensitivity toward oxygen and cyanide.[107,108] The reduction of a variety of nonactivated carboxylic acids requires a lower pH value, as well as a very negative electron carrier such as tetramethyl viologen ($E'_0 = -550$ mV) reduced by CO. The tungsten-containing enzyme of *M. thermoacetica* contains up to three subunits (64, 14, and 43 kDa), corresponding to an AOR, an Fe/S protein, and the latter 43-kDa subunit, which acts as a viologen-accepting reduced pyridine nucleotide oxidoreductase, probably due to the FAD content of that subunit.[109] The aldehyde dehydrogenase/carboxylic acid reductase ratio, as measured with methyl viologen, is about 20 to 1, but reduced pyridine nucleotides were unable to reduce nonactivated carboxylic acids.[110] Upstream of the AOR gene of *E. acidaminophilum* is a sigma-54 promoter and a gene encoding a regulator (Fig. 3).[102] The deduced amino acid composition of AOR exhibits

57% identity to the AOR of *P. furiosus*, differing, for example, in that it lacks an iron binding site, which means it is monomeric.

Due to the similarity of the nomenclature of aldehyde oxidoreductase (AOR) and aldehyde oxidase concomitant with their common ability to use both aldehydes as principal substrates, we were at first surprised that both enzymes are very different, as they are a tungsto- or molybdoenzyme with a different domain structure in *D. gigas*.[111] The crystal structure of the tungstopterin enzyme AOR is quite different[112] from that known for the classical molybdopterin enzyme aldehyde oxidase, belonging to the xanthine oxidase family.[26,77] Molybdenum or vanadium is unable to replace tungsten in one of the AOR's of *P. furiosus*.[113] This might be due to the catalysis at a very low redox potential and the preference to transfer two electrons by the family of only tungsten-containing AOR enzymes.[114]

The Cytoplasmic Molbindin Protein Mop Binds Both Tungstate and Molybdate

Labeling studies of *E. acidaminophilum* with ^{185}W indicate the presence of a 40-kDa and a 5.5-kDa protein when there are higher concentrations of tungstate or molybdate, as stated previously. Screening for proteins involved in putative W or Mo homeostasis in *E. acidaminophilum* led to the identification of a *mop* gene, using one of the three *mop* genes of *Colostridium pasteurianum* as a probe.[115,116] Such a protein or its derived domain has been termed "molbindin" to emphasize its molybdate-binding ability.[117] The *mop* gene encodes a 69 amino acid–containing protein of 7.3 kDa, forming a hexamer. The deduced amino acid sequence exhibits the highest similarity (84–70%) to the Mop proteins from *C. pasteurianum*, *Haemophilus influenzae*, and *Sporomusa ovata*. The gene of *E. acidaminophilum* is cotranscribed with those of hypothetical proteins.[115] *Rhodobacter capsulatus* contains a molybdate-activated promoter within the intergenic region upstream of its *mop* gene. The latter organism seems to be highly sophisticated in its molybdenum metabolism, for it contains different nitrogenases as well as different MPT-containing enzymes. Thus, its genome contains two similar, but partly different, molybdenum-dependent regulator proteins, MopA and MopB, with functions similar to ModE of *E. coli* and even two different ABC-type transporters for molybdate.[36,118] Using the protein gel-shift assay, molybdate and tungstate are bound with about the same affinity by the Mop protein of *E. acidaminophilum*, showing a small shift for chromate as a member of the common sixth group of transition elements. A different shift behavior is obtained after site-directed mutagenesis within the conserved N-terminal SARN motif (R6K or R6L), or even no shift is seen (R6E). Due to the absence of aromatic amino acids in this Mop protein, no fluorescence studies are possible. Isothermal titration calorimetric (ITC) studies specify for Mop of *H. influenzae* a stoichiometry of 8 mol of molybdate per Mop hexamer, binding at two different sites with different affinities (K_D of 1.1 to 2.7×10^{-8} M).[119] Similar results are obtained for Mop of *S. ovata* and ModG of *A. vinelandii*, where the hexameric structure is arranged as a trimer of dimers or in the case of the di-mop domain of ModG as a trimer, respectively, creating in both cases two different binding sites at the interphase of the subunits and domains.[36,116] So far, no Mop protein is known to be able to discriminate between molybdate and tungstate. Thus, it is not a selective donor to deliver molybdenum or tungsten to a given protein. However, it can be important for the homeostasis of these tetrahedral oxyanions.

A different and new type of storage protein for polynuclear molybdenum oxides is present in *A. vinelandii* (and also similar genes in *R. capsulatus*) that might store up to 90 atoms of molybdenum per octameric arrangement of the two different, but related, subunits of about 29 kDa each. This protein can also bind tungsten, and its exchange with free molybdate requires ATP, whereas Mo release from this storage protein is ATP-independent, but is pH regulated.[120]

The Selective Transport of Tungstate into the Cytoplasma of Prokaryotes is Accomplished by Two Different ABC Transporter Systems: TupA and WtpA Are the Highly Selective Binding Proteins

The first example of a protein able to discriminate between the highly similar tetrahedral oxyanions of tungstate and molybdate is reported for TupA (Tup stands for *t*ungstate *up*take) of *E. acidaminophilum*, already showing a clear and selective difference in the protein gel-shift assay.[93] The genes for this ABC transporter are within the cluster of genes involved in MPT cofactor biosynthesis, close to those encoding the tungsten-containing FDH 1, and adjacent to two genes of *moeA* (FIG. 3). MoeA is responsible for the final incorporation of Mo or W into the pyranopterin cofactor.[38,121] The cleavage products observed before are no longer present after using the intein system for overexpression of TupA in *E. coli* and also avoiding its natural signal sequence for a genuine lipoprotein, as it is present in this gram-positive bacterium. The K_D-value for tungstate is lower than 0.5 μM according to the protein gel-shift assay and is not influenced by a 1000-fold molar excess of molybdate. TupA seems to bind tungstate in a ratio of about 1 to 1.[93]

The presence of two tryptophan residues in TupA allows the application of fluorescence spectroscopy. Although a titration with molybdate shows a higher quench of about twice that observed for tungstate, the decrease in fluorescence is exactly within a 1-to-1 ratio for tungstate and TupA, whereas at least a 100-fold excess of molybdate is required to obtain a saturation of TupA. Adding molybdate first to TupA, tungstate is able to compete for the same binding site and replaces molybdate. The K_D-constant is about 1 μM for molybdate, but lower than 10 nM for tungstate (Rauh *et al.*, in preparation). Very similar results are obtained for TupA proteins obtained from the fluorescence studies of three other, very different organisms: *M. thermoacetica*, the gram-positive organism that started our research on tungsten, the gram-negative betaproteobacterium *Ralstonia metallidurans*, encoding a putative tungsten-containing FDH, and the archaeon *Methanosarcina mazei* (Makdessi *et al.*, in preparation). The latter is one of the few archaea that does not contain the Wtp (*W-transport*) system.[122] The genes of TupA homologues can be found in the genomes of many prokaryotes that do not live under anaerobic conditions like *Ralstonia*. As indicated before,[93] the TTTS motif is one of the typical signatures indicating a selective binding of tungstate, as was also revealed by X-ray analysis of TupA crystals from *E. acidaminophilum* (Rauh *et al.*, in preparation).

The ITC is a very sensitive method for monitoring the interaction of different components by the heat release initiated by binding, for example, tungstate. The enthalpy value for tungstate was twice that of molybdate. The K_D value determined for tungstate is about 0.2 nM, an extremely low value for an ABC transporter (Rauh *et al.*, in preparation), representing a concentration that should be present in most natural environments. This extremely high affinity might explain our inability to observe any dependence on tungstate during the growth of *E. acidaminophilum* on formate/betaine (in contrast to that of selenite), although the selenocysteine-containing tungstoenzyme FDH is an essential enzyme. During competetive ITC experiments, TupA is first saturated with molybdate before tungstate is added by titration. These experiments, like the fluorescence studies, demonstrated that both ions compete for a common binding site. The much lower enthalpy value determined by this procedure for tungstate indicates that the energy required to remove molybdate from the common binding site will be compensated by the energy released by binding tungstate. Again, a K_D value for tungstate of about 1 nM was obtained by the competitive ITC. The K_D-value for molybdate is about 1.4 μM (Rauh *et al.*, in preparation), which is at least three orders of magnitude higher than that observed for tungstate. This indicates the ability of TupA to discriminate very sharply between these two similar anions. This has never been observed for any ModA protein analyzed, and shows a K_D for molybdate of 20 nM at its best, but all ModA proteins bind tungstate as well.[123]

Quite recently, a new class of tungstate binding protein was found to be present mainly in archaea (except in, e.g., *M. mazei* and *Pyrobaculum aerophilum*) and a few bacteria, such as *Desulfotalea psychrophila*. The latter bacterium contains all three binding proteins, TupA, ModA, and WtpA.[122] These three proteins do not contain a common binding motif, so they should be recognized quite easily. Although not listed by Bevers *et al.*,[122] *Archaeoglobus fulgidus* also contains, along with the Wtp system, the ABC transporter ModABC for molybdate, whose general structure could be resolved very recently for the first time, especially the membrane-spanning regions with two gates within the 12 transmembrane helices of the $ModB_2C_2$ complex.[124] ITC studies with WtpA indicate a K_D value for molybdate of 11 nM, which is the lowest value measured so far for molybdate. The K_D value for tungstate is below 1 nM, and again the limiting datasets do not allow a precise value to be given. According to competitive ITC studies, the K_D value of tungsten is somewhat higher, about 15 nM, but if it is calculated by the given formula, and due to the low K_D value observed for molybdate, the apparent binding constant for tungstate is about 17 pM, which is the lowest K_D value ever reported for an inbound ABC transporter system.[122]

Crystallography of TupA of *E. acidaminophilum*

Crystals obtained from TupA proved to be quite fragile, especially in the presence of tungstate. Thus, only the analysis of crystals without tungstate became possible, giving a structure with up to 1.9 Å resolution and good statistics (Rauh *et al.*, in preparation). TupA's overall structure is very similar to the binding proteins analyzed before, showing two different domains connected by a flexible-hinge region of two beta strands. The amino acids detected in this flexible cleft represent the potential binding site at the two coils of both domains. The alignment of TupAs from different prokaryotes reveals common characteristic motifs that are located in about the same area as observed for molybdate in the two prokaryotic ModA proteins analyzed.[36,93] The newly discovered vanadate binding protein, VupA, of *Anabaena variabilis*[125] exhibits the highest similarity to TupA of *E. acidaminophilum,* and the corresponding phylogenetic tree of the tetrahedral oxyanion binding proteins that are specific for tungstate, molybdate, sulfate/ thiosulfate, and phosphate[93,122] clearly show its affiliation with the TupA cluster (Rauh *et al.*, in preparation). Site-directed mutagenesis of, for example, an extending positively charged residue of arginine to lysine diminishes strongly the specific binding of tungstate (Makdessi *et al.*, in preparation). Although the buffer citrate was found at the proposed tungstate binding site, tungstate can be modeled to fit into this site.

Conclusions

The data obtained for TupA from *E. acidaminophilum* by different methods, such as protein gel-shift, fluorescence quench, ITC, and site-directed mutagenesis, clearly indicate that TupA is actually the binding protein of an ABC uptake system that specifies this unusually high selectivity for tungstate compared to molybdate. The molecular mechanism by which this is accomplished is still unknown. It might be due partly to the different pK_a-values of molybdate and tungstate, for the latter has a K_a-value of 4.7 compared to 3.8 of molybdate. One sequence motif of TupA exhibits a close similarity to phosphate-binding proteins as well as to VupA for vanadate.[93,125] A conserved histidine residue might be quite important for a selective protonation of tungstate, thus differentiating it from molybdate. Due to the preference of TupA for tungstate over molybdate by at least three orders of magnitude, and an observed lack of *modA* gene in the genome of *E. acidaminophilum*, the preferred incorporation of tungsten into the MPT-cofactor can be anticipated to occur. The cytoplasmic Mop protein should not be the direct receptor due to the generally observed nondiscrimination between molybdate and tungstate. The energizing TupC protein of *E. acidaminophilum* does not contain a molbindin domain, as observed for ModC of *E. coli* or *A. fulgidus*,[36,124] which might immediately chelate the transported oxyanion.

E. acidaminophilum contains two genes of the *moeA* type.[93] MoeA is responsible for the incorporation of molybdate into the adenylated MPT catalyzed by MogA to form the final active cofactor.[121] The *mogA* gene is present in *E. acidaminophilum*, but often missing in archaea, where its function might become replaced by MoaB, a protein absent in most bacteria.[38] The shorter MoeA (43.1 kDa) is coexpressed with TupC, the larger MoeA-1 (68.7 kDa) with MoaA and MoaC.[93] The presence of two *moeA* genes is found in some bacteria and archaea; however, no data are published for an additional function residing on the C-terminal extension of MoeA-1. The gene(s) and mechanism of forming the bis-pyranopterin cofactor form are still unknown. This cofactor form is generally present in the enzymes of the DMSO family like FDH and of the AOR family, and thus present in all of the known tungstoenzymes of *E. acidaminophilum.* The four oxygen atoms originally present in tungstate and molybdate must be replaced by the two dithiolene moieties of the bis-cofactor form, which is first chelated by a copper moiety.[126,127] However, no experimental proof is given for the exchange of the oxygen atoms by sulfur. Perhaps the proposed intermediate adenosine 5′phosphomolybdate (APMo) is first bound in a hydrophobic pocket of MoeA in order to avoid an instantly occurring hydrolysis of the APMo.[128] It has been suggested that APMo is formed from MPT-AMP by MoeA.[38,121,127] Replacement of the four oxygen atoms of molybdate or tungstate by the four sulfur atoms of the bis-dithiolene–containing cofactor might be accompanied by a successively occurring change in the redox state of the metal and removal of an oxygen atom as water [perhaps analogous to the reduction of sulfate via adenosine 5′-phosphosulfate (APS) to sulfite and water, or otherwise like selenate via a molybdenum-containing selenate reductase[129]]. Changes in the coordination numbers or geometries during such redox changes might also be important. In addition, the intermediate formation of thiotungstates might be anticipated, especially for tungstate.[7,11] In any case, the intermediates and the final Mo- or W-cofactors are quite labile and oxygen-sensitive.

At least, the finally loaded forms of both labile cofactors can become protected and a potential surplus can be stored. A cofactor-carrier protein of 64 kDa, a tetramer of 16.5 kDa subunits, is analyzed in the case of the algae *Chlamydomonas reinhardtii* and protects both, the final Mo-cofactor, and an artificially obtained W-cofactor, but not the metal-free MPT-form.[130] Similar deduced proteins are found in genomes of cyanobacteria, some other bacteria, and archaea. Although the pathway of the Mo-/W-pterin cofactor biosynthesis is explored in its general outline and shown to be surprisingly similar in pro- and eukaryotes,[38] there are further questions to be solved about the involvement of general or specialized chaperones during the final assembly of the often quite complex proteins, such as dissimilatory nitrate reductase[131] or FDH, and their potential transport into the periplasma of gram-negative bacteria (most probably via the Tat system) or their anchoring at the outside of the cytoplasmic membrane of gram-positive bacteria. Due to the widespread occurrence of *tupA* or *wtpA* genes in different prokaryotes, there will be more interesting enzymes discovered that should also contain tungsten, besides the known tungstoenzymes acetylene hydratase[132] and formylmethanofuran dehydrogenases[133] or enzymes that can be active with both molybdenum and tungsten, like some DMSO and trimethylamine N-oxide (TMAO) reductases.[2]

To conclude, the classic acetogenic model organism for the Wood–Ljundahl pathway of autotrophic CO_2 fixation, *M. thermoacetica*, represents the first clear example of the positive action and the requirement of tungsten, as became experimentally shown for its FDH.[19] The prejudice of humans is a primary obstacle to accepting any unexpected results presented by nature. Tungsten is one such example. In the end, Bortels[13,14] was at least correct in pointing out that tungsten might catalyze part of a biological reaction because the W-substituted nitrogenase of *R. capsulatus* is catalytically active in at least one general reaction catalyzed by nitrogenases, the production of hydrogen.[33] One can be confident that more will be learned about tungstoenzymes in the years to come.

Acknowledgments

Financial support by the Deutsche Forschungsgemeinschaft is gratefully acknowledged. We thank David Rauh and Anett Heinrich for providing unpublished data.

Conflict of Interest

The authors declare no conflicts of interest.

References

1. BOLL, M. *et al.* 2005. Novel bacterial molybdenum and tungsten enzymes: three-dimensional structure, spectroscopy, and reaction mechanism. Biol. Chem. **386:** 999–1006.
2. GARNER, C.D. & L.J. STEWART. 2002. Tungsten-substituted molybdenum enzymes. *In* Metal Ions Biol. Syst., Vol. 39. A. Sigel & H. Sigel, Eds.: 699–726. New York: Marcel Dekker.
3. HILLE, R. 2002. Molybdenum and tungsten in biology. Trends Biochem. Sci. **27:** 360–367.
4. CALLIS, G.E. & R.A. WENTWORTH. 1977. Tungsten vs. Molybdenum in models for biological systems. Bioinorg. Chem. **7:** 57–70.
5. HAGEN, W.R. & A.F. ARENDSEN. 1998. The bio-inorganic chemistry of tungsten. Struct. Bond. **90:** 161–191.
6. L'VOV, N.P., A.N. NOSIKOV & A.N. ANTIPOV. 2002. Tungsten-containing enzymes. Biochemistry (Moscow) **67:** 196–200.
7. STIEFEL, E.I. 2002. The biogeochemistry of molybdenum and tungsten. *In* Metal Ions Biol. Syst., Vol. 39. A. Sigel & H. Sigel, Eds.: 1–29. New York: Marcel Dekker.
8. WENTWORTH, R.A.D. 1976. Mechanism for the reactions of molybdenum in enzymes. Coord. Chem. Rev. **18:** 1–27.
9. POPE, M.T. & A. MÜLLER. 1991. Polyoxometalate chemistry: an old field with new dimensions in several disciplines. Angew. Chem. **103:** 56–70.
10. TYTKO, K.-H. & O. GLEMSER. 1976. Isopolymolybdates and isopolytungstates. Adv. Inorg. Chem. Radiochem. **19:** 239–215.
11. MÜLLER, A. *et al.* 1981. Transition metal thiometalates: properties and significance in complex and bioinorganic chemistry. Angew. Chem. **93:** 957–977.
12. ARNOLD, G.L. *et al.* 2004. Molybdenum isotope evidence for widespread anoxia in mid-Proterozoic oceans. Science **304:** 87–90.
13. BORTELS, H. 1930. Molybdän als Katalysator bei der biologischen Stickstoffbindung. Arch. Mikrobiol. **1:** 333–342.
14. BORTELS, H. 1936. Weitere Untersuchungen über die Bedeutung von Molybdän, Vanadium, Wolfram und andere Erdaschenstoffe für stickstoffbindene und andere Mikroorganismen. Centralbl. Bakt. II **95:** 193–218.
15. NICHOLAS, D.J. & A. NASON. 1954. Molybdenum and nitrate reductase. II. Molybdenum as a constituent of nitrate reductase. J. Biol. Chem. **207:** 353–360.
16. PINSENT, J. 1954. The need for selenite and molybdate in the formation of formic dehydrogenase by members of the *coli-aerogenes* group of bacteria. Biochem. J. **57:** 10–16.
17. HIGGINS, E.S., D.A. RICHERT & W.W. WESTERFELD. 1956. Molybdenum deficiency and tungstate inhibition studies. J. Nutrition. **59:** 539–559.
18. JOHNSON, J.L., K.V. RAJAGOPALAN & H.J. COHEN. 1974. Molecular basis of the biological function of molybdenum. Effect of tungsten on xanthine oxidase and sulfite oxidase in the rat. J. Biol. Chem. **249:** 859–866.
19. LJUNGDAHL, L.G. 1976. Tungsten, a biologically active metal. Trends Biochem. Sci. **1:** 63–65.
20. MUKUND, S. & M.W. ADAMS. 1991. The novel tungsten-iron-sulfur protein of the hyperthermophilic archaebacterium, *Pyrococcus furiosus*, is an aldehyde ferredoxin oxidoreductase. Evidence for its participation in a unique glycolytic pathway. J. Biol. Chem. **266:** 14208–14216.
21. SCHMITZ, R.A., S.P. ALBRACHT & R.K. THAUER. 1992. A molybdenum and a tungsten isoenzyme of formylmethanofuran dehydrogenase in the thermophilic archaeon *Methanobacterium wolfei*. Eur. J. Biochem. **209:** 1013–1018.
22. ADAMS, M.W.W. 1992. Novel iron-sulfur clusters in metalloenzymes and redox proteins from extremely thermophilic bacteria. Adv. Inorg. Chem. **38:** 341–396.
23. FRAUSTO DA SILVA, J.J.R. & R.J.P. WILLIAMS. 1991. The Biological Chemistry of the Elements. The Inorganic Chemistry of Life. Oxford: Clarendon Press.

24. SIGEL, A. & H. SIGEL. 2002. Molybdenum and Tungsten: Their Roles in Biological Processes. New York: Marcel Dekker.
25. JOHNSON, M.K., D.C. REES & M.W. ADAMS. 1996. Tungstoenzymes. Chem. Rev. **96:** 2817–2839.
26. KISKER, C., H. SCHINDELIN & D.C. REES. 1997. Molybdenum-cofactor-containing enzymes: structure and mechanism. Annu. Rev. Biochem. **66:** 233–267.
27. KLETZIN, A. & M.W. ADAMS. 1996. Tungsten in biological systems. FEMS Microbiol. Rev. **18:** 5–63.
28. MCMASTER, J. & J.H. ENEMARK. 1998. The active sites of molybdenum- and tungsten-containing enzymes. Curr. Opin. Chem. Biol. **2:** 201–207.
29. ROMÃO, M.J. *et al.* 1997. Structure and function of molybdopterin containing enzymes. Prog. Biophys. Mol. Biol. **68:** 121–144.
30. SCHINDELIN, H., C. KISKER & K.V. RAJAGOPALAN. 2001. Molybdopterin from molybdenum and tungsten enzymes. Adv. Protein. Chem. **58:** 47–94.
31. RUBIO, L.M. & P.W. LUDDEN. 2005. Maturation of nitrogenase: a biochemical puzzle. J. Bacteriol. **187:** 405–414.
32. RÜTTIMANN-JOHNSON, C. *et al.* 2003. VnfY is required for full activity of the vanadium-containing dinitrogenase in *Azotobacter vinelandii*. J. Bacteriol. **185:** 2383–2386.
33. SIEMANN, S. *et al.* 2003. Characterization of a tungsten-substituted nitrogenase isolated from *Rhodobacter capsulatus*. Biochemistry **42:** 3846–3857.
34. GRUNDEN, A.M. & K.T. SHANMUGAM. 1997. Molybdate transport and regulation in bacteria. Arch. Microbiol. **168:** 345–354.
35. PAU, R.N., W. KLIPP & S. LEIMKÜHLER. 1997. Molybdenum transport, processing and gene regulation. *In* Transition Metals in Microbial Metabolism. G. Winkelmann & C.J. Carrano, Eds.: 217–234. Amsterdam, The Netherlands. Harwood Academic Publishers.
36. PAU, R.N. & D.M. LAWSON. 2002. Transport, homeostasis, regulation, and binding of molybdate and tungstate to proteins. *In* Met. Ions Biol. Syst., Vol. 39. A. Sigel & H. Sigel, Eds.: 31–74. Marcel Dekker. New York.
37. SELF, W.T. *et al.* 2001. Molybdate transport. Res. Microbiol. **152:** 311–321.
38. SCHWARZ, G., P.L. HAGEDOORN & K. FISCHER. 2007. Molybdate and tungstate: uptake, homeostasis, cofactors, and enzymes. *In* Molecular Microbiology of Heavy Metals. D.H. Nies & S. Silver, Eds.: 421–451. Berlin: Springer-Verlag.
39. ANTIPOV, A.N. *et al.* 2003. New enzyme belonging to the family of molybdenum-free nitrate reductases. Biochem. J. **369:** 185–189.
40. AFSHAR, S. *et al.* 1998. Effect of tungstate on nitrate reduction by the hyperthermophilic archaeon *Pyrobaculum aerophilum*. Appl. Environ. Microbiol. **64:** 3004–3008.
41. HAGEDOQRN, P.L. *et al.* 2005. Purification and characterization of the tungsten enzyme aldehyde:ferredoxin oxidoreductase from the hyperthermophilic denitrifier *Pyrobaculum aerophilum*. J. Biol. Inorg. Chem. **10:** 259–269.
42. ANDREESEN, J.R. & L.G. LJUNGDAHL. 1973. Formate dehydrogenase of *Clostridium thermoaceticum*: incorporation of selenium-75, and the effects of selenite, molybdate, and tungstate on the enzyme. J. Bacteriol. **116:** 867–873.
43. ANDREESEN, J.R. *et al.* 1973. Fermentation of glucose, fructose, and xylose by *Clostridium thermoaceticum*: effect of metals on growth yield, enzymes, and the synthesis of acetate from CO_2. J. Bacteriol. **114:** 743–751.
44. ENOCH, H.G. & R.L. LESTER. 1972. Effects of molybdate, tungstate, and selenium compounds on formate dehydrogenase and other enzyme systems in *Escherichia coli*. J. Bacteriol. **110:** 1032–1040.
45. ANDREESEN, J.R. 1980. Role of selenium, molybdenum and tungsten in anaerobes. *In* Anaerobes and Anaerobic Infections. G. Gottschalk, N. Pfennig & H. Werner, Eds.: 31–40. Stuttgart: Gustav Fischer Verlag.
46. DRAKE, H.L. 1994. Acetogenesis. New York: Chapman & Hall.
47. LI, L.F., L. LJUNGDAHL & H.G. WOOD. 1966. Properties of nicotinamide adenine dinucleotide phosphate-dependent formate dehydrogenase from *Clostridium thermoaceticum*. J. Bacteriol. **92:** 405–412.
48. ANDREESEN, J.R. & L.G. LJUNGDAHL. 1974. Nicotinamide adenine dinucleotide phosphate-dependent formate dehydrogenase from *Clostridium thermoaceticum*: purification and properties. J. Bacteriol. **120:** 6–14.
49. LJUNGDAHL, L.G. & J.R. ANDREESEN. 1975. Tungsten, a component of active formate dehydrogenase from *Clostridium thermoacetium*. FEBS Lett. **54:** 279–282.
50. LJUNGDAHL, L.G. & J.R. ANDREESEN. 1976. Reduction of CO_2 to acetate in homoacetate fermenting clostridia and the involvment of tungsten in formate dehydrogenase. *In* Symposium on Microbial Production and Utilization of Gases (H_2, CH_4, CO). H.G. Schlegel, G. Gottschalk & N. Pfennig, Eds.: 163–172. Göttingen: E. Goltze.
51. LJUNGDAHL, L.G. & J.R. ANDREESEN. 1978. Formate dehydrogenase, a selenium–tungsten enzyme from *Clostridium thermoaceticum*. Methods Enzymol. **53:** 360–372.
52. SAIKI, T., G. SHACKLEFORD & L.G. LJUNGDAHL. 1980. Composition of tungsten:selenium-containing formate dehydrogenase from *Clostridium thermoaceticum*. *In* Selenium in Biology and Medicine. J.E. Spallholz, J.L. Martin & H.E. Ganther, Eds.: 220–229. Westport, CT: Avi Publishing Coompany.
53. YAMAMOTO, I. *et al.* 1983. Purification and properties of NADP-dependent formate dehydrogenase from *Clostridium thermoaceticum*, a tungsten-selenium-iron protein. J. Biol. Chem. **258:** 1826–1832.
54. GOTTWALD, M. *et al.* 1975. Presence of cytochrome and menaquinone in *Clostridium formicoaceticum* and *Clostridium thermoaceticum*. J. Bacteriol. **122:** 325–328.
55. ARENDSEN, A.F., M.Q. SOLIMAN & S.W. RAGSDALE. 1999. Nitrate-dependent regulation of acetate biosynthesis and nitrate respiration by *Clostridium thermoaceticum*. J. Bacteriol. **181:** 1489–1495.
56. GRAENTZDOERFFER, A. *et al.* 2003. Molecular and biochemical characterization of two tungsten- and selenium-containing formate dehydrogenases from *Eubacterium acidaminophilum* that are associated with components of an iron-only hydrogenase. Arch. Microbiol. **179:** 116–130.
57. THAUER, R.K. 1972. CO_2 reduction to formate by NADPH. The initial step in the total synthesis of acetate from CO_2 in *Clostridium thermoaceticum*. FEBS Lett. **27:** 111–115.

58. THAUER, R.K. 1973. CO_2 reduction to formate in *Clostridium acidiurici*. J. Bacteriol. **114:** 443–444.

59. HAGEDOORN, P.-L. 2002. Metalloproteins containing iron and tungsten: biocatalytic links between organic and inorganic redox chemistry. Ph.D. Thesis. Delft University of Technology, Delft, the Netherlands.

60. ANDREESEN, J.R., G. GOTTSCHALK & H.G. SCHLEGEL. 1970. *Clostridium formicoaceticum* nov. spec. isolation, description and distinction from *C. aceticum* and *C. thermoaceticum*. Arch. Microbiol. **72:** 154–174.

61. BRAUN, M., F. MAYER & G. GOTTSCHALK. 1981. *Clostridium aceticum* (Wieringa), a microorganism producing acetic acid from molecular hydrogen and carbon dioxide. Arch. Microbiol. **128:** 288–293.

62. ANDREESEN, J.R., E. EL GHAZZAWI & G. GOTTSCHALK. 1974. The effect of ferrous ions, tungstate and selenite on the level of formate dehydrogenase in *Clostridium formicoaceticum* and formate synthesis from CO_2 during pyruvate fermentation. Arch. Microbiol. **96:** 103–118.

63. LEONHARDT, U. & J.R. ANDREESEN. 1977. Some properties of formate dehydrogenase, accumulation and incorporation of 185W-tungsten into proteins of *Clostridium formicoaceticum*. Arch. Microbiol. **115:** 277–284.

64. WHITE, H. *et al.* 1991. Purification and some properties of the tungsten-containing carboxylic acid reductase from *Clostridium formicoaceticum*. Biol. Chem. Hoppe-Seyler **372:** 999–1005.

65. BALCH, W.E. *et al.* 1977. *Acetobacterium*, a new genus of hydrogen-oxidizing, carbon dioxide-reducing, anaerobic bacteria. Int. J. Syst. Bacteriol. **27:** 355–361.

66. SCHOBERTH, S. 1977. Acetic acid from H_2 and CO_2. Formation of acetate by cell extracts of *Acetobacterium woodii*. Arch. Microbiol. **114:** 143–148.

67. ANDREESEN, J.R. & S. FETZNER. 2002. The molybdenum-containing hydroxylases of nicotinate, isonicotinate, and nicotine. *In* Metal Ions Biol. Syst., Vol. 39. A. Sigel & H. Sigel, Eds.: 405–430. New York: Marcel Dekker.

68. DILWORTH, G.L. 1982. Properties of the selenium-containing moiety of nicotinic acid hydroxylase from *Clostridium barkeri*. Arch. Biochem. Biophys. **219:** 30–38.

69. GLADYSHEV, V.N., S.V. KHANGULOV & T.C. STADTMAN. 1996. Properties of the selenium- and molybdenum-containing nicotinic acid hydroxylase from *Clostridium barkeri*. Biochemistry **35:** 212–223.

70. IMHOFF, D. & J.R. ANDREESEN. 1979. Nicotinic acid hydroxylase from *Clostridium barkeri*: selenium-dependent formation of active enzyme. FEMS Microbiol. Lett. **5:** 155–158.

71. SCHRÄDER, T., A. RIENHÖFER & J.R. ANDREESEN. 1999. Selenium-containing xanthine dehydrogenase from *Eubacterium barkeri*. Eur. J. Biochem. **264:** 862–871.

72. WAGNER, R. & J.R. ANDREESEN. 1977. Differentiation between *Clostridium acidiurici* and *Clostridium cylindrosporum* on the basis of specific metal requirements for formate dehydrogenase formation. Arch. Microbiol. **114:** 219–224.

73. WAGNER, R. & J.R. ANDREESEN. 1979. Selenium requirement for active xanthine dehydrogenase from *Clostridium acidiurici* and *Clostridium cylindrosporum*. Arch. Microbiol. **121:** 255–260.

74. WAGNER, R., R. CAMMACK & J.R. ANDREESEN. 1984. Purification and properties of xanthine dehydrogenase from *Clostridium acidiurici* grown in the presence of selenium. Biochim. Biophys. Acta **791:** 63–74.

75. BRAY, R.C. & J.C. SWANN. 1972. Molybdenum-containing enzymes. Struct. Bonding **11:** 107–144.

76. CHAMPION, A.B. & J.C. RABINOWITZ. 1977. Ferredoxin and formyltetrahydrofolate synthetase: comparative studies with *Clostridium acidiurici*, *Clostridium cylindrosporum*, and newly isolated anaerobic uric acid-fermenting strains. J. Bacteriol. **132:** 1003–1020.

77. COUGHLAN, M.P. 1980. Aldehyde oxidase, xanthine oxidase and xanthine dehydrogenase: hydroxylases containing molybdenum, iron-sulphur and flavin. *In* Molybdenum and Molybdenum-containing Enzymes. M.P. Coughlan, Ed.: 119–185. Oxford: Pergamon Press.

78. VOGELS, G.D. & C. VAN DER DRIFT. 1976. Degradation of purines and pyrimidines by microorganisms. Bacteriol. Rev. **40:** 403–468.

79. DÜRRE, P., W. ANDERSCH & J.R. ANDREESEN. 1981. Isolation and characterization of an adenine utilizing, anaerobic sporeformer, *Clostridium purinolyticum* nov. spec. Int. J. Syst. Bacteriol. **31:** 184–194.

80. DÜRRE, P. & J.R. ANDREESEN. 1983. Purine and glycine metabolism by purinolytic clostridia. J. Bacteriol. **154:** 192–199.

81. SCHIEFER-ULLRICH, H. *et al.* 1984. Comparative studies on physiology and taxonomy of obligately purinolytic clostridia. Arch. Microbiol. **138:** 345–353.

82. WAGNER, R. & J.R. ANDREESEN. 1987. Accumulation and incorporation of 185W-tungsten into proteins of *Clostridium acidiurici* and *Clostridium cylindrosporum*. Arch. Microbiol. **147:** 195–199.

83. BEUSCHER, H.U. & J.R. ANDREESEN. 1984. *Eubacterium angustum* sp. nov., a gram-positive, anaerobic, nonsporeforming, obligate purine fermenting organism. Arch. Microbiol. **140:** 2–8.

84. SCHIEFER-ULLRICH, H. & J.R. ANDREESEN. 1985. *Peptostreptococcus barnesae* sp. nov., a gram-positive, anaerobic, obligately purine utilizing coccus from chicken feces. Arch. Microbiol. **143:** 26–31.

85. TZIAKA, C. 1987. Untersuchungen zur Charakterisierung von Stämmen der Gattungen *Peptococcus* und *Peptostreptococcus*. Ph.D. Thesis. University of Göttingen, Göttingen.

86. DÜRRE, P., R. SPAHR & J.R. ANDREESEN. 1983. Glycine fermentation via glycine reductase in *Peptococcus glycinophilus* and *Peptococcus magnus*. Arch. Microbiol. **134:** 127–135.

87. LEBERTZ, H. 1984. Selenabhängiger Glycin-Stoffwechsel bei anaeroben Bakterien und vergleichende Untersuchungen zur Glycin-Reduktase und zur Glycin-Decarboxylase. Ph.D. Thesis. University of Göttingen, Göttingen.

88. ZINDEL, U. *et al.* 1988. *Eubacterium acidaminophilum* sp. nov., a versatile amino acid-degrading anaerobe producing or utilzing H_2 or formate. Description and enzymatic studies. Arch. Microbiol. **150:** 254–266.

89. ANDREESEN, J.R. 1994. Glycine metabolism in anaerobes. Antonie Leeuwenhoek **66:** 223–237.

90. ANDREESEN, J.R. 2004. Glycine reductase mechanism. Curr. Opin. Chem. Biol. **8:** 454–461.
91. GRANDERATH, K. 1993. Charakterisierung der Formiat-Dehydrogenase und Aldehyd-Dehydrogenase als wolframhaltige Proteine von *Eubacterium acidaminophilum*. Ph.D. Thesis. University of Göttingen, Göttingen.
92. MEYER, M., K. GRANDERATH & J.R. ANDREESEN. 1995. Purification and characterization of protein PB of betaine reductase and relationship to the corresponding proteins of glycine reductase and sarcosine reductase from *Eubacterium acidaminophilum*. Eur. J. Biochem. **234:** 184–191.
93. MAKDESSI, K., J.R. ANDREESEN & A. PICH. 2001. Tungstate uptake by a highly specific ABC transporter in *Eubacterium acidaminophilum*. J. Biol. Chem. **276:** 24557–24564.
94. SOBOH, B., D. LINDER & R. HEDDERICH. 2004. A multisubunit membrane-bound [NiFe] hydrogenase and an NADH-dependent Fe-only hydrogenase in the fermenting bacterium *Thermoanaerobacter tengcongensis*. Microbiology **150:** 2451–2463.
95. DE BOK, F.A. *et al.* 2003. Two W-containing formate dehydrogenases (CO2-reductases) involved in syntrophic propionate oxidation by *Syntrophobacter fumaroxidans*. Eur. J. Biochem. **270:** 2476–2485.
96. RAAIJMAKERS, H. *et al.* 2002. Gene sequence and the 1.8 Å crystal structure of the tungsten-containing formate dehydrogenase from *Desulfovibrio gigas*. Structure **10:** 1261–1272.
97. LAUKEL, M. *et al.* 2003. The tungsten-containing formate dehydrogenase from *Methylobacterium extorquens* AM1: purification and properties. Eur. J. Biochem. **270:** 325–333.
98. POHLMANN, A. *et al.* 2006. Genome sequence of the bioplastic-producing "Knallgas" bacterium *Ralstonia eutropha* H16. Nat. Biotechnol. **24:** 1257–1262.
99. OH, J.I. & B. BOWIEN. 1998. Structural analysis of the fds operon encoding the NAD+-linked formate dehydrogenase of *Ralstonia eutropha*. J. Biol. Chem. **273:** 26349–26360.
100. BURGDORF, T., D. BÖMMER & B. BOWIEN. 2001. Involvement of an unusual *mol* operon in molybdopterin cofactor biosynthesis in *Ralstonia eutropha*. J. Mol. Microbiol. Biotechnol. **3:** 619–629.
101. MOURA, J.J. *et al.* 2004. Mo and W bis-MGD enzymes: nitrate reductases and formate dehydrogenases. J. Biol. Inorg. Chem. **9:** 791–799.
102. RAUH, D. *et al.* 2004. Tungsten-containing aldehyde oxidoreductase of *Eubacterium acidaminophilum*. Eur. J. Biochem. **271:** 212–219.
103. BEVERS, L.E. *et al.* 2005. WOR5, a novel tungsten-containing aldehyde oxidoreductase from *Pyrococcus furiosus* with a broad substrate specificity. J. Bacteriol. **187:** 7056–7061.
104. MUKUND, S. & M.W. ADAMS. 1993. Characterization of a novel tungsten-containing formaldehyde ferredoxin oxidoreductase from the hyperthermophilic archaeon, *Thermococcus litoralis*. A role for tungsten in peptide catabolism. J. Biol. Chem. **268:** 13592–13600.
105. ROY, R. & M.W. ADAMS. 2002. Characterization of a fourth tungsten-containing enzyme from the hyperthermophilic archaeon *Pyrococcus furiosus*. J. Bacteriol. **184:** 6952–6956.
106. MA, K. *et al.* 1997. Pyruvate ferredoxin oxidoreductase from the hyperthermophilic archaeon, *Pyrococcus furiosus*, functions as a CoA-dependent pyruvate decarboxylase. Proc. Natl. Acad. Sci. USA **94:** 9608–9613.
107. HUBER, C. *et al.* 1994. Further characterization of two different, reversible aldehyde oxidoreductases from *Clostridium formicoaceticum*, one containing tungsten and the other molybdenum. Arch. Microbiol. **162:** 303–309.
108. WHITE, H., C. HUBER & H. SIMON. 1993. On a reversible molybdenum-containing aldehyde oxidoreductase from *Clostridium formicoacteticum*. Arch. Microbiol. **159:** 244–249.
109. STROBL, G. *et al.* 1992. The tungsten-containing aldehyde oxidoreductase from *Clostridium thermoaceticum* and its complex with a viologen-accepting NADPH oxidoreductase. Biol. Chem. Hoppe Seyler. **373:** 123–132.
110. WHITE, H. *et al.* 1989. Carboxylic acid reductase: a new tungsten enzyme catalyses the reduction of non-activated carboxylic acids to aldehydes. Eur. J. Biochem. **184:** 89–96.
111. HENSGENS, C.M., W.R. HAGEN & T.A. HANSEN. 1995. Purification and characterization of a benzylviologen-linked, tungsten-containing aldehyde oxidoreductase from *Desulfovibrio gigas*. J. Bacteriol. **177:** 6195–6200.
112. CHAN, M.K. *et al.* 1995. Structure of a hyperthermophilic tungstopterin enzyme, aldehyde ferredoxin oxidoreductase. Science **267:** 1463–1469.
113. MUKUND, S. & M.W. ADAMS. 1996. Molybdenum and vanadium do not replace tungsten in the catalytically active forms of the three tungstoenzymes in the hyperthermophilic archaeon *Pyrococcus furiosus*. J. Bacteriol. **178:** 163–167.
114. BOL, E. *et al.* 2006. Redox chemistry of tungsten and iron-sulfur prosthetic groups in *Pyrococcus furiosus* formaldehyde ferredoxin oxidoreductase. J. Biol. Inorg. Chem. **11:** 999–1006.
115. MAKDESSI, K. *et al.* 2004. Identification and characterization of the cytoplasmic tungstate/molybdate-binding protein (Mop) from *Eubacterium acidaminophilum*. Arch. Microbiol. **181:** 45–51.
116. SCHÜTTELKOPF, A.W. *et al.* 2002. Passive acquisition of ligand by the MopII molbindin from *Clostridium pasteurianum*: structures of apo and oxyanion-bound forms. J. Biol. Chem. **277:** 15013–15020.
117. LAWSON, D.M. *et al.* 1997. Protein ligands for molybdate. Specificity and charge stabilisation at the anion-binding sites of periplasmic and intracellular molybdate-binding proteins of *Azotobacter vinelandii*. J. Chem. Soc. Dalton Trans. 3981–3984.
118. WIETHAUS, J. *et al.* 2006. Overlapping and specialized functions of the molybdenum-dependent regulators MopA and MopB in *Rhodobacter capsulatus*. J. Bacteriol. **188:** 8441–8451.
119. MASTERS, S.L., G.J. HOWLETT & R.N. PAU. 2005. The molybdate binding protein Mop from *Haemophilus influenzae*–biochemical and thermodynamic characterisation. Arch. Biochem. Biophys. **439:** 105–112.

120. FENSKE, D. *et al.* 2005. A new type of metalloprotein: the Mo storage protein from *Azotobacter vinelandii* contains a polynuclear molybdenum-oxide cluster. ChemBioChem **6:** 405–413.
121. LLAMAS, A. *et al.* 2006. The mechanism of nucleotide-assisted molybdenum insertion into molybdopterin. A novel route toward metal cofactor assembly. J. Biol. Chem. **281:** 18343–18350.
122. BEVERS, L.E. *et al.* 2006. Tungsten transport protein A (WtpA) in *Pyrococcus furiosus*: the first member of a new class of tungstate and molybdate transporters. J. Bacteriol. **188:** 6498–6505.
123. IMPERIAL, J., M. HADI & N.K. AMY. 1998. Molybdate binding by ModA, the periplasmic component of the *Escherichia coli* mod molybdate transport system. Biochim. Biophys. Acta. **1370:** 337–346.
124. HOLLENSTEIN, K., D.C. FREI & K.P. LOCHER. 2007. Structure of an ABC transporter in complex with its binding protein. Nature **446:** 213–216.
125. PRATTE, B.S. & T. THIEL. 2006. High-affinity vanadate transport system in the cyanobacterium *Anabaena variabilis* ATCC 29413. J. Bacteriol. **188:** 464–468.
126. KUPER, J. *et al.* 2004. Structure of the molybdopterin-bound Cnx1G domain links molybdenum and copper metabolism. Nature **430:** 803–806.
127. LLAMAS, A., R.R. MENDEL & G. SCHWARZ. 2004. Synthesis of adenylated molybdopterin: an essential step for molybdenum insertion. J. Biol. Chem. **279:** 55241–55246.
128. WILSON, L.G. & R.S. BANDURSKI. 1958. Enzymatic reactions involving sulfate, sulfite, selenate, and molybdate. J. Biol. Chem. **233:** 975–981.
129. STOLZ, J.F. *et al.* 2006. Arsenic and selenium in microbial metabolism. Annu. Rev. Microbiol. **60:** 107–130.
130. FISCHER, K. *et al.* 2006. Function and structure of the molybdenum cofactor carrier protein from *Chlamydomonas reinhardtii*. J. Biol. Chem. **281:** 30186–30194.
131. VERGNES, A. *et al.* 2004. Involvement of the molybdenum cofactor biosynthetic machinery in the maturation of the *Escherichia coli* nitrate reductase A. J. Biol. Chem. **279:** 41398–41403.
132. SEIFFERT, G.B. *et al.* 2007. Structure of the non-redox-active tungsten/[4Fe:4S] enzyme acetylene hydratase. Proc. Natl. Acad. Sci. USA **104:** 3073–3077.
133. VORHOLT, J.A. & R.K. THAUER. 2002. Molybdenum and tungsten enzymes in C1 metabolism. *In* Metal Ions Biol. Syst., Vol. 39. A. Sigel & H. Sigel, Eds.: 571–619. New York: Marcel Dekker.

Transformation of Inorganic and Organic Arsenic by Alkaliphilus oremlandii *sp. nov. Strain OhILAs*

EDWARD FISHER,[a] ASIA M. DAWSON,[a] GANNA POLSHYNA,[b] JOY LISAK,[a] BRYAN CRABLE,[a] ERANDA PERERA,[b] MRUNALNI RANGANATHAN,[a] MIRUNALNI THANGAVELU,[a] PARTHA BASU,[b] AND JOHN F. STOLZ[a]

[a]*Departments of Biology, and*
[b]*Chemistry and Biochemistry, Duquesne University, Pittsburgh, Pennsylvania, USA*

Alkaliphilus oremlandii **sp. nov. strain OhILAs is a mesophilic, spore-forming, motile, low mole%GC gram positive. It was enriched from Ohio River sediments on a basal medium with 20 mM lactate and 5 mM arsenate and isolated through passage on medium with increased arsenic concentration (10 and 20 mM), tindalization, and serial dilution. The pH optimal for growth was 8.4 and 16S rRNA gene sequence analysis indicated it is most closely related to species in the genus *Alkaliphilus* (*A. crotonoxidans* 95%, *A. auruminator* 95%, *A. metalliredigens*, 94%). A strict anaerobe, it can ferment lactate via the acrylate pathway as well as fructose and glycerol. *A. oremlandii* also has respiratory capability, as it is able to use arsenate and thiosulfate as terminal electron acceptors with acetate, pyruvate, formate, lactate, fumarate, glycerol, or fructose as the electron donor. A respiratory arsenate reductase, which is constitutively expressed, has been identified through biochemical and Western blot analyses and confirmed by cloning and sequencing of the gene encoding the structural subunit *arrA*. The entire *arr* operon as well as the *ars* operon have also been identified in the fully annotated genome. *A. oremlandii* also transforms the organoarsenical 3-nitro-4-hydroxy benzene arsonic acid (roxarsone). Growth experiments and genomic analysis suggest that it couples the reduction of the nitro group of the organoarsenical to the oxidation of either lactate or fructose in a dissimilatory manner, generating ATP via a sodium dependent ATP synthase.**

Key words: **arsenic; roxarsone; organoarsenical; 3-amino-4-hydroxybenzene arsonic acid**

Introduction

Despite an overall low crustal abundance (0.00018%), arsenic may be present in certain environments at high enough concentrations to be a significant environmental factor affecting the local ecology.[1] Its predominant oxidation states are arsine (−3), elemental As (0), arsenate (+5), and arsenite (+3). Arsenate ($HAsO_4^{2-}$ and $H_2AsO_4^{1-}$) and arsenite (H_3AsO^0 and $H_2AsO_3^{2-}$) are the more common species in soil and water and are typically associated with oxic and anoxic conditions, respectively.[1] The toxicology of arsenic is well known; arsenate is an analog of phosphate and can disrupt oxidative phosphorylation, while arsenite binds to thiol groups and can inhibit crucial respiratory enzymes (e.g., pyruvate dehydrogenase complex).[2] Nevertheless, over the last decade an increasing number of prokaryotes have been described that metabolize arsenic.[3] Most striking is the discovery of dissimilatory arsenate-reducing bacteria and chemolithoautotrophic arsenite–oxidizing bacteria, organisms that use these oxyanions for energy generation. Many new species of arsenic-resistant species have been identified as well, many of which have a chromosomal locus for the resistance genes. Progress has also been made on understanding the process of methylation.[4] Most recently, microorganisms have been implicated in the transformation of organoarsenicals.[5–7] Ecologically, studies of environments with elevated concentrations of arsenic revealed diverse arsenic metabolizing microbial communities and a robust biogeochemical cycle.[8–10] More importantly, the activity of these organisms have been shown to impact the speciation and mobilization of arsenic in the environment.

Naturally occurring arsenic-containing minerals, such as arsenopyrite (and other arsenosulfides), iron

Address for correspondence: John F. Stolz, Department of Biological Sciences, Duquesne University, Pittsburgh, PA 15282. Voice: 412 396 6333; fax: 412 396 5907.
stolz@duq.edu

hydroxides, and aluminosilicates (e.g., clays), as well as geothermal sources, are the major contributors of arsenic in the environment.[11] Anthropogenic inputs, however, may also be considerable. While the use of inorganic arsenicals as pesticides and in wood preservation has been banned or voluntarily discontinued, organoarsenicals, such as roxarsone (3-nitro-4-hydroxy benzene arsonic acid) and monosodium and disodium methyl arsonate (MSMA and DSMA), continue to see widespread use. Roxarsone is used in the poultry industry as a feed additive to prevent coccidiosis, but has the added benefit of stimulating growth and improving pigmentation.[12,13] Of the more than 8 billion broiler chickens raised in the United States annually, a significant percentage (~70%) have been fed roxarsone, resulting in the release of an estimated 9×10^5 kg of the organoarsenical into the environment.[14–16] MSMA and DSMA are used in the production of cotton and for the maintenance of golf courses, with almost 3000 tons of MSMA and DSMA applied to cotton, citrus, and sod annually.[17] More importantly, the application may be confined to certain geographic locations, resulting in localized "hot spots."

Arsenic in water is a major environmental problem in many areas in the world including South America, India, China, Vietnam, and Cambodia.[18,19] In Bangladesh, it is estimated that over 33 million people use arsenic-contaminated water, and arsenicosis (poisoning due to chronic exposure) has reached epidemic proportions. Arsenic contamination is also a concern in the United States, as groundwater is the source of about 50% of the drinking water. A recent study of 1600 public and private water sources in six New England states indicated that close to a third had levels of arsenic exceeding 10 ppb.[20] Different processes can result in arsenic-contaminated aquifers, including (1) the oxidation of arsenic-containing pyrites, (2) the release of As(V) from reduction of iron oxides by autochthonous organic matter, (3) the reduction of iron oxides by allochthonous organic matter from dissolved organics in recharging waters, (4) the exchange of adsorbed As(V) with fertilizer phosphates, and (5) recharge with water contaminated with inorganic arsenic derived from wood preservatives, pesticides (e.g., calcium arsenate), and organoarsenicals (e.g., roxarsone, monomethyl arsonic acid), or contamination as a result of the production of these arsenicals.[1,18] Microbial activities are either directly involved or enhance these processes. For example, arsenic bound to ferrioxyhdroxides can be liberated by arsenate respiring bacteria.[21,22] Studies on aquifer sediments in Bangladesh[23–25] and lake sediments in Massachusetts[26] have indicated that microbial iron reduction, arsenate reduction, and arsenite oxidation affect solubility. Thus, specific processes may be directly associated with specific microbial species or communities.

Microbes metabolize inorganic arsenic primarily for resistance or energy generation. Organoarsenicals may also be used in energy generation or as a carbon source. Depending on the form (i.e., inorganic or organic) and the purpose, the transformation may involve oxidation, reduction, methylation, or demethylation. Bacterial oxidation of arsenite to arsenate may be used for both detoxification and energy generation. There are over 30 strains representing at least nine genera of arsenite oxidizing prokaryotes, including alpha-, beta-, and gamma- proteobacteria, Deinocci (i.e., *Thermus*), and Crenarchaeota.[27] Physiologically diverse, they include heterotrophic and chemolithoautotrophic species.[28–32] Microbial arsenite oxidation has been known for almost a century, first being reported in 1918.[33] It has been only recently, however, that chemolithoautotrophic species have been identified. Aerobic species such as strain *Rhizobium* sp. NT-26 oxidize arsenite with oxygen as the teminal electron acceptor.[31] Anaerobic species such as *Alkalilimnicola ehrlichii* strain MLHE-1, *Azoarcus* sp. strain DAO-1, and *Sinorhizobium* sp. strain DAO-10 use nitrate as the terminal electron acceptor instead.[34,35] Both strains DAO-1 and DAO-10 are capable of full denitrification, whereas in *A. ehrlichii* the end product is nitrite. Interestingly, analysis of the complete genome of *A. ehrlichii* indicates that it does have operons for nitric oxide (*norDQBC*) and nitrous oxide (*nosLYDZR*) reductases. However, neither *nirK* (copper-containing nitrite reductase) nor *nirS* (cd_1 nitrite reductase) homologs were found, indicating the absence of the enzyme that catalyzes a critical step in denitrification (the reduction of nitrite to nitric oxide).[34] Whether the oxidation of arsenite is coupled to energy generation or not, in most species the process involves the same enzyme, arsenite oxidase (AoxAB or AroBA). The enzyme is a member of the dimethyl sulfoxide reductase (DMSOR) family of molybdoenzymes, as the catalytic subunit AoxB contains a molybdopterin cofactor and an iron–sulfur cluster. The smaller electron-transfer subunit AoxA is unique among this family, as it is a Rieske-type iron–sulfur cluster protein (2Fe-2S) and contains the leader sequence that targets the complex to the periplasm.[36] The genes encoding the catalytic subunit (*aoxB*) and Rieske-type iron–sulfur protein (*aoxA*) are highly conserved and have even been identified in green phototrophic bacteria (e.g., *Chloroflexus aurantiacus*, *Chlorobium limnicola*).[3] Additional genes, including those encoding regulatory elements, may be present. The same gene cluster has been identified in *Rhizobium* sp. strain NT-26 and *Agrobacterium tumefaciens* (*aoxRSABCmoeA*), where *aoxR* and *aoxS* are part of a two-component regulatory system and *aoxC* encodes a *c*-type cytochrome.[3,37] Several studies have used primers designed from the gene for the catalytic subunit (*aoxB* or *aroA*) to investigate the occurrence and distribution of arsenite oxidizing species in natural environments.[38,39] The absence of genes encoding an AoxAB homolog in the genome of *A. ehrlichii* and the failure to detect *aroB* in strain DOA-1, however, suggest that there may be other forms of arsenite oxidase.[34,35]

The dissimilatory reduction of arsenate by a bacterium was first reported in 1994 and subsequently an additional 20 species of prokaryotes were shown to have the ability to use arsenate as a terminal electron acceptor.[40–43] They include representatives from the Crenarchaeota, thermophilic bacteria, gamma-, delta-, and epsilon-proteobacteria, and low and high mole%GC gram-positive bacteria.[1,44–47] Metabolically diverse, most are capable of using other terminal electron acceptors.[3] Acetate, lactate, pyruvate, formate, aromatics (syringic acid, ferulic acid, phenol, benzoate, toluene) as well hydrogen can serve as electron donors.[3] Searles Lake strain SLAS-1 and Mono Lake strain MLMS-1, a chemolithoautotroph, have been shown to couple the reduction of arsenate to the oxidation of hydrogen sulfide.[9,48] The respiratory arsenate reductase (Arr) has been purified from two species, *Chrysiogenes arsenatis* and *Bacillus selenitireducens*.[49,50] The enzyme, a heterodimer (ArrAB), is also a member of the DMSOR family of molybdoenzymes, but differs significantly from arsenite oxidase. Phylogenetic analysis of the protein sequence places the catalytic subunit, ArrA, closer to thiosulfate and polysulfide reductase.[3] The amino acids that constitute the catalytic pocket are also highly conserved and are different than those in AoxB.[3] The smaller electron-transfer protein subunit, ArrB, contains four iron–sulfur clusters and does not have the leader sequence. Although the genes encoding ArrA and ArrB are highly conserved, additional subunits (e.g., a membrane-anchoring protein ArrC, and a chaperon-like protein ArrD) as well as regulatory components may be present, but they vary from species to species.[3] Several research groups have used the *arrA* gene to design primers to investigate the occurrence and distribution of arsenate reducing bacteria.[19,51,52] Although these studies have been successful in amplifying *arrA* genes from various environments, not all organisms are recognized by any given set of primers (e.g., there is no "universal" set of primers). Thus, multiple sets of primers must be employed.

Many more microorganisms are known to be resistant to arsenic via the ars system.[53] Originally, the genes for arsenic resistance were discovered on plasmids.[54] More recently, however, chromosomal loci have been identified in over 50 organisms, including archaea, bacteria, yeasts, and protoctists.[55] Three different but comparable systems (e.g., *Escherichia coli*, *Staphylococcus*, *Saccharomyces*) have evolved through convergent evolution.[54] They are composed of an arsenate reductase (ArsC, ACR2), which catalyzes the reduction of arsenate to arsenite, and an arsenite-specific efflux pump (ArsB, ACR3). Additional components include an ATPase (ArsA) (which, when present, forms the ArsAB complex), and regulatory elements (ArsR, ArsD). Those organisms possessing the ATP-independent resistance system, such as *Staphylococcus aureus*, use thioredoxin as the source of reducing equivalents, while those that are ATP-dependent (ArsAB), like *E. coli*, use thioredoxin.[53] In some species, multiple *arsC* genes may be present, increasing the level of resistance.[56] Many species of arsenite-oxidizing and arsenate-reducing bacteria also have the arsC resistance system, perhaps aiding the cell in maintaining the internal concentration of arsenic. Mutants of *A. tumefaciens* deficient in arsenite oxidase still have an active arsC resistance system and express an arsenate-reducing phenotype.[37] This presents the interesting possibility that the organism may simultaneously oxidize arsenite in the periplasm while reducing arsenate in the cytoplasm.

Far less is known about methylation and demethylation. Demethylation has been demonstrated in organisms that can use methylated arsenicals as a carbon source.[57] Little is known, however, about the mechanism involved.[3] In methylation, the reduction of arsenate is followed by oxidative addition of a methyl group and may result in the formation of methylarsenite, dimethylarsenate, dimethylarsenite, and trimethylarsine oxide.[3] Methylated arsenicals can be generated by different processes with a different source of the methyl group (*S*-adenosylmethionine (SAM), methylcobalamin) and transfer reaction. Recently, a SAM-dependent methylase, ArsM, has been characterized in *Rhodopseudomonas palustris*. The cells methylated arsenite to dimethylarsenate (DMA(V)) and trimethylarsine oxide (TMAO). The purified enzyme converted arsenite to DMA(V) or DMA(V) to TMAO when glutathione and SAM were present in the reaction mixture.[4] Homologs were subsequently identified in 125 bacteria and 16 archaea, suggesting this is a common mechanism for arsenic detoxification and may possibly play a role in biogeochemical cycling.[4]

As part of our ongoing investigation of arsenic in microbial metabolism we have isolated a low mole% GC gram-positive organism that is capable of both anaerobic respiration using arsenate as the electron acceptor and transforming the organoarsenical roxarsone. We describe here the general characterization of the organism, describe its ability to metabolize arsenicals, and propose the name *Alkaliphilus oremlandii* sp. nov. strain OhILAs.

Materials and Methods

Enrichment and Isolation

The basal medium used in enrichment and pure cultures was essentially the same as used in Reference 58 with the exception that ammonium chloride and magnesium chloride were substitute for $(NH_4)_2SO_4$ and $MgSO_4$, the pH was adjusted to 7.5, and the donor and acceptors were added separately. The medium was dispensed into Wheaton bottles (either 50 or 125 mL), and degassed with 80:20 N_2:CO_2 (5 min for the liquid, 2 min for the headspace). Bottles were sealed with butyl rubber stoppers and capped with aluminum crimp tops before autoclaving. Amendments (e.g., electron donors and acceptors) were added using syringes degassed with 80:20 N_2:CO_2.

Sediment samples of the Ohio River were obtained with a ponar dredge sampler and were frozen on site. A slurry was made by injecting approximately 5 grams of sediment into 100 mL of anoxic basal media (in a 125-mL Wheaton bottle) that did not contain electron donor or acceptor. One milliliter of the slurry was then used as an inoculum for bottles of medium that contained 20 mM lactate as the electron donor and 5 mM sodium arsenate. The bottles were incubated at 37°C in the dark and monitored daily for arsenic trisulfide (orpiment) production. Enrichments that showed copious production of orpiment were transferred weekly.

In an attempt to isolate a single strain, enrichment cultures that were positive for orpiment production were transferred to medium containing higher concentrations of arsenate and subjected to tindalization and dilution to extinction. Initially, the cultures (1 mL inoculum) were inoculated into 10 mL of fresh media (in 20-mL tubes) and allowed to grow for 24 h. The tubes were then incubated in a water bath at 60°C for 10 min, followed by incubation for 24 h at 37°C. These cultures were then transferred to medium containing 20 mM arsenate. After 24 h growth, the tubes were incubated at 80°C for 10 min, followed by serial dilution. These tubes were then incubated at 37°C overnight. In this manner a spore-forming rod was isolated (strain OhILAs). The isolate was maintained on the basal medium with lactate as the electron donor and sodium arsenate (10 mM) as the terminal electron acceptor. Cells (2–3 mL innoculum per 100 mL of fresh medium) were transferred once a week. The transfer syringes were degassed with 80:20 N_2:CO_2. Cultures were incubated at ambient temperature.

Growth Experiments

For growth experiments, basal medium was prepared in 20-mL volumes in 50-mL Wheaton bottles.

Cultures were transferred six times in each medium before evaluation. Growth was determined as optical density at 600 nm on a Perkin-Elmer Lamda 2 dual-beam spectrophotometer. To determine the optimal concentration for arsenate as the terminal electron acceptor, cultures were grown on 20 mM acetate and different arsenate concentrations (1, 5, 10, 20, 30, and 40 mM). Other electron acceptors tested (at 20 mM) were sodium arsenite, sodium nitrate, sodium sulfate, sodium thiosulfate, sodium selenate, sodium selenite, and dimethylsulfoxide. To determine the suite of electron donors, cultures were grown on medium containing 10 mM arsenate and either 20 mM acetate, pyruvate, formate, lactate, fumarate, glycerol, or no donor (control). The pH optimum was determined using basal medium containing 10 mM arsenate and 20 mM acetate at different pH. The pH was measured after autoclaving (4.3, 5.4, 6.9, 7.6, 8.1, 8.25, 8.35, 8.44, and 8.8). The same medium with different concentrations of NaCl (0, 0.1, 0.5, 1.0, 2.5, 5.0, 7.5, 10, 15, and 20 g/L) was used to determine the salinity optimum. Cultures grown in basal medium containing 5 mM arsenate and 10 mM formate incubated at 4°, 10°, 22°, 32°, 37°, 4°, or 50°C were used to determine the temperature optimum. To examine the affect of arsenite concentration, cultures were grown in basal medium with 10 mM thiosulfate and 10 mM acetate containing different concentrations of arsenite (0, 0.1, 0.5, 1, 5, 7, 8, 9, and 10 mM). All growth experiments were done in triplicate with the standard deviation calculated using Prism (GraphPad Software, San Diego, CA).

Transformation of 3-Nitro-4-Hydroxy Benzene Arsonic Acid (Roxarsone)

To determine the ability of *Alkaliphilus* sp. strain OhILAs to use roxarsone as a growth substrate, the basal medium was amended with either 20 mM lactate or fructose and 1 mM roxarsone. Controls contained lactate, fructose, or roxarsone alone. The media were prepared as described earlier, in 50-mL volumes in 125-mL Wheaton bottles. Absorbance at 600 nm was determined to monitor growth and the measured optical densities were calibrated by direct cell counts.

Morphology

Light micrographs were taken using a Nikon Microphot SA equipped with a Kodak DC290 digital camera. Transmission electron microscopy was done using the methods described in Reference 59, with thin sections observed on the JEOL 100CX transmission electron microscope at 60 kV. Images were captured using an SIA-7C digital camera.

Cloning of the 16S rRNA and arrA Genes, and Phylogenetic Analysis

The 16S rRNA gene was cloned and sequenced using the methods described in Reference 60 and assigned the GenBank accession number DQ250645. The sequences of the closest relatives based on a BLAST search were obtained from GenBank. Sequence alignments were done using ClustalX and a phylogenetic tree constructed with maximum parsimony using PAUP* 4.0b (Sinauer Associates Inc., Sunderland, MA). Mole% GC was determined through total genome analysis courtesy of the DOE Joint Genome Institute (http://genome.jgi-psf.org/draft_microbes/clo_o/clo_o.home.html).

An 852-bp fragment of the *arrA* gene was amplified using degenerate primers (ArrAUF1 TGTCAAGGHTGTACBDCHTGG, ArrAUR3 GCW GCC CAY TCV GGN GT) designed by M. Berekaa (University of Alexandria, Egypt). The product was amplified using the RedMix Plus (PGC Scientific Corporation, Gaithersburg, MD) with the following parameters: denaturing at 95°C, annealing at 60°C, extension at 72°C. The amplicon was cloned into the pCR-4-TOPO sequencing vector (Invitrogen, Carlsbad, CA). Sequencing was done using an ABI 300 capillary sequencer at Duquesne University using the BigDye kit (Applied Biosystems, Foster City, CA). The inferred amino acid sequence data were determined using the NCBI Orf finder program. The sequence was also used for BLAST searches (www.ncbi.nlm.nih.gov"blastn," "blastp," "blastx") to look for close matches in the database.

E. Chemical, Biochemical and Western Blot Analyses

As(V), As(III), roxarsone, 3-amino-4-hydroxybenzene arsonic acid, 4-hydroxybenzene arsonic acid, 2-nitrophenol, 3-aminophenol, lactate, acetate, and proprionate were detected using the methods described in Reference 7. Respiratory arsenate reductase activity was detected by both spectrophotometric and in gel assays with reduced methyl viologen as the artificial electron donor using the methods described in Reference 50. Western blot analysis, using polyclonal antibodies raised against a 15 amino acid peptide fragment from ArrA (the catalytic subunit of the respiratory arsenate reductase), was done using the methods in Reference 61.

Results

General Characterization

Strain OhILAs occurs as single motile rods or chains measuring 0.5 μm in diameter and up to 2 μm in length (FIG. 1A). Spores, when they occurred, were terminally located. Cells stained gram positive and ultrastructural analysis indicated the presence of a gram-positive cell wall (FIG. 1B). Strain OhILAs exhibited metabolic versatility, as it could couple the reduction of As(V) to the oxidation of a variety of electron donors. It was also capable of fermentative growth. Growth on acetate, formate, lactate, pyruvate, fumarate, glycerol, and fructose was enhanced with the presence of As(V) when compared to the bottles containing only the donor. Growth on pyruvate, lactate, and fructose also occurred fermentatively. Cultures grown with medium containing pyruvate but no As(V) exhibited greater growth than pyruvate and As(V). Conversely, growth on glycerol and As(V) was greater than glycerol alone; however, a subsequent transfer from the glycerol/As(V) medium to medium containing glycerol alone exhibited growth for five successive transfers at 1/100-mL dilutions. In addition to As(V), strain OhILAs also used thiosulfate as a terminal electron acceptor, but not nitrate, sulfate, fumarate, selenate, or selenite.

The growth kinetics of strain OhILAs was determined using medium containing 20 mM sodium arsenate and 10 mM sodium acetate. Cell growth was concomitant with the reduction of arsenate with equimolar amounts of As(V) reduced to As(III) (FIG. 2A). Stationary phase occurred after 42 h, which coincided with the cessation of As(III) production. The calculated doubling time was 4.7 h.

The optimal temperature for the growth of strain OhILAs was 37°C (FIG. 2B). Growth occurred at 32°C and 44°C, but was inhibited at temperatures below 22°C and above 50°C.

The optimal concentration of NaCl was in the range between 0.1 g/L and 2.5 g/L with significant growth occurring up to 5 g/L (FIG. 2C). Growth was inhibited at NaCl concentrations above 10 g NaCl/L. The optimum pH was approximately 8.4, with growth occurring in between 8.0 and 8.8 (FIG. 2D). No growth was found below pH 7.5 in the medium in which the pH was tested.

Phylogenetic analysis using the 16S rRNA gene sequence indicated that strain OhILAs is most closely

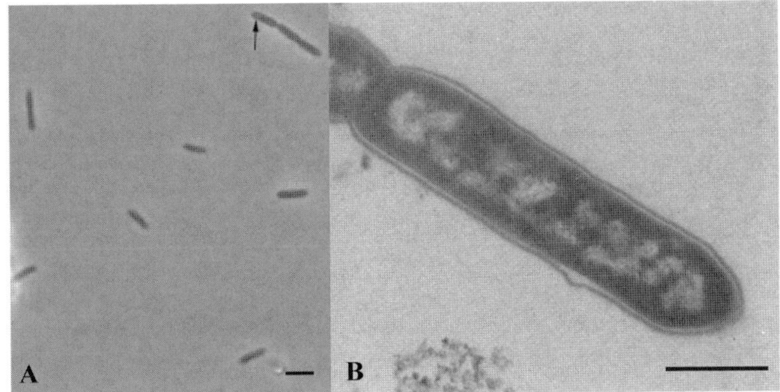

FIGURE 1. Morphology of *Alkaliphilus* sp. strain OhILAs. (**A**) Phase contrast, *arrow* denotes terminal spore. (Bar 2 μm), (**B**) Transmission electron micrograph. (Bar 0.5 μm.)

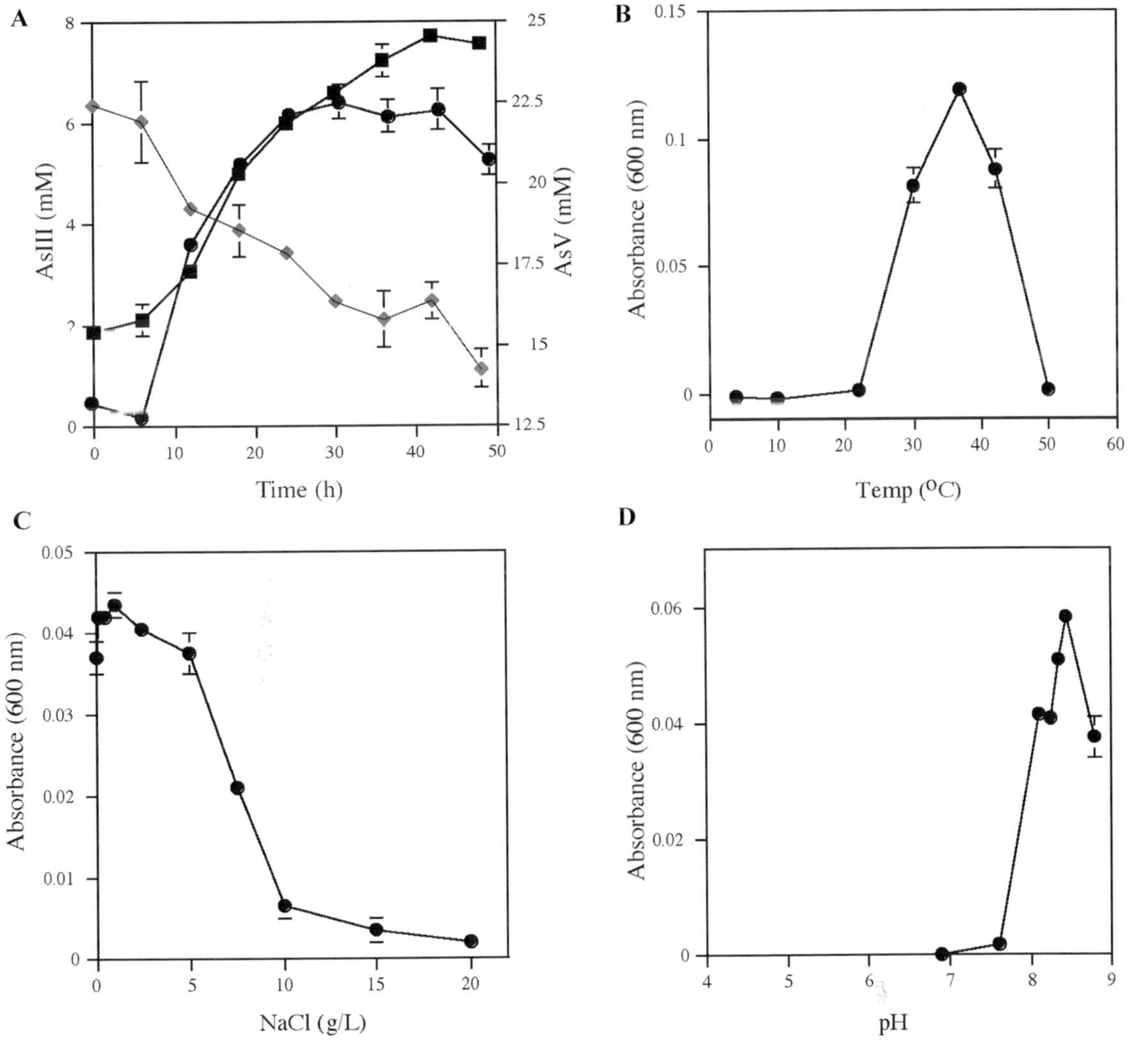

FIGURE 2. Arsenate reduction and growth optimum of *Alkaliphilus* sp. strain OhILAs (**A**) cell growth (●) with arsenate consumption (◆) and arsenite production (■). The axis for cell growth is not shown, but cell numbers increased from 1.5×10^4 (inoculum) to 7.5×10^5 cells/mL, (**B**) temperature optimum, (**C**) sodium chloride optimum, (**D**) pH optimum.

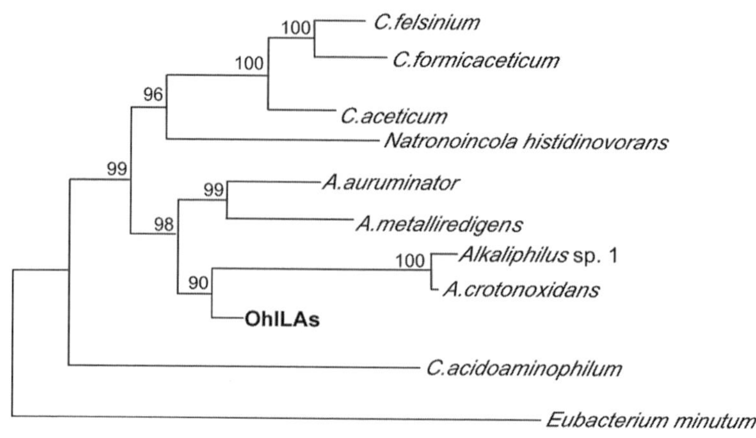

FIGURE 3. Phylogenetic tree (neighbor joining) showing the relatedness of *Alkaliphilus* sp. strain OhILAs to *Alkaliphilus* species.

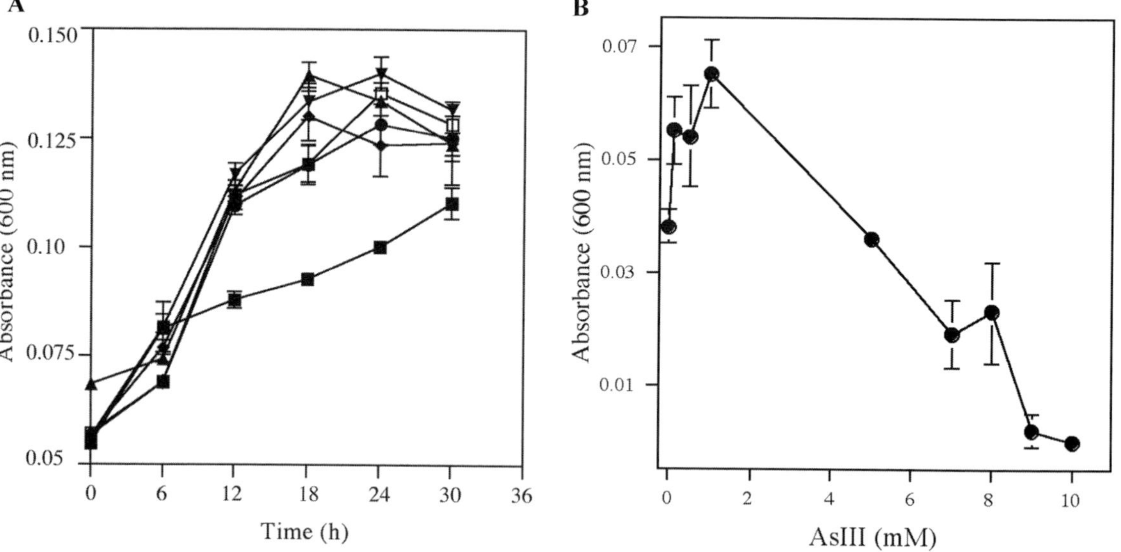

FIGURE 4. Affects of arsenate and arsenite concentration on the growth of *Alkaliphilus* sp. strain OhILAs. (**A**) arsenate concentrations 1 (■), 5 (▲), 10 (▼), 20 (●), 30 (◆), 40 (□), (**B**) arsenite.

related to *Alkaliphilus* species, with sequence identities of 95% for *A. crotonoxidans* and *A. auruminator*, and 94% for *A. metalliredigens* (FIG. 3). The closest clostridial species were *C. aceticum* (92%), *C. felsineum* (92%), and *C. formicoaceticum* (91.7%).

Inorganic Arsenic Metabolism

Strain OhILAs exhibited growth at all concentrations of As(V) tested (1, 5, 10, 20, 30, 40 mM) (FIG. 4A). The highest cell yields were obtained at concentrations of 5 mM and 10 mM arsenate, while the significantly lower yield with 1 mM arsenate suggested that the concentration was not sufficient to sustain robust respiratory growth. The growth curves for all other concentrations tested were similar, as exponential growth occurred from hour 6 to hour 24. Maximum cell yield and growth rate occurred with 5 mM As(V), suggesting that this was the optimal concentration for growth.

Regardless of whether the cultures were grown with 10, 20, 30, or 40 mM As(V), only a maximum of 10 mM was consumed. Thus, the affect on growth of As(III) concentration was investigated (FIG. 4B). Growth of strain OhILAs was completely inhibited by 10 mM As(III). Interestingly, growth seemed to be enhanced at low concentrations of As(III), including 0.1, 0.5, and 1 mM. Acidification of the cultures with HCl allowed

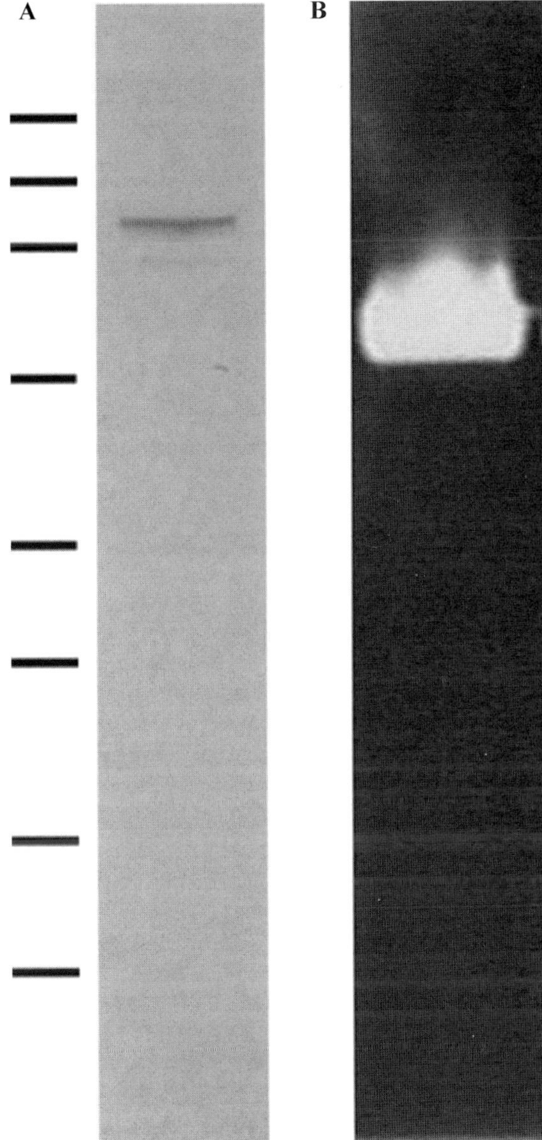

FIGURE 5. The respiratory arsenate reductase of *Alkaliphilus* sp. strain OhILAs. (**A**) Western blot analysis using anti-ArrA of cell lysate with a 12% polyacrylamide gel. Molecular-weight standards 206, 115, 98, 54, 37, 29, 20, 7 kD. (**B**) In gel activity assay showing active arsenate reductase in nondenaturing polyacrylamide gel. The active complex migrates further into the 12% polyacrylamide gel.

for the formation of orpiment in sample sets containing concentrations of 0.5 mM As(III) and higher, providing evidence that the thiosulfate had been reduced to sulfide (data not shown).

In an effort to further establish that strain OhILAs was capable of arsenate respiration, an attempt was made to identify the respiratory arsenate reductase.

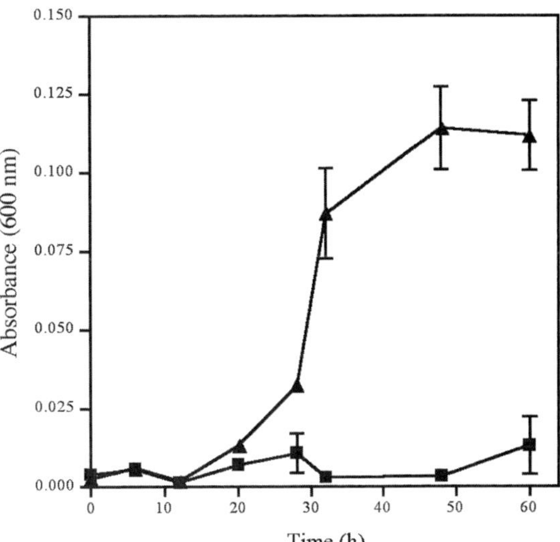

FIGURE 6. Growth of *Alkaliphilus* sp. strain OhILAs with lactate and roxarsone (▲) or lactate alone (■).

Arsenate reductase activity was detected in cell lysates prepared from cultures grown on 20 mM lactate and 10 mM As(V), both spectrophotometrically (data not shown), and in gel assays (FIG. 5A). Additional evidence was provided by Western blot analysis, which showed that the protein band exhibiting arsenate reductase activity also cross-reacted with the anti-ArrA polyclonal antisera (FIG. 5B). An 852-bp portion of the 5′ end of the gene encoding the catalytic subunit (*arrA*) was cloned and sequenced, yielding an unambiguous inferred amino acid sequence of 252 residues. The inferred amino acid sequence shares a 63% identity and 77% similarity with ArrA from two other arsenate-respiring gram-positive bacteria *Bacillus arseniciselenatis* and *Desulfitobacterium hafniense*.[50,62]

Organic Arsenic Metabolism

Strain OhILAs rapidly transformed roxarsone when either lactate or fructose (FIG. 6) was provided in the medium. The cell yields were typically ~10-fold greater on either lactate or fructose (FIG. 6) when roxarsone was provided, as compared to growth on lactate or fructose alone. There was, however, a noticeable, prolonged lag phase in the cultures grown on roxarsone (FIG. 6) that may be attributed to the cells acclimating to the medium (i.e., the cells used for the inoculum were grown under fermentative conditions with lactate or fructose). Little roxarsone transformation occurred in the basal medium without the addition of carbon source or additional yeast extract (3 g/L). For either lactate or fructose, roxarsone was completely transformed with the primary end-products detected being

3A4HBAA and inorganic arsenic (As(V)). Attempts to grow strain OhILAs with 3A4HBAA or 4HBAA resulted in cell yields comparable to lactate or fructose alone, and no detectable transformation of either 3A4HBAA or 4HBAA (data not shown).

Discussion

The investigation into the physiology of a new species of *Alkaliphilus*, strain OhILAs, has demonstrated that it is a metabolically versatile anaerobe. It can ferment lactate, fructose, glycerol, and yeast, and is capable of anaerobic respiration with arsenate or thiosulfate as terminal electron acceptors with acetate, pyruvate, formate, lactate, fumarate, glycerol, or fructose as the electron donor. Its temperature optimum is 37°C, although it is not pathogenic, while the salinity (1 g/L NaCl) and pH optimum (8.4) are indicative of an organism from a lacustrine environment (Fig. 2).

The recent completion of the fully annotated genome of *Alkaliphilus* sp. strain OhILAs has allowed for further validation of its metabolic capability. All known components of the acrylate pathway (e.g., lactate dehydrogenase, pyruvate-ferredoxin oxidoreductase, phosphotransacetylase, acetate kinase) are present. Conversely, components of the succinate-propionate pathway (e.g., malate dehydrogenase, fumarate hydratase) are missing. An almost complete glycolytic pathway is also present; however, glucokinase and glucose-6-phosphate isomerase are missing thus explaining why *Alkaliphilus* sp. strain OhILAs can utilize fructose but not glucose. Furthermore, an ability for Stickland fermentation is suggested based on the presence of a glycine reductase. Genomic analysis also confirms the ability to respire sulfur oxyanions with the identification of the tetrathionate reductase (Ttr) and dissimilatory sulfite reductase (Dsr). Interestingly, the annotation suggests the presence of a selenocysteine in the catalytic subunit of Ttr, which if verified through biochemical analysis, would be a first.

Although the sediments from which it was isolated typically have low levels of arsenic (4 ppb in the water),[63] *Alkaliphilus* sp. strain OhILAs is capable of metabolizing both inorganic and organic arsenic. While resistant to elevated concentration of arsenate (at least 40 mM), it is sensitive to elevated levels of arsenite (Fig. 4). When grown in medium containing lactate and arsenate, equal molar amounts of arsenate were converted to arsenite concomitant with cell growth (Fig. 1). Arsenate reductase activity and positive Western blot analysis with anti-ArrA indicate the presence of a respiratory arsenate reductase (Fig. 5), further confirmed through the identification of the *arr* operon (*arrCAB*) in the genome. The organism should also be resistant, as it is also possesses an *ars* operon (*arsDARCB*). Interestingly, an *arsM* homolog is also present, but as part of the *arr* operon (immediately upstream of *arrC*) and not the *ars* operon as reported for *R. palustris*.[4]

The ability of *Alkaliphilus* strain OhILAs to transform the organoarsenical roxarsone was of particular importance, as clostridial species are common in both the chicken gastrointestinal tract as well as the litter.[64,65] In separate experiments, roxarsone was provided as either the electron donor/carbon source (with thiosulfate as the electron acceptor) or electron acceptor with lactate or fructose (as the electron donor and carbon source). Strain OhILAs was found to rapidly convert roxarsone to 3A4HBAA with the concomitant oxidation of lactate.[7] The data suggest the following reaction stoichiometry:

$$\begin{aligned} &2\ \text{roxarsone} + 3\ \text{lactate} \rightarrow 2\ \text{3A4HBAA} \\ &\quad + 3\ \text{acetate} + 3\,CO_2 + H_2O \end{aligned} \tag{1}$$

The reduction of nitro groups on aromatic compounds by clostridia is well known.[66,67] Two homologs of Pentaerythritol tetranitrate reductase (NemA), three nitroreductases, carbon monoxide dehydrogenase, and hydrogenase, all implicated in the reduction of nitro groups on aromatic compounds, have been identified in the genome. Inorganic arsenic was also detected in the cultures subsequent to the generation of 3A4HBAA; however, the mechanism by which it was generated (whether metabolically or abiotic) is not as yet known.[7] Interestingly, cell yields in medium with roxarsone with lactate (as in Ref. 7) or fructose (Fig. 6) were ~10-fold greater than medium without roxarsone. The recent description of the reduction of the phenylacrylate ester caffeate coupled to a Na^+ dependent F_1/F_0 ATPase suggests a possible mechanism.[68] *Alkaliphilus* sp. strain OhILAs does have homologs of Na^+-dependent F_1/F_0 ATPase, NADH dehydrogenase, and hydrogenase; however, confirmation awaits further investigation.

Phylogenetic analysis clearly places the organism within Group XI of clostridia (Fig. 3).[69] Although it was isolated from freshwater river sediments, it clusters with alkaliphilic strains. To date, three species of *Alkaliphilus* have been described and the range and optimum for growth varies significantly. The type species of the genus, *A. transvaalensis* strain SAGM1, has a pH range of 8.5 to 12.5 and optimum of 10.[69] The pH optimum (and range) for *A. crotonaoxidans* strain B11-2 and *A. metalliredigens* strain QYMF are 7.5 (5.5–9) and 9.5 (7.5–11), respectively.[70,71] The pH range (8–9) and optimum (8.4) for strain OhILAs falls well within this range. A comparison between the fully annotated genome of *A. metalliredigens* and strain OhILAs reveals a significant difference in genome size (2.7 MB for strain OhILAs compared with 4.9 MB for *A. metalliredigenes*). Of the 2741 predicted open reading frames in the strain OhILAs genome, only 77% have homologs in *A. metalliredigens*. The 36.1% mole GC content compares with 30.6 for *A. crotonaoxidans*, 36.4 for *A. transvaalensis*, and 36.8 for *A. metalliredigens*. Based on phylogenetic affiliation, ability to grow at alkaline pH, and the genomic comparison, strain OhILAs warrants the assignment to the genus *Alkaliphilus* and the designation of a new species. We propose *Alkaliphilus oremlandii*. The strain has been deposited with the American Type Culture Collection and assigned the accession number of BAA1360.

Species Description

Alkaliphilus oremlandii (or.em.lan'di.i. N.L. masc. gen. n. *oremlandii* of Oremland, in honor of Ronald S. Oremland of the U.S. Geological Survey, Menlo Park, CA). Spore-forming, motile, gram-positive rod. 36.1% mole GC. Cells are 0.5 μm in diameter and up to 2 μm in length, often forming chains. Ferments glycerol, lactate, pyruvate, fructose, and yeast. Anaerobic respiration with arsenate, thiosulfate, and 3-nitro-4-hydroxybenzene arsonic acid as electron acceptors, acetate, lactate, and fructose as electron donors. Strict anaerobe isolated from sediments from the Ohio River, PA. The type strain is OhILAs^T (ATCC BAA1360^T).

Acknowledgments

This work was funded in part by the USGS National Water Resources Institute. Thanks to Jonathan Franks for electron microscopy, Mahmoud Berekaa for primer design, Lars Ljundal for helpful discussion, and Aharon Oren for etymology.

Conflict of Interest

The authors declare no conflicts of interest.

References

1. OREMLAND, R.S. & J.F. STOLZ. 2003. Ecology of arsenic. Science **300:** 939–944.
2. NATIONAL RESEARCH COUNCIL. 1999. Arsenic in Drinking Water. Washington, D.C.: National Academy Press.
3. STOLZ, J.F. *et al.* 2006. Arsenic and selenium in microbial metabolism. Ann. Rev. Microbiol. **60:** 107–130.
4. QIN, J. *et al.* 2006. Arsenic detoxification and evolution of trimethylarsine gas by a microbial arsenite S-adenosylmethionine methyltransferase. Proc. Natl. Acad. Sci. USA **103:** 2075–2080.
5. SIERRA-ALVAREZ, R. *et al.* 2006. Anaerobic biotransformation of organoarsenical pesticides monomethylarsonic acid and dimethylarsinic acid. J. Agric. Food Chem. **54:** 3959–3966.
6. CORTINAS, I. *et al.* 2006. Anaerobic biotransformation of roxarsone and related N-substituted phenylarsonic acids. Environ. Sci. Technol. **40:** 2951–2957.
7. STOLZ, J.F. *et al.* 2006. Biotransformation of 3-nitro-4-hydroxybenzene arsonic acid (roxarsone) and release of inorganic arsenic by clostridium species. Environ. Sci. Technol. **41:** 818–823.
8. OREMLAND, R.S. *et al.* 2004. The microbial arsenic cycle in Mono Lake, California. FEMS Microbiol. Ecol. **48:** 14–27.
9. OREMLAND, R.S. *et al.* 2005. A microbial arsenic cycle in a salt-saturated, extreme environment. Science **308:** 1305–1308.
10. KULP, T.R. *et al.* 2006. Dissimilatory arsenate and sulfate reduction in sedements of two hypersaline, arsenic-rich soda lakes: Mono and Searles Lakes, California. Appl. Environ. Microbiol. **72:** 6514–6526.
11. WELCH, A.H. *et al.* 2000. Arsenic in groundwater of the United States: occurrence and geochemistry. Groundwater **38:** 589–604.
12. CZARNECKI, G.L. & D.H. BAKER. 1982. Roxarsone toxicity in the chick as influenced by dietary cysteine and copper and by experimental infection with *Eimeria acervulina*. Poult. Sci. **61:** 516–523.
13. CHAPMAN, H.D. & Z.B. JOHNSON. 2002. Use of antibiotics and roxarsone in broiler chickens in the USA: analysis for the years 1995 to 2000. Poult. Sci. **81:** 356–364.
14. GARBARINO, J.R. *et al.* 2003. Environmental fate of roxarsone in poultry litter. I. Degrading of roxarsone during composting. Environ. Sci. Technol. **37:** 1509–1514.
15. RUTHERFORD, D.W. *et al.* 2003. Environmental fate of roxarsone in poultry litter. II. Mobility of arsenic in soils amended with poultry litter. Environ. Sci. Technol. **37:** 1515–1520.
16. CHRISTEN, K. 2001. Chickens, manure, and arsenic. Environ. Sci. Tech. **35:** 184A–185A.
17. NATIONAL CENTER FOR FOOD AND AGRICULTURAL POLICY. 2003. http://www.ncfap.org/database/default.htm. (accessed December 17, 2007).
18. OREMLAND, R.S. & J.F. STOLZ. 2005. Arsenic, microbes and contaminated aquifers. Trends Microbiol. **13:** 45–49.
19. LEAR, G. *et al.* 2007. Molecular analysis of arsenate-reducing bacteria within Cambodian sediments following amendment with acetate. Appl. Environ. Microbiol. **73:** 1041–1048.
20. AYOTTE, J.D. *et al.* 2003. Arsenic groundwater in eastern New England: occurrence, controls, and human health implications. Environ. Sci. Technol. **37:** 2075–2083.

21. ZOBRIST, J. *et al.* 2000. Mobilization of arsenite by dissimilatory reduction of adsorbed arsenate. Environ. Sci. Technol. **34:** 4747–4753.

22. ISLAM, F.S. *et al.* 2004. Direct evidence of arsenic release from Bengali sediments due to metal-reducing bacteria. Nature **430:** 68–71.

23. HARVEY, C.F. *et al.* 2002. Arsenic mobility and groundwater extraction in Bangladesh. Science **298:** 1602–1606.

24. HORNEMAN, A. *et al.* 2004. Decoupling of As and Fe release to Bangladesh groundwater under reducing conditions. Part I: Evidence from sediment profiles. Geochim. Cosmochim. Acta **68:** 3459–3473.

25. VAN GEEN, A. *et al.* 2004. Decoupling of As and Fe release to Bangladesh groundwater under reducing conditions. Part II: Evidence from sediment incubations. Geochim. Cosmochim. Acta **68:** 3475–3486.

26. SENN, D.B. & H.F. HEMMOND. 2002. Nitrate controls on iron and arsenic in a suburban lake. Science **296:** 2373–2376.

27. STOLZ, J.F. *et al.* 2002. Microbial transformation of elements: the case of arsenic and selenium. Int. Microbiol. **5:** 201–207.

28. GIHRING, T.M. *et al.* 2001. Rapid arsenite oxidation by *Thermus aquaticus* and *Thermus thermophilus*: field and laboratory observations. Environ. Sci. Technol. **35:** 3857–3862.

29. GIHRING, T.M. & J.F. BANFIELD. 2001. Arsenite oxidation and arsenate respiration by a new *Thermus* isolate. FEMS Microbiol. Letts. **204:** 335–340.

30. SALMASSI, T.M. *et al.* 2002. Oxidation of arsenite by *Agrobacterium albertimagni*, AOL15, sp. nov. from Hot Creek, California. Geomicrobiol. J. **19:** 53–66.

31. SANTINI, J.M. *et al.* 2000. A new chemolithoautotrophic arsenite-oxidizing bacterium isolated from a gold mine: phylogenetic, physiological, and preliminary biochemical studies. Appl. Environ. Microbiol. **66:** 92–97.

32. SANTINI, J.M. *et al.* 2002. New arsenite-oxidizing bacteria isolated from Australian gold mining environments—phylogenetic relationships. Geomicrobiol. J. **19:** 67–76.

33. GREEN, H.H. 1918. Description of a bacterium which oxidizes arsenite to arsenate and one of which reduces arsenate to arsenite, isolated from a cattle-dipping tank. S. Afr. J. Sci. **14:** 465–467.

34. HOEFT, S.E. *et al.* 2007. *Alkalilimnicola ehrlichii* sp. nov., a novel arsenite-oxidizing haloalkaliphilic gammaproteobacterium capable of chemoautotrophic or heterotrophic growth with nitrate or oxygen as the electron acceptor. Int. J. Syst. Evol. Microbiol. **57:** 504–512.

35. RHINE, E.D. *et al.* 2006. Anaerobic arsenite oxidation by novel denitrifying isolates. Environ. Microbiol. **8:** 899–908.

36. ANDERSON, G.L. *et al.* 1992. The purification and characterization of arsenite oxidase from *Alcaligenes faecalis*, a molybdenum-containing hydroxylase. J. Biol. Chem. **267:** 23674–23682.

37. KASHYAP, D.R. *et al.* 2006. Complex regulation of arsenite oxidation in *Agrobacterium tumefaciens*. J. Bacteriol. **188:** 1081–1088.

38. RHINE, E.D., *et al.* 2007. The arsenite oxidase genes (aroAB) in novel chemoautotrophic arsenite oxidizers. Biochem. Biophys. Res. Comm. **354:** 662–667.

39. INSKEEP, W.P. *et al.* 2007. Detection, diversity and expression of aerobic bacterial arsenite oxidase genes. Environ. Microbiol. **9:** 934–943.

40. AHMANN, D. *et al.* 1994. Microbe grows by reducing arsenic. Nature **371:** 750.

41. LAVERMAN, A.M. *et al.* 1995. Growth of strain SES-3 with arsenate and other diverse electron acceptors. Appl. Environ. Microbiol. **61:** 3556–3561.

42. MACY, J.M. *et al.* 1996. *Chrysiogenes arsenatis*, gen. nov. sp. nov., a new arsenate-respiring bacterium isolated from gold mine wastewater. Int. J. Syst. Bacteriol. **46:** 1153–1157

43. NEWMAN, D.K. *et al.* 1997. Dissimilatory arsenate and sulfate reduction in *Desulfotomaculum auripigmentum* sp. nov. Arch. Microbiol. **168:** 380–388.

44. OREMLAND, R.S. *et al.* 2001. Bacterial respiration of arsenate and its significance in the environment. *In* Environmental Chemistry of Arsenic. W.T. FRANKENBERGER, Jr., Ed.: 273–295. New York: Marcel Dekker.

45. OREMLAND, R.S. & J.F. STOLZ. 2000. Dissimilatory reduction of selenate and arsenate in nature. *In* Environmental Metal-Microbe Interaction. D.R. Lovley, Ed.: 199–224. Washington, D.C.: ASM Press.

46. STOLZ, J.F. & R.S. OREMLAND. 1999. Bacterial arsenate and selenate reduction. FEMS Microbiol. Rev. **23:** 615–627.

47. MACY, J.M. *et al.* 2000. Two new arsenate/sulfate reducing bacteria: mechanisms of arsenate reduction. Arch. Microbiol. **173:** 49–57.

48. HOEFT, S.E. *et al.* 2004. Dissimilatory arsenate reduction with sulfide as the electron donor: experiments with Mono Lake water and isolation of strain MLMS-1, a chemoautotrophic arsenate respirer. Appl. Environ. Microbiol. **70:** 2741–2747.

49. KRAFFT, T. & J.M. MACY. 1998. Purification and characterization of the respiratory arsenate reductase of Chrysiogenes arsenatis. Eur. J. Biochem. **255:** 647–53.

50. AFKAR, E. *et al.* 2003. The respiratory arsenate reductase from *Bacillus selenitireducens* strain MLS10. FEMS Microbiol. Letts. **226:** 107–112.

51. MALASARN, D. *et al.* 2004. *arrA* is a reliable marker for As(V)-respiration in the environment. Science **306:** 455.

52. PEREZ-JIMENEZ, J.R. *et al.* 2005. Arsenate respiratory reductase gene (*arrA*) for *Desulfosporosinus* sp. strain Y5. Biochem. Biophys. Res. Commun. **338:** 825–829.

53. SILVER, S. & L.T. PHUNG. 2005. Genes and enzymes involved in bacterial oxidation and reduction of inorganic arsenic. Appl. Environ. Microbiol. **71:** 599–608.

54. MUKHOPADHYAY, R. *et al.* 2002. Microbial arsenic: from geocycles to genes and enzymes. FEMS Microbiol. Rev. **26:** 311–325.

55. JACKSON C.R. & S.L. DUGAS. 2003. Phylogenetic analysis of bacterial and archaeal arsC gene sequences suggests an ancient, common origin for arsenate reductase. BMC Evol. Biol. **3:** 18–27.

56. LI, X. & L.R. KRUMHOLZ. 2007. Regulation of arsenate resistance in *Desulfovibrio desulfuricans* G20 by an *arsRBCC* operon and an *arsC* gene. J. Bacteriol. **189:** 3705–3711.

57. MAKI, T. *et al.* 2005. Classification for dimethylarsenate-decomposing bacteria using a restriction fragment length polymorphism analysis of 16S rRNA genes. Anal. Sci. **20:** 1–8.

58. STOLZ, J.F. *et al.* 1997. Differential cytochrome content and reductase activity in *Geospirillum barnesii* strain SeS3. Arch. Microbiol. **167:** 1–5.

59. SWITZER BLUM, J. *et al.* 1998. *Bacillus arsenicoselenatis* sp. nov., and *Bacillus selenitireducens* sp. nov.: two haloalkaliphiles from Mono Lake, California, which respire oxyanions of selenium and arsenic. Arch. Microbiol. **171:** 19–30.

60. STOLZ, J.F. *et al.* 1999. *Sulfurospirillum barnesii* sp. nov., *Sulfurospirillum arsenophilus* sp. nov., and the *Sulfurospirillum* clade in the epsilon proteobacteria. Int. J. Syst. Bacteriol. **49:** 1177–1180.

61. THANGAVELU, M. 2004. Development of a Biochemical Probe for Arsenate Respiring Bacteria using *Bacillus selenitireducens* strain MLS10. Masters Thesis. Duquesne University, Pittsburgh, PA.

62. NIGGEMYER, A. *et al.* 2001. Isolation and characterization of a novel As(V)-reducing bacterium: implications for arsenic mobilization and the genus *Desulfitobacterium*. Appl. Environ. Microbiol. **67:** 5568–5580.

63. ANDERSON, R.M. *et al.* 2000. Water Quality in the Allegheny and Monongahela River Basins Pennsylvania, West Virginia, New York, and Maryland, 1996–98: U.S. Geological Survey Circular 1202, 32 p., on-line at http://pubs.water.usgs.gov/circ1202/ (accessed December 17, 2007).

64. LU, J. *et al.* 2003. Diversity and succession of the intestinal bacterial community of the maturing broiler chicken. Appl. Environ. Microbiol. **69:** 6816–6824.

65. LU, J. *et al.* 2003. Evalulation of broiler litter with reference to the microbial composition as assessed by using 16S rRNA and functional gene markers. Appl. Environ. Microbiol. **69:** 901–908.

66. SPAIN, J.C. 1995. Biodegradation of nitroaromatic compounds. Ann. Rev. Microbiol. **49:** 523–555.

67. ESTEVE-NUNEZ, A., A. CABALLERO & J.L. RAMOS. 2001. Biological Degradation of 2,4,6-trinitrotoluene. Microbiol. Mol. Biol. Revs. **65:** 335–352.

68. IMKAMP, F. & V. MUELLER. 2002. Chemiosmotic energy conservation with Na+ as the coupling ion during hydrogen-dependent caffeate reduction by *Acetobacterium woodii*. J. Bacteriol. **184:** 1947–1951.

69. TAKAI, K. *et al.* 2001. *Alkaliphilus transvaalensis* gen. nov. sp. nov., an extremely alkaliphilic bacterium isolated from a deep South African gold mine. Int. J. Syst. Evol. Microbiol. **51:** 1245–1256.

70. COA, X., X. LIU & X. DONG. 2003. *Alkaliphilus crotonaoxidans* sp. nov., a strictly anaerobic, crotonate-dismutating bacterium isolated from a methanogenic environment. Int. J. Syst. Evol. Microbiol. **53:** 971–975.

71. YE, Q. *et al.* 2004. Alkaline anaerobic respiration: isolation and characterization of a novel alkaliphilic and metal-reducing bacterium. Appl. Environ. Microbiol. **70:** 5595–5602.

Hydrogen and Nickel Metabolism in Helicobacter *Species*

STÉPHANE L. BENOIT AND ROBERT J. MAIER

Department of Microbiology, University of Georgia, Athens, Georgia, USA

Anaerobic microorganisms (such as clostridia) present in the large intestine of animals generate molecular hydrogen (H_2) by fermentation using "H_2-evolving" hydrogenases. The gas can also be detected in other tissues in mice, including the stomach, liver, spleen, or small intestine. It is established that this available H_2 can in turn be used as a source of energy by some pathogenic bacteria, including *Helicobacter* species like *H. pylori* and *H. hepaticus*. Both species possess one hydrogenase, which has been studied for H_2 oxidation characteristics and for its role in conferring animal colonization. On the basis of available annotated gene sequences, other *Helicobacter* species also appear to have one well-conserved respiratory, membrane-bound, nickel-iron-containing [NiFe] hydrogenase. Although *H. pylori* has been well-studied, many other (poorly studied) *Helicobacter* species likely represent a spectrum of emerging pathogens. The important role of hydrogenases in *Helicobacter* species is discussed, and the hydrogenases, their maturation/accessory factors, their regulation, as well as nickel transport and metabolism among the different species are compared.

Key words: **hydrogen; nickel; *Helicobacter***

Microbial Reactions Producing Biological Hydrogen

It is well established that anaerobic fermentative reactions carried out by microbes result in the production of molecular hydrogen, H_2.[1] The physiological basis for this H_2 production are "H_2-evolving" hydrogenase enzymes facilitating disposal of excess electrons that usually originate from the anaerobic glycolytic breakdown of sugars. Formate or a reduced electron carrier (i.e., ferredoxin) can serve as the direct reductant for H_2-evolving hydrogenases. Although these reactions are being optimized *in vitro* for possible use in generating harvestable levels of H_2,[1] the fermentative metabolism, especially by clostridia, occur naturally in the intestinal tract of animals.[2] The sugars left undigested by the host are fermented by colonic anaerobes, which show remarkable ability to utilize the complex sugars not adsorbed by the host. The colonically generated H_2 can be transported through the bloodstream of the host animal via cross-epithelial diffusion from the colon.[3–5] As a result, H_2 can be detected on the breath of animals (including humans). The colonic H_2-producing fermentations typified by clostridial metabolism also produce acetate, butyrate, and propionate as additional products; it is proposed these short-chain fatty-acid products can even be used as carbon sources by the host animal.[5]

Colonic flora such as clostridia ferment different carbon sources with differing efficiency. It is therefore not surprising that the amount of H_2 produced in animals (including in humans), depends on the digestibility of carbohydrates ingested.[6–8] The highest breath H_2 levels in humans are associated with ingestion of nondigestible (beta-linked) oligosaccharides,[3,9] and breath H_2 can be used diagnostically to assess digestibility of carbohydrates or inefficiencies in digestion by the host.[2] Interestingly, most colonically produced H_2 (in humans) was attributed to clostridial species, and the highest clostridial numbers, and breath H_2 levels were associated with ingestion of pectin rather than cellulose by the human host. Up to 20% of the colonically produced H_2 is estimated to reach the lungs of humans, to be lost as expirant. That H_2-utilizing pathogens use H_2 as a growth substrate while in the host was first shown for the gastric pathogen *Helicobacter pylori* just a few years ago.[10]

Helicobacter Species

The genus *Helicobacter* is constituted of nonspore-forming, gram-negative bacteria, which can be curved,

Address for correspondence: Robert J. Maier, 815 Biological Sciences Building, Department of Microbiology, University of Georgia, Athens, GA 30602.

rmaier@uga.edu

spiral, or fusiform. *H. pylori*, the original and best studied *Helicobacter* species, is a gastric pathogen that solely colonizes the mucosal surfaces of the human (or primate) stomach; the result is peptic ulceration, gastritis, and a predisposition for types of gastric carcinomas. The highly inflammatory nature of the infection undoubtedly leads to further host-cell destruction or malfunction.[11] The most severe manifestations of infection are attributed to the persistent nature of the pathogen, and many virulence factors have been identified; these either support bacterial persistence/survival or facilitate tissue destruction. The latter pathology typically occurs in long-term infections. The bacterium is considered to be a microaerophile, but it has terminal respiratory enzymes that could be involved in anaerobic metabolism. Respiration via dehydrogenases, quinones, and cytochromes is terminated via a high O_2 affinity ccoNOQP terminal oxidase.[12] The bacterium will oxidize organic acids, such as formate, lactate, succinate, and pyruvate readily, but glucose and other sugars are poorly utilized. *H. pylori* has a branched incomplete citric acid cycle[13]; the amount of oxygen uptake is insufficient to account for the complete oxidation of any carbon substrates tested. In addition, the (unusual) lack of allosteric inhibition of *H. pylori* citrate synthase by reduced nicotinamide adenine dinucleotide is taken to mean the cycle is poised for biosynthesis rather than for energy generation.

Helicobacter hepaticus is classified as an enterohepatic *Helicobacter* closely related to *H. pylori*, but it is bile resistant and has different colony morphology than *H. pylori*. It is even more O_2 sensitive than *H. pylori*, and it has been shown to be able to grow anaerobically, presumably via nitrate respiration.[14] *H. hepaticus* has not been isolated from humans. Still, enterohepatic *Helicobacter* DNA has been associated with liver cirrhosis and cancers in humans, but whether a *Helicobacter* species is the causative agent of these diseases (in humans) cannot yet be predicted. The bacterium is primarily found in the lower intestinal tract of many mouse lines, but it can be routinely recovered (at lower numbers) from the liver, too. The latter association is most often observed in mice with hepatitis symptoms, but *H. hepaticus*–induced colitis and colon carcinoma have been studied in mouse models. Considering their *in vivo* niches, it would be expected that *H. hepaticus* encounters a lower O_2 level than *H. pylori*, and one that is much more competitive in terms of encountering toxins and the (poor) availability of nutrients. As such, a comparison of the species metabolism is interesting and also important for understanding animal colonization by bacteria. Besides *H. pylori* and *H. hepaticus*, other *Helicobacter* species that have been characterized over the years include the gastric ferret pathogen *H. mustelae*[15]; the cat or dog pathogen *H. felis*[16]; *H. acinonychis*, a species specific to large felines, including cheetahs, lions, and tigers[17]; the mouse pathogen *H. bilis*[18]; and the dog pathogen *H. canis*.[19]

TABLE 1. Hydrogen levels in tissues and *Helicobacter* species (known to have hydrogenases)

Tissue	Hydrogen levels in μM (mouse)	Pathogen (host)
Stomach	43 ± 16	*H. pylori* (human)
		H. acinonychis (feline)
		H. mustelae (ferret)
Liver	53 ± 18	*H. hepaticus* (mouse)
Spleen	48 ± 19	(*enterobacteriaceae*)[a]
Small intestine	168 ± 43	*H. hepaticus*

[a]Includes some species of *Salmonella*, *Shigella*, and *Yersinia*.

H_2 Levels within Tissues Colonized by Pathogens

Although H_2 excretion from live animals and breath H_2 levels from human volunteers have been studied for many years, levels of H_2 within specific vertebrate tissues were only recently assessed (TABLE 1). The levels were first measured in the gastric mucosa of live mice, due to the observed attenuated virulence of a *H. pylori* strain that was unable to use H_2.[10] The possibility that H_2 was available at the site of infection by the gastric pathogen was investigated. Subsequently it was determined that the mucus lining of the stomach, all the lobes of the liver, the spleen, and the heart all had H_2 levels of 40–55 μM, and the small intestine of live mice contained about three times the level determined in these organs.[3] It is significant that these levels are much greater than the half-saturation affinity of the H_2 binding enzymes, including the hydrogenases from *Helicobacter* species.[3] It was concluded that the attenuated colonization abilities (compared to the wild-type parents) of both *H. pylori* (stomach lining) and *H. hepaticus* (liver tissue) *hyd* mutants was due to their inability to use H_2 within the animal.[10,20]

Roles of H_2 Oxidation in *Helicobacter* Species

H_2 oxidation is thought to provide energy (adenosine triphosphate, or ATP) or reductant for use in anabolic reactions while the bacterium is growing, or perhaps alternatively for maintenance energy purposes. Use of H_2 by a bacterium that is otherwise metabolically deficient in carbon source catabolism (*Helicobacter* species use sugars poorly or not at all) may enable *Helicobacter* species to capitalize on a readily available simple energy source, while obtaining carbon sources via peptides and amino acids.[21–23] *Helicobacter* species also suffer considerable oxidative stress *in vivo*, and some of the oxygen radical related detoxification enzymes require reducing power.[24] Perhaps a further role for H_2 oxidation is to facilitate production of such reductant. For *H. hepaticus*, H_2 oxidation was implicated in providing energy for amino acid transport.[20] The amino acid uptake (from a labeled mixture) was sevenfold greater in a 2-h period for *H. hepaticus* wild-type cells incubated in H_2 versus the control atmosphere containing Argon instead. Similarly, the amino acid uptake of wild-type cells was sixfold greater than the *hyd* mutant strain, when both were in an H_2-containing atmosphere. This H_2-supported uptake is more pronounced when cells are incubated in low serum as opposed to the higher 5% serum medium. The results indicate that H_2 oxidation is most important for transport when the bacterium is limiting in nutrients, and perhaps even more important when the bacterium needs to depend on amino acids as their sole carbon sources. The maximum whole-cell velocity (analogous to a V_{max} in enzymology) achieved for amino acid uptake was 2.2-fold greater when cells were in H_2 versus Ar and the half-saturation affinity for the amino acids was unaffected by H_2, consistent with the proposal that H_2 oxidation is providing energy directly for the transport process, and not for synthesis of new transporters or for enzymes in reactions incorporating the amino acids as substrates. There is at least one precedent for H_2 oxidation-mediated carbon-source transport, and this also is for a nonchemoautotroph; *Azotobacter vinelandii* mannose uptake was stimulated by H_2, and this was at the whole-cell V_{max} level rather than affecting the K_m for the carbon substrate.[25]

Helicobacter Species and H_2 Oxidizing Hydrogenases

Until now, five genomes of *Helicobacter* species have been completed and published: genomes of three different strains of *H. pylori* (26695, J99, and AG1), *H. hepaticus* strain 51449, and *H. acinonychis* strain "Sheeba" are now available.[17,26–29] *H. acinonychis* is specific for large felines, including cheetahs, lions, and tigers.[17] In addition, the finished but nonannotated genome sequence of the ferret gastric pathogen *Helicobacter mustelae* is available on-line at http://www.sanger.ac.uk/Projects/H_mustelae, allowing for comparative genomics between these six *Helicobacter* species. Analysis of their genomes reveals that all six species are likely to possess only one hydrogenase (TABLE 2). While many microorganisms have two or more hydrogenases (for instance, the well-studied *Escherichia coli* bacterium has four [NiFe] hydrogenase isoenzymes), a single H_2-oxidizing hydrogenase seems to be a hallmark of *Helicobacter* species. Hydrogenases within both *H. pylori* and *H. hepaticus* have been studied for H_2 oxidation characteristics (see TABLE 2), and for their roles in conferring animal colonization, but the enzymes have not been purified. For *H. pylori*, H_2 activation via hydrogenase in membranes results in cytochrome reduction as well, indicating that H_2 oxidation is linked to a respiratory energy–generating electron-transport chain. Both *H. pylori* and *H. hepaticus* membranes could couple H_2 oxidation to a variety of positive redox potential components, indicating that the enzyme is poised to function primarily in the uptake direction. Even with reduced redox dyes, no H_2 evolution could be observed for either system. Like many H_2 uptake enzymes, the *Helicobacter* enzymes are most active when they are first reduced (such as with dithionite).

While the minimum content of [NiFe] hydrogenases in all organisms is two subunits (a large subunit containing the [NiFe] site and a small subunit containing multiple [Fe-S] clusters), hydrogenases from *Helicobacter* appear to be organized as heterotrimeric complexes. The *hydA* (*hyaA*) gene encodes for the small (β) subunit, while *hydB* (*hyaB*) encodes the large (α) subunit, and the third gene, *hydC* (*hyaC*) encodes the membrane-anchored cytochrome *b* (γ) subunit. These genes are always clustered together on the same locus. Two other genes (*hydD* and *hydE*) are usually found downstream on the same operon; the specific role of these two genes remains hypothetical; however, their involvement as structural hydrogenase genes seems unlikely. Rather they probably can be considered as accessory proteins needed to complete Ni-enzyme maturation (see below and TABLE 4).

Based on previous works on hydrogenases from other microorganisms, it is hypothesized that the small β subunit possesses [Fe-S] clusters, while the large α subunit contains Ni and Fe (the latter is usually coordinated to carbon monoxide (CO) and cyanide (CN) ligands), which are needed for the catalytic activity of the enzyme. The cytochrome *b* subunit is involved in transferring electrons from H_2 to the pool of quinones. The amino acid sequences and the length of the small and large subunits are well conserved between *Helicobacter*

TABLE 2. Hydrogenase structural proteins in sequenced *Helicobacter* species

Characteristics	*Helicobacter* species					
	H. pylori			*H. hepaticus*	*H. mustelae*[a]	*H. acinonychis*
Strain	26695	J99	AG1	51449	43772	Sheeba
Number of hydrogenases/strain	1	1	1	1	1	1
Number of subunits/hydrogenase	3	3	3	3	3	3
Small subunit (β)						
Gene name	*hydA*	*hyaA*	*hydA*	*hyaA*	NA	*hyaA*
Gene number	*0631*	*0574*	*0614*	*0056*	NA	*0744*
Amino acids	384	384	384	386	381	385
Molecular weight (Da)	42,354	42,396	42,255	41,966	41,426	42,509
Large subunit (α)						
Gene name	*hydB*	*hyaB*	*hydB*	*hyaB*	NA	*hyaB*
Gene number	*0632*	*0575*	*0615*	*0057*	NA	*0745*
Amino acids	578	578	578	576	578	578
Molecular weight (kDa)	64,397	64,391	64,394	63,961	64,429	64,459
Cytochrome *b* subunit (γ)						
Gene name	*hydC*	*hyaC*	*hydC*	*hyaC*	NA	*hyaC*
Gene number	*0633*	*0576*	*0616*	*0058*	NA	*0746*
Amino acids	224	224	224	254	220	222
Molecular weight (kDa)	25,824	26,066	25,897	29,104	25,440	25,843
Membrane associated	Yes	Yes	Yes	Unknown	Unknown	Unknown
Tat motif (β subunit)	Yes	Yes	Yes	Yes	Yes	Yes
Affinity ($K_{0.5S}$) for H_2 (μM)	1.8	ND	ND	2.5	ND	ND
Whole-cell activity (O_2)	33 ± 4	ND	ND	3.2 ± 0.2	ND	ND
Whole-cell activity (MB, no O_2)	176 ± 15	ND	ND	25 ± 2	ND	ND

The gene numbers for *H. pylori*, *H. hepaticus*, or *H. acinonychis* are from TIGR. Theoretical molecular weights are from the Swiss-Prot database.

[a]The genome of *H. mustelae* has not yet been annotated.

ABBREVIATIONS: MB = methylene blue; ND = not determined; NA = not available.

species, while there is usually less similarity between sequences of the cytochrome *b* subunit (TABLES 2 and 3). The homology between hydrogenases from *H. pylori* strains and *H. acinonychis* strain Sheeba is striking, and confirms the evolutionary link between both *Helicobacter* species, as suggested by Eppinger and coworkers.[17] The non-*Helicobacter* hydrogenase with the closest homology to *Helicobacter* hydrogenases is that from *Wolinella succinogenes* (67%, 68%, and 51% identity between *H. pylori* 26695 and *W. succinogenes* HydA, HydB, and HydC, respectively).

Hydrogenases from *H. pylori* ATCC strains 43504 or 26695 or from *H. hepaticus* strain 51449 have been shown to be membrane-associated,[30,31] and it is hypothesized that their homologs in *H. mustelae* or *H. acinonychis* are also membrane-bound. Indeed, not only the cytochrome *b* subunit is an anchor for the binding of the uptake hydrogenase to the membrane, but the presence of a conserved –RRxFxK– twin arginine motif (Tat) in the N-terminal peptide sequence of each β subunit strongly suggests that the completely folded αβ heterodimer will be targeted to the cytoplasmic membrane by the Tat translocation pathway.[32]

Accessory Proteins Needed for Hydrogenase Maturation

Genome analysis of the six sequenced strains indicates that each of them possesses all the genes needed for maturation of NiFe hydrogenases, that is, *hypABCDEF* (the term *hyp* is for hydrogenase pleiotropic) (TABLE 4). In most H_2-utilizing bacteria, these genes are clustered together in one operon.[33] In *Helicobacter* species, however, *hyp* genes are located on three different loci of the chromosome, and this distribution seems to be conserved among the different species for which genome sequences are available: the *hypA* gene is always alone, *hypBCD* are clustered together, while *hypE* and *hypF* can be found on the same locus at a third location (TABLE 4). While *hypE* is located directly upstream of *hypF* in *H. pylori* and *H. acinonychis*, the organization is different in *H. hepaticus* (see gene numbers in TABLE 3) and possibly in *H. mustelae*. Mutants have been generated in each of the six *hyp* genes in *H. pylori*; disruption of these genes abolished the hydrogenase activity and addition of nickel could partially restore the activity in some of the mutants.[34,35] Likewise,

TABLE 3. Sequence homology between *Helicobacter* hydrogenase structural proteins

	Helicobacter species					
	H. pylori			*H. hepaticus*	*H. mustelae*	*H. acinonychis*
Strain	26695	J99	AG1	51449	43772	Sheeba
Amino acid identity/similarity (%)						
H. pylori str. 26695						
HydA	____	96/97	99/99	70/85	73/83	97/98
HydB	____	98/99	99/99	69/84	77 /89	96/98
HydC	____	94/96	97/99	48/63	56/69	87/93
H. hepaticus 51449						
HydA	____	____	____	____	75/88	71/84
HydB	____	____	____	____	72/84	69/84
HydC	____	____	____	____	53/65	48/63
H. mustelae 43772						
HydA	____	____	____	____	____	73/83
HydB	____	____	____	____	____	78/89
HydC	____	____	____	____	____	55/68

disruption of *H. hepaticus hypA, hypB,* or *hypC* led to negligible hydrogenase activity; however, in contrast to *H. pylori*, Ni supplementation to the growth medium did not lead to partial restoration of the hydrogenase activity.[36] Homologs of each accessory protein have been extensively studied in *E. coli* (for a recent review, see Ref. 37); therefore, at least working models for the currently assigned roles for *Helicobacter* Hyp proteins (see TABLE 4) can be derived from these *E. coli* studies. Nevertheless, putative roles for *H. pylori* HypA or HypB have been confirmed after both proteins were overexpressed, purified, and biochemically characterized.[38] HypA was shown to form dimers in solution, with each dimer being able to bind two atoms of Ni; a role for HypB as a GTPase was confirmed. Besides, site-directed mutagenesis studies revealed that the His2 residue of HypA was crucial for both Ni-binding and hydrogenase activity, while the Lys59 residue of HypB was shown to be required for both GTPase activity and hydrogenase activity.[38,39] Finally, cross-linking of the purified proteins followed by immunoblotting revealed that HypA and HypB interact with each other in a 1:1 molar ratio.

Interestingly, HypA and HypB proteins also play a role in the nickel maturation of the urease, another important Ni-binding protein present in most *Helicobacter* species. The pleiotropic role of HypA and HypB for Ni-dependent maturation of both Ni-enzymes has been shown in two *Helicobacter* species, *H. pylori* and *H. hepaticus*.[35,36] It is likely that these proteins play a similar role in the maturation of both the hydrogenase and the urease in *H. mustelae* or *H. acinonychis*.

In addition to the six Hyp proteins, HydD and HydE probably can be considered as "accessory" proteins. Based on sequence homologies with *W. succinogenes hydD* and *hydE*, HydD could play a role as protease involved in the maturation of the large subunit, while HydE could be involved in the maturation, processing, or assembly of hydrogenase, in a role similar to those played by other Hyp proteins.[40] Disruption of both genes in *H. pylori* abolished hydrogenase activity in the gastric pathogen, therefore confirming their involvement at some point in the hydrogenase activation process.[34]

Proteins Involved in Nickel Transport, Metabolism, or Regulation

Helicobacter species have evolved different systems for importing nickel ions (Ni^{2+}) into the cell (TABLE 5). The most characterized system is an integral cytoplasmic membrane protein, the Ni-specific permease termed NixA. The *nixA* gene was first isolated as a gene-enhancing urease activity in an *E. coli* strain expressing plasmid-borne urease structural genes.[41] While *H. pylori nixA* mutants usually show a decrease in urease activity of up to 50% in some strains,[42,43] it is not known whether disruption of the *nixA* gene has an effect on hydrogenase activity in *H. pylori*. *nixA* mutants have decreased colonization ability, and this phenotype is probably linked to decreased urease activity.[42] *H. pylori* NixA has a high affinity for nickel, with an estimated K_d of 11.3 nM,[41] thus making the gastric pathogen a good nickel scavenger in the human body. This high affinity is needed because Ni concentrations have been estimated to range between only 2 and 11 nM in the human serum, according to independent studies.[44,45] On the other side of the spectrum, since high concentrations of Ni are toxic to *H. pylori*, the import of nickel inside the cell has to be regulated; this process is

TABLE 4. Hydrogenase accessory genes in sequenced *Helicobacter* species

Gene	*Helicobacter* species						Proposed function
	H. pylori			*H. hepaticus*	*H. mustelae*	*H. acinonychis*	
Strain	26695	J99	AG1	51449	43772	Sheeba	
hypA	0869	0803	0852	0808	Present	1232	Ni-insertion/maturation
hypB	0900	0837	0880	0325	Present	1288	GTPase
hypC	0899	0836	0879	0324	Present	1287	Chaperone
hypD	0898	0835	0878	0322	Present	1286	Fe/S protein maturation
hypE	0047	0040	0043	0320	Present	0082	Maturation
hypF	0048	0041	0044	0326	Present	0083	CO and CN synthesis
hydD	0634	0577	0617	0059	Present	0747	Protease
hydE	0635	0578	0618	0060	Present	0748	Maturation

mediated via the nickel-responsive regulator NikR (see later in the chapter), which represses the transcription of *nixA* in the presence of increasing nickel concentrations.[46,47] Genome sequence analysis reveals that *nixA* is also present in *H. mustelae* and *H. acinonychis*, but not in the mouse liver pathogen. Instead, *H. hepaticus* seems to rely on the NikABDE transport system, which displays homologies to the *E. coli* NikABCDE nickel transport system.[48] Based on sequence comparisons with the *E. coli* ABC transport system, NikA is expected to be a periplasmic nickel-binding protein; NikB is thought to be a cytoplasmic nickel permease; NikD and NikE are predicted to provide the energy to the transport system as ATPases.[49]

Because *H. pylori nixA* mutants retain some urease activity and are still able to colonize the mouse,[42] the existence of alternate Ni transport systems has been suggested, such as AbcCD or Dpp, but none of them was confirmed to play a role in nickel transport.[50,51] However, the mechanism by which nickel traverses the outer membrane (OM) of the gram-negative bacterium has recently started to be unraveled: it is now known that TonB-dependent proteins are involved in nickel transport across the OM. Indeed, two independent studies have shown that two TonB-dependent OM proteins (HP1400 [FecA3] and HP1512 [FrpB3 or FrpB4]) were regulated by nickel and NikR.[52,53] In addition, Schauer *et al.* have shown that the TonB/ExbB/ExbD (HP1341/1339/1340) machinery as well as *hp1512* was involved in nickel accumulation at low pH.[54] Whether mutagenesis of TonB, FecA3, or FrpB4 has an effect on hydrogenase activity in *H. pylori* is not known at this time. Homologs of *tonB*, *exbB*, and *exbD* can be found in every *Helicobacter* species for which genome sequences are available, whereas homologs of OM receptors genes *hp1400* or *hp1512* can only be found in *H. pylori* and *H. acinonychis*; the product of *hh0418* could play a similar role in *H. hepaticus*. Although Belzer and co-workers report the presence of up to three copies of an *hh0418* homolog in the genome of *H. mustelae*,[55] our own genomic sequence analysis could not verify this hypothesis.

Two proteins rich in histidine residues have been identified in *H. pylori*: Hpn (60 amino acid residues) and Hpn-like (72 amino acid residues), which possess 47% and 25% histidine residues in their sequence, respectively.[26] Hpn, first identified in 1995, was named after the protein; it was identified in *H. pylori* and had affinity for nickel.[56] Disruption of the *hpn* gene leads to decreased tolerance to nickel concentrations.[57,58] The *H. pylori* Hpn protein has been recently biochemically characterized: it is mostly found as a multimer in solution, with each monomer of 7 kDa reversibly binding five nickel ions at pH 7.4; nickel can be released by a decrease in pH or in the presence of nickel-chelating, agents such as ethylenediaminetetraacetic acid.[59] Although previous studies suggested that there was no difference in urease activity in the *hpn* mutant as compared to the wild type,[56,59] recent data from Seshadri *et al.* actually indicated that urease activity was up to eight-fold higher in the *hpn* mutant with respect to the wild type when cells were grown in non-Ni-supplemented medium.[57] The same effect was observed in an *hpn-like* mutant.[57] The phenotype of the mutants was consistent with a nickel storage role for both proteins. Although disruption of either (*hpn* or *hpn*-like) gene in *H. pylori* was shown to have no effect on hydrogenase activity, it is possible that Hpn and Hpn-like might be involved in hydrogenase nickel-maturation under certain conditions. While both genes can be found in all sequenced *H. pylori* strains, sequence analysis did not reveal any homolog in *H. hepaticus* or *H. acinonychis*. The presence of Hpn was shown in *H. mustelae*,[56] but it is not known whether the ferret pathogen also possesses Hpn-like. Perhaps Ni-storage/detoxification is most important in the gastric environment where Ni levels would be expected to fluctuate widely.

TABLE 5. Genes involved in nickel transport/metabolism/regulation in sequenced *Helicobacter* species

Gene	*Helicobacter* species						Proposed function
	H. pylori			*H. hepaticus*	*H. mustelae*	*H. acinonychis*	
Strain	26695	J99	AG1	51449	43772	Sheeba	
fecA3	1400	1426	1469	-	-	1303	Outer membrane receptor
frpB4	1512	1405	1400	-	-	0072	Outer membrane receptor
hh0418	-	-	-	0418	?	-	Outer membrane receptor
tonB/exbD /*exbB*	1341–1339	1260–1258	1288–1286	0355–0353	Present	0271–0269	Inner membrane complex
hpn	1427	1320	1352	-	Present	-	Nickel-storage/detoxification
hpn-like	1432	1321	1357	-	-	-	Nickel-storage/detoxification
hspA	0011	0009	0011	-	Present	1697	Nickel-binding chaperone
nikR	1338	1257	1285	0352	Present	0267	Nickel-responsive regulator
nixA	1077	0348	0370	-	Present	1190	Inner membrane Ni transporter
nikABDE	-	-	-	0417–0414	-	-	Nickel ABC transporter
cznABC	0969–0971	0903–0905	0950–0952	0625–0623	Present	1054–1056	Nickel efflux system

Another protein displaying interesting nickel-binding capabilities is HspA, a heat-shock protein that belongs to the GroES family of proteins.[60,61] The N terminus of the protein is homologous to other GroES proteins, but the C terminus of HspA is rich in histidine residues; the protein showed the greatest specificity for nickel among a series of divalent cations tested in the study; a purified MBP-HspA was able to bind 2 Ni^{2+} per molecule with an apparent Kd of 1.8 μM. Expression of HspA in *E. coli* cells expressing plasmid-borne urease genes led to a marked increase in urease activity, which suggests a role of HspA in urease maturation. Homologs of *H. pylori hspA* can be found in the genome of *H. mustelae* or *H. acinonychis*, but not in *H. hepaticus*. The involvement of HspA in hydrogenase maturation/activity in *Helicobacter* species has yet to be demonstrated.

Since nickel is toxic to *Helicobacter* at high concentration (millimolar range), the excess metal levels have to be either sequestered internally or excreted out of the cell. The latter is accomplished through the CznABC system, which has been recently discovered in *H. pylori*.[62] Mutants in any of the three genes were unable to colonize the stomach of Mongolian gerbils, highlighting the importance of metal homeostasis for the gastric pathogen.[62] Also, urease activity was significantly increased in *cznA* and *cznC* mutants, and all three mutants were more sensitive to nickel, as well as to cadmium and zinc.[62] Homologs of the *cznABC* genes are found in each *Helicobacter* species (TABLE 5). The effect of mutations in these genes on hydrogenase activity is not known.

Although transcriptional regulation of hydrogenase genes has not been investigated *per se* in any *Helicobacter* species, genetic profiling of various *H. pylori* mutants has given us some insight about the regulation of the H_2-oxidizing enzyme in this bacterium. For instance, studies of the *H. pylori nikR* mutant in the presence of excess nickel showed that all three hydrogenase structural genes *hydA*, *hydB*, and *hydC* are repressed by NikR.[63] The nickel-responding regulator NikR from *H. pylori* and *E. coli* has been well characterized. In *E. coli*, it was first described as a repressor of the *nikABCDE* nickel uptake operon.[64] In *H. pylori*, it has been shown to have multiple roles, including as transcriptional activator of the urease operon,[65,66] and as transcriptional repressor of the *nixA* gene.[47,67] The ferric uptake regulator, Fur, also seems to be involved in regulation of hydrogenase in *H. pylori*. Indeed, studies with the *fur* mutant showed that hydrogenases are repressed by the apo-Fur protein,[68] suggesting that hydrogenase subunits are synthesized only when the iron cofactor is present. In an independent study, Merrell and co-workers showed that the *H. pylori hydB* and *hydC* genes are both iron-regulated: they appeared to be repressed under iron-starved conditions and activated again when iron was added back to the medium.[69] In addition, hydrogenase expression increased when cells entered the stationary phase or in aging cultures,[70] suggesting an increased need for H_2 oxidation under these conditions: when the nutrients in the culture medium are depleted, H_2 might be used as an additional source of energy. Finally, using a reporter gene system, Olson and Maier showed that the transcription of plasmid-borne *hydA* was enhanced fourfold when *H. pylori* cells were grown in the presence of H_2.[10] The situation seems different in *H. hepaticus*, where hydrogenase activity (and by extension probably hydrogenase expression) appears to be constitutive, regardless of the level of H_2 in the medium.[20] Whether these differences are due to different regulatory factors or due to indirect physiological differences in the two organisms (i.e., nickel transport or storage) is not known.

Acknowledgments

The sequence data for *Helicobacter mustelae* were produced by the *H. mustelae* Sequencing Group at the Sanger Institute and can be obtained from http://www.sanger.ac.uk/Projects/H_mustelae. We thank Dr. Paul O'Toole and Dr. Julian Parkhill for access to this unpublished genome sequence. The sequencing of the *H. mustelae* genome is funded by the Wellcome Trust.

Conflict of Interest

The authors declare no conflicts of interest.

References

1. HALLENBECK, P.C. 2005. Fundamentals of the fermentative production of hydrogen. Water Sci. Technol. **52:** 21–29.
2. CHINDA, D. *et al.* 2004. The fermentation of different dietary fibers is associated with fecal clostridia levels in men. J. Nutr. **134:** 1881–1886.
3. MAIER, R.J. 2005. Use of molecular hydrogen as an energy substrate by human pathogenic bacteria. Biochem. Soc. Trans. **33:** 83–85.
4. CUMMINGS, J.H., G.T. MACFARLANE & H.N. ENGLYST. 2001. Prebiotic digestion and fermentation. Am. J. Clin. Nutr. **73**: 415S–420S.
5. MILLER, T.L. & M.J. WOLIN. 1996. Pathways of acetate, propionate, and butyrate formation by the human fecal microbial flora. Appl. Environ. Microbiol. **62:** 1589–1592.
6. TADESSE, K & M.A. EASTWOOD. 1978. Metabolism of dietary fibre components in man assessed by breath hydrogen and methane. Br. J. Nutr. **40:** 393–396.
7. MUIR, J.G. *et al.* 1995. Resistant starch in the diet increases breath hydrogen and serum acetate in human subjects. Am. J. Clin. Nutr. **61:** 792–799.
8. HALLFRISCH, J. & K.M. BEHALL. 1999. Breath hydrogen and methane responses of men and women to breads made with white flour or whole wheat flours of different particle sizes. J. Am. Coll. Nutr. **18:** 296–302.
9. NAKAMURA, S., T. OKU & M. ICHINOSE. 2004. Bioavailability of cellobiose by tolerance test and breath hydrogen excretion in humans. Nutrition **20:** 979–983.
10. OLSON, J.W. & R.J. MAIER. 2002. Molecular hydrogen as an energy source for *Helicobacter pylori*. Science **298:** 1788–1790.
11. FOX, J.G. & T.C. WANG. 2007. Inflammation, atrophy, and gastric cancer. J. Clin. Invest. **117:** 60–69.
12. KELLY, D.J., N.J. HUGHES & R.K. POOLE. 2001. Microaerobic physiology: aerobic respiration, anaerobic respiration and carbon dioxide metabolism. *In* Helicobacter: Physiology and Genetics. H.L. MOBLEY, G.L. MENDZ & S.L. HAZELL, Eds.: 113–124. Washington, DC: ASM Press.
13. KELLY, D.J. & N.J. HUGHES. 2001. The citric acid cycle and fatty acid biosynthesis. *In* Helicobacter: Physiology and Genetics. H.L. Mobley, G.L. Mendz & S.L. Hazell, Eds.: 135–146. Washington, DC: ASM Press.
14. FOX, J.G. *et al.* 1994. *Helicobacter hepaticus* sp. nov., a microaerophilic bacterium isolated from livers and intestinal mucosal scrapings from mice. J. Clin. Microbiol. **32:** 1238–1245.
15. FOX, J.G. *et al.* 1988. Gastric colonization by *Campylobacter pylori* subsp. *mustelae* in ferrets. Infect. Immun. **56:** 2994–2996.
16. PASTER, B.J. *et al.* 1991. Phylogeny of *Helicobacter felis* sp. nov., *Helicobacter mustelae*, and related bacteria. Int. J. Syst. Bacteriol. **41:** 31–38.
17. EPPINGER, M. *et al.* 2006. Who ate whom? Adaptive *Helicobacter* genomic changes that accompanied a host jump from early humans to large felines. PLoS Genet. **2:** e120.
18. FOX, J.G. *et al.* 1995. *Helicobacter bilis* sp. nov., a novel *Helicobacter* species isolated from bile, livers, and intestines of aged, inbred mice. J. Clin. Microbiol. **33:** 445–454.
19. STANLEY, J. *et al.* 1993. *Helicobacter canis* sp. nov., a new species from dogs: an integrated study of phenotype and genotype. J. Gen. Microbiol. **139:** 2495–2504.
20. MEHTA, N.S. *et al.* 2005. *Helicobacter hepaticus* hydrogenase mutants are deficient in hydrogen-supported amino acid uptake and in causing liver lesions in A/J mice. Infect. Immun. **73:** 5311–5318.
21. WEINBERG, M.V. & R.J. MAIER. 2007. Peptide transport in *Helicobacter pylori*: roles of *dpp* and *opp* systems and evidence for additional peptide transporters. J. Bacteriol. **189:** 3392–3402.
22. DOIG, P. *et al.* 1999. *Helicobacter pylori* physiology predicted from genomic comparison of two strains. Microbiol. Mol. Biol. Rev. **63:** 675–707.
23. TESTERMAN, T.L. *et al.* 2006. Nutritional requirements and antibiotic resistance patterns of *Helicobacter* species in chemically defined media. J. Clin. Microbiol. **44:** 1650–1658.
24. WANG, G., P. ALAMURI & R.J. MAIER. 2006. The diverse antioxidant systems of *Helicobacter pylori*. Mol. Microbiol. **61:** 847–860.
25. MAIER, R.J. & J. PROSSER. 1988. Hydrogen-mediated mannose uptake in *Azotobacter vinelandii*. J. Bacteriol. **170:** 1986–1989.
26. TOMB, J.F. *et al.* 1997. The complete genome sequence of the gastric pathogen *Helicobacter pylori*. Nature **388:** 539–547.
27. ALM, R.A. *et al.* 1999. Genomic-sequence comparison of two unrelated isolates of the human gastric pathogen *Helicobacter pylori*. Nature **397:** 176–180.
28. OH, J.D. *et al.* 2006. The complete genome sequence of a chronic atrophic gastritis *Helicobacter pylori* strain: evolution during disease progression. Proc. Natl. Acad. Sci. USA **103:** 9999–10004.
29. SUERBAUM, S. *et al.* 2003. The complete genome sequence of the carcinogenic bacterium *Helicobacter hepaticus*. Proc. Natl. Acad. Sci. USA **100:** 7901–7906.
30. MAIER, R.J. *et al.* 1996. Hydrogen uptake hydrogenase in *Helicobacter pylori*. FEMS Microbiol. Lett. **141:** 71–76.

31. Maier, R.J., J. Olson & A. Olczak. 2003. Hydrogen-oxidizing capabilities of *Helicobacter hepaticus* and in vivo availability of the substrate. J. Bacteriol. **185:** 2680–2682.
32. Berks, B.C., T. Palmer & F. Sargent. 2005. Protein targeting by the bacterial twin-arginine translocation (Tat) pathway. Curr. Opin. Microbiol. **8:** 174–181.
33. Vignais, P.M. & A. Colbeau. 2004. Molecular biology of microbial hydrogenases. Curr. Issues Mol. Biol. **6:** 159–188.
34. Benoit, S. *et al.* 2004. Requirement of *hydD, hydE, hypC* and *hypE* genes for hydrogenase activity in *Helicobacter pylori*. Microb. Pathog. **36:** 153–157.
35. Olson, J.W., N.S. Mehta & R.J. Maier. 2001. Requirement of nickel metabolism proteins HypA and HypB for full activity of both hydrogenase and urease in *Helicobacter pylori*. Mol. Microbiol. **39:** 176–182.
36. Benoit, S., A. Zbell & R.J. Maier. 2007. Nickel enzyme maturation in *Helicobacter hepaticus*: role of accessory proteins in hydrogenase and urease maturation. Microbiology **153:** 3748–3756.
37. Forzi, L. & R.G. Sawers. 2007. Maturation of [NiFe]-hydrogenases in *Escherichia coli*. Biometals **20:** 565–578.
38. Mehta, N., J.W. Olson & R.J. Maier. 2003. Characterization of *Helicobacter pylori* nickel metabolism accessory proteins needed for maturation of both urease and hydrogenase. J. Bacteriol. **185:** 726–734.
39. Mehta, N., S. Benoit & R.J. Maier. 2003. Roles of conserved nucleotide-binding domains in accessory proteins, HypB and UreG, in the maturation of nickel-enzymes required for efficient *Helicobacter pylori* colonization. Microb. Pathog. **35:** 229–234.
40. Gross, R. *et al.* 1998. Two membrane anchors of Wolinella succinogenes hydrogenase and their function in fumarate and polysulfide respiration. Arch. Microbiol. **170:** 50–58.
41. Mobley, H.L., R.M. Garner & P. Bauerfeind. 1995. *Helicobacter pylori* nickel-transport gene nixA: synthesis of catalytically active urease in *Escherichia coli* independent of growth conditions. Mol. Microbiol. **16:** 97–109.
42. Nolan, K.J. *et al.* 2002. In vivo behavior of a *Helicobacter pylori* SS1 nixA mutant with reduced urease activity. Infect. Immun. **70:** 685–691.
43. Bauerfeind, P., R.M. Garner & L.T. Mobley. 1996. Allelic exchange mutagenesis of nixA in *Helicobacter pylori* results in reduced nickel transport and urease activity. Infect. Immun. **64:** 2877–2880.
44. Sundermann, F.W., Jr. 1993. Biological monitoring of nickel in humans. Scand. J. Work Environ. Health **19:** 34–38.
45. Christensen, J.M. *et al.* 1999. Nickel concentrations in serum and urine of patients with nickel eczema. Toxicol Lett. **108:** 185–189.
46. van Vliet, A.H., F.D. Ernst & J.G. Kusters. 2004. NikR-mediated regulation of *Helicobacter pylori* acid adaptation. Trends Microbiol. **12:** 489–494.
47. Wolfram, L., E. Haas & P. Bauerfeind. 2006. Nickel represses the synthesis of the nickel permease NixA of *Helicobacter pylori*. J Bacteriol. **188:** 1245–1250.
48. Beckwith, C.S. *et al.* 2001. Cloning, expression, and catalytic activity of *Helicobacter hepaticus* urease. Infect Immun. **69:** 5914–5920.
49. Navarro, C., L.F. Wu & M.A. Mandrand-Berthelot. 1993. The nik operon of *Escherichia coli* encodes a periplasmic binding-protein-dependent transport system for nickel. Mol. Microbiol. **9:** 1181–1191.
50. Davis, G.S. & H.L. Mobley. 2005. Contribution of *dppA* to urease activity in *Helicobacter pylori* 26695. Helicobacter **10:** 416–423.
51. Hendricks, J.K. & H.L. Mobley. 1997. *Helicobacter pylori* ABC transporter: effect of allelic exchange mutagenesis on urease activity. J. Bacteriol. **179:** 5892–5902.
52. Ernst, F.D. *et al.* 2006. NikR mediates nickel-responsive transcriptional repression of the *Helicobacter pylori* outer membrane proteins FecA3 (HP1400) and FrpB4 (HP1512). Infect. Immun. **74:** 6821–6828.
53. Davis, G.S., E.L. Flannery & H.L. Mobley. 2006. *Helicobacter pylori* HP1512 is a nickel-responsive NikR-regulated outer membrane protein. Infect. Immun. **74:** 6811–6820.
54. Schauer, K. *et al.* 2007. Novel nickel transport mechanism across the bacterial outer membrane energized by the TonB/ExbB/ExbD machinery. Mol. Microbiol. **63:** 1054–1068.
55. Belzer, C., J. Stoof & A.H. van Vliet. 2007. Metal-responsive gene regulation and metal transport in *Helicobacter* species. Biometals **20:** 417–429.
56. Gilbert, J.V. *et al.* 1995. Protein Hpn: cloning and characterization of a histidine-rich metal-binding polypeptide in *Helicobacter pylori* and *Helicobacter mustelae*. Infect. Immun. **63:** 2682–2688.
57. Seshadri, S., S.L. Benoit & R.J. Maier. 2007. Roles of His-rich Hpn and Hpn-like proteins in *Helicobacter pylori* nickel physiology. J. Bacteriol. **189:** 4120–4126.
58. Mobley, H.L. *et al.* 1999. Role of Hpn and NixA of *Helicobacter pylori* in susceptibility and resistance to bismuth and other metal ions. Helicobacter **4:** 162–169.
59. Ge, R. *et al.* 2006. Expression and characterization of a histidine-rich protein, Hpn: potential for Ni^{2+} storage in *Helicobacter pylori*. Biochem. J. **393:** 285–293.
60. Kansau, I. *et al.* 1996. Nickel binding and immunological properties of the C-terminal domain of the *Helicobacter pylori* GroES homologue (HspA). Mol. Microbiol. **22:** 1013–1023.
61. Kansau, I. & A. Labigne. 1996. Heat shock proteins of *Helicobacter pylori*. Aliment. Pharmacol. Ther. **10**(Suppl 1): 51–56.
62. Stahler, F.N. *et al.* 2006. The novel *Helicobacter pylori* CznABC metal efflux pump is required for cadmium, zinc, and nickel resistance, urease modulation, and gastric colonization. Infect. Immun. **74:** 3845–3852.
63. Contreras, M. *et al.* 2003. Characterization of the roles of NikR, a nickel-responsive pleiotropic autoregulator of *Helicobacter pylori*. Mol. Microbiol. **49:** 947–963.
64. De Pina, K. *et al.* 1999. Isolation and characterization of the *nikR* gene encoding a nickel-responsive regulator in *Escherichia coli*. J. Bacteriol. **181:** 670–674.

65. DELANY, I. *et al.* 2005. In vitro analysis of protein-operator interactions of the NikR and *fur* metal-responsive regulators of coregulated genes in *Helicobacter pylori*. J. Bacteriol. **187:** 7703–7715.
66. VAN VLIET, A.H. *et al.* 2002. NikR mediates nickel-responsive transcriptional induction of urease expression in *Helicobacter pylori*. Infect. Immun. **70:** 2846–2852.
67. ERNST, F.D. *et al.* 2005. The nickel-responsive regulator NikR controls activation and repression of gene transcription in *Helicobacter pylori*. Infect. Immun. **73:** 7252–7258.
68. ERNST, F.D. *et al.* 2005. Transcriptional profiling of *Helicobacter pylori* Fur- and iron-regulated gene expression. Microbiology **151:** 533–546.
69. MERRELL, D.S. *et al.* 2003. Growth phase-dependent response of *Helicobacter pylori* to iron starvation. Infect Immun. **71:** 6510–6525.
70. THOMPSON, L.J. *et al.* 2003. Gene expression profiling of *Helicobacter pylori* reveals a growth-phase-dependent switch in virulence gene expression. Infect. Immun. **71:** 2643–2655.

Hydrogenases of the Model Hyperthermophiles

FRANCIS E. JENNEY, JR. AND MICHAEL W. W. ADAMS

Department of Biochemistry and Molecular Biology, University of Georgia, Athens, Georgia, USA

Hydrogenases are enzymes found in all domains of life that catalyze a remarkably simple chemistry, the reversible oxidation of molecular hydrogen to protons and electrons. In order to perform this chemistry, cells have evolved, several different times, intricate organometal complexes built around a binuclear Ni-Fe or Fe-Fe center, with bound CO and CN^- groups, as well as multiple FeS centers. These complicated enzymes have been an area of intense study for many decades, with interest peaking on the occasions of major increases in national energy costs. Interest in biologically generated hydrogen as a potential substitute for fossil fuels is again at the forefront, and the new tools of the postgenomic world available for manipulating these enzymes make it a truly viable possibility. Hydrogenases from hyperthermophilic microorganisms such as *Pyrococcus furiosus* and *Thermotoga maritima*, with optimal growth temperatures near 100°C, are of particular interest and promise for elucidating and manipulating these enzymatic mechanisms.

Key words: **bioenergy; hydrogen; hydrogenase; hyperthermophile; NiFe; Fe-only**

Introduction

Biological production and utilization of hydrogen has been investigated experimentally for many decades.[1,2] These enzymes catalyze the reversible oxidation of hydrogen

$$H_2 \leftrightarrow 2H^+ + 2e^- \quad (1)$$

and have been found in all domains of life. While most enzymes catalyze a reversible reaction *in vitro*, *in vivo* they will typically catalyze only H_2 oxidation (as a source of energy) or reduction of H^+ (as an electron sink), depending on the metabolic pathway being utilized.[2] This chemistry is used throughout the spectrum of metabolic strategies on earth, from anaerobic autotrophs and heterotrophs to both oxygenic and anoxygenic photosynthesis.[3,4] Hydrogenases are sensitive to oxygen, and thus in facultatively aerobic or microaerophilic microorganisms, they are typically only expressed under low oxygen conditions.[5,6] Some microorganisms, such as *Ralstonia eutropha*, have evolved strategies to make hydrogenase less O_2-sensitive, such as hydrophobic cavities, too small to allow O_2 passage, which act to channel H_2 to the active site.[7,8] This hypothesis is supported by mutagenesis experiments in the green algae *Chlamydomonas reinhardtii*, which make the hydrogenase less oxygen sensitive.[9] Nevertheless, hydrogenase enzymes from aerobic and anaerobic microorganisms can be classified in similar ways. Hydrogenases can be divided into three major classes based on their catalytic metal center; the [NiFe]-hydrogenases,[10,11] the [FeFe]-hydrogenases,[12] and the smaller class of iron–sulfur cluster-free hydrogenases.[13] Hydrogenases have long been of interest both from a chemical/industrial perspective and from the biological perspective, and their structure, function, assembly, and physiological roles have been extensively reviewed).[2,7,12,14–25] As energy prices increase, so does interest in hydrogen as a potential replacement,[26] and biofuels, such as ethanol, butanol, and hydrogen, become fruitful areas for research.[27] Currently, H_2 is not a viable alternative to fossil fuels, as economical production of H_2 is based on steam reforming of fossil fuels.[28] Elucidation of the still incompletely understood mechanisms by which living microorganisms harness H_2 as an energy source via the enzymatic action of hydrogenases will hopefully lead to, at a minimum, design of better and more inexpensive chemical catalysts for H_2 production, and at best, biological generation of economically viable amounts of H_2, ultimately using the megatons of agricultural biomass and animal waste produced in the United States alone each year.[27]

The purpose of this review is to focus on the metabolic role(s) hydrogenases play in anaerobic,

Address for correspondence: Michael W. W. Adams, Department of Biochemistry and Molecular Biology, Life Sciences Bldg. University of Georgia, Athens, GA 30602-7229.
adams@bmb.uga.edu

Ann. N.Y. Acad. Sci. 1125: 252–266 (2008).
doi: 10.1196/annals.1419.013

TABLE 1. Hyperthermophilic genera and their metabolisms

Genus	T_{max}	Metabolism	Substrates	Acceptors
S^0-Dependent Archaea				
Thermofilum (c)	100°	hetero	Pep	S^0, H^+
Staphylothermus (d/m)	98°	hetero	Pep	S^0, H^+
Thermodiscus (d/c)	98°	hetero	Pep	S^0, H^+
Desulfurococcus (d/c)	90°	hetero	Pep	S^0, H^+
Thermoproteus (c)	92°	hetero (auto)	Pep, CBH (H_2)	S^0, H^+
Pyrodictium (d/m)	110°	hetero (auto)	Pep, CBH (H_2)	S^0, H^+
Pyrococcus (d/m)	105°	hetero	Pep, CBH	$\pm S^0$, H^+
Thermococcus (d/m)	97°	hetero	Pep, CBH	$\pm S^0$, H^+
Hyperthermus (m)	110°	hetero	Pep (H_2)	$\pm S^0$, H^+
Stetteria (m)	103°	hetero	Pep + H_2	S^0, $S_2O_3^{2-}$
Pyrobaculum (d/c)	102°	hetero (auto)	Pep (H_2)	$\pm S^0$, O_2, NO_3^-
Acidianus (m/c)	96°	auto	S^0, H_2	S^0, O_2
Palaeococcus (d)	90°	hetero	Pep, Fe^{2+}	S^0, H^+, O_2
Ignicoccus (d/m)	98°	auto	H_2	S^0
Vulcanisaeta (c)	90°	hetero	Pep	S^0, $S_2O_3^{2-}$
Caldivirga (c)	92°	hetero	Pep, CBH	S^0, $S_2O_3^{2-}$, SO_4^{2-}, O_2
S^0-Independent Archaea				
Sulfophobococcus (c)	95°	hetero	Pep	-
Pyrolobus (d)	?121°	auto	H_2	$S_2O_3^{2-}$, O_2, NO_3^-
Aeropyrum (m)	100°	hetero	Pep	O_2
Thermosphaera (c)	90°	hetero	Pep	H^+
Sulfate-reducing Archaea				
Archaeoglobus (d/m)	95°	hetero (auto)	CBH, H_2	$S_2O_3^{2-}$, SO_4^{2-}
Iron-oxidizing Archaea				
Ferroglobus (m)	95°	auto	Fe^{2+}, H_2, S_2^-	$S_2O_3^{2-}$, NO_3^-
Iron-reducing Archaea				
Geoglobus (d)	90°	hetero (auto)	Fe^{3+}, H_2	Fe_2O_3
Methanogenic Archaea				
Methanococcus (d/c)	91°	auto	H_2	CO_2
Methanothermus (c)	97°	auto	H_2	CO_2
Methanopyrus (d/m)	110°	auto	H_2	CO_2
Bacteria				
Thermotoga (d/m)	90°	hetero	Pep, CBH	S^0, H^+
Aquifex (m)	95°	auto	S^0, H_2	O_2, NO_3^-
Thermocrinis (c)	90°	auto	CBH	H^+, O_2

Six of the archaea and two of the bacteria genera are either facultative anaerobes or microaerophiles, as indicated by O_2 in the electron acceptor column; the rest are strictly anaerobic.[29,30,63,78,82,167–169]

ABBREVIATIONS: m = shallow marine; d = deep sea; c = continental; Pep = peptides; CBH = carbohydrates.

hyperthermophilic microorganisms, using a specific example from the Bacterial domain and one from the Archaeal domain. Hyperthermophilic microorganisms were discovered 25 years ago[29] and are defined as having optimal growth temperatures at or above 80°C. While there is still some debate, hyperthermophiles are considered to represent the earliest organisms on earth, adapted as they are to conditions very similar to those thought to exist on the early earth, and they are the deepest branching representatives of both the Archaeal and Bacterial 16S rRNA–based phylogenetic trees.[29–33] There are examples of hyperthermophiles that grow both autotrophically, using electrons from H_2 (as well as other substrates), and heterotrophically, reducing protons as terminal electron acceptors (TABLE 1). Many of these microorganisms can also utilize elemental sulfur as a terminal electron acceptor (some prefer it to H^+). For example, the archaeon *Pyrococcus furiosus* immediately shifts its metabolism from H_2 production to sulfur utilization, based on microarray and biochemical data, upon addition of S^0 to a growing culture.[34,35] Hydrogenases represent a family of enzymes that are relatively well-characterized biochemically in general terms, as a number have been purified and characterized both structurally and functionally.[2,7,12,19,36] However, the rapidly accelerating number of completed geneome sequencing projects from both aerobic and anaerobic (and unknown) microorganisms will doubtlessly provide new insights into the function of these enzymes.

The Structural Hydrogenase Families

The first active hydrogenase enzymes purified were all shown to be iron–sulfur center (FeS)-containing enzymes; however, subsequent work has demonstrated that there are three particular types of catalytic centers containing different metal coordination. The [NiFe]-hydrogenases contain a binuclear site containing one Ni atom and one Fe atom with bound carbon monoxide (CO) and cyanide $(CN)^-$ ligands, in addition to several FeS centers (see later in the chapter for more detail).[11,37,38] The Ni can vary in oxidation state (I, II, or III), while the Fe atom is found as Fe^{+2}.[18] A subclass of these enzymes contains a [NiFeSe] center with a selenocysteine coordinated at the active site.[39] The [FeFe]-hydrogenases are phylogenetically distinct proteins and contain a binuclear Fe center also coordinating CO and CN^- ligands in the active site, a fascinating example of convergent evolution.[20,40] These two types of hydrogenases are structurally similar in the sense that they contain at least one [4Fe4S] center associated with the active site. A third, phylogenetically distinct type of hydrogenase is found only in the archaea, in methane-producing methanogens. This type was originally believed to have an active site based on an organic cofactor with no metals involved, but was ultimately demonstrated to contain a mononuclear Fe.[13,41] It is known as the H_2-forming methylenetetrahydromethanopterin dehydrogenase (Hmd), or iron–sulfur cluster-free hydrogenase.[13] This enzyme lacks acid-labile sulfur, and while the Fe originally was considered to play a structural role, it has been shown to play a key role in activating H_2[13] In addition to at least one Fe atom, all three types of hydrogenase also contain a CO ligand to the Fe in their active sites, as discussed further later in the chapter.

Distribution and Functions of Hydrogenases

The [FeFe]-hydrogenases are found only in a limited number of anaerobic Bacteria, and in some anaerobically adapted Eukarya,[2,12,42] while [NiFe]-hydrogenases are found extensively in Bacteria and Archaea, but not in Eukarya. In most cases, the [FeFe]-hydrogenases are monomeric enzymes, and appear to function *in vivo* primarily to eliminate electrons by reducing protons to H_2.[2,20,43] Location also generally correlates to function, as typically cytoplasmic hydrogenases evolve H_2, and membrane-bound or periplasmically located hydrogenases are typically uptake hydrogenases.[2] Many prokaryotes contain multiple hydrogenases located in different cell compartments, presumably to allow them to respond to varying environmental conditions.[2] The [NiFe]-hydrogenases all contain at least two different subunits and can be broken into four groups. The first are the respiratory enzymes, which couple the oxidation of H_2 via a number of different electron carriers, such as cytochromes, to reduction of terminal electron acceptors, such as SO_4^{2-} or NO_3^- anaerobically, or O_2 in aerobic microorganisms.[7] These are found as both membrane-bound multisubunit complexes, as well as periplasmic enzymes. For example, those found in *Desulfovibrio* sp. couple H_2 oxidation to a proton motive force via a high-molecular-weight cytochrome.[2] A second class is the cytoplasmic hydrogenases that function as H_2 sensors[44] and activate the expression of the metabolically active [NiFe]-hydrogenases.[7,45] A third class of [NiFe]-hydrogenase is the bidirectional cytoplasmic enzymes, which contain multiple subunits able to interact with electron carriers such as nicotinamide adenine dinucleotide phosphate (NADP). These are readily reversible *in vitro*, but *in vivo* probably only oxidize H_2. They are found in many Archaea, for example, *P. furiosus*,[46] and bacteria.[7] The fourth class of [NiFe]-hydrogenase are membrane-bound, energy-conserving multienzyme complexes, which are found in Bacteria,[47] as well as in many Archaea, for example *P. furiosus*.[48] The phylogenetically distinct "iron–sulfur-free" type of hydrogenases are specific to methanogenic Archaea and rather than reversibly oxidize H_2, they catalyze direct reduction of a specialized cofactor.[13]

Hyperthermophiles and Hydrogen

Hydrogen gas is found at significant levels in most hydrothermal systems due to volcanic outgassing[3,49–52] or abiotic production, for example, by oxidation of Fe^{2+} ions by water.[3,53] Low concentrations of H_2 are often associated with methanogens (as well as increased $[CH_4]$), which utilize H_2 produced locally by fermentation of organics, for example, as seen in waste-treatment sludge.[51,54–57] It has been estimated that (hyper)thermophilic anaerobes represent more than half the microbial biomass on earth, in deep underground biomes.[51,58,59] A large source of energy for such microorganisms may well be abiotically produced H_2.[3,60]

and examples of both obligate and facultative relationships between microorganisms living in partnership and exchanging reductant in the form of H_2 to complete metabolic pathways can be found.[3,61,62] Hyperthermophiles are typically found in geothermally heated environments where H_2 is available from both biological and abiological sources.[29,30,36,63–66] These biomes are of further interest because they likely represent the types of emerging life on the Early Earth, as well as extraterrestrial life that potentially exists in the solar system.[66] There are a handful of hyperthermophilic members of the Bacteria; however, the majority are members of the domain Archaea. Hydrogenases can be found in virtually all of these microorganisms (TABLE 1). Of the hyperthermophilic bacteria, *Thermotoga* and *Thermocrinis* species can use protons as an electron acceptor to produce H_2, while *Aquifex* sp. use H_2 as a source of electrons for reducing such acceptors as oxygen or nitrate.[64,67–69] *Persephonella* species (members of the order Aquificales), which grow optimally near 70°C, are just at the edge of the definition of hyperthermophiles ($T_{opt} = 80°C$), and show a similar ability to utilize H_2 as a source of electrons.[70]

The Archaea can generally be divided into those that utilize elemental sulfur and those that do not (TABLE 1). The S^0-independent microorganisms utilize molecular hydrogen as an electron source, ultimately reducing, for example, oxygen, nitrate, sulfate, sulfite, iron, or in the case of the methanogens, carbon dioxide. The methanogenic archaea have been well-studied and in fact the genome of *Methanococcus jannaschii* (now *Methanocaldococcus*) was the first completed archaeal genome.[71–73] Genetic systems are now available for these microorganisms, particularly *Methanococcus maripaludis*.[74,75] It is the methanogens that contain the specialized hmd-type hydrogenase that reduces an organic cofactor rather than catalyzing the reversible oxidation of H_2,[13] though they also contain [NiFe] hydrogenases.[2,76] Essentially, nothing is known of the iron-oxidizing *Ferroglobus* sp. and iron-reducing *Geoglobus* sp. hydrogenase enzymes, but both have been shown to utilize H_2 as an electron source,[77,78] as has the sulfate-reducer *Archaeoglobus fulgidus*.[79] Of the other S^0-independent hyperthermophiles, *Sulfophobococcus* and *Aeropyrum* sp. do not apparently utilize H_2, *Pyrolobus* can utilize H_2 as a source of electrons, and *Thermosphaera* can utilize H^+ as an electron sink, though little is known of the enzymes that do this.[80,81] The S^0-dependent microorganisms can either evolve H_2 or use it as an energy source for S^0 respiration. In a few cases, hydrogenase activity has been measured in cell extracts,[82,83] and in several cases they have been purified[84–87]; however, in most cases information on hydrogenases in these microorganisms comes only from growth experiments using H_2 as an electron donor or detecting H_2 as a metabolic product, or else is predicted from genome analyses. As noted earlier, the hydrogenases catalyzing these reactions are [FeFe] for the bacteria, and [NiFe] for the archaea, and the few specific examples where the native enzyme(s) have been most thoroughly examined are discussed later in the chapter.

[FeFe] Hydrogenase of *Thermotoga maritima*

The [FeFe] hydrogenases are typically single subunit, soluble enzymes found in some bacteria and eukaryotes (but not archaea), and their evolution, structure, and function have been recently and extensively reviewed.[12] Their active sites bear a striking resemblance to that of the [NiFe]-hydrogenases described later in the chapter, particularly the functionally conserved CO and CN^- ligands to the Fe in the active site.[12,20,88] As stated earlier, however, they are phylogenetically distinct, and are typically a single subunit (~40 kDa) containing the so-called "H cluster," which coordinates both the binuclear FeFe site bridged to a [4Fe4S] center by only a single protein (cysteine) ligand.[12,88] The majority of structural and functional information comes from mesophilic representatives, but one hyperthermophilic [FeFe] hydrogenase has been extensively characterized. *Thermotoga maritima* (Tm) is a heterotrophic, hyperthermophilic bacterium that ferments various carbohydrates, ultimately reducing elemental sulfur to H_2S or protons to H_2. The biology of this microorganism was thoroughly reviewed just recently.[64] Tm contains an [FeFe] hydrogenase that has been purified to homogeneity.[12,89–91] This was the first Fe-only enzyme characterized with more than two subunits.[90] It is extremely oxygen-sensitive ($t_{1/2}$ in air ~2 min for the purified enzyme) and is a heterotrimeric protein encoded by the open reading frames (ORFs) TM1424–TM1426. Biochemical analyses of the Tm hydrogenase and of native (α) and recombinant (γ) forms of some of its subunits indicate that the α-subunit contains the catalytic H-cluster, three [4Fe-4S], and one [2Fe-2S] cluster, while the γ-subunit contains a single [2Fe-2S] cluster.[90,92] Based on sequence analyses, additional FeS centers are predicted in the α-subunit ([2Fe-2S] cluster), and the β-subunit is predicted to contain three [4Fe-4S] clusters and one [2Fe-2S] cluster, although biochemical analyses indicate the presence of only one [4Fe-4S] cluster and two [2Fe-2S] clusters for this subunit.[89,90] The β-subunit contains putative nicotinamide adenine dinucleotide (NAD) and flavin mononucleotide (FMN) -binding domains, indicating an ability to utilize reduced NAD (NADH) as an electron donor *in vivo*.[90] The β-subunit also contains a NuoF-like domain, a flavin-binding domain in the NADH dehydrogenase of *Escherichia coli*,[93] while all three hydrogenase subunits contain a domain similar to *E. coli* NuoE, which contains a [2Fe2S] cluster,[93] consistent with the proposed evolutionary relationship between Fe-hydrogenases and NADH dehydrogenases of respiratory complexes.[2] The genome sequence of *T. maritima*[94] shows that the genes encoding the three subunits of the Tm hydrogenase are part of an operon that contains eight ORFs (TM1420–TM1427), in which the other genes are presumably involved in the synthesis, assembly, regulation, or stability of the holoenzyme. TM1420, TM1421, and TM1422 contain cysteine motifs, suggesting that they are FeS cluster-containing proteins. The sequence of TM1423 has no obvious cysteine motif, but it contains nine cysteines residues, as well as a conserved serine/threonine phosphatase domain. The last ORF in the operon (TM1427) is a conserved hypothetical protein with homology to a redox-sensing transcriptional repressor protein (rex) from *Streptomyces coelicolor*.[95] The roles of these five additional ORFs are unknown. They might be involved in the electron-transfer pathway for H_2 production. For example, the glucose fermentation pathway of *T. maritima* generates NADH and reduced ferredoxin, but the purified hydrogenase does not use either as an electron donor for H_2 production *in vitro*.[90] What provides reductant to the enzyme for H_2 production is a major unresolved issue. Since the electron carrier for the hydrogenase is uncharacterized, one or more of the ORFs in the hydrogenase gene cluster may encode proteins that function in electron transfer to this [FeFe]-hydrogenase, with the last ORF perhaps acting as a redox sensor/regulator. Several other ORFs, hydE (TM1269), hydF (TM0445), and hydG (TM1267), have recently been demonstrated to be involved in maturation of the holoprotein in *Thermotoga*.[96,97] HydF has been shown to contain a [4Fe-4S] center and have guanosine triphosphate (GTP) -ase activity,[96] and by analogy to the role of HypB in GTP-dependent insertion of Ni atoms in the assembly of the [NiFe] hydrogenases,[16] it has been proposed to play a similar role in assembly of this cognate [FeFe] center.[96] HydE and HydG are both radical *S*-adenosylmathionine (SAM) enzymes They contain [4Fe-4S] clusters and cleave *S*-adenosyl methionine.[97] Their function remains unclear, but it is proposed that they are involved in assembly of the [FeFe] active center, possibly by acting as a source of sulfur for the bridging organic molecule, thought to be di(thiomethyl)amine or dithiopropane.[97–100] Like the [NiFe] hydrogenases, there are multiple [4Fe4S] clusters that connect the active site to the surface of the proteins, and are appropriately spaced for electron transfer.[20] The active-site H cluster is made up of a standard [4Fe4S] cluster bridged to the binuclear iron center [FeFe] by a cysteine sulfur. The two Fe atoms of the binuclear center each have two CO/CN^- ligands, and are connected as well by two sulfur atoms that are part of a small, as yet unknown, organic molecule as mentioned earlier.[12,20,98]

[NiFe] Hydrogenases and *Pyrococcus furiosus*

The NiFe-hydrogenases have been extensively characterized and several crystal structures are available,[101–103] although none have been obtained from a hyperthermophilic hydrogenase. As described earlier, all of the NiFe-enzymes are composed of a minimum of two subunits, the large subunit (or LSU, typically ~50 kDa), which contains the NiFe-catalytic site, and the small subunit (or SSU, typically ~20 kDa), which contains three iron–sulfur (FeS) clusters. These clusters transfer electrons to or from the electron carrier for the enzyme (which interacts with the small subunit) to and from the NiFe site in the large subunit. The Ni atom is bound to four cysteinyl residues of the large subunit, two of which are near the N terminus and two near the C terminus. Two of the Cys (one from the C and one from the N terminus) also bind a single Fe atom, which is also coordinated by organic ligands, one carbon monoxide (CO) and two cyanide (CN).[104] These diatomic ligands serve to activate the iron atom (maintaining it in the low spin state) facilitating catalysis. Very recent studies have shown that one [NiFe]-hydrogenase can contain additional CN^- ligands to the Ni[105] as well as to the Fe,[106] and these reduce the sensitivity of the enzyme-to-oxygen inactivation. The [NiFe]-hydrogenases are typically purified under aerobic conditions (even from anaerobic microorganisms) and as such are catalytically inactive (and exhibit the so-called Ni-A EPR-active state). The enzymes can be reactivated over several hours by treatment with H_2 under reducing conditions (to give an enzyme that exhibits the "ready" or Ni-B EPR-active state). The fully-active enzyme further exhibits another electron paramagnetic resonance (EPR) active state (Ni-C).[107–111] The inactive forms are thought to have an oxygen species, probably peroxide or hydroxide, bridging the Ni and Fe atoms at the active site,[37,110,112,113] although the nature of the different inactive oxidized states is not yet clear.[110] During reductive activation with H_2, the bridging O ligand is replaced by a hydride intermediate. Several complex mechanisms have been proposed for both the activation and catalytic reactions,[11,19,102,114] and it is generally accepted that the iron atom remains in the low spin ferrous form throughout the catalytic cycle, while the nickel switches between diamagnetic Ni(II) and paramagnetic Ni(III). Interestingly, H_2 does not appear to freely diffuse within the protein, but is transferred via hydrophobic channels that lead from the exterior of the protein to the active site.[102,115]

The hyperthermophilic archaeon *P. furiosus* is a strictly anaerobic archaeon that grows optimally at 100°C.[116] It ferments carbohydrates and peptides as sources of energy and carbon for growth, and produces organic acids, CO_2, and H_2 as end-products. *P. furiosus* can also dispose of the excess reductant generated during growth by reducing elemental sulfur (S^0) to H_2S, rather than producing H_2, and in fact appears to prefer S^0 to protons.[35] It contains three distinct NiFe-hydrogenases,[117–119] all three of which are expressed and functional in cells grown in the absence of elemental sulfur. Their expression is downregulated in the presence of S^0.[35,120] All three enzymes have been purified under strictly anaerobic conditions from *P. furiosus* biomass. Consequently, they are obtained in their catalytically active states and do not require activation. Two of the them are soluble, cytoplasmic enzymes and use NAD(P)(H) as the electron carrier.[46,117] They are referred to as hydrogenase I or SH-I and hydrogenase II or SH-II (S for soluble hydrogenase), respectively. The third hydrogenase is located in the membrane[119] and is referred to as MBH (for membrane-bound hydrogenase). It uses the cytoplasmic redox protein ferredoxin[121] as the electron donor and is the hydrogenase primarily responsible for producing H_2 during growth. In *P. furiosus*, reduced ferredoxin is generated directly by the unusual pathway for glucose oxidation in this microorganism,[122] where the oxidation steps are catalyzed by glyceraldehyde ferredoxin oxidoreductase (GAPOR),[123] and pyruvate ferredoxin oxidoreductase (POR),[124] It was demonstrated using *P. furiosus* membrane vesicles that when the oxidation of ferredoxin is coupled to H_2 production by MBH, protons are pumped across the membrane and energy is conserved in the form of adenosine triphosphate (ATP) by a membrane-bound ATPase.[48] The *in vivo* functions of the soluble hydrogenases I and II, however, are not known, even though they are among the most abundant proteins in the cell.[118] They are able *in vitro* to reduce S^0 to H_2S with H_2 or NADPH, but this is

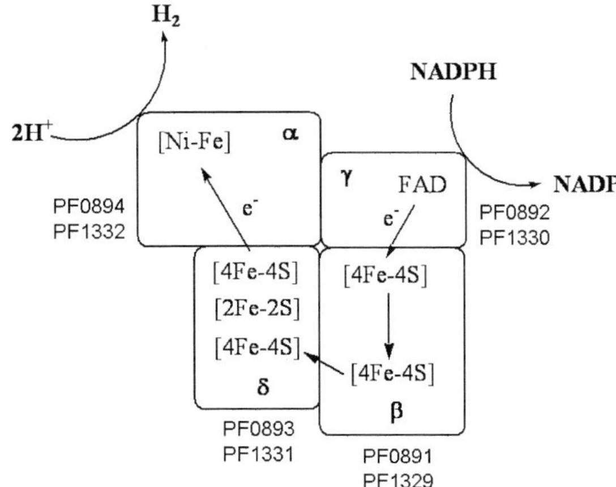

FIGURE 1. Proposed structures, cofactor contents, and pathway of electron flow in the two cytoplasmic hydrogenases of *P. furiosus*. Hydrogenase I and hydrogenase II are represented by PF0891-PF0894 and PF1329-PF1330, respectively. (Adapted from Ma and Adams.[46])

not their physiological role. Enzyme assays,[125] DNA microarray analyses,[34] and growth studies[35] demonstrate that the biosynthesis of the three hydrogenases (SH-I, SH-II, and the MBH) does not occur if S^0 is present in the growth medium. The two soluble hydrogenases may play roles in recycling H_2 to provide NADPH for biosynthesis, although why two enzymes, whose regulation is not significantly different under the conditions examined to date, are needed is not obvious.

The complete genome sequence of *P. furiosus* is known[126] and the genes encoding the three hydrogenases have been identified using sequence information from the purified proteins.[117,118] The two soluble hydrogenases each contain four distinct subunits, which are encoded by two operons, PF0891–PF0894 and PF1329–PF1332, respectively. Thus, these two enzymes each contain two subunits in addition to the catalytic NiFe-containing large subunit (PF0894 and PF1332) and the FeS-cluster-containing small subunit (PF0893 and PF1331). One of these additional subunits (PF0892 and PF1330) contains flavin adenine dinucleotide (FAD) and interacts with NADP(H), while the other (PF0891 and PF1329) contains yet more FeS centers, presumably to transfer electrons from the flavin-containing subunit to the small subunit. A model showing the proposed functions of the four subunits in soluble hydrogenase I and soluble hydrogenase II is shown in FIGURE 1. The two enzymes are predicted to be extremely similar structurally, with their respective subunits showing 55–63% sequence similarity at the amino acid level. They are also extremely thermostable; for example, pure SH-I shows a loss of activity with $t_{1/2} \approx 12$ hours at 80°C. Where they differ is in their catalytic properties. SH-I is an extremely active enzyme, with high turnover rates for H_2 evolution using the artificial electron donor methyl viologen (42,900 s^{-1}), and in H_2 oxidation using NADP as the electron acceptor (11,250 s^{-1}).[46] By comparison, soluble hydrogenase II is much less active in all assays, by an order of magnitude or more. In addition, in the standard H_2 evolution and H_2 oxidation assays using the artificial electron carriers methyl viologen (MV) and methylene blue (MeB), virtually all other purified hydrogenases, of which there are now several dozen,[7,27,127] preferentially catalyze H_2 oxidation. *P. furiosus* SH-I, however, catalyzes H_2 production with MV at about 10-fold the rate of H_2 oxidation with MeB, and it is reasonably oxygen resistant, with a half-life in air of about 6 h at 23°C.[117] Thus, *P. furiosus*-soluble hydrogenase I is an extremely thermostable, highly active enzyme that preferentially catalyzes H_2 evolution. Of course, what determines catalytic prowess and preferred direction (evolution or oxidation), either *in vivo* or *in vitro,* is not understood at any level. The third hydrogenase in *P. furiosus*, the membrane-bound enzyme MBH, it uses the redox protein ferredoxin as the electron donor, and does not use NAD(P)H.[48] The ferredoxin-dependent activity is lost when the enzyme is solubilized from the membrane using detergents.[119] Quite remarkably, MBH is almost exclusively unidirectional, as the only significant catalytic activity is H_2 evolution from artificial electron carriers. The solubilized enzyme has a specific activity for producing H_2 that is 250-fold greater than the rate of H_2 oxidation that it catalyzes (using redox dyes).[119] In fact, in intact membranes, ATP hydrolysis can be used to drive H_2 oxidation (FIG. 2).

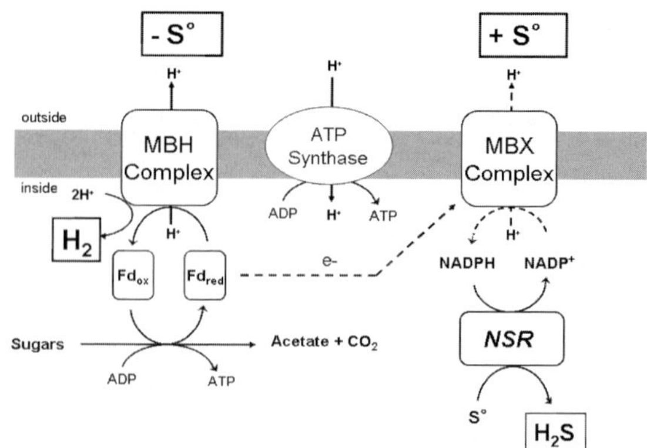

FIGURE 2. Proposed pathway of electron flow in the metabolism of *P. furiosus*. Reduced ferredoxin (Fd_{red}) produced by metabolism of sugars must be reoxidized. In the presence of elemental sulfur (S^0) the newly described NADPH:sulfur oxidoreductase (NSR) transfers electrons ultimately to produce H_2S. However, while the mechanism invokes an interaction between Fd and the MBX membrane-bound complex, which is induced under these conditions, it has not yet been demonstrated.[35] In the absence of elemental sulfur (S^0), the membrane-bound hydrogenase (MBH) can oxidize Fd_{red}, additionally generating a proton gradient for ATP production.[48]

The genes encoding MBH are more complex than those encoding the soluble enzymes (TABLE 2). The purified form of MBH contained only two subunits (of 17 and 48 kDa), which are encoded by genes PF1433 (termed *mbhK*) and PF1434 (*mbhL*). These two genes are part of a large 14-gene operon encoded by PF1423–PF1436 (*mbhA–mbhN*). One of them (PF1434, *mbhL*) corresponds to the large catalytic subunit of MBH. As stated earlier, MBH is a respiratory enzyme and both reduces protons to form H_2 and also pumps protons as a means of conserving energy.[48] The enzyme was subsequently recognized as a new subclass of so-called energy-converting hydrogenases that function as ion pumps.[24,25] Members of this class contain six conserved genes (*mbhHJKLMN* in *P. furiosus*) and include part of the hyc operon of *E. coli* (*hycCDEFG*), which encodes hydrogenase 3. These hydrogenases serve to either couple the oxidation of low potential carriers, such as ferredoxin, with proton reduction and conserve energy by pumping protons (like the *P. furiosus* enzyme), or they use H_2 to reduce ferredoxin in a reaction driven by reverse electron transport. Note that the second subunit of solubilized *P. furiosus* MBH (PF1433, *mbhK*) shows sequence similarity to the N-terminal region of the *E. coli* HycE protein. Hence, the purified *P. furiosus* enzyme (PF1433, PF1434) is equivalent to the single large subunit (HycE) of the *E. coli* hydrogenase 3 complex. It is not clear, however, if the *E. coli* HYD3 enzyme conserves energy during H_2 production.[24] Indeed, it is important to point out that the *E. coli* enzyme is very unstable and has yet to be purified.[24]

The six *P. furiosus* genes conserved in the core of the energy-converting hydrogenases also have close homologs in the NADH:quinone oxidoreductase (complex I) of the aerobic respiratory chain (TABLE 2). In fact, an evolutionary relationship between the large and small subunits of hydrogenase and NADH:quinone oxidoreductase is well established,[128] where the latter is thought to have evolved from hydrogenase, at least in part, by conversion of the NiFe site to a quinone-binding site.[129] Several of these energy-converting hydrogenases, including the *P. furiosus* enzyme, contain, in addition to the six genes encoding the "core" hydrogenase subunits, eight or more genes and these are predicted to encode small membrane-bound

TABLE 2. Components of the predicted MBH operon of *Pyrococcus furiosus*

PF #	Gene name	Location[1]	Size (kDa)	Metallo centers?	Homologous subunit (PFAM/TIGRFAM)[170,171]
PF1423	MbhA	Mem(2–3)	19	No	MnhE (Na+/H+ antiporter)
PF1424	MbhB	Mem(3)	9	No	MrpF (Na+/H+ antiporter)
PF1425	MbhC	Mem(3)	14	No	MnhG (Na+/H+ antiporter)
PF1426	MbhD	Mem(3)	10	No	-
PF1427	MbhE	Mem(0–2)?	11	No	-
PF1428	MbhF	Mem(3–4)	16	No	MnhB (Na+/H+ antiporter)
PF1429	MbhG	Mem(2–3)	13	No	HyfE
PF1430	MbhH	Mem(12–13)	55	No	HycC
PF1431	MbhI	Mem(1–2)	13	No	-
PF1432	MbhJ	Cyto(0)	18	[FeS]?	HycG (SSU)
PF1433	MbhK	Cyto(0)	20	No	HycE (LSU N-term)
PF1434	MbhL	Cyto(0)	48	[NiFe]	HycE (LSU)
PF1435	MbhM	Mem(5–8)	35	No	HycD
PF1436	MbhN	Cyto(0)	15	2x[4Fe-4S]	HycF

[1] Based on predicted transmembrane domains.[172]
ABBREVIATIONS: Mem = membrane bound; Cyto = cytoplasmic.

proteins.[24,25] These other eight genes in the *P. furiosus* MBH operon (*mbhABCDEFGI*) are similar to genes encoding ion transporters, such as the Na^+/H^+ transporter in *Bacillus.*[119] It is possible, therefore, that this half of the MBH operon encodes a pump that serves to convert a primary Na^+ ion gradient into a secondary proton gradient,[24,25] and recent data are consistent with this idea.[130] In other words, does the core hydrogenase of the *Pyrococcus* MBH (*mbhHJKLMN*) pump protons, or does it pump Na^+ ions? The nature and likely function of the 14 genes encoding the MBH operon remain unclear. There is another operon in *P. furiosus*, designated MBX, similar to the MBH which is upregulated in the presence of elemental sulfur just as MBH expression is down-regulated (FIG. 2).[35] The role of this putative MBX protein complex is, as yet, unclear, but is very likely involved in somehow oxidizing reduced ferredoxin to ultimately transfer these electrons to elemental sulfur.

The genome sequence of *P. furiosus* also contains obvious homologs of seven of the eight accessory genes involved in the biosynthesis of the NiFe-containing large subunit of hydrogenases that have been well described in *E. coli.*[16] An area of intense study in the hydrogenase field has been, and continues to be, how the NiFe catalytic site of hydrogenase is synthesized. Elegant studies by Böck and co-workers using one of the hydrogenases of *E. coli* (HYD3) as the model system have established that the assembly of the [NiFe] catalytic site within the large subunit requires the participation of eight accessory proteins.[16,131–134] These are designated HypA, HypB, HypC, HypD and HypE (which in *E. coli* are organized in a single operon), HypF (which is part of a bicistronic operon with a gene of unknown function), HycI, and SlyD. The first step is the assembly of the $Fe(CN)_2(CO)$ group on a HypC-HypD complex in an ATP-dependent process. HypF and HypE provide the CN ligands, and possibly also the CO ligand, using carbamoyl phosphate as the source of the CN units. HypF functions as a carbamoyl transferase and transfers the carbamoyl group (from carbamoyladenylate) to the C-terminal Cys of the HypE protein. HypE catalyzes the ATP-dependent dehydration of the *S*-carbamoyl group to give an enzyme-bound thiocyanate, which releases cyanide to the iron. It has very recently been demonstrated with isotopic labeling studies in *E. coli* and *R. eutropha* that carbamoyl phosphate serves as the source of the CN^- ligands, but not the CO ligand, which comes from an as yet undefined source.[135,136] After the $Fe(CN)_2(CO)$ group is transferred to the hydrogenase apoprotein, nickel insertion occurs in a GTP-dependent manner. It requires the concerted action of HypA and HypB, with HypA functioning as a metallochaperone and HypB as a regulator controlling the interaction of HypA with the large subunit. Very recent experiments have indicated that an additional protein, SlyD, is involved in the nickel insertion step.[137] SlyD has a histidine-rich C terminus that could be used for nickel binding, and the protein forms a complex with HypB. The final step of hydrogenase maturation involves a specific peptidase (HycI) that removes the C-terminal "assembly peptide" of 20 or so amino acids from the large subunit, which leads to protein folding around the Ni-Fe active site. The homologs for these genes in *P. furiosus* are (with proposed function) termed HypA (Ni insertion, PF0615), HypC and HypD (Fe insertion, PF0548 and PF0549), HypE and HypF (CO/CN synthesis, PF0604 and PF0559), HycI (proteolysis, PF0617), and SlyD (PF1401). There is no obvious homolog for HypB (a GTPase); however, the ORF located in an operon with HypA and HycI has homology to ATPases and could encode a protein that functions as a HypB. The presence of a homolog of the HycI protease suggests that one or perhaps all three of the large subunits of the *P. furiosus* hydrogenases undergo C-terminal processing. Only one example of each of these accessory genes is present in the *P. furiosus* genome, so it appears that each is responsible for the maturation of the catalytic subunits of all three hydrogenases, MBH, SH-I, and SH-II.

In summary, while the [FeFe]-hydrogenase of *T. maritima* and the three [NiFe]-hydrogenases of *P. furiosus* are very well-characterized examples of hyperthermophilic hydrogenases, both of which are similar in many ways to mesophilic counterparts, the roles and *in vivo* functions of the three cytoplasmic enzymes remain to a large extent a mystery, although these are reasonably well-defined for the membrane-bound enzyme in *Pyrococcus*.

Biotransformations and Biofuel Production

Economical access to feed stocks, and production of fossil fuels, an essential capability for all national economies, becomes increasingly difficult for both practical and political reasons every year.[138] Interest in the processing of renewable resources into chemicals and alternative fuels, such as ethanol[139] and H_2,[17,140] to augment energy reserves in the United States, has surfaced from time to time over the past several decades in response to geopolitical events.[26,27] Despite its promise, as of yet many bioenergy conversion technologies remain unproven or prohibitively expensive to implement. In the postgenomics era, however, an assortment of powerful new tools is now available that could address current limitations of bioenergy conversion processes.[141] The challenge is to determine how to use these tools strategically to achieve technologically and economically sound routes for extracting energy from renewable resources. Biomass available for energy extraction typically includes plant-based starch, lignocellulosics, and other agricultural renewables.[138]

Part of the challenge in converting biomass into versatile liquid (ethanol) and gaseous (H_2) fuels continues to be the heterogeneity of the material. There is a broad spectrum of chemical contexts within biomass so that converting such material into alternative fuels requires a series of integrated biotransformations.[142] Assuming the economics are favorable, biorefineries will ultimately require both *in vitro* (biocatalysts) and *in vivo* (microbial systems) biotransformations to convert biomass into alternative fuels.[143,144] The challenge is to identify, develop, and strategically blend enzymes and microorganisms (containing the appropriate metabolic pathways) to achieve economically attractive routes to bioenergy production.[145] Directed evolution and other recombinant technologies hold great promise for optimizing the properties of individual biocatalysts, and even multistep pathways, that are needed for biomass conversion.[146] These approaches work best, however, when biochemical and biophysical properties of the naturally occurring biocatalyst are already favorable. Metabolic engineering will certainly play an important role in developing optimal microbial systems for biomass conversion.[147–150] Nevertheless, concerns about release of recombinant microorganisms from large-scale bioprocesses make identification and characterization of naturally occurring, wild-type microorganisms an important pursuit for bioenergy production, particularly as a near-term critical need. Relevant biotransformation strategies used by such strains will no doubt ultimately inspire metabolic engineering efforts to develop optimal biomass conversion processes.

Hyperthermophilic Microorganisms and Their Enzymes for Bioenergy Conversion Processes

There are many factors to suggest that bioprocesses converting biomass into alternative fuels might be best done partly or entirely at elevated temperatures using hyperthermophilic biocatalysts and microorganisms.[151–153] Mechanisms underlying the thermophilicity and thermoactivity of proteins have been extensively investigated.[154–161] While known hyperthermophiles (see TABLE 1) do not seem to produce significant amounts of ethanol or other alcohols, they do produce considerable amounts of H_2 from sugars. Several hyperthermophilic archaea can convert glucan-based substrates to H_2, and hyperthermophilic bacteria of the genus *Thermotoga* are capable of fermenting a wider variety of mono- and polysaccharides to H_2.[64,162] There are several potential advantages to H_2 production at elevated temperatures. The solubility of substrates is enhanced, and the spectrum of fermentation products appears to be more limited with increasing temperature.[152,163] Differences in the glycolytic pathway for hyperthermophiles compared to mesophiles may also lead to improved biohydrogen yields. For example, *P. furiosus* reduces ferredoxin during sugar oxidation rather than NAD^+ as a consequence of the dehydrogenation of glyceraldehyde-3-phosphate and formation of phosphoglycerate, which occurs without ATP generation; this may allow ferredoxin reduction to proceed in the face of elevated pH_2.[17] *Thermotoga* species have also been shown to produce considerable amounts of H_2, a process that is strongly dependent on the type of sugar fermented.[164] One member of this genus, *T. neapolitana*, has been reported to produce H_2 under microaerophilic conditions, suggesting that an O_2-insensitive, and perhaps technologically important, evolution hydrogenase might be involved.[164] As discussed in the Introduction, some hydrogenases can be rendered more insensitive to oxygen by mutagenesis[9]; while all hydrogenases are O_2-sensitive to some degree (in particular the [FeFe]-hydrogenases), some are less so than others. Harnessing photosynthetic microorganisms to split water and provide these electrons to hydrogenases for H_2 generation is also possible, although this process would generate O_2, and thus requires engineering of the cell to spatially or functionally isolate the hydrogenase from O_2. Growth at higher temperatures reduces the solubility of gases, which would help remove contaminating O_2, and anaerobic degradation of biomass can provide the reducing power in the absence of light, another advantage of thermophiles.

Since bioenergy conversion plants will likely operate continuously, risk of contamination from nonsterile biomass feed stocks represents a bioprocessing challenge. This can potentially upset the metabolic balance in the resident microbial population, and could also give rise to contamination from H_2-oxidizing anaerobes, such as methanogens (generating methane) and acetogens (generating acetate); acetate production would lower the overall yield of H_2, and methane is a less desirable energy product. Continuous production of H_2 at high levels in chemostat culture has been demonstrated for several hyperthermophiles, including *P. furiosus*,[165] *T. maritima*,[166] and both *Thermococcus litoralis*[166] and *Thermococcus kodakaraensis* KOD1.[152] Elevated operating temperatures would preclude intrusion by mesophilic contaminants, resulting in less process variability.

Another important challenge to address for biohydrogen production by either mesophilic or thermophilic cultures is formation of metabolic side products, especially organic acids in addition to acetate, in the face of product inhibition by H_2. Metabolic engineering approaches may be used to avert this, and recent advances in the genetic manipulation of hyperthermophiles[152] indicates that such a strategy is applicable at least in principle to these microorganisms. Many questions remain concerning the efficacy of using hyperthermophilic microorganisms, and enzymes derived from them, for bioenergy conversion processes, but the application of the newly developed genomic-based tools of the last few years are being increasingly utilized to address these issues. A clearer understanding of the mechanisms of these highly complex hydrogenase enzymes, which ironically catalyze the simplest of chemical reactions, will lead ultimately to economical production of molecular hydrogen, which can be utilized for the energy requirements of a rapidly increasing world population.

Acknowledgments

The work described here, carried out in the authors' laboratory, was funded by the Department of Energy and the National Science Foundation.

Conflict of Interest

The authors declare no conflicts of interest.

References

1. Adams, M.W., L.E. Mortenson & J.S. Chen. 1980. Hydrogenase. Biochim. Biophys. Acta **594:** 105–176.
2. Vignais, P.M., B. Billoud & J. Meyer. 2001. Classification and phylogeny of hydrogenases. FEMS Microbiol. Rev. **25:** 455–501.
3. Hoehler, T.M. 2005. Biogeochemistry of dihydrogen (H_2). Metal Ions Biol. Syst. **43:** 9–48.
4. Dutta, D. *et al.* 2005. Hydrogen production by Cyanobacteria. Microb. Cell Factories **4:** 36–47.
5. Happe, T. *et al.* 2002. Hydrogenases in green algae: do they save the algae's life and solve our energy problems? Trends Plant Sci. **7:** 246–250.
6. Happe, T. & A. Kaminski. 2002. Differential regulation of the Fe-hydrogenase during anaerobic adaptation in the green alga *Chlamydomonas reinhardtii*. Eur. J. Biochem. **269:** 1022–1032.
7. Vignais, P.M. & A. Colbeau. 2004. Molecular biology of microbial hydrogenases. Curr. Issues Mol. Biol. **6:** 159–188.
8. Volbeda, A. *et al.* 2002. High-resolution crystallographic analysis of *Desulfovibrio fructosovorans* [NiFe] hydrogenase. Int. J. Hydrogen Energy. **27:** 1449–1461.
9. Flynn, T., M.L. Ghirardi & M. Seibert. 2002. Accumulation of O_2-tolerant phenotypes in H_2-producing strains

of *Chlamydomonas reinhardtii* by sequential applications of chemical mutagenesis and selection. Int. J. Hydrogen Energy **27:** 1421–1430.

10. CASALOT, L. & M. ROUSSET. 2001. Maturation of the [NiFe] hydrogenases. Trends Microbiol. **9:** 228–237.
11. GARCIN, E. *et al.* 1998. Structural bases for the catalytic mechanism of [NiFe] hydrogenases. Biochem. Soc. Trans. **26:** 396–401.
12. MEYER, J. 2007. [FeFe] hydrogenases and their evolution: a genomic perspective. Cell. Mol. Life Sci. **64:** 1063–1084.
13. SHIMA, S. & R.K. THAUER. 2007. A third type of hydrogenase catalyzing H(2) activation. Chem. Rec. **7:** 37–46.
14. ROBSON, R. 2001. Biodiversity of hydrogenases. *In* Hydrogen as a Fuel: Learning from Nature. R. CAMMACK, M. FREY & R. ROBSON, Eds.: pp. 9–32. London: Taylor & Francis.
15. GHIRARDI, M.L. *et al.* 2007. Hydrogenases and hydrogen photoproduction in oxygenic photosynthetic organisms. Annu. Rev. Plant Biol. **58:** 71–91.
16. BOCK, A. *et al.* 2006. Maturation of hydrogenases. Adv. Microb. Physiol. **51:** 1–71.
17. HALLENBECK, P.C. 2005. Fundamentals of the fermentative production of hydrogen. Water Sci. Technol. **52:** 21–29.
18. ARMSTRONG, F.A. 2004. Hydrogenases: active site puzzles and progress. Curr. Opin. Chem. Biol. **8:** 133–140.
19. EVANS, D.J. & C.J. PICKETT. 2003. Chemistry and the hydrogenases. Chem. Soc. Rev. **32:** 268–275.
20. NICOLET, Y., C. CAVAZZA & J.C. FONTECILLA-CAMPS. 2002. Fe-Only hydrogenases: structure, function and evolution. J. Inorg. Biochem. **91:** 1–8.
21. HORNER, D.S. *et al.* 2002. Iron hydrogenases—Ancient enzymes in modern eukaryotes. Trends Biochem. Sci. **27:** 148–153.
22. RAGSDALE, S.W. 2000. Nickel containing CO dehydrogenases and hydrogenases. Subcell. Biochem. **35:** 487–518.
23. NANDI, R. & S. SENGUPTA. 1998. Microbial production of hydrogen: an overview. Crit. Rev. Microbiol. **24:** 61–84.
24. HEDDERICH, R. 2004. Energy-converting [NiFe] hydrogenases from archaea and extremophiles: ancestors of complex I. J. Bioenerg. Biomembr. **36:** 65–75.
25. HEDDERICH, R. & L. FORZI. 2005. Energy-converting [NiFe] hydrogenases: more than just H_2 activation. J. Mol. Microbiol. Biotechnol. **10:** 92–104.
26. DICKSON, E.M., J.W. RYAN & M.H. SMULYAN. 1977. The Hydrogen Energy Economy: A Realistic Appraisal of Prospects and Impacts. New York: Praeger.
27. CAMMACK, R., M. FREY & R. ROBSON, Eds. 2001. Hydrogen as a Fuel: Learning from Nature. London: Taylor & Francis.
28. RAO, K.K. & R. CAMMACK. 2001. Producing hydrogen as a fuel. *In* Hydrogen as a Fuel: Learning from Nature. R. CAMMACK, M. FREY & R.L. ROBSON, Eds.: 201–230. London: Taylor & Francis.
29. STETTER, K.O. 2006. History of discovery of the first hyperthermophiles. Extremophiles **10:** 357–362.
30. KLENK, H.P. *et al.* 2004. Phylogenomics of hyperthermophilic Archaea and Bacteria. Biochem. Soc. Trans. **32:** 175–178.
31. SNEL, B., M.A. HUYNEN & B.E. DUTILH. 2005. Genome trees and the nature of genome evolution. Annu. Rev. Microbiol. **59:** 191–209.
32. GRIBALDO, S. & C. BROCHIER-ARMANET. 2006. The origin and evolution of Archaea: a state of the art. Philos. Trans. R. Soc. Lond. B. Biol. Sci. **361:** 1007–1022.
33. SCHWARTZMAN, D.W. & C.H. LINEWEAVER. 2004. The hyperthermophilic origin of life revisited. Biochem. Soc. Trans. **32:** 168–171.
34. SCHUT, G.J., J. ZHOU & M.W. ADAMS. 2001. DNA microarray analysis of the hyperthermophilic archaeon *Pyrococcus furiosus*: evidence for a new type of sulfur-reducing enzyme complex. J. Bacteriol. **183:** 7027–7036.
35. SCHUT, G.J., S.L. BRIDGER & M.W. ADAMS. 2007. Insights into the metabolism of elemental sulfur by the hyperthermophilic archaeon *Pyrococcus furiosus*: characterization of a coenzyme A-dependent NAD(P)H sulfur oxidoreductase. J. Bacteriol. **189:** 4431–4441.
36. MIROSHNICHENKO, M.L. & E.A. BONCH-OSMOLOVSKAYA. 2006. Recent developments in the thermophilic microbiology of deep-sea hydrothermal vents. Extremophiles **10:** 85–96.
37. VOLBEDA, A. *et al.* 1996. Structure of the [NiFe] hydrogenase active site: Evidence for biologically uncommon Fe ligands. J. Am. Chem. Soc. **118:** 12989–12996.
38. VOLBEDA, A. *et al.* 1995. Crystal-structure of the nickel-iron hydrogenase from *Desulfovibrio gigas*. Nature **373:** 580–587.
39. GARCIN, E. *et al.* 1999. The crystal structure of a reduced [NiFeSe] hydrogenase provides an image of the activated catalytic center. Structure **7:** 557–566.
40. PETERS, J.W. *et al.* 1998. X-Ray crystal structure of the Fe-only hydrogenase (CpI) from *Clostridium pasteurianum* to 1.8 angstrom resolution. Science **282:** 1853–1858.
41. ZIRNGIBL, C. *et al.* 1992. H_2-forming methylenetetrahydromethanopterin dehydrogenase, a novel type of hydrogenase without iron-sulfur clusters in methanogenic archaea. Eur. J. Biochem. **208:** 511–520.
42. PUTZ, S. *et al.* 2006. Fe-Hydrogenase maturases in the hydrogenosomes of *Trichomonas vaginalis*. Eukaryot. Cell. **5:** 579–586.
43. ADAMS, M.W.W. 1990. The structure and mechanism of iron-hydrogenases. Biochim. Biophys. Acta **1020:** 115–145.
44. GEBLER, A. *et al.* 2007. Impact of alterations near the [NiFe] active site on the function of the H(2) sensor from *Ralstonia eutropha*. FEBS J. **274:** 74–85.
45. LUDWIG, M., R. SCHULZ-FRIEDRICH & J. APPEL. 2006. Occurrence of hydrogenases in cyanobacteria and anoxygenic photosynthetic bacteria: implications for the phylogenetic origin of cyanobacterial and algal hydrogenases. J. Mol. Evol. **63:** 758–768.
46. MA, K. & M.W. ADAMS. 2001. Hydrogenases I and II from *Pyrococcus furiosus*. Methods Enzymol. **331:** 208–216.
47. MNATSAKANYAN, N., K. BAGRAMYAN & A. TRCHOUNIAN. 2004. Hydrogenase 3 but not hydrogenase 4 is major in hydrogen gas production by *Escherichia coli* formate hydrogenlyase at acidic pH and in the presence of external formate. Cell Biochem. Biophys. **41:** 357–365.

48. SAPRA, R., K. BAGRAMYAN & M.W. ADAMS. 2003. A simple energy-conserving system: proton reduction coupled to proton translocation. Proc. Natl. Acad. Sci. USA **100:** 7545–7550.
49. AMEND, J.P. & E.L. SHOCK. 2001. Energetics of overall metabolic reactions of thermophilic and hyperthermophilic Archaea and bacteria. FEMS Microbiol. Rev. **25:** 175–243.
50. CONRAD, R. 1996. Soil microorganisms as controllers of atmospheric trace gases (H_2, CO, CH_4, OCS, N_2O, and NO). Microbiol. Rev. **60:** 609–640.
51. KELLEY, D.S., J.A. BAROSS & J.R. DELANEY. 2002. Volcanoes, fluids, and life at mid-ocean ridge spreading centers. Annu. Rev. Earth Planetary Sci. **30:** 385–491.
52. ELDERFIELD, H. & A. SCHULTZ. 1996. Mid-ocean ridge hydrothermal fluxes and the chemical composition of the ocean. Annu. Rev. Earth Planetary Sci. **24:** 191–224.
53. MCCOLLOM, T.M. & J.S. SEEWALD. 2007. Abiotic synthesis of organic compounds in deep-sea hydrothermal environments. Chem. Rev. **107:** 382–401.
54. JOHNSON, M.R. *et al.* 2006. The *Thermotoga maritima* phenotype is impacted by syntrophic interaction with *Methanococcus jannaschii* in hyperthermophilic coculture. Appl. Environ. Microbiol. **72:** 811–818.
55. BONCHOSMOLOVSKAYA, E.A. & K.O. STETTER. 1991. Interspecies hydrogen transfer in cocultures of thermophilic Archaea. Syst. Appl. Microbiol. **14:** 205–208.
56. ISHII, S. *et al.* 2005. Coaggregation facilitates interspecies hydrogen transfer between *Pelotomaculum thermopropionicum* and *Methanothermobacter thermautotrophicus*. Appl. Environ. Microbiol. **71:** 7838–7845.
57. IMACHI, H. *et al.* 2000. Cultivation and in situ detection of a thermophilic bacterium capable of oxidizing propionate in syntrophic association with hydrogenotrophic methanogens in a thermophilic methanogenic granular sludge. Appl. Environ. Microbiol. **66:** 3608–3615.
58. SCHIPPERS, A. *et al.* 2005. Prokaryotic cells of the deep subseafloor biosphere identified as living bacteria. Nature **433:** 861–864.
59. WHITMAN, W.B., D.C. COLEMAN & W.J. WIEBE. 1998. Prokaryotes: the unseen majority. Proc. Natl. Acad. Sci. USA **95:** 6578–6583.
60. CHAPELLE, F.H. *et al.* 2002. A hydrogen-based subsurface microbial community dominated by methanogens. Nature **415:** 312–315.
61. TRAORE, A.S. *et al.* 1983. Energetics of growth of a defined mixed culture of *Desulfovibrio vulgaris* and *Methanosarcina barkeri*: interspecies hydrogen transfer in batch and continuous cultures. Appl. Environ. Microbiol. **46:** 1152–1156.
62. VALENTINE, D.L., W.S. REEBURGH & D.C. BLANTON. 2000. A culture apparatus for maintaining H_2 at subnanomolar concentrations. J. Microbiol. Methods **39:** 243–251.
63. CHABAN, B., S.Y. NG & K.F. JARRELL. 2006. Archaeal habitats—From the extreme to the ordinary. Can. J. Microbiol. **52:** 73–116.
64. CONNERS, S.B. *et al.* 2006. Microbial biochemistry, physiology, and biotechnology of hyperthermophilic *Thermotoga* species. FEMS Microbiol. Rev. **30:** 872–905.
65. STETTER, K.O. 2006. Hyperthermophiles in the history of life. Philos. Trans. R. Soc. Lond. B. Biol. Sci. **361:** 1837–1842; discussion 1842–1843.
66. TAKAI, K. *et al.* 2004. Geochemical and microbiological evidence for a hydrogen-based, hyperthermophilic subsurface lithoautotrophic microbial ecosystem (HyperSLiME) beneath an active deep-sea hydrothermal field. Extremophiles **8:** 269–282.
67. GUIRAL, M., C. AUBERT & M.T. GIUDICI-ORTICONI. 2005. Hydrogen metabolism in the hyperthermophilic bacterium *Aquifex aeolicus*. Biochem. Soc. Trans. **33:** 22–24.
68. GUIRAL, M. *et al.* 2006. Hyperthermostable and oxygen resistant hydrogenases from a hyperthermophilic bacterium *Aquifex aeolicus*: physicochemical properties. Int. J. Hydrogen Energy **31:** 1424–1431.
69. EDER, W. & R. HUBER. 2002. New isolates and physiological properties of the Aquificales and description of *Thermocrinis albus* sp. nov. Extremophiles **6:** 309–318.
70. GOTZ, D. *et al.* 2002. *Persephonella marina* gen. nov., sp. nov. and *Persephonella guaymasensis* sp. nov., two novel, thermophilic, hydrogen-oxidizing microaerophiles from deep-sea hydrothermal vents. Int. J. Syst. Evol. Microbiol. **52:** 1349–1359.
71. BULT, C.J. *et al.* 1996. Complete genome sequence of the methanogenic archaeon, *Methanococcus jannaschii*. Science **273:** 1058–1073.
72. SLESAREV, A.I. *et al.* 2002. The complete genome of hyperthermophile *Methanopyrus kandleri* AV19 and monophyly of archaeal methanogens. Proc. Natl. Acad. Sci. USA **99:** 4644–4649.
73. SMITH, D.R. *et al.* 1997. Complete genome sequence of *Methanobacterium thermoautotrophicum* deltaH: functional analysis and comparative genomics. J. Bacteriol. **179:** 7135–7155.
74. ROTHER, M. & W.W. METCALF. 2005. Genetic technologies for Archaea. Curr. Opin. Microbiol. **8:** 745–751.
75. TUMBULA, D.L. & W.B. WHITMAN. 1999. Genetics of *Methanococcus*: possibilities for functional genomics in Archaea. Mol. Microbiol. **33:** 1–7.
76. SORGENFREI, O. *et al.* 1997. The [NiFe] hydrogenases of *Methanococcus voltae*: genes, enzymes and regulation. Arch. Microbiol. **167:** 189–195.
77. HAFENBRADL, D. *et al.* 1996. *Ferroglobus placidus* gen. nov., sp. nov., a novel hyperthermophilic archaeum that oxidizes Fe^{2+} at neutral pH under anoxic conditions. Arch. Microbiol. **166:** 308–314.
78. KASHEFI, K. *et al.* 2003. Thermophily in the Geobacteraceae: *Geothermobacter ehrlichii* gen. nov., sp. nov., a novel thermophilic member of the Geobacteraceae from the "Bag City" hydrothermal vent. Appl. Environ. Microbiol. **69:** 2985–2993.
79. KLENK, H.P. *et al.* 1997. The complete genome sequence of the hyperthermophilic, sulphate-reducing archaeon *Archaeoglobus fulgidus*. Nature **390:** 364–370.
80. BLOCHL, E. *et al.* 1997. *Pyrolobus fumarii*, gen. and sp. nov., represents a novel group of archaea, extending the upper temperature limit for life to 113 degrees C. Extremophiles **1:** 14–21.
81. HUBER, R. *et al.* 1998. Sulfur-inhibited *Thermosphaera aggregans* sp. nov., a new genus of hyperthermophilic archaea

isolated after its prediction from environmentally derived 16S rRNA sequences. Int. J. Syst. Bacteriol. **48:** 31–38.

82. HAO, X. & K. MA. 2003. Minimal sulfur requirement for growth and sulfur-dependent metabolism of the hyperthermophilic archaeon *Staphylothermus marinus*. Archaea **1:** 191–197.
83. LASKA, S. & A. KLETZIN. 2000. Improved purification of the membrane-bound hydrogenase-sulfur-reductase complex from thermophilic archaea using epsilon-aminocaproic acid-containing chromatography buffers. J. Chromatogr. B Biomed. Sci. Appl. **737:** 151–160.
84. PIHL, T.D. & R.J. MAIER. 1991. Purification and characterization of the hydrogen uptake hydrogenase from the hyperthermophilic archaebacterium *Pyrodictium brockii*. J. Bacteriol. **173:** 1839–1844.
85. KANAI, T., S. ITO & T. IMANAKA. 2003. Characterization of a cytosolic NiFe-hydrogenase from the hyperthermophilic archaeon *Thermococcus kodakaraensis* KOD1. J. Bacteriol. **185:** 1705–1711.
86. SMITH, E.T. *et al.* 2001. Direct electrochemical characterization of hyperthermophilic *Thermococcus celer* metalloenzymes involved in hydrogen production from pyruvate. J. Biol. Inorg. Chem. **6:** 227–231.
87. LASKA, S., F. LOTTSPEICH & A. KLETZIN. 2003. Membrane-bound hydrogenase and sulfur reductase of the hyperthermophilic and acidophilic archaeon *Acidianus ambivalens*. Microbiology **149:** 2357–2371.
88. ADAMS, M.W.W. & E.I. STIEFEL. 2000. Organometallic iron: the key to biological hydrogen metabolism. Curr. Opin. Chem. Biol. **4:** 214–220.
89. VERHAGEN, M.F. & M.W. ADAMS. 2001. Fe-Only hydrogenase from *Thermotoga maritima*. Methods Enzymol. **331:** 216–226.
90. VERHAGEN, M.F., T. O'ROURKE & M.W. ADAMS. 1999. The hyperthermophilic bacterium, *Thermotoga maritima*, contains an unusually complex iron-hydrogenase: amino acid sequence analyses versus biochemical characterization. Biochim. Biophys. Acta **1412:** 212–229.
91. PAN, G., A.L. MENON & M.W. ADAMS. 2003. Characterization of a [2Fe-2S] protein encoded in the iron-hydrogenase operon of *Thermotoga maritima*. J. Biol. Inorg. Chem. **8:** 469–474.
92. VERHAGEN, M.F. *et al.* 2001. Heterologous expression and properties of the gamma-subunit of the Fe-only hydrogenase from *Thermotoga maritima*. Biochim. Biophys. Acta **1505:** 209–219.
93. WEIDNER, U. *et al.* 1993. The gene locus of the proton-translocating NADH: ubiquinone oxidoreductase in *Escherichia coli*. Organization of the 14 genes and relationship between the derived proteins and subunits of mitochondrial complex I. J. Mol. Biol. **233:** 109–122.
94. NELSON, K.E., J.A. EISEN & C.M. FRASER. 2001. Genome of *Thermotoga maritima* MSB8. Methods Enzymol. **330:** 169–180.
95. BREKASIS, D. & M.S. PAGET. 2003. A novel sensor of NADH/NAD+ redox poise in *Streptomyces coelicolor* A3(2). EMBO J. **22:** 4856–4865.
96. BRAZZOLOTTO, X. *et al.* 2006. The [Fe-Fe]-hydrogenase maturation protein HydF from *Thermotoga maritima* is a GTPase with an iron-sulfur cluster. J. Biol. Chem. **281:** 769–774.
97. RUBACH, J.K. *et al.* 2005. Biochemical characterization of the HydE and HydG iron-only hydrogenase maturation enzymes from *Thermatoga maritima*. FEBS Lett. **579:** 5055–5060.
98. PETERS, J.W. *et al.* 2006. A radical solution for the biosynthesis of the H-cluster of hydrogenase. FEBS Lett. **580:** 363–367.
99. NICOLET, Y. & C.L. DRENNAN. 2004. AdoMet radical proteins—From structure to evolution—Alignment of divergent protein sequences reveals strong secondary structure element conservation. Nucleic Acids Res. **32:** 4015–4025.
100. POSEWITZ, M.C. *et al.* 2004. Discovery of two novel radical S-adenosylmethionine proteins required for the assembly of an active [Fe] hydrogenase. J. Biol Chem. **279:** 25711–25720.
101. HIGUCHI, Y. *et al.* 1999. Removal of the bridging ligand atom at the Ni-Fe active site of [NiFe] hydrogenase upon reduction with H-2, as revealed by X-ray structure analysis at 1.4 angstrom resolution. Structure **7:** 549–556.
102. VOLBEDA, A. & J.C. FONTECILLA-CAMPS. 2003. The active site and catalytic mechanism of NiFe hydrogenases. Dalton Trans. 4030–4038.
103. VOLBEDA, A. *et al.* 2005. Structural differences between the ready and unready oxidized states of [NiFe] hydrogenases. J. Biol. Inorg. Chem. **10:** 239–249.
104. PIERIK, A.J. *et al.* 1999. Carbon monoxide and cyanide as intrinsic ligands to iron in the active site of [NiFe]-hydrogenases–NiFe(CN)(2)CO, biology's way to activate H-2. J. Biol. Chem. **274:** 3331–3337.
105. BLEIJLEVENS, B. *et al.* 2004. The auxiliary protein HypX provides oxygen tolerance to the soluble [NiFe]-hydrogenase of *Ralstonia eutropha* H16 by way of a cyanide ligand to nickel. J. Biol. Chem. **279:** 46686–46691.
106. VAN DER LINDEN, E. *et al.* 2004. The soluble [NiFe]-hydrogenase from *Ralstonia eutropha* contains four cyanides in its active site, one of which is responsible for the insensitivity towards oxygen. J. Biol. Inorg. Chem. **9:** 616–626.
107. CAMMACK, R., C. BAGYINKA & K.L. KOVACS. 1989. Spectroscopic characterization of the nickel and iron-sulphur clusters of hydrogenase from the purple photosynthetic bacterium *Thiocapsa roseopersicina*. 1. Electron spin resonance spectroscopy. Eur. J. Biochem. **182:** 357–362.
108. FOERSTER, S. *et al.* 2005. An orientation-selected ENDOR and HYSCORE study of the Ni-C active state of *Desulfovibrio vulgaris* Miyazaki F hydrogenase. J. Biol. Inorg. Chem. **10:** 51–62.
109. JONES, A.K. *et al.* 2003. Enzyme electrokinetics: electrochemical studies of the anaerobic interconversions between active and inactive states of *Allochromatium vinosum* [NiFe]-hydrogenase. J. Am. Chem. Soc. **125:** 8505–8514.
110. LAMLE, S.E., S.P. ALBRACHT & F.A. ARMSTRONG. 2004. Electrochemical potential-step investigations of the aerobic interconversions of [NiFe]-hydrogenase from *Allochromatium vinosum*: insights into the puzzling difference between unready and ready oxidized inactive states. J. Am. Chem. Soc. **126:** 14899–14909.

111. TROFANCHUK, O. *et al.* 2000. Single crystal EPR studies of the oxidized active site of [NiFe] hydrogenase from *Desulfovibrio vulgaris* Miyazaki F. J. Biol. Inorg. Chem. **5:** 36–44.
112. BAGLEY, K.A. *et al.* 1995. Infrared-detectable groups sense changes in charge-density on the nickel center in hydrogenase from *Chromatium vinosum*. Biochemistry **34:** 5527–5535.
113. HAPPE, R.P. *et al.* 1997. Biological activation of hydrogen. Nature **385:** 126.
114. FREY, M. 2002. Hydrogenases: hydrogen-activating enzymes. ChembioChem **3:** 153–160.
115. MONTET, Y. *et al.* 1997. Gas access to the active site of Ni-Fe hydrogenases probed by X-ray crystallography and molecular dynamics. Nature Struct. Biol. **4:** 523–526.
116. FIALA, G. & K.O. STETTER. 1986. *Pyrococcus furiosus* sp. nov. represents a novel genus of marine heterotrophic Archaebacteria growing optimally at 100 degrees C. Arch. Microbiol. **145:** 56–61.
117. BRYANT, F.O. & M.W. ADAMS. 1989. Characterization of hydrogenase from the hyperthermophilic archaebacterium, *Pyrococcus furiosus*. J. Biol. Chem. **264:** 5070–5079.
118. MA, K., R. WEISS & M.W. ADAMS. 2000. Characterization of hydrogenase II from the hyperthermophilic archaeon *Pyrococcus furiosus* and assessment of its role in sulfur reduction. J. Bacteriol. **182:** 1864–1871.
119. SAPRA, R., M.F. VERHAGEN & M.W. ADAMS. 2000. Purification and characterization of a membrane-bound hydrogenase from the hyperthermophilic archaeon *Pyrococcus furiosus*. J. Bacteriol. **182:** 3423–3428.
120. SCHUT, G.J. *et al.* 2003. Whole-genome DNA microarray analysis of a hyperthermophile and an archaeon: *Pyrococcus furiosus* grown on carbohydrates or peptides. J. Bacteriol. **185:** 3935–3947.
121. AONO, S., F.O. BRYANT & M.W. ADAMS. 1989. A novel and remarkably thermostable ferredoxin from the hyperthermophilic archaebacterium *Pyrococcus furiosus*. J. Bacteriol. **171:** 3433–3439.
122. VERHEES, C.H. *et al.* 2003. The unique features of glycolytic pathways in Archaea. Biochem. J. **375:** 231–246.
123. MUKUND, S. & M.W.W. ADAMS. 1995. Glyceraldehyde-3-phosphate ferredoxin oxidoreductase, a novel tungsten-containing enzyme with a potential glycolytic role in the hyperthermophilic archaeon *Pyrococcus furiosus*. J. Biol. Chem. **270:** 8389–8392.
124. BLAMEY, J.M. & M.W.W. ADAMS. 1993. Purification and characterization of pyruvate ferredoxin oxidoreductase from the hyperthermophilic archaeon *Pyrococcus furiosus*. Biochim. Biophys. Acta **1161:** 19–27.
125. ADAMS, M.W. *et al.* 2001. Key role for sulfur in peptide metabolism and in regulation of three hydrogenases in the hyperthermophilic archaeon *Pyrococcus furiosus*. J. Bacteriol. **183:** 716–724.
126. ROBB, F.T. *et al.* 2001. Genomic sequence of hyperthermophile, *Pyrococcus furiosus*: implications for physiology and enzymology. Hyperthermophilic Enzymes, A. **330:** 134–157.
127. FRIEDRICH, B. & E. SCHWARTZ. 1993. Molecular-biology of hydrogen utilization in aerobic chemolithotrophs. Annu. Rev. Microbiol. **47:** 351–383.
128. ALBRACHT, S.P.J. 1993. Intimate-relationships of the large and the small subunits of all nickel hydrogenases with 2 nuclear-encoded subunits of mitochondrial NADH—Ubiquinone oxidoreductase. Biochim Biophys. Acta **1144:** 221–224.
129. KASHANI-POOR, N. *et al.* 2001. A central functional role for the 49-kDa subunit within the catalytic core of mitochondrial complex I. J. Biol. Chem. **276:** 24082–24087.
130. PISA, K.Y. *et al.* 2007. A sodium ion-dependent A(1)A(O) ATP synthase from the hyperthermophilic archaeon *Pyrococcus furiosus*. FEBS J. **274:** 3928–3938.
131. BLOKESCH, M. *et al.* 2004. The complex between hydrogenase-maturation proteins HypC and HypD is an intermediate in the supply of cyanide to the active site iron of [NiFe]-hydrogenases. J. Mol. Biol. **344:** 155–167.
132. BLOKESCH, M. & A. BOCK. 2002. Maturation of [NiFe]-hydrogenases in *Escherichia coli*: the HypC cycle. J. Mol. Biol. **324:** 287–296.
133. BLOKESCH, M. *et al.* 2002. Metal insertion into NiFe-hydrogenases. Biochem. Soc. Trans. **30:** 674–680.
134. BLOKESCH, M. *et al.* 2004. HybF, a zinc-containing protein involved in NiFe. hydrogenase maturation. J. Bacteriol. **186:** 2603–2611.
135. FORZI, L. *et al.* 2007. The CO and CN(-) ligands to the active site Fe in [NiFe]-hydrogenase of *Escherichia coli* have different metabolic origins. FEBS Lett. **581:** 3317–3321.
136. LENZ, O. *et al.* 2007. Carbamoylphosphate serves as the source of CN(-), but not of the intrinsic CO in the active site of the regulatory [NiFe]-hydrogenase from *Ralstonia eutropha*. FEBS Lett. **581:** 3322–3326.
137. ZHANG, J.W. *et al.* 2005. A role for SlyD in the *Escherichia coli* hydrogenase biosynthetic pathway. J. Biol. Chem. **280:** 4360–4366.
138. BOHLMANN, G.M. 2005. Biorefinery—Process economics. Chem. Eng. Prog. **101:** 37–43.
139. LIN, Y. & S. TANAKA. 2006. Ethanol fermentation from biomass resources: current state and prospects. Appl. Microbiol. Biotechnol. **69:** 627–642.
140. MERTENS, R. & A. LIESE. 2004. Biotechnological applications of hydrogenases. Curr. Opin. Biotechnol. **15:** 343–348.
141. MORRISSEY, S.R. 2005. Employment genomics and clean energy. Chem. Eng. News **83:** 39–41.
142. SAHA, B.C. 2003. Hemicellulose bioconversion. J. Ind. Microbiol. Biotechnol. **30:** 279–291.
143. LYND, L. *et al.* 2006. Energy returns on ethanol production. Science **312:** 1746–1748; author reply, 1746–1748.
144. LYND, L.R. *et al.* 2005. Consolidated bioprocessing of cellulosic biomass: an update. Curr. Opin. Biotechnol. **16:** 577–583.
145. LAWFORD, H.G. & J.D. ROUSSEAU. 2003. Cellulosic fuel ethanol: alternative fermentation process designs with wild-type and recombinant *Zymomonas mobilis*. Appl. Biochem. Biotechnol. **105-108:** 457–469.
146. WANG, T.W. *et al.* 2006. Mutant library construction in directed molecular evolution: casting a wider net. Mol. Biotechnol. **34:** 55–68.

147. ARISTIDOU, A. & M. PENTTILA. 2000. Metabolic engineering applications to renewable resource utilization. Curr. Opin. Biotechnol. **11:** 187–198.

148. DEMAIN, A.L., M. NEWCOMB & J.H. WU. 2005. Cellulase, clostridia, and ethanol. Microbiol. Mol. Biol. Rev. **69:** 124–154.

149. INGRAM, L.O. *et al.* 1998. Metabolic engineering of bacteria for ethanol production. Biotechnol. Bioeng. **58:** 204–214.

150. SWARTZ, J. 2006. Developing cell-free biology for industrial applications. J. Ind. Microbiol. Biotechnol. **33:** 476–485.

151. SCHIRALDI, C., M. GIULIANO & M. DE ROSA. 2002. Perspectives on biotechnological applications of archaea. Archaea **1:** 75–86.

152. KANAI, T. *et al.* 2005. Continuous hydrogen production by the hyperthermophilic archaeon, *Thermococcus kodakaraensis* KOD1. J. Biotechnol. **116:** 271–282.

153. ATOMI, H. 2005. Recent progress towards the application of hyperthermophiles and their enzymes. Curr. Opin. Chem. Biol. **9:** 166–173.

154. ADAMS, M.W. 1993. Enzymes and proteins from organisms that grow near and above 100 degrees C. Annu. Rev. Microbiol. **47:** 627–658.

155. ADAMS, M.W.W. & R.M. KELLY. 2001. Hyperthermophilic Enzymes, Pt A. San Diego, CA: Academic Press.

156. ADAMS, M.W.W. & R.M. KELLY. 2001. Hyperthermophilic Enzymes, Pt B. San Diego, CA: Academic Press.

157. ADAMS, M.W.W. & R.M. KELLY. 2001. Hyperthermophilic Enzymes, Pt C. San Diego, CA: Academic Press.

158. FARIAS, S.T. *et al.* 2004. Thermo-search: lifestyle and thermostability analysis. In Silico Biol. **4:** 377–380.

159. FUJIWARA, S. 2002. Extremophiles: developments of their special functions and potential resources. J. Biosci. Bioeng. **94:** 518–525.

160. STERNER, R. & W. LIEBL. 2001. Thermophilic adaptation of proteins. Crit. Rev. Biochem. Mol. Biol. **36:** 39–106.

161. VIEILLE, C. & G.J. ZEIKUS. 2001. Hyperthermophilic enzymes: sources, uses, and molecular mechanisms for thermostability. Microbiol. Mol. Biol. Rev. **65:** 1–43.

162. DRISKILL, L.E., M.W. BAUER & R.M. KELLY. 1999. Synergistic interactions among beta-laminarinase, beta-1,4-glucanase, and beta-glucosidase from the hyperthermophilic archaeon *Pyrococcus furiosus* during hydrolysis of beta-1,4-, beta-1,3-, and mixed-linked polysaccharides. Biotechnol. Bioeng. **66:** 51–60.

163. VAN NIEL, E.W., P.A. CLAASSEN & A.J. STAMS. 2003. Substrate and product inhibition of hydrogen production by the extreme thermophile, *Caldicellulosiruptor saccharolyticus*. Biotechnol. Bioeng. **81:** 255–262.

164. VAN OOTEGHEM, S.A., S.K. BEER & P.C. YUE. 2002. Hydrogen production by the thermophilic bacterium *Thermotoga neapolitana*. Appl. Biochem. Biotechnol. **98-100:** 177–189.

165. SCHICHO, R.N. *et al.* 1993. Bioenergetics of sulfur reduction in the hyperthermophilic archaeon *Pyrococcus furiosus*. J. Bacteriol. **175:** 1823–1830.

166. RINKER, K.D. & R.M. KELLY. 1996. Growth physiology of the hyperthermophilic archaeon *Thermococcus litoralis*: development of a sulfur-free defined medium, characterization of an exopolysaccharide, and evidence of biofilm formation. Appl. Environ. Microbiol. **62:** 4478–4485.

167. COWAN, D.A. 2004. The upper temperature for life—Where do we draw the line? Trends Microbiol. **12:** 58–60.

168. STETTER, K.O. 1996. Hyperthermophilic procaryotes. FEMS Microbiol. Rev. **18:** 149–158.

169. JOHNSON, M.R. *et al.* 2004. Functional genomics-based studies of the microbial ecology of hyperthermophilic micro-organisms. Biochem. Soc. Trans. **32:** 188–192.

170. FINN, R.D. *et al.* 2006. Pfam: clans, web tools and services. Nucleic Acids Res. **34**: D247–D251.

171. HAFT, D.H., J.D. SELENGUT & O. WHITE. 2003. The TIGRFAMs database of protein families. Nucleic Acids Res. **31:** 371–373.

172. HOLDEN, J.F. *et al.* 2001. Identification of membrane proteins in the hyperthermophilic archaeon *Pyrococcus furiosus* using proteomics and prediction programs. Comp. Funct. Genomics **2:** 275–288.

Cellulases of Mesophilic Microorganisms

Cellulosome and Noncellulosome Producers

Roy H. Doi

Section of Molecular and Cellular Biology, University of California, Davis, California, USA

The cellulolytic activity of mesophilic bacteria and fungi is described, with special emphasis on the large extracellular enzyme complex called the cellulosome. The cellulosome is composed of a scaffolding protein, which is attached to various cellulolytic and hemicellulolytic enzymes, and this complex allows the organisms to degrade plant cell walls very efficently. The enzymes include a variety of cellulases, hemicellulases, and pectinases that work synergistically to degrade complex cell-wall molecules.

Key words: **cellulosomes; cellulases; hemicellulases; Clostridia; mesophiles**

Introduction

Microorganisms that grow and thrive at temperatures around 10°C and 45°C are considered to be mesophiles. There are many bacteria, fungi, and plants that grow at these temperatures, and they play a major role in the carbon cycle on earth. The plants actively fix CO_2 into plant cell walls, particularly as cellulose, hemicellulose, and lignin. These compounds, which make up the bulk of the biomass on earth, are relatively stable forms of biomass polymers, but are degraded by the action of microorganisms that are capable of hydrolyzing cellulose and hemicellulose into hexoses and pentoses that are used metabolically for their growth and sustenance. There is increasing interest in these organisms since they and the enzymes they produce may play a significant role in the conversion of plant biomass into biofuels.

There is a large number of bacteria and fungi that are capable of degrading plant cell walls, and some of these are listed in Tables 1–3. In addition, the sugars found in plant cell walls are listed below:

α-D-Glucose
α-D-Xylose
α-D-Mannose
α-D-Galactose
α-L-Rhamnose
β-L-Arabinose
α-L-Fucose
α-D-Apiose
α-D-Glucuronic acid
α-D-Galacturonic acid
α-D-Mannuronic acid

Both aerobic and anaerobic forms of microorganisms are involved in the degradation of plant cell walls (Schwarz: http://www.wzw.tum.de/mbiotec/cellmo.htm). The anaerobic bacteria are found in the soil, on decaying plant materials, in rumens, in sewage sludge, in termite gut, in wood-chip piles, in compost piles, and at paper mills and wood processing plants (Table 1). Most of these bacteria occur in natural habitats such as soil and decaying plant materials, but some are enriched by human activities, such as in compost piles, in sewage plants, and at wood processing plants. Other natural habitats include the anaerobic rumen of various ruminants and the gut of termites, where they process plant materials for the host organism's nutrition. The aerobic bacteria are usually found in the soil, in water, on plant materials, in humus, animal feces, sugar cane fields, and leaf litter (Table 2).

The aerobic fungi play a major role in the degradation of plant materials and are found on decomposing wood and plants, in the soil, and on agricultural wastes (Table 3). One of the most studied and industrially important aerobic fungi is *Trichoderma reesei*.

The anaerobic fungi are found in the intestinal tract of large herbivorous animals, such as the rumen of cows and sheep, and the hindgut of elephants and horses (Table 3). There are six recognized genera of anaerobic fungi, namely, *Anaeromyces*, *Caecomyces*, *Cyllamyces*, *Neocallimastix*, *Orpinomyces*, and *Piromyces*.[48] Unlike aerobic fungi, anaerobic fungi produce large, multienzyme cellulase-hemicellulase complexes similar to bacterial cellulosomes.[45,49] The enzyme complexes are

Address for correspondence: Roy H. Doi, Section of Molecular and Cellular Biology, University of California, One Shields Avenue, Davis, CA 95616. Voice: (530) 752-3191.
rhdoi@ucdavis.edu

Ann. N.Y. Acad. Sci. 1125: 267–279 (2008).
doi: 10.1196/annals.1419.002

TABLE 1. Mesophillic anaerobic bacteria with active cellulolytic systems

Microorganism	Habitat
Acetivibrio cellulolyticus	Sewage sludge[1,2]
Bacteroides cellulosolvens	Sewage sludge[3]
Butyrivibrio fibrisolvens	Bovine rumen[4]
Clostridium acetobutylicum	Soil[5]
Clostridium aldrichii	Wood digester[6]
Clostridium cellobioparum	Soil[7]
Clostridium cellulofermentans	Dairy farm soil[8]
Clostridium cellulolyticum	Decayed grass[9]
Clostridium cellulovorans	Wood chips[10]
Clostridium herbivorans	Pig intestine[11]
Clostridium hungatei	Soil[12]
Clostridium josui	Compost[13]
Clostridium papyrosolvens	Paper mill[14]
Fibrobacter succinogenes	Rumen[15]
Ruminococcus albus	Rumen[16]
Ruminococcus flavefaciens	Rumen[17]

TABLE 2. Some mesophilic aerobic cellulolytic bacteria

Microorganism	Habitat
Bacillus megaterium	Soil[18]
Bacillus pumilus	Soil, dead plant[19]
Cellulomonas fimi	Soil[20]
Cellulomonas flavigena	Soil, leaf litter[21]
Cellulomonas gelida	Soil[22]
Cellulomonas iranensis	Forest humus soils[23]
Cellulomonas persica	Forest humus soils[23]
Cellulomonas uda	Sugar cane field[24]
Cellvibrio gilvus	Bovine feces[25]
Cellvibrio mixtus	Soil[26]
Pseudomonas fluorescens	Soil, water[27]
Streptomyces antibioticus	Soil[28]
Streptomyces cellulolyticus	Soil[29]
Streptomyces lividans	Soil[30]
Streptomyces reticuli	Soil[31]

extremely active and can degrade both amorphous and crystalline cellulose.

The target for the enzymes of these various microorganisms is the plant cell wall, which comprises many different polysaccharides, proteins, and aromatic substances arranged as fibers with cross-linkers. The plant cell wall composition and structure varies between plant species, between tissues of a single species, and even among individual cells. The complex structure of the plant cell wall consists of cellulose fibers linked with hemicellulose, pectin, and lignin. The enzymes that are produced by the various microorganisms are classified as cellulases, hemicellulases, and ligninases. The composition and complexity of plant cell wall structure requires a multitude of enzymes for its degradation. In nature it is therefore likely that cooperative action also

TABLE 3. Some mesophilic aerobic and anaerobic cellulolytic fungi

Aerobic fungi
Habitat: soil, decomposing wood, agricultural waste
Aspergillus niger[32]
Phanerochaete chrysosporium[33]
Piptoporus betulinus[34]
Pycnoporus cinnabarinus[35]
Rhizopus stolonifer[36]
Serpula lacrymans[37]
Sporotrichum pulverulentum[38]
Trichoderma reesei (Hypocrea jecorina)[39]
Anaerobic Fungi
Habitat: rumen and intestinal tract of large herbivorous animals
Anaeromyces mucronatus 543[40]
Caecomyces communis[41]
Cyllamyces aberensis[42]
Neocallimastix frontalis[43]
Orpinomyces sp.[44]
Piromyces sp.[45]
Piromyces equi[46]
Piromyces sp. strain E2[47]

occurs between a number of different microorganisms to degrade plant cell wall materials.

This review will be concerned primarily with mesophilic cellulosomes and cellulases and hemicellulases whose actions result in the production of glucose and xylose, and also other hexoses and pentoses (see list preseted earlier in this section). The properties of the degradative systems of mesophilic anaerobic microorganisms are quite similar to those of thermophilic anaerobic microorganisms, with some variations in genetic organization and enzyme stability. Generally speaking, two types of systems occur in regard to plant cell wall degradation by microorganisms. In one type, the organism produces a set of free enzymes that work synergistically to degrade plant cell walls. In the second type, the degradative enzymes are organized into an enzyme complex called the cellulosome. This complex is very effective in degrading plant cell walls. The emphasis of this review will be on the cellulosomes.[50]

The Cellulosome System of Mesophiles

The cellulosome system of anaerobic microorganisms has been studied extensively, and a number of reviews have been written recently.[51–59] The occurrence of a cellulosome was first observed with the thermophilic bacterium, *Clostridium thermocellum*.[60,61] The cellulosome has now been described in a number of mesophilic anaerobic bacteria (TABLE 4) and with some anaerobic fungi, particularly *Piromyces* sp. strain E2 (TABLE 3).[74]

TABLE 4. Anaerobic mesophilic bacteria that produce cellulosomes

Species	Source
Acetivibrio cellulolyticus	Sewage sludge[1,62]
Bacterioides cellulosolvens	Sewage sludge[63–65]
Butyrivibrio fibrisolvens	Bovine rumen[4,66]
Clostridium acetobutylicum	Soil[5,67]
Clostridium cellulovorans	Wood-chip pile[10,54]
Clostridium cellobioparum	Bovine rumen[69,70]
Clostridium cellulolyticum	Decayed grass[9,71]
Clostridium josui	Compost pile[13,72]
Clostridium papyrosolvens	Paper mill[14,73]
Ruminococcus albus	Rumen[16,74]
Ruminococcus flavefaciens	Rumen[75,76]

The cellulosome is characterized by the presence of two general components: (1) the nonenzymatic scaffolding protein(s) with enzyme binding sites called cohesions, and (2) a variety of cellulosomal enzymes with dockerins, which interact with the cohesins in the scaffolding protein. In the simplest system, there is a single scaffolding protein (scaffoldin) with a number of cohesins and a cellulose binding domain (CBD). The enzymatic subunits are bound to the scaffolding through the interaction of the cohesins and dockerins to form the cellulosome (FIG. 1).

The scaffolding protein usually has a variable number of cohesions, depending on the bacterial species and a CBD more generally called a carbohydrate binding module (CBM). The CBD or CBM binds the cellulosome tightly to the substrate and concentrates the enzymes to a particular site of the substrate. A more complex type of cellulosome structure, in which there are multiple interacting scaffolding proteins that form a more complex structure that allows the binding of as many as 96 enzymes, has been revealed more recently.[62]

The cellulosomal enzymes include a wide variety of glycoside hydrolases and these have been listed in several reviews.[53,56,75] They include cellulases, xylanases, pectinases, mannanases, esterases, debranching glycosidases, and lichenases. The cellulosomal enzymes are characterized by the presence of duplicated sequences called dockerins, which interact with the cohesins of scaffolding proteins.

Why Is the Cellulosome Such an Efficient System for Degrading Plant Cell Walls?

First, the cellulosomes contain substrate binding sites or CBM, which bind the cellulosome tightly to the substrate and concentrate the hydrolytic enzymes to specific sites. Second, the presence of multiple cohesins on the scaffolding proteins and the presence of large number of cellulosomal enzymes results in a large variety of combinations of enzymes in the cellulosome and results in a functional variety of cellulosomes capable of attacking different types of plant cell wall materials. Third, there exists synergistic action of the cellulosomal enzymes which results in more efficient degradation of the complex plant cell wall substrate. Fourth, the cellulosomes act in concert with noncellulosomal glycosidic hydrolases, probably in sequential and/or nonsequential fashion, thus amplifying the whole degradative process. In essence the cellulosome system may exceed the potential of noncellulosomal degradative systems because of structural organization, efficient binding to the substrate, the large number of different types of hydrolytic enzymes, a high degree of enzymatic synergy, and cooperative action with noncellulosomal enzymes.

The Cellulosomal System of *Clostridium cellulovorans*

The *Clostridium cellulovorans* cellulosome system has been studied extensively for the last 20 years and has resulted in providing basic information about mesophilic cellulosomes. The cellulosome system of *C. cellulovorans* will be described here as a model for the simple type of cellulosomes containing a single scaffolding protein. This organism was isolated from a wood-chip pile and is an anaerobic spore–forming bacterium whose optimal growth temperature is 37°C.[68] It has the ability to utilize cellulose, xylan, pectin, cellobiose, glucose, fructose, galactose and mannose as carbon sources for growth. Its fermentation products include H_2, CO_2, acetate, butyrate, formate, lactate and ethanol. When grown in the presence of cellulose, electron micrographs have shown that large protuberances are present on its cell surface,[76] while little or no protuberances are evident when cells are grown in the presence of glucose or cellobiose.[77] The protuberances contain a large number of cellulosomes whose molecular mass is about 1000 kDa.[78]

An analysis of the *C. cellulovorans* cellulosome revealed a large nonenzymatic scaffolding protein (CbpA) that was essential for cellulosomal cellulolytic activity[78] and contained several functional domains[79] (FIG. 2). The CbpA domains included a CBD,[80] four hydrophilic domains,[81,82] and nine cohesins.[79] The CBD contained a planar configuration that interacted with the cellulose and involved the amino acids

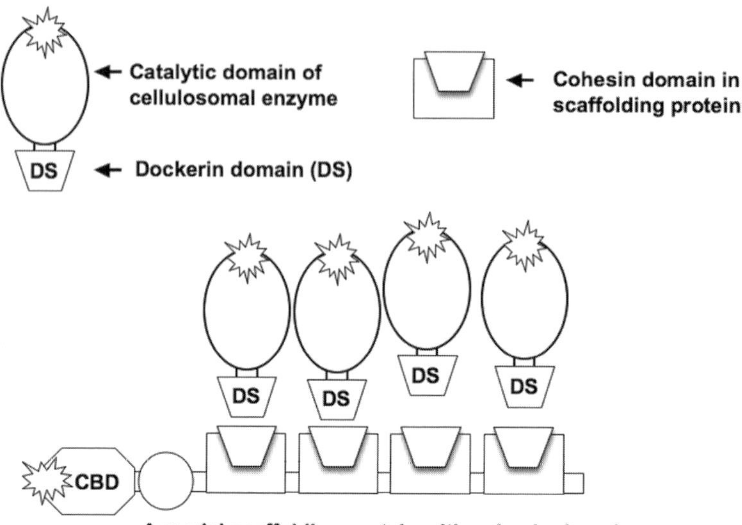

FIGURE 1. Model of the scaffolding protein CbpA of *Clostridium cellulovorans*.

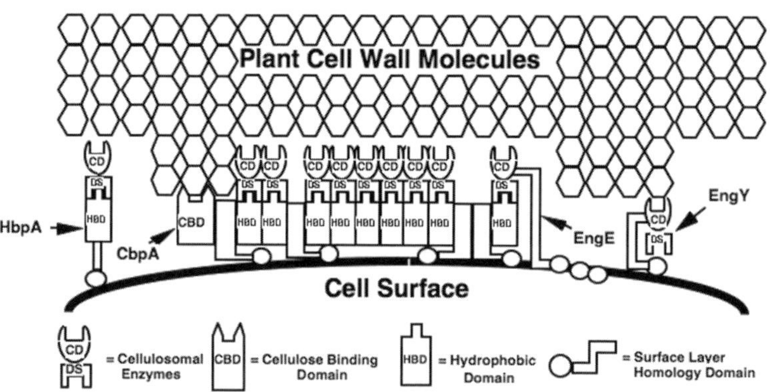

FIGURE 2. Model of a cellulosome attached to its substrate and cell surface.

tryptophan, aspartic acid, histidine, tyrosine, and arginine in binding the scaffolding protein to cellulose.[82] The four hydrophilic domains had a dual role in binding the scaffolding protein to the substrate and to the cell surface.[81] The nine cohesin domains were shown to bind the dockerin domains of the cellulosomal enzymes.[83]

The *C. cellulovorans* cellulosomal enzymes that have been identified to date include a large gene cluster that encodes the genes for CbpA-ExgS-EngH-EngK-HbpA-EngL-ManA-EngM-EngN,[84–88] and genes for endoglucanases EngB,[89] EngE,[87] pectate lyase A,[90] XynA,[91] and XynB[92] that are dispersed throughout the genome. XynA not only exhibited xylanase activity but also acetyl xylan esterase activity and released acetate from acetylated xylan.[91] Thus the cellulosomal enzymes can degrade cellulose, xylose, mannan, and pectin. It is likely, based on the analysis of the *C. thermocellum* genome, that the *C. cellulovorans* genome also encodes some 50–60 cellulosomal enzymes. It is this collection of enzymes that makes the cellulosome system so versatile in attacking plant cell walls and also that results in a large diverse functional population of cellulosomes that allows the system to attack a variety of plant materials with different compositions.

Regulation of the expression of the cellulosomal genes is evident at the transcriptional level. Coordinate expression of cellulase and hemicellulase genes was observed in the presence of cellulose as the carbon source, as well as catabolite repression when cells were grown in glucose or cellobiose.[93] It was also shown that the presence of xylan or pectin as the carbon source enhanced the expression of cellulosomal xylanase and pectate lyase, as well as several noncellulosomal enzymes.[94] In fact, pectin-grown cells produced enzymes that were most effective in converting plant cells into protoplasts.[95] In addition, mixed carbon substrates induced a wider variety of enzymes than a single carbon source, such as cellobiose, pectin, and xylan. Therefore it is evident that the expression of cellulosomal genes can be modified during growth on different carbon substrates such that optimal levels of certain enzymes will be attained.

The compositional difference between cellulosomes has been illustrated by fractionation of the cellulosome into subpopulations followed by analysis of the enzymatic subunits present in the subpopulations.[94,96,97] These studies have indicated that a large number of different types of cellulosomes are present at any one time and that the enzymatic composition of the cellulosomes depends on the carbon source that is used for growth. For instance, the types and compositions of cellulosomes are different if the cells are grown on cellobiose, cellulose, xylan, pectin, or a mixture of the substrates.[97]

Synergy of *Clostridium cellulovorans* Enzymes for the Degradation of Plant Cell Wall Materials

The presence of a large variety of cellulosomal enzymes allows the cellulosome to degrade a wide variety of lignocellulosic materials. However, in addition to the multiplicity of different enzymes, the synergistic action of the enzymes further enhances their capabilities. Synergism is described as the activity observed when two or more enzymes are used simultaneously or sequentially and the activity is greater than the additive activity of the individual enzymes. This has been illustrated in the case of cellulosomal cellulases, in which synergy was observed between three cellulases: endoglucanases EngE and EngH, and exoglucanase ExgS.[98] Optimal synergy with these three enzymes was observed when a proper ratio of the enzymes was present. Also it was observed that sequential action of the enzymes enhanced synergy, for example, addition of EngE or EngH followed by the addition of ExgS increased the synergistic action, suggesting that the initial reactions of EngE and EngH promoted effective degradation by ExgS.

Synergism was also observed between cellulosomal cellulases and xylanase when a substrate, such as corn cell wall, was used. Minicellulosomes containing either XynA, EngE, EngH, or ExgS were used to degrade corn cell wall. The mixture containing XynA minicellulosome and all three cellulase minicellulosomes at a 1:2 ratio (XynA:3 cellulases) showed the highest synergy.[99] In this case, simultaneous addition of the enzymes, but not sequential addition, showed the highest degree of synergy.

Besides synergistic action between cellulosomal enzymes, which might be expected since enzymes within the cellulosome are colocated to cooperate in their actions, synergism was found to occur between specific cellulosomal enzymes and noncellulosomal enzymes. This is exemplified by the action of cellulosomal hemicellulase XynA and noncellulosomal hemicellulases ArfA and BgaA.[100] When corn-stem powder was used as the substrate, synergy was observed when either XynA and ArfA (2.7) or XynA and BgaA (1.2) were used together, and the greatest degree of synergy was observed when XynA, ArfA and BgaA (2.9) were used together. Thus the action of noncellulosomal ArfA and BgaA, which would remove certain groups from xylan, also enhances the activity of XynA, probably by allowing better access of xylanase to the substrate.

When cellulosomes and crude noncellulosomal fractions of *C. cellulovorans* were mixed and tested simultaneously for activity on cellulose-arabinoxylan (CAX), a high degree of synergy was observed when the preparations were from cells grown on cellulose (3.86), but somewhat less with enzymes from cells grown on cellulose–pectin (1.04), cellulose–xylan (1.83), or cellulose–pectin–xylan (1.71).[94] In another approach, when enzymes were added sequentially and tested for activity on CAX, an initial treatment with XynA followed by EngL resulted in synergy, but EngL followed by XynA did not.[101] Furthermore, if mini-CbpA was used to form minicellulosomes in these experiments, greater activity always occurred with the minicellulosomes than with the free enzymes.

As further evidence for the cooperative action of cellulosomes with the noncellulosomal fraction, when cellulosomes were used simultaneously with noncellulosomal β-glucan glucohydrolase (BglA),[102] cellulose was degraded to glucose more rapidly than with cellulosomes alone. BglA was found to degrade the cellooligosaccharides produced by cellulosomes to glucose more rapidly than it was able to degrade cellobiose. Thus, BglA plays an important role in the conversion of cellulose to glucose.

Thus, the overall picture of plant cell wall degradation by *C. cellulovorans*appears to involve a sequential action of enzymes: initially a battery of noncellulosomal enzymes removes groups, such as arabinose, glucuronic acid, ferulic acid, and acetate, from xylan, followed by endoxylanase and β-xylosidase releasing xylobiose and xylose, and finally after xylan has been loosened by the hemicellulases, cellulose is attacked by noncellulosomal endoglucanases and exoglucanases and cellulosomes, resulting in the release of cellobiose and glucose. The synergy observed between individual enzymes of the cellulosome and between the cellulosome and noncellulosomal enzymes is a result of the sequential and simultaneous actions of these enzymes.

The Cellulosomal System of *Clostridium cellulolyticum*

The other mesophilic system that has been studied in great detail involves *Clostridium cellulolyticum*.[52] This organism has a cellulosome that is quite similar to that of *C. cellulovorans* with a large scaffolding protein, CipC, with a CBM at the N terminus followed by hydrophilic X2 module, seven cohesins, a second hydrophilic S2 module, and a terminal cohesin. Thus, the CipC is able to bind eight cellulosomal enzymes. The cohesins are highly conserved sequences, with most showing greater than 60% identity. Enzyme binding studies have indicated that all the cohesins bind with equal affinity to cellulosomal enzymes.[103]

A large gene cluster for cellulosomal genes is also found with *C. cellulolyticum* coding for CipC-Cel48F-Cel8C-Cel9G-Cel9E-orfX-Cel9H-Cel9J-Man5K-Cel9M-Rgl11Y-Cel5N.[52,103] This large gene cluster is 26 kb long and codes for seven cellulases, one pectinase, one mannanase, and one rhamnogalacturonan lyase. The core proteins for the cellulosome are CipC, Cel48F, and Cel9E, which are similar to the core proteins of the *C. cellulovorans* cellulosome, CbpA, ExgS, and EngE. The modular structure of the cellulosomal enzymes varies and usually includes the catalytic and dockerin domains, and depending on the enzyme, other domains, such as CBMs and Ig-like domains. Several cellulosomal enzymes whose genes are not linked to the large gene cluster have been found and include Cel5A, Cel5D, and Cel44O. There are at least 22 dockerin-bearing proteins.[52]

The multiplicity of cellulases is again reasonable, since they appear to have different properties, such as endo- or exoglucanase activity, substrate specificity, partial activity on xylan and cellulose, sensitivity or lack of sensitivity to cellobiose inhibition, endo-processive activity, cellobiohydrolase activity, and different pH and temperature optima. Thus, there is a large diversity of enzyme types among the cellulases, which again could allow the organism to degrade a variety of substrates under a variety of environmental conditions.

A particularly fruitful approach with the *C. cellulolyticum* system has been in the construction of designer minicellulosomes with specific functions. This has been accomplished by using a miniscaffolding protein containing cohesins that would interact specifically with designated enzymes containing cognate dockerins.[104–106] This approach has allowed the study of enzyme activity based on colocation of enzymes in the minicellulosome, alternate position of enzymes in the minicellulosome, and specific enzyme functions. The application of this approach should lead to the development of highly active, highly specific cellulosomes that would be useful in practical applications.[107]

The Cellulosomal System of *Clostridium josui*

The cellulosome system of *Clostridium josui* is similar to that found in *C. cellulolyticum* and *C. cellulovorans*. The scaffolding protein CipA of *C. josui* contains a CBM and six cohesin modules, and is therefore smaller than the CipA of *C. cellulolyticum* and the CbpA of *C. cellulovorans*. A large cluster of cellulosomal genes also is present in *C. josui* that contains *cipA-cel48A-cel8A-cel9E-cel9F-orfX-cel9G-cel9H-man5A*.[108] A somewhat unusual cellulosomal enzyme, α-galactosidase Aga27, was found that belonged to family 27.[109] The enzyme was found to be very active on guar gum, and it was proposed that this enzyme attacks galactomannan that is present in plant cell walls and allows the cellulosome to attack the cellulose chains buried in galactomannan.

In a study of the properties of CipA cohesins, the binding of dockerins from Aga27A and Cel8A with four of the cohesins was analyzed. These studies indicated that the binding affinity of the cohesins varied and that there was an eight-fold difference in binding of Aga27A dockerin with the cohesins and a 34-fold difference in binding of Cel8A dockerin with the cohesins.[110] Although past studies on cohesin–dockerin interactions suggested that the binding was relatively constant, the sensitivity of these studies indicated that indeed differences in binding occurred. These results indicate that the enzymatic composition of cellulosomes could be affected by competitive binding of enzymes and that this competition could ultimately determine the enzymatic activity and specificity of the cellulosome population.

The Lack of a Functional Cellulosomal System in *Clostridium acetobutylicum*

The absence of a functional cellulosomal in *Clostridium acetobutylicum* is somewhat puzzling, since genomic analysis has revealed the presence of cellulosomal genes.[111] It is apparent that the expression of these genes has been partially blocked either by the structure or absence of promoters for these genes, the presence of a repression system that is not inducible, the absence of a suitable transcription apparatus, or such a low expression of the genes that cellulosomes are not synthesized or assembled.

However, it has been reported that the cells do produce cellulase[112] and an inactive cellulosome.[5]

The CipA scaffolding protein of *C. acetobutylicum* has been cloned and sequenced. CipA contains family 3a CBM, five type I cohesin domains, and six hydrophilc domains.[113] The cohesin domains are separated by hydrophilic domains, which makes the organization of the cohesin domains different from that found in other mesophilic *Clostridia*. The genome does contain a large cluster of genes involved in cellulolytic functions encoding a scaffolding protein CipA followed by eight genes encoding glycosyl hydrolases of families 5, 9, and 48.[111]

Since *C. acetobutylicum* is a well-known solventogenic bacterium that converts sugars and polysaccharides into solvents and acids, there is much interest in converting this organism into a cellulosome producer and utilizer of cellulosic biomass (see chapter by P. Duerre, this issue). In this regard, pilot studies have been carried out successfully to produce minicellulosomes from *C. acetobutylicum*.[113,114] Both from the basic and applied aspects, solving the reasons for the production of inactive cellulosomes by this organism is intriguing.

Cellulosomes with Multiple Scaffolding Proteins

Besides the cellulosomes with relatively simple structure, that is, a single scaffolding protein and enzymes, more complex cellulosomal structures have been observed in which there are multiple scaffolding proteins and accessory proteins for binding the cellulosome to the cell surface. These include the cellulosomes from *Acetovibrio cellulolyticus*,[2,61] *Bacteroides cellulosolvens*,[3] and *Ruminococcus flavefaciens*.[115,116]

As an example of a more complex cellulosome, the cellulosome from *A. cellulolyticus* contains three nonenzymatic scaffolding proteins, ScaA, ScaB, and ScaC.[117] ScaA is considered the primary scaffoldin, ScaB an adaptor scaffoldin, and ScaC an anchoring scaffoldin. These scaffoldins interact through cohesin–dockerin-type interactions to form a large multiscaffolding structure capable of binding up to 96 enzymes. This large structure is held to the cell surface by the anchoring scaffoldin. It is possible that these cellulosomes with a potential mass of 5 mDa and the 7 mDa polycellulosomes from *C. thermocellum* are the largest extracellular enzyme complexes produced by any cell.[118]

It is likely that further diversity of cellulosome structure will be observed as more cellulolytic organisms are discovered and studied.

Noncellulosomal Systems

This review has emphasized the properties of cellulosome-producing mesophilic organisms. However, the noncellulosomal mesophilic systems that have been studied most actively are the cellulases from the fungus *T. reesei*[119] and the aerobic bacterium *Cellulomonas fimi*.[120] Much of the early work on cellulases was directed toward the cellulolytic enzymes produced by *T. reesei*, since they also had commercial value. The basic difference between the enzyme systems is the formation of cellulosomes and noncellulosomal enzymes by the cellulosome producers and the production of free interacting enzymes of noncellulosome producers. Also the cellulosome producers appear to have many more genes for lignocellulolytic activity than the noncellulosome producers.

The number and types of enzymes produced by *T. reesei* are shown in TABLE 5. They consist of endoglucanases, exoglucanases, and a family of hemicellulases. In this regard, it appears that the hemicellulases may attack xylans of plant cell walls initially to clear the way for the cellulases to attack the cellulose core. The aerobic soil bacterium, *C. fimi*, also produces a number of endoglucanases, exoglucanases, and hemicellulases (TABLE 6).

A thorough review of carbohydrate-degrading enzymes by various microbial organisms is presented by Warren[120] and is highly recommended to obtain a wider view of lignocellulolytic activity of aerobic and anaerobic microorganisms as well as thermophilic microorganisms.

Thoughts on Lignocellulolytic Microorganisms and Possible Future Applications

The efficient degradation of lignocellulose in nature appears to be related to the properties of the enzymes,

TABLE 5. ***Trichoderma reesei*** **cellulolytic and hemicellulolytic enzymes**

Cellulases
β-1,4-Cellobiohydrolase (CBH I & CBH II)
β-1,4-Endoglucanases (EG I, EG II, EG III, EG IV, EG V)
Hemicellulases
Xylanase (XYN I, XYN II, XYN III)
β-Glucosidase
α-L-Arabinofuranosidase
Acetyl xylan esterase
β-Mannanase
α-Glucuronidase
α-1,2-Mannosidase

Source: Claeyssens[119] and Warren.[120]

TABLE 6. ***Cellulomonas fimi*** **enzymes**

CenA	*endo*-β-1,4-Glucanase[121]
CenB	*endo*-β-1,4-Glucanase[121]
CenC	*endo*-β-1,4-Glucanase; *exo*-Glucanase[121]
CenD	*endo*-β-1,4-Glucanase[122] [TR4]
Cex	β-1,4-Exoglucanase/xylanase[123]
CbhA	Cellobiohydrolase A[124]
CbhB	Cellobiohydrolase B[125]
Man26A	Mannanase 26A[126]
Man2A	β-Mannosidase 2A[126]
XynD	Xylanase D; deacetylase[127]

the multiplicity of enzyme types, and the synergistic activity of the enzymes. There is efficient binding of the enzymes and enzyme complexes to their substrates through their CBMs. The large number of enzyme types with varying biochemical properties allows the microorganisms to attack various lignocellulosic materials under a variety of environmental conditions, for example, various temperatures, pHs, and ionic conditions. The common occurrence of synergy between the enzymes facilitates the degradation of recalcitrant substrates. The three factors—efficient binding of substrate, multiplicity of enzyme types, and synergy between enzymes—create a common theme for efficient degradation of plant cell walls.

The practical use of mesophilic enzymes will require further research and development of the enzyme systems. These improvement would include the following:

1. Genetic engineering or directed evolution of enzymes to be more efficient in degrading cellulose and hemicellulose. This could lead to the development of very efficient cellulosomes and organisms that could produce large numbers of cellulosomes and noncellulosomal enzymes.
2. Use of mesophilic enzyme genes in the construction of transgenic plants that will "deconstruct" the plant cell wall. For instance, promoters controlling cellulases and hemicellulases would be turned on after growth of the plant, or after harvesting of the product (corn, rice, etc.).
3. Use of mesophilic enzyme genes to transform an aerobic microorganism, such as *Bacillus subtilis*, to be an efficient cellulosome producer. In this regard, preliminary results indicate minicellulosomes can be produced by *B. subtilis*.[128,129]
4. Genetically engineer industrially important microorganisms such as *Corynebacterium* and *Streptomyces* to be cellulose degraders so that they could grow on plant biomass.
5. Not only the enzyme system should be improved or modified so that renewable biomass can be utilized, but one might consider the use of the ocean as a farm for producing biomass. Land and water is limited for growing plants on earth, but the ocean covers two-thirds of the earth's surface. It would seem that marine plants may be a good source for biomass in the future.

Acknowledgments

The author thanks the U.S. Department of Energy and the RITE Institute of Kyoto, Japan, for support of the research carried out in the author's laboratory. He also expresses his appreciation to the students and postdoctoral fellows for their contributions, and to Helen Chan for her technical support.

Conflict of Interest

The author declares no conflicts of interest.

References

1. Patel, G. B., A.W. Khan, B.J. Agnew & J.R. Colvin. 1980. Isolation and characterization of an anaerobic, cellulolytic microorganism, *Acetivibrio cellulolyticus* gen. nov., sp. nov. Int. J. Syst. Bacteriol. **30:** 179–185.
2. Ding, S.-Y., E.A. Bayer, D. Steiner, *et al.* 1999. A novel cellulosomal scaffoldin from *Acetivibrio cellulolyticus* that contains a family 9 glycosyl hydrolase. J. Bacteriol. **181:** 6720–6729.
3. Ding, S.-Y., E.A. Bayer, D. Steiner, *et al.* 2000. A scaffoldin of the *Bacteroides cellulosolvens* cellulosome that contains 11 Type II cohesins. J. Bacteriol. **182:** 4915–4925.
4. Lin, L.-L. & J.A. Thomson. 1991. An analysis of the extracellular xylanases and cellulases of *Butyrivibrio fibrisolvens* H17c. FEMS Microbiol. Lett. **84:** 197–204.
5. Sabathe, F., A. Belaich & P. Soucaille. 2002. Characterization of the cellulolytic complex (cellulosome) of *Clostridium acetobutylicum*. FEMS Microbiol. Lett. **217:** 15–22.

6. Yang, J.C., D.P. Chynoweth, D.S. Williams & A. Li. 1990. *Clostridium aldrichii* sp. nov., a cellulolytic mesophile inhabiting a wood-fermenting anaerobic digester. Int. J. Syst. Bacteriol. **40:** 268–272.
7. Hungate, R.E. 1944. Studies on cellulose fermentation. I. The culture and physiology of an anaerobic cellulose-digesting bacterium. J. Bacteriol. **48:** 499–513.
8. He, Y.L., Y.F. Ding & Y.Q. Long. 1991. Two cellulolytic *Clostridium* species: *Clostridium cellulosi* sp. nov. and *Clostridium cellulofermentans* sp. nov. Int. J. Syst. Bacteriol. **41:** 306–309.
9. Gal, L., S. Pages, C. Gaudin, *et al.* 1997. Characterization of the cellulolytic complex (cellulosome) produced by *Clostridium cellulolyticum*. Appl. Environ. Microbiol. **63:** 903–909.
10. Doi, R.H. & Y. Tamaru. 2001. *The Clostridium cellulovorans* cellulosome: an enzyme complex with plant cell wall degrading activity. Chem. Rec. **1:** 24–32.
11. Varel, V.H., R.S. Tanner & C.R. Woese. 1995. *Clostridium herbivorans* sp. nov., a cellulolytic anaerobe from the pig intestine. Int. J. Syst. Bacteriol. **45:** 490–494.
12. Monserrate, E., S.B. Leschine & E. Canale-Parola. 2001. *Clostridium hungatei* sp. nov., a mesophilic, N2-fixing cellulolytic bacterium isolated from soil. Int. J. Syst. Evol. Microbiol. **51:** 123–132.
13. Kakiuchi, M., A. Isui, K. Suzuki, *et al.* 1998. Cloning and DNA sequencing of the genes encoding *Clostridium josui* scaffolding protein CipA and cellulase CelD and identification of their gene products as major components of the cellulosome. J. Bacteriol. **180:** 1303–4308.
14. Pohlschroder, M., S.B. Leschine & E. Canale-Parola. 1994. Multicomplex cellulase-xylanase system of *Clostridium papyrosolvens* C7. J. Bacteriol. **176:** 70–76.
15. Forsberg, C.W., T.J. Beveridge & A. Hellstrom. 1981. Cellulase and xylanase release from *Bacteriodes succinogenes* and its importance in the rumen environment. Appl. Environ. Microbiol. **42:** 886–896.
16. Ohara, H., S. Karita, T. Kimura, *et al.* 2000. Characterization of the cellulolytic complex (cellulosome) from *Ruminococcus albus*. Biosci. Biotechnol. Biochem. **64:** 254–260.
17. Aurilia, V., J.C. Martin, S.I. McCrae, *et al.* 2000. Three multidomain esterases from the cellulolytic rumen anaerobe *Ruminococcus flavefaciens* 17 that carry divergent dockerin sequences. Microbiology **146:** 1391–1397.
18. Beukes, N. & B.I. Pletschke. 2006. Effect of sulfur-containing compounds on *Bacillus cellulosome*–associated 'CMCase' and 'Avicelase'activities. FEMS Microbiol. Lett. **264:** 226–231.
19. Kotchoni, O.S., O.O. Shonukan & W.E. Gachomo. 2003. *Bacillus pumilus* BpCRI6, a promising candidate for cellulase production under conditions of catabolite repression. Afr. J. Biotechnol. **2:** 140–146.
20. Langsford, M.L., N.R. Gilkes, W.W. Wakarchuk, *et al.* 1984. The cellulase system of *Cellulomonas fimi*. J. Gen. Microbiol. **130:** 1367–1376.
21. Pérez-Avalos, O., T. Ponce-Noyola, I. Magaña-Plaza & M. de la Torre. 2004. Induction of xylanase and beta-xylosidase in *Cellulomonas flavigena* growing on different carbon sources. Appl. Microbiol. Biotechnol. **46:** 405–409.
22. Stackebrandt, E. & O. Kandler. 1979. Taxonomy of the genus *Cellulomonas*, based on phenotypic characters and deoxyribonucleic acid-deoxyribonucleic acid homology, and proposal of seven neotype strains. Int. J. Syst. Bacteriol. **29:** 273–282.
23. Elberson, M.A., F. Malekzadeh, M.T. Yazdi, *et al.* 2000. *Cellulomonas persica* sp. nov. and *Cellulomonas iranensis* sp. nov., mesophilic cellulose-degrading bacteria isolated from forest soils. Int. J. Syst. Evol. Microbiol. **50:** 993–996.
24. Stoppok, W., P. Rapp & F. Wagner. 1982. Formation, location, and regulation of endo-1,4-β-glucanases and β-glucosidases from *Cellulomonas uda*. Appl. Environ. Microbiol. **44:** 44–53.
25. Storwick, W.O. & K.W. King. 1960. The complexity and mode of action of the cellulase system of *Cellvibrio gilvus*. J. Biol. Chem. **235:** 303–307.
26. Blackall, L.L., A.C. Hayward & L.I. Sly. 1985. Cellulolytic and dextranolytic gram-negative bacteria: revival of the genus *Cellvibrio*. J. Appl. Bacteriol. **59:** 81–97.
27. Hazlewood, G.P., J.I. Laurie, L.M. Ferreira & H.J. Gilbert. 1992. *Pseudomonas fluorescens* subsp. *cellulosa*: an alternative model for bacterial cellulase. J. Appl. Bacteriol. **72:** 244–251.
28. Enger, M.D. & B.P. Sleeper. 1965. Multiple cellulase system from *Streptomyces antibioticus*. J. Bacteriol. **89:** 23–27.
29. Li, X.-Z. 1997. *Streptomyces cellulolyticus* sp. nov., a new cellulolytic member of the genus *Streptomyces*. Int. J. Syst. Bacteriol. **47:** 443–445.
30. Kluepfel, D., F. Shareck, F. Mondou & R. Morosoli. 1986. Characterization of cellulase and xylanase activities of *Streptomyces lividans*. Appl. Microbiol. Biotechnol. **24:** 230–234.
31. Schlochtermeier, A., S. Walter, J. Schroder, *et al.* 1992. The gene encoding the cellulase (Avicelase) Cel1 from *Streptomyces reticuli* and analysis of the protein domains. Mol. Microbiol. **6:** 3611–3621.
32. Gielkens, M.M.C., E. Dekkers, J. Visser & L.H. de Graaff. 1999. Two cellobiohydrolase-encoding genes from *Aspergillus niger* require D-xylose and the xylanolytic transcriptional activator XlnR for their expression. Appl. Environ. Microbiol. **65:** 4340–4345.
33. Covert, S.F., J. Bolduc & D. Cullen. 1992. Genomic organization of a cellulase gene family in *Phanerochaete chrysosporium*. Curr. Genet. **22:** 407–413.
34. Valaskova, V. & P. Baldrian. 2006. Degradation of cellulose and hemicelluloses by he brown rot fungus *Piptoporus betulinus*–production of extracellular enzymes and characterization of the major cellulases. Microbiology **152:** 3613–3622.
35. Sigoillot, C., A. Lomascolo, E. Record, *et al.* 2002. Lignocellulolytic and hemicellulolytic system of *Pycnoporus cinnabarinus*: isolation and characterization of a cellobiose dehydrogenase and a new xylanase. Enzyme Microb. Technol. **31:** 876–883.
36. Pothiraj, C., P. Balaji & M. Eyini. 2006. Enhanced production of cellulases by various fungal cultures in solid

state fermentation of cassava waste. Afr. J. Biotechnol. **5:** 1882–1885.

37. HASTRUP, A.C.S., B. JENSEN, C. CLAUSEN & F. GREEN III. 2006. The effect of CaCl2 on growth rate, wood decay and oxalic acid accumulaion in *Serpula lacrymans* and related brown-rot fungi. Holzforshung **60:** 339–345.
38. ERIKSSON, K.E. & B. PETTERSSON. 1975. Extracellular enzyme system utilized by the fungus *Sporotrichum pulverulentum* (*Chrysosporium lignorum*) for the breakdown of cellulose. 1. Separation, purification and physico-chemical characterization of five endo-1,4-beta-glucanases. Eur. J. Biochem. **51:** 193–206.
39. KUHLS, K., E. LIECKFELDT, G. J. SAMUELSDAGGER, *et al.* 1996. Molecular evidence that the asexual industrial fungus *Trichoderma reesei* is a clonal derivative of the ascomycete *Hypocrea jecorina*. Proc. Natl. Acad. Sci. USA **93:** 7755–7760.
40. LEE, S.S., J.K. HA & K.-J. CHENG. 2001. The effects of sequential inoculation of mixed rumen protozoa on the degradation of orchard grass cell walls by anaerobic fungus *Anaeromyces mucronatus* 543. Can. J. Microbiol. **47:** 754–760.
41. GERBI, C., J. BATA, A. BRETON & G. PRENSIER. 1996. Glycoside and polysaccharide hydrolase activity of the rumen anaerobic fungus *Caecomyces communis* (*Sphaeromonas communis* SENSU ORPIN) at early and final stages of the developmental cycle. Curr. Microbiol. **32:** 256–259.
42. OZKOSE, E., B. J. THOMAS, D.R. DAVIES, *et al.* 2001. *Cyllamyces aberensis* gen.nov. sp.nov., a new anaerobic gut fungus with branched sporangiophores isolated from cattle. Can. J. Bot. **79:** 666–673.
43. LI, X. & R. CALZA. 1991. Fractionation of cellulases from the ruminal *fungus Neocallimastix frontalis* EB188. Appl. Environ. Microbiol. **57:** 3331–3336.
44. CHEN, H., X.L. LI, D.L. BLUM & L.G. LJUNGDAHL. 1998. Two genes of the anaerobic fungus *Orpinomyces* sp. strain PC-2 encoding cellulases with endoglucanase activities may have arisen by gene duplication. FEMS Microbiol. Lett. **159:** 63–68.
45. ALI, B.R.S., L. ZHOU, F.M. GRAVES, *et al.* 1995. Cellulases and hemicellulases of the anaerobic fungus *Piromyces* constitute a multiprotein cellulose-binding complex and are encoded by multigene families. FEMS Microbiol. Lett. **125:** 15–21.
46. FILLINGHAM, I., P. KROON, G. WILLIAMSON & G. HAZLEWOOD. 1999. A modular cinnamoyl ester hydrolase from the anaerobic fungus *Piromyces equi* acts synergistically with xylanase and is part of a multiprotein cellulose-binding cellulase-hemicellulase complex. Biochem. J. **343:** 215–224.
47. STEENBAKKERS, P.J.M., W. UBHAYASEKERA, H. J.A.M. GOOSSEN, *et al.* 2002. An intron-containing glycoside hydrolase family 9 cellulase gene encodes the dominant 90 kDa component of the cellulosome of the anaerobic fungus *Piromyces* sp. strain E2. Biochem. J. **365:** 193–204.
48. NICHOLSON, M.J., M.K. THEODOROU & J.L. BROOKMAN. 2005. Molecular analysis of the anaerobic rumen fungus *Orpinomyces*—Insights into an AT-rich genome. Microbiology **151:** 121–133.
49. EBERHARDT, R.Y., H.J. GILBERT & G.P. HAZLEWOOD. 2000. Primary sequence and enzymic properties of two modular endoglucanases, Cel5A and Cel45A, from the anaerobic fungus *Piromyces equi*. Microbiology **146:** 1999–2008.
50. UVERSKY, V. & I.A. KATAEVA (Eds.). 2006. Cellulosome. New York: Nova Science Publishers.
51. DOI, R.H. 2006. The *Clostridium cellulovorans* cellulosome. *In* Cellulosome. V. Uversky & I.A. Kataeva, Eds.: 153–168. New York: Nova Science Publishers.
52. TARDIF, C., A. BELAICH, H.-P. FIEROBE, *et al.* 2006. *Clostridium cellulolyticum*: Cellulosomes and cellulases. *In* Cellulosome. V. Uversky & I.A. Kataeva, Eds.: 231–259. New York: Nova Science Publishers.
53. ZERLOV, V.V. & W.H. SCHWARZ. 2006. The *C. thermocellum* cellulosome: novel components and insights from the genomic sequence. *In* Cellulosome. V. Uversky & I.A. Kataeva, Eds.: 119–151. New York: Nova Science Publishers.
54. DEMAIN, A.L., M. NEWCOMB & J.H. WU. 2005. Cellulase, clostridia and ethanol. Microbiol. Mol. Biol. Rev. **69:** 124–154.
55. DESVAUX, M. 2005. *Clostridium cellulolyticum*: model organism of mesophilic cellulolytic clostridia. FEMS Microbiol. Rev. **29:** 741–764.
56. DOI, R.H. & A. KOSUGI. 2004. Cellulosomes: plant cell wall degrading enzyme complexes. Nat. Rev. Microbiol. **2:** 541–551.
57. LYND, L.R., P.J. WEIMER, W.H. VANZYL & I.S. PRETORIUS. 2002. Microbial cellulose utilization: fundamentals and biotechnology. Microbiol. Mol. Biol. Rev. **66:** 506–577.
58. BAYER, E.A., J. P. BELAICH, Y. SHOHAM & R. LAMED. 2004. The cellulosomes: multienzyme machines for degradation of plant cell wall polysaccharides. Annu. Rev. Microbiol. **58:** 521–554.
59. SCHWARZ, W.H. 2001. The cellulosome and cellulose degradation by anaerobic bacteria. Appl. Microbiol. Biotechnol. **56:** 634–649.
60. LAMED, R., J. NAIMARK, E. MORGENSTERN & E.A. BAYER. 1987. Specialized surface structures in cellulolytic bacteria. J. Bacteriol. **169:** 3792–3800.
61. WIEGEL, J. & M. DYKSTRA. 1984. *Clostridium thermocellum*: adhesion and sporulation while adhered to cellulose and hemicellulose. Eur. J. Appl. Microbiol. Biotechnol. **20:** 59–65.
62. XU, Q., W. GAO, S.-Y. DING, *et al.* 2003. The cellulosome system of *Acetivibrio cellulolyticus* includes a novel type of adaptor protein and a cell surface anchoring protein. J. Bacteriol. **185:** 4548–4557
63. MURRAY, W.D., L.C. SOWDEN & J.R. COLVIN. 1984. *Bacteroides cellulosolvens* sp. nov., a cellulolytic species from sewage sludge. Int. J. Syst. Bacteriol. **34:** 185–187.
64. LAMED, R., E. MORAG, O. MORYOSEF & E.A. BAYER. 1991. Cellulosome-like entities in *Bacteroides cellulosolvens*. Curr. Microbiol. **22:** 27–34.
65. XU, Q., E.A. BAYER, M. GOLDMAN, *et al.* 2004. Architecture of the *Bacteroides cellulosolvens* cellulosome: description of a cell surface-anchoring scaffolding and a family 48 cellulase. J. Bacteriol. **186:** 968–977.

66. Bryant, M.P. 1984. *Butyrivibrio*. *In* Bergey's manual of systemic bacteriology, 9th ed., vol. 1. N.R. Krieg, Ed.: 641–643. Baltimore, MD: The Williams & Wilkins Company.

67. McCoy, E., E.B. Fred, W.H. Peterson & E.G. Hastings. 1926. A cultural study of the acetone butyl alcohol organisms. J. Infect. Dis. **39:** 457–483.

68. Sleat, R., R.A. Mah & R. Robinson. 1984. Isolation and characterization of an anaerobic, cellulolytic bacterium, *Clostridium cellulovorans* sp. nov. Appl. Environ. Microbiol. **48:** 88–91.

69. Kelly, W.J., R.V. Asmundson & D.H. Hopcroft. 1987. Isolation and characterization of a strictly anaerobic cellulolytic spore former: *Clostridium chartatabidum* sp. nov. Arch. Microbiol. **147:** 169–173.

70. Felix, C.R. & L.G. Ljungdahl. 1993. The cellulosome: the exocellular organelle of *Clostridium*. Annu. Rev. Microbiol. **47:** 791–819.

71. Petitdemange, E., F. Caillet, J. Giallo & C. Gaudin. 1984. *Clostridium cellulolyticum* sp. nov., a cellulolytic mesophilic species from decayed grass. Int. J. Syst. Bacteriol. **34:** 155–159.

72. Sukhumavasi, J., K. Ohmiya, S. Shimizu & K. Ueno. 1988. *Clostridium josui sp. nov.*, a cellulolytic, moderate thermophilic species from Thai compost. Int. J. Syst. Bacteriol. **38:** 179–182.

73. Madden, R.H., M.J. Bryder & N.J. Poole. 1982. Isolation and characterization of an anaerobic, cellulolytic bacterium, *Clostridium papyrosolvens* sp. nov. Int. J. Syst. Bacteriol. **32:** 87–91.

74. Ljungdahl, L.G., H.J.M. Op den Camp, H.J. Gilbert, *et al.* 2006. Cellulosomes of anaerobic fungi. *In* Cellulosome. V. Uversky & I.A. Kataeva, Eds.: 271–303. New York: Nova Science Publishers.

75. Doi, R.H., A. Kosugi, K. Murashima, *et al.* 2003. Cellulosomes from mesophilic bacteria. J. Bacteriol. **185:** 5907–5914.

76. Blair, B.G. & K.L. Anderson. 1998. Comparison of staining techniques for scanning electron microscopic dtetection of ultrastructural protuberances on cellulolytic bacteria. Biotech. Histochem. **73:** 107–113.

77. Blair, B.G. & K.L. Anderson. 1999. Regulation of cellulose-inducible structures of *Clostridium cellulovorans*. Can. J. Microbiol. **45:** 242–249.

78. Shoseyov, O. & R.H. Doi. 1990. Essential 170 kDa subunit for degradation of crystalline cellulose by *Clostridium cellulovorans* cellulase. Proc. Natl. Acad. Sci. USA **87:** 2192–2195.

79. Shoseyov, O., M. Takagi, M. Goldstein & R.H. Doi. 1992. Primary sequence analysis of *Clostridium cellulovorans* cellulose binding protein A (CbpA). Proc. Natl. Acad. Sci. USA **89:** 3483–3487.

80. Goldstein, M., M. Takagi, S. Hashida, *et al.* 1993. Characterization of the cellulose binding domain of the *Clostridium cellulovorans* cellulose binding protein A (CbpA). J. Bacteriol. **175:** 5762–5768.

81. Kosugi, A., Y. Amano, K. Murashima & R.H. Doi. 2004. Hydrophilic domains of scaffolding protein CbpA promote glycosyl hydrolase activity and localization of cellulosomes to the cell surface of *Clostridium cellulovorans*. J. Bacteriol. **186:** 6351–6359.

82. Murashima, K., A. Kosugi & R.H. Doi. 2005. Site-directed mutagenesis and expression of the soluble form of the family IIIa cellulose binding domain from the cellulosomal scaffolding protein of *Clostridium cellulovorans*. J. Bacteriol. **187:** 7146–7149.

83. Park, J.-S., Y. Matano & R.H. Doi. 2001. Cohesin-dockerin interactions of cellulosomal subunits of *Clostridium cellulovorans*. J. Bacteriol. **183:** 5431–5435.

84. Tamaru, Y., S. Karita, A. Ibrahim, *et al.* 2000. A large gene cluster for the *Clostridium cellulovorans* cellulosome. J. Bacteriol. **182:** 5906–5910.

85. Foong, F., T. Hamamoto, O. Shoseyov & R.H. Doi. 1991. Nucleotide sequence and characteristics of endoglucanase gene *engB* from *Clostridium cellulovorans*. J. Gen. Microbiol. **137:** 1729–1736.

86. Liu, C.-C. & R.H. Doi. 1998. Properties of *exgS*, a gene for a major subunit of the *Clostridium cellulovorans* cellulosome. Gene **211:** 39–47.

87. Tamaru, Y. & R.H. Doi. 1999. Three surface layer homology domains at the N terminus of the *Clostridium cellulovorans* major cellulosomal subunit EngE. J. Bacteriol. **181:** 3270–3276.

88. Tamaru, Y. & R.H. Doi. 2000. The *engL* gene cluster of *Clostridium cellulovorans* contains a gene for cellulosomal ManA. J. Bacteriol. **182:** 244–247.

89. Foong, F. C.-F. & R.H. Doi. 1992. Characterization and comparison of *Clostridium cellulovorans* endoglucanases-xylanases EngB and EngD expressed in *Escherichia coli*. J. Bacteriol. **174:** 1403–1409.

90. Tamaru, Y. & R.H. Doi. 2001. Pectate lyase A, an enzymatic subunit of the *Clostridium cellulovorans* cellulosome. Proc. Natl. Acad. Sci. USA **98:** 4125–4129.

91. Kosugi, A., K. Murashima & R.H. Doi. 2002. Xylanase and acetyl xylan esterase activities of XynA, a key subunit of the *Clostridium cellulovorans* cellulosome for xylan degradation. Appl. Environ. Microbiol. **68:** 6399–6402.

92. Han, S.-O., H. Yukawa, M. Inui & R.H. Doi. 2004. Isolation and expression of the *xynB* gene and its product, XynB, a consistent component of the *Clostridium cellulovorans* cellulosome. J. Bacteriol. **186:** 8347–8355.

93. Han, S.-O., H. Yukawa, M. Inui & R.H. Doi. 2003. Regulation of expression of cellulosomal cellulase and hemicellulase genes in *Clostridium cellulovorans*. J. Bacteriol. **1185:** 6067–6075.

94. Han, S.-O., H.-Y. Cho, H. Yukawa, *et al.* 2004. Regulation of expression of cellulosomes and noncellulosomal (hemi) cellulolytic enzymes in *Clostridium cellulovorans* during growth on different carbon sources. J. Bacteriol. **1186:** 4218–4227.

95. Tamaru, Y., S. Ui, K. Murashima, *et al.* 2002. Formation of protoplasts from cultured tobacco cells and *Arabidopsis thaliana* by the action of cellulosomes and pectate lyase from *Clostridium cellulovorans*. Appl. Environ. Microbiol. **68:** 2614–2618.

96. Murashima, K., A. Kosugi & R.H. Doi. 2002. Determination of subunit composition of *Clostridium cellulovorans* cellulosomes that degrade plant cell walls. Appl. Environ. Microbiol. **68:** 1610–1615.

97. HAN, S.-O., H. YUKAWA, M. INUI & R.H. DOI. 2005. Effect of carbon source on the cellulosomal subpopulations of *Clostridium cellulovorans*. Microbiology **151:** 1491–1497.

98. MURASHIMA, K., A. KOSUGI & R.H. DOI. 2002. Synergistic effects on crystalline degradation between cellulosomal cellulases from *Clostridium cellulovorans*. J. Bacteriol. **184:** 5088–5095.

99. MURASHIMA, K., A. KOSUGI & R.H. DOI. 2003. Synergistic effects of cellulosomal xylanase and cellulases from *Clostridium cellulovorans* on plant cell wall degradation. J. Bacteriol. **185:** 1518–1524.

100. KOSUGI, A., K. MURASHIMA & R.H. DOI. 2002. Characterization of two noncellulosomal subunits, ArfA and BgaA, from *Clostridium cellulovorans* that cooperate with the cellulosome in plant cell wall degradation. J. Bacteriol. **184:** 6859–6865.

101. KOUKIEKOLO, R., H.-Y. CHO, A. KOSUGI, *et al.* 2005. Degradation of corn fiber by *Clostridium cellulovorans* cellulases and hemicellulases and contribution of scaffolding protein CbpA. Appl. Environ. Microbiol. **71:** 3504–3511.

102. KOSUGI, A., T. ARAI & R.H. DOI. 2006. Degradation of cellulosome-produced cello-oligosaccharides by an extracellular non-cellulosomal β-glucan glucohydrolase, BglA, from *Clostridium cellulovorans*. Biochem. Biophys. Res. Commun. **349:** 20–23.

103. PAGES, S.A., A. BELAICH, H.P. FIEROBE, *et al.* 1999. Sequence analysis of scaffolding protein CipC and ORFXp, a new cohesin-containing protein in *Clostridium cellulolyticum*: comparison of various cohesin domains and subcellular localization of ORFXp. J. Bacteriol. **181:** 1801–1810.

104. FIEROBE, H.-P., E.A. BAYER, C. TARDIF, *et al.* 2002. Degradation of cellulose substrates by cellulosome chimerase: substrate targeting versus proximity of enzyme components. J. Biol. Chem. **277:** 49621–49630.

105. FIEROBE, H.-P., A. MECHALY, C. TARDIF, *et al.* 2001. Design and production of active cellulosome chimeras: selective incorporation of dockerin-containing enzymes into defined functional complexes. J. Biol. Chem. **276:** 21257–21261.

106. FIEROBE, H.-P., F. MINGARDON, A. MECHALY, *et al.* 2005. Action of designer cellulosomes on homogeneous versus complex substrates: controlled incorporation of three distinct enzymes into a defined tri-functional scaffolding. J. Biol. Chem. **280:** 16325–16334.

107. FIEROBE, H.-P., S. PAGES, C. TARDIF, *et al.* 2006. Cellulosome chimeras: a powerful tool to investigate the functioning of cellulosomes with potential biotechnological applications. *In* Cellulosome. V. Uversky & I.A. Kataeva, Eds.: 191–209. New York: Nova Science Publishers.

108. KARITA, S., S. JINDOU, A. SHIMAMURA, *et al.* 2006. The cellulosomes of *Clostridium josui* and *Ruminococcus albus*. *In* Cellulosome. V. Uversky & I.A. Kataeva, Eds.: 169–189. New York: Nova Science Publishers.

109. JINDOU, S., S. KARITA, E. FUJINO, *et al.* 2002. α-Galactosidase Aga27A, an enzymatic component of the *Clostridium josui* cellulosome. J. Bacteriol. **184:** 600–604.

110. JINDOU, S., A. SODA, S. KARITA, *et al.* 2004. Cohesin-dockerin interactions within and between *Clostridium josui* and *Clostridium thermocellum*. J. Biol. Chem. **279:** 9867–9874.

111. NOLLING, J., G. BRETON, M.V. OMELCHENKO, *et al.* 2001. Genome sequence and comparative analysis of the solvent-producing bacterium *Clostridium acetobutylicum*. J. Bacteriol. **183:** 4823–4838.

112. LEE, S.F., C.W. FORSBERG & L.N. GIBBINS. 1985. Cellulolytic activity *of Clostridium acetobutylicum*. Appl. Environ. Microbiol. **50:** 220–228.

113. SABATHE, F. & P. SOUCAILLE. 2003. Characterization of the CipA scaffolding protein and *in vivo* production of a minicellulosome in *Clostridium acetobutylicum*. J. Bacvteriol. **185:** 1092–1096.

114. PERRET, S., L. CASALOT, H.-P, FIEROBE, *et al.* 2004. Production of heterologous and chimeric scaffoldins by *Clostridium acetobutylicum* ATCC 824. J. Bacteriol. **186:** 253–257.

115. DING, S.Y., M.T. RINCON, R. LAMED, *et al.* 2001. Cellulosomal scaffolding-like proteins from *Ruminococcus flavefaciens*. J. Bacteriol. **183:** 1945–1953.

116. FLINT, H.J. & M.T. RINCON. 2006. Cellulosome organization in *Ruminococcus flavefaciens*. *In* Cellulosome. V. Uversky & I.A. Kataeva, Eds.: 211–229. New York: Nova Science Publishers.

117. BAYER, E.A. & R. LAMED. 2006. The cellulosome saga: early history. Flint, H.J. and M.T. Rincon. 2006. Cellulosome organization in *Ruminococcus flavefaciens*. *In* Cellulosome. V. Uversky & I.A. Kataeva, Eds.: 11–46. New York: Nova Science Publishers.

118. NORRIS, V., T. DEN BLAAUWEN, A. CABIN-FLAMAN, *et al.* 2007. Functional taxonomy of bacterial hyperstructures. Microb. Mol. Biol. Rev. **71:** 230–253.

119. CLAEYSSENS, M., W. NERINCKX & K. PIENS (Eds.). 1998. Carbohydrases from *Trichoderma reesei* and Other Microorganisms: Structures, Biochemistry, Genetics and Applications. Cambridge: The Royal Society of Chemistry.

120. WARREN, R.A.J. 1996. Microbial hydrolysis of polysaccharides. Annu. Rev. Microbiol. **50:** 183–212.

121. TOMME, P., E. KWAN, N.R. GILKES, *et al.* 1996. Characterization of CenC, an enzyme from *Cellulomonas fimi* with both endo- and exoglucanase activities. J. Bacteriol. **178:** 4216–4223.

122. MEINKE, A., N.R. GILKES, D.G. KILBURN, *et al.* 1993. Cellulose-binding polypeptides from *Cellulomonas fimi*: endoglucanse D (CenD), a family A beta-1,4-glucanase. J. Bacteriol. **175:** 1910–1918.

123. GILKES, N.R., B. HENRISSAT, D.G. KILBURN, *et al.* 1991. Domains in microbial beta-1,4-glycanases: sequence conservation, function, and enzyme families. Microbiol. Rev. 55, 303–315.

124. MEINKE, A., N.R. GILKES, E. KWAN, *et al.* 1994. Cellobiohydrolase A (CbhA) from the cellulolytic bacterium *Cellulomonas fimi* is a -1,4-exocellobiohydrolase analogous to *Trichoderma reesei* CBHII. Mol. Microbiol. **12:** 413–442.

125. SHEN, H., N.R. GILKES, D.G. KILBURN, *et al.* 1995. Cellobiohydrolase B, a second exocellobiohydrolase from the cellulolytic bacterium *Cellulomonas fimi*. Biochem. J. **311:** 67–74.

126. STOLL, D., H. STALBRAND & R.A.J. WARREN. 1999. Mannan-degrading enzymes from *Cellulomonas fimi*. Appl. Environ. Microbiol. **65:** 2598–2605.

127. LAURIE, J.I., J.H. CLARKE, A. CIRUELA, *et al.* 1997. The NodB domain of a multidomain xylanase from *Cellulomonas fimi* deacetylates acetylxylan. FEMS Microbiol. Lett. **148:** 261–264.

128. CHO, H.-Y., H. YUKAWA, M. INUI, *et al.* 2004. Production of minicellulosomes from *Clostridium cellulovorans* in *Bacillus subtilis* WB800. Appl. Environ. Microbiol. **70:** 5704–5707.

129. ARAI, T., S. MATSUOKA, H.-Y. CHO, *et al.* 2007. Synthesis of *Clostridium cellulovorans* minicellulosomes by intercellular complementation. Proc. Natl. Acad. Sci. USA **104:** 1456–1460.

Plant Cell Wall Breakdown by Anaerobic Microorganisms from the Mammalian Digestive Tract

HARRY J. FLINT[a] AND EDWARD A. BAYER[b]

[a]*Microbial Ecology Group, Rowett Research Institute, Greenburn Road, Bucksburn, Aberdeen, United Kingdom*

[b]*Department of Biological Chemistry, The Weizmann Institute of Science, Rehovot, Israel*

Degradation of lignocellulosic plant material in the mammalian digestive tract is accomplished by communities of anaerobic microorganisms that exist in symbiotic association with the host. Catalytic domains and substrate-binding modules concerned with plant polysaccharide degradation are found in a variety of anaerobic bacteria, fungi, and protozoa from the mammalian gut. The organization of plant cell wall–degrading enzymes, however, varies widely. The cellulolytic gram-positive bacterium *Ruminococcus flavefaciens* produces an elaborate cellulosomal enzyme complex that is anchored to the bacterial cell wall; assembly of the complex involves at least five different dockerin:cohesin specificities, and the *R. flavefaciens* genome encodes at least 180 dockerin-containing proteins that encompass a wide array of catalytic and binding activities. On the other hand, in the cellulolytic protozoan, *Polyplastron multivesiculatum*, individual plant cell wall–degrading enzymes appear to be secreted into food vacuoles, while the gram-negative bacterium *Prevotella bryantii* appears to possess a sequestration-type system for the utilization of soluble xylans. The system that is employed for polysaccharide utilization must play a major role in defining the ecological niche that each organism occupies within a complex gut community. 16S rRNA analyses are also revealing uncultured bacterial species closely adherent to fibrous substrates in the rumen and in the large intestine of animals and humans. The true complexity, both at a single organism and community level, of the microbial enzyme systems that allow animals to digest plant material is beginning to become apparent.

Key words: **lignocellulose; rumen; cellulosome; *Ruminococcus*; *Prevotella*; large intestine; microbial ecology**

Introduction

Lignocellulose-rich plant material is a major component of the diet of herbivorous and omnivorous mammals. This material is, however, largely undegradable by mammalian digestive enzymes, and its breakdown within the gut is accomplished by communities of resident microorganisms that live in symbiotic or mutualistic association with the host. Large herbivores devote some 70% of their gut volume to microbial fermentation, either pregastrically, in the case of ruminants, or in the colon and cecum, in the case of hindgut fermentors.[1] Conditions in these gut compartments result in the development of microbial communities of very high cell density (more than 10^{11} cells/g) in which oxygen becomes depleted, and anaerobic microorganisms are predominant.[2] The major products of anaerobic fermentation are short-chain fatty acids that are in turn utilized by the host as energy sources and can contribute up to 70% of total dietary energy in the case of large herbivores.

By comparison with other environments, such as soils and sediments, the main feature of microbial lignocellulose breakdown in the gut is the constant, relatively rapid, throughput of substrate. This argues that resident gut microorganisms should collectively possess a high capacity for fast and efficient substrate degradation if they are to gain energy from the large amount of potentially available carbohydrate that is passing through the gut. In their favor are the relatively high temperature (37–39°C) and relatively stable conditions of pH and ionic environment, plus the potential for synergy between different members of the gut community. Against them, however, is the intractable nature

Address for correspondence: Harry J. Flint, Microbial Ecology Group, Rowett Research Institute, Greenburn Road, Bucksburn, Aberdeen, AB21 9SB, UK.
H.Flint@rowett.ac.uk

TABLE 1. Colonization of insoluble substrates by mixed human fecal bacteria in *in vitro* continuous culture

		Substrates		
		Wheat bran	Hylon starch	Pig mucin
Bacterial group	Species		No. 16S rRNA sequences	
Bacteroidetes	*Bacteroides vulgatus*	1	-	12
	Other *Bacteroides* sp.	15	2	-
Clostridia (XIVa)	*Roseburia/E. rectale*	21	14	2
	Ruminococcus spp.	2	1	21
	C. hathewayi-related	28	-	-
	Other	6	5	7
Clostridia (IV)	*Ruminococcus bromii*	-	19	-
	Other cluster IV	3	1	-
Bifidobacteria	*B. adolescentis*	-	18	-
	B. bifidum	-	-	18
	Other Bifids	1	15	10

Residual substrate was recovered following 24-h incubation and washed extensively. Adherent bacteria were identified by PCR amplification and sequencing of 16S rRNA genes (data of McWilliam Leitch *et al.*[14]).

of insoluble plant cell wall material, and only a few specialized gut microorganisms appear to be capable of degrading this material in pure culture.[3] We are only just beginning to appreciate the remarkable complexity of the enzyme systems that have evolved to allow certain gut microorganisms to degrade lignocellulose.

Microbial Ecology of Plant Fiber Breakdown in the Gut

Recent molecular surveys based on amplified 16S rRNA sequences have revealed the extent of bacterial diversity in gut microbial communities in man[4–7] and in farm animals.[8–11] Although relatively few phyla are represented, with most sequences belonging to the Bacteroidetes or Firmicutes, many hundreds of phylotypes are typically detected within single samples of gut contents or feces. The gram-negative Bacteroidetes are represented mainly by *Bacteroides* spp. in the human large intestine and by *Prevotella* spp. in the rumen community. The Firmicutes include low% G + C content gram-positive bacteria, and among these, representatives of Clostridial 16S rRNA sequence clusters XIVa, IV, and IX appear particularly numerous in gut communities.[2,7,11]

So far, rather few bacterial phylotypes or species have been shown to be capable of degrading lignocellulose-rich plant cell wall material. In the rumen, the major cellulolytic species identified by early isolation studies were *Fibrobacter succinogenes,* which belongs to a discrete phylum of gram-negative bacteria, and two species of gram-positive Firmicutes, *Ruminococcus flavefaciens* and *Ruminococcus albus* that belong to the Clostridial 16S rRNA cluster IV. It is possible, and indeed likely, that important cellulolytic gut bacteria have remained uncultured to date; culture-independent molecular community analyses are helping to reveal which phylogenetic groups contribute to fiber breakdown.[12] Fractionation of rumen samples into solid and liquid reveals some broad differences between the planktonic and tightly adherent bacterial communities analyzed using 16S rRNA sequencing, with fewer Bacteroidetes and more Firmicutes associated with the insoluble substrate.[0] Similar analyses of the human large intestinal microbiota also indicated higher proportions of *Firmicutes* species among the fiber-associated community, with bacterial phylotypes related to cluster IV ruminococci showing the most significant association with fiber.[13] The preferences of different phylotypes for colonization of particular insoluble substrates were investigated recently using a simple continuous-flow fermentor system in which mixed human fecal bacteria were allowed to colonize wheat bran, resistant starch, or porcine mucin. 16S rRNA sequence analysis of bacteria found tightly attached after 24 h revealed that each substrate supported a different attached community, including several apparently uncultured, novel groups.[14] Sequences related to cultured species, in particular *Ruminococcus bromii* and *Bifidobacterium adolescentis*, for example, accounted for many of the sequences attached to resistant starch (TABLE 1).

Molecular studies on the rumen community have confirmed significant populations of the cultured cellulolytic species *F. succinogenes*, *R. flavefaciens,* and *R. albus*, and these isolates continue to provide the best and most obvious route, via detailed functional and genetic

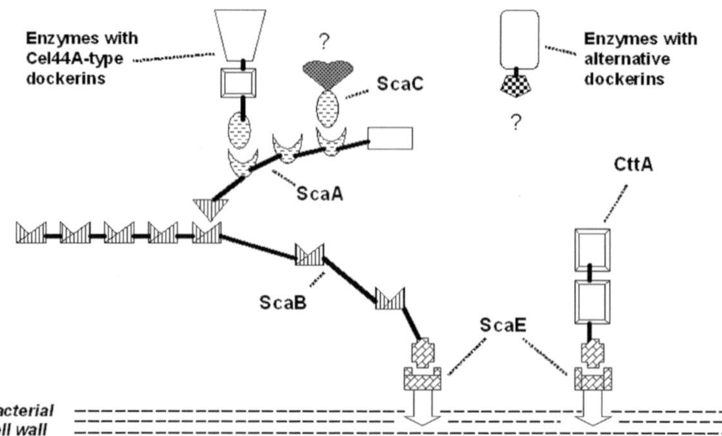

FIGURE 1. The cellulosome system of *Ruminococcus flavefaciens*. The different dockerin-cohesin pairings involved in the assembly of the *R. flavefaciens* cellulosome complex are distinguished by different *cross-hatchings*. The complex is connected to the cell surface through the pairing between the structural proteins ScaB and ScaE, with ScaE anchored to the bacterial cell surface via a sortase-mediated mechanism.[26] ScaE also accommodates the abundant cellulose binding protein CttA.[28] *Question marks* indicate dockerins and cohesins for which the partner is currently unknown.

studies, toward understanding the process of plant cell wall degradation by gut bacteria. *R. flavefaciens* was also well represented among sequences recovered from human fecal fiber,[13] and related strains have been isolated from the human GI tract.[15]

In the rumen, and in the large intestine, of large herbivores it is clear that eukaryotic microorganisms also play a significant role in the degradation of plant fiber. Anaerobic fungi are considered to be important in the initial degradation of large substrate particles, and their activities probably enhance degradation by bacteria by decreasing particle size and increasing access to the plant cell wall. Protozoa are able to engulf substrate particles as well as other gut microorganisms, and hence have multiple effects on rumen fiber digestion.[1]

Plant Cell Wall Breakdown by the Cellulolytic Bacterium *Ruminococcus flavefaciens*

R. flavefaciens was first isolated from the rumen over 50 years ago.[16] Many strains show high activity against crystalline cellulose, although individual strains show some variation in their ability to degrade different types of plant cell wall and different forms of cellulose.[17] The range of sugars utilized for growth is typically narrow, often restricted to cellobiose and cello-oligosaccharides, although many strains are also able to utilize xylo-oligosaccharides.[18] The failure to grow on glucose is assumed to be explained by the energetic advantage of possessing a cellobiose phosphorylase.[19] Phosphorylytic cleavage of cellobiose, and possibly also of cello-oligosaccharides, during transport into the cell yields glucose and glucose-1-phosphate; the latter is converted to glucose-6-phosphate, which goes directly into glycolysis without an adenosine triphosphate (ATP)-consuming kinase step.

Enzymes from *R. flavefaciens* 17 were shown to have complex multidomain structures often involving more than one catalytic domain, together with substrate-binding modules and dockerin sequences.[20,21] Analysis of the draft genome of the related strain *R. flavefaciens* FD1 has now shown that dockerins are present in at least 180 gene products.[22] Evidence for cellulosomal enzyme complexes,[23] whose organization depends on specific dockerin:cohesin interactions, on the bacterial cell surface has now been obtained for both of these strains. In particular, work on *R. flavefaciens* 17 revealed a cluster of five genes, four of whose products carry one or more cohesin domains capable of pairing with dockerins.[24,25] One of these, ScaE, is covalently attached to the cell wall and provides the point of attachment for the largest structural protein, ScaB[26] (FIG. 1). The structural protein, ScaA is assembled onto ScaB, and together with ScaC, these proteins provide cohesins to which a wide variety of enzyme subunits, and also many uncharacterized proteins, can bind. In *R. flavefaciens* 17, the ScaB protein carries seven cohesins that bind to ScaA only, while the ScaA protein carries three cohesins that bind a major group of enzyme dockerins.[25] In *R. flavefaciens* FD1, the ScaB protein homologue differs in carrying two types of cohesin, one, present in four copies, that binds directly to enzyme-borne dockerins, and another, present in five copies, that binds the C-terminal dockerin of ScaA.[27] ScaA from *R. flavefaciens* FD1 carries only two identified enzyme-binding cohesins, as opposed to the three that are present in ScaA from strain 17. Otherwise, the homologous *sca* gene clusters from the two strains show remarkable conservation of gene order and organization of their gene products.

The role of the fifth gene within the *sca* gene, designated *cttA*, has only recently become apparent. The C terminus resembles that of ScaB, and this protein is also able to bind to the ScaE protein, thereby becoming attached to the bacterial cell surface.[28] Interestingly, CttA is the most abundant protein that remains attached to residual cellulose in 7- to 10-day cultures of *R. flavefaciens* 17, thus suggesting that CttA would account for a major part of the adhesion of the bacterium to the substrate.[28] The N-terminal regions of CttA represent novel protein sequences. Purified recombinant protein fragments carrying these regions bind avidly to crystalline cellulose, and provide evidence that two distinct cellulose binding modules (CBMs) are present that act in concert.[28] The appearance of the CBMs on a separate cell surface protein component appears to compensate for the curious lack of a more conventional scaffoldin-associated CBM that has characterized all previously described cellulosome systems.

Earlier work by Stewart *et al.*[29,30] showed that another *R. flavefaciens* strain, 007, could lose its ability to degrade cotton cellulose when cultured repeatedly on cellobiose, although it retained its ability to degrade other forms of cellulose, including Avicel. The amount of CttA protein attached to residual Avicel cellulose was severely reduced or absent in cultures of the strain that had lost its cotton-degrading activity.[28] This suggests that CttA may play a critical role in the degradation of cotton cellulose, either because of a synergistic interaction with cellulolytic enzymes, or because this protein is indispensable for binding to this substrate.

It is not clear whether other rumen cellulolytic species, such as the related *R. albus,* or the unrelated *F. succinogenes,* display cellulosome organization. Several prominent cellulases from *R. albus* 8, including the potentially important GH48 enzyme,[31] do not carry dockerin sequences,[32] although the homologous GH48 cellulase from *R. flavefaciens* 17 does appear to carry a dockerin.[28] It also has been proposed that cell surface pili may be involved in attachment to cellulose by *R. albus*.[33] Thus it would appear that several different solutions to the problem of cellulose breakdown among rumen cellulolytic bacteria may have emerged.

Alternative Microbial Strategies for Gaining Energy from Plant Fiber

Cellulolytic Eukaryotes—Rumen Protozoa

As noted earlier, enzyme systems for the breakdown of plant cell wall polymers are not limited to cellulolytic prokaryotes. The remarkable enzyme systems and fiber-degrading activities of the rumen anaerobic fungi have, of course, been of special interest to Lars Ljungdahl over the years, and are considered elsewhere in this volume by Lars and his coauthors. Another remarkable group of anaerobic eukaryotes, however, are the rumen protozoa. For many decades these organisms were considered to be likely contributors to plant fiber breakdown, but they also engulf and predate bacteria. This makes it difficult to establish whether they produce their own fiber-degrading enzymes, or whether any apparent activity is due to ingested cellulolytic bacteria. This question has now been resolved by the cloning of reverse-transcribed polyA+ mRNA recovered from rumen contents of monofaunated sheep that carry a single protozoal species. Analysis of cDNA libraries has shown that *Polyplastron multivesiculatum* encodes its own xylanases,[34,35] and this conclusion has now been extended to cellulases and xylanases from a variety of rumen protozoa.[36] Enzyme structures so far reported lack the complex multimodular organization seen in anaerobic cellulolytic bacteria and fungi of gut origin, although substrate-binding and catalytic domains have been found in the same polypeptide.[35] Interestingly, some amino acid sequences for protozoal xylanases and cellulases are rather closely related to those from enzymes of the same family from rumen bacteria.[34] This suggests strongly that there has been gene acquisition through horizontal transfer between prokaryotes and eukaryotes, perhaps facilitated by the engulfment and digestion of bacterial cells resulting in DNA release within protozoal food vacuoles. Codon usage is highly biased in the rumen protozoa, which have exceptionally genomic low% G + C contents, so that such gene acquisition events must presumably be followed by extensive codon changes.

Protozoal cellulases and xylanases are assumed to be secreted into the food vacuoles of the protozoa rather than being expressed on the surface of the cell, as in anaerobic fungi and bacteria. This may have consequences for the modular organization of these enzymes. Although unusual in anaerobic microorganisms, the possibility exists that the protozoal enzymes function as a free soluble enzyme system, since the food vacuole will tend to limit the loss of enzymes and degradation products from the cell. There is no indication so far of dockerin:cohesin interactions that might be involved in membrane-anchoring or in enzyme-complex formation, and these features may indeed be absent from protozoal enzymes. It is also possible that the protozoal enzymes have evolved particularly to act in concert with the more elaborate enzyme complexes, discussed earlier, that are present on ingested bacterial cells. The recently reported protozoal plant cell wall–degrading enzymes therefore present a number of intriguing questions that call for further investigation.

Noncellulolytic Bacteria—Prevotella bryantii

In all anaerobic gut communities involved in fiber breakdown, a large number of species exist by metabolic cross-feeding, gaining energy either from the breakdown of carbohydrate substrates that are released by the action of the primary degrading species, or from the further conversion of fermentation products.[37–39] Thus, many highly abundant species of gut bacteria that are unable to degrade intact plant cell walls, nevertheless produce multiple polysaccharidases that are able to degrade structural plant polysaccharides, including pectins, xylans, and amorphous (noncrystalline) forms of cellulose. Such species include members of the dominant gram-negative CFB (*Cytophaga–Flavobacteria–Bacteroidetes*) phylum,[40] as well as gram-positive *Firmicutes* related to *Butyrivibrio fibrisolvens,*[41] *Roseburia,* and *Eubacterium rectale.*[42] It seems that these organisms, in addition to utilizing soluble components already present in the diet, take advantage of solubilized products of plant cell wall breakdown.[38,39] Furthermore, removal of degradation products and nutritional cross-feeding may enhance the rate of breakdown of insoluble cell-wall material by the primary cellulolytic species.

The noncellulolytic rumen CFB bacterium *P. bryantii*[43,44] produces multiple xylanases and mannanases, and a carboxymethylcellulase/mixed linkage β-glucanase.[45–47] The ability to utilize xylans is, however, strictly limited to soluble molecules[48] (FIG. 2A). Furthermore, xylanase activity was found to be at least six fold greater when assayed on freshly harvested whole cells compared with cells that have been disrupted, for example, by sonication[48] (FIG. 2B). Fractionation studies indicate that the bulk of xylanase activity is located in the periplasm or membranes, rather than being on the cell surface.[48] This recalls the organization of amylases in the related human colonic species *Bacteroides thetaiotaomicron* which, it is proposed, promotes sequestration of soluble polysaccharides by initial binding to Sus proteins on the outer membrane.[49,50]

One gene cluster identified in *P. bryantii* B_14 that is concerned with xylan utilization encodes a GH family 10 endoxylanase, and a GH43 β-xylosidase that is very unusual in exhibiting sensitivity to oxygen.[47] A homologous gene cluster is found in the human-gut relative *Bacteroides ovatus.*[51] These and other linked genes in *P. bryantii* are under the control of an adjacent two-component regulator that acts at a transcriptional level, and responds to large xylo-oligosaccharides[52,53] (FIG. 2C). A separately encoded GH10 xylanase (Xyn10C) includes an unusual large peptide insertion.[54] It is hypothesized, but not proven, that Xyn10C accounts for extracellular xylanase activity, whereas the other known genes may encode periplasmic activities. An additional gene cluster identified in this strain encodes GH5 and GH26 glucanases and mannanases,[55] and inserts carrying variations on this arrangement have recently been recovered with surprisingly high frequency from a rumen metagenome library, accounting for 4 out of 16 CMCase or β-glucanase positive lambda ZAP inserts analyzed.[56] This may indicate that multiple arrangements of this cluster exist within individual genomes, which is perhaps to be expected following analysis of the *B. thetaiotaomicron* genome[57] that revealed a multiplicity of operons concerned with utilization of dietary polysaccharidases. In addition, *Prevotella* DNA must have been highly abundant within the metagenome library. The functional significance of these many intriguing features of *Prevotella* polysaccharidases should become more apparent when genome sequence is available.

Metagenomics

Metagenomic approaches offer the potential to examine the diversity of polysaccharidase genes in gut communities without the constraints imposed by cultivability. Several metagenome libraries constructed from rumen-extracted DNA have now been screened for inserts that encode polysaccharidase activities.[56,58] Thus far, few, if any, genes have been reported that correspond to those found in cultured cellulolytic rumen bacteria, that is, that demonstrate complex multidomain organization, including potential substrate- or protein-binding domains as well as catalytic domains. This might imply that the available cultured cellulolytic bacteria are representative of minor components of the community. On the other hand, the technical difficulties of recovering DNA of a size suitable for metagenome library construction

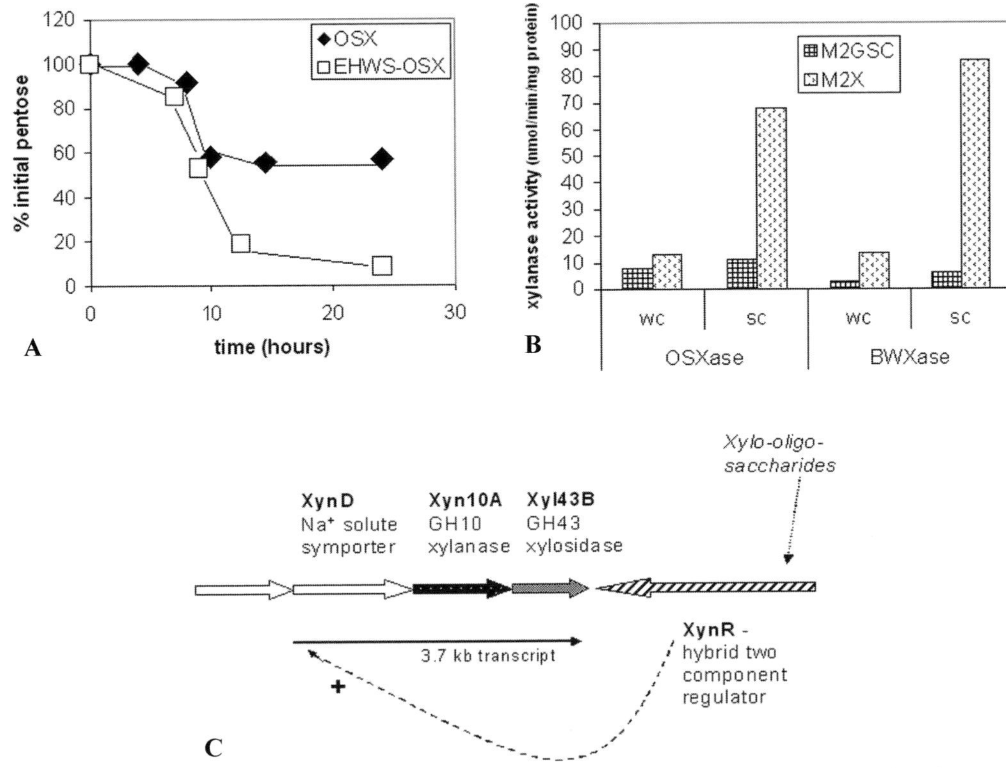

FIGURE 2. Xylan utilization in *Prevotella bryantii*. (**A**) Preferential utilization of a partially hydrolyzed, water-soluble fraction of oat spelt xylan (EHWS-OSX) compared with unfractionated oat spelt xylan (OSX) by cultures of *P. bryantii* B_14.[48] (**B**) Induction of cell-associated xylanase activity (OSXase = oat spelt xylanase, BWXase = birch wood xylanase) in *P. bryantii* grown on xylan medium (M2X) compared with medium containing glucose, starch, and cellobiose (M2GSC). Results are shown for freshly harvested whole cells (wc) and for sonicated cells (sc).[48] (**C**) Induction of the major xylanase gene cluster in *P. bryantii* B_14.[47,52,53] The genes that encode Xyn10A (*xynA*) and Xyl43B (*xynB*), together with the *xynD* transporter gene, are induced by the two-component regulator XynR, in response to xylo-oligosaccharide signal. Xyl43B appears to make a major contribution to inducible BWXase activity, and Xyl10A to inducible OSXase activity (part B).[47,52]

from bacteria that are tightly adherent to the substrate have not really been addressed. It seems more likely therefore that most of the genes that have been recovered derive from planktonic phase bacteria rather than tightly attached cellulolytic species, as suggested by the high representation of putative *Prevotella*-derived gene clusters mentioned earlier. Nevertheless, the screening studies undertaken so far have yielded novel sequences, for example, among rumen cellulases, that appear to represent new enzyme families or subfamilies. Other metagenomic approaches rely on large-scale sequencing,[59] but the information available on gut communities from this approach is currently limited relative to the complexity of the community.

Conclusions

Gut microorganisms, acting as complex consortia, have a remarkable ability to break down complex structural plant structures and to recover energy from them. Understanding how this process occurs, and harnessing components that can help to create alternatives to fossil fuels as energy sources, has never been more important.

Molecular ecological analysis of gut microbial communities is helping to identify those members of the complex community that are primarily involved in plant cell wall degradation, and is in general confirming the importance of the cellulolytic organisms that were isolated by the pioneers of anaerobic microbiology.[60] Genome sequence information on these organisms is clearly a priority, and is eagerly awaited, as this should reveal the range of strategies employed by microorganisms to extract energy from these substrates, and also allow comparison with cellulolytic microorganisms from non-gut sources. Engineered enzyme complexes, isolated anaerobic microbes, and complex consortia should all have roles to play in future biotechnological applications aimed at recovering energy from biomass.

Acknowledgments

The Rowett Institute receives support from the Scottish Government Rural and Environment Research and Analysis Directorate. One of the authors (E.A.B.) is the incumbent of The Maynard I. and Elaine Wishner Chair of Bio-organic Chemistry at The Weizmann Institute of Science and appreciates support (Research Grant 422/05) from the Israel Science Foundation (Jerusalem).

Conflict of Interest

The authors declare no conflicts of interest.

References

1. Van Soest, P.J. 1984. Nutritional Ecology of the Ruminant. 2nd ed. Itaca, NY: Cornell Univtrsity Press.
2. Flint, H.J. 2004. Polysaccharide breakdown by anaerobic micro-organisms inhabiting the mammalian gut. Adv. Appl. Microbiol. **56:** 89–120.
3. Dehority, B.A. & H.W. Scott. 1967. Extent of cellulose and hemicellulose digestion in various forages by pure cultures of rumen bacteria. J. Dairy Sci. **50:** 1136–1141.
4. Suau A., R. Bonnet, M. Sutren, *et al.* 1999. Direct analysis of genes encoding 16S rRNA from complex communities reveals many novel molecular species within the human gut. Appl. Environ. Microbiol. **24:** 4799–4807.
5. Hold, G. L., S.E. Pryde, V.J. Russell, *et al.* 2002. Assessment of microbial diversity in human colonic samples by 16S rDNA sequence analysis. FEMS Microbiol. Ecol. **39:** 33–39.
6. Hayashi, H., M. Sakamoto, & Y. Benno. 2002. Phylogenetic analysis of the human gut microbiota using 16S rDNA clone libraries and strictly anaerobic culture-based methods. Microbiol. Immunol. **46:** 535–548.
7. Eckburg, P.B., E.M. Bik, C.N. Bernstein, *et al.* 2005. Diversity of the human intestinal microbial flora. Science **308:** 1635–1638.
8. Tajima, K., R. I. Aminov, T. Nagamine, *et al.* 1999. Rumen bacterial diversity as determined by sequence analysis of 16S rDNA libraries. FEMS Microbiol. Ecol. **29:** 159–169.
9. Pryde, S.E., A.J. Richardson, C.S. Stewart, *et al.* 1999. Molecular analysis of the microbial diversity present in the colonic wall, colonic lumen, and cecal lumen of a pig. Appl. Environ. Microbiol. **65:** 5372–5377.
10. Daly, K., C.S. Stewart, H.J. Flint, *et al.* 2001. Bacterial diversity within the equine large intestine as revealed by molecular analysis of cloned 16S rRNA genes. FEMS Microbiol. Ecol. **38:** 141–151.
11. Leser, T.D., J.Z. Amenuvor, T.K. Jensen, *et al.* 2002. Culture independent analysis of gut bacteria: the pig gastrointestinal tract revisited. Appl. Environ. Microbiol. **68:** 673–690.
12. Larue, R., Z. Yu, V.A. Parisi, *et al.* 2005. Novel microbial diversity adherent to plant biomass in the herbivore gastrointestinal tract, as revealed by ribosomal intergenic spacer analysis and *rrs* gene sequencing. Environ. Microbiol. **7:** 530–543.
13. Walker, A.W. 2005. Influence of substrate and environmental factors on human gut microbial ecology and metabolism. PhD thesis. University of Aberdeen, Aberdeen, Scotland.
14. McWilliam Leitch, E.C., A.W. Walker, S.H. Duncan, *et al.* 2007. Selective colonisation of insoluble substrates by human faecal bacteria. Environ. Microbiol. **72:** 667–679.
15. Robert, C. & A. Bernalier-Donadille. 2003. The cellulolytic microflora of the human colon: evidence of microcrystalline cellulose-degrading bacteria in methane excreting subjects. FEMS Microbiol. Ecol. **46:** 81–89.
16. Sijpestein, A.K. 1951. On *Ruminococcus flavefaciens*, a cellulose decomposing bacterium from the rumen of sheep. J. Gen. Microbiol. **5:** 317–325.
17. Krause, D.O., R.J. Bunch, W.J.M. Smith, *et al.* 1999. Diversity of *Ruminococcus* strains: a survey of genetic polymorphisms and plant digestibility. J. Appl. Microbiol. **86:** 487–495.
18. Aurilia, V., J.C. Martin, C.A. Munro, *et al.* 2000. Organisation and strain distribution of genes responsible for the utilization of xylans by the rumen cellulolytic bacterium *Ruminococcus flavefaciens*. Anaerobe **6:** 333–340.
19. Ayers, W.A. 1959. Phosphorolysis and synthesis of cellobiose by cell extracts of *Ruminococcus flavefaciens*. J. Biol. Chem. **234:** 2819–2822.
20. Kirby, J., J.C. Martin, A.S. Daniel, *et al.* 1997. Dockerin-like sequences in cellulases and xylanases from the rumen cellulolytic bacterium *Ruminococcus flavefaciens*. FEMS Microbiol. Letts. **149:** 213–219.
21. Aurilia, V., J.C. Martin, S.I. McCrae, *et al.* 2000. Three multidomain esterases belonging to the plant cell wall degrading enzyme system of the rumen cellulolytic bacterium *Ruminococcus flavefaciens* 17 carry divergent dockerin sequences. Microbiology **146:** 1391–1397.
22. Rincon, M.T., I. Borovok, M.E. Berg, *et al.* 2006. Novel structural and catalytic elements in the *Ruminococcus flavefaciens* cellulosome revealed by genome analysis. Reprod. Nutr. Dev. **46**(Suppl. 1): S57.
23. Bayer, E.A., L.J. Shimon, Y. Shoham, *et al.* 1998. Cellulosomes-structure and ultrastructure. J. Struct. Biol. **124:** 221–234.
24. Ding, S.Y., M.T. Rincon, R. Lamed, *et al.* 2001. Cellulosomal scaffoldin-like proteins from *Ruminococcus flavefaciens*. J. Bacteriol. **183:** 1945–1953.
25. Rincon, M.T., S.-Y. Ding, S.I. McCrae, *et al.* 2003. Novel organisation and divergent dockerin specificities in the cellulosome system of *Ruminococcus flavefaciens*. J. Bacteriol. **185:** 703–713.
26. Rincon, M.T., T. Cepeljnik, J.C. Martin, *et al.* 2005. Unconventional mode of attachment of the *Ruminococcus flavefaciens* cellulosome to the cell surface. J. Bacteriol. **187:** 7569–7578.

27. Jindou S., I. Borovok, M.T. Rincon, *et al.* 2006. Conservation and divergence in cellulosome architecture between two strains of *Ruminococcus flavefaciens*. J. Bacteriol. **188:** 7971–7976.

28. Rincon, M.T., T. Cepelnik, J.C. Martin, *et al.* 2007. A novel cell-surface anchored cellulose-binding protein encoded by the *sca* gene cluster of *Ruminococcus flavefaciens*. J. Bacteriol. **189:** 4774–4783.

29. Stewart, C.S., S.H. Duncan, C.A. McPherson, *et al.* 1990. The implications of the loss & regain of cotton degrading activity for the degradation of straw by *Ruminococcus flavefaciens* 007. J. Appl. Bacteriol. **68:** 349–356.

30. Stewart, C.S., S.H. Duncan & H.J. Flint. 1990. The properties of forms of *Ruminococcus flavefaciens* which differ in their ability to degrade cotton cellulose. FEMS Microbiol. Letts. **72:** 47–50.

31. Bourne, Y. & B. Henrissat. 2001. Glycoside hydrolases and glycosyltransferases: families and functional modules. Curr. Opin. Struct. Biol. **11:** 593–600.

32. Devillard, E., D.B. Goodheart, K.R. Sanjay, *et al.* 2004. *Ruminococcus albus* 8 mutants defective in cellulose degradation are deficient in two processive endocellulases, Cel48A and Cel9B, both of which possess a novel modular architecture. J. Bacteriol. **186:** 136–145.

33. Rakotoarivinina, H., M.A. Larson, M. Morrison, *et al.* 2005. The *Ruminococcus albus* pilA1-pilA2 locus: expression and putative role of two adjacent pil genes in pilus formation and bacterial adhesion to cellulose. Microbiology **151:** 1291–1299.

34. Devillard, E., C.J. Newbold, K.P. Scott, *et al.* 1999. A family 11 xylanase from the ruminal protozoan *Polyplastron multivesiculatum*. FEMS Microbiol. Letts. **181:** 145–152.

35. Devillard, E., C. Bera-Maillet, H.J. Flint, *et al.* 2003. Characterisation of XynB, a modular xylanase from the ruminal protozoan *Polyplastron multivesiculatum*, bearing a family 22 carbohydrate binding module that binds to cellulose. Biochem. J. **373:** 495–503.

36. Ricard, G., N.R. McEwan, B.A. Dulith, *et al.* 2006. Horizontal gene transfer from bacteria to rumen ciliates indicates adaptation to their anaerobic, carbohydrate-rich environment. BMC Genomics **7:** Art. No. 22.

37. Belenguer, A., S.H. Duncan, G. Calder *et al.* 2006. Two routes of metabolic cross-feeding between bifidobacteria and butyrate-producing anaerobes from the human gut. Appl. Environ. Microbiol. **72:** 3593–3599.

38. Flint, H.J., S.H. Duncan, K.P. Scott, *et al.* 2007. Interactions and competition within the microbial community of the human large intestine: links between diet and health. Environ. Microbiol. **9:** 1101–1111.

39. Osborne, J.M. & B.A. Dehority. 1989. Synergism in degradation and utilization of intact forage cellulose, hemicellulose and pectin by three pure cultures of rumen bacteria. Appl. Environ, Microbiol. **55:** 2247–2250.

40. Sonnenburg, E.D., J. Xu, D.D. Leip, *et al.* 2005. Glycan foraging in vivo by an intestine-adapted bacterial symbiont. Science **307:** 1955–1959.

41. Hespell, R.B. & M.A. Cotta. 1995. Degradation and utilization by *Butyrivibrio fibrisolvens* H17c of xylans with different chemical and physical properties. Appl. Environ. Microbiol. **61:** 3042–3050.

42. Aminov, R.I., A.W. Walker, S.H. Duncan, *et al.* 2006. Molecular diversity, cultivation, and improved detection by fluorescent *in situ* hybridization of a dominant group of human gut bacteria related to *Roseburia* spp. or *Eubacterium rectale*. Appl. Environ. Microbiol. **72:** 6371–6376.

43. Avgustin, G., R.J. Wallace & H.J. Flint. 1997. Phenotypic diversity among rumen isolates of *Prevotella ruminicola*: proposal for redefinition of *Prevotella ruminicola* and the creation of *Prevotella brevis* sp. nov., *Prevotella bryantii* sp. nov. and *Prevotella albensis* sp. nov. Int. J. Syst. Bacteriol. **47:** 284–288.

44. Ramsak, A., M. Peterka, K. Tajima, *et al.* 2000. Unravelling the genetic diversity of ruminal bacteria belonging to the CFB phylum. FEMS Microbiol. Ecol. **33:** 69–79.

45. Fields, M.W., J.B. Russell & D.B. Wilson. 1998. The role of ruminal carboxymethyl cellulases in the degradation of beta-glucans from cereal grain. FEMS Microbiol. Ecol. **27:** 261–268.

46. Gasparic, A., R. Marinsek-Logar, J. Martin, *et al.* 1995. Isolation of genes encoding xylanase, β-D-xylosidase and α-L-arabinofuranosidase activities from the rumen bacterium *Prevotella ruminicola* B14. FEMS Microbiol. Lett. **125:** 135–142.

47. Gasparic, A., A. Daniel, J. Martin, *et al.* 1995. A xylan hydrolase gene cluster from *Prevotella ruminicola*: sequence relationships, oxygen sensitivity and synergistic interactions of a novel exoxylanase. Appl. Environ. Microbiol. **61:** 2958–2964.

48. Miyazaki, K., J.C. Martin, R. Marinsek-Logar, *et al.* 1997. Degradation and utilization of xylans by the rumen anaerobe *Prevotella bryantii* (formerly *P. ruminicola* subsp. *brevis* B14). Anaerobe **3:** 373–381.

49. Anderson, K.L. & A.A. Salyers. 1989. Biochemical evidence that starch breakdown by *Bacteroides thetaiotaomicron* involves outer-membrane starch-binding sites and periplasmic starch-degrading enzymes. J. Bacteriol. **171:** 3192–3198.

50. Reeves, A.R., G.R. Wang & A.A. Salyers. 1997. Characterization of four outer membrane proteins that play a role in utilization of starch by *Bacteroides thetaiotaomicron*. J. Bacteriol. **179:** 643–649.

51. Weaver, J., T.R. Whitehead, M.A. Cotta, *et al.* 1992. Genetic analysis of a locus on the *Bacteroides ovatus* chromosome which contains xylan utilization genes. Appl. Environ. Microbiol. **58:** 2764–2770.

52. Miyazaki, K., H. Miyamoto, D.K. Mercer, *et al.* 2003. Involvement of the two component regulatory protein XynR in positive control of xylanase gene expression in the ruminal anaerobe *Prevotella bryantii* B14. J. Bacteriol. **185:** 2219–2226.

53. Miyazaki, K., T. Hirase, Y. Kojima, *et al.* 2005. Medium to large sized xylo-oligosaccharides are responsible for xylanase induction in *Prevotella bryantii* B14. Microbiology **151:** 4121–4125.

54. Flint, H.J., T.R. Whitehead, J.C. Martin, *et al.* 1997. Interrupted domain structures in xylanases from two distantly related strains of *Prevotella ruminicola*. Biochim. Biophys. Acta **1337:** 161–165.

55. GARDNER, R.G., J.E. WELLS, M.W. FIELDS, *et al.* 1997. A *Prevotella ruminicola* B14 operon encoding extracellular polysaccharide hydrolases. Curr. Microbiol. **35:** 274–277.
56. RINCON, M.T., D. HENDERSON, J.C. MARTIN, *et al.* 2006. Polysaccharidases isolated from the rumen metagenome. Reprod. Nutr. Dev. **46**(Suppl 1): S57.
57. XU, J., M.K. BJURSELL, J. HIMROD, *et al.* 2003. A genomic view of the human-*Bacteroides thetaiotaomicron* symbiosis. Science **299**(5615): 2074–2076.
58. FERRER, M., O.V. GOLYSHINA, T.N. CHERMIKOVA, *et al.* 2005. Novel hydrolase diversity retrieved from a metagenome library of bovine rumen microflora. Environ. Microbiol. **7:** 1996–2010.
59. GILL, S.R., M. POP, R.T. DEBOY, *et al.* 2006. Metagenomic analysis of the human distal gut microbiome. Science **312:** 1355–1359.
60. BRYANT, M.P. 1972. Commentary on the Hungate technique for cultivation of anaerobic bacteria. Am. J. Clin. Nutr. **25:** 1324–1328.

Three Microbial Strategies for Plant Cell Wall Degradation

DAVID B. WILSON

Department of Molecular Biology and Genetics, Cornell University, Ithaca, New York, USA

Cellulolytic bacteria and fungi have been shown to use two different approaches to degrade cellulose. Most aerobic microbes secrete sets of individual cellulases, many of which contain a carbohydrate binding molecule (CBM), which act synergistically on native cellulose. Most anaerobic microorganisms produce large multienzyme complexes called cellulosomes, which are usually attached to the outer surface of the microorganism. Most of the cellulosomal enzymes lack a CBM, but the cohesin subunit, to which they are bound, does contain a CBM. The cellulases present in each class show considerable overlap in their catalytic domains, and processive cellulases (exocellulases and processive endocellulases) are the most abundant components of both the sets of free enzymes and of the cellulosomal cellulases. Analysis of the genomic sequences of two cellulolytic bacteria, *Cytophaga hutchinsonii*, an aerobe, and *Fibrobacter succinogenes*, an anaerobe, suggest that these organisms must use a third mechanism. This is because neither of these organisms, encodes processive cellulases and most of their many endocellulase genes do not encode CBMs. Furthermore, neither organism appears to encode the dockerin and cohesin domains that are key components of cellulosomes.

Key words: **cellulase; hemicellulose; synergism; cellulosome; bacteria; fungi**

Cellulose is the most abundant form of fixed carbon, with 10^{11} tons produced in cell walls by plants each year.[1] Most of the terrestrial plant cell wall material, especially cellulose, is degraded by living organisms, although some plant cell wall material is recycled by fire and by photodegradation.[2] Despite its abundance, only a small fraction of organisms can degrade cellulose, since it is very recalcitrant due to its insolubility and the presence of crystalline regions in which the adjacent cellulose molecules have strong interactions, such as hydrogen bonds and hydrophobic stacking. In fact, it is estimated that the half-life of crystalline cellulose at neutral pH in the absence of organisms is about a hundred million years and it requires concentrated acid at high temperature to chemically hydrolyze crystalline cellulose. Cellulolytic microorganisms, primarily fungi and bacteria, are responsible for much of the cellulose degradation in soils, although some insects and mollusks produce their own cellulases and degrade cellulose.[3,4] Termites are very important in cellulose degradation, especially in tropical regions, and most cellulose-degrading insects contain symbiotic cellulolytic microorganisms.[5,6] Some highly cellulolytic termites utilize aerobic symbiotic cellulolytic fungi to breakdown plant material in their nests and then eat the fungi and residual plant material.[7] Ruminants, such as cows, sheep, and deer, also are important cellulose-degrading organisms, but all of the cellulose that they utilize is degraded by symbiotic rumen microorganisms, primarily bacteria. The rumen is an extremely anaerobic environment and this area has been reviewed recently.[8,9]

Cellulose degradation is catalyzed by enzymes called cellulases, which hydrolyze the β-1–4 linkages that are present in cellulose. Cellulases are named by the name of the organism that produces them, followed by the family number associated with their catalytic domain (CD), followed by a capital letter that is assigned based on the order in which family members were discovered in that organism, with A being used for the first, that is, *Tricoderma reesei* CBHI is *T. reesei* Cel7A.[10]

Most cellulytic bacteria and fungi secrete their cellulases outside their cell wall, as these organisms are unable to transport insoluble materials, like cellulose, across their cell membrane. The soluble sugars produced by cellulase digestion are transported inside the cell and metabolized. The specific activities of individual cellulases are much lower than those of most enzymes. However, in terms of catalytic enhancement, cellulases are very active enzymes, due to the very long half-life of crystalline cellulose. When cellulases are assayed on low molecular-weight soluble substrates, they

Address for correspondence: David B. Wilson Department of Molecular Biology and Genetics, Cornell University, Ithaca, NY 14853. dbw3@cornell.edu

Ann. N.Y. Acad. Sci. 1125: 289–297 (2008).
doi: 10.1196/annals.1419.026

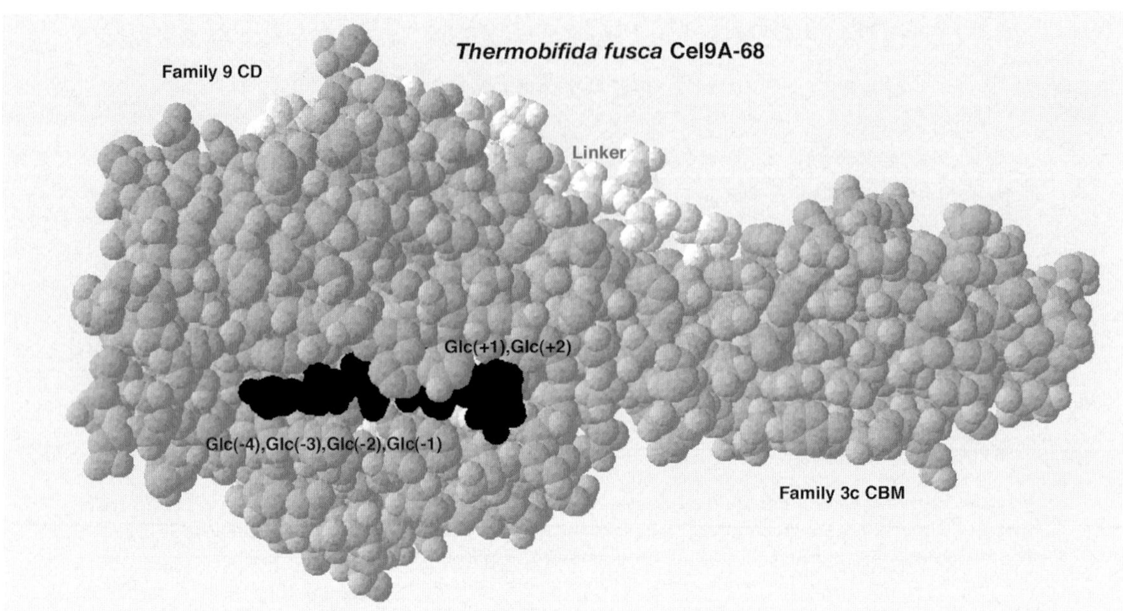

FIGURE 1. Structure of the processive endoglucanase, *T. fusca* Cel9A-68, with two molecules of cellobiose bound in the active site.

follow Michaelis–Menten kinetics, and some have respectable specific activities, showing that they can function like normal enzymes. However, when cellulases are assayed with insoluble cellulose, the enzyme properties are very different, because of the heterogeneity of insoluble cellulose. Most cellulase assays are nonlinear both with time and with the amount of enzyme. In addition, although activity does increase with an increase in substrate concentration, activity on insoluble cellulose does not show Michaelis–Menten kinetics.[11] Studies of different enzymes have produced several explanations for this behavior, but for *Thermobifida fusca* endocellulase, Cel6A, its nonlinearity was caused by substrate heterogeneity.[12]

All cellulases use one of two mechanisms to degrade cellulose: hydrolysis with retention of the stereochemistry of the anomeric hydroxyl group, or hydrolysis with inversion of the anomeric hydroxyl group.[13] One important difference between these mechanisms is that most retaining enzymes can catalyze both transglycosylation and hydrolysis, since a covalent intermediate is produced on the enzyme, while no known inverting enzyme catalyzes transglycosylation.[14]

There are three functionally different types of cellulases: endocellulases, also called endoglucanases, exocellulases, also called cellobiohydrolases, and processive endocellulases, which were discovered later.[15] To completely hydrolyze cellulose to glucose, a fourth enzyme, β glucosidase, is required, which hydrolyzes the soluble oligosaccharides produced by the cellulases to glucose. Many aerobic fungi secrete a β glucosidase as part of their crude cellulase, while most cellulolytic aerobic bacteria do not, and their β glucosidases are usually cytoplasmic. Some organisms, mainly anaerobic bacteria, contain cellobiose phosphorylase, also called dextrin phosphorylase, which converts cellobiose and soluble dextrins to glucose and glucose-1-phosphate, conserving the energy in the cellobiose linkage.[16]

All endocellulase CDs, whose structures have been determined, have an open active site, as would be expected, since they are able to bind to the interior of long cellulose molecules.[17] In contrast, all exocellulases have their active sites in a tunnel, consistent with their processive activity, where they sequentially cleave cellobiose residues from a cellulose molecule (FIG. 1).[18] In the case of glycosyl hydrolase family GH-48 enzymes, only part of the active site is in the tunnel, but these enzymes are just as processive as family GH-7 enzymes, where the entire active site is in the tunnel.[19] There are two classes of exocellulases[20]: one class attacks the nonreducing end of a cellulose molecule and all known members of this class are in family GH-6. Members of the other class attack the reducing end of a cellulose chain and all aerobic fungal members of this class are in family GH-7, while the bacterial members are in family GH-48. It is interesting that the anaerobic fungal members of this class are in family GH-48, rather than in family GH-7.[21] It has been claimed that the *T.*

reesei exocellulase, Cel6A, can act as an endocellulase and that is the reason it can synergize with *T. reesei* exocellulase, Cel7A[22]; however, it has been shown that all of the hydrolysis in a synergistic mixture of these two enzymes results from exocellulolytic activity.[23]

Most aerobic cellulolytic microorganisms degrade cellulose by secreting a set of individual cellulases, each of which contains a carbohydrate binding module (CBM) joined by a flexible linker peptide to the CD, and additional domains are often present. In some cellulases, the CBM is N-terminal to the CD, while in others it is C-terminal and the location probably does not affect its function.

Most anaerobic microorganisms digest cellulose differently from the "free cellulase" strategy described earlier that is used by many aerobic microorganisms, as they produce large (>1 million MW) multienzyme complexes, called cellulosomes, which are usually bound to the outer surface of the microorganism.[24] Only a few of the enzymes in cellulosomes contain a CBM, but the scaffoldin protein to which they are attached does contain a CBM, which binds the complex to cellulose. Some anaerobic bacteria also produce some free cellulases, and it is not known if and how they function in cellulose degradation.[25] A key unanswered question about cellulosomes is: How are they able to degrade cellulose, as or more effectively than a mixture of free cellulases? Cellulosomes have much less access to the cellulose surface then free cellulases, due to their large size, which should prevent them from entering the extensive pores of cellulose particles that contain most of the surface and are accessible to free cellulases.

There are two cellulolytic bacteria that appear to use a third strategy to degrade cellulose. *Cytophaga hutchinsonii* is an aerobic cellulolytic bacterium, and the DOE Joint Genome Institute determined the DNA sequence of its genome (http://genome.jgi-psf.org/finished_microbes/cythu/cythu.home.html). While *C. hutchinsonii* codes for a number of cellulase genes, most of them do not encode a CBM, none of them encode a dockerin domain, and all of them appear to code for endoglucanases.[26] Thus, this organism does not code for any processive cellulases, which are major components in most other cellulolytic microorganisms. These results clearly distinguish *C. hutchinsonii* from most other well-studied aerobic and anaerobic cellulolytic microorganisms.

The genome sequence of the anaerobic rumen bacterium, *Fibrobacter succinogenes,* which was determined by The Institute of Genomic Research (TIGR) funded by a USDA grant to the North American Consortium for Genomics of Fibrolytic Ruminal Bacteria, also does not code for any known processive cellulases and only one of the many endocellulases that have been cloned and sequenced appears to bind to cellulose.[27–29] *F. succinogenes* does not appear to encode dockerin domains and a scaffoldin gene has not been identified. Thus, *F. succinogenes* also appears to use a novel strategy for degrading cellulose. *F. succinogenes* grows very rapidly on cellulose, so that its cellulose-degrading mechanism is very efficient.[30] One possible mechanism for cellulose degradation by these organisms is the one proposed for starch degradation by *Bacteroides thetaiotaomicron*.[31] In this mechanism, starch is bound to a complex present in the outer membrane and individual molecules are transported into the periplasmic space, where they are degraded by starch-degrading enzymes. This mechanism would not require processive cellulases, as individual cellulose molecules would be readily degraded by endoglucanases. If this is the process by which cellulose is degraded, it will be very interesting to determine the mechanism by which the outer membrane proteins are able to bind and transport individual cellulose molecules. It is possible that this information would allow the design of new cellulases or cellulose-modifying proteins, which would be able to increase the rate of cellulose degradation by free cellulases.

In both aerobic and anaerobic organisms, certain cellulases can act synergistically on crystalline cellulose, with the specific activity of some cellulase mixtures containing four to six enzymes being up to 15 times that of any single cellulase in the mixture.[23] Synergism is usually only seen in the digestion of substrates that contain crystalline cellulose, probably because there are only a few regions in this substrate that are accessible to each cellulase. It seems likely that synergism occurs only when two cellulases attack different regions of the cellulose microfibril and each cellulase creates new sites of attack for the other enzymes in the mixture. There is no evidence that synergism requires interactions between the synergizing cellulases, as cellulases from unrelated organisms show similar synergism to those from the same organism. All good endocellulases are able to show synergism with any exocellulase, but most endocellulases do not show synergism with each other. Exocellulases show synergism with other exocellulases, but only if they attack different ends of the cellulose chains. Processive endocellulases give synergy with endocellulases and with exocellulases.[32] It is interesting that cellulose pretreated with an endocellulase is a better substrate for exocellulases than is untreated cellulose, but the reverse is not true.[33] In synergistic mixtures of an exocellulase and an endocellulase, endocellulase activity is increased as much in the mixture as is exocellulase activity.[23] It seems likely

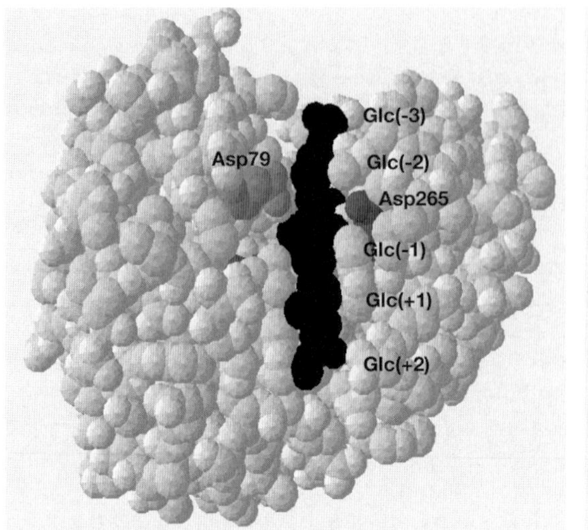

Thermobifida fusca Cel6A D117A
(unpublished data, Anna Larssen)

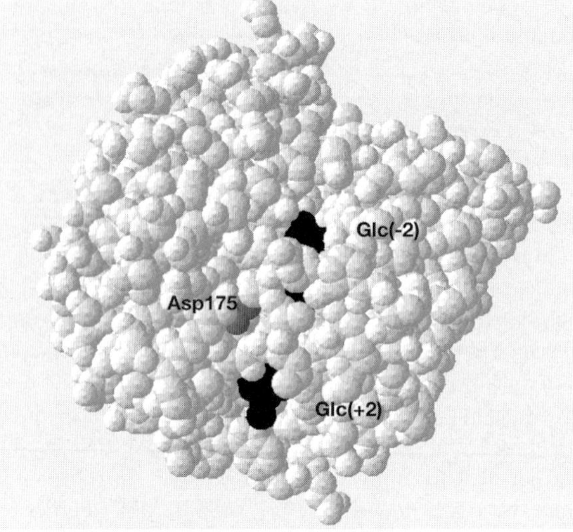

Trichoderma reesei Cel6A
(pdb code 1QK2)

FIGURE 2. Structure of the catalytic domain of the inactive mutant *T. fusca* exocellulase Cel6A D117A with cellopentose bound in the active site along with the structure of the CD of *T. reesei* exocellulase Cel6A with two cellobiose residues bound in the active site.

that when an exocellulase degrades a cellulose chain, it disrupts some regions in adjacent cellulose molecules, making them more available to an endocellulase. Over time, however, the molecules in the disrupted regions may reform their interactions with adjacent chains, so that when an endocellulase is added after exocellulase treatment, the endocellulase is not able to utilize the transiently disrupted chains.

Even though cellulose is a homopolymer of glucose, with only a single type of linkage (β-1–4), and with the disaccharide cellobiose being the repeating unit, cellulases are very diverse in their structures, mechanisms, and sequences. There are 14 cellulase families listed on the CAZy Web site: (http://afmb.cnrsmrs.fr/CAZY/fam/acc_GH.html). Several of these families (10, 26, 51, 74) mainly contain other types of glycosyl hydrolases, with only a few members having cellulase activity, but even if these families are excluded, there are still 10 cellulase families. The enzymes in any given family show significant sequence homology with some or all of the other family members. All members of a glycosyl hydrolase family have the same basic protein fold and utilize the same catalytic mechanism, but their substrate specificities can be quite different. Because there is a wide range of amino acid sequences that can give the same protein fold, several families share the same fold, even though the sets of sequences in each family show little similarity between families.

There are nearly twice as many cellulase families as there are in the next largest group of hydrolases, the seven xylanase families. Furthermore, there are seven different protein folds found for known cellulase structures, and a structure is not yet known for one family. There are two possible reasons why cellulases are so diverse. One reason for diversity is that the actual substrate of most cellulases is not pure cellulose but rather plant cell walls, which are extremely diverse both among different plants and in different types of cell walls in a single plant species. This complexity is due to the many other compounds present in cell walls, some of which are bound to the cellulose fibrils.[34] The other reason is that cellulose itself is quite complex with both crystalline and amorphous regions. It appears that cellulases are under positive selection, as when the DNA and protein sequences of two related cellulase genes were compared there were nearly as many DNA changes that caused an amino acid change (nonsynonymous) as there were DNA changes that did not change the amino acid (synonymous).[35]

All well-documented processive endocellulases are in family G-9, which is the largest cellulase family and includes most plant cellulases, animal cellulases, many bacterial cellulases, and surprisingly, very few fungal cellulases. Processive endoglucanases have an open active site cleft like all endocellulases, but in addition they contain a family 3 CBM, which is rigidly attached to the C terminus of the CD (FIG. 2).[36] The two domains are oriented so that a cellulose chain can bind simultaneously to both domains. The family 3c CBMs, that are present in processive endocellulases, differ from family 3a and 3b CBMs, in that they lack the conserved aromatic residues, which cause the high affinity for cellulose. Although the 3c CBMs bind very weakly to cellulose, it has been shown that they are necessary for the processive activity of these enzymes.[15]

There do not appear to be major differences between the CD families of the cellulases present in cellulosomes and the families of cellulases secreted by aerobic microorganisms, as most cellulase families contain cellulases from both types of microorganisms. However, all known GH-7 cellulases are from aerobic fungi or termites, and there are no known GH-6 cellulases produced by anaerobic bacteria. Furthermore, all GH-12 cellulases appear to be produced by aerobic microorganisms, but this is currently a small family that only contains endocellulases. At this time, all known GH-48 cellulases are exocellulases and this is the only cellulase family that does not contain endocellulases.

CBMs play an important role in the ability of cellulases to degrade crystalline cellulose,[37] but they have little or no effect on the activity of most cellulases on a soluble cellulose derivative, carboxymethylcellulose, amorphous cellulose, or oligosaccharides.[38] One role of a CBM is to anchor the cellulase to the insoluble cellulose, so that the CD remains close to the substrate. The flexible linker that separates the CBM from the CD allows the CD to access regions of the cellulose, adjacent to the bound CBM. Some workers have proposed that a CBM can also disrupt the structure of cellulose, making it more accessible to the CD,[39,40] but this is still controversial.[41] This activity would be equivalent to the Cx activity proposed by Reese in his early discussion of the nature of cellulases.[42] CBMs also have been reported to target cellulases to specific regions of cellulose, presumably to the regions where they will be most active.[43] It has been shown that family 2 CBMs can diffuse on the surface of cellulose without dissociation, giving them the ability to readily access new regions of a cellulose particle after the CD has hydrolyzed cellulose near the original site of binding.[44] Other CBMs that bind to crystalline cellulose may also have this ability, since they appear to bind in a way similar to that of the family 2 CBMs.

There are many CBM families (45 listed on the CAZy Web site) (http://afmb.cnrs-mrs.fr/CAZY/-fam/acc_CBM.html). Not all CBMs bind cellulose, as many families contain chitin binding domains, xylan binding domains, or mannose binding domains. Some CBMs can bind to several polymers, while others are specific for only one. Labeled CBMs are being used to stain plant materials, and different members of a given family can give very different staining, showing that there is an even greater diversity of binding specificity than is seen with pure substrates.[45] Almost all known enzymes that have high activity on insoluble substrates contain a substrate-binding domain, in addition to a CD, so that the presence of such a substrate-binding domain is a general property of this type of enzyme.

All fungal cellulase CBMs are in family I and they are small, containing about 30 amino acids. Most aerobic bacterial cellulase CBMs are in family 2 and they are larger, containing about 120 amino acids. The CBMs on cellulosomal scaffoldins are in family 3. Most of the CBMs in these three families bind to crystalline cellulose and have a relatively flat binding surface that usually contains three aromatic residues spaced so they can bind to three adjacent glucose residues in a cellulose molecule. They also contain a number of residues, which can hydrogen bond to the cellulose chain, but site-directed mutagenesis has shown that the aromatic residues are essential for high-affinity binding, while the other residues play a secondary role.[46] The CBMs in families 4 and 6 bind to single cellulose molecules and their binding sites are in a groove.[47] A number of cellulases contain multiple CBMs and in some cases they are from the same family, and in other cases they are from different families. It has been shown that the affinity of a protein containing two domains can be significantly higher than one with only one domain.[48] Atomic force microscopy of the binding of a family I CBM to cellulose found that the bound CBM was present in aggregates, not as single domains.[49]

The most studied aerobic cellulolytic microorganism is the fungus, *Hypocrea jecorina,* originally called *Trichoderma reesei.* It was isolated and studied by Drs. Reese and Mandels at the Army Quartermaster Lab in Natick, Massachusetts, during the Second World War, because it was degrading the cotton fabrics used by the army for tents, gun straps, and so forth, on islands in the Pacific Ocean.[50] The original goal of this work was to find cellulase inhibitors, which was not achieved, as only the toxic ions, Hg and Ag, are good inhibitors. However, this group also carried out many studies on the organism and its crude cellulase, which then led to the development of high-producing mutant strains by Dr. Eveleigh; these strains were used to develop the strains used for industrial cellulase production by several companies.[51] Recently the *T. reesei* genome was sequenced by the DOE Joint Genome Institute: http://genome.jgi-psf.org/Trire2/Trire2.home.html.

The most abundant cellulase produced by *T. reesei* is the reducing end-specific exocellulase, Cel7A (cellobiohydrolase I), which makes up about 70% of the cellulase protein secreted by *T. reesei*.[52] The next most abundant cellulase is Cel6A (CBH II), which makes up a further 10% of *T. reesei* secreted cellulase. *T. reesei* crude cellulase contains seven endoglucanases, of which Cel7B (EGLI) is the most abundant. In addition, Cel5A (EGLII), Cel12A (EGLIII), Cel61A (EGLIV) and Cel45A (EGLV), Cel5B, and Cel61B are also present in *T. reesei*–secreted cellulase. Most of the *T. reesei* cellulases contain a family I CBM, except Cel5B, Cel12A, and Cel61B. It is not clear why *T. reesei* produces so many endocellulases, but Cel12A was shown to have expansin activity as well as cellulase activity.[52] Expansin is present in plants and it appears to disrupt the hydrogen bonds that bind different carbohydrate chains together in plant cell walls, so that it may make the chains more accessible to hydrolytic enzymes.[53] The least studied endocellulase is Cel61A, which has extremely low cellulase activity.[54] It is quite surprising that when a set of thermophilic fungal cellulases were screened for the ability to stimulate the activity of *T. reesei* crude cellulase, a number of them were able to increase it about threefold, and the component that was most active in giving this stimulation was a family 61 enzyme.[55] Little is known about the role of Cel45A in cellulose degradation.[56] Another protein secreted by *T. reesei* is swollenin, which is a low-molecular-weight protein that has no catalytic activity, but appears to disrupt the structure of cellulose microfibers, possibly by breaking hydrogen bonds.[57]

Most of the *T. reesei* cellulases are glycosylated, and glycosylation appears to protect the cellulases from proteolysis.[58] The linker peptide is particularly susceptible to proteolysis and *T. reesei* secretes proteases,[59] so that protection from proteolysis may be an important role for the O-linked glycosylation found on the linker peptide.[60] The role of the N-linked glycosylation on the CD is unclear at this time, but glycosylation has been shown to enhance stability in some fungal hydrolases. There is a great deal of heterogeneity in the glycosylation of any given cellulose, and this can cause heterogeneity of each enzyme during gel electrophoresis and column chromatography.[61]

Another organism used for industrial cellulase production is *Humicola insolens*, which seems to produce the same set of cellulases as *T. reesei*, except that it produces a family GH-6 endoglucanase.[62] These two organisms are not closely related, however, even though they are both brown rot fungi, which do not degrade lignin. *Phanerochaete chrysosporium* is a white-rot fungus that degrades lignin, while not degrading much cellulose, despite containing a set of cellulase genes.[63] A surprising finding is that it contains seven CBH I genes that are differentially regulated, in contrast to the single CBH I gene in *T. reesei*. The cellulases produced by *Aspergillus aculeatus* have been extensively studied, and it produces nine cellulases, of which three have been sequenced: Cel7A, Cel12A, and Cel5A. From these limited data, it seems that its cellulases may be similar to those of *T. reesei*.[64] Another fungus whose cellulases have been studied is *Talaromyces emersonii*, which produces two exocellulases: Cel7A and Cel6A, which have been extensively studied, and several endocellulases, of which only Cel5A has been studied.[65] *Chrysosporium lucknowense* cellulases also have been studied and seem to resemble those of *T. reesei*.[66] There are many other cellulolytic fungi whose cellulases have had some research, including *Agaricus bisporus*, several *Aspergillus* species, *Aureobasidium pullulans*, *Cochlibolus carbonum*, several *Fusarium* species, several *Penicillium* species, *Pleurotus ostreatus*, and *Thermoascus aurantiacus*, but the total cellulase system has not been determined for any of these organisms at this time. All of these aerobic fungi produce a set of individual cellulases; however, an aerobic *Chaetomium* strain has been reported to produce a large cellulase complex.[67] At this time it is not known what cellulases are present in the complex and if the complex is assembled using the cohesin–dockerin binding seen in cellulosomes. This system clearly should be studied further.

Two aerobic cellulolytic bacteria whose cellulases have been well characterized are *Cellulomonas fimi* and *T. fusca*. The sets of six cellulases produced by these two organisms are similar in their activity, CD families, and CBMs, but differ significantly in their sequences and domain order. These data suggest that these organisms did not obtain their cellulase genes from a common ancestor, but rather each set is the result of convergent evolution. In five of the six corresponding cellulase pairs from each species, the family 2 CBM is at the opposite end of the enzyme.[11] These organisms are both actinomycetes and they are both found in soils, but *C. fimi* is mesophillic with an optimum growth temperature near 30°C, while *T. fusca* is moderately thermophilic with an optimum growth temperature of 50°C. An interdisciplinary group at the University of British Columbia has studied *C. fimi* cellulases, and this group has made many important contributions to cellulase research.[68]

T. fusca is often found in compost piles, rotting hay, and manure piles. The genome sequence of *T. fusca* was determined in 2000 by the DOE Joint Genome Institute and the finished 3.7 mb sequence is available at: http://genome.jgi-psf.org/finished_microbes/thefu/thefu.home.html.[69]

These two organisms are not closely related, as *C. fimi* is in the suborder Micrococcineae while *T. fusca* is in the suborder Streptosporangineae. It is interesting that *Streptomyces lividans,* which is a relative of *T. fusca,* contains five cellulase genes that are similar to those present in *T. fusca* and, in every one, the family 2 CBM is in the same location as it is in *T. fusca*, suggesting that these sets of genes did come from a common ancestor.

Saccharophagus degradans is an aerobic marine plant cell wall–degrading organism, whose genome was recently sequenced. Preliminary analysis of its 13 cellulase genes shows that it contains 10 family GH-5 enzymes, which are most similar to endoglucanases, and most of them also contain cellulose binding domains. It also contains two family GH-9 endoglucanase genes, one of which encodes both a family 10 and a family 2 CBM.[70] It also encodes a family 6 exocellulase, but not a known reducing end-specific exocellulase. This is a rather unusual set of cellulases, as it lacks a reducing end attacking exocellulase and includes so many family 5 endocellulases.

The best-studied anaerobic cellulolytic microorganism is *Clostridium thermocellum*, where cellulosomes were first identified.[71] The genome sequence of *C. thermocellum* has been determined by the DOE Joint Genome Institute: http://genome.jgi-psf.org/finished_microbes/cloth/cloth.home.html. There are more than 70 genes in *C. thermocellum,* which encode open reading frames that include a dockerin sequence, showing that they are cellulosomal proteins. Most of these proteins appear to encode cellulases, hemicellulases, or pectin-degrading enzymes, but a few have other functions. Extensive research is continuing on this organism and past work is reviewed in detail in a 2004 review of anaerobic cellulolytic microorganisms.[72]

Conflict of Interest

The author declares no conflicts of interest.

References

1. Malhi, Y. 2002. Carbon in the atmosphere and terrestrial biosphere in the 21st century. Philos. Trans., Ser. A, Math. Phys. Eng. Sci. **360:** 2925–2945.
2. Falkowski, P., R.J. Scholes, E. Boyle, *et al.* 2000. The global carbon cycle: a test of our knowledge of earth as a system. Science **290:** 291–296.
3. Breznak, J.A. 1982. Intestinal microbiota of termites and other xylophagous insects. Annu. Rev. Microbiol. **36:** 323–343.
4. Ohkuma, M. 2003. Termite symbiotic systems: efficient biorecycling of lignocellulose. Appl. Microbiol. Biotech. **61:** 1–9.
5. Watanabe, H. & G. Tokuda. 2001. Animal cellulases. Cell. Mol. Life Sci. **58:** 1167–1178.
6. Kunieda, T., T. Fujiyuki, R. Kucharski, *et al.* 2006. Carbohydrate metabolism genes and pathways in insects: insights from the honey bee genome. Insect Mol. Biol. **15:** 563–576.
7. Shinzato, N., M. Muramatsu, Y. Watanabe & T. Matsui. 2005. Termite-regulated fungal monoculture in fungus combs of a macrotermitine termite *Odontotermes formosanus*. Zool. Sci. **22:** 917–922.
8. Lynd, L.R., P.J. Weimer, W.H. van Zyl & I.S. Pretorius. 2002. Microbial cellulose utilization: fundamentals and biotechnology. Microbiol. Molec. Biol. Rev. **66:** 506–577.
9. Desvaux, M. 2006. Unraveling carbon metabolism in anaerobic cellulolytic bacteria. Biotech. Prog. **22:** 1229–1238.
10. Henrissat, B., T.T. Teeri & R.A.J. Warren. 1998. A scheme for designating enzymes that hydrolyze the polysaccharides in the cell walls of plants. FEBS Lett. **425:** 352–354.
11. Wilson, D.B. & D. Irwin. 1999. Genetics and properties of cellulases. *In* Advances in Biochemical Engineering/Biotechnology: Recent Progress in Bioconversion of Lignocellulosics. G.T. Tsao & T. Scheper, Eds., Vol. 65: 1–21. Berlin: Springer-Verlag.
12. Zhang, S., D. Wolfgang & D.B. Wilson. 1999. Substrate heterogeneity causes the non-linear kinetics of insoluble cellulose hydrolysis. Biotechnol. Bioeng. **66:** 35–41.
13. McCarter, J.D. & S.G. Withers. 1994. Mechanisms of enzymatic glycoside hydrolysis. Curr. Opin. Struc. Biol. **4:** 885–892.
14. Blanchard, J.E. & S.G. Withers. 2001. Rapid screening of the aglycone specificity of glycosidases: applications to enzymatic synthesis of oligosaccharides. Chem. Biol. **8:** 627–633.
15. Irwin, D., D.-H. Shin, S. Zhang, *et al.* 1998. Roles of the catalytic domain and two cellulose binding domains of *Thermomonospora fusca* E4 in cellulose hydrolysis. J. Bacteriol. **180:** 1709–1714.
16. Kim, Y.K., M. Kitaoka, M. Krishnareddy, *et al.* 2002. Kinetic studies of a recombinant cellobiose phosphorylase (CBP) of the *Clostridium thermocellum* YM4 strain expressed in *Escherichia coli*. J. Biochem. (Tokyo) **132:** 197–203.
17. Juy, M., A.G. Amit, P.M. Alzari, *et al.* 1992. Three-dimensional structure of a thermostable bacterial cellulase. Nature **357:** 89–91.
18. Rouvinen, J., T. Bergfors, T. Teeri, *et al.* 1990. Three-dimensional structure of cellobiohydrolase II from *Trichoderma reesei*. Science **249:** 380–386.
19. Parsiegla, G., M. Juy, C. Reverbel-Leroy, *et al.* 1998. The crystal structure of the processive endocellulase CelF of *Clostridium cellulolyticum* in complex with a thiooligosaccharide inhibitor at 2.0Å resolution. EMBO J. **17:** 5551–5562.
20. Barr, B.K., Y.L. Hsieh, B. Ganem & D.B. Wilson. 1996. Identification of two functionally different classes of exocellulases. Biochemistry **35:** 586–592.

21. TEUNISSEN, M.J. & H.J. OP DEN CAMP. 1993. Anaerobic fungi and their cellulolytic and xylanolytic enzymes. Antonie Leeuwenhoek **63:** 63–76.

22. BOISSET, C., C. FRASCHINI, M. SCHULEIN, *et al.* 2000. Imaging the enzymatic digestion of bacterial cellulose ribbons reveals the endo character of the cellobiohydrolase Cel6A from *Humicola insolens* and its mode of synergy with cellobiohydrolase Cel7A. Appl. Environ. Microbiol. **66:** 1444–1452.

23. IRWIN, D.C., M. SPEZIO, L.P. WALKER & D.B. WILSON. 1993. Activity studies of eight purified cellulases: specificity, synergism, and binding domain effects. Biotechol. Bioeng. **42:** 1002–1013.

24. BAYER, E.A., J.P. BELAICH, Y. SHOHAM & R. LAMED. 2004. The cellulosomes: multienzyme machines for degradation of plant cell wall polysaccharides. Annu. Rev. Microbiol. **58:** 521–554.

25. DOI, R.H. & A. KOSUGI. 2004. Cellulosomes: plant-cell-wall-degrading enzyme complexes. Nat. Rev. Microbiol. **2:** 541–551.

26. XIE, G., D.C. BRUCE, J.F. CHALLACOMBE, *et al.* 2007. Genome sequence of the cellulolytic gliding bacterium *Cytophaga hutchinsonii*. Appl. Environ. Microbiol. **73:** 3536–346.

27. IYO, A.H. & C.W. FORSBERG. 1996. Endoglucanase G from *Fibrobacter succinogenes* S85 belongs to a class of enzymes characterized by a basic C-terminal domain. Can. J. Microbiol. **42:** 934–943.

28. MALBURG, S.R., L.M. MALBURG, JR., T. LIU, *et al.* 1997. Catalytic properties of the cellulose-binding endoglucanase F from *Fibrobacter succinogenes* S85. Appl. Environ. Microbiol. **63:** 2449–2453.

29. JUN, H.S., M. QI, J. GONG, *et al.* 2007. Outer membrane proteins of *Fibrobacter succinogenes* with potential roles in adhesion to cellulose and in cellulose digestion. J. Bacteriol. In press.

30. FIELDS, M.W., S. MALLIK & J.B. RUSSELL. 2000. *Fibrobacter succinogenes* S85 ferments ball-milled cellulose as fast as cellobiose until cellulose surface area is limiting. Appl. Microbiol. Biotechnol. **54:** 570–574.

31. CHO, K.H. & A.A. SALYERS. 2001. Biochemical analysis of interactions between outer membrane proteins that contribute to starch utilization by *Bacteroides thetaiotaomicron*. J. Bacteriol. **183:** 7224–7230.

32. WILSON, D.B. 2004. Studies of *Thermobifida fusca* plant cell wall degrading enzymes. Chem. Rec. **4:** 72–82.

33. NIDETZKY, B., W. STEINER, M. HAYN & M. CLAEYSSENS. 1994. Cellulose hydrolysis by the cellulases from *Trichoderma reesei*: a new model for synergistic interaction. Biochem. J. **298:** 705–710.

34. *In* The Plant Cell Wall. 2003. Annual Plant Reviews. J.K.C. ROSE, Ed., Vol. 8. Oxford: Blackwell Publishing.

35. LAO, G., G.S. GHANGAS, E.D. JUNG & D.B. WILSON. 1991. DNA sequence of three β-1,4-endoglucanase genes from *Thermomonospora fusca*. J. Bacteriol. **173:** 3397–3407.

36. SAKON, J., D. IRWIN, D.B. WILSON & P.A. KARPLUS. 1997. Structure and mechanism of endo/exocellulase E4 from *Thermomonospora fusca*. Nature Struc. Biol. **4:** 810–818.

37. SHOSEYOV, O., Z. SHANI & I. LEVY. 2006. Carbohydrate binding modules: biochemical properties and novel applications. Microbiol. Mol. Biol. Rev. **70:** 283–295.

38. GILKES, N.R., R.A. WARREN, R.C. MILLER, JR. & D.G. KILBURN. 1988. Precise excision of the cellulose binding domains from two *Cellulomonas fimi* cellulases by a homologous protease and the effect on catalysis. J. Biol. Chem. **263:** 10401–10407.

39. DIN, N., N.R. GILKES, B. TEKANT, *et al.* 1991. Non-hydrolytic disruption of cellulose fibres by the binding domain of a bacterial cellulase. Bio/Technology **9:** 1096–1099.

40. ESTEGHLALIAN, A.R., V. SRIVASTAVA, N.R. GILKES, *et al.* 2001. Do cellulose binding domains increase substrate accessibility? Appl. Biochem. Biotechnol. **91–93:** 575–592.

41. BOLAM, D.N., A. CIRUELA, S. MCQUEEN-MASON, *et al.* 1998. *Pseudomonas* cellulose binding domains mediate their effects by increasing enzyme substrate proximity. Biochem. J. **331:** 775–781.

42. REESE, E.T., R.G.H. SUI & H.S. LEVINSON. 1950. The biological degradation of soluble cellulose derivatives and its relationship to the mechanism of cellulose hydrolysis. J. Bacteriol. **59:** 485–497.

43. BORASTON, A.B., D.N. BOLAM, H.J. GILBERT & G.J. DAVIES. 2004. Carbohydrate binding modules: fine-tuning polysaccharide recognition. Biochem. J. **382:** 769–781.

44. JERVIS, E.J., C.A. HAYNES & D.G. KILBURN. 1997. Surface diffusion of cellulases and their isolated binding domains on cellulose. J. Biol. Chem. **272:** 24016–24023.

45. MCCARTNEY, L., A.W. BLAKE, J. FLINT, *et al.* 2006. Differential recognition of plant cell walls by microbial xylan-specific carbohydrate-binding modules. Proc. Natl. Acad. Sci. USA **103:** 4765–4770.

46. DIN, N., I.J. FORSYTHE, L.D. BURTNICK, *et al.* 1994. The cellulose-binding domain of endoglucanase A (CenA) from *Cellulomonas fimi*: evidence for the involvement of tryptophan residues in binding. Mol. Microbiol. **11:** 747–755.

47. KORMOS, J., P.E. JOHNSON, E. BRUN, *et al.* 2000. Binding site analysis of cellulose binding domain CBD(N1) from endoglucanse C of *Cellulomonas fimi* by site-directed mutagenesis. Biochemistry **39:** 8844–8852.

48. LINDER, M., I. SALOVUORI, L. RUOHONEN & T.T. TEERI. 1996. Characterization of a double cellulose-binding domain. Synergistic high affinity binding to crystalline cellulose. J. Biol. Chem. **271:** 21268–21272.

49. NIGMATULLIN, R., R. LOVITT, C. WRIGHT, *et al.* 2004. Atomic force microscopy study of cellulose surface interaction controlled by cellulose binding domains. Colloids Sur. B, Biointerfaces **35:** 125–135.

50. REESE, E.T. 1976. History of the Cellulose Program at the U.S. Army Natick Development Center. Biotechnology and Bioengineering Symposium, Vol. 6: 9–20. New York: Wiley Interscience.

51. GHOSH, A., B.K. GHOSH, H. TRIMINO-VAZQUEZ, *et al.* 1984. Cellulase secretion from a hypercellulolytic mutant of *Trichoderma reesei* RUT-C-30. Arch. Microbiol. **140:** 126–133.

52. MARKOV, A.V., A.V. GUSAKOV, E.G. KONDRATYEVA, *et al.* 2005. New effective method for analysis of the component

composition of enzyme complexes from *Trichoderma reesei*. Biochemistry (Moscow) **70:** 657–663.

53. YUAN, S., Y. WU & D.J. COSGROVE. 2001. A fungal endoglucanase with plant cell wall extension activity. Plant Physiol. **127:** 324–333.
54. KARLSSON, J., M. SALOHEIMO, M. SIIKA-AHO, *et al.* 2001. Homologous expression and characterization of Cel61A (EG IV) of *Trichoderma reesei*. Eur. J. Biochem. **268:** 6498–6507.
55. ROSGAARD, L., S. PEDERSEN, J.R. CHERRY, *et al.* 2006. Efficiency of new fungal cellulase systems in boosting enzymatic degradation of barley straw lignocellulose. Biotechnol. Prog. **22:** 493–498.
56. SALOHEIMO, M., M. PALOHEIMO, S. HAKOLA, *et al.* 2002. Swollenin, a *Trichoderma reesei* protein with sequence similarity to the plant expansins, exhibits disruption activity on cellulosic materials. Eur. J. Biochem. **269:** 4202–4211.
57. SALOHEIMO, A., B. HENRISSAT, A.M. HOFFREN, *et al.* 1994. A novel, small endoglucanase gene, egl5, from *Trichoderma reesei* isolated by expression in yeast. Mol. Microbiol. **13:** 219–228.
58. HU, J.P., P. LANTHIER, T.C. WHITE, *et al.* 2001. Characterization of cellobiohydrolase I (Cel7A) glycoforms from extracts of *Trichoderma reesei* using capillary isoelectric focusing and electrospray mass spectrometry. J. Chromatogr. B, Biomed. Sci. Appl. **752:** 349–368.
59. KREDICS, L., Z. ANTAL, A. SZEKERES, *et al.* 2005. Extracellular proteases of *Trichoderma* species. A review. Acta Microbiol. Immunol. Hung. **52:** 169–184.
60. HARRISON, M.J., A.S. NOUWENS, D.R. JARDINE, *et al.* 1998. Modified glycosylation of cellobiohydrolase I from a high cellulase-producing mutant strain of *Trichoderma reesei*. Eur. J. Biochem. **256:** 119–127.
61. HU, J.P., T.C. WHITE & P. THIBAULT. 2002. Identification of glycan structure and glycosylation sites in cellobiohydrolase II and endoglucanases I and II from *Trichoderma reesei*. Glycobiology **12:** 837–849.
62. SCHULEIN, M. 1997. Enzymatic properties of cellulases from *Humicola insolens*. J. Biotechnol. **57:** 71–81.
63. MUNOZ, I.G., W. UBHAYASEKERA, H. HENRIKSSON, *et al.* 2001. Family 7 cellobiohydrolases from *Phanerochaete chrysosporium*: crystal structure of the catalytic module of Cel7D (CBH58) at 1.32Å resolution and homology models of the isozymes. J. Mol. Biol. **314:** 1097–1111.
64. TAKADA, G., M. KAWASAKI, M. KITAWAKI, *et al.* 2002. Cloning and transcription analysis of the *Aspergillus aculeatus* No. F-50 endoglucanase 2 (cmc2) gene. J. Biosci. Bioeng. **94:** 482–485.
65. GRASSICK, A., P.G. MURRAY, R. THOMPSON, *et al.* 2004. Three-dimensional structure of a thermostable native cellobiohydrolase, CBH IB, and molecular characterization of the cel7 gene from the filamentous fungus, *Talaromyces emersonii*. Eur. J. Biochem. **271:** 4495–4506.
66. BUKHTOJAROV, F.E., B.B. USTINOV, T.N. SALANOVICH, *et al.* 2004. Cellulase complex of the fungus *Chrysosporium lucknowense:* isolation and characterization of endoglucanases and cellobiohydrolases. Biochemistry (Moscow) **69:** 542–551.
67. OHTSUKI, T., SUYANTO, S. YAZAKI, *et al.* 2005. Production of large multienzyme complex by aerobic thermophilic fungus Chaetomium sp. nov. MS-017 grown on palm oil mill fibre. Lett. Appl. Microbiol. **40:** 111–116.
68. TOMME, P., E. KWAN, N.R. GILKES, *et al.* 1996. Characterization of CenC, an enzyme from *Cellulomonas fimi* with both endo- and exoglucanase activities. J. Bacteriol. **178:** 4216–4223.
69. LYKIDIS, A., K. MAVROMATIS, N. IVANOVA, *et al.* 2007. Genome sequence and analysis of the soil cellulolytic actinomycete *Thermobifida fusca* YX. J. Bacteriol. **189:** 2477–2486.
70. TAYLOR, L.E. 2ND, B. HENRISSAT, P.M. COUTINHO, *et al.* Complete cellulase system in the marine bacterium *Saccharophagus degradans* strain 2–40T. J. Bacteriol. **188:** 3849–3861.
71. LAMED, R., E. SETTER & E.A. BAYER. 1983. Characterization of a cellulose-binding, cellulase-containing complex in Clostridium thermocellum. J. Bacteriol. **156:** 828–836.
72. DESVAUX, M. 2006. Unravelling carbon metabolism in anaerobic cellulolytic bacteria. Biotechnol. Prog. **22:** 1229–1238.

Bacterial Cellulose Hydrolysis in Anaerobic Environmental Subsystems—Clostridium thermocellum *and* Clostridium stercorarium, *Thermophilic Plant-fiber Degraders*

VLADIMIR V. ZVERLOV AND WOLFGANG H. SCHWARZ

Department of Microbiology, Technische Universität München, Freising, Germany

Cellulose degradation is a rare trait in bacteria. However, the truly cellulolytic bacteria are extremely efficient hydrolyzers of plant cell wall polysaccharides, especially those in thermophilic anaerobic ecosystems. *Clostridium stercorarium*, a thermophilic ubiquitous soil dweller, has a simple cellulose hydrolyzing enzyme system of only two cellulases. However, it seems to be better suited for the hydrolysis of a wide range of hemicelluloses. *Clostridium thermocellum*, an ubiquitous thermophilic gram-type positive bacterium, is one of the most successful cellulose degraders known. Its extracellular enzyme complex, the cellulosome, was prepared from *C. thermocellum* cultures grown on cellulose, cellobiose, barley β-1,3-1,4-glucan, or a mixture of xylan and cellulose. The single proteins were identified by peptide chromatography and MALDI-TOF-TOF. Eight cellulosomal proteins could be found in all eight preparations, 32 proteins occur in at least one preparation. A number of enzymatic components had not been identified previously. The proportion of components changes if *C. thermocellum* is grown on different substrates. Mutants of *C. thermocellum,* devoid of scaffoldin CipA, that now allow new types of experiments with *in vitro* cellulosome reassembly and a role in cellulose hydrolysis are described. The characteristics of these mutants provide strong evidence of the positive effect of complex (cellulosome) formation on hydrolysis of crystalline cellulose.

Key words: **bacterial; cellulose; hydrolysis; anaerobic; thermophilic; cellulosome; mutant; scaffoldin; enzyme; complex**

Introduction

Climate is changing. One indication for this change is the large number of strong hurricanes in the United States in 2005, including Katrina and Rita, the hot summer in 2006, and the early and extremely dry spring in Europe in 2007. A long time indicator for global warming is the melting of the alpine glaciers, which can be observed by a larger number of people, in contrast to the less familiar melting of the Arctic and Antarctic ice shields and other indicators.[1] The Furtwängler glacier on mount Kilimandjaro (Tansania, Africa) has supposedly existed for 11,700 years but will disappear completely within 20 years.[1] The publication of the Intergovernmental Panel on Climate Change (IPCC) report,[2] the first parts of which present the scientific evidence for global warming and the human impact. The Stern report[3] made clear that immediate measures for the reduction of CO_2 emissions would be economically more efficient than subsequent repair measures and damage.

With the growing prosperity of an increasing number of people, associated with an advancing dependence on technology, the demand for energy is growing steadily. More than 90% of our energy is currently made from stored natural resources that are rapidly depleted. The most important of those sources is crude (fossil) oil, which is burned with the release of CO_2, a climate-active greenhouse gas. This leads to a global increase in CO_2 content of the atmosphere. The reduction of CO_2 release is a prime issue and calls the scientific world to present solutions for the production of sustainable energy.[4] New sustainably produced liquid and gaseous fuels for the energy sector could contribute considerably to a CO_2-neutral energy usage. Even more effective measures would be a more rational use of energy combined with energy saving.

At present, only about 1% of the world's energy need is met by biofuels. New biofuels are to be produced in

Address for correspondence: Wolfgang H. Schwarz, Department of Microbiology, TUM, Am Hochanger 4, D-85350 Freising, Germany. +49-8161-715445.
wschwarz@wzw.tum.de

Ann. N.Y. Acad. Sci. 1125: 298–307 (2008).
doi: 10.1196/annals.1419.008

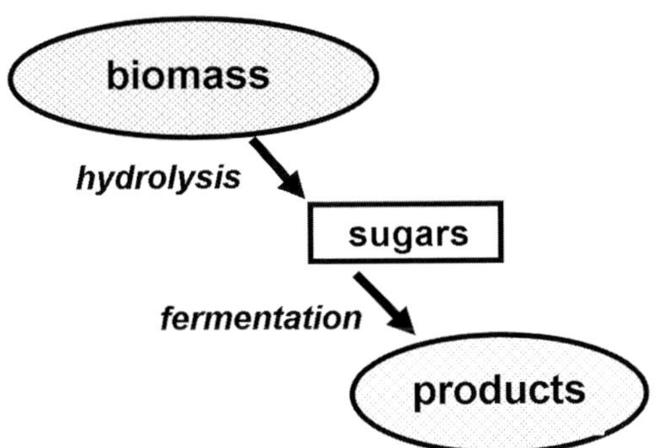

FIGURE 1. The double role of microbes in biofuel production.

greater amounts. This has to include the production of new types of fuels that are optimized for conversion to energy with the existing and even more with advanced and specially adapted technology. Both achievements have to go hand in hand with utilization of new and more abundant substrates, such as cellulose and hemicellulose, from plant biomass.

Microbiology can contribute to the CO_2-neutral production of biofuels from renewable biomass in two major fields: (1) the supply of efficient polysaccharide hydrolyzing enzymes, especially for the cheap and abundant polysaccharides cellulose and hemicellulose, and (2) the fermentation of the resulting sugars to solvents and fuels (FIG. 1). However, the hydrolysis technology for the production of sugary fermentation substrates is lagging behind and commercial conversion processes for biomass-to-sugar plants are still in their infancy. New and more efficient cellulase and hemicellulase preparations have to be developed.

Clostridium stercorarium produces a variety of plant cell wall hydrolyzing enzymes, especially for the hydrolysis of hemicellulose. However, one of the most efficient cellulose-degrading microorganisms known so far is *Clostridium thermocellum*.[5,6] The cellulosome, an extracellular, remarkably efficient cellulase complex of this anaerobic, thermophilic bacterium, is one of the targets of this overview.[7] The understanding of the detailed cellulosome function could provide insight into the mechanism of degradation enzymes for crystalline (= native) cellulose. This chapter gives a short introduction and summarizes new findings in the enzymatic polysaccharide hydrolysis and the possible role of the thermophilic bacteria *C. stercorarium* and *C. thermocellum* therein.

The Role of Biomass Hydrolysis

Bacteria seem to be the major degraders of organic biomass in various ecosystems, either in the environment, for example, in biofilms, or in symbiosis with uni- or multicellular organisms, for example, the intestine of termites or ruminants. For this task they produce a multiplicity of enzymes. Especially the hydrolysis of refractory material, such as crystalline cellulose or the heterogeneous hemicellulose, is dependent on the simultaneous presence of a large number of interacting (synergistic) enzymes in high local concentration on the target site. The heterogeneity of hemicellulose obviously requires a variety of enzymes to cleave the different chemical bonds.[8] In contrast, the chemically homogeneous cellulose has to be hydrolyzed by enzymes that cleave all the same type of chemical bond (β-1,4-glycosidic). However, they attack in a distinct mode of action, such as progressively from one end (from the reducing or the nonreducing end) or nonprogressively in an endomode, by splitting off cellobiose or cellotetraose units, in amorphic or crystalline regions of the cellulose molecule bundle, in Iα- or Iβ-type cellulose, and so on.[9–11]

Cellulases are distinguished by two cleavage mechanisms, retaining or inverting cleavage. Besides these mechanisms there are only the above mentioned two ways to cut an otherwise uniform chemical bond, that is, the cellulases belong to a number of different structural enzyme classes, called glycosyl hydrolase (GH) families.[12] However, beyond the retaining–inverting modus, the mode of action is not defined by the GH family. Most GH families contain nonprogressive endoglucanases as well as progressive cellobiohydrolases or progressive endoglucanases (CAZY database). The difference between those two action modes is generally dependent on the absence or presence of a lid of protruding amino acid residues over the active site pocket that forms a tunnel for the substrate.[13,14]

Another way of influencing the mode of cellulase activity is the attachment of noncatalytic modules, many of which are carbohydrate binding modules (CBMs). These modules bind to single, long or short, cellulose molecules, or to crystalline structures of a different type in cellulose.[15] Some thread a single molecule into the active site pocket, as was shown for the *Thermonospora fusca* cellulase.[16] The direction of activity in progressive enzymes (from the reducing or nonreducing end) may also be influenced by those modules. At present such enzyme mechanisms cannot be predicted by bioinformatics (yet?); they have to be determined experimentally in the laboratory.

Anaerobic hydrolytic bacteria play a major role in the first step of biomass degradation cascades, such as in compost heaps, plant-eating animals (especially in ruminants), or biogas-producing plants. The sugars produced by enzymatic degradation are metabolized

FIGURE 2. A compost plant for household biowaste with forced aeration in Germany: temperatures of up to 75°C are reached. (Photograph provided by W. Hiegl.)

by the cellulolytic microorganisms either aerobically (by fungi and some bacteria) or anaerobically (mostly by bacteria). The anaerobic conversion of biomass results in energy-rich fatty acids and alcohols (plus CO_2 and H_2) that accumulate in some cases to considerable amounts.

The formation of ethyl- or butyl-alcohol by anaerobic fermentation has been exploited for the production of biofuels.[4] Although no commercially viable process is yet available to directly ferment cellulosic biomass into ethanol, a pilot plant with a combination of weak acid treatment and enzymatic hydrolysis is under construction in Japan.[17] The process uses separate conversion of the pentose and hexose sugars formed and involves metabolically engineered bacteria for the fermentation step. In contrast, *C. thermocellum* would be able to directly ferment cellulose and hemicellulose to ethanol in coculture with *Clostridium thermohydrosulfuricum* as has been suggested by Ng *et al.*[18] Various *C. stercorarium* strains were applied during the early 1990s at a pilot plant of Takara Shuzo Inc., Japan, for the direct fermentation of biomass to ethanol.[19]

For drawing out single cellulose molecules from the crystal surface, a rise of temperature should be energetically advantageous for cellulolysis. In fact, at least some thermophilic cellulolytic bacteria seem to be very efficient cellulose degraders.[5] This is in agreement with the increasing number of thermophilic processes for fiber hydrolysis. Thermophilic processes are developed for composting, that is, *in vitro* enzymatic digestion of plant material or biogas formation. Thermophilic treatment plants for biological household and agricultural waste are meanwhile state of the art (FIG. 2).

TABLE 1. Comparison of hydrolytic activities in culture supernatants of type strains of *Clostridium thermocellum* and *Clostridium stercorarium* grown in cellobiose medium

Substrate	*C. thermocellum* (mu/mL)	*C. stercorarium* (mu/mL)
Microcrist. cellulose	2	2
Phosphoric acid swollen cellulose	13	10
Carboxymethylcellulose (CMC)	140	120
1,3-1,4-β-glucan (lichenan)	6,500	12,000
Arabino-xylan	3,000	20,000
*p*NP-β-glucopyranoside	1.3	7
*p*NP-β-cellobioside	12	1.7
*p*NP-β-xylopyranoside	0.3	2
*p*NP-α-arabinofuranoside	1.3	21

Cell-free culture supernatants were incubated with the substrates indicated and activity was determined by estimating reducing sugars (upper part of the list) with dinitrosalicylic acid and *p*-nitrophenol, respectively (lower part) released.

Activity in cell-free culture fluid (grown on cellobiose).

Polysaccharide Degradation by *Clostridium stercorarium*

For a process of direct conversion of lignocellulosic plant biomass to ethanol, new thermophilic bacterial strains had been isolated by Takara Shuzo Inc.[19] Three of those strains have been investigated and found to be very similar to *C. stercorarium* in the pattern of genomic restriction fragments and the hybridization of cellulase genes to macrorestriction fragments of the same size.[19] The degradation of crystalline cellulose in these strains was relatively modest as was the case with the type strain *C. stercorarium*. However, the new strains as well as the *C. stercorarium* type strain were very good degraders of a wide range of hemicellulosic model substrates.[20]

When culture supernatants of cellobiose-grown type strains of *C. stercorarium* and *C. thermocellum* were compared with different substrates, it became clear that both strains can degrade the crystalline or amorphic cellulose, the mixed-linkage glucan, the xylan, and the arylglycosides (TABLE 1). But apparently the activity of *C. stercorarium* was slower with the hydrolysis of cellulosic substrates, whereas it was faster, and in many cases very much faster, with the β-glucan, the xylan, and the arylglycosides. An exception is *p*NP-β-cellobioside, where *C. thermocellum* has an enzyme system that is better suited to degrade the aryl bond between the *p*-nitrophenyl residue and the cellobiose (cellobiohydrolase). This agrees well with the better cellulose degrading ability, which is also obvious with the shorter

TABLE 2. Screening of a genomic library of *Clostridium stercorarium* DNA in *E. coli*

Substrate	Isolates	Genes isolated	Alternative substrates	Ref.
Cellulose (CMC, Avicel)	2	***celY, celZ***		38, 39
Xylan	17	***xynA***		8, 40
Lichenan	18	***xynB, xynC***	xylan	8, 40
PNP-β-xyloside (PNPX)	18	***bxlA***		8, 40
PNP-α-arabinoside (PNPAf)	4	***arfB***		8, 40, 41, 42
PNPX + PNPAf	4	***arfA, bxlB***		8, 40
PNP-β-glucoside	4	***bglZ***		8
PNP-α-rhamnoside	7	***ramA***	6 on PNPX	43
PNP-α-fucoside	2			
PNP-α-galactoside	2			
PNP-β-galactoside	3			
PNP-maltoside	3			
PNP-β-glucuronoside	2		PNPX	
PNP-β-galacturonoside	4		PNP-β-glucuronoside	

Source: Schwarz *et al.*[40]

In all, cell-free extracts from 1139 clones were prepared and assayed on the substrates indicated. Not all clones were tested on all substrates and not all identified active clones were characterized. Designation of identified genes is indicated, together with the references for the publication of sequence analysis, methods, and biochemical characterization. Some activities found with a clone were also active on other substrates (alternative substrates).

time in which Whatman filter paper is completely hydrolyzed during growth (unpublished observation). *C. stercorarium* can thus be regarded as a specialist for hydrolysis of hemicellulose, whereas *C. thermocellum* is a specialist for cellulose.

This fact persuaded us to screen hydrolytic genes from a genomic library of *C. stercorarium* DNA.[21] A library of 1139 clones were obtained by the partial digest of genomic DNA with different restriction enzymes. Cell-free extracts were prepared and screened with 14 different substrates indicative for enzyme activities involved in cellulose and hemicellulose hydrolysis (TABLE 2). Substrates containing, for example, α- and β-linkages, hexoses and pentoses, or uronic acids, were applied. As soon as a handful of active clones on a substrate were identified, the screening with that substrate was stopped. TABLE 2 thus cannot be used for a statistical calculation of genes present. On any substrate tested, a number of active clones could be found. For those enzyme activities with the greatest importance for cellulose or hemicellulose hydrolysis, the genes expressing activity were subcloned, the DNA restriction maps were compared, and the genes responsible for the activity were sequenced (TABLE 2). The biochemical characteristics of the single enzymes were investigated.

Only two genes of *C. stercorarium* could be identified with activity on cellulose and soluble β-1,4-glucans: Cel9Z and Cel48Y. The two enzymes produced were also the only enzymes isolated from the culture supernatant. They were found to be responsible for cellulose degradation in a synergistic manner.[22] They were the only cellulase genes of that bacterium identified so far in a number of screenings of different genomic libraries and in protein isolations: it can be assumed that they are indeed the only cellulases of that organism, although this cannot be said with certainty until the genome is sequenced.

A number of genes expressing enzyme activities related to hemicellulose hydrolysis were sequenced (TABLE 2). A combination of recombinant xylanase XynA, arabinofuranosidase ArfB, and β-xylosidase BxlB was sufficient to degrade arabinoxylan completely to the monosaccharides xylose and arabinose with a mass ratio of about 10:1.[8] Other genes expressing similar activities (such as ArfA and BxlA) were not able to substitute these enzymes on arabinoxylan because they were arylhydrolases and not active on oligosaccharides (alkyl compounds). Their function is probably the detoxification of arylglycosides.

In contrast to *C. thermocellum*, *C. stercorarium* thus degrades a wide range of polysaccharides (especially hemicellulose) and ferments pentoses more readily. Many strains of *C. thermocellum* do not metabolize pentoses and even have a long lag phase when glucose is provided as the only carbon source. However, there are differences between the isolates (see, e.g., Ref. 23).

The *Clostridium thermocellum* Cellulosome

C. thermocellum, an anaerobic thermophilic ubiquitous bacterium frequently isolated from decaying

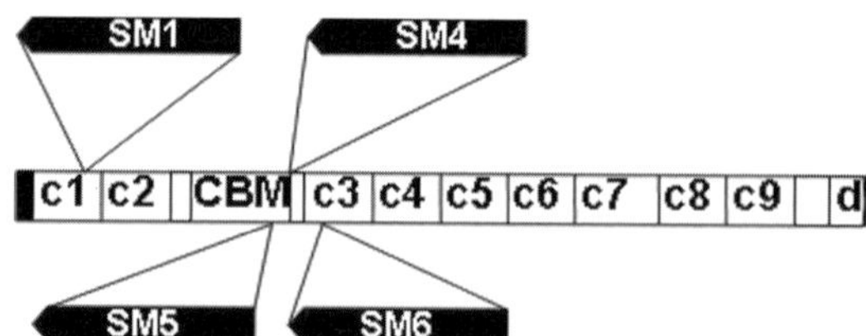

FIGURE 3. Map of the CipA protein of *C. thermocellum*, the scaffoldin protein. c1 to c9, cohesin modules; d, dockerin module of type 2; carbohydrate binding module (CBM) family III. The leader peptide is indicated by a *black box*. The insertion points, translation direction, and approximate size of IS 1447 in four different mutants are indicated by *black arrow boxes*.

biomass and soil,[5] was reported to be one of the most efficient bacteria in degradation of native cellulose. It can also be found in high numbers in habitats that are not "hot" (thermophilic). Recently, *C. thermocellum* was shown to be present in a thermophilic biomass digester that was run at an elevated temperature, as was the more moderately thermophilic bacterium, *C. stercorarium*[24] which is closely related to *C. thermocellum*, but has a much simpler cellulase system.[25] These data were confirmed in our laboratory when various enrichments of plant fiber degrading bacteria at 55 or 60°C were investigated (Hiegl, personal communication).

The natural substrate of *C. thermocellum* catabolism are the hydrolysis products of cellulose hydrolysis, the cellodextrins, which are degraded by the consecutive action of intracellular cellodextrin- and cellobiose-phosphorylases, as was shown for the type strain ATCC 27405.[6,26] This results in a much more energy-efficient metabolism, which makes up for the costly production of the large amount of extracellular cellulase complex needed for the degradation of the refractory substrate.[6]

For the degradation of crystalline cellulose, *C. thermocellum* synthesizes a large extracellular protein particle. It contains a core protein called scaffoldin (the CipA protein).[15,27,28] CipA is a noncatalytic structural protein of intricate structure, and, among other modules, it consists of nine binding modules (FIG. 3) that are called cohesins. One enzyme component, probably randomly selected, is firmly attached to each cohesin by virtue of its dockerin module. The dockerins consist of highly conserved dual 24 aa repeats of amino acids. CipA is connected to the bacterial cell wall by an S-layer homologous module; it is also connected to the cellulosic substrate via a CBM.[27]

The assembly of the cellulosome is a highly demanding task to discover. The sequence of events and the assembly mechanism have not been uncovered to date. The cellulosome formation may begin with the secretion of all components to the cell surface, and is succeeded by the assembly into the enzyme complex on the cell surface. The steps in between are a black box. The present working hypothesis is that the exact composition of a single particle cannot be predicted. The particles are apparently composed at random, according to the concentration of single components secreted through the cell membrane. It can be speculated that the most frequent (and most important) enzyme components, such as Cel48S, may have a defined cohesin where they are preferentially bound. To find proof of this would need much more sophisticated experiments than have been performed to date.

The number of cellulosome components is surprisingly large. An analysis of the closed genome sequence of *C. thermocellum* (GenBank No. CP000568, DOE JGI as of February 16, 2007) detected 74 reading frames containing a dockerin duplet (unpublished data and Ref. 29). This unequivocally indicates cellulosomal components. The high number of components makes working with an artificial cellulosome difficult. However, not all components are present in the same molar amount (TABLE 3 and Ref. 29). It can be assumed that components less frequently represented in the cellulosome are less important for cellulose hydrolysis. Moreover, a part of the most frequent components are xylanases. This reduces the number of components for *in vitro* reconstruction of the cellulosome.

Besides the cell-bound cellulosome complex, the *C. thermocellum* genomic sequence also revealed the presence of a putative second cellulase system. This system consists of the cellulases Cel9I and Cel48Y (GHF9 and GHF48, respectively). Both enzymes cooperate synergistically.[30] This soluble cellulase system resembles the one in *C. stercorarium* that consists of the two cellulases, Cel9Z and Cel48Y,[30] the only cellulases found in *C. stercorarium*. Cs-Cel9Z resembles Ct-Cel9I, and Cs-Cel48Y resembles Ct-Cel48Y in structure and mode of activity. These soluble cellulases of *C. thermocellum* possibly assist the cellulosomal cellulose hydrolysis, whereas the cellulosomes are cell-bound, the soluble enzymes putatively act on different subsites of the cellulose crystals, and at a greater distance to the producer cell.

The cellulosomes contain not only cellulases but also a large number of xylanases and glycosidases (in addition to a chitinase and to pectinases). These enzymes are obviously removing covering hemicellulosic material from the surface of the cellulose crystals that are the substrate for the cellulases and for the host bacterium. To test the ability of different *C. thermocellum* strains, such as the various DSMZ strains and strain VKPM2203, to degrade plant cell wall, the bacteria were inoculated to ground plant fibers (2 mm) and

TABLE 3. Proteins identified in cellulosomes from *Clostridium thermocellum* produced on different substrates

Protein (GHF)	Locus_tag	C	Cb	X + C	βG	C2D
CipA	Cthe_3077	■	■	■	■	■
CelS (48)	Cthe_2089	■	■	■	■	■
CelG (5)	Cthe_2872	■	■	■	■	■
CelK (9)	Cthe_0412	■	■	■	■	■
CbhA (9)	Cthe_0413	■	■	■	■	■
CelR (9)	Cthe_0578	■	■	■	■	■
CelA (8)	Cthe_0269	■	■	■	■	■
XynC (10)	Cthe_1838	■	■	■	■	■
- (9)	Cthe_0433	■	■	■	■	
CelJ (9,44)	Cthe_0624	■	■	■	■	
CelQ (9)	Cthe_0625	■	■	■	■	
CelF (9)	Cthe_0543	■	■	■	■	
XynY (10)	Cthe_0912	■	■	■	■	
CelV (9)	Cthe_2760	■	■	■		
CprA	Cthe_3136	■	■	■		
XynZ (10)	Cthe_1963	■			■	■
XghA (74)	Cthe_1398		■	■		■
XynD (10)	Cthe_2590		■	■		■
ChiA (18)	Cthe_0270		■			■
CelT (9)	Cthe_2812	■	■			
CelB (5)	Cthe_0536		■	■		
CelU (9)	Cthe_2360		■	■		
- (PL11)	Cthe_0246		■	■		
- (28)	Cthe_2038		■	■		
- (9)	Cthe_2761		■	■		
CelN (9)	Cthe_0043					■
- (5)	Cthe_2193		■			
CelL (5)	Cthe_0405		■			
CelW (9)	Cthe_0745			■		
- (53)	Cthe_1400		■			
- (UN)	Cthe_3132		■			
CseP (UN)	Cthe_0044		■			

The glycosyl hydrolase family is indicated in the first column in brackets; in one case this is a pectate lyase (PL); unknown module (UN); no number in brackets is indicating a protein without depolymerizing function (or a putative functional protein). The presence of a protein in the protein identification of culture supernatants is indicated by a *black box*. The proteins are sorted according to their frequency of detection. ABBREVIATIONS: C, cellulose; Cb, cellobiose; X + C, xylane + cellulose; βG, barley β-glucan; C2D, cellulose experiment 2 (proteins identified with 2D-gel electrophoresis).

incubated in GS2 medium.[31] After 5 days incubation at 60°C, the residual insoluble fibers were washed, dried, and weighed. Although no drastic volume degradation was apparent in the medium (the sedimented material at the bottom of the example bottles in FIG. 4), the reduction in dry weight was up to 80% compared to the control with no bacteria in it. A mixture of the gases CO_2 and H_2 was formed and cell growth was observed (turbidity of the medium in FIG. 4); furthermore, short-chain fatty acids (acetic acid), lactate, and ethanol were formed (data not shown; Hiegl, personal communication). This result shows that many *C. thermocellum* strains are able to degrade native plant fibers well and to a considerable degree.

Differential Expression of Cellulosome Components

C. thermocellum is usually cultivated on cellobiose as a carbon source. Cellulosomes are produced well under these conditions, although they are not necessary for substrate degradation; cellobiose is a good substrate for cellulosome expression.[44] The protein complexes can be isolated from the culture supernatant relatively easily.[32] Although it is difficult to detach cellulosomes in natural form from insoluble cellulose particles, cellulosomes can be isolated in a sufficient amount and purity if such substrates are digested to completion. Otherwise, the cellulosomes would be found in the pellet fraction, together with the substrate particles and the cells.

Cellulosome preparations were obtained from *C. thermocellum* ATCC27405 (type strain) grown on four different substrates, crystalline cellulose MN300, cellobiose, xylan plus cellulose, and barley β-1,3-1,4-D-glucane.[32] Growth on xylan alone was too poor to obtain a sufficient amount of cellulosomes. Soluble barley β-glucan was a good but expensive substrate for *C. thermocellum.* The isolated cellulosomes were analyzed by ICPL peptide mapping and N terminal sequencing (J. Kellermann, personal communication). The C2D fraction of cellulose-grown cellulosomes was analyzed by MALDI-TOF-TOF from 2D-gel electrophoresis.[29]

Combining the five preparations investigated with two different methods, 32 different proteins were identified in the cellulosomes (TABLE 1):

- 4 potentially structured, noncatalytic proteins, including CipA,
- 19 β-glucanases, including 1 each of GHF8 and GHF48, 4 of GHF5, and 13 of GHF9,
- 6 xylanases,
- 7 proteins from not yet cloned genes and not yet studied enzymes.

All proteins (except CipA) contain a module coding for a dockerin type I, indicating unambiguously their localization on the cohesin modules of CipA. This is corroborated by the fact that they were identified in the cellulosome complex. Contrarily, all proteins identified in the cellulosome are containing dockerin modules, which emphasizes that the major and maybe the

FIGURE 4. Ground maize silage incubated with *C. thermocellum*. Anaerobic bottles are filled with 10% (w/v dry mass) in GS2 medium and stoppered under anaerobic conditions. The left bottle without inoculation (negative control), the right bottle five days after inoculation with *C. thermocellum* ATCC 27405. Incubation was at 65°C. (Photograph provided by W. Hiegl.)

only principle for integration in the cellulosome is the dockerin-cohesin interaction. Even putative structural proteins (besides CipA itself) are bound by dockerin-cohesin (type I) interactions. The dockerin module of CipA is of type II and responsible for binding the complex to the cell surface or to multicomplex integrating proteins, such as OlpB.

Interestingly, almost 50% of the genes (32/72) containing a dockerin module (Ref. 29 and TABLE 3) are actually expressed, secreted, and assembled as components of the cellulosome. Some components may be minor proteins in the cellulosome and have not been detected yet. The peptides, which were most frequently picked up, belonged to cellulases of GHF9, a hydrolase family containing nonprogressive as well as progressive β-glucanases. GHF5, GHF8, and GHF48 cellulases appear to be less represented. However, given the uncertainty of quantification by the method, this cannot be stated with certainty. Nevertheless, the number of peptide counts coincides with the density of the corresponding spots in 2D-gel electrophoresis.

The lower part of the list in TABLE 3 represents the proteins that presumably occur less frequently in cellulosomes; they were not identified in each preparation. No conclusive picture concerning the substrate-depending regulation can be drawn, such as the up-regulation of xylanase genes when xylan is used as a substrate. This ambiguous result may be due to the presence of cellulose in the xylan or the inoculum added to the culture medium. The question of differential expression of cellulosomal components cannot be answered conclusively with the data currently available. Influence on the expression of cellulase genes by the conditions of the culture were shown earlier, but with mRNA and not with protein.[6,33,34]

Knockout Mutants of the *Clostridium thermocellum cipA* Gene

For identification of the components necessary for crystalline cellulose hydrolysis, mutants with reduced ability to hydrolyze crystalline cellulose were isolated. They were created by mutagenesis with ethyl-methane-sulfonate (EMS) treatment of growing cultures of *C. thermocellum*. Mutagenized cells were plated in turbid cellulose plates and screened for colonies without a clear halo around them due to reduced cellulose degradation. Analysis of the cellulosomal proteins should eventually reveal the components missing in mutants reduced in cellulase activity.

Six almost completely cellulose-defective mutants were purified by single-colony isolation and subjected to SDS-PAGE of their culture supernatant proteins. The protein pattern was almost identical to that of cellulosomal proteins, but the most prominent band of cellulosomes, the scaffoldin band (CipA), was missing.[35] Size exclusion chromatography indicated that the mutants were defective in the formation of the cellulosome complexes that were absent in cleared culture supernatants. Sequencing of the amplified CipA region revealed the insertion of a new insertion element, called IS1447 (gene bank account numbers AM491039 to AM491042), into the N terminal region of the *cipA* gene, with opposite transcription direction. This should lead to reading-frame interruption, and thus to seriously shortened CipA fragments missing, for example, all cohesin modules (as in mutant SM1; FIG. 3). The location of the insertions of four mutants is indicated in FIGURE 3.

The lack of a functional scaffoldin has the consequence that the *C. thermocellum* culture supernatant contains all cellulosomal proteins except CipA. The components of the cellulosome are not integrated in a large protein complex. They are not bound to the cell wall because the cell wall binding of the complex is also mediated by the now missing CipA. This particle-free mixture of cellulosomal components will be used for follow-up experiments that will reveal the mechanism of the assembly of the cellulosome and the nature of the cellulase synergism in the complex.

Preliminary experiments have shown that CipA mutants show similar activity compared to the wild type on soluble β-glucans, such as carboxymethylcellulose (CMC) or barley β-glucan. However they have lost most of their activity (up to 90%) on crystalline cellulose.[35] It has always been a paradigm of cellulase research that the complex formation, the combination of several cellulases in the cellulosome, is responsible for the outstanding effectiveness of the "true cellulase" activity in *C. thermocellum* and other anaerobic bacteria. The generally lower specific activity of cellulase systems produced by other bacteria that do not form a cellulosome complex, such as *C. stercorarium*[25] or most aerobic cellulolytic bacteria can perhaps be explained by the findings on the key role of CipA in maintaining the cellulosome.

These results—together with the genomic DNA sequence completed recently, the possibility for metabolic engineering through gene transfer,[36] the remarkable cellulase system, and the earlier reports on cocultures of *C. thermocellum* with other thermophilic clostridia for ethanol production directly from biomass[37]—open chances to make use of improved *C. thermocellum* strains for the production of chemicals and biofuels in future white biotechnology and biorefinery applications. *C. thermocellum* with its two independent cellulase systems, the soluble and the insoluble (cellulosomal), is a good candidate for further developments in novel cellulases for industrial applications. *C. stercorarium* may have an important role for the production of new hemicellulases for the saccharification of plant biomass.

Acknowledgment

This work was supported by a grant from the Deutsche Forschungsgemeinschaft DFG. Thanks to J. Kellermann for providing the MALDI-TOF results, and M. Klupp for preparing the *Clostridium thermocellum* cultures on different substrates. Experimental results from E. Berger, S. Freiding, W. Hiegl, J. Krauss, and M. Hösl (all Inst. Microbiology, TUM) were used for preparation of this manuscript.

Conflict of Interest

The authors declare no conflicts of interest.

References

1. MINARCEK, A. 2003. Mount Kilimanjaro's Glacier Is Crumbling. National Geographic Adventure. URL: http://news.nationalgeographic.com/news/2003/09/0923_030923_kilimanjaroglaciers.html. Pictures of other glaciers can be visited at the homepage of Gesellschaft für ökologische Forschung e.V., URL: http://www.gletscherarchiv.de/.
2. IPCC. 2007. Climate Change 2007: The Physical Science Basis. *Contribution of Working Group I to the Fourth Assessment Report of the Intergovernmental Panel on Climate Change (IPCC)*. ISBN 13: 9780521705967. http://ipcc-wg1.ucar.edu/wg1/wg1-report.html Cambridge: Cambridge University Press, in press.
3. STERN, N. 2006. The Economics of Climate Change. The Stern Review. Cabinet Office–HM Treasury. Cambridge: Cambridge University Press. http://www.hm-treasury.gov.uk/independent_reviews/stern_review_economics_climate_change/stern_review_report.cfm
4. ANTONI, D., V.V. ZVERLOV & W.H. SCHWARZ. 2007. Biofuels from microbes. Appl. Microbiol. Biotechnol. **77:** 23–35.
5. LYND, L.R., P.J. WEIMER, W.H. VAN ZYL & I.S. PRETORIUS. 2002. Microbial cellulose utilization: fundamentals and biotechnology. Microbiol. Molec. Biol. Rev. **66:** 506–577.
6. ZHANG, Y.H. & L.R. LYND. 2005. Cellulose utilization by *Clostridium thermocellum*: bioenergetics and hydrolysis product assimilation. PNAS **102:** 7321–7325.
7. COUGHLAN, M.P., K. HON-NAMI, H. HON-NAMI, *et al.* 1985. The cellulolytic enzyme complex of *Clostridium thermocellum* is very large. Biochem. Biophys. Res. Commun. **130:** 904–909.
8. ADELSBERGER, H., C. HERTEL, E. GLAWISCHNIG, *et al.* 2004. Enzyme system of *Clostridium stercorarium* for hydrolysis of arabinoxylan: reconstitution of the in vivo system from recombinant enzymes. Microbiol. **150:** 2257–2266.
9. SCHWARZ, W.H. 2001. The cellulosome and cellulose degradation by anaerobic bacteria. Appl. Microbiol. Biotechnol. **56:** 634–649.

10. SCHWARZ, W.H. 2004. Cellulose—Struktur ohne Ende. Naturwiss. Rundsch. **8:** 443–445.

11. BAYER, E.A., Y. SHOHAM & R. LAMED. 2000. Cellulose-decomposing bacteria and their enzyme systems. *In* M. Dworkin, S. Falkow, E. Rosenberg, *et al.*, Eds.: The Prokaryotes: An Evolving Electronic Resource for the Microbiological Community, 3rd ed. New York: Springer Verlag.

12. HENRISSAT, B. & G.J. DAVIES. 1997. Structural and sequence-based classification of glycoside hydrolases. Curr. Opin. Struct. Biol. **7:** 637–644.

13. TEERI, T.T. 1997. Crystalline cellulose degradation: new insights into the function of cellobiohydrolases. TibTech **15:** 160–166.

14. ZVERLOV, V.V., G.A. VELIKODVORSKAYA & W.H. SCHWARZ. 2003. Two new cellulosome components encoded downstream of *celI* in the genome of *Clostridium thermocellum*: the non-processive endoglucanase CelN and the possibly structural protein CseP. Microbiol. **149:** 515–524.

15. BAYER, E.A., E. MORAG, R. LAMED, *et al.* 1998. Cellulosome structure: four-pronged attack using biochemistry, molecular biology, crystallography and bioinformatics. *In* Carbohydrases from *Trichoderma reesei* and Other Microorganisms. M. Claeyssens, W. Nerinckx & K. Piens, Eds.: 39–65. London. The Royal Society of Chemistry.

16. IRWIN, D., D.H. SHIN, S. ZHANG, *et al.* 1998. Roles of the catalytic domain and two cellulose binding domains of *Thermomonospora fusca* E4 in cellulose hydrolysis. J. Bacteriol. **180:** 1709–1714.

17. INGRAM, L.O. 2008. Ethanol production by genetically engineered *Escherichia coli*. Ann. N.Y. Acad. Sci. Incredible Anaerobes: From Physiology to Genomics to Fuels. In Press.

18. NG, T.K., A. BEN-BASSAT & J.G. ZEIKUS. 1981. Ethanol production by thermophilic bacteria: fermentation of cellulosic substrates by cocultures of *Clostridium thermocellum* and *Clostridium thermohydrosulfuricum*. Appl. Environ. Microbiol. **41:** 1337–1343.

19. SCHWARZ, W.H., K. BRONNENMEIER, B. LANDMANN, *et al.* 1995. Molecular characterization of four strains of the cellulolytic thermophile *Clostridium stercorarium*. Biosci. Biotechnol. Biochem. **59:** 1661–1665.

20. BRONNENMEIER, K., C. EBENBICHLER & W.L. STAUDENBAUER. 1990. Separation of the cellulolytic and xylanolytic enzymes of *Clostridium stercorarium*. J. Chromatogr. **521:** 301–310.

21. SCHWARZ, W.H., S. JAURIS, M. KOUBA, *et al.* 1989. Cloning and expression of *Clostridium stercorarium* cellulase genes in *Escherichia coli*. Biotechnol. Lett. **11:** 461–466.

22. RIEDEL, K. & K. BRONNENMEIER. 1998. Intramolecular synergism in an engineered exo-endo-1,4-β-glucanase fusion protein. Mol. Microbiol. **28:** 767–775.

23. FREIER, D., C.P. MOTHERSHED & J. WIEGEL. 1988. Characterization of *Clostridium thermocellum* JW20. Appl. Environ. Microbiol. **54:** 204–211.

24. BURRELL, P.C., C.O. O'SULLIVAN, H. SONG, *et al.* 2004. Identification, detection, and spatial resolution of *Clostridium* populations responsible for cellulose degradation in a methanogenic landfill leachate bioreactor. Appl. Environ. Microbiol. **70:** 2414–2419.

25. BRONNENMEIER, K., H. ADELSBERGER, F. LOTTSPEICH & W.L. STAUDENBAUER. 1996. Affinity purification of cellulose-binding enzymes of *Clostridium stercorarium*. Bioseparation **6:** 41–45.

26. REICHENBECHER, M., F. LOTTSPEICH & K. BRONNENMEIER. 1997. Purification and properties of a cellobiose phosphorylase (CepA) and a cellodextrin phosphorylase (CepB) from the cellulolytic thermophile *Clostridium stercorarium*. Eur. J. Biochem. **247:** 262–267.

27. SCHWARZ, W.H., V.V. ZVERLOV & H. BAHL. 2004. Extracellular glycosyl hydrolases from clostridia. Adv. Appl. Microbiol. **56:** 215–261.

28. BAYER, E.A., J.P. BELAICH, Y. SHOHAM & R. LAMED. 2004. The cellulosomes: multienzyme machines for degradation of plant cell wall polysaccharides. Annu. Rev. Microbiol. **58:** 521–554.

29. ZVERLOV, V.V., J. KELLERMANN & W.H. SCHWARZ. 2005. Functional subgenomics of *Clostridium thermocellum* cellulosomal genes: identification of the major catalytic components in the extracellular complex and detection of three new enzymes. Proteomics **5:** 3646–3653.

30. BERGER, E., D. ZHANG, V.V. ZVERLOV & W.H. SCHWARZ. 2007. Two noncellulosomal cellulases of *Clostridium thermocellum*, Cel9I and Cel48Y, hydrolyse crystalline cellulose synergistically. FEMS Microbiol. Lett. **268:** 194–201.

31. JOHNSON, E.A., A. MADIA & A.L. DEMAIN. 1982. Chemically defined minimal medium for growth of the anaerobic cellulolytic thermophile *Clostridium thermocellum*. Appl. Environ. Microbiol. **41:** 1060–1062.

32. MORAG, E., I. HARVEY, E.A. BAYER & R. LAMED. 1991. Isolation and properties of a major cellobiohydrolases from the cellulosome of *Clostridium thermocellum*. J. Bacteriol. **173:** 4155–4162.

33. DROR, T.W., A. ROLIDER, E.A. BAYER, *et al.* 2005. Regulation of major cellulosomal endoglucanases of Clostridium thermocellum differs from that of a prominent cellulosomal xylanase. J. Bacteriol. **187:** 2261–2266.

34. STEVENSON, D.M. & P.J. WEIMER. 2005. Expression of 17 genes in *Clostridium thermocellum* ATCC 27405 during fermentation of cellulose or cellobiose in continuous culture. Appl. Environ. Microbiol. **71:** 4672–4678.

35. ZVERLOV, V.V., M. KLUPP, J. KRAUSS & W.H. SCHWARZ. Mutants in the scaffoldin gene *cipA* of *Clostridium thermocellum* with impaired cellulosome formation and cellulose hydrolysis: insertions of a new IS-element, IS1447, and implications for the assembly of the genomic sequence. In preparation.

36. TYURIN, M.V., S.G. DESAI & L.R. LYND. 2004. Electrotransformation of *Clostridium thermocellum*. Appl. Environ. Microbiol. **70:** 883–890.

37. MIELENZ, J.R. 2001. Ethanol production from biomass: technology and commercialization status. Curr. Opin. Microbiol. **4:** 324–329.

38. BRONNENMEIER, K., K. KUNDT, K. RIEDEL, *et al.* 1997. Structure of the *Clostridium stercorarium* gene *celY* encoding the exo-1,4-β-glucanase Avicelase II. Microbiology **143:** 891–898.

39. JAURIS, S., K.P. RÜCKNAGEL, W.H. SCHWARZ, *et al.* 1990. Sequence analysis of the *Clostridium stercorarium* *celZ* gene encoding a thermoactive cellulase (Avicelase I): identification of catalytic and cellulose-binding domains. Mol. Gen. Genet. **223:** 258–267.

40. SCHWARZ, W.H., H. ADELSBERGER, S. JAURIS, *et al.* 1990. Xylan degradation by the thermophile *Clostridium stercorarium*: cloning and expression of xylanase, β-D-xylosidase, and alpha-L-arabinofuranosidase genes in *Escherichia coli*. BBRC **170:** 368–374.
41. ZVERLOV, V.V., W. LIEBL, M. BACHLEITNER & W.H. SCHWARZ. 1998. Nucleotide sequence of *arfB* of *Clostridium stercorarium*, and prediction of catalytic residues of alpha-L-arabinofuranosidases based on local similarity with several families of glycosyl hydrolases. FEMS Microbiol. Lett. **164:** 337–343.
42. SCHWARZ, W.H., K. BRONNENMEIER, B. KRAUSE, *et al.* 1995. Debranching of arabinoxylan: properties of the thermoactive recombinant alpha-L-arabinofuranosidase from *Clostridium stercorarium* (ArfB). Appl. Microbiol. Biotechnol. **43:** 856–860.
43. ZVERLOV, V.V., C. HERTEL, K. BRONNENMEIER, *et al.* 2000. The thermostable alpha-L-rhamnosidase RamA of *Clostridium stercorarium*: biochemical characterization and primary structure of a bacterial alpha-L-rhamnoside hydrolyse, a new type of inverting glycoside hydrolase. Molec. Microbiol. **35:** 173–179.
44. BHAT, S., P.W. GOODENOUGH, E. OWEN & M.K. BHAT. 1993. Cellobiose: a true inducer of cellulosome in different strains of *Clostridium thermocellum*. FEMS Microbiol. Lett. **111:** 73–78.

The Cellulase/Hemicellulase System of the Anaerobic Fungus Orpinomyces *PC-2 and Aspects of Its Applied Use*

LARS G. LJUNGDAHL

Department of Biochemistry and Molecular Biology, The University of Georgia, Athens, Georgia, USA

Anaerobic fungi, first described in 1975 by Orpin, live in close contact with bacteria and other microorganisms in the rumen and caecum of herbivorous animals, where they digest ingested plant food. Seventeen distinct anaerobic fungi belonging to five different genera have been described. They have been found in at least 50 different herbivorous animals. Anaerobic fungi do not possess mitochondria, but instead have hydrogenosomes, which form hydrogen and carbon dioxide from pyruvate and malate during fermentation of carbohydrates. In addition, they are very oxygen- and temperature-sensitive, and their DNA has an unusually high AT content of from 72 to 87 mol%. My initial reason for studying anaerobic fungi was because they solubilize lignocellulose and produce all enzymes needed to efficiently hydrolyze cellulose and hemicelluloses. Although some of these enzymes are found free in the medium, most of them are associated with cellulosomal and polycellulosomal complexes, in which the enzymes are attached through fungal dockerins to scaffolding proteins; this is similar to what has been found for cellulosomes from anaerobic bacteria. Although cellulosomes from anaerobic fungi share many properties with cellulosomes of anaerobic cellulolytic bacteria and have comparable structures, their structures differ in their amino acid sequences. I discuss some features of the cellulosome of the anaerobic fungus *Orpinomyces* sp. PC-2 and some possible uses of its enzymes in industrial settings.

Key words: **acetyl xylan esterase; anaerobic fungi; cellulases; cyclophilin; feruloyl and coumaroyl esterases; fungal cellulosomes; glucanase; glucosidase; hemicellulases; herbivorous animals; mannanase; *Orpinomyces* PC-2; xylanases**

Introduction

Anaerobic fungi, first described in 1975 by Orpin,[1] are present in the gastrointestinal tract, and especially in the rumen and caecum, of herbivorous animals. At present at least 17 different anaerobic fungi isolated from over 50 different herbivorous animals have been described. They are placed in five genera depending on ultrastructural characteristics, including monocentric or polycentric growth and uni-, bi-, or polyflagellated zoospores:[2–4] *Caecomyces*, *Neocallimastix*, and *Piromyces* species are monocentric, whereas *Anaeromyces* and *Orpinomyces* species are polycentric. Although anaerobic fungi genera differ from one another morphologically, they appear to constitute a homogeneous group as indicated by analyses of 18S ribosomal RNA, high homology of genes encoding enzymes, and very low GC content of the DNA, ranging from 13% to 22%.[3,5–9] The anaerobic fungi differ from aerobic fungi in that they have hydrogenosomes instead of mitochondria; like mitochondria, hydrogenosomes are involved in electron transport and energy generation.[10] It is likely that the hydrogenosome has a mitochondrial origin;[11,12] this is suggested by the observation that the two organelles have homologous enzymes.[13,14] The anaerobic fungi are very sensitive to oxygen and temperature and generally do not survive in conditions other than those found in the intestinal tracts of animals.[15,16]

Plant materials, such as grass and straw, are rather rapidly hydrolyzed by anaerobic fungi, yielding sugars and other plant products. It has been suggested that, in herbivorous animals, anaerobic fungi are the initial colonizers of lignocellulose and play an essential role in fiber digestion.[17–19] They ferment the sugars, yielding formate, acetate, ethanol, lactate, CO_2, and H_2 as the main end products.[20,21] Knowledge about anaerobic fungi is rapidly accumulating, especially pertaining to their ability to produce plant hydrolyzing enzymes,

Address for correspondence: Lars G. Ljungdahl, Department of Biochemistry and Molecular Biology, Fred C. Davison Life Sciences Complex, University of Georgia, Athens, GA 30602-7229. Voice: +1-706-542-1334 or +1-706-548-5190; fax: +1-706-542-1738.

Larsljd@bmb.uga.edu

Ann. N.Y. Acad. Sci. 1125: 308–321 (2008).
doi: 10.1196/annals.1419.030

including cellulases, β-glucosidases, xylanases, β-glucanases/lichenases, mannanases, cinnamoyl esterases/feruloyl- and *p*-coumaroyl esterases, acetyl xylan esterase, and (4-0-methyl)-D-glucuronidase/α-glucuronidase.[21–23] Excellent information about anaerobic fungi can be found in the book edited by Mountford and Orpin.[24] Work to be discussed here has been performed with the polycentric fungus *Orpinomyces* PC-2, which is one of five different fungi isolated from the rumen of a cow.[21]

The Cellulosome, an Extracellular Organelle

In the natural environment, cellulose and other carbohydrates are degraded almost exclusively by microorganisms. In the presence of oxygen, aerobic bacteria and fungi oxidize carbohydrates to carbon dioxide, whereas in anaerobic environments, carbohydrates are converted to methane and carbon dioxide.[25] Cellulolytic enzyme systems of aerobic fungi and bacteria seem to differ from those of anaerobic bacteria and fungi. Aerobic cellulolytic microorganisms, which are exemplified by the fungus *Trichoderma reesei* (anamorph *Hypocrea jecorina*) and the aerobic bacteria *Cellulomonas fimi* and *Thermomonospora fusca*, produce and secrete separate enzymes that act synergistically to degrade crystalline cellulose.[26,27] In contrast, many anaerobic bacteria produce extracellular multienzyme complexes called cellulosomes, which effectively hydrolyze cellulose and hemicelluloses.[27] The discovery of the cellulosome was made with *Clostridium thermocellum*.[28,29] Cellulosomes can be considered exocellular organelles of about 3 million Da, which also exist as polycellulosomes of up to 100 million Da containing 20 or more different carbohydrate hydrolytic enzymes bound together by noncatalytic scaffolding proteins.[27,30] All polypeptides of cellulosomal complexes are modular. The enzymatically active subunits have, in addition to the catalytic sites, modules termed *dockerins*, which bind to cohesin modules of scaffolding proteins, thus forming the cellulosomal complexes.[27,31,32] Other modules found in polypeptides of cellulosomes include carbohydrate-binding modules (CBMs), immunoglobulin-like modules, and fibronectin type 3-like modules, which are also designated *X-modules*. Finally, many glycoside hydrolases (GH) contain two or several enzymatically active modules, which can be of similar activity or have different catalytic activities, making the enzymes bifunctional.[33] Several excellent reviews of cellulosomes are found in the book *Cellulosome*,[34] and in works by Bayer *et al.*,[27] and Doi *et al.*[35]

Cellulosomes from Anaerobic Fungi

Anaerobic fungi efficiently hydrolyze several types of polysaccharides and they produce both cellulosomes and free separate glycolytic enzymes. Wilson and Wood were the first to isolate a cellulosome-type complex from *Neocallimastix frontalis*.[36] It consisted of at least six different polypeptides with very high activity against cotton fiber. Cellulosome-type complexes with endoglucanase, xylanase, mannanase, and β-glucosidase activities containing at least 10 polypeptides have been found in *Piromyces*.[37–39] Fanutti *et al.*[40] concluded from the amino acid sequences of xylanase A and mannanase A (MANA) from *Piromyces*, that these enzymes contained, in addition to the catalytic modules (CMs), reiterated noncatalytic sequences. These bound specifically to two polypeptides of 97 and 116 kDa components of the glycolytic complexes from *Piromyces* and *Neocallimastix patriciarum*, respectively. These noncatalytic sequences properties indicated that the reiterated sequences function as noncatalytic docking domains (here to be called *fungal dockerin domain*, or FDD, replacing noncatalytic docking domain) similar to dockerins of enzymatic subunits found in cellulosomes from *C. thermocellum* and other anaerobic bacteria. Consequently, the 97 and 116 kDa subunits function as scaffolding polypeptides. Similar results were obtained with cellulosomes from *Orpinomyces* sp. PC-2.[41] However, in this case, the FDD of CelC from *Orpinomyces* bound to four polypeptides of 64, 66, 95, and 130 kDa, indicating the possibility of four scaffolding proteins in the cellulosome of *Orpinomyces* PC-2. So far no scaffolding polypeptide has been isolated from a cellulosome from an anaerobic fungus, and there is no detailed knowledge regarding cohesins in scaffolding peptides from anaerobic fungi. Much more is known about FDDs, or docking domains, which are required modules in enzymes found in cellulosomes from anaerobic fungi. Although FDDs function in a manner similar to that of dockerins of *C. thermocellum* and other anaerobic bacteria, I prefer to refer to them as FDDs because their amino acid sequences are very different from those of bacterial dockerins.[41,42]

Most enzymes of anaerobic fungi associated with cellulosomes contain two copies of an FDD, each consisting of about 40 highly conserved amino acids linked together by a short novel linker sequence. FDDs are cysteine-rich and, depending on the number and position of the cysteines, FDDs can be divided into three types or subfamilies:[41] Types 1 and 3 have six cysteines and Type 2 has four. All 50 known FDDs except one contain vicinal cysteines. These cysteines are important for the structure of FDDs, as demonstrated for the N-terminal FDD of Cel45A from *Piromyces equi*,

TABLE 1. GH, esterases, and cyclophilin from the anaerobic fungus *Orpinomyces* sp. strain PC-2[a]

Enzyme	Family	No. residues	Size, Da	Special domain	Accession	Reference
CelA	GH6	459	50,560	FDD[3]	U63837	55
CelB	GH5	471	53,103	FDD	U57818	60
CelC	GH6	449	49,390	FDD	U63838	55
CelD	GH6	455	50,279	FDD	AAC090661	Unpubl.
CelE	GH5	477	53,635	FDD	U97153	61
CelF	GH6	432	46,736	CBD	U97154	61
CelG[1]	GH5	193	21,150	FDD	U97155	Unpubl.
CelH	GH6	491	53,956	FDD	AAL01211.1	57
CelI	GH6	490	54,051	FDD	AAL01212.1	57
CelJ[1]	GH5	229	24,907	FDD&CBD	AF177207	Unpubl.
BglA	GH1	663	75,228	None	AF016864	72
XynA	GH11	362	39,542	FDD	U57819	60
LicA	GH16	245	27,929	None	U63813	96
ManA	GH5	579	64,425	CBD&FDD	AF177206	94
AxeA	CE6	313	34,845	None	AF001178	91
FaeA	CE1	530	59,013	None	AF164351	81
CypB	–	203	21,969	None	U17900	48, 50

[a]GH, glycoside hydrolase; CE, carbohydrate esterase; FDD, fungal dockerin domain; CBD, carbohydrate-binding domain; [1]fragment.

for which the three-dimensional structure has been determined.[42] This is a Type 2 FDD; it is held together in a cradle-like structure by disulfide bridges between the vicinal cysteines and the other cysteines located at the N and C termina, respectively. Reduction of the disulfides results in the loss of the structure. Three amino acids located in the cradle—tyrosine (Y8), tryptophan (W35), and aspartic acid (D23)—are essential for docking to the cellulosomes. These amino acids are present in all known FDDs.[41]

Cellulosomes from *Orpinomyces* sp. PC-2 have been isolated from cultures grown for 3–4 days on Avicel or coastal Bermuda grass.[43] Cellulosomes are found in the supernatant of the culture, but most of them are bound to the residual cellulosic substrate particles, from which they can be recovered by extraction with distilled water using methods previously established for the isolation of cellulosomes from cultures of the bacterium *C. thermocellum*.[44] The cellulosomes from the *Orpinomyces* have many properties similar to those of cellulosomes from *C. thermocellum*. Gel filtration of the cellulosomal distilled water fraction separates the cellulosomes into two distinct particles of 25 and 80 nm, the sizes of which were determined using electron microscopy as described by Mayer *et al.*[30] If the particles are spherical, they should have molecular masses of 3000 and 80,000 kDa, respectively. These sizes correspond to those of the cellulosome and polycellulosome of *C. thermocellum*.[30] The *Orpinomyces*' cellulosomes bind to cellulose and are located at the surface of the tip of the fungal mycelium. Thus, the cellulosomes mediate the attachment of the fungal mycelium to plant tissue. This arrangement enables the enzymes of the cellulosome to act on the hemicellulose/lignin layer of plant tissue, thus facilitating its penetration by the mycelia, as has been observed.[45] SDS-PAGE analyses of the *Orpinomyces*' cellulosome and polycellulosome showed identical protein profiles; they contain at least 20 different polypeptides with molecular masses from 30 to 190 kDa. These results correspond to those obtained with the cellulosomal complex from *C. thermocellum*. Additional details including electron microscope pictures of cellulosomes from *Orpinomyces* can be found elsewhere.[46]

Enzymes of the Anaerobic Fungus *Orpinomyces* PC-2

Anaerobic fungi effectively hydrolyze cellulose, hemicelluloses, and other plant materials, and they produce a full spectrum of enzymes needed to degrade plant tissue, which include enzymes breaking bonds between lignin and hemicellulose.[21,45–47] Most of these enzymes are associated with cellulosomes, but anaerobic fungi also produce free individual enzymes. Genes for 17 enzymes (TABLE 1) have been isolated from an *Orpinomyces* sp. PC-2 cDNA library in phage λZAPII using mRNA from cells grown on Avicel and xylan.[48] Three approaches were used to isolate the genes, including screening for enzyme-producing plaques on agar plates with substrates for the enzymes, construction of oligonucleotide probes based on peptide sequences, and specifically selecting genes encoding FDD docking sequences. Of the 17 enzymes listed in TABLE 1, 10 are cellulases belonging to GH family 6 (CelA, CelC, CelD, CelF, CelH and CelI), four are cellulases of family 5 (CelB, CelE, CelG, and CelJ), and one is a β-glucosidase (BglA) of GH family 1. Five other enzymes involved in the hydrolysis of hemicelluloses are xylanase XynA GH11, lichenase or β-glucanase LicA GH16, mannanase ManA GH5, acetyl xylan esterase AxeA, and feruloyl esterase FaeA.

An additional enzyme listed in TABLE 1, cyclophilin B (CypB), is a peptidyl-prolyl *cis-trans* isomerase—an important enzyme for protein folding and a determinant of protein structure.[49] CypB was first recognized as a protein binding tightly to β-glucosidase and was isolated directly from the mycelium of *Orpinomyces*. Its nucleotide sequence was determined using the cDNA library in phage λZAPII.[48] Surprisingly, the deduced amino acid sequence of the *Orpinomyces* CypB is almost identical to that of CypB from humans and it has higher identity with CypB from vertebrate animals (65%–70%) and plants (55%–60%) than with that of other fungi and bacteria (30%–35%). The cyclophilin gene from *Orpinomyces* has been cloned into *Escherichia coli*, where it is expressed to a level of 20% of the soluble protein.[50] The recombinant enzyme has properties identical to those of the native enzyme and very similar to those of CypB from humans and cattle. Its three-dimensional structure, as predicted using the Swiss-Model protein modeling server, is also homologous to that of human CypB. Genomic analysis of *cypB* from *Orpinomyces* showed that it is present as a single copy and that it contains two introns. Introns have been observed in genes of anaerobic fungi, but they are rare; their presence has been considered to indicate that the gene has eukaryotic origin.[12,51] Because of the symbiotic relationship between anaerobic fungi and herbivorous animals, and considering that CypBs from *Orpinomyces*, humans, and other animals are highly homologous, it seems possible that a lateral transfer of the cyclophilin gene has occurred between the anaerobic fungus and its host animal.[52]

Enzymes hydrolyzing cellulases, hemicellulases, and other plant wall polysaccharides constitute the large group of GH. These and related enzymes and associated modules have been classified by Henrissat and associates into families based on amino acid sequence similarities and hydrophobic cluster analysis.[53,54] It has long been recognized that, to completely hydrolyze cellulose to glucose, three types of GH are required: endoglucanase, which cleaves the cellulose inside the glucan chain; exoglucanase or cellobiohydrolase, which cleaves cellobiose units, the building blocks of cellulose, from the ends of the cellulose chain; and cellobiase or β-glucosidase, which cleaves cellobiose, yielding glucose. It has been demonstrated that endo- and exoglucanases work in synergy. Cellobiose is an inhibitor of the glucanases. β-glucosidase, by hydrolyzing cellobiose, removes the inhibition, and thus stimulates the hydrolysis of the cellulose. This is, however, a very simplistic view of enzymatic hydrolysis of cellulose. As mentioned above, many cellulases, in addition to their catalytic sites, have additional active sites as well as modules with different properties, all of which may affect the activity of the enzymes. In addition, the cleavage of the β-1-4 bonds of the glucan chain may involve different types of stereochemistry. The mechanism can be either inverting, leading to a product with α-configuration, or retaining, yielding a product with maintained β-configuration. Furthermore, exocellulases may act on the reducing or on the nonreducing end of the glucan chain. Finally, some cellulases degrade soluble or amorphic cellulose more efficiently than crystalline cellulose, whereas others seem to prefer the crystalline substrate. Excellent discussions of the properties of enzymes that degrade plant cell wall polysaccharides are available.[26,27]

Cellulolytic Enzymes of *Orpinomyces* PC-2

As mentioned above, 10 cellulases from *Orpinomyces* sp. PC-2 have been sequenced and characterized (TABLE 1), and there is evidence for additional cellulases in this fungus. It is rather remarkable that one microorganism has acquired so many cellulases; why does it need so many to hydrolyze the β-1-4 bonds of glucan chains? *Orpinomyces* is not alone in having acquired many cellulases. Many cellulolytic fungi and bacteria have batteries of cellulases, especially those that produce cellulosomes. An example is *C. thermocellum*, which has 15 known cellulases. Of these, 13 are in the cellulosome and two are secreted free into the medium.[27] Considering, first, the structure of the cellulosome with the scaffolding protein binding several cellulases through the dockerin and cohesin interactions and, second, that these cellulases are lined up along the cellulose chain, the cellulases may then cut the cellulose chain simultaneously at several places in a multicutting event.[30] This could explain why preparations of cellulosomes seem to have a higher rate of cellulolytic activity than cellulase preparations of free noncellulosomal cellulases.[36]

Cellulases within the same family have similar catalytic mechanisms, which include the stereochemistry of cleavage of the β-1-4 bond, but there are differences

TABLE 2. Homology between the CDs of the GH family 6 cellulases of *Orpinomyces* PC-2

Cellulase[a]	CelA	CelC	CelD	CelF	CelH	CelI
CelA	–	75.6[b]	56.4[b]	64.8[b]	40.4[b]	38.5[b]
CelC	81.0[c]	–	57.1[b]	63.6[b]	40.3[b]	35.0[b]
CelD	66.7[c]	67.7[c]	–	61.2[b]	37.7[b]	36.2[b]
CelF	71.8[c]	71.5[c]	67.7[c]	–	38.7[b]	34.5[b]
CelH	63.1[c]	62.5[c]	59.6[c]	61.0[c]	–	91.4[b]
CelI	62.2[c]	64.8[c]	62.0[c]	60.4[c]	93.1[c]	–

[a]Numbers are expressed as identity percentage; [b]amino acid residue identity; [c]nucleotide identity. Reprinted from Reference 57.

among enzymes of the same family. *Orpinomyces* has six cellulases belonging to family 6, which can be divided into two groups.[55–57] The CMs of CelA, CelC, CelD, and CelF have from 56% to 76% sequence identity but only 34% and 40% identity with CelH and CelI, respectively. The identity between CelH and CelI is over 90% (TABLE 2). A phylogenetic tree (FIG. 1) of the catalytic domains (CDs) of family 6 cellulases suggests two subfamilies.[57] CelH and CelI differ from the other cellulases in size as they contain 491 and 490 amino acid residues, respectively, whereas the other cellulases have from 449 to 459 residues. All the cellulases, except CelF, have the same modular or domain composition; this includes an N-terminal signal peptide, two copies of FDD, a linker, and the CD at the C terminus. The fact that they contain FDDs indicates that these cellulases are cellulosome-associated. CelF differs from the other cellulases by having a family 1 carbohydrate-binding domain (CBD) replacing the FDDs, which suggests that CelF is a free enzyme.[56] CelF also differs from the other cellulases in that its gene has an intron of 111 base pairs located in the CBD region. Splicing boundaries for the intron are GT and TA, which correspond to consensus sequences for introns of filamentous fungi. The finding of an intron in CelF is good evidence for a fungal origin of the gene. It has been proposed that the family 6 cellulases of *Orpinomyces* PC-2 have a eukaryotic origin similar to those of family 6 cellulase from other fungal sources.[56] Family 6 cellulases harbor both cellobiohydrolases, as exemplified by CBHII from *T. reesei*,[58] and endoglucanases, as shown for E2 from *T. fusca*.[59] The three-dimensional structures of these two enzymes have been determined, and the catalytic sites are similar in the two enzymes. In CBHII, the catalytic site is covered by amino acids forming two loops and resulting in a tunnel, which can accommodate the ends of cellulose chains to be hydrolyzed, yielding cellobiose, but prevents binding of long cellulose chains. In E2, one of the loops is missing and the second is pulled away; this allows long cellulose chains to bind and be cut in an endo-fashion way. The family 6 *Orpinomyces* cellulases seem to contain loops different from those found in CBHII and E2 because they lack, or have alternative, amino acid residues; these cellulases have both endo- and exo-cellulase activities.[55]

The genes for four family 5 cellulases (CelB, CelE, CelG, and CelJ) have been cloned and sequenced from *Orpinomyces* PC-2.[41,60,61] They are modular like the family 6 cellulases. However, their CDs are N-terminal, in contrast to the CDs of the family 6 enzymes where they are C-terminal. Thus, the N-terminal signal peptide is followed by the N-terminal family 5 CD, which is followed by a linker and two copies of FDD at the C terminus. The sequence identities between CelE and the CelB from *Orpinomyces* PC-2 is 72.3%, and with CelB from *N. patriciarum* 67.9%.[62] Three family 5 cellulases, CelA, CelB2, and CelB29, have been characterized from *Orpinomyces joynii*.[63,64] It has been noted that the gene for the CelB29 enzyme from *O. joynii* has 99% sequence identity with CelB from *Orpinomyces* PC-2; they differ in only six amino acid residues. None of the family 5 cellulases from the anaerobic fungi have an intron. They hydrolyze carboxymethylcellulose and cello-oligosaccharides in patterns consistent with endoglucanases. Their CDs show homology with those of GH from several anaerobic bacteria, and it has been suggested that the gene for the family 5 cellulases of anaerobic fungi was originally transferred from a rumen bacterium and subsequently underwent gene duplication, as has been postulated for other genes.[51,65]

As mentioned above, cellobiose is the main product of the hydrolysis of cellulose by endo- and exoglucanases and it inhibits these enzymes. Thus, to efficiently hydrolyze cellulose, a β-glucosidase is needed to hydrolyze the cellobiose to glucose. β-Glucosidases have been found in several anaerobic fungi, including *N. frontalis*,[66–68] *Piromyces* E2,[69,70] and *Orpinomyces*.[71] Most of these enzymes are found free in the cultural fluid and in the cytoplasm. They are N-glycosylated to varying degrees. The genes of BglA of *Orpinomyces* and Cel1A of *Piromyces* have been sequenced. They belong to GH family 1 β-glucosidases, have molecular masses of 75,227 and 75,800 Da, respectively, and have a sequence identity of 72%.[69,72] With the exception of a short signal peptide and the CDs, they have no obvious modules. Thus, they lack FDDs and, consequently, they are not associated with cellulosomes. They have significant homology with several bacterial β-glucosidases and are considered to have a bacterial origin. Several three-dimensional structures are known for family 1 β-glucosidases, and it has been recognized that conserved amino acid residues are involved in

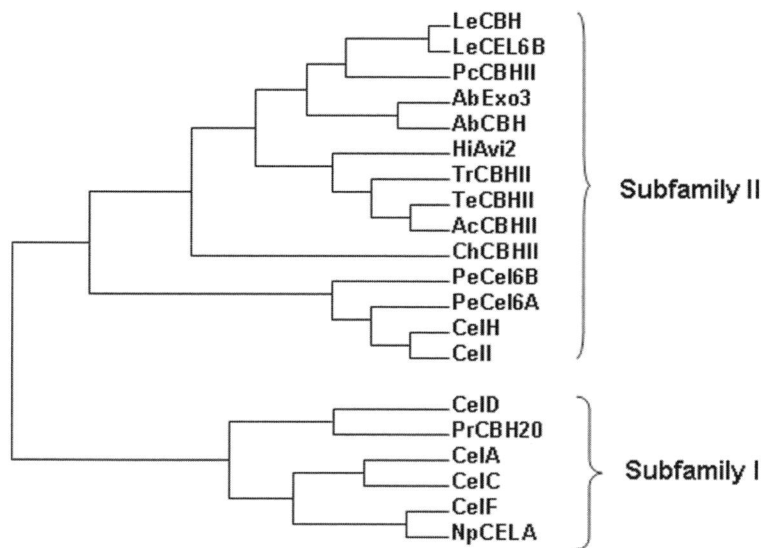

FIGURE 1. A phylogenetic tree for the CD regions of fungal cellulases belonging to GH family 6. Amino acid sequences from GenBank include *Orpinomyces* PC-2 CelA (AAC09228.1), CelC (AAB92670.1), CelD (AAC09066.1), CelF (AAC09228.1), CelH (AAL01211.1), CelI (AAL01212.1), *N. patriciarum* NpCELA (AAC94315.1), *Piromyces rhizinflata* PrCBH20 (AAD51054.2), *P. equi* PeCel6A (AAL92497.1), PeCel6B (AAP30749.1), *Cochliobolus heterostrophus* ChCBHII (AAM76664.1), *Talaromyces emersonii* TeCBHII (AAL78165.2), *Humicola insolens* HiAvi2 (BAB39154.1), *Agaricus bisporus* AbExo3 (AAA50607.1), AbCBH (AAA50608.1), *Acremonium cellulolyticus* AcCBHII (BAA74458.1), *Phanerochaete chrysosporium* PcCBHII (AAB32942.1), *Lentinula edodes* LeCBH (AAK28357.1), LeCEL6B (AAK5564.1), and *T. reesei* TrCBHII (AAA34210.1). Reprinted from Reference 57.

catalysis and in binding of substrates.[74,75] Thus, in *Orpinomyces*, BglA Glu-250 and Glu-523 seem to be involved in the catalytic reaction, and Gln-82, His-260, Tyr-433, Glu-523, and Tyr-607 recognize and bind the substrate.[72] Cel1A from *Piromyces* has transglycosylation activity[69]; the relevance of this observation is uncertain, but it is possible that the formation of some oligosaccharides can trigger induction of cellulolytic enzymes. Attempts to express the genes of BglA and Cel1A in *E. coli* were not successful, but active enzymes were obtained for Cel1A with *Pichia pastoris*, and for BglA with *Saccharomyces cerevisiae*.[70,72] The addition of the recombinant *Orpinomyces* BglA to cellulase preparations of *T. reesei* greatly enhanced the saccharification of Avicel. With the *T. reesei* enzyme, glucose, cellobiose, and cellotriose were produced, whereas in the presence of CelA, essentially only glucose was formed. The amount of glucose produced was higher than the total amount expected from the hydrolysis of the cellobiose and the cellotriose formed with the *T. reesei* enzyme alone, indicating that the inhibition of the *T. reesei* cellulase was removed and that it was more active.[72,73]

A second β-glucosidase, Cel3A, has been identified in *Piromyces* sp. Strain E2.[70] It is a rather large protein of 867 amino acid residues and a calculated molecular mass of 93,588 kDa. It differs from the above mentioned family 1 β-glucosidases in that it has, after the N-terminal signal peptide, a CD resembling fungal β-glucosidases of GH family 3. The CD is followed by an auxiliary domain, a 25-residue linker, and, at the C terminus, three copies of FDD, which indicates that Cel3A is cellulosome-associated. It is the first β-glucosidase to be identified with a fungal cellulosome and, although it is a minor component of the cellulosome from *Piromyces* E2, it is responsible for the formation of glucose from cellobiose by the cellulosomal complex during hydrolysis of cellulose.[70] A family 3 β-glucosidase has not been found in the cellulosome from *Orpinomyces* PC-2.

Hemicellulases and Associated Enzymes from *Orpinomyces* PC-2

Anaerobic fungi have the enzymes needed for the complete and efficient degradation of most types of hemicelluloses including β-glucans, mannans, and xylans, which are major constituents of plant cell walls. It was first noticed with *N. frontalis* species that xylan was rapidly degraded and that culture media had high levels of xylanase activity.[17,76,77] Five anaerobic fungal isolates from the rumen of a cow, representing *Piromyces*, *Neocallimastix*, and *Orpinomyces* species and growing on coastal Bermuda grass, had from five to seven times higher xylanase than cellulase activity in culture media. In addition, accessory enzymes were found for hemicellulose degradation, including acetyl xylan esterase, feruloyl- and *p*-coumaroyl esterases, and α-(4-O-methyl)-glucoronidase.[19,20] Two feruloyl esterases (FAE-I and FAE-II), and one *p*-coumaroyl esterase (CAE) have been purified from culture media of *Neocallimastix* strain MC-2.[78,79] These esterases function synergetically together with commercial xylanase preparations from *Trichoderma viride* (Sigma-Aldrich Corp., St. Louis, MO) and *Basidiomycetes* (Driselase; Sigma).[80] For example, 10 mg of coastal Bermuda grass (finely ground, 80 μm) incubated for 5 h at 40°C with one unit of xylanase released 38 μg of reducing sugar and 13 ng of feruloylate. Incubation under the same conditions with 2.5 mU of FAE-II generated 11 μg of reducing sugar and 463 ng of feruloylate, whereas a combination of the xylanase and the esterase produced 69 μg of reducing sugar and 1002 ng of feruloylate. Similar results were obtained with the other esterases and Driselase. A gene, *faeA* (GenBank accession no. AF164351), coding for a feruloyl esterase has been isolated and sequenced from *Orpinomyces* PC-2. Sequence analysis indicated homology between the FaeA sequence and sequences of unknown domains of xylanases XynY and XynZ of the cellulosome of *C. thermocellum*.[81] These unknown domains were subsequently shown to have feruloyl esterase activity. The three-dimensional structures of the two Fae domains of XynY and XynZ have been determined.[82,83]

Several different xylanases have been found in anaerobic fungi. *N. patriciarum* contains three—XYLA, XylB, and XynC—which have been characterized.[84–86] XYLA has a molecular mass of 66.2 kDa and contains FDD and two almost identical CDs, indicating gene duplication. Recombinant forms of the two CDs have been produced separately and both of them were enzymatically active. XynC has one CM similar to those of XYLA, a C-terminal CBM of family 1, and FDD. The CDs of XYLA and XylC belong to GH family 11, and they are in the cellulosome. XylB has a family 10 CM, a long unusual domain, which contains the amino acid sequence TLPG repeated 57 times, and a domain related to a CBM of family 1. Like *N. patriciarum*, *N. frontalis* has several xylanases belonging to either GH family 10 or family 11.[87,88] XYN3 has a molecular mass of 66 kDa and consists of 607 amino acid residues. It has two homologous CMs of GH family 11 with endoxylanase activity. It also has an FDD at the C terminus, indicating that it is associated with the cellulosome.[89] A complex-bound and a free β-1,4-endoxylanase have been obtained from the cultural fluid of *Piromyces* sp. E2.[37] These enzymes seem to be identical based on physical and kinetic data: they have the same molecular mass of 12.5 kDa; the same response to temperature and pH; and K_m and V_{max} values of 3 mg/mL and 2600 IU/mg, respectively, with oat spelt xylan as substrate. A gene has been isolated from *P. equi* that encodes a larger xylanase, XYLA, which has a molecular mass of 68 kDa and contains 625 amino acid residues.[40] Like XYN3 from *N. frontalis* and XYLA from *N. patriciarum* it has two CMs of GH family 11. It differs, however, from XYN3 and XYLA, in that the CMs are separated by an FDD and that only one, the C-terminal, yields an active recombinant form after cloning. The location of the FDD is unusual as these are generally found at the C terminus or, less commonly, at the N terminus.

Only one xylanase, XynA, has been characterized from *Orpinomyces* PS-2.[60] XynA has a mass of 39.5 kDa and contains 362 amino acid residues. It has an N-terminal signal peptide, a GH family 11 CM, a linker, and an FDD at the C terminus. XynA is definitely a xylanase associated with the cellulosome. It should be noted that the molecular mass of the *Orpinomyces* PS-2 xylanase is about half those of XYN3 and XYLA from the *Neocallimastix* sp. and XYLA of the *P. equi*. This is clearly related to the fact that the *Orpinomyces* enzyme has only one CM, whereas the xylanases from the other fungi have two. A comparison of the deduced amino sequences of the CDs of the different xylanases from the anaerobic fungi revealed that the CD of XynA from *Orpinomyces* PS-2 has about 88% identity with the CDs of XYN3 and XYLA from the *Neocallimastix* species, but only 34% with the CDs of XYLA from *P. equi*. On the other hand, the FDDs from the different xylanases are highly identical and all belong to FDD type 1, which is the type found in most hemicellulolytic enzymes.[41] A case may be made that xylanases XYN3 and XYLA of the *Neocallimastix* species and XynA from *Orpinomyces* PC-2 have a similar ancestral origin, which is different from that of XYLA from *P. equi*. However, the FDDs in all of the fungal xylanases are very similar, indicating a common origin. Clearly, anaerobic fungi have been able to selectively combine domains from different origins to compose their complete modular enzymes. In addition, the results obtained with the xylanases and cellulases demonstrate that horizontal gene transfer and gene duplication occur among the anaerobic fungi.

Xylose moieties in arabinoxylan may be acetylated at the O-2 or O-3 position from 20% to 50%; this hinders the hydrolysis of xylan by xylanases.[90] Thus, an important enzyme for the degradation of xylans is acetyl xylan esterase, which hydrolyzes the ester bonds, removing the acetyl group from arabinoxylan. Acetyl xylan esterases are present in many bacteria and fungi. A gene designated *axeA*, coding for an acetyl xylan esterase (AxeA), has been isolated from *Orpinomyces* PC-2. It contains 313 amino acid residues having a mass of 34.8 kDa,[91] and has 56% amino acid sequence identity with an acetyl xylan esterase BnaA from *N. patriciarum.*[92] No other protein with an identity over 20% was found in the database. It has been suggested that AxeA and BnaA constitute a new family of hydrolases.[91,92] *N. patriciarum* has two additional acetyl xylan esterases, BnaB and BnaC, which do not have sequence homology with BnaA. This suggests that there are at least two different acetyl xylan esterases in anaerobic fungi. The sequence similarity of AxeA and BnaA pertains only to the CD. They differ in that BnaA has a C-terminal FDD, which is attached to the CD with a linker, and is cellulosome associated.[92] In contrast, AxeA from *Orpinomyces* is devoid of an FDD and is considered a free enzyme. AxeA hydrolyzes *p*-nitrophenyl acetate (*p*NA), glucose pentaacetate, xylose tetraacetate, tri-*O*-acetyl-D-galactal, and acetylated birchwood xylan, but not FAXX, a substrate for feruloyl esterases and cellulose acetate. As discussed earlier, feruloyl and *p*-coumaroyl esterases work synergetically with xylanses to degrade coastal Bermuda grass. Similarly, recombinant BnaA and XynA from *N. patriciarum* and AxeA and XynA from *Orpinomyces*, when combined, substantially increase the rate of hydrolysis of acetylated xylan and other hemicelluloses.[91,93] In the case of the enzymes from *Orpinomyces*, the rate of hydrolysis of acetylated xylan triples.

A gene encoding a mannanase, Man A, has been isolated from *Orpinomyces* PR-2,[94] and genes for three mannanases from *P. equi*—MANA, MANB, and MANC—have been sequenced.[38,40,65] All four mannanases are modular. The three *P. equi* enzymes are very similar and they have, starting with the N terminus, a signal peptide, a CD of GH family 26, a linker, and an FDD. MANB and MANC have 571 and 569 amino acid residues, respectively. MANA is somewhat larger with 606 amino acid residues; this is because MANA has three copies of the FDD sequence instead of the two found in MANB, MANC, and in most cellulosome-associated enzymes. It has been suggested that the *P. equi* mannanases, which appear to have bacterial origin, have arisen through gene duplication.[65] ManA from *Orpinomyces* is different from the *P. equi* mannanases. In particular, ManA from *Orpinomyces* contains 579 amino acids and is of the same size as the *P. equi* enzymes. It has an N-terminal signal peptide, but this is followed by a CD belonging to GH family 5, a linker, a family 1 CBM, a second linker, and FDD.[94] The CBM and the CD both seem to have an aerobic fungal origin, whereas the FDD has been found only in anaerobic fungi. ManA is an example of a protein in which modules from different origins have been combined. Active ManA was not produced in *E. coli* containing a plasmid with the full length of ManA. However, active *ManA* was found produced both intercellularly and secreted by the yeast *S. cerevisia* harboring a plasmid containing DNA coding for the signal peptide, CD, linker, and CBM of ManA. ManA purified from the culture medium hydrolyzes locust bean and ivory nut mannans; it also hydrolyzes oat spelt xylan, but at a lower rate. No activity has been observed with barley β-glucan or carboxymethyl cellulose. With locust bean mannan, a K_m of 3.4 mg/mL and a V_{max} of 179.1 U/mg have been obtained. The gene encoding the CBM of ManA has been separately expressed in *E. coli*, and the recombinant protein has been purified. This recombinant CBM bound strongly to insoluble polysaccharides, such as acid swollen cellulose, insoluble oat spelt xylan, and chitin, but bound less strongly to bacterial crystalline cellulose and dewaxed cotton; it bound weakly to Avicel and filter paper and did not bind to soluble polysaccharides. CBMs are generally not present in enzymes associated with cellulosomes from anaerobic bacteria because the scaffolding polypeptides have CBMs.[27] It is possible that the CBM of ManA is involved in the binding of the fungal cellulosome to lignocellulose. A scaffoldin has not yet been isolated from an anaerobic fungus, thus it is not known if scaffoldins from anaerobic fungi have CBMs. A novel CBM-containing protein, NCP1, which is associated with the cellulosome, has been found in *P. equi.*[95] It has three copies of FDDs binding it to the cellulosome and two C-terminal copies of CBMs of family 29, which primarily bind to mannan, cellulose, and glucomannan. NCP1 seems to be an ideal protein for anchoring the cellulosome to its substrates.

Orpinomyces PC-2 has a gene designated *licA*, which encodes a protein with 245 amino acid residues and a mass of 27.9 kDa.[96] This protein, LicA, has 1,3-1,4-β-D-glucan 4-glucano hydrolase (β-glucanase; lichenase) activity, and contains a 29-residue signal peptide, which is followed by a CD of GH family 16. LicA does not have an FDD and is not connected with the cellulosome. It is homologous with β-glucanases from several mesophilic, thermophilic, and ruminal anaerobic bacteria. It has insignificant homology with β-glucanases from plants. Recombinant LicA has been produced in *E. coli*, and a major part of the enzyme is found in the extracellular supernatant. That β-glucanases from other sources are secreted when cloned into *E. coli* has been previously observed, and it has been shown that β-glucanases alter the outer membrane of *E.coli*, making it permeable and allowing periplasmic enzymes to enter the supernatant.[97] Recombinant LicA purified from the *E. coli* extracellular supernatant hydrolyzes lichenin and barley β-glucan at pH 6.0 and 40°C with V_{max} values of 3790 and 5320 μmol/min/mg, respectively. These rates are quite respectable and demonstrate that several GH from anaerobic fungi have higher specific activities than similar enzymes from other sources. LicA does not hydrolyze laminarin, carboxymethylcellulose, pustulan, or xylan.

Orpinomyces Enzymes—Some Applied Work

An upcoming energy crisis is predicted to result from global warming, a higher demand for energy, and an expanding population.[98] We simply must find renewable resources for energy and food production to sustain our world and life. Microorganisms may help us in this effort.[99] Many examples in this Proceedings, covering "Incredible Anaerobes: From Physiology to Genomics to Fuels," demonstrate how this might occur. Anaerobic fungi, in particular, may play a key role in such endeavors. They produce cellulosomal enzyme systems, which degrade cotton fiber at rates higher than any other cellulase or cellulase component so far.[36] They are responsible for most of the plant cell wall degradation in the rumen ecosystem, where they seem to interact with rumen bacteria.[19,45,100] As pointed out above, individual enzymes from anaerobic fungi isolated or obtained by cloning generally have higher specific activities than similar enzymes from other sources. These properties seem to be of value when considering enzymes from anaerobic fungi for industrial uses, which may involve the bioconversion of lignocellulose to sugars fermentable to ethanol for fuel or the production of other industrial feed stock chemicals.[101] Cellulases, hemicellulases, and related enzymes are already used in the food, brewery, animal feed, textile, and pulp and paper industries.[102] It can be predicted that more and more enzymes from a variety of sources will be used in industrial settings as they become available and are produced economically.

The xylanase, XynA, from *Orpinomyces* PC-2 has potential uses in the pulp and paper, animal feed, and bakery industries. In one study, the part of the XynA gene coding for the active site with the FDD removed was cloned into plasmid pET28b and transformed into *E. coli* HMS174(DE3) to obtain large quantities of XynA. The recombinant form of the CD of $XynA_{CD}$ yielded a xylanase with a specific activity of 4000 μmol/min/mg of protein at 50°C and pH 6.0 with birch wood xylan as substrate. This activity is about seven times higher than that of xylanases from other sources; however, only a low amount of the recombinant enzyme was generated (75 U/mL of growth culture). Fermentation process development was carried out using a bank of four 5 L New Brunswick Scientific (New Brunswick, NJ) Bioflo 3000 fermentors equipped with New Brunswick Scientific BioCommand control software. Initial fed-batch cultures yielded 4950 U/mL of $XynA_{CD}$. Further improvement was obtained by optimizing the culture conditions, such as stirring rate, temperature, and oxygen supply. A yield of 13,400 U/mL was reached. Finally, by optimizing the carbon source and the inducer, a production of 25,800 U/mL was achieved.[103] This is a 520-fold increase in the yield of $XynA_{CD}$ that corresponded to an enzyme yield of 6.5 g/L. A production cost at commercial scale was estimated to be $0.109 per million enzyme units. The process has been scaled up using a 500-L fermentor. Using similar procedures, high yields of *Orpinomyces* CelE and of the CD of the FaeA sequence of XynZ of the *C. thermocellum* cellulosome have also been produced.[81] Recently, the recombinant form of the CD of XynA has been produced in *H. jecorina* (anamorph *T. reesei*), demonstrating that genes encoding for GH from anaerobic fungi can be expressed in this organism.[104]

$XynA_{CD}$ has been tested as an agent for bleaching of oxygen-bleached hardwood kraft pulp.[105] Pulp at 5% consistency in 0.05 M phosphate buffer was incubated with 2 U of xylanase per gram of pulp at pH values 6, 7, and 8 at 40°C for 120 min. The best results were obtained at pH 8. The Kappa number (a measure of lignin content) was decreased by 25% and the brightness increased by 2.7% ISO. The xylanase treatment also positively affected further bleaching with ozone, which resulted in 90% ISO brightness, a number unattainable without the xylanase treatment. The xylanase treament released both lignin and reducing sugars, but did not change the strength properties of the paper.

The poultry industry is of global importance. It is well established that addition of GH to barley or wheat diets increases the growth of broiler chicks. Available enzyme preparations often contain multiple enzyme activities. In a recent trial, the addition of 1000 U/kg of pure $XynA_{CD}$ or 3000 U/kg of β-glucanase (LicA) obtained from *Orpinomyces* to a basal barley diet was found to enhance the growth of broiler chicks; further, the effects of the individual enzymes were additive.[106] The basic barley diet had 3000 kcal/kg and contained 1.3% lysine. The growth of the chicks was monitored for 10 days. Addition of β-glucanase to the basic barley diet resulted in a 25% increase in growth of the chicks compared with the basic diet without enzyme addition. Addition of xylanase alone resulted in a growth increase of 19%. With β-glucanase and xylanase combined, the increase was 41%. Clearly, the fungal enzymes have potential utility in industrial processes.

Summary

Seventeen morphologically distinct anaerobic fungi representing five genera have been found in the intestinal tracts of more than 50 different herbivorous animals. These fungi differ from aerobic fungi in that they have hydrogenosomes instead of mitochondria. They are very sensitive to oxygen and to temperatures below 30°C. Although morphologically different, they form a homogenous group, as shown by analyses of 18S ribosomal RNA, low GC content from 13% to 22%, and high similarity among genes. These fungi are considered to be the initial colonizers of lignocellulosic feed and are essential for fiber digestion. They hydrolyze plant carbohydrates, forming soluble sugars, which they ferment to formate, acetate, ethanol, lactate, carbon dioxide, and hydrogen. They produce a full battery of glycosyl hydrolases and auxiliary enzymes for the hydrolysis of hemicelluloses. Some of the enzymes are secreted free in the growth medium, but most are found in an extracellular cellulase/hemicellulase complex similar to the cellulosome of anaerobic bacteria. The enzymes of the fungal cellulosome are all modular, each having at least, in addition to the CD, an FDD, which attaches the enzyme to the cellulosome. The molecular mass of the fungal cellulosome is about 2000 kDa, and the cellulosome forms polycellulosomes of close to 100 million Da. It contains at least 20 different hydrolytic enzymes, and perhaps four scaffolding nonenzymatic polypeptides, which hold the complex together.

Comparisons of the sequences of the enzymes from anaerobic fungi with those of the equivalent enzymes from aerobic fungi and bacteria have led to the conclusion that some enzymes of anaerobic fungi have origins similar to those of bacteria, whereas other enzymes seem to have aerobic fungal and even animal origins. These observations indicate horizontal transfer of genes from different sources to anaerobic fungi. The occurrence of several similar enzymes having high homology with each other suggests gene duplication. The finding that some enzymes have two or more catalytically active domains demonstrates internal gene duplication within a gene. In addition, combining domains with specific functions to form complete and novel genes is also common in anaerobic fungi. Examples are the combinations of FDDs or carbohydrate binding domains with various CDs. These results also indicate that microorganisms living in close contact with each other, as they do in the intestinal tracts of herbivorous animals, have a high likelihood of exchanging genetic material.

Acknowledgments

Initial work involving the isolation of the anaerobic fungi and studies of feruloyl and coumaroyl esterases was in cooperation with Danny E. Akin, William S. Borneman, Roy D. Hartley, and David S. Himmelsbach at the Richard B. Russell Research Center, Agricultural Research Center, U.S. Department of Agriculture in Athens, Georgia. The work on the cellulosome, sequencing and characterizing enzymes from *Orpinomyces* in the Department of Biochemistry and Molecular Biology at The University of Georgia, was in cooperation with David L. Blum, Huizhong Chen, Timothy Davies, Carlos Felix, Yi He, Irina Kateva, Xin-Liang Li, and Eduardo A. Ximenes, each of whom performed this work expertly and contributed to pleasant work conditions in the lab. Support for a Georgia Power Distinguished Professorship in Biotechnology, and from the U.S. Department of Energy, Consortium for Plant Biomass Research, Georgia Research Alliance, and Aureozyme Inc. is gratefully acknowledged.

Conflict of Interest

The author declares no conflicts of interest.

References

1. Orpin, C.G. 1975. Studies on the rumen flagellate *Neocallimastix frontalis*. J. Gen. Microbiol. **91:** 249–262.
2. Trinci, A.P.J., D.R. Davies, K. Gull, *et al.* 1994. Anaerobic fungi in herbivorous animals. Mycol. Res. **96:** 129–152.
3. Barr, D.J.S., L.J. Yanke, H.D. Bae, *et al.* 1995. Contributions on the morphology and taxonomy of some rumen fungi from Canada. Mycotaxon LIV: 203–214.
4. Wubah, D.A. 2004. Anaerobic zoosporic fungi associated with animals. *In* Biodiversity of Fungi: Inventory and Monotoring Methods. G.M. Mueller, G.F. Bills & M.S. Foster, Eds.: 501–510. Elsevier Academic Press. Burlington, MA.

5. Li, J. & I.B. Heath. 1992. The phylogenetic relationships of the anaerobic chytridiomycetous gut fungi (*Neocallimasticaceae)* and the *Chytridiomycota.* I. Cladistic analysis of rRNA sequences. Can. J. Bot. **70:** 1738–1746.
6. Li, J., I.B. Heath & L. Packer. 1993. The phylogenetic relationships of the anaerobic chytridiomycetous gut fungi (*Neocallimasticaceae)* and the *Chytridiomycota.* II. Cladistic analysis of structural data and description of *Neocallimasticales* ord. nov. Can. J. Bot. **71:** 393–407.
7. Brownlee, A.G. 1989. Remarkable AT-rich genomic DNA from the anaerobic fungus *Neocallimastix.* Nucl. Acid Res. **17:** 1327–1335.
8. Billon-Grand, G., J.B. Fiol, A. Breton, *et al.* 1991. DNA of some anaerobic rumen fungi: G-C content determination. FEMS Microbiol. Lett. **82:** 267–270.
9. Chen, H., S.L. Hopper, X.-L. Li, *et al.* 2006. Large scale isolation of extremely AT-rich genomic DNA and analysis of genes encoding carbohydrate-degrading enzymes from the anaerobic fungus *Orpinomyces* Strain PC-2. Curr. Microbiol. **53:** 396–400.
10. Marvin-Sikkema, F.D., T.M.P. Gomes, J.-P. Grivet, *et al.* 1993. Characterization of hydrogenosomes and their role in glucose metabolism of *Neocallimastix* sp. L2. Arch. Microbiol. **160:** 388–396.
11. Marvin-Sikkema, F.D., A.J.M. Driessen, J.C. Gottschal & R.A. Prins. 1994. Metabolic energy generation in hydrogenosomes of the anaerobic fungus *Neocallimastix*: evidence for a functional relationship with mitochondria. Mycol. Res. **98:** 205–212.
12. Akhmanova, A., F.G.J. Voncken, K.M. Hosea, *et al.* 1999. A hydrogenosome with pyruvate formate-lyase: anaerobic chytrid fungi use an alternative route for pyruvate catabolism. Molec. Microbiol. **32:** 1103–1114.
13. Akhmanova, A., F.G.J. Voncken, H. Harangi, *et al.* 1999. Cytosolic enzymes with a mitochondrial ancestry from anaerobic chytrid *Piromyces* sp.E2. Molec. Microbiol. **30:** 1017–1027.
14. Hackstein, J.F.P., A. Akhmanova, B. Boxma, *et al.* 1999. Hydrogenosomes: eukaryotic adaptations to anaerobic environments. Trends Microbiol. **7:** 441–446.
15. Milne, A., M.K. Theodorou, M.G.C. Jordan, *et al.* 1989. Survival of anaerobic fungi in faeces, in saliva, and in pure culture. Exp. Mycol. **13:** 27–37.
16. Trinci, A.P.J., S.E. Lowe, A. Milne & M.K. Theodorou. 1988. Growth and survival of rumen fungi. Biosystems **21:** 357–363.
17. Orpin, C.G. 1977. Invasion of plant tissue in the rumen by the flagellate *Neocallimastix frontalis*. J. Gen. Microbiol. **98:** 423–430.
18. Bauchop, T. 1979. Rumen anaerobic fungi of cattle and sheep. Appl. Environ. Microbiol. **38:** 148–158.
19. Akin, D.E., G.L.R. Gordon & J.P. Hogan. 1983. Rumen bacterial and fungal degradation of *Digitaria pentzii* grown with or without sulfur. Appl. Environ. Microbiol. **46:** 738–748.
20. Bauchop, T. & D.O. Mountford. 1981. Cellulose fermentation by a rumen anaerobic fungus in both the absence and the presence of rumen bacteria. Appl. Environ. Microbiol. **42:** 1103–1110.
21. Borneman, W.S., D.E. Akin & L.G. Ljungdahl. 1989. Fermentation products and plant cell wall-degrading enzymes produced by monocentric and polycentric anaerobic ruminal fungi. Appl. Environ. Microbial. **55:** 1066–1073.
22. Mountford, D.O., R.A. Asher & T. Bauchop. 1982. Fermentation of cellulose to methane and carbon dioxide by a rumen anaerobic fungus in a triculture with *Methanobrievibacter* sp. Strain RA-1 and *Methanosarcina barkeri*. Appl. Environ. Microbiol. **44:** 128–134.
23. Theodorou, M.K., A.C. Longland, M.S. Dhanoa, *et al.* 1989. Growth of *Neocallimastix* sp. Strain R1: on Italian ryegrass hay: removal of neutral sugars from plant cell walls. Appl. Environ. Microbiol. **55:** 1363–1367.
24. Mountford, D.O. & C.G. Orpin, Eds. 1994. Anaerobic Fungi: Biology, Ecology & Function. Marcel Dekker, Inc. New York, NY.
25. Ljungdahl, L.G. & K.-E. Eriksson. 1985. Ecology of microbial cellulose degradation. *In* Advances in Microbial Ecology, Vol. 8. K.C. Marshall, Ed.: 237–299. Plenum Publishing Corporation. New York, New York.
26. Wilson, D.B. & D.C. Irwin. 1999. Genetics and properties of cellulases. *In* Advances in Biochemical Engineering/Biotechnology, Vol. 65. T. Scheper, Ed.: 1–21. Springer Verlag. Berlin, Germany.
27. Bayer, E.A., J.-P. Belaich, Y. Shoham & R. Lamed. 2004. The Cellulosomes: multienzyme machines for degradation of plant cell wall poly saccharides. Annu. Rev. Microbiol. **58:** 521–554.
28. Lamed, R., E. Setter & E.A. Bayer. 1983. Characterization of a cellulose-binding, cellulase-containing complex in *Clostridium thermocellum*. J. Bacteriol. **156:** 828–836.
29. Lamed, R., E. Setter, R. Kenig & E.A. Bayer. 1983. The cellulosome—a discrete cell surface organelle of *Clostridium thermocellum* which exhibits separate antigenic, cellulose-binding and various cellulolytic activities. Biotechnol. Bioeng. Symp. **13:** 163–181.
30. Mayer, F., M.P. Coughlan, Y. Mori & L.G. Ljungdahl. 1987. Macromolecular organization of the cellulolytic enzyme complex of *Clostridium thermocellum* as revealed by electron microscopy. Appl. Environ. Microbiol. **53:** 2785–2792.
31. Shimon, L.J.W., E.A. Bayer, E. Morag, *et al.* 1997. A cohesin domain from *Clostridium thermocellum*: the crystal structure provides new insights into cellulosome assembly. Structure **5:** 381–390.
32. Carvalho, A.L., F.M.V. Dias, J.A.M. Prates, *et al.* 2003. Cellulosome assembly revealed by the crystal structure of the cohesin-dockerin complex. Proc. Natl. Acad. Sci. USA **100:** 13809–13814.
33. Ljungdahl, L.G., I.A. Kataeva & V.N. Uversky. 2008. Contribution of domain interactions and calcium binding to the stability of carbohydrate active enzymes. *In* Bioenergy. J. Wall, C. Harwood & A.L. Demain, Eds. ASM Press. Washington, DC. In press.
34. Uversky, V. & I.A. Kataeva, Eds. 2006. Cellulosome. Nova Science Publishers, Inc. New York, NY.
35. Doi, R.H., A. Kosugi, K. Murashima, *et al.* 2003. Cellulosomes from mesophilic bacteria. J. Bacteriol. **185:** 5907–5914.

36. Wilson, C.A. & T.M. Wood. 1992. The anaerobic fungus *Neocallimastix frontalis:* isolation and properties of a cellulosome-type enzyme fraction with the capacity to solubilize hydrogen-bond-ordered cellulose. Appl. Microbiol. Biotechnol. **37:** 125–129.
37. Teunissen, M.J., J.M.H. Hermans, J.H.J. Huis in't Veld & G.D. Vogels. 1993. Purification and characterization of a complex-bound and free β-1,4-endoxylanase from the culture fluid of the anaerobic fungus *Piromyces* sp. Strain E2. Arch. Microbiol. **159:** 265–271.
38. Ali, B.R.S., L. Zhou, F.M. Graves, *et al.* 1995. Cellulases and hemicellulases of the anaerobic fungus *Piromyces* constitute a multiprotein cellulose-binding complex and are encoded by multigene families. FEMS Microbiol. Lett. **125:** 15–22.
39. Dijkerman, R., H.J.M. Op den Camp, C. Van Der Drift & G.D. Vogels. 1997. The role of the cellulolytic high molecular mass (HMM) complex of the anaerobic fungus *Piromyces* sp. strain E2 in the hydrolysis of microcrystalline cellulose. Arch. Microbiol. **167:** 137–142.
40. Fanutti, C., T. Ponyi, G.W. Black, *et al.* 1995. The conserved noncatalytic 40-residue sequence in cellulases and hemicellulases from anaerobic fungi functions as a protein docking domain. J. Biol. Chem. **270:** 29314–29322.
41. Steenbakkers, P.J.M., X.-L. Li, E.A. Ximenes, *et al.* 2001. Noncatalytic docking domains of cellulosomes of anaerobic bacteria. J. Bacteriol. **183:** 5325–5333.
42. Raghothama, S., R.Y. Eberhardt, P. Simpson, *et al.* 2001. Characterization of a cellulosome dockerin domain from the anaerobic fungus *Piromyces equi*. Nature Struct. Biol. **8:** 775–778.
43. Li, X.-L., H. Chen, Y. He, *et al.* 1997. High molecular weight cellulase/hemicellulase complexes of anaerobic fungi. Annu. Meeting Am. Soc. Microbiol. Abstr. O-31, p. 424.
44. Ljungdahl, L.G., M.P. Coughlan, F. Mayer, *et al.* 1988. Macrocellulase complexes and yellow affinity substance from *Clostridium thermocellum*. Methods Enzymol. **160:** 483–500.
45. Akin, D.E., C.E. Lyon, W.R. Windham & L.L. Rigsby. 1989. Physical degradation of lignified stem tissue by ruminal fungi. Appl. Environ. Microbiol. **55:** 611–616.
46. Ljungdahl, L.G., H.J.M. Op den Camp, H.J. Gilbert, *et al.* 2006 Cellulosomes of anaerobic fungi. *In* Cellulosome. V. Uversky & I.A. Kataeva, Eds.: 271–303. Nova Science Publishers, Inc. New York, NY.
47. Selinger, L.B., C.W. Forsberg & K.-J. Cheng. 1996. The rumen: a unique source of enzymes for enhancing livestock production. Anaerobe **2:** 263–284.
48. Chen, H., X.-L Li & L.G. Ljungdahl. 1995. A cylophilin from the polycentric anaerobic rumen fungus *Orpinomyces* sp. Strain PC-2 is highly homologous to vertebrate cyclophilin B. Proc. Natl. Acad. Sci. USA **92:** 2587–2591.
49. Reimer, U. & G. Fischer. 2002. Local structure changes caused by peptidyl-prolyl cis/trans isomerization in the native state of proteins. Biophys. Chem. **96:** 203–212.
50. Chen, H., X.-L. Li, H. Xu, *et al.* 2006. High level expression and characterization of the cyclophilin B gene from the anaerobic fungus *Orpinomyces* sp. Strain PC-2. Protein Peptide Lett. **13:** 727–732.
51. Steenbakkers, P.J.M., W. Ubhayasekera, H.J.A.M. Goossen, *et al.* 2002. An intron-containing glycoside hydrolase family 9 cellulase gene encodes the dominant 90 kDa component of the cellulosome of the anaerobic fungus *Piromyces* sp. Strain E2. Biochem. J. **365:** 193–204.
52. Ljungdahl, L.G., X.-L. Li & H. Chen. 1998. Evidence in anaerobic fungi of transfer of genes between them and from aerobic fungi, bacteria and animal hosts. *In* Thermophiles: The Keys to Molecular Evolution and the Origin of Life. J. Wiegel & M.W.W. Adams, Eds.: 187–197. Taylor and Francis, Inc. Philadelphia, PA.
53. Henrissat, B. 1991. A classification of glycosyl hydrolases based on amino acid similarities. Biochem. J. **280:** 309–316.
54. Coutinho, P.M. & B. Henrissat. 1999. Carbohydrate-active enzymes and associated modular organization server (CAZyModO Website). http://afmb.cnrs.fr/~pedro/DB/db.html.
55. Li, X.-L., H. Chen & L.G. Ljungdahl. 1997. Two cellulases, CelA and CelC, from the polycentric anaerobic fungus *Orpinomyces* strain PC-2 contain N-terminal docking domains for a cellulase-hemicellulase complex. Appl. Environ. Microbiol. **63:** 4721–4728.
56. Chen, H., X.-L. Li, D.L. Blum, *et al.* 2003. CelF of *Orpinomyces* PC-2 has an intron and encodes a cellulase (CelF) containing a carbohydrate-binding module. Appl. Biochem. Biotech. **105–108:** 775–785.
57. Li, X.-L., E.A. Ximenes, H. Chen, *et al.* 2004. Cloning and sequencing of two highly homologous cellulase genes, CelH and CelI from the anaerobic fungus *Orpinomyces* strain PC-2. *In* Biotechnology of Lignocellulose Degradation and Biomass Utilization. K. Ohmiya, K. Sakka, S. Karita, *et al.* Eds.: 638–641. Uni Publishers Co., LTD. Tokyo, Japan.
58. Rouvinen, J., T. Bergfors, T. Teeri, *et al.* 1990. Three-dimensional structure of cellobiohydrolase II from *Trichoderma reesei*. Science **249:** 380–386.
59. Spezio, M., D.B. Wilson & P.A. Karplus. 1993. Crystal structure of the catalytic domain of a thermophilic endocellulase. Biochemistry **32:** 9906–9916.
60. Li, X.-L., H. Chen & L.G. Ljungdahl. 1997. Monocentric and polycentric anaerobic fungi produce related cellulases and xylanases. Appl. Environ. Microbiol. **63:** 628–635.
61. Chen, H., X.-L. Li, D.L. Blum & L.G. Ljungdahl 1998. Two genes of the anaerobic fungus *Orpinomyces* sp. Strain PC-2 encoding cellulases with endoglucanase activities may have arisen by gene duplication. FEMS Microbiol. Lett. **159:** 63–68.
62. Zhou, L., G.-P. Xue, C.G. Orpin, *et al.* 1994. Intronless *celB* from the anaerobic fungus *Neocallimastix patriciarum* encodes a modular family A endoglucanase. Biochem. J. **297:** 359–364.
63. Lui, J.-H., L.B. Selinger, Y.-J. Hu, *et al.* 1997. An endoglucanase from the anaerobic fungus *Orpinomyces joyonii*: characterization of the gene and its product. Can. J. Microbiol. **43:** 477–485.
64. Qui, X., B. Selinger, L.-J. Yanke & K.-J. Cheng. 2000. Isolation and analysis of two cellulase cDNAs from *Orpinomyces joyonii*. Gene **245:** 119–126.

65. MILLWARD-SADLER, S.J., J. HALL, G.W. BLACK, *et al.* 1996. Evidence that the *Piromyces* gene family encoding endo-1,4-mannanases arose through gene duplication. FEMS Microbiol. Lett. **141:** 183–188.
66. LI, X.O & R.E. CALZA. 1991. Kinetic study of a cellobiase purified from *Neocallimastix frontalis* EB188. Biochim.Biophys. Acta **1080:** 148–154.
67. HEBRAUD, M. & M. FÈVRE. 1990. Purification and characterization of an aspecific glycoside hydrolase from the anaerobic fungus *Neocallimastix frontalis*. Appl. Environ. Microbiol. **56:** 3164–3169.
68. WILSON, C.A., S.I. MCCRAE & T.M. WOOD. 1994. Characterization of a β-D-glucosidase from the anaerobic fungus *Neocallimastix frontalis* with particular reference to attack on cello-oligosaccharides. J. Biotech. **37:** 217–227.
69. HARANGI, H.R., P.J.M. STEENBAKKERS, A. AKHMANOVA, *et al.* 2002. A highly expressed family 1 β-glucosidase with transglycosidation capacity from the anaerobic fungus *Piromyces* sp. E2. Biochim. Biophys. Acta **1574:** 293–303.
70. STEENBAKKERS, P.J.M., H.R. HARANGI, M.W. BOSSCHER, *et al.* 2003. β-Glucosidase in cellulosome of the anaerobic fungus *Piromyces* sp. Strain E2 is a family 3 glycoside hydrolase. Biochem. J. **370:** 963–970.
71. CHEN, H., X.-L. LI & L.G. LJUNGDAHL. 1994. Isolation and properties of an extracellular β-glucosidase from the rumen fungus *Orpinomyces* sp. Strain PC-2. Appl. Environ. Microbiol. **60:** 64–70.
72. LI, X.-L., L.G. LJUNGDAHL, E.A. XIMENES, *et al.* 2004. Properties of a recombinant β-glucosidase from the polycentric anaerobic fungus *Orpinomyces* PC-2 and its application for cellulose hydrolysis. Appl. Biochem. Biotech. **113:** 233–250.
73. CHEN, H., E.A. XIMENES, X.-L. LI & L.G. LJUNGDAHL. 1999. Glucose production by *Trichoderma reesei* cellulase is enhanced by a recombinant *Orpinomyces* beta-glucosidase. *In* Genetics, Biochemistry and Ecology of Cellulose Degradation. K. Ohmiya, K. Sakka, S. Karita, *et al.* Eds.: 173–181. Uni Publishers Co., LTD. Tokyo, Japan.
74. SANZ-APARICIO, J., J.A. HERMOSO, M. MARTINEZ-RIPOLI, *et al.* 1998. Crystal structure of beta-glucosidase A from *Bacillus polymyxa*: insight into the catalytic activity in family 1 glycosyl hydrolases. J. Mol. Biol. **275:** 491–502.
75. ISORNA P., J. POLAINA, L. LATORRE-GARCIA, *et al.* 2007. Crystal structures of *Paenibacillus polymyxa* β-glucosidase B complexes reveal the molecular basis of substrate specificity and give new insights into the catalytic machinery of family 1 glycosidases. J. Mol. Biol. **371:** 1204–1218.
76. ORPIN, C.G. & A.J. LESTER. 1979. Utilization of cellulose, starch, xylan, and other hemicelluloses for growth by the rumen phycomycete *Neocallimastix frontalis*. Curr. Microbiol. **3:** 121–124.
77. LOWE, S.E., M.K. THEODOROU & A.P.J. TRINCI. 1987. Growth and fermentation of an anaerobic rumen fungus on various carbon sources and effect of temperature on development. Appl. Environ. Microbiol. **53:** 1210–1215.
78. BORNEMAN, S. & D.E. AKIN. 1994. The nature of anaerobic fungi and their polysaccharide degrading enzymes. Mycoscience **35:** 199–211.
79. BORNEMAN, W.S., L.G. LJUNGDAHL, R.D. HARTLEY & D.E. AKIN. 1991. Isolation and characterization of *p*-coumaroyl esterase from the anaerobic fungus *Neocallimastix* strain MC-2. Appl. Environ. Microbiol. **57:** 2337–2344.
80. BORNEMAN, W.S., L.G. LJUNGDAHL, R.D. HARTLEY & D.E. AKIN. 1993. Feruloyl and *p*-coumaroyl esterases from the anaerobic fungus *Neocallimastix* strain MC-2: properties and functions in plant cell wall degradation. *In* Hemicellulose and Hemicellulases. M.P. Coughlan & G.P. Hazlewood, Eds.: 85–102. Portland Press. London, UK.
81. BLUM, D.L., I.A. KATAEVA, X.-L. LI & L.G. LJUNGDAHL. 2000. Feruloyl esterase activity of the *Clostridium thermocellum* cellulosome can be attributed to previously unknown domains of XynY and XynZ. J. Bacteriol. **182:** 1346–1351.
82. SCHUBOT, F.D., I.A. KATAEVA, D.L. BLUM, *et al.* 2001. Structural basis for the substrate specificity of the feruloyl esterase domain of the cellulosomal xylanase from *Clostridium thermocellum*. Biochemistry **40:** 12524–12532.
83. PRATES, J.A.M., N. TARBOUREICH, S.J. CHARNOCK, *et al.* 2001. The structure of the feruloyl esterase module of xylanase 10B from *Clostridium thermocellum* provides insights into substrate recognition. Structure **9:** 1183–1190.
84. GILBERT, H.J., G.P. HAZLEWOOD, J.I. LAURIE, *et al.* 1992. Homologues catalytic domains in a rumen fungal xylanase: evidence for gene duplication and prokaryotic origin. Mol. Microbiol. **59:** 2065–2072.
85. BLACK, G.W., G.P. HAZLEWOOD, G.-P. XUE, *et al.* 1994. Xylanase B from *Neocallimastix patriciarum* contains a non-catalytic 455-residue linker sequence comprised of 57 repeats of an octapeptide. Biochem. J. **299:** 381–387.
86. LUI, J.H., L.B. SELINGER, C.F. TSAI & K.J. CHENG. 1999. Characterization of a *Neocallimastix patriciarum* xylanase gene and its product. Can J. Microbiol. **45:** 970–974.
87. GOMEZ DE SEGURA, B.G. & M. FÈVRE. 1993. Purification and characterization of two 1,4-β-xylan endohydrolases from the rumen fungus *Neocallimastix frontalis*. Appl. Environ. Microbiol. **59:** 3654–3660.
88. GOMEZ DE SEGURA, B., R. DURAND & M. FÈVRE. 1998. Multiplicity and expression of xylanases in the rumen fungus of *Neocallimastix frontalis*. FEMS Microbiol. Lett. **164:** 47–53.
89. DURAND, R., C. RASCLE & M FÈVRE. 1996. Molecular characterization of *xyn3*, a member of the endoxylanase multigene family of the rumen anaerobic fungus *Neocallimastix frontalis*. Curr. Genetics **30:** 531–540.
90. CHESSON, A. & C.W. FORSBERG. 1988. Polysaccharide degradation by rumen microorganisms. *In* The Rumen Microbial Ecosystem. P.N. Hobson, Ed.: 251–284. Elsevier Applied Science. New York, NY.
91. BLUM D.L., X.-L. LI, H. CHEN & L.G. LJUNGDAHL. 1999. Characterization of an acetyl xylan esterase from the anaerobic fungus *Orpinomyces* sp. Strain PC-2. Appl. Environ. Microbiol. **65:** 3990–3995.
92. DALRYMPLE, B.P., D.H. CYBINSKI, I. LAYTON, *et al.* 1997. Three *Neocallimastix patriciarum* esterases associated with the degradation of complex polysaccharides are members of a new family of hydrolases. Microbiology **143:** 2605–2614.
93. CYBINSKI, D.H., L. LAYTON, J.B. LOWRY & B.P. DALRYMPLE. 1999. An acetylxylan esterase and a xylanase expressed

from genes cloned from the ruminal fungus *Neocallimastix patriciarum* act synergistically to degrade acetylated xylans. Appl. Microbiol. Biotechnol. **52:** 221–225.

94. Ximenes, E.A., H. Chen, I.A, Kataeva, *et al.* 2005. A mannanase, ManA, of the polycentric anaerobic fungus *Orpinomyces* sp. Strain PC-2 has carbohydrate binding and docking modules. Can. J. Microbiol. **51:** 559–568.
95. Freelove A.C.J., D.N. Bolam, P. White, *et al.* 2001. A novel carbohydrate-binding protein is a component of the plant cell wall-degrading complex of *Piromyces equi*. J. Biol. Chem. **276:** 43010–43017.
96. Chen, H., X.-L. Li & L.G. Ljungdahl. 1997. Sequencing of a 1,3-1,4-β-D-glucanase (lichenase) from the anaerobic fungus *Orpinomyces* strain PC-2: properties of the enzyme expressed in *Escherichia coli* and evidence that the gene has a bacterial origin. J. Bacteriol. **179:** 6028–6034.
97. Cantwell, B.A., P.M. Sharp, E. Gormley & D.L. McConnell. 1988. Molecular cloning of *Bacillus* β-glucanases. *In* Biochemistry and Genetics of Cellulose Degradation. J.-P. Aubert, P. Béguin & J. Millet, Eds.: 181–201, Academic Press. London, UK.
98. Speth J.G. 2004. Red Sky at Morning. Yale University Press. New Haven, CT.
99. Buckley M. & J. Wall. 2006. Microbial Energy Production. Colloquium Report, American Academy of Microbiology. Washington, DC.
100. Lee, S.S., J.K. Ha & K.-J. Cheng. 2000. Relative contributions of bacteria, protozoa, and fungi to in vitro degradation of orchard grass cell walls and their interactions. Appl. Environ. Microbiol. **66:** 3807–3813.
101. Himmel, M.E., M.F. Ruth & C.H. Wyman. 1999. Cellulase for commodity products from cellulosic biomass. Curr. Opinion Biotech. **10:** 358–364.
102. Bhat, M.K. 2000. Cellulases and related enzymes in biotechnology. Biotech. Adv. **18:** 355–383.
103. Davies, E.T., X.-L. Li & L.G. Ljungdahl. 2002. Optimization of recombinant protein production in high cell density fed-batch *Escherichia coli* cultures. Abstract PC43, Society for Industrial Microbiology, Annual Meeting, Philadelphia, PA.
104. Li, X.-L., C.D. Skory, E.A. Ximenes, *et al.* 2007. Expression of an AT-rich xylanase from the anaerobic fungus *Orpinomyces* sp. Strain PC-2 and secretion of the heterologous enzyme by *Hypocrea jecorina*. Appl. Microbiol. Biotechnol. **74:** 1264–1275.
105. Shah, A.K., A. Cooper, R.B. Adolphson, *et al.* 2000. Use of an extremely high specific activity xylanase in ECF and TCF pulp bleaching. Tappi J. **83:** 1–12.
106. Azain, M.J., X.-L. Li, A.K. Shah & T.E. Davies. 2002. Separation of the effects of xylanase and β-glucanase addition on performance of broiler chicks fed barley based diets. Int. Poultry Sci. Forum Abstr. 46.

Polysaccharide Degradation and Synthesis by Extremely Thermophilic Anaerobes

AMY L. VANFOSSEN, DERRICK L. LEWIS, JASON D. NICHOLS, AND ROBERT M. KELLY

Department of Chemical and Biomolecular Engineering, North Carolina State University, Raleigh, North Carolina, USA

Extremely thermophilic fermentative anaerobes (growth $T_{opt} \geq 70°C$) have the capacity to use a variety of carbohydrates as carbon and energy sources. As such, a wide variety of glycoside hydrolases and transferases have been identified in these microorganisms. The genomes of three model extreme thermophiles—an archaeon *Pyrococcus furiosus* ($T_{opt} = 98°C$), and two bacteria, *Thermotoga maritima* ($T_{opt} = 80°C$) and *Caldicellulosiruptor saccharolyticus* ($T_{opt} = 70°C$)—encode numerous carbohydrate-active enzymes, many of which have been characterized biochemically in their native or recombinant forms. In addition to their voracious appetite for polysaccharide degradation, polysaccharide production has also been noted for extremely thermophilic fermentative anaerobes; *T. maritima* generates exopolysaccharides that aid in biofilm formation, a process that appears to be driven by intraspecies and interspecies interactions.

Key words: ***Pyrococcus furiosus*; *Thermotoga maritima*; *Caldicellulosiruptor saccharolyticus*; extremely thermophilic organism; thermophile; glycoside hydrolases; exopolysaccharides**

Introduction

Thermophilic ("heat loving") microorganisms can be found in every phyla of bacteria and archaea so far described.[1] In the latter parts of the twentieth century, the established upper temperature for life was revised frequently as geothermal and hydrothermal habitats gave rise to more and more thermophilic bacteria and archaea. This led to a still-evolving nomenclature that currently classifies moderate thermophiles (50–70°C) and extreme thermophiles (>70°C) based upon optimal growth temperatures;[2] those extreme thermophiles that grow optimally at 80°C and above are often referred to as "hyperthermophiles." Extreme thermophiles inhabit a wide range of biotopes that include terrestrial hot springs and solfataric fields, shallow submarine hydrothermal systems, abyssal hot-vent environments, hot coal-refuse piles, and geothermally heated oil reservoirs.[2] The low solubility of O_2 at high temperatures and the presence of reducing gases renders many of these biotopes anoxic, appropriate only for anaerobic or microaerophilic microorganisms.

As with mesophilic organisms, extremely thermophilic anaerobes obtain their nutrient and energy requirements by a variety of mechanisms. Despite their stark, biologically restrictive habitats, many microorganisms in this group still rely upon simple and complex carbohydrates for their carbon and energy sources. A survey of genome sequences for several anaerobic extreme thermophiles (TABLE 1) reveals a surprisingly diverse set of encoded proteins and enzymes that are devoted to the uptake, processing, and synthesis of carbohydrates. As one would expect, the heterotrophic thermophiles possess an array of glycoside hydrolases (GHs) and transferases (GTs), enzymes that are central to carbohydrate mobilization and metabolism in all the three domains of life.[3] GHs specifically hydrolyze the glycosidic bond between two or more carbohydrates or between a carbohydrate and a noncarbohydrate moiety and function during the mobilization of complex carbohydrates for subsequent metabolism. GTs catalyze the transfer of a saccharide unit from a donor molecule to a specific acceptor molecule, forming a new glycosidic bond, and are required for the production of energy storage biopolymers and for the synthesis of cellular material. While GHs are generally only found in a significantly large number in heterotrophic organisms, even autotrophic extreme thermophile (e.g., *Methanococcus jannaschii*) genomes can encode a wide array of carbohydrate-active enzymes that appear to function in the cytosolic cycling of storage polymers, such as glycogen.[4]

Address for correspondence: Robert M. Kelly, Department of Chemical and Biomolecular Engineering, North Carolina State University, Raleigh, NC 27695-7905.
rmkelly@eos.ncsu.edu

Ann. N.Y. Acad. Sci. 1125: 322–337 (2008).
doi: 10.1196/annals.1419.017

TABLE 1. Anaerobic extreme thermophiles with sequenced genomes

Microorganism	Domain	Size (Mb)	Growth T (°C)	Metabolism	Reference
Caldicellulosiruptor saccharolyticus	Bacteria	2.97	70	Heterotroph	http://genome.ornl.gov/microbial/csac
Thermoanaerobacter tengcongensis	Bacteria	2.68	75	Heterotroph	114
Thermotoga maritima MSB8	Bacteria	1.86	80	Heterotroph	16
Thermotoga neapolitana	Bacteria	~1.8	80	Heterotroph	Nelson et al., unpublished data
Thermotoga petrophila	Bacteria	1.82	80	Heterotroph	http://genome.ornl.gov/microbial/tpet/
Archaeoglobus fulgidus	Archaea	2.17	83	Heterotroph / Autotroph	115
Methanococcus jannaschii	Archaea	1.73	85	Autotroph	4
Pyrococcus horikoshii	Archaea	1.73	95	Heterotroph	116
Hyperthermus butylicus	Archaea	1.67	95–106	Heterotroph	117
Methanopyrus kandleri	Archaea	1.69	98	Autotroph	118
Staphylothermus marinus	Archaea	1.57	98	Heterotroph	http://genome.ornl.gov/microbial/smar/
Pyrococcus abyssi	Archaea	1.76	100	Heterotroph	http://www.genoscope.cns.fr/Pab/
Pyrococcus furiosus	Archaea	1.90	100	Heterotroph	17
Pyrobaculum islandicum	Archaea	1.83	103	Heterotroph	http://genome.ornl.gov/microbial/pisl

As genome sequence data for anaerobic extreme thermophiles have become increasingly available, the capacity to gain broader (i.e., whole-organism or community) insights into the biology and ecology of extreme habitats has grown immensely. The use of functional genomics-based approaches has provided unprecedented insights into the various mechanisms employed by these microorganisms' carbohydrate assimilation and metabolism,[5] and has helped to identify the specific genes and operons involved.[6–8] For example, genomewide transcriptional response experiments for the hyperthermophiles *Pyrococcus furiosus* and *Thermotoga maritima* have revealed that these organisms respond in complex ways to the availability of different carbohydrates.[9–12] The results from these transcriptional analyses allowed the assignment of substrates to transporters and enzymes, revealed the association of unannotated or poorly annotated genes to operons and operons to pathways, and provided whole-organism views of the complex (and at times temporal) interplay of various proteins/enzymes/pathways, many previously unannotated and/or seemingly unaffiliated with these specific metabolic strategies, used by the organism for substrate acquisition and metabolism.

This chapter will provide the reader with an overview of carbohydrate mobilization and utilization for three model anaerobic extreme thermophiles: the archaeon *P. furiosus*,[13] the bacterium *T. maritima*,[14] and *Caldicellulosiruptor saccharolyticus*.[15] While there are many thermophiles that have been studied to one extent or another, these three organisms have an unusually high capacity to degrade and ferment a myriad of simple and complex carbohydrates. Note that, while the genome sequences of *T. maritima* and *P. furiosus* have been available since the beginning of this decade,[16,17] the sequence for *C. saccharolyticus* was recently completed (http://genome.ornl.gov/microbial/csac). As sequence data are generated faster than the definitive annotation of putative open reading frames (ORFs) can be accomplished (through biochemical characterization and bioinformatic methods) hence, many ORFs in sequenced genomes, even those that have been available for quite some time, as yet have either no defined function or have a biochemically characterized function with no clear associated physiological function. Thus, the story of how anaerobic extreme thermophiles break down and create glycosidic linkages to meet their nutritional, bioenergetic and ecological needs is not yet complete.

Pyrococcus furiosus

P. furiosus is a hyperthermophilic archaeon of the order Thermococcales,[18] originally isolated from a geothermally heated shallow sea vent near Vulcano Island, Italy.[13] *P. furiosus* is a motile, flagellated coccus, 0.8–2.5 μm in diameter, growing between 70 and 105°C, with an optimum temperature of 98–100°C.[13] Each cell has approximately 50 monopolar polytrichous flagella, each measuring about 7 nm in width and 7 μm in length.[13] The *P. furiosus* genome consists of 2065 protein-encoding ORFs in 1.9 Mb, approximately half of which were assigned a functional role in the initial annotation.[17]

In addition to peptides, *P. furiosus* can ferment a variety of glucosides with various linkages, including starch, pullulan, glycogen, maltose, cellobiose, laminaran, barley glucan, and chitin.[13,19–21] Little or no growth is observed with the monosaccharides glucose,

TABLE 2. α-Specific glycoside hydrolases of *Pyrococcus furiosus*

Enzyme	GH family	MW (kDa)	3D Structure	SwissProt	Locus	Reference
α-Amylase	57	76.3	NA	P49067	PF0272	29
α-Amylase	13	52.9	NA	O08452	PF0477	27
Amylopullulanase	57	124.4	NA	Q8TZQ1	PF1935*[a]	12, 119
Neopullulanase (α-amylase II)	13	76.1	NA	Q8TZP8	PF1939	35
α-Galactosidase	57	41.5	NA	Q9HHB5	PF0444	120
Isomaltase (α-glucosidase)	New	54.8	NA	Q8U4F6	PF0132	26, 31, 32

[a]Sequencing error noted by Lee et al.[12] revealed that PF1934 and PF1935 were one ORF, now termed PF1935*.

ribose, or xylose, possibly due to their thermal lability at the growth temperature of this archaeon.[19] Following polysaccharide or oligosaccharide hydrolysis, the resulting glucose is metabolized via a modified, adenosine diphosphate (ADP)–dependent, version of the Embden–Meyerhof–Parnas pathway (EMP), generating acetate, alanine, CO_2, and H_2 as the major metabolic by-products.[22] As with many H_2 producers, H_2 inhibits *P. furiosus* growth to some extent,[23] but such inhibition can be minimized by employing cultivation conditions that maintain a low partial pressure of H_2[21] or by coculturing *P. furiosus* with a hydrogenotrophic hyperthermophilic methanogen.[24] The presence of elemental sulfur also removes the H_2 inhibition effect as it allows *P. furiosus* to divert e^- into H_2S which is not inhibitory to *P. furiosus*.[23]

Pyrococcus furiosus *α-Specific Glycoside Hydrolases*

Starch, comprising of amylose and amylopectin, requires the action of several enzymes for complete hydrolysis; α- and β-amylases, glucoamylase, and debranching enzymes (pullulanases) reduce it to oligosaccharides that are subsequently cleaved to release monosaccharides by α-glucosidases. *P. furiosus* grows quite well on α-1,4- and α-1,6-linked glucosides, including maltose and starch. In fact, the first hyperthermophilic GHs identified were α-specific GHs from *P. furiosus*.[25–29] Biochemically characterized α-specific GHs from *P. furiosus* are listed in TABLE 2. The majority of the amylolytic enzymes listed have been cloned and produced recombinantly in heterologous hosts. The amylopullulanase PF1935* hydrolyzes α-1,4 linkages in starch as well as α-1,6 linkages in pullulan at optimal temperatures above 120°C, and is dramatically upregulated when starch or maltose is the carbon source.[12,25] Pullulanases are used in industrial starch processes for the production of maltose syrups and high-purity glucose and fructose.[30] The transcriptional data suggest that PF1935* is a major component of the multienzyme system for starch hydrolysis. Other *P. furiosus* enzymes believed to be involved are two α-amylases, one intracellular (PF0272), and one extracellular (PF0477); both have optimal activities for starch hydrolysis at ~100°C.[27,29] However, of the two, only PF0272 is transcriptionally upregulated to a significant extent during growth on starch and maltose.[27] Lee *et al.*[12] showed that this ORF is transcribed at high levels on maltose and proposed that it acts as an intracellular 4-α-glucanotransferase, generating maltodextrins and glucose. The large number of intracellular oligosaccharide-active enzymes encoded in the *P. furiosus* genome demonstrates the level of polysaccharide hydrolysis achieved prior to intracellular transport. Biochemical characterization of the intracellular isomaltase/α-glucosidase PF0132 showed that this enzyme hydrolyzes maltose and isomaltose with similar catalytic efficiencies[26,31] at 105–115°C.[32] This ORF is transcribed at high levels on a number of α-glucans, but is highest on pullulan, suggesting that its key role is in the processing of this α-glucan.[12] The capacity of the enzymes in TABLE 2 to hydrolyze starch and related α-glucosides indicate that these have significant biotechnological potential.[33,34]

The often imprecise nature of gene annotation makes it difficult, when insufficient bench-top validation of the gene in question, or of any high similarity orthologs, is available, to provide a detailed and accurate annotation. Genomic analysis showed that PF1939 has homology to various cyclodextrin-hydrolyzing enzymes.[35] Biochemical characterization of this putative amylolytic enzyme revealed it to be, in fact, a novel neopullulanase, able to hydrolyze α-amylase-resistant pullulan and β-cyclodextrin.[35] Despite the homology, it is distinct from other cyclodextrin-hydrolyzing enzymes, producing maltooligosaccharides instead of maltose and lacking any transglycosylation activity.[35] The cellular function of PF1939 is unknown, however; transcriptional response experiments by Lee *et al.*[12] found no induction by a variety of α-glucans, raising questions about its direct involvement in starch/pullulan degradation.

TABLE 3. β-Specific glycoside hydrolases of *Pyrococcus furiosus*

Enzyme	GH Family	MW (kDa)	3D Structure	SwissProt	Locus	Reference
β-Glycosidase$_4$ (CelB)	1	54.7	NA	Q51723	PF0073	121
β-1,3-Glucanase (LamA)	16	33.8	NA	O73951	PF0076	122
endo-1,4-Glucanase A (EglA)	12	36.0	NA	Q9V2T0	PF0854	36
β-Mannosidase$_4$ (BglB)	1	56.3	NA	Q8U3U9	PF0356	41
β-Glucosidase (BglA)	1	49.8	NA	Q9HHB3	PF0442	40
β-Mannosidase$_4$ (BmaA)	1	59.1	NA	Q51733	PF1208	123
Chitinase (ChiB)	18	78.6	2DSK	Q8U1H5	PF1233	20
Chitinase (ChiA)	18	39.7	NA	Q8U1H4	PF1234	20

Pyrococcus furiosus *β-specific Glycoside Hydrolases*

Growth of *P. furiosus* on various β-glucans revealed a preference for β-1,3- and β-1,3/β-1,4 mixed-linkage saccharides.[19] The breakdown of β-glucans involves synergistic action of (at least) three enzymes: endo-1,4-glucanase (EglA),[36] laminarinase (LamA), and β-glucosidase (CelB).[37,38] These enzymes demonstrated much higher tolerance for glucose than their mesophilic counterparts, with little or no inhibition of activity.[39] In addition to CelB, other β-specific exo-acting enzymes that have been identified include a β-glucosidase (BglA, PF0442), with optimal activity against *p*NP-β-D-glucopyranoside at 100°C,[40] and a β-mannosidase (BglB, PF0356) which is 88% identical at the amino acid level with PH0501, a biochemically characterized β-mannosidase from *Pyrococcus horikoshii*[41] (TABLE 3).

Chitin can also serve as a growth substrate for *P. furiosus*. Chitin is processed by two different chitinases (ChiA and ChiB) and β-*N*-acetylglucosaminidases.[20] ChiA is an endochitinase that is only active on chitooligomers, while ChiB is a chitobiosidase that releases chitobiose from chitooligomers.[20] Synergism was noted for these two enzymes on colloidal chitin.[20] The only other hyperthermophilic archaeon known to utilize chitin is *Thermococcus kodakaraensis*, which produces several chitin-degrading enzymes related to those in *P. furiosus*.[42]

Carbohydrate Uptake by Pyrococcus furiosus

In bacteria, cellobiose enters the cell either by a phosphoenolpyruvate-dependent phosphotransferase system (PTS) or by a binding protein-dependent adenosine triphosphate (ATP)–binding cassette (ABC) transporter. Analysis of the completed genome sequences of a variety of members of Archaea demonstrates that PTSs are absent from these organisms.[43] Growth studies have shown *P. furiosus* reaches higher cell densities when grown on complex oligosaccharides, suggesting that its ABC transporters, many of which had/have putative annotations as oligopeptide transporters, have a preference for oligosaccharides containing two to six sugar units.[44]

Cellobiose/β-glucoside, trehalose/maltose, and maltodextrin uptake by *P. furiosus* ABC transporters has been examined with respect to substrate identification, binding specificity/affinity and genomic locus.[12,43,45] The cellobiose transport system is homologous to the Opp (for Oligopeptide) family transporters, and has affinity not just for cellobiose but also cellotriose, cellotetraose, cellopentaose, laminaribiose, laminaritriose, and sophorose.[43] Lee *et al.*[12] found that the CUT1 (for Carbohydrate Uptake) family Mal-I transporter (PF1739–1741 and PF1744), which recognizes and transports maltose and trehalose, but not maltooligosaccharides, was upregulated with growth upon maltose, starch, laminarin, cellobiose, and chitin. In contrast, the Mal-II transporter (PF1933 and PF1936–1938) is maltooligosaccharide-specific and is upregulated only on maltose and starch. Both of these transporters are regulated by the repressor, TrmB. Repression of Mal-I is relieved in the presence of maltose or trehalose, while the repression of Mal-II is relieved by sucrose and maltodextrins.[46] Genome analysis revealed 16 other putative ABC transporters, three of which were annotated as carbohydrate-specific.[47] Transcriptional analysis of maltose and starch-grown cells also revealed an upregulation of two transmembrane permeases, PF1748 and PF1749, proximal to the Mal-I transporter operon.[12,47] While originally annotated as an iron/thiamin ABC transporter-family set of permeases, transcriptional data suggest that they may also (or additionally) function in starch and maltose assimilation.[12]

Other Pyrococcus *Species*

A comparison of the *P. furiosus* genome with the genomes of two closely related species: *P. abyssi* (http://www.genoscope.cns.fr/Pab) and *P. horikoshii* (http://www.bio.nite.go.jp/ngac/e/ot3-e.html) with 1765 and 2061 predicted ORFs, respectively, revealed

198 ORFs unique to *P. furiosus* and 240 ORFs unique to *P. abyssi*, while *P. horikoshii* has 352 ORFs not found in *P. abyssi* or *P. furiosus*.[48,49] As with *P. furiosus*, examination of the genomes of these two organisms reveals numerous GHs that appear to function in the degradation of α- and β-linked glucosides, though this assessment is based primarily upon bioinformatic annotation, as only a few of these enzymes have as yet been biochemically characterized. One GH that has been studied *in vitro* is a *P. horikoshii* β-1,4-endoglucanase (PH1171).[50] Interestingly, this enzyme reportedly hydrolyzes crystalline cellulose, in contrast to the homologous enzyme from *P. furiosus* (PF0854).[50,51]

Thermotoga maritima

Representatives of the bacterial order Thermotogales have been found in a variety of hydrothermal environments worldwide.[10] *Thermotoga maritima* MSB8, the most studied member of this order, was isolated from geothermally heated sea floors near Vulcano Island, Italy. *T. maritima* grows between 55 and 90°C, with an optimum at 80°C; befitting its marine habitat, salt concentrations of 2.7% (w/v) were found to support optimal growth.[14] The rod-shaped (2 μm by 0.6 μm) gram-negative bacterium is surrounded by a sheath-like structure, resembling a toga that balloons at the ends. *T. maritima*'s genome of 1.86 Mb encodes 1877 ORFs (covering 95%) with a GC content of 46%; 54% of the coding regions were assigned a function in the initial annotation.[16] A re-evaluation of the genome annotation by Kyrpides *et al.* increased the number of functional assignments for ORFs in the genome from 1014 to 1178 (63%).[52] *T. maritima* has a high percentage of archaeal-like genes (~24%), suggesting a significant amount of lateral gene transfer in its evolutionary history.[16]

T. maritima is a strictly anaerobic fermentative heterotroph that grows on a variety of simple and complex sugars, including ribose, xylose, glucose, galactose, sucrose, maltose, lactose, starch, and glycogen.[21] Unlike *P. furiosus*, *T. maritima* is able to ferment non-glucose-based sugars; however, the bacterium is unable to grow in the absence of sugars with peptides as the sole carbon and energy source. *T. maritima* has a preference for complex carbohydrates, as growth in the presence of monosaccharides is slower than growth in the presence of oligo/polysaccharides.[9] However, as with *P. furiosus*, the increased thermal lability of the monosaccharide makes it impossible to describe this as a wholly biological preference. *T. maritima* ferments carbohydrates primarily to acetate, lactate, CO_2, and H_2 through conventional EMP and Entner-Doudoroff (ED) glycolytic pathways with a flux ratio of ~17:3 between the two.[53] While primarily a fermenter, *T. maritima* is able to grow in a respiratory mode using Fe(III) as an electron acceptor.[54] When growing in a fermentative mode, *T. maritima* is even more susceptible to H_2-induced growth inhibition than *P. furiosus*. As with *P. furiosus*, the addition of sulfur eliminates this effect; however, H_2S production does not appear to be enzymatically facilitated energy dissimilation or conservation.[14,55] *T. maritima* also does not produce ATP from S^0 or thiosulphate reduction.[14,16]

Polysaccharide Degradation by Thermotoga maritima

T. maritima is adapted for growth on sugars; ~7% of the predicted ORFs are involved in the breakdown and metabolism of simple and complex carbohydrates.[16] The genome encodes for the largest number of GHs of any sequenced thermophile, consistent with its capacity to grow on a wide range of polysaccharides and simple sugars.[9] The immense library of thermostable and thermoactive enzymes produced by this organism has generated great interest from the biotechnology sector; these enzymes are excellent catalysts for the hydrolysis (under industrially relevant conditions) of both simple and complex plant polymers that have potential as renewable energy sources.[10]

Thermotoga maritima *α-specific Glycoside Hydrolases*

Several α-specific GHs from *T. maritima* have been biochemically characterized and whole-genome transcriptional analysis has provided insights into the roles of specific GHs (TABLE 4). When *T. maritima* was grown on starch, genes coding for a membrane-bound, endo-acting α-amylase, TM1840, and an extracellular debranching pullulanase, TM1845, were stimulated, signifying their cellular role in processing this polysaccharide.[9] TM1840 hydrolyzes amylose/starch but is less active on highly-branched polysaccharides, such as glycogen and pullulan.[56] TM1845 is a type I pullulanase, hydrolyzing pullulan efficiently but showing no significant activity on amylose.[57]

Full hydrolysis of starch is completed intracellularly; malto-oligosaccharides are imported and hydrolyzed by the α-amylases TM1438 and TM1650.[58,59] Also involved is TM1834 (AglA), an intracellular α-glucosidase, that liberates glucose from maltose and maltotriose.[60] This enzymes requires Mn^{2+}, oxidized nicotinamide adenine dinucleotide (NAD^+), and dithiothreitol (DTT) for full activity, similar to other members of family 4 GHs.[60] Ca^{2+} dependence for

TABLE 4. α-Specific glycoside hydrolases of *Thermotoga maritima*

Enzyme	GH family	MW (kDa)	3D structure	SwissProt	Locus	Reference
Pectinase (PelA)	Lyase	40.6	NA	Q9WYR4	TM0433	61
Pectinase (PelB)	28	50.5	NA	Q9WYR8	TM0437	62
α-Glucosidase	4	55.0	1OBB	O33830	TM1834	124
4-α-Glucanotransferase	13	51.9	1LWH	Q60035	TM0364	125
α-Amylase	13	50.2	NA	Q9X1Y3	TM1650	59
Cyclomaltodextrinase	13	55.2	NA	Q9X2F4	TM1835	126
Maltodextrin glycosyltransferase	13	73.9	NA	Q9S5X2	TM0767	127
α-Amylase	13	64.7	NA	P96107	TM1840	56
α-Amylase	57	62.6	2B5D	NA	TM1438	58
Pullulanase	13	96.3	NA	O33840	TM1845	60

α-amylases is common (PF0477, PF1935*, TM1840), but a Mn^{2+} requirement is unusual for GHs.[60] Surprisingly, transcriptional response experiments have shown that TM1650 (α-amylase) and TM1834 (α-glucosidase) are constitutively produced and unaffected by the presence of starch, suggesting that their primary role may not be in starch degradation.[9]

Pectin, a heteropolymer primarily comprised of α-1,4-linked galacturonic acid, degradation is important in fruit-juice production and in textile and paper treatment. *T. maritima* is one of the few extreme thermophiles that can use pectin as a carbon and energy source.[61] TM0433 (PelA), an extracellular pectate lyase that cleaves the backbone by nonhydrolytic β-elimination, is implicated in the growth of *T. maritima* on pectin.[61] The enzyme is exo-acting, cleaving one to three sugars off the reducing end of the polymer.[61] Another enzyme, the exopolygalacturonase, PelB (TM0437), can hydrolyze the α-1,4-linkages between two α-galacturonic acid residues and can remove galacturonate units from the nonreducing ends of the molecule.[62] Both pectin-degrading enzymes are highly thermostable with optimal activity at 90°C.[61,62]

Thermotoga maritima β-specific Glycoside Hydrolases

T. maritima can also process a variety of β-linked biopolymers, although there is no evidence of a complex, multisubunit cellulosic degradation system akin to the cellulosome (TABLE 5). An intracellular endo-β-1,4-glucanase (TM1524) and an extracellular exo-β-1,4-glucanase (TM1525) have been isolated from *T. maritima* grown on cellobiose[63] and shown to be active on a variety of β-1,4-linked glucans. Transcriptional analysis revealed that both genes were upregulated upon growth of *T. maritima* on carboxymethyl cellulose (CMC) as well as β-1,3/1,4 barley glucan.[64,65] An extracellular endo-β-1,4-glucanase (TM0305) was also induced by CMC, along with a variety of other polysaccharides (albeit to a lesser extent). TM0305 hydrolyzes β-glucans into oligosaccharides, which can then be transported into the cell for subsequent degradation by TM1524.[65] When the substrate was switched from β-1,4 to β-1,3 glucans, transcription of TM0024, an extracellular laminarinase,[63] and TM0025, an intracellular exo-glycosidase,[64] was noted. These two enzymes can completely hydrolyze laminarin to glucose *in vitro* and, hence, are believed to enable *T. maritima* to grow on β-1,3-glucans.[10]

T. maritima can also utilize mannan and xylan hemicelluloses. Growth on mannose and mannan polysaccharides stimulates the transcription of several GH genes: TM1227, TM1751, and TM1752.[64] Recombinant TM1227, an extracellular mannanase, and TM1751, an intracellular β-specific endoglucanase,[64] were examined biochemically in order to properly assign a cellular function. TM1227 hydrolyzed mannan-based substrates, but not glucans,[65] while TM1751, homologous to some cellulosome-associated endoglucanases, was active on both mannan-based and glucan-based polysaccharides, signifying a major role in the breakdown of glucomannans.[64]

Thermostable xylanases are of great importance for the pulp and paper industries and increasingly in the biofuels sector, where high temperatures (55–70°C) are encountered in bioprocessing operations, such that xylanases from *T. maritima* are highly attractive candidates for the breakdown of hemicellulose feedstocks.[66–69] Transcriptional response experiments with *T. maritima* grown on birchwood xylan revealed that two extracellular endo-acting xylanases (TM0061 and TM0070) and an intracellular exo-acting β-xylosidase (TM0076) were transcribed at high levels.[9] A debranching α-glucuronidase (TM0055), an acetyl xylan esterase (TM0077), and an α-glucosidase (TM0752) were all upregulated on xylan, though not to the same level as the other enzymes.[9] Another α-glucuronidase, TM0434, is also

TABLE 5. β-Specific glycoside hydrolases of *Thermotoga maritima*

Enzyme	GH family	MW (kDa)	3D structure	SwissProt	Locus	Reference
endo-1,4-Glucanase	Not known	79.5	NA	Q9WYE1	TM0305	64
β-Glucosidase (laminaribiase)	3	81.1	NA	Q9WXN2	TM0025	76
endo-Glucanase	5	37.4	NA	Q9X273	TM1751	64
endo-Glucanase	5	39.3	1VJZ	Q9X274	TM1752	64
endo-1,4-Glucanase A	12	29.7	NA	Q60032	TM1524	128
exo-1,4-Glucanase B	12	31.7	NA	Q60033	TM1525	128
endo-1,3-β-Glucanase (laminarinase)	16	72.5	NA	Q9WXN1	TM0024	63
α-L-Fucosidase	29	52.2	1HL8	Q9WYE2	TM0306	129
endo-β-1,4-D-Galactanase	53	68.6	NA	Q9X0S8	TM1201	130
β-Glucosidase (bglT)	4	47.6	1UP4	Q9X108	TM1281	131
β-Galactosidase	2	127.6	NA	Q56307	TM1193	132
β-Galactosidase	42	51.9	NA	Q56306	TM1195	132
α-Galactosidase	36	63.7	1ZY9	O33835	TM1192	133
β-Mannosidase	2	92.4	NA	Q9X1V9	TM1624	134
endo-1,4-β-Mannosidase	5	76.9	NA	Q9X0V4	TM1227	135
cytosolic α-Mannosidase	38	117.9	NA	Q9X2G6	TM1851	136
β-Xylosidase	3	86.8	NA	Q9WXT1	TM0076	137
α-glucuronidase	4	54.6	NA	Q9WYR5	TM0434	70
α-Glucuronidase	4	55.4	1VJT	Q9WZL1	TM0752	68
α-Glucuronidase	67	78.6	NA	P96105	TM0055	67
endo-1,4-β-Xylanase B	10	40.7	1VBR	Q9WXS5	TM0070	69
endo-1,4-β-Xylanase A	10	119.6	NA	Q60037	TM0061	138
β-Glucuronidase	2	65.7	NA	Q9X0F2	TM1062	139

upregulated in xylan-grown cells, but is oddly much less thermostable than most other *T. maritima* enzymes, with an optimal temperature of only 60°C. TM0434 also has an unusual cofactor dependence—Mn^{2+} and 2-mercaptoethanol.[70] DTT can substitute as the reducing agent, but Mn^{2+} is strictly required, similar to the α-glucosidase TM1834.

Carbohydrate Uptake by Thermotoga maritima

The general and specific components of a PTS are missing from the *T. maritima* genome.[16] As such, the uptake of most carbohydrates likely occurs via binding protein-dependent ABC transporters; in fact, 84% of the identifiable transport proteins in the *T. maritima* genome are of the ABC-type.[16] Many of these transporters were initially annotated as putative oligopeptide transporters; however, biochemical studies coupled with a thorough transcriptional and subsequent *in vitro* analysis resulted in the re-annotation of many as sugar-specific transporters.[9–11,71] Perhaps surprisingly, the binding proteins for extremophilic ABC transporters appear to retain the canonical "Venus flytrap" structure typical of their mesophilic counterparts.[47,72] Substrate specificities were assigned for many of the *T. maritima* transporters based upon transcriptional analysis; however, these assignments have also mostly been verified and detailed by *in vitro* binding studies.[10,11,71] A maltose-binding protein [myelin basic protein (MBP), TM1839], for instance, was found to bind maltose, maltotriose, and trehalose, with binding efficiency increasing with temperature.[71,73]

Other Thermotogales

T. maritima is not the only member of the Thermotogales with a variety of carbohydrate-degrading enzymes. *Thermotoga neapolitana* ($T_{opt} = 80°C$), isolated from a hydrothermal vent at Lucrino near Naples, Italy,[74] is also able to grow on a wide variety of α- and β-linked glycans through the action of thermostable and thermoactive GHs. *T. neapolitana* enzymes, with orthologs in *T. maritima*, that have been studied *in vitro* include an α-galactosidase,[75] a laminarinase,[76] and two cellulases (endo-1,4-β-glucanases), all of which are optimally active above 100°C.[75–77] *T. neapolitana* produces several unique GHs as well. One example is a 1,4-β-D-glucan glucohydrolase (GghA), which hydrolyzes cellotetraose, cellotriose, cellobiose, and lactose.[10,78] A draft sequence of the *T. neapolitana* genome is available in GenBank (GenBank accession NC-06811).

The beginning of this decade saw the isolation and description of still other Thermotogales, including *Thermotoga petrophila* and *Thermotoga naphthophila*.[79] The genome of *T. petrophila*, isolated from the Kubiki oil reservoir in Japan, has recently been sequenced (http://genome.ornl.gov/microbial/tpet/). It differs slightly from *T. maritima* and *T. neapolitana* with respect to utilization of certain sugars, growing weakly on cellulose and not at all on xylose as sole carbon sources.[79] Given that the organism came from an oil well, attempts were made to grow it on hydrocarbons, including kerosene, light-oil, A-heavy oil and crude oil, but without success.[79] It has also become evident that there is immense genetic diversity even between different strains of *T. maritima*. A comparison of strain MSB8, corresponding to the published genome sequence and the standard laboratory *T. maritima* strain, and strain RQ2 by suppressive subtractive hybridization revealed over 300 genes in strain RQ2 that are distinct from strain MSB8.[80] Many of these genes are associated with sugar transport and polysaccharide degradation, suggesting that strain RQ2 is even more dependent on the degradation of polysaccharides than MSB8.

Caldicellulosiruptor saccharolyticus

Previously, the most extensively studied anaerobic cellulolytic thermophiles were the rod-shaped, spore-forming members of the bacterial genus *Clostridium,* especially *Clostridium thermocellum*. This organism produces a multisubunit complex, the cellulosome, to degrade cellulose and hemicellulose, and ferments the resulting sugars to ethanol and short-chain aliphatic acids.[81] Other anaerobic thermophilic bacteria have been reported recently, including cellulolytic species from New Zealand thermal springs.[81] While none of these isolates exceed the cellulolytic capacity of *C. thermocellum*, some are able to produce enzymes with optimal activities 10 to 15°C above those from *Clostridia* species.[82] The strain that produces the most active and stable cellulolytic enzymes, TP8T 6331, was initially named *Caldocellum saccharolyticum*, but was later renamed *Caldicellulosiruptor saccharolyticus*.[15]

The original isolate of *C. saccharolyticus* was taken from a piece of wood downstream from a 78°C pool in the Tokaanu thermal area of New Zealand.[81] *C. saccharolyticus* is a nonmotile, nonflagellated, oval-ended straight rod,[82] 0.4–0.6 μm by 3.0–4.0 μm, occurring singly and in pairs.[15] It is a strict, fermentative anaerobe, growing between 45 and 80°C ($T_{opt} = 70$°C), and monosaccharides, disaccharides, and polysaccharides can all serve as carbon sources for this organism.[15,82] Previous studies with *C. saccharolyticus* have shown the strain to be phenotypically and genetically distinct from cellulose-degrading, sporogenous *Clostridium* species.[15] The major phenotypic differences between *C. saccharolyticus* and *clostridia* is its ability to degrade a wide range of polysaccharides, despite the absence of a cellulosome complex, at temperatures greater than 70°C, and the fact that it is asporogenous.[81] The strain is also different from other asporogenous, thermophilic organisms in the genera *Thermoanaerobacter*, *Thermoanaerobacterium,* and *Thermoanaerobium*, as it is capable of utilizing a wide range of cellulose forms, including microcrystalline cellulose.[15]

C. saccharolyticus is gram-positive, with a GC content of 35.2%. Its 2.97-Mb genome has only recently been completed (http://genome.ornl.gov/microbial/csac), and its initial annotation revealed 2695 ORFs; approximately 20% of the ORFs are involved in carbohydrate degradation, transport, and metabolism. This fact, in addition to the wide variety of carbohydrates that support its growth, suggests that *C. saccharolyticus* will yield a plethora of industrially relevant GHs.

Carbohydrate Utilization by Caldicellulosiruptor saccharolyticus

C. saccharolyticus has the ability to use a variety of polysaccharides, including xylan, cellulose, pullulan, pectin, mannan, and starch for growth.[15,83] Fermentation of sugars by the organism in nitrogen-rich medium results in the production of hydrogen, acetate, lactate, and traces of ethanol.[15,84] Experiments with ^{13}C-labeled glucose showed the canonical EMP pathway to be the main route for glycolysis.[85] As with *T. maritima* and *P. furiosus*, H_2 and acetate inhibit the growth of the microorganism, though it is much more tolerant of H_2 than other thermophiles, including *T. maritima* and *P. furiosus*; H_2 levels that are inhibitory to *P. furiosus* grown on pyruvate and *T. maritima* grown on glucose (1.6 kPa and 2 kPa, respectively), were observed to diminish *C. saccharolyticus* growth on sucrose by only 7%.[86]

Because the biofuels revolution is getting more attention for political and economic reasons, interest in H_2 as a possible nonhydrocarbon fuel of the future is growing. H_2 production will need to be, at least in the short term, a decentralized process, due to significant storage and transportation problems. Therefore, biological production of large amounts of H_2 is a very attractive option. Dark fermentation to produce H_2, while not the only biological means for its synthesis, is doubly attractive, as it will take advantage of the enormous technological advancements that will undoubtedly be made as part of the Department of Energy's biofuels mission (http://genomics-gtl.energy.gov/biofuels/index.shtml). *C. saccharolyticus* is receiving a great deal of attention, as it provides an immense variety of cellulolytic and hemicellulolytic enzymes that can be utilized for upstream processing of cellulosic feedstocks in biofuels processes, and it is

TABLE 6. Glycoside hydrolases of *Caldicellulosiruptor saccharolyticus*

Enzyme	GH family	MW (kDa)	3D structure	SwissProt	Locus	Reference
Pullulanase (pulA)	13	95.7	NA	Q59319	Csac0671	87
β-Mannanase	5 and 44	146.9	NA	P22533	Csac1077	89
Cellulase (CelA)	9 and 48	193.7	NA	P22534	Csac1076	140
Cellulase (CelB)	10 and 5	117.6	NA	P10474	Csac1078	141
β-Xylosidase (XynB)	39	56.4	NA	P23552	Csac2404	90
endo-1,4-β-Xylanase	10	36.5	NA	P23557	Csac2405	90
Acetyl esterase (ORF2)		30.6	NA	P23553	Csac2407	90
Xylanase (XynA)	10	40.5	NA	P23556	Csac2408	90

an excellent candidate for an industrial H_2 producer. Previous studies on the production of H_2 by *C. saccharolyticus* growing on paper-sludge hydrolysate indicated that both glucose and xylose were cofermented, with xylose consumption higher than glucose; however, less H_2 was produced from xylose than from glucose.[84] The cofermentation of these two sugars is an unusual but biotechnologically very desirable microbial characteristic, raising questions about the regulation of carbohydrate utilization in *C. saccharolyticus*. Catabolite repression generally prevents mesophiles from cofermenting sugars, while yeast, an organism employed for bioethanol production, does not naturally possess the enzymatic repertoire to process non-glucan hemicellulosic substrates.

Caldicellulosiruptor saccharolyticus α-specific Glycoside Hydrolases

A short list of the *C. saccharolyticus* GHs previously studied is provided in TABLE 6. Unfortunately, the lack, until recently, of a genome sequence, and by extension transcriptional data, has precluded biochemical analysis of the *C. saccharolyticus* enzymatic machinery to the same depth as *P. furiosus* and *T. maritima*. Several groups have, however, strived to provide a glimpse of the mechanisms by which *C. saccharolyticus* degrades cellulosic materials. While more β-specific than α-specific GHs have been identified (prior to release and annotation of the genome), a type I extracellular pullulanase from *C. saccharolyticus* was identified using a genomic library screen.[87] It was highly active on pullulan, hydrolyzing α-1,6-bonds, but not α-1,4-bonds, producing maltotriose as the only end-product.

Caldicellulosiruptor saccharolyticus β-specific Glycoside Hydrolases

Initial attempts to purify native cellulolytic enzymes from *C. saccharolyticus* were complicated by the multiplicity of glycosidic activities.[88] To address this issue, genes encoding cellulases, mannanases, and xylanases were identified by screening *C. saccharolyticus* genomic libraries in *Escherichia coli* hosts.[89,90] Even without genome sequence data, these studies demonstrated that, though lacking a full cellulosome, *C. saccharolyticus* possesses a multitude of large, multidomain enzymes. There are more than 60 GHs apparent in the final genome sequence of *C. saccharolyticus*. We have identified other putative xylanases and mannanases, which underlies the capability of this bacterium to hydrolyze paper-sludge hydrolysate.[84] A three-domain, cellulolytic enzyme, CelB (Csac1078), found screening the library contains both exo-1,4-β-glucanase activity, akin to a xylanase cellobiohydrolase, in the first domain, and endo-1,4-β-glucanase activity in its third domain. A second enzyme discovered through the same technique (Csac1076) was described as a cellulase, though only one of its four domains showed (endo) activity toward cellulose.[140] A mannanase gene was found encoded between the two cellulase genes. This enzyme comprises four domains, with domain 1 hydrolyzing only mannan, and domain 4 hydrolyzing CMC, oat spelt xylan, and lichenin.[89]

While screening for xylan-degrading enzymes, five related ORFs (Csac2404–2408) were identified. One of these, *xynA*, had xylanase/β-xylosidase activity, while another, *xynB*, was not active on xylan, but did have β-xylosidase activity.[90] Further characterization of XynA showed that its optimum temperature and thermal stability were among the highest reported for xylanases from any microorganism.[91] The importance of thermally active and hyperstable hemicellulases for industrial purposes suggests that other as yet uncharacterized *C. saccharolyticus* xylanases will prove attractive targets for enzyme discovery efforts.

Carbohydrate Uptake by Caldicellulosiruptor saccharolyticus

In contrast to *T. maritima* and *P. furiosus*, components of a putative PTS (Csac2437–2440) have been identified from the newly available genome sequence. This locus comprises general EI and HPr proteins (Csac2437 and Csac2438) and sugar-specific components EIIc and EIIa (Csac2439 and Csac2440). The

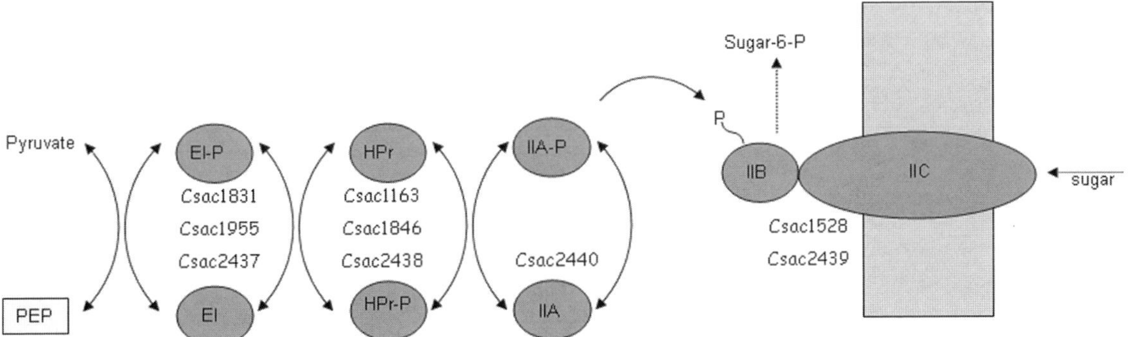

FIGURE 1. Phosphotransferase system–related proteins found in the *Caldicellulosiruptor saccharolyticus* genome.

presence of these and other putative PTS components, shown in FIGURE 1, would suggest that this constitutes a minor part (if not the major part) of solute transport in *C. saccharolyticus*. Conversely, approximately 160 ABC transporter ORFs were also identified, many of which are found in close proximity to genes for GHs and carbohydrate-responsive regulator family proteins. The relative importance of PTS compared to ABC transporters for carbohydrate uptake by *C. saccharolyticus* remains to be determined.

Polysaccharide Formation by Thermophilic Anaerobes

Thermophilic microorganisms produce polysaccharides to facilitate a variety of cellular functions, and this property is by no means restricted to the anaerobes. Various thermoacidophilic archaea, including members of the genera *Thermococcus* and *Sulfolobus*, have been observed to accumulate intracellular polysaccharides, such as glycogen, when grown with sucrose as a carbon source.[92] In addition, *Sulfolobus*,[93] *Thermococcus*,[94] and *Thermotoga*[95] species produce exopolysaccharides (EPS) that can act indirectly as extracellular storage polymers, and therefore as the predominant subsequent substrates for the various GHs they possess, for which no other source of its substrate has been identified in their natural extreme environments. In addition to cellular functions, EPS mediates adhesion, allowing cells to attach to other cells and solid surfaces, forming a matrix of cells, with the EPS commonly referred to as a biofilm. Pure cultures of *T. maritima*,[95] *Archaeoglobus fulgidus*,[96] and *Thermococcus litoralis*[94] formed significant biofilms under a variety of experimental conditions. Cocultures of *T. maritima* and the H_2-consuming methanogen *Methanococcus jannaschii* also developed significant biofilms,[97,98] with polysaccharides comprising ~5% of the dry biomass. Analysis of the coculture polysaccharide found ~91.2% glucose, 5.2% ribose, and ~2.7% mannose.[98]

The basics of thermophilic biofilm formation appears to proceed in much the same way as for mesophiles and involves the initial attachment of cells to a solid support, production of EPS, early biofilm development, mature biofilm development, and the subsequent detachment of cells.[99] Analysis of biofilm formation for a variety of mesophilic bacteria has indicated that the expression of select genes is required for transition between sessile and planktonic cells.[100,101] These key genes were found to be involved in chemotaxis, motility, EPS synthesis, and stress response. The expression of these specific genes has been linked to quorum-sensing behavior in a number of bacterial species. Quorum sensing describes the coordination of microbial communitywide events, such as biofilm formation, through the use of chemical signaling molecules called autoinducers. The ability of cells to participate in quorum sensing allows for rapid response to changes in environmental conditions, including fluctuations in availability of nutrients, defense against other microorganisms, and the avoidance of toxic compounds.[102,103]

Although hyperthermophiles have been shown to form biofilms, the pathways utilized and role of quorum sensing had not been addressed.[95,104,105] The cocultivation of *T. maritima* and *M. jannaschii* resulted in a fivefold increase in *T. maritima* cell densities when compared to monocultures, as well as an increase in EPS formation.[98] Increased biomass production from the cocultivation of methanogens and fermentative anaerobes is due to the methanogen serving as a hydrogen sink, thereby eliminating inhibitory levels of hydrogen.[97] In addition, studies indicate that microbial interactions, such as heterotroph:methanogen syntrophy, are enhanced by the formation of biofilms held together or engulfed by EPS.[106]

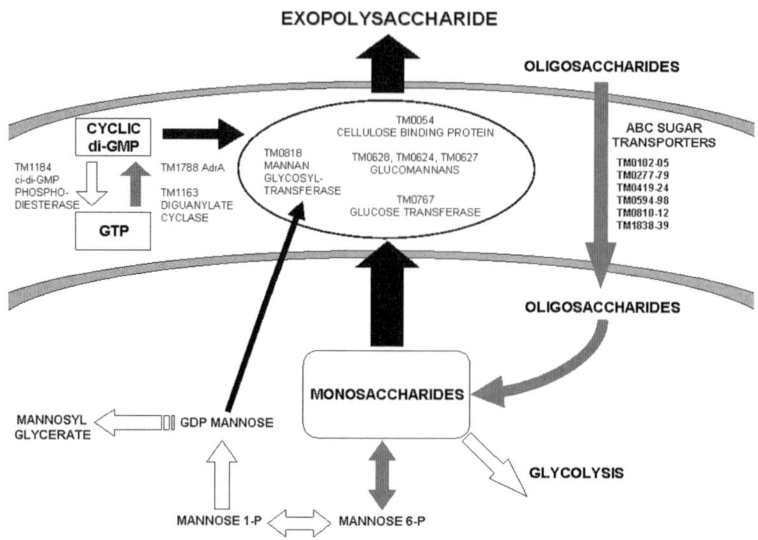

FIGURE 2. Proposed pathway for exopolysaccharide production in *Thermotoga maritima*.[98] *Shaded arrows* indicate a pathway is upregulated, while *nonshaded arrows* indicate a pathway is downregulated.

Transcriptional response data from the cocultivation experiments, in conjunction with previous studies of EPS production in mesophilic bacteria, provided the basis for a proposed EPS production pathway in *T. maritima,* as shown in FIGURE 2. As expected, a large number of ABC transporters were upregulated. In addition, TM1788 and TM1163, an AdrA regulator and diguanylate cyclase, respectively, were also upregulated in the high cell density cocultures of *T. maritima* and *M. jannaschii*, while TM1184, a cyclic di-guanosine monophosphate (di-GMP) phosphodiesterase, was downregulated under the same conditions. The diguanylate cyclase and cyclic di-GMP phosphodiesterase act to cycle between cyclic di-GMP, an allosteric activator of cellulose synthase, and GTP.[107] Because cyclic di-GMP serves as a second messenger for many extracellular signals and has been linked to a host of cell density–dependent responses, it is believed to be associated with quorum-sensing responses, although no definitive link has been made.[108]

The transcriptional analysis revealed a strong differential regulation of the TM0504 gene in the cocultivation of *T. maritima* and *M. jannaschii* at high cell densities compared to low cell density pure cultures. Examination of the TM0504 gene, which codes for a 42 amino acid peptide, revealed a double Gly motif similar to the cleavage site of immature peptide autoinducers found in *Streptococcus pneumoniae* and lactic acid bacteria, giving rise to the possibility of peptide-based quorum-sensing mechanisms in *T. maritima*.[109] The genome locus also includes a putative oligopeptide ABC transporter (TM0500–503); the lack of a periplasmic-binding protein suggested that this transporter may function as an exporter for the mature form of this peptide. A synthetic version of TM0504, lacking the first 14 amino acids and corresponding to the predicted mature length peptide, dosed into pure cultures of *T. maritima* during early log phase elicited a similar, albeit diminished response to that observed with the coculture. EPS production was observed 30 min after dosing, but not 10 min after dosing, suggesting that a period of time was necessary for uptake of the peptide and the subsequent response.[98] Transcriptional analysis revealed four GTs were upregulated in both the cocultivation with *M. jannaschii* and in pure cultures dosed with the peptide.[98]

The identification of TM0504 as a possible signaling peptide involved in the production of EPS brings into question the role of other small peptides that may play a role in a variety of cellular functions. Further examination of microbial genomes sequenced to date reveals a large proportion of small ORFs lack functional annotation due in part to the lack of comparative sequence information. Studies have shown that a number of these peptides, despite lacking readily identifiable structural features, are involved in inter- and intraspecies signaling, antimicrobial behavior, surface adhesion, and substrate acquisition.[110–112] Understanding the details of EPS formation in hyperthermophilic microorganisms will require determining what signals expression of key genes and how cells respond to such signals. The most likely system is a two-component regulatory system consisting of a histidine kinase and response regulator, which are present in both *T. maritima*, and *C. saccharolyticus*.[113]

Summary

Extreme thermophiles are evolutionary slow organisms that are not only interesting from a phylogenetical point of view, but their requirement for high temperatures also makes their enzymes ideal candidates for use in industrial processes. At present, many GHs from *P. furiosus* and *T. maritima* have been characterized biochemically and genetically. *C. saccharolyticus*, while less studied, also has great potential as a source of important glycolytic enzymes. The organism also serves as a model for comparison to the novel signaling system identified in *T. maritima* and linked to EPS production. Transcriptional response analysis can facilitate genomewide exploration of issues involving carbohydrate metabolism and complement *in vitro* biochemical studies in *C. saccharolyticus*, as it has previously in *P. furiosus* and *T. maritima*. Though much has been done, detailed understanding of these organisms' metabolism in order to fully exploit their enzymatic and metabolic capacity to process carbohydrates is still an ongoing process.

Acknowledgments

This work was supported in part by grants from the Natural Science Foundation (Biotechnology Program), the Department of Energy (Energy Bioscience Program), and the NASA Exobiology Program. A.L.V. and D.L.L. acknowledge support from the Department of Education GAANN Fellowship.

Conflict of Interest

The authors declare no conflicts of interest.

References

1. Miyazaki, J. *et al.* 2001. Ancestral residues stabilizing 3-isopropylmalate dehydrogenase of an extreme thermophile: experimental evidence supporting the thermophilic common ancestor hypothesis. J. Biochem. (Tokyo) **129:** 777–782.
2. Blochl, E. *et al.* 1995. Isolation, taxonomy and phylogeny of hyperthermophilic microorganisms. World J. Microbiol. Biotechnol. **11:** 9–16.
3. Henrissat, B. 1991. A classification of glycosyl hydrolases based on amino acid sequence similarities. Biochem. J. **280:** 309–316.
4. Bult, C.J. *et al.* 1996. Complete genome sequence of the methanogenic archaeon, *Methanococcus jannaschii*. Science **273:** 1058–1073.
5. Meibom, K.L. *et al.* 2004. The *Vibrio cholerae* chitin utilization program. Proc. Natl. Acad. Sci. USA **101:** 2524–2529.
6. Barrangou, R. *et al.* 2003. Functional and comparative genomic analyses of an operon involved in fructooligosaccharide utilization by *Lactobacillus acidophilus*. Proc. Natl. Acad. Sci. USA **100:** 8957–8962.
7. Bertram, R. *et al.* 2004. In silico and transcriptional analysis of carbohydrate uptake systems of *Streptomyces coelicolor A3(2)*. J. Bacteriol. **186:** 1362–1373.
8. Nguyen, T.N. *et al.* 2004. Whole-genome expression profiling of *Thermotoga maritima* in response to growth on sugars in a chemostat. J. Bacteriol. **186:** 4824–4828.
9. Chhabra, S.R. *et al.* 2003. Carbohydrate-induced differential gene expression patterns in the hyperthermophilic bacterium *Thermotoga maritima*. J. Biol. Chem. **278:** 7540–7552.
10. Conners, S.B. *et al.* 2006. Microbial biochemistry, physiology, and biotechnology of hyperthermophilic *Thermotoga* species. FEMS Microbiol. Rev. **30:** 872–905.
11. Conners, S.B. *et al.* 2005. An expression-driven approach to the prediction of carbohydrate transport and utilization regulons in the hyperthermophilic bacterium *Thermotoga maritima*. J. Bacteriol. **187:** 7267–7282.
12. Lee, H.S. *et al.* 2006. Transcriptional and biochemical analysis of starch metabolism in the hyperthermophilic archaeon *Pyrococcus furiosus*. J. Bacteriol. **188:** 2115–2125.
13. Fiala, G. & K.O. Stetter. 1986. *Pyrococcus furiosus*, new species represents a novel genus of marine heterotrophic archaebacteria growing optimally at 100°C. Arch. Microbiol. **145:** 56–60.
14. Huber, R. *et al.* 1986. *Thermotoga maritima* sp. nov. represents a new genus of unique extremely thermophilic eubacteria growing up to 90°C. Arch. Microbiol. **144:** 324–333.
15. Rainey, F.A. *et al.* 1994. Description of *Caldicellulosiruptor saccharolyticus* gen. nov., sp. nov: an obligately anaerobic, extremely thermophilic, cellulolytic bacterium. FEMS Microbiol. Lett. **120:** 263–266.
16. Nelson, K.E. *et al.* 1999. Evidence for lateral gene transfer between Archaea and bacteria from genome sequence of *Thermotoga maritima*. Nature **399:** 323–329.
17. Robb, F.T. *et al.* 2001. Genomic sequence of hyperthermophile, *Pyrococcus furiosus*: implications for physiology and enzymology. Methods Enzymol. **330:** 134–157.
18. Maeder, D.L. *et al.* 1999. Divergence of the hyperthermophilic archaea *Pyrococcus furiosus* and *P. horikoshii* inferred from complete genomic sequences. Genetics **152:** 1299–1305.
19. Driskill, L.E. *et al.* 1999. Relationship between glycosyl hydrolase inventory and growth physiology of the hyperthermophile *Pyrococcus furiosus* on carbohydrate-based media. Appl. Environ. Microbiol. **65:** 893–897.
20. Gao, J. *et al.* 2003. Growth of hyperthermophilic archaeon *Pyrococcus furiosus* on chitin involves two family 18 chitinases. Appl. Environ. Microbiol. **69:** 3119–3128.
21. Schonheit, P.a.T.S. 1995. Metabolism of hyperthermophiles. World J. Microbiol. Biotechnol. **11:** 26–57.

22. VERHAGEN, M.F. *et al.* 2001. *Pyrococcus furiosus*: large-scale cultivation and enzyme purification. Methods Enzymol. **330:** 25–30.

23. SCHICHO, R.N. *et al.* 1993. Bioenergetics of sulfur reduction in the hyperthermophilic archaeon *Pyrococcus furiosus*. J. Bacteriol. **175:** 1823–1830.

24. BONCH-OSMOLOVSKAYA, E.A. & K.O. STETTER. 1991. Interspecies hydrogen transfer in cocultures of thermophilic *Archaea*. Syst. Appl. Microbiol. **14:** 205–208.

25. BROWN, S.H. & R.M. KELLY. 1993. Characterization of amylolytic enzymes, having both alpha-1,4 and alpha-1,6 hydrolytic activity, from the thermophilic archaea *Pyrococcus furiosus* and *Thermococcus litoralis*. Appl. Environ. Microbiol. **59:** 2614–2621.

26. COSTANTINO, H.R., S.H. BROWN & R.M. KELLY. 1990. Purification and characterization of an alpha-glucosidase from a hyperthermophilic archaebacterium, *Pyrococcus furiosus*, exhibiting a temperature optimum of 105 to 115 degrees C. J. Bacteriol. **172:** 3654–3660.

27. DONG, G. *et al.* 1997. Cloning, sequencing, and expression of the gene encoding extracellular alpha-amylase from *Pyrococcus furiosus* and biochemical characterization of the recombinant enzyme. Appl. Environ. Microbiol. **63:** 3569–3576.

28. DONG, G., C. VIEILLE & J.G. ZEIKUS. 1997. Cloning, sequencing, and expression of the gene encoding amylopullulanase from *Pyrococcus furiosus* and biochemical characterization of the recombinant enzyme. Appl. Environ. Microbiol. **63:** 3577–3584.

29. LADERMAN, K.A. *et al.* 1993. The purification and characterization of an extremely thermostable alpha-amylase from the hyperthermophilic archaebacterium *Pyrococcus furiosus*. J. Biol. Chem. **268:** 24394–24401.

30. DOMAN-PYTKA, M. & J. Bardowski. 2004. Pullulan degrading enzymes of bacterial origin. Crit. Rev. Microbiol. **30**(2): 107–121.

31. COMFORT, D.A. 2006. Discovery, functional genomics and biochemical characterization of alpha-specific glycosyl hydrolases from hyperthermophilic microorganisms. *In* Chemical and Biomdecular Engineering. Raleigh, NC: North Carolina State University.

32. CHANG, S.T. *et al.* 2001. Alpha-glucosidase from *Pyrococcus furiosus*. Methods Enzymol. **330:** 260–269.

33. BERTOLDO, C. & G. ANTRANIKIAN. 2001. Amylolytic enzymes from hyperthermophiles. Methods Enzymol. **330:** 269–289.

34. ANTRANIKIAN, G., C.E. VORGIAS & C. BERTOLDO. 2005. Extreme environments as a resource for microorganisms and novel biocatalysts. Adv. Biochem. Eng. Biotechnol. **96:** 219–262.

35. YANG, S.J. *et al.* 2004. Enzymatic analysis of an amylolytic enzyme from the hyperthermophilic archaeon *Pyrococcus furiosus* reveals its novel catalytic properties as both an alpha-amylase and a cyclodextrin-hydrolyzing enzyme. Appl. Environ. Microbiol. **70:** 5988–5995.

36. BAUER, M.W. *et al.* 1999. An endoglucanase, EglA, from the hyperthermophilic archaeon *Pyrococcus furiosus* hydrolyzes beta-1,4 bonds in mixed-linkage (1→3),(1→4)-beta-D-glucans and cellulose. J. Bacteriol. **181:** 284–290.

37. LEBBINK, J.H. *et al.* 2001. Beta-glucosidase CelB from *Pyrococcus furiosus*: production by *Escherichia coli*, purification, and *in vitro* evolution. Methods Enzymol. **330:** 364–379.

38. CADY, S.G. *et al.* 2001. Beta-endoglucanase from *Pyrococcus furiosus*. Methods Enzymol. **330:** 346–354.

39. DRISKILL, L.E., M.W. BAUER & R.M. KELLY. 1999. Synergistic interactions among beta-laminarinase, beta-1,4-glucanase, and beta-glucosidase from the hyperthermophilic archaeon *Pyrococcus furiosus* during hydrolysis of beta-1,4-, beta-1,3-, and mixed-linked polysaccharides. Biotechnol. Bioeng. **66:** 51–60.

40. KAPER, T. *et al.* 2001. Characterization of beta-glycosylhydrolases from *Pyrococcus furiosus*. Methods Enzymol. **330:** 329–346.

41. KAPER, T. *et al.* 2002. Substrate specificity engineering of beta-mannosidase and beta-glucosidase from *Pyrococcus* by exchange of unique active site residues. Biochemistry **41:** 4147–4155.

42. TANAKA, T. *et al.* 2003. Characterization of an exo-beta-D-glucosaminidase involved in a novel chitinolytic pathway from the hyperthermophilic archaeon *Thermococcus kodakaraensis KOD1*. J. Bacteriol. **185:** 5175–5181.

43. KONING, S.M. *et al.* 2001. Cellobiose uptake in the hyperthermophilic archaeon *Pyrococcus furiosus* is mediated by an inducible, high-affinity ABC transporter. J. Bacteriol. **183:** 4979–4984.

44. KENGEN, S.W.M., A.J.M. STAMS & W.M. DE VOS. 1996. Sugar metabolism of hyperthermophiles. FEMS Microbiol. Rev. **18:** 119–137.

45. KONING, S.M., W.N. KONINGS & A.J. DRIESSEN. 2002. Biochemical evidence for the presence of two alpha-glucoside ABC-transport systems in the hyperthermophilic archaeon *Pyrococcus furiosus*. Archaea **1:** 19–25.

46. LEE, S.J. *et al.* 2005. TrmB, a sugar sensing regulator of ABC transporter genes in *Pyrococcus furiosus* exhibits dual promoter specificity and is controlled by different inducers. Mol. Microbiol. **57:** 1797–1807.

47. ALBERS, S.V. *et al.* 2004. Insights into ABC transport in archaea. J. Bioenerg. Biomembr. **36:** 5–15.

48. FUKUI, T. *et al.* 2005. Complete genome sequence of the hyperthermophilic archaeon *Thermococcus kodakaraensis KOD1* and comparison with *Pyrococcus* genomes. Genome Res. **15:** 352–363.

49. LECOMPTE, O. *et al.* 2001. Genome evolution at the genus level: comparison of three complete genomes of hyperthermophilic archaea. Genome Res. **11:** 981–993.

50. ANDO, S. *et al.* 2002. Hyperthermostable endoglucanase from *Pyrococcus horikoshii*. Appl. Environ. Microbiol. **68:** 430–433.

51. KASHIMA, Y. *et al.* 2005. Analysis of the function of a hyperthermophilic endoglucanase from *Pyrococcus horikoshii* that hydrolyzes crystalline cellulose. Extremophiles **9:** 37–43.

52. KYRPIDES, N.C. *et al.* 2000. Analysis of the *Thermotoga maritima* genome combining a variety of sequence similarity and genome context tools. Nucleic Acids Res. **28:** 4573–4576.

53. SELIG, M. *et al.* 1997. Comparative analysis of Embden-Meyerhof and Entner-Doudoroff glycolytic pathways in hyperthermophilic archaea and the bacterium *Thermotoga*. Arch. Microbiol. **167:** 217–232.

54. VARGAS, M. *et al.* 1998. Microbiological evidence for Fe(III) reduction on early Earth. Nature **395:** 65–67.
55. SCHRODER, C., M. SELIG & P. SCHONHEIT. 1994. Glucose fermentation to acetate, CO_2 and H_2 in the anaerobic hyperthermophilic eubacterium *Thermotoga maritima*: involvement of the Embden-Meyerhof pathway. Arch. Microbiol. **161:** 460–470.
56. LIEBL, W., I. STEMPLINGER & P. RUILE. 1997. Properties and gene structure of the *Thermotoga maritima* alpha-amylase AmyA, a putative lipoprotein of a hyperthermophilic bacterium. J. Bacteriol. **179:** 941–948.
57. KRIEGSHAUSER, G. & W. LIEBL. 2000. Pullulanase from the hyperthermophilic bacterium *Thermotoga maritima*: purification of beta-cyclodextrin affinity chromatography. J. Chromatogr. B Biomed. Sci. Appl. **737:** 245–251.
58. BALLSCHMITER, M., O. FUTTERER & W. LIEBL. 2006. Identification and characterization of a novel intracellular alkaline alpha-amylase from the hyperthermophilic bacterium *Thermotoga maritima MSB8*. Appl. Environ. Microbiol. **72:** 2206–2211.
59. LIM, W.J. *et al.* 2003. Cloning and characterization of a thermostable intracellular alpha-amylase gene from the hyperthermophilic bacterium *Thermotoga maritima MSB8*. Res. Microbiol. **154:** 681–687.
60. BIBEL, M. *et al.* 1998. Isolation and analysis of genes for amylolytic enzymes of the hyperthermophilic bacterium *Thermotoga maritima*. FEMS Microbiol. Lett. **158:** 9–15.
61. KLUSKENS, L.D. *et al.* 2003. Molecular and biochemical characterization of the thermoactive family 1 pectate lyase from the hyperthermophilic bacterium *Thermotoga maritima*. Biochem. J. **370**(Pt. 2): 651–659.
62. PARISOT, J. *et al.* 2003. Cloning expression and characterization of a thermostable exopolygalacturonase from *Thermotoga maritima*. Carbohydr. Res. **338:** 1333–1337.
63. BRONNENMEIER, K. *et al.* 1995. Purification of *Thermotoga maritima* enzymes for the degradation of cellulosic materials. Appl. Environ. Microbiol. **61:** 1399–1407.
64. CHHABRA, S.R. *et al.* 2002. Regulation of endo-acting glycosyl hydrolases in the hyperthermophilic bacterium *Thermotoga maritima* grown on glucan- and mannan-based polysaccharides. Appl. Environ. Microbiol. **68:** 545–554.
65. CHHABRA, S.R. & R.M. KELLY. 2002. Biochemical characterization of *Thermotoga maritima* endoglucanase Cel74 with and without a carbohydrate binding module (CBM). FEBS Lett. **531:** 375–380.
66. COLLINS, T., C. GERDAY & G. FELLER. 2005. Xylanases, xylanase families and extremophilic xylanases. FEMS Microbiol. Rev. **29:** 3–23.
67. RUILE, P., C. WINTERHALTER & W. LIEBL. 1997. Isolation and analysis of a gene encoding alpha-glucuronidase, an enzyme with a novel primary structure involved in the breakdown of xylan. Mol. Microbiol. **23:** 267–279.
68. SURESH, C., M. KITAOKA & K. HAYASHI. 2003. A thermostable non-xylanolytic alpha-glucuronidase of *Thermotoga maritima MSB8*. Biosci. Biotechnol. Biochem. **67:** 2359–2364.
69. ZHENGQIANG, J. *et al.* 2001. Characterization of a thermostable family 10 endo-xylanase (XynB) from *Thermotoga maritima* that cleaves p-nitrophenyl-beta-D-xyloside. J. Biosci. Bioeng. **92:** 423–428.
70. SURESH, C. *et al.* 2002. Evidence that the putative alpha-glucosidase of *Thermotoga maritima MSB8* is a pNP alpha-D-glucuronopyranoside hydrolyzing alpha-glucuronidase. FEBS Lett. **517:** 159–162.
71. NANAVATI, D.M., K. THIRANGOON & K.M. NOLL. 2006. Several archaeal homologs of putative oligopeptide-binding proteins encoded by *Thermotoga maritima* bind sugars. Appl. Environ. Microbiol. **72:** 1336–1345.
72. KONING, S.M. 2003. Sugar transport in (hyper) thermophilic archaea. Res. Microbiol. **153:** 61–67.
73. WASSENBERG, D., W. LIEBL & R. JAENICKE. 2000. Maltose-binding protein from the hyperthermophilic bacterium *Thermotoga maritima*: stability and binding properties. J. Mol. Biol. **295:** 279–288.
74. BELKIN, S., C.O. WIRSEN & H.W. JANNASCH. 1986. A new sulfur-reducing, extremely thermophilic eubacterium from a submarine thermal vent. Appl. Environ. Microbiol. **51:** 1180–1185.
75. KING, M.R. *et al.* 1998. Thermostable alpha-galactosidase from *Thermotoga neapolitana*: cloning, sequencing and expression. FEMS Microbiol. Lett. **163:** 37–42.
76. ZVERLOV, V.V. *et al.* 1997. Highly thermostable endo-1,3-beta-glucanase (laminarinase) LamA from *Thermotoga neapolitana*: nucleotide sequence of the gene and characterization of the recombinant gene product. Microbiology **143**(Pt. 5): 1701–1708.
77. BOK, J.D., D.A. YERNOOL & D.E. EVELEIGH. 1998. Purification, characterization, and molecular analysis of thermostable cellulases CelA and CelB from *Thermotoga neapolitana*. Appl. Environ. Microbiol. **64:** 4774–4781.
78. MCCARTHY, J.K. *et al.* 2004. Improved catalytic efficiency and active site modification of 1,4-beta-D-glucan glucohydrolase A from *Thermotoga neapolitana* by directed evolution. J. Biol. Chem. **279:** 11495–11502.
79. TAKAHATA, Y. *et al.* 2001. *Thermotoga petrophila* sp. nov. and *Thermotoga naphthophila* sp. nov., two hyperthermophilic bacteria from the Kubiki oil reservoir in Niigata, Japan. Int. J. Syst. Evol. Microbiol. **51**(Pt. 5): 1901–1909.
80. NESBO, C.L., K.E. NELSON & W.F. DOOLITTLE. 2002. Suppressive subtractive hybridization detects extensive genomic diversity in *Thermotoga maritima*. J. Bacteriol. **184:** 4475–4488.
81. SISSONS, C.H. *et al.* 1987. Isolation of cellulolytic anaerobic extreme thermophiles from New Zealand thermal sites. Appl. Environ. Microbiol. **53:** 832–838.
82. REYNOLDS, P.H. *et al.* 1986. Comparison of cellulolytic activities in *Clostridium thermocellum* and three thermophilic, cellulolytic anaerobes. Appl. Environ. Microbiol. **51:** 12–17.
83. DONNISON, A.M., C.M. BROCKELSBY, H.W. MORGAN & R.M. DANIEL. 1989. The degradation of lignocellulosics by extremely thermophilic microorganisms. Biotechnol. Bioeng. **33:** 1495–1499.
84. KADAR, Z. *et al.* 2003. Hydrogen production from paper sludge hydrolysate. Appl. Biochem. Biotechnol. **105–108**: 557–566.
85. DE VRIJE, T. *et al.* 2007. Glycolytic pathway and hydrogen yield studies of the extreme thermophile *Caldicellulosiruptor saccharolyticus*. Appl. Microbiol. Biotechnol. **74**: 1358–1367.

86. van Niel, E.W., P.A. Claassen & A.J. Stams. 2003. Substrate and product inhibition of hydrogen production by the extreme thermophile, *Caldicellulosiruptor saccharolyticus*. Biotechnol. Bioeng. **81:** 255–262.
87. Albertson, G.D. *et al.* 1997. Cloning and sequence of a type I pullulanase from an extremely thermophilic anaerobic bacterium, *Caldicellulosiruptor saccharolyticus*. Biochim. Biophys. Acta **1354:** 35–39.
88. Schofield, L.R. & R.M. Daniel. 1993. Purification and properties of a beta-1,4-xylanase from a cellulolytic extreme thermophile expressed in *Escherichia coli*. Int. J. Biochem. **25:** 609–617.
89. Gibbs, M.D. *et al.* 1992. The beta-mannanase from "*Caldocellum saccharolyticum*" is part of a multidomain enzyme. Appl. Environ. Microbiol. **58:** 3864–3867.
90. Luthi, E. *et al.* 1990. Cloning, sequence analysis, and expression of genes encoding xylan-degrading enzymes from the thermophile "*Caldocellum saccharolyticum*". Appl. Environ. Microbiol. **56:** 1017–1024.
91. Luthi, E., N.B. Jasmat & P.L. Bergquist. 1990. Xylanase from the extremely thermophilic bacterium "*Caldocellum saccharolyticum*": overexpression of the gene in *Escherichia coli* and characterization of the gene product. Appl. Environ. Microbiol. **56:** 2677–2683.
92. Konig, H., R. Skorko, W. Zillig & W.D. Reiter. 1982. Glycogen in thermoacidophilic archaebacteria of the genera *Sulfolobus, Thermoproteus, Desulfurococcus*, and *Thermococcus*. Arch. Microbiol. **132:** 297–303.
93. Nicolaus, B., M.C. Manca, I. Romano & L. Lama. 1993. Production of an exopolysaccharide from two thermophilic archaea belonging to the genus *Sulfolobus*. FEMS Microbiol. Lett. **109:** 203–206.
94. Rinker, K.D. & R.M. Kelly. 1996. Growth physiology of the hyperthermophilic archaeon *Thermococcus litoralis*: development of a sulfur-free defined medium, characterization of an exopolysaccharide, and evidence of biofilm formation. Appl. Environ. Microbiol. **62:** 4478–4485.
95. Rinker, K.D. & R.M. Kelly. 2000. Effect of carbon and nitrogen sources on growth dynamics and exopolysaccharide production for the hyperthermophilic archaeon *Thermococcus litoralis* and bacterium *Thermotoga maritima*. Biotechnol. Bioeng. **69:** 537–547.
96. Hartzell, P.L., J. Millstein & C. LaPaglia. 1999. Biofilm formation in hyperthermophilic Archaea. Methods Enzymol. **310:** 335–349.
97. Muralidharan, V. *et al.* 1997. Hydrogen transfer between methanogens and fermentative heterotrophs in hyperthermophilic cocultures. Biotechnol. Bioeng. **56:** 268–278.
98. Johnson, M.R. *et al.* 2005. Population density-dependent regulation of exopolysaccharide formation in the hyperthermophilic bacterium *Thermotoga maritima*. Mol. Microbiol. **55:** 664–674.
99. Kolter, R. & R. Losick. 1998. One for all and all for one. Science **280:** 226–227.
100. Davies, D.G. *et al.* 1998. The involvement of cell-to-cell signals in the development of a bacterial biofilm. Science **280:** 295–298.
101. Islam, M.S. *et al.* 2007. Biofilm acts as a microenvironment for plankton-associated *Vibrio cholerae* in the aquatic environment of Bangladesh. Microbiol. Immunol. **51:** 369–379.
102. Armitage, J.P. 1999. Bacterial tactic responses. Adv. Microb. Physiol. **41:** 229–289.
103. Miller, M.B. & B.L. Bassler. 2001. Quorum sensing in bacteria. Annu. Rev. Microbiol. **55:** 165–199.
104. Lapaglia, C. & P.L. Hartzell. 1997. Stress-induced production of biofilm in the hyperthermophile *Archaeoglobus fulgidus*. Appl. Environ. Microbiol. **63:** 3158–3163.
105. Pysz, M.A. *et al.* 2004. Transcriptional analysis of biofilm formation processes in the anaerobic, hyperthermophilic bacterium *Thermotoga maritima*. Appl. Environ. Microbiol. **70:** 6098–6112.
106. Costerton, J.W. 1995. Overview of microbial biofilms. J. Ind. Microbiol. **15:** 137–140.
107. Romling, U., M. Gomelsky & M.Y. Galperin. 2005. C-di-GMP: the dawning of a novel bacterial signalling system. Mol. Microbiol. **57:** 629–639.
108. Camilli, A. & B.L. Bassler. 2006. Bacterial small-molecule signaling pathways. Science **311:** 1113–1136.
109. Havarstein, L.S., D.B. Diep & I.F. Nes. 1995. A family of bacteriocin ABC transporters carry out proteolytic processing of their substrates concomitant with export. Mol. Microbiol. **16:** 229–240.
110. Bassler, B.L. 2002. Small talk. Cell-to-cell communication in bacteria. Cell **109:** 421–424.
111. Fuqua, C., S.C. Winans & E.P. Greenberg. 1996. Census and consensus in bacterial ecosystems: the LuxR-LuxI family of quorum-sensing transcriptional regulators. Annu. Rev. Microbiol. **50:** 727–751.
112. Lazazzera, B.A. & A.D. Grossman. 1998. The ins and outs of peptide signaling. Trends Microbiol. **6:** 288–294.
113. Galperin, M.Y. 2005. A census of membrane-bound and intracellular signal transduction proteins in bacteria: bacterial IQ, extroverts and introverts. BMC Microbiol. **5:** 35.
114. Bao, Q. *et al.* 2002. A complete sequence of the *T. tengcongensis* genome. Genome Res. **12:** 689–700.
115. Klenk, H.P. *et al.* 1997. The complete genome sequence of the hyperthermophilic, sulphate-reducing archaeon *Archaeoglobus fulgidus*. Nature **390:** 364–370.
116. Kawarabayasi, Y. *et al.* 1998. Complete sequence and gene organization of the genome of a hyper-thermophilic archaebacterium, *Pyrococcus horikoshii OT3*. DNA Res. **5:** 55–76.
117. Zillig, W. *et al.* 1990. *Hyperthermus butylicus*, a hyperthermophilic sulfur-reducing archaebacterium that ferments peptides. J. Bacteriol. **172:** 3959–3965.
118. Slesarev, A.I. *et al.* 2002. The complete genome of hyperthermophile *Methanopyrus kandleri AV19* and monophyly of archaeal methanogens. Proc. Natl. Acad. Sci. USA **99:** 4644–4649.
119. Brown, S.H., H.R. Costantino & R.M. Kelly. 1990. Characterization of amylolytic enzyme activities associated with the hyperthermophilic archaebacterium *Pyrococcus furiosus*. Appl. Environ. Microbiol. **56:** 1985–1991.
120. Van Lieshout, J.F.T. *et al.* 2003. Identification and molecular characterization of a novel type of alpha-galactosidase from *Pyrococcus furiosus*. Biocat. Biotransfor. **21:** 243–252.

121. VOORHORST, W.G. *et al.* 1995. Characterization of the celB gene coding for beta-glucosidase from the hyperthermophilic archaeon *Pyrococcus furiosus* and its expression and site-directed mutation in *Escherichia coli*. J. Bacteriol. **177:** 7105–7111.

122. GUEGUEN, Y. *et al.* 1997. Molecular and biochemical characterization of an endo-beta-1,3-glucanase of the hyperthermophilic archaeon *Pyrococcus furiosus*. J. Biol. Chem. **272:** 31258–31264.

123. BAUER, M.W. *et al.* 1996. Comparison of a beta-glucosidase and a beta-mannosidase from the hyperthermophilic archaeon *Pyrococcus furiosus*. Purification, characterization, gene cloning, and sequence analysis. J. Biol. Chem. **271:** 23749–23755.

124. RAASCH C., S.W., J. SCHANZER, M. BIBEL, *et al.* 2000. *Thermotoga maritima* AglA, an extremely thermostable NAD+-, Mn2+-, and thiol-dependent alpha-glucosidase." Extremophiles **4:** 189–200.

125. LIEBL, W. *et al.* 1992. Purification and characterization of a novel thermostable 4-alpha-glucanotransferase of *Thermotoga maritima* cloned in *Escherichia coli*. Eur. J. Biochem. **207:** 81–88.

126. LEE, M.H. *et al.* 2002. A novel amylolytic enzyme from *Thermotoga maritima*, resembling cyclodextrinase and alpha-glucosidase, that liberates glucose from the reducing end of the substrates. Biochem. Biophys. Res. Commun. **295:** 818–825.

127. MEISSNER, H. & W. LIEBL. 1998. *Thermotoga maritima* maltosyltransferase, a novel type of maltodextrin glycosyltransferase acting on starch and malto-oligosaccharides. Eur. J. Biochem. **258:** 1050–1058.

128. LIEBL, W. *et al.* 1996. Analysis of a *Thermotoga maritima* DNA fragment encoding two similar thermostable cellulases, CelA and CelB, and characterization of the recombinant enzymes. Microbiology **142**(Pt. 9): 2533–2542.

129. TARLING, C.A. *et al.* 2003. Identification of the catalytic nucleophile of the family 29 alpha-L-fucosidase from *Thermotoga maritima* through trapping of a covalent glycosyl-enzyme intermediate and mutagenesis. J. Biol. Chem. **278:** 47394–47399.

130. YANG, H. *et al.* 2006. Characterization of a thermostable endo-beta-1,4-D-galactanase from the hyperthermophile *Thermotoga maritima*. Biosci. Biotechnol. Biochem. **70:** 538–541.

131. YIP, V.L. *et al.* 2004. An unusual mechanism of glycoside hydrolysis involving redox and elimination steps by a family 4 beta-glycosidase from *Thermotoga maritima*. J. Am. Chem. Soc. **126:** 8354–8355.

132. MOORE, J.B., P. MARKIEWICZ & J.H. MILLER. 1994. Identification and sequencing of the *Thermotoga maritima* lacZ gene, part of a divergently transcribed operon. Gene **147:** 101–106.

133. LIEBL, W., B. WAGNER & J. SCHELLHASE. 1998. Properties of an alpha-galactosidase, and structure of its gene galA, within an alpha- and beta- galactoside utilization gene cluster of the hyperthermophilic bacterium *Thermotoga maritima*. Syst. Appl. Microbiol. **21:** 1–11.

134. PARKER, K.N. *et al.* 2001. Galactomannanases Man2 and Man5 from *Thermotoga* species: growth physiology on galactomannans, gene sequence analysis, and biochemical properties of recombinant enzymes. Biotechnol. Bioeng. **75:** 322–333.

135. DUFFAUD, G.D. *et al.* 1997. Purification and characterization of extremely thermostable beta-mannanase, beta-mannosidase, and alpha-galactosidase from the hyperthermophilic eubacterium *Thermotoga neapolitana 5068*. Appl. Environ. Microbiol. **63:** 169–177.

136. NAKAJIMA, M. *et al.* 2003. Unique metal dependency of cytosolic alpha-mannosidase from *Thermotoga maritima*, a hyperthermophilic bacterium. Arch. Biochem. Biophys. **415:** 87–93.

137. XUE, Y. & W. SHAO. 2004. Expression and characterization of a thermostable beta-xylosidase from the hyperthermophile, *Thermotoga maritima*. Biotechnol. Lett. **26:** 1511–1515.

138. WINTERHALTER, C. *et al.* 1995. Identification of a novel cellulose-binding domain within the multidomain 120 kDa xylanase XynA of the hyperthermophilic bacterium *Thermotoga maritima*. Mol. Microbiol. **15:** 431–444.

139. SALLEH, H.M. *et al.* 2006. Cloning and characterization of *Thermotoga maritima* beta-glucuronidase. Carbohydr. Res. **341:** 49–59.

140. TE'O, V.S., D.J. SAUL & P.L. BERGQUIST. 1995. celA, another gene coding for a multidomain cellulase from the extreme thermophile *Caldocellum saccharolyticum*. Appl. Microbiol. Biotechnol. **43:** 291–296.

141. SAUL, D.J. *et al.* 1990. celB, a gene coding for a bifunctional cellulase from the extreme thermophile "*Caldocellum saccharolyticum*". Appl. Environ. Microbiol. **56:** 3117–3124.

A Preliminary Analysis of Microbial and Biochemical Properties of High-Temperature Compost

TAIRO OSHIMA[a,b] AND TOSHIYUKI MORIYA[b]

[a]*Institute of Environmental Microbiology, Kyowa-kako Co., Machida, Tokyo, Japan*

[b]*Tokyo University of Pharmacy and Life Science, Hachioji, Tokyo, Japan*

We investigated the microbial community of a high-temperature compost process exhibiting an internal temperature exceeding 90°C. The waste pile was crosscut and samples were collected from the bottom to the top of the refuse pile. PCR–denaturing gradient gel electrophoresis analysis suggested that the microbial community of the high-temperature compost is heterogeneous and differs from one locality to another. Heat-stable collagenases and amylases were extracted directly from the compost pile. Collagenases were located in the upper half of the pile, whereas amylases were detected mainly in the lower parts. Several extremely thermophilic strains were isolated at 80°C; these strains were aerobes. Based on 16S rRNA sequence analysis, the isolates clustered together and represent one or two closely related species. We propose that these thermophilic isolates belong to a novel genus, *Caldaterra*, gen. nov.

Key words: **PCR-DGGE; compost pile; zymography; collagenase; amylase; microbial ecology of compost; thermophile**

Introduction

Composting is a traditional way to decompose various organic wastes, such as those from agriculture, cattle farming, food industries, and city life. Composting has many advantages over burning; in particular, it is economical, poses fewer human health risks, and is environmentally safer. Unlike burning, composting does not require extra oil or sophisticated equipment; it also does not produce harmful gaseous products, such as nitrogen oxides (NO_x), sulfur oxides (SO_x), and dioxins. The final products can be used as fertilizers and thus contribute to the production of food resources. Because burning is now generally prohibited in Japan and some other countries, composting has become more important for decomposing organic wastes.

Heat emitted from the fermentation processes of microorganisms causes the inside of compost piles to become hot, creating an ideal environment for isolating thermophiles.[1] According to the literature, the inside temperature reaches up to 75°C–80°C.[2,3] The isolation of many moderate thermophiles belonging to the genera *Geobacillus*, *Bacillus*, *Clostridium*, and related genera has been reported.

Recently, the Sanyu Company in the city of Kagoshima, Japan, invented a high-temperature compost process in which the internal temperature exceeds 95°C or even 100°C. This high-temperature composting rapidly degrades waste from cattle farming, including bones, skin, and even carcasses.[4] Because of the rapid degradation of organic wastes at these high temperatures, we became interested in describing the aerobic and anaerobic bacterial community and enzymes causing degradation in the high-temperature compost. We have attempted to analyze the bacterial and biochemical nature of high-temperature composting and report the preliminary results of our analytical studies in this paper.

High-Temperature Compost Pile Construction

The compost pile analyzed was 10 m wide, 8 m deep, and 280 cm high (FIG. 1), and sampling was performed 28 days after the pile was constructed. The high-temperature compost process was started by mixing 2/3 organic wastes (water content adjusted with polymer coagulant) with 1/3 fully fermented compost as inoculum. Air was continuously supplied from the

Address for correspondence: Tairo Oshima, Institute of Environmental Microbiology, Kyowa-kako Co., 2-15-5 Tadao, Machida, Tokyo 194-0035, Japan.

tairo.oshima@kyowa-kako.co.jp

Ann. N.Y. Acad. Sci. 1125: 338–344 (2008). © 2008 New York Academy of Sciences.
doi: 10.1196/annals.1419.012

FIGURE 1. High-temperature compost. Photograph taken at Sanyu Company, Kagoshima, Japan.

bottom floor using a compressor through three pipes buried in the floor. Small holes every 38 cm in the pipe provided a conduit for air to be supplied to the inside of the waste pile. The whole pile was mixed thoroughly every 7–8 days. The waste used in this study was raw sludge carried through sewage lines from houses in the city of Kagoshima and consisted mainly of food wastes from kitchens, waste from toilets including papers, and active sludge. The front half of the compost pile was removed to crosscut the pile. Samples were taken every 20 cm from the bottom to the top of the exposed surface of the crosscut and were immediately frozen in a dry ice bath.

Determination of Physical and Chemical Properties

The samples (2.0 g) were suspended in 10 mL of water and vigorously stirred. The suspension was centrifuged and the pH of the supernatant was determined. The pH was slightly alkaline in the range of 7.2–7.8 and did not change significantly from the bottom to the top surface.

The temperature was measured by inserting a thermometer 1 m deep from the crosscut surface into the pile. The temperature change is shown in FIGURE 2. The highest temperature was recorded about 160 cm from the bottom. The temperature at 5 cm above the floor and at 5 cm below the top surface was 68°C, suggesting that the waste pile is a good thermal insulator and little heat can escape from the inside of the waste pile.

Total carbon content and total nitrogen content were determined using a carbon and nitrogen content analyzer (CN Corder MT-700; Yanaco, Tokyo, Japan). The carbon content was decreased when the temperature was higher, and was the lowest at 160–180 cm from the bottom where temperature was the highest. These results suggest that organic carbon compounds were most actively degraded by extreme thermophiles rather than moderate thermophiles or mesophiles, as had been reported for common lower-temperature compost (FIG. 2).

Variations in total nitrogen content were similar to those of carbon content: the lowest value was recorded 160 cm from the bottom—the region with the highest temperature—whereas the higher nitrogen content was recorded in the cooler sites. Because changes in nitrogen content paralleled those of carbon content, the ratio of carbon to nitrogen was almost constant throughout the compost pile.

The water content of the compost material is also shown in FIGURE 2. Contrary to our expectations, generally, the higher the temperature, the more water existed in the area. The highest water content (around 25%) was in the region around 180 cm from the bottom. The bottom and the top surface were drier than the inside.

Microbial Community

We have attempted to isolate extreme thermophiles from the compost materials. The medium used consisted of 0.3% NZ-case (Wako Pure Chemicals, Osaka,

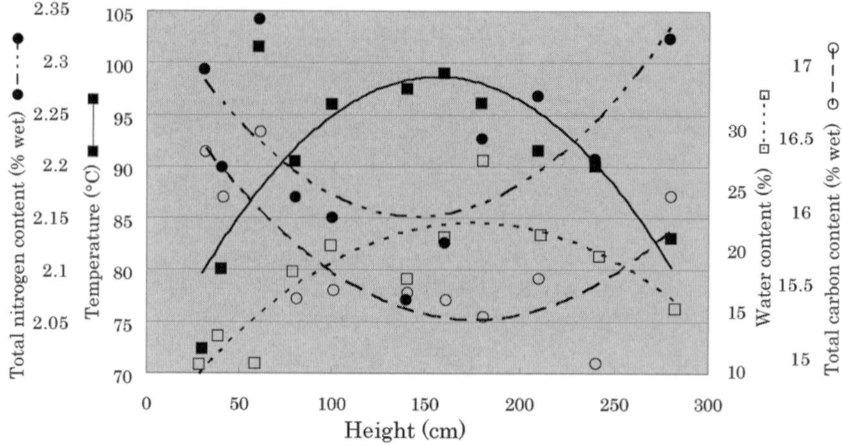

FIGURE 2. Changes in temperature, water content, total carbon content, and total nitrogen content depending on the height from the bottom.

Japan), 0.2% Difco yeast extract (Becton, Dickinson and Company, Sparks, MD), 0.1% soluble starch, 3% NaCl, 0.0125% $MgCl_2$, 0.0025% $CaCl_2$, and 0.001% $FeSO_4.7H_2O$. The pH of the medium was adjusted to 7.5 at room temperature.

Many moderate thermophiles belonging to *Geobacillus stearothermophilus* (formerly *Bacillus stearothermophilus*) or related species have been isolated at 70°C from samples throughout the waste pile including from the hottest spot. This suggests the presence of spores of moderate thermophiles distributed throughout the compost pile and present even at the hottest site.

Some extreme thermophiles were isolated at 80°C. However, so far all isolates at 80°C belong to only one or two closely related aerobic species. No bacterial growth has been observed when isolation was attempted at 85°C or higher temperatures. This suggests that most of the extreme thermophiles responsible for the degradation of organic substances in the hotter sites of the high-temperature compost could not be cultured using our current conditions.

Strain YMO81 is the typical aerobic isolate at 80°C and is a Gram staining negative, nonspore-forming, rod-shaped bacterium. The highest growth temperature of this isolate was 83°C. We investigated proteolytic activities of a cell-free extract of YMO81 using different proteins, such as casein, as substrates. YMO81 cannot digest collagen, although zymography for collagenases (FIG. 4) showed that enzyme activity exists even in the hottest site of the waste pile. This observation also supports our prediction that as yet uncultured thermophiles exist at least in sites hotter than 80°C. More work is needed to determine the source of collagenase activity in the hotter sites in the compost pile.

The G+C content of the chromosomal DNA of strain YMO81 is about 70%. Although the isolate is nonspore-forming and Gram staining negative, small subunit rRNA gene sequence homology indicated that the species most similar to YMO81 is *G. stearothermophilus*. The sequence homology between the base sequence of a small subunit rRNA gene of strain YMO81 and that of *G. stearothermophilus* is 85%, indicating that YMO81 can be classified into a different genus. Using 16S rRNA gene sequence comparison, strain YMO81 belongs to group I of the Firmicute. However, the G+C mol% content of the strain is high, and the cell envelope structure of the isolate more closely resembles that of a Gram staining negative bacterium and consists of a cytoplasmic membrane, cell wall, and outer membrane.

Based on these observations, we concluded that YMO81 represents the first species (and thus the type species) of a novel genus. We are proposing to name isolate YMO81 *Caldaterra satsumae* spec. nov. (manuscript in preparation). The name *satsumae* is derived from the ancient name of the city of Kagoshima from which the sample material for the isolation was obtained.

Because it appears that the majority of the extreme thermophiles responsible for fermentation at sites hotter than 80°C cannot be cultured using our current techniques, we extracted DNA from the sample soils of the high-temperature compost. The extracted DNA was analyzed by PCR-denaturing gradient gel electrophoresis (PCR-DGGE) techniques.

DNA was extracted from each sample using a DNA extraction kit, ISOIL (Nippon Gene Co., LTD., Tokyo, Japan). PCR was performed using Ampli Taq Gold according to the manufacturer's instruction manual (Perkin Elmer Japan, Applied Biosystems Division,

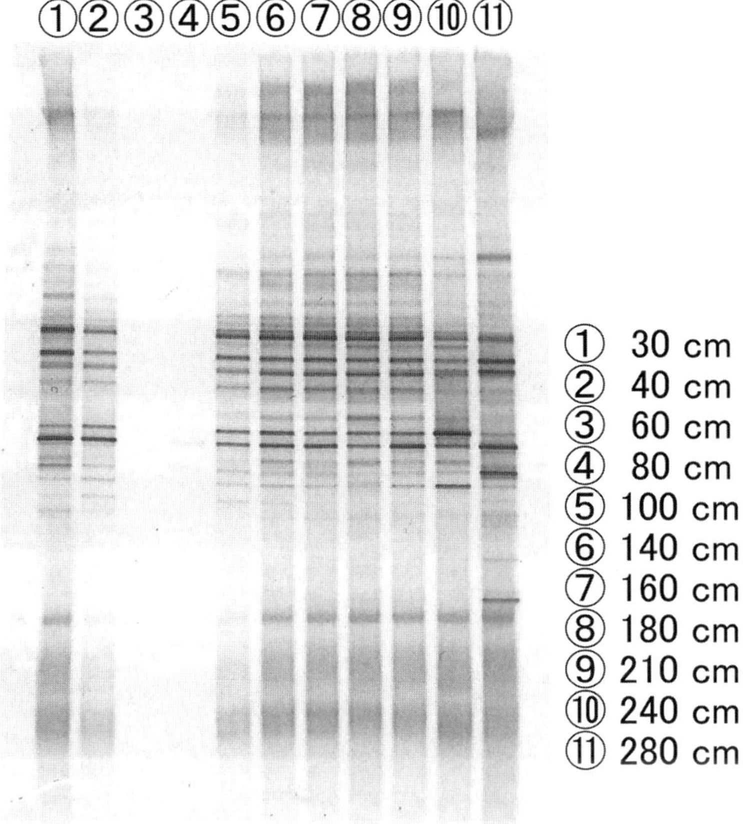

FIGURE 3. DGGE analysis of DNA extracted from selected sites in the compost pile. DNA extracted from the compost pile was amplified by PCR using bacterial domain primers. No detectable amount of DNA was extracted from samples taken from 60 cm and 80 cm from the bottom for reasons that are unclear. Samples were taken from 30 cm (lane 1), 40 cm (lane 2), 60 cm (lane 3), 80 cm (lane 4), 100 cm (lane 5), 140 cm (lane 6), 160 cm (lane 7), 180 cm (lane 8), 210 cm (lane 9), 240 cm (lane 10), and 280 cm (lane 11) from the bottom of the compost pile.

Chiba, Japan). Primers used for the reactions were 357F-GC, 5′-CGC CCG CCG CGC GCG GCG GGC GGG GCG GGG GCA CGG GGG GCC TAC GGG AGG CAG CAG-3′ and 517R, 5′-ATT ACC GCG GCT GCT GG-5′.[5] The temperature cycle was 30 s denaturation at 93°C, 30 s annealing at 65°C for the first 10 cycles, 60°C for the second 10 cycles, and 55°C for the last 10 cycles, and finally 1 min primer extension at 72°C using TaKaRa Thermal Cycler Dice TP600 (Takara Bio Co., Otsu, Japan). The products of PCR amplification were examined by electrophoresis on 2% agarose gel plates prior to DGGE analysis.

DGGE analyses of the PCR products were evaluated by the method reported by Muyzer *et al.*[5] with the DCode Universal Mutation Detection System (Bio-Rad, Hercules, CA). Denaturing gradient gel, 1 mm thickness and 160 × 160 mm, was prepared in 20 mM Tris buffer, pH 7.4, containing 0.5 mM ethylenediamine tetraacetic acid (EDTA) and 10 mM sodium acetate. The concentration gradient used was 6%–12% polyacrylamide and 1.4–4.9 M urea with 8%–28% formamide. Electrophoresis was carried out at 200 V for 3 h in 20 mM Tris buffer, pH 8.0, at 60°C. After electrophoresis, the gel was stained with SYBR Green I (Invitrogen Japan, Tokyo, Japan).

A representative DGGE gel is shown in FIGURE 3. Many bacterial rRNA gene amplicons were generated via PCR amplification and some of them were sequenced. For the sequence determination, a band detected on the gel was excised from the gel plate with a razor blade. The gel piece was washed with ethanol and then soaked in 0.5 M ammonium acetate containing 10 mM magnesium acetate, 1 mM EDTA,

TABLE 1. Results of 16S rRNA gene sequence comparison

Species[a]	rRNA gene sequence homology (%)	Presence or absence (+ or −) at three locations (height from the bottom)		
		30 cm	160 cm	280 cm
Bacillus licheniformis	95	±	±	+
Bacillus niabensis	96	+	+	+
Planifilum yunnanesis	95	+	+	+
Bacillus eolicus	99	−	+	+
Bacillus pumilus	97	+	+	±
Bacillus thermocloacae	97	+	±	+
Anoxybacillus toebii	96	+	+	+
Thermoactinomyces dichotomicus	98	+	−	+
Sulfobacillus sp.	90	+	−	+

[a]The species closest to the rRNA gene sequence detected in DGGE analysis shown in FIGURE 3 are listed.

and 0.1% SDS (pH was adjusted to 8.0). DNA was extracted by incubations at 50°C for 1 h and then obtained by using the QIAEXII gel extraction kits (Qiagen Company, Tokyo, Japan). PCR amplification was carried out with a 517R and 357F primer that had deleted GC clamp from 357F-GC. The amplified products were again analyzed using DGGE to confirm that the preparation was identical to the detected DNA on the original DGGE gel as shown in FIGURE 3. Sequences were determined by Macrogen Inc. (Seoul, Korea). The sequence data were phylogenetically analyzed using the DDBJ Blast program provided by DNA Data Bank of Japan (Mishima, Japan). If the sequence homology was higher than 98% to the 16S rRNA gene sequences of any known species, we tentatively concluded that the detected DNA originated from an isolate belonging to that species.[6] Some of the results are summarized in TABLE 1, in which the taxonomic names of the closest species and the sequence homology to the 16S rRNA gene sequences of the closest species are listed.

Microbial diversity was slightly reduced as the temperature increased, as evidenced by the decrease in number of amplicons generated in the sample material taken from the hottest spot. Many 16S rRNA gene sequence bands detected on the DGGE electrophoreogram can be assigned to *Bacillus* species or related microorganisms, including *Thermoactinomyces* species. Several 16S rRNA gene sequences representing anaerobic bacteria were also detected. One example is the presence of the 16S rRNA gene sequences homologous to that of *Anoxybacillus* species, as shown in TABLE 1. This suggests that anaerobes could be involved in the biodegradation process. Further isolations and characterization of the anaerobic community has to be done in the near future.

Meta-enzymology of the High-Temperature Compost

Compost samples (1 g wet weight) were suspended in 10 mL of 10 mM potassium phosphate buffer, pH 7.0, and then centrifuged. Enzymatic activities were analyzed by zymography. The gel for collagenase activity consists of 5% (w/v) polyacrylamide containing 1.6% w/v gelatin.[7] The electrophoresis was performed for 3 h at 100 V. The gel was incubated at 70°C overnight in 10 mM potassium phosphate buffer, pH 7.0. After the incubation overnight, the gel was stained with Coomassie Blue. For the analysis of amylase activities, the extracts from the waste pile were subjected to gel electrophoresis. Conditions were the same as those mentioned above except that gelatin was not added. After electrophoresis, the gel was incubated in 10 mM potassium phosphate buffer, pH 7.0, containing 1% soluble starch at 70°C for 3 h. To visualize the enzymatic activity—that is, the degradation of starch—the gel was stained with an iodine solution.[8] The results are shown in FIGURES 4 and 5. Collagenases were detected only in samples taken from 140 cm or higher from the bottom. At least four different collagenases exist in the high-temperature compost. These collagenases were heat-resistant up to at least 80°C (data not shown). In contrast, amylase activities were found only in the lower parts of the pile. The zymography suggested the presence of only a limited number of amylases.

We are currently attempting to purify the enzymes from samples taken from varying sites in the compost pile. The presence of a variety of organic substances, especially deeply colored substances bound tightly with proteins, perhaps lignins, complicates this procedure.

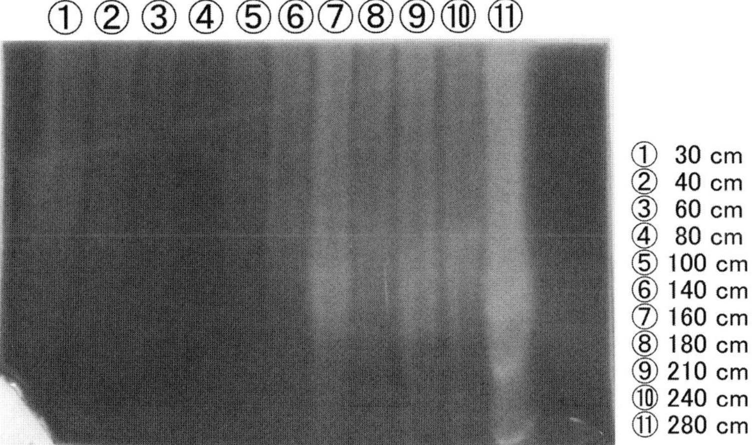

FIGURE 4. Gelatin zymography. Extracts of the samples taken from the compost were applied on a gel. After electrophoresis, the gel was incubated at 70°C overnight and then stained using Coomassie Blue. Samples were taken at 30 cm (lane 1), 40 cm (lane 2), 60 cm (lane 3), 80 cm (lane 4), 100 cm (lane 5), 140 cm (lane 6), 160 cm (lane 7), 180 cm (lane 8), 210 cm (lane 9), 240 cm (lane 10), and 280 cm (lane 11) from the bottom of the compost pile.

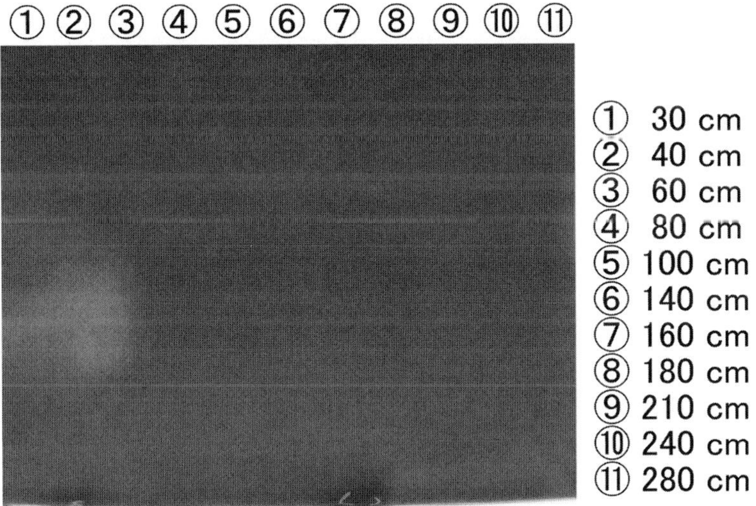

FIGURE 5. Starch zymography. Activity was detected using iodine staining. Samples were taken at 30 cm (lane 1), 40 cm (lane 2), 60 cm (lane 3), 80 cm (lane 4), 100 cm (lane 5), 140 cm (lane 6), 160 cm (lane 7), 180 cm (lane 8), 210 cm (lane 9), 240 cm (lane 10), and 280 cm (lane 11) from the bottom of the compost pile.

Conclusions

The fermentation processes of the high-temperature compost are heterogeneous and consist of combinations of many different micro-environments. With the exception of the top surface and the bottom, the temperature varied from 70°C to 100°C. However many mesophiles and moderately thermophilic microorganisms were isolated from samples removed throughout the compost pile, including the hottest spot, probably because of the presence of heat-resistant spores. Most of the isolated bacteria belonged to *Bacillus* species or related genera. From the hottest spot, only a limited number of extreme thermophiles were isolated, suggesting that most of the extreme thermophiles present in the high-temperature compost were not cultured under the growth conditions used in the present study.

The bacterial community also differed from sample to sample. Although high-temperature composting involves a continuous supply of air, DGGE analyses showed the presence of anaerobes, suggesting that anaerobic processes also take place, probably in micro-environments. One example is the presence of facultative aerobic *Anoxybacillus* species. The 16S rRNA gene sequences belonging to bacteria clustering with anaerobic species were detected anywhere from the bottom to the top surface of the compost pile.

Likewise, enzymatic activities are localized and are not homogeneous. Starch degradation was observed only in samples taken from the lower one-fourth of the pile. In contrast, collagen-degrading activity was located in the upper half of the pile. Oxidation activity of organic substances seems to be highest in the hottest spot and parallels the temperature. Where the temperature is higher, the total carbon and nitrogen content is lower (FIG. 2).

High-temperature compost appears to be a promising source for isolation of new thermophilic organisms as well as heat-stable enzymes of industrial value. We isolated a new extreme thermophile that is capable of growing at 83°C. Its 16S rRNA gene sequence showed low homology to any of the 16S rRNA gene sequences deposited in the ribosomal data bank. The DNA base sequence comparison showed that the closest organism to the new isolate, "*C. satsumae*," is *G. stearothermophilus*, a Gram staining positive and spore-forming thermophile, although the isolate is Gram staining negative and nonspore-forming, The sequence homology between base sequences of 16S rRNA genes of these two microorganisms is 85%, suggesting that the thermophilic isolate can be classified into a novel genus.

Conflict of Interest

The authors declare no conflicts of interest.

References

1. NORMAN, A.G., L.A. RICHARDS & R.E. CARLYL. 1941. Microbial thermogenesis in the decomposition of plant materials. Part 1. An adiabatic fermentation apparatus. J. Bacteriol. **41:** 689–697.
2. ZINDER, S.H. 1986. Thermophilic waste treatment systems. *In* Thermophiles: General, Molecular, and Applied Microbiology. T.D. Brock, Ed.: 257–277. John Wiley & Sons. New York, NY.
3. SAIKI, T., T. BEPPU, K. ARIMA, *et al.* 1978. Changes in microbial flora, including thermophiles, during composting of animal manure. *In* Biochemistry of Thermophily. S.M. Friedman, Ed.: 103–115. Academic Press. New York, NY.
4. KANAZAWA, S., Y. ISHIKAWA, K. TOMITA-YOKOTANI, *et al.* 2008. Space agriculture for habitation on Mars with hyperthermophilic aerobic composting bacteria. Adv. Space Res. **41:** 696–700.
5. MUYZER, G., E.C.D. WAAL & A.G. UITTERLINDEN. 1993. Profiling of complex microbial populations by denaturing gradient gel electrophoresis analysis of polymerase chain reaction-amplified genes coding for 16S rRNA. Appl. Environ. Microbiol. **59:** 695–700.
6. STACKEBRANDT E. & B.M. GOEBEL. 1994. Taxonomic note: a place for DNA-DNA reassociation and 16S rRNA sequence analysis in the present species definition in bacteriology. Int. J. Syst. Bacteriol. **44:** 846–849.
7. BLUMENTALS, I.I., A.S. ROBINSON & R.M. KELLY. 1990. Characterization of sodium dodecyl sulfate-resistant proteolytic activity in the hyperthermophilic archaebacterium *Pyrococcus furiosus*. Appl. Environ. Microbiol. **56:** 1992–1998.
8. NAKAMURA, K., S. HARUTA, H.L. NGUYEN, *et al.* 2004. Enzyme production-based approach for determining the functions of microorganisms within a community. Appl. Environ. Microbiol. **70:** 3329–3337.

Microbiology to Help Solve Our Energy Needs

Methanogenesis from Oil and the Impact of Nitrate on the Oil-field Sulfur Cycle

ALEXANDER GRIGORYAN AND GERRIT VOORDOUW

Department of Biological Sciences, University of Calgary, Calgary, Alberta, Canada

Our society depends greatly on fossil fuels, and the environmental consequences of this are well known and include significant increases of the CO_2 concentration in the earth's atmosphere. Although microbiology has traditionally played only a minor role in fossil-fuel extraction, two novel key discoveries indicate that this may change. First, the realization that oil components can be converted to methane and CO_2 by methanogenic consortia in the absence of electron acceptors (oxygen, nitrate, sulfate) explains how much of the world's oil has been biodegraded *in situ*. In addition to inorganic nutrients, only water is needed for these methanogenic conversions. Hence, continued methanogenic biodegradation may have shaped the heavy-oil reservoirs that are so prevalent today. The potential to exploit these reactions, for example, by *in situ* gasification, is currently being actively investigated. Second, injection of nitrate in oil and gas fields can lower sulfide concentrations. High sulfide concentrations, caused by the action of sulfate-reducing bacteria (SRB), are associated with increased risk of corrosion, reservoir plugging (through precipitated sulfides), and human safety. Nitrate injection into an oil field stimulates subsurface heterotrophic nitrate-reducing bacteria (hNRB) and nitrate-reducing, sulfide-oxidizing bacteria (NR-SOB). Nitrite, formed by these NRB by partial reduction of nitrate, is a strong and specific SRB inhibitor. Nitrate injection has, therefore, promise in positively controlling the oil-field sulfur cycle. There is now more interest in and potential to apply petroleum microbiology than there has been in the past, allowing microbiologists to contribute to a sustainable energy future.

Key words: **oil field; sulfur cycle; sulfate-reducing bacteria; nitrate-reducing bacteria; souring; methanogenesis; sulfide**

Introduction

This article is based on a lecture presented at the Symposium on Incredible Anaerobes: From Physiology to Genomics to Fuels, to honor Lars Ljungdahl on the occasion of his 80th birthday. Lars contributed greatly to uncovering the metabolism of anaerobic bacteria and fungi, including those that can hydrolyze polymeric cellulose to monomeric glucose. These microorganisms ferment the liberated glucose to a variety of products, including hydrogen, alcohols, and ketones. Because the energy yield of these fermentations is small, allowing synthesis of only 2–4 adenosine 5′-triphosphate (ATP) per glucose fermented, these anaerobes must process a large amount of glucose to synthesize biomass. Hence, their specific growth yield (Y_{xg}), defined as grams of biomass formed per gram of glucose fermented is low, $Y_{xg} = 0.10$ being a typical value. In contrast, aerobic microorganisms access the full free energy available in the glucose molecule ($C_6H_{12}O_6$) by catalyzing reaction 1 (TABLE 1), allowing the synthesis of more than 30 ATP per glucose oxidized. As a consequence their specific growth yield is much higher, with $Y_{xg} = 0.55$ being a typical value. Hence, in order to grow as fast as an aerobe, an anaerobe must process more glucose molecules to synthesize a similar number of ATP molecules per unit time. This high glucose flux gives rise to large amounts of fermentation products, which are more conspicuous then the smaller amounts of biomass formed. The anaerobes that catalyze fermentations are oftentimes named after the main products, as in *Clostridium acetobutylicum* and *Clostridium propionicum*.

There is currently enormous interest in the question as to what extent conversion of plant polymers, such as cellulose, into fermentation products like ethanol can meet the growing energy needs of the world. The use of biomass is considered green-house-gas neutral, because CO_2, liberated when the ethanol is burned,

Address for correspondence: Department of Biological Sciences, University of Calgary, 2500 University Dr. NW, Calgary, Alberta, T2N 1N4, Canada.

voordouw@ucalgary.ca

Ann. N.Y. Acad. Sci. 1125: 345–352 (2008). © 2008 New York Academy of Sciences.
doi: 10.1196/annals.1419.004

was only recently sequestered when the biomass was grown. In contrast, because the CO_2, liberated when fossil fuel is burned, was sequestered tens of millions of years ago, the massive burning of fossil fuels today contributes to increasing atmospheric CO_2 concentrations with associated global warming. Although this would tend to give an edge to biomass-derived fuels, it has been stated that the ratio of energy returned on energy invested (EROI) is low for these fuels.[1] In contrast, this ratio is high for fossil fuels, for example, 100 to 1 for Saudi Arabian crude. The fact that meeting the energy needs of our society requires an abundant energy source with a high EROI presents a huge challenge for using alternative forms of energy. Key steps toward a sustainable energy future will therefore be to (1) reduce per capita energy consumption, (2) have renewables contribute a larger fraction of our energy supply, and (3) make extraction and use of fossil fuels as efficient and green as possible. Microbiology contributes to step 2 through fermentation processes, but can also contribute to step 3, as will be discussed in this chapter.

TABLE 1. Reactions as used and as explained in the text

Reaction 1	$C_6H_{12}O_6 + 6O_2 \rightarrow 6H_2O + 6CO_2$
Reaction 2	$6CO_2 + 6H_2O + hv \rightarrow C_6H_{12}O6 + 6O_2$
Reaction 3	$4C_{16}H_{34} + 40H_2O \rightarrow 49CH_4 + 15CO_2$
Reaction 4	$4C_{16}H_{34} + 64H_2O \rightarrow 32CH_3COO^- + 32H^+ + 68H_2$
Reaction 5	$32CH_3COO^- + 32H^+ \rightarrow 32CH_4 + 32CO_2$
Reaction 6	$68H_2 + 17CO_2 \rightarrow 17CH_4$
Reaction 7	$FeCO_3 + HS^- \rightarrow FeS + HCO_3^-$
Reaction 8	$HS^- + 4NO_3^- \rightarrow H^+ + SO_4^{2-} + 4NO_2^-$
Reaction 9	$7H^+ + 5HS^- + 2NO_3^- \rightarrow 5S^0 + N_2 + 6H_2O$

Mechanism of Anaerobic Oil Degradation *in Situ*

Oil is ultimately derived from photosynthesis. Light-driven fixation of CO_2 into carbohydrate (TABLE 1: reaction 2, where hν symbolizes light energy) and subsequently into biomass of photosynthesizing plants, algae, and microbes provides a net flux of fixed carbon into ecosystems, referred to as their net primary productivity. The net primary productivity of ecosystems varies from 100 to 5000 g of dry organic matter per m^2 and per year.[2] This net amount of photosynthetic biomass produced is available to heterotrophic consumers, which recycle the fixed carbon to CO_2, by reaction 1 (TABLE 1). However, because of the absence of oxygen or the recalcitrance of some biomass fractions (e.g., the lignin component of plant cell walls), some of the produced biomass escapes recycling by hetrotrophic consumers. Recycling this fraction can require millions of years and can involve burial into sedimentary rock, downward movement increasing temperature and pressure, and finally, when temperatures reach values well over 100°C, thermal decomposition into oil components and ultimately into CO_2 and methane.[3,4] During this process, a low concentration of oil is formed in sedimentary source rock over a large area. Because it is poorly miscible with water and lighter (i.e., having a density at room temperature of $\rho < 1.0\,g\,cm^{-3}$), it moves upward where it can be concentrated underneath dome-shaped, impermeable geological layers.[3] This upward and lateral movement concentrates oil hydrocarbons from a large area of source rock into spatially more limited fields, from which oil can be produced commercially. Hence, an oil field is created over tens of millions of years by biological fermentation and chemical degradation reactions of ample net produced biomass, along with a geological fractionation process that leads to concentration of the energy source. Overall this concentrated energy source (the oil and gas) requires little energy input to produce (and hence the favorable EROI). When viewed as a whole, the process resembles converting photosynthetic biomass into biofuel, such as ethanol. The main differences are in the dilute nature of the biomass resource and in the desire for a short time scale of the process in which it is used (1 year versus the millions of years required to create a subsurface oil field). This increases the required energy input due to the need to grow, collect, pretreat, ferment, and fractionate in an ordered, action-driven manner (i.e., to produce ethanol) rather than in a random, opportunity-driven fashion (i.e., to produce oil). The increased energy input associated with biomass to ethanol processes reduces their EROI and thereby the likelihood that they will contribute significantly to covering the world's energy needs.

A subsurface oil field, once formed, is not forever stable, but subject to *in situ* biodegradation. The nature of the microbial process causing subsurface oil biodegradation has long been a mystery. Earlier dogma was that oil biodegradation requires an electron acceptor, either oxygen, nitrate, Fe(III), or sulfate. However, because these are always in short supply in the oil-bearing subsurface, extensive biodegradation could not be explained. The recent demonstration that methanogenic consortia can catalyze the reaction of selected oil components with water[5] has changed our perspective on how oil can be biodegraded *in situ*. Zengler *et al.*[5] discussed and demonstrated how a consortium of

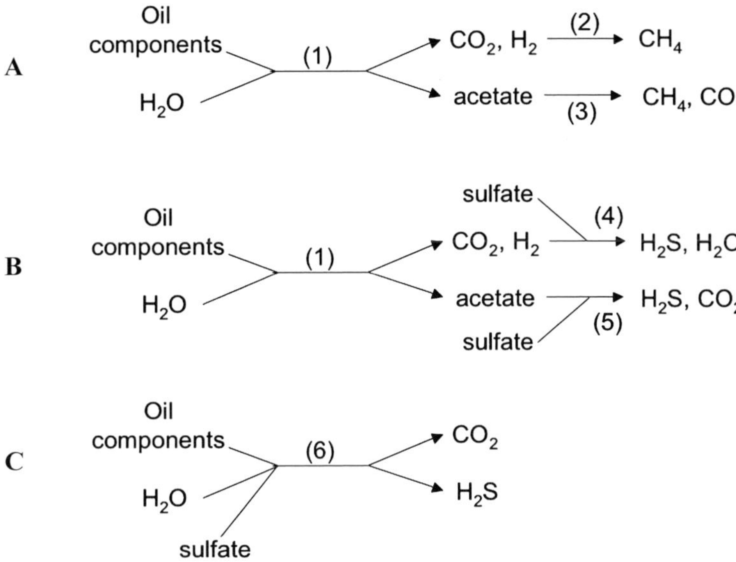

FIGURE 1. Mechanism of *in situ* anaerobic oil degradation in oil reservoirs. (**A**) In the absence of sulfate, oil is attacked by (1) syntrophic bacteria forming CO_2, H_2, and acetate. These products are converted to methane and CO_2 by (2) CO_2-reducing methanogens, and (3) acetotrophic methanogens. (**B**) In the presence of sulfate the H_2 and acetate produced by (1) the syntrophic bacteria can potentially be used by SRB of (4) the genus *Desulfovibrio* and (5) of the genus *Desulfobacter*. (**C**) Oil fields also harbor SRB (6), which directly oxidize selected oil components with sulfate.

bacteria can catalyze the overall conversion of hexadecane into methane and CO_2 by reaction 3 (TABLE 1, $\Delta G = -1600$ kJ). In this consortium, syntrophs catalyze the initial attack, producing acetate and hydrogen intermediates (TABLE 1, reaction 4), which are used by acetotrophic methanogens (TABLE 1, reaction 5) and methanogens (TABLE 1, reaction 6). The latter two groups serve to keep the concentration of hydrogen and acetate intermediates low, allowing the oil hydrocarbon attack by the syntrophs to proceed. Therefore, in addition to the inorganic nutrients required for microbial growth (e.g., phosphate, ammonium), the only ingredient needed for continued oil degradation is water, which is generally present in abundance. The terminal electron acceptors listed earlier in this chapter are not required. It had been previously demonstrated that other oil hydrocarbons, for example, low molecular-weight aromatics, such as toluene, are also converted by anaerobic, methanogenic consortia into methane and CO_2.[6] A schematic of these reactions is provided in FIGURE 1A. Hence, when methanogenic consortia are present, active oil degradation may be expected to occur at the bottom of the trapped oil column in the zone of oil–water contact (OWC),[7] causing the oil column to be topped by a methane gas cap.[3]

The methanogenic consortia (FIG. 1A) are selective in the oil hydrocarbons that they use, preferring the light-end, lower molecular-weight aromatic and aliphatic hydrocarbons, while presumably ignoring the more complex heavier resin and asphaltene fractions. Hence, methanogenic conversion increases the content of resins and asphaltenes, while decreasing the content of lighter aliphatic and aromatic fractions (TABLE 2).[7] In this process, light oil, with a density of $\rho < 1.0\,\mathrm{g\,cm^{-3}}$ and low viscosity is converted to heavy oil and eventually to tar ($\rho > 1.0\,\mathrm{g\,cm^{-3}}$) with very high viscosities (TABLE 2). Physical weathering of spilled oil can cause similar transformations. Light oil on the water surface ($\rho < 1.0\,\mathrm{g\,cm^{-3}}$) is converted by evaporation and

TABLE 2. Effect of biodegradation of specific oil fractions in the subsurface on oil composition and physical properties[1]. The arrows indicate the natural progression towards oils that are more viscous and have lower aliphatic content

	light oil →	heavy oil →	tar
Density (g/cc)	0.8-0.9	1.0-1.004	1.02
Viscosity (centipoise)	10^3-10^4	10^4-10^5	10^5-10^7
Aliphatics	40%	22%	17%
Aromatics	30%	20%	18%
Resins	20%	41%	44%
Asphaltenes	10%	17%	17%

[1]The values given could differ substantially for individual oils.

(bio)degradation of its light components to tarry pellets that sink ($\rho > 1.0\,g\,cm^{-3}$) and may wash up on the beach. Subsurface oil is subject to methanogenic transformation at the base of the oil column in the OWC zone.[8] Hence, the properties of an oil undergoing biodegradation *in situ* differ as a function of vertical distance. At the bottom near the OWC, the oil may be heavy, viscous, and devoid of alkanes, whereas at the top, it may have the physical and chemical characteristics of light oil (TABLE 2). Complete biodegradation requires some form of oil convection or diffusion to bring oil in the middle or at the top of the oil column in contact with water.[8] The higher viscosities in the OWC would slow down such diffusion or convection processes. The scale of methanogenic transformations can be appreciated when one considers that the Alberta tar sands, spanning about 100,000 km^2 and holding about 2.5×10^{12} barrels of bitumen (TABLE 2: tar), are thought to have resulted from such a methanogenic transformation.[7] Hence, large amounts of aliphatic and aromatic hydrocarbons of the order of 10^{12} barrels, 1 barrel being 159 L, were converted into methane and CO_2 (TABLE 1, reaction 3) over geological time, converting a reservoir holding light oil into the highly viscous bitumen deposits that are being mined today.

Most subsurface oil is produced by water injection and over time the fraction of water (referred to as the water cut) increases at the expense of the fraction of oil. A focus of petroleum engineering is to understand and potentially manipulate reservoir permeability in order to prevent or slow down this increase in water cut. Acceleration of the methanogenic transformation of oil *in situ* is being considered as an alternative for producing residual oil that cannot be economically produced by water injection or other conventional means, as methane (TABLE 1, reaction 3). Technologies based on such a microbially based process would have to be able to deal with the increases in viscosity that will result from these methanogenic transformations. Thermophilic methanogenic consortia operating at $\sim 50°C$ have been demonstrated,[9] and one objective of these technologies could be to identify and use consortia that can operate at an even higher temperature (80°C), where oil viscosities are substantially lower, creating more favorable conditions for microbial metabolism.

Impact of Nitrate on the Oil-field Sulfur Cycle

When water is injected to maintain the reservoir pressure required to keep the flow of oil, the oil–water mixtures emerging at producing wells provide a unique opportunity to obtain samples from the oil-bearing subsurface and determine which bacterial groups are present. These groups include sulfate-reducing bacteria and archaea, which we will collectively refer to as SRB, the methanogenic archaea, which were already discussed, as well as fermentative, iron-reducing, and nitrate-reducing bacteria (NRB). These communities are indigenous and are generally distinct from those present in the injection source water.[10] Among bacteria present in oil fields, the SRB are the most studied, primarily because their metabolic activity increases the H_2S concentrations in the produced oil, water, and gas. This process is referred to as souring[11] and is unwanted because high sulfide concentrations can cause metal corrosion, reservoir plugging, as well as health and safety concerns. These negative effects explain the preoccupation of the oil and gas industry with preventing or inhibiting SRB activity.[10,11,34] Isolation and characterization of oil-field SRB has yielded a wide variety of organisms. The taxonomic affiliation and properties of some mesophilic SRB (optimum growth temperature 20–37°C), isolated from a variety of oil-field settings in the world,[12–22] are summarized in TABLE 3. All except *Desulfobacter vibrioformis* can use hydrogen, whereas the latter organism is the only one listed that uses acetate.

It would be helpful if we understood which oil organics serve as an electron donor for sulfate reduction in oil fields *in situ*. Formation waters (waters in contact with the oil) often contain volatile fatty acids (VFA, a mixture of acetate, propionate, and butyrate), which would allow growth of the *Desulfobulbus* and *Desulfobacter*, but not of the *Desulfovibrio* spp. (TABLE 3). SRB capable of using hexadecane or toluene directly as an electron donor for sulfate reduction (FIG. 1C) have been isolated.[23,24] These organisms may be important in oxidizing selected components of light oils. The role of *Desulfovibrio* spp., which tend to oxidize substrates that are not abundant in the oil-field environment, is somewhat enigmatic. Perhaps these bacteria grow in association with syntrophs, that is, replace the methanogens from syntrophic consortia, whenever sulfate is present (FIG. 1B). *Desulfovibrio* spp. are excellent hydrogen oxidizers and could stimulate syntrophic degradation of hydrocarbon by keeping the hydrogen concentrations low. Because we know little about the substrate specificity and activity of syntrophs, the effective concentration of electron donors available for sulfate reduction in an oil field is hard to estimate. This, in turn, makes it difficult to predict the extent to which sulfate present in the injection water may be converted into sulfide, that is, the degree to which the field may sour (see later in this chapter).

TABLE 3. Some characteristics of mesophilic sulfate-reducing bacteria isolated from oil fields

Organism, taxa	Source	Salinity[a]	t (°C)[a]	Donors[b]	Other acceptors[c]	Ref.
Desulfovibrio desulfuricans	Oil-field formation waters, Azerbaijan and Guinea	2	20	H_2, lactate, malate		12, 13
Desulfovibrio africanus	Oil-field formation waters, Azerbaijan	2	35	H_2, lactate		12
Desulfovibrio bastinii	Oil-field formation waters, Congo	4	37	H_2, formate, lactate, malate, fumarate, ethanol, butanol	S^0	14
Desulfovibrio gabonensis	African oil pipeline	5.5	30	H_2, formate, lactate, malate, fumarate, succinate ethanol, butanol	S^0	15
Desulfovibrio gracilis	Oil-field formation waters, Congo	5.5	37	H_2, lactate, fumarate, and acetate	S^0, fumarate	14
Desulfovibrio vietnamensis	Oil-field production waters, Vietnam	5	37	H_2, formate, lactate, malate, fumarate, ethanol, glycerol	fumarate	16
Desulfomicrobium apsheronum	Oil-field formation waters, Azerbaijan	1	30	H_2, formate, lactate, alcohols		17
Desulfobacterium cetonicum	Oil-field formation waters, Azerbaijan	1	30	H_2, C_1-C_{16} fatty acids, lactate, alcohols, ketones	S^0	18
Desulfobacter vibrioformis	Water–oil separation system, North Sea oil platform	3	33	Acetate		19
Desulfobulbus rhabdoformis	Water–oil separation system, North Sea oil platform	2	31	H_2, lactate, propionate, ethanol, propanol, fumarate, malate		20
Desulfovermiculus halophilus	Oil-field formation waters	9	37	H_2, formate, propionate, butyrate, ethanol, lactate, fumarate, malate	S^0	21
Desulfotomaculum halophilum	Oil-field formation waters, Paris Basin, France	5	35	H_2, formate, lactate, malate, ethanol, butanol		22

[a]Optimal salinity (NaCl, % w/v) and optimal growth temperature (°C).
[b]Electron donors used. All strains using lactate also use pyruvate. Strains using H_2 require CO_2 or CO_2 and acetate as the carbon source for growth. Note that the list is not necessarily exhaustive, that is, *D. africanus* likely also uses formate.
[c]Electron acceptors other than sulfate, sulfite, and thiosulfate, which are used by all.

Souring of an oil field is a frequent consequence of the injection of sulfate-containing water. Although some fields are naturally sour, most acquire high sulfide concentrations as the result of SRB activity. Consider, for instance, an offshore oil field into which 10,000 m^3 of seawater is injected per day, producing 8000 m^3 of oil and 8000 m^3 of water per day. Because of the high sulfate concentration of seawater (~1 g/L of sulfate sulfur), 10 tons of sulfate sulfur are injected daily. If 10% of this is reduced by SRB oxidizing oil organics, the daily sulfide production would approximate 1 ton/day. The produced sulfide could remain in the reservoir if this is rich in iron minerals, for example, through reaction with siderite ($FeCO_3$) producing precipitated iron sulfide and soluble bicarbonate (TABLE 1, reaction 7).[11] However, eventually this buffering capacity of the reservoir will be exhausted and soluble sulfide will start to emerge in the oil, aqueous, and gas phases. Daily production of 1 ton of sulfide together with 16,000 m^3 oil–water would result in an overall sulfide concentration of 60 ppm, which is considered sour and in need of mitigation. This theoretical example may be compared with an actual case of sulfide production in the Skjold field in the Danish sector of the North Sea, where seawater injection started in 1985 and total H_2S production in all three phases increased from 100 to 1000 kg per day from 1994 to 2000, respectively.[25] The resulting high sulfide concentrations need to be removed from the oil and gas phases before shipment by pipeline to shore in order to minimize potential corrosion problems. This can be done by several methods, including using H_2S-scavenging chemicals postproduction, by trying to reduce SRB activity as much as possible by the application of biocides, or by the fieldwide application of nitrate to prevent the production of H_2S.

The use of nitrate to remediate H_2S has a long history. SRB (*Spirillum desulfuricans*) were discovered by Beyerinck in 1895[26] to be the causative agent of the horrible H_2S smell that plagued the citizens of Amsterdam every summer when warming temperatures allowed the rapid formation of H_2S from the reaction of sewage organic carbon with seawater sulfate in city canals. Nitrates have been used to remove sulfides from sewage systems since the early 1900s and have been applied in remediating sulfides from produced oil and gas since the 1990s. Inclusion of nitrate in waters, injected into oil fields to produce the oil, stimulates nitrate-reducing, sulfide oxidizing bacteria (NR-SOB). These catalyze the oxidation of sulfide with nitrate to form sulfur and/or sulfate, as well as nitrite and nitrogen or nitrite and ammonia. One of the best-characterized oil field NR-SOB is the *Thiomicrospira* sp. strain CVO, isolated from the Coleville field in western Canada.[27] This mesophilic organism produces sulfur and/or sulfate or nitrite and/or nitrogen, depending on the nitrate-to-sulfide ratio.[28] At a high nitrate-to-sulfide ratio, the main products are nitrite and sulfate (TABLE 1, reaction 8), whereas at a low nitrate-to-sulfide ratio, the main products are nitrogen and sulfur (TABLE 1, reaction 9). Reaction 8 is the preferable way to remove sulfide, because nitrite is a strong SRB inhibitor[28,29] and sulfur is considered corrosive toward iron. Hence, if possible, conditions favoring reaction 9 should be avoided. Nitrite specifically inhibits dissimilatory sulfite reductase (Dsr), the enzyme responsible for H_2S production by SRB.[29,30] Although nitrite also has general bacteriostatic properties, it can be considered a magic bullet in the case of SRB, inhibiting precisely the reaction one would target to be inhibited. Because of this specific inhibition, many (but not all) mesophilic SRB have a periplasmic, membrane-associated nitrite reductase, which reduces the incoming nitrite to ammonia, preventing inhibition of Dsr. Once all nitrite is reduced, sulfate reduction starts again.[28,29] The nitrite concentration needed to permanently inhibit mid-log phase cultures of mesophilic SRB with nitrite reductase is in excess of 10 mM, whereas mid-log phase cultures of mesophilic SRB lacking nitrite reductase are inhibited by as little as 0.5 mM nitrite.[28] It appears that thermophilic SRB from oil fields, growing at temperatures of 60°C or higher, lack nitrite reductase and these SRB are therefore inhibited by low nitrite concentrations.[32]

When nitrate is injected into an oil field at a moderate depth (e.g., 1000 m) and, as a consequence, at a moderate subsurface temperature (30°C), it also reacts with organic carbon through the action of heterotrophic nitrate-reducing bacteria (hNRB). These bacteria couple oxidation of organic carbon to CO_2 (or acetate and CO_2, in the case of incomplete oxidizers such as the oil-field *Sulfurospirillum* spp.[31]) to the reduction of nitrate to nitrite and then to either nitrogen or ammonia. Some hNRB have neither of these last two activities, and therefore produce nitrite as an end-product. It has been stated that hNRB contribute to souring control by using organic electron donors that would otherwise be used by SRB, thereby preventing sulfide production.[31] However, from the point of view of sulfide remediation, reactions of nitrate with organic carbon can be considered as competitive with reactions of nitrate with sulfide, that is, the hNRB reactions, or the SRB-mediated reaction of organic carbon with nitrite, increase the nitrate dose needed to convert a given amount of sulfide to sulfur and sulfate. Although some bacteria have both NR-SOB and hNRB activities, these activities are usually quite distinct, for example, *Thiomicrospira* sp, strain CVO is an autotroph that uses only CO_2 as the carbon source. Hence, this organism lacks hNRB activity.

The mechanism of sulfide remediation in higher-temperature fields harboring thermophilic SRB (tSRB) consortia may be quite different from the scenario described previously, because tNR-SOB have never been isolated and may be absent altogether. However, nitrite reacts chemically with sulfide, producing sulfur, polysulfide, and sulfate, while nitrite is reduced to ammonia.[32] High-temperature sulfide remediation may therefore involve (1) tNRB-mediated reaction of injected nitrate with organic carbon to form nitrite, (2) inhibition of tSRB by nitrite,[32] and (3) chemical reaction of nitrite with sulfide. The tNRB-mediated reduction of nitrite to either nitrogen or ammonium using organic carbon as an electron donor is competitive with the remediation of sulfide.

Successful fieldwide sulfide remediation by nitrate injection has been demonstrated for landlocked reservoirs of moderate subsurface temperature, from which oil is obtained by produced-water reinjection (PWRI), for example, in the Coleville field, injection of 300 ppm of nitrate over 5 weeks reduced sulfide concentrations in the produced water by 40–100%.[33] The mesophilic NR-SOB *Thiomicrospira* sp. strain CVO became a major component of the reservoir microbial community under these conditions.[33] Nitrate amendment of fields subjected to seawater flooding has been very successful, for example, fieldwide injection of typically 100 ppm sodium nitrate into the Gullfaks and Halfdan fields yielded a permanent reduction of produced sulfide.[34] At a typical injection rate of 10,000 m^3 of seawater/day, this amounts to a dose of one ton of sodium nitrate per day. Although the costs of a continuous fieldwide nitrate injection at this rate are in the millions of dollars per year, this is considered worthwhile when compared with the potential costs of corrosion or other sulfide-mediated equipment failures. The use of nitrate has been so successful that it is now also injected prophylactically. To prevent the emergence of sulfide, Shell has, since seawater injection was started, continuously injected 45 ppm of nitrate into the offshore Bonga field in Nigeria.[35]

Conclusions

Several new developments, namely, understanding *in situ* oil biodegradation by methanogenic consortia and understanding the impact of nitrate on the oilfield sulfur cycle to reduce souring and corrosion, have made petroleum microbiology an important field that may spur development of novel, advanced techniques to reduce costs and improve production. Such approaches are expected to become more important with time. Future developments in this area require a better understanding of key oil-field microbes, especially the syntrophs postulated to attack the oil in FIGURE 1A B. Because these organisms are very hard to culture, such information may best be obtained through a well-constructed, targeted metagenomics project, aimed at broadly characterizing oil-field microbial communities by large-scale sequencing.

Acknowledgments

This research has been supported through a Strategic Grant from the Natural Science and Engineering Research Council of Canada (NSERC) to one of the authors (G.V.).

Conflict of Interest

The authors declare no conflicts of interest.

References

1. HALL, C., P. THARAKAN, J. HALLOCK, *et al.* 2003. Hydrocarbons and the evolution of human culture. Nature **426:** 318–322.
2. ATLAS, R.M. & R. BARTHA. 1993. Microbial Ecology. Redwood City, CA. The Benjamin Cummins Publishing Company.
3. PHILP, R.P. 1986. Geochemistry in search of oil. Chem. Eng. News, February 10, 28–43.
4. SHOCK, E.L. 1988. Organic acid metastability in sedimentary basins. Geology **16:** 886–890.
5. ZENGLER, K., H.H. RICHNOW, R. ROSSELLO-MORA, *et al.* 1999. Methane formation from long-chain alkanes by anaerobic microorganisms. Nature **401:** 266–269.
6. EDWARDS, E.A. & D. GRBIC-GALIC. 1994. Anaerobic degradation of toluene and *o*-xylene by a methanogenic consortium. Appl. Environ. Microbiol. **60:** 313–322.
7. HEAD, I.M., D.M. JONES & S. LARTER. 2003. Biological activity in the deep subsurface and the origin of heavy oil. Nature **426:** 344–352.
8. HUANG, H. & S. LARTER. 2005. Biodegradation of petroleum in subsurface geological reservoirs. *In* Petroleum Microbiology. B. Ollivier & M. Magot, Eds.: 91–121. Washington, D.C.: ASM Press.
9. KASTER, K.M. & G. VOORDOUW. 2006. Effect of nitrite on a thermophilic, methanogenic consortium from an oil storage tank. Appl. Microbiol. Biotechnol. **72:** 1308–1315.
10. MAGOT, M., B. OLLIVIER & B.K.C. PATEL. 2000. Microbiology of petroleum reservoirs. Antonie Leeuwenhoek **77:** 103–116.
11. VANCE, I. & D.R. THRASHER. 2005. Reservoir souring: mechanisms and prevention. *In* Petroleum Microbiology. B. Ollivier & M. Magot, Eds.: 123–142. Washington, D.C.: ASM Press.
12. ROZANOVA, E.P. & S.I. KUZNETSOV. 1974. Microflora of Oil Fields. Moscow: Nauka (in Russian).
13. TARDY-JACQUENOD, C., P. CAUMETTE, R. MATHERON, *et al.* 1996. Characterization of sulfate-reducing bacteria isolated from oil-field waters. Can. J. Microbiol. **42:** 259–266.
14. MAGOT, M., O. BASSO, C. TARDY-JACQUENOD & P. CAUMETTE. 2004. *Desulfovibrio bastinii* sp. nov. and *Desulfovibrio gracilis* sp. nov., moderately halophilic, sulfate-reducing bacteria isolated from deep subsurface oilfield water. Int. J. Syst. Evol. Microbiol. **54:** 1693–1697.
15. TARDY-JACQUENOD, C., M. MAGOT, F. LAIGRET, *et al.* 1996. *Desulfovibrio gabonensis* sp. nov., a new moderately halophilic sulfate-reducing bacterium isolated from an oil pipeline. Int. J. Syst. Bacteriol. **46:** 710–715.
16. NGA, D.P., D.T. CAM HA, L.T. HIEN & H. STAN-LOTTER. 1996. *Desulfovibrio vietnamensis* sp. nov., a halophilic sulfate-reducing bacterium from Vietnamese oil fields. Anaerobe **2:** 385–392.
17. ROZANOVA E.P., T.N. NAZINA & A.S. GALUSHKO. 1988. Isolation of a new genus of sulfate-reducing bacteria and description of a new species of this genus, *Desulfomicrobium apsheronum* gen. nov., sp. nov. Mikrobiologiya **57:** 634–641.
18. GALUSHKO A.S. & E.P. ROZANOVA. 1991. *Desulfobacterium cetonicum* sp. nov.: a sulfate-reducing bacterium which oxidizes fatty acids and ketones. Mikrobiologiya **60:** 102–107.
19. LIEN T. & J. BEEDER. 1997. *Desulfobacter vibrioformis* sp. nov., a sulfate reducer from a water-oil separation system. Int. J. Syst. Bacteriol. **47:** 1124–1128.
20. LIEN T., M. MADSEN, I.H. STEEN & K. GJERDEVIK. 1998. *Desulfobulbus rhabdoformis* sp. nov., a sulfate reducer from a water-oil separation system. Int. J. Syst. Bacteriol. **48:** 469–474.
21. BELYAKOVA, E.V., E.P. ROZANOVA, I.A. BORZENKOV, *et al.* 2006. The new facultatively chemolithoautotrophic, moderately halophilic, sulfate-reducing bacterium

Desulfovermiculus halophilus gen. nov., sp. nov., isolated from an oil field. Microbiologiya **75:** 161–171.

22. Tardy-Jacquenod C., M. Magot, B.K.C. Patel, *et al.* 1998. *Desulfotomaculum halophilum* sp. nov., a halophilic sulfate-reducing bacterium isolated from oil-production facilities. Int. J. Syst. Bacteriol. **48:** 333–338.
23. Aeckersberg, F., F. Bak & F. Widdel. 1991. Anaerobic oxidation of saturated hydrocarbons to CO_2 by a new type of sulfate-reducing bacterium. Arch. Microbiol. **156:** 5–14.
24. Rabus, R., R. Nordhaus, W. Ludwig & F. Widdel. 1993. Complete oxidation of toluene under strictly anoxic conditions by a new sulfate-reducing bacterium. Appl. Environ. Microbiol. **59:** 1444–1451.
25. Larsen, J. 2002. Downhole nitrate applications to control sulfate reducing bacteria activity and reservoir souring. Corrosion 2002. Paper 02025, NACE International, Houston, TX.
26. Beyerinck, W.M. 1895. Ueber *Spirillum desulfuricans* als ursache von sulfat-reduction. Zentralbl. Bakteriol. Parasitenkd. **1:** 1–9.
27. Gevertz, D., A.J. Telang, G. Voordouw & G.E. Jenneman. 2000. Isolation and characterization of strains CVO and FWKO B: Two novel nitrate-reducing, sulfide-oxidizing bacteria isolated from oil field brine. Appl. Environ. Microbiol. **66:** 2491–2501.
28. Greene, E.A., C. Hubert, M. Nemati, *et al.* 2003. Nitrite reductase activity of sulfate-reducing bacteria prevents their inhibition by nitrate-reducing, sulfide-oxidizing bacteria. Environ. Microbiol. **5:** 607–617.
29. Haveman, S.A., E.A. Greene, C.P. Stilwell, *et al.* 2004. Physiological and gene expression analysis of inhibition of *Desulfovibrio vulgaris* Hildenborough by nitrite. J. Bacteriol. **186:** 7944–7950.
30. He, Q., K.H. Huang, Z. He, *et al.* 2006. Energetic consequences of nitrite stress in *Desulfovibrio vulgaris* Hildenborough, inferred from global transcriptional analysis. Appl. Environ. Microbiol. **72:** 4370–4381.
31. Hubert, C. & G. Voordouw. 2007. Oil field souring control by nitrate-reducing *Sulfurospirillum* spp. that outcompete sulfate-reducing bacteria for organic electron donors. Appl. Environ. Microbiol. **73:** 2644–2652.
32. Kaster, K.M., A. Grigoriyan, G. Jenneman & G. Voordouw. 2007. Effect of nitrate and nitrite on two thermophilic, sulfate-reducing enrichments from an oil field in the North Sea. Appl. Microbiol. Biotechnol. **75:** 195–203.
33. Telang, A.J., S. Ebert, J.M. Foght, *et al.* 1997. The effect of nitrate injection on the microbial community in an oil field as monitored by reverse sample genome probing. Appl. Environ. Microbiol. **63:** 1785–1793.
34. Sunde, E. & T. Torsvik. 2005. Microbial control of hydrogen sulfide production in oil reservoirs. *In* Petroleum Microbiology. B. Ollivier & M. Magot, Eds.: 201–213. Washington, D.C.: ASM Press.
35. Kuijvenhoven, C., J. Noirot, P. Hubbard & L, Oduola. 2007. 1 year experience with the injection of nitrate to control souring on Bonga deepwater development offshore nigeria. Proceedings of the SPE International Symposium on Oilfield Chemistry, Paper SPE 105784, pp. 1–9, Society of Petroleum Engineers, Richardson, Texas.

Fermentative Butanol Production

Bulk Chemical and Biofuel

PETER DÜRRE

Institut für Mikrobiologie und Biotechnologie, Universität Ulm, Ulm, Germany

Clostridium acetobutylicum **is an anaerobic, spore-forming bacterium with the ability to ferment starch and sugars into solvents. In the past, it has been used for industrial production of acetone and butanol, until cheap crude oil rendered petrochemical synthesis more economically feasible. Both economic (price of crude oil) and environmental aspects (carbon dioxide emissions) have caused the pendulum to swing back again. Molecular biology has allowed a detailed understanding of genes and enzymes, required for solventogenesis. Thus, construction of strains with improved fermentation ability is now possible. Advances in continuous culture technology and improved downstream processing also add to economic advantages of a new biotechnological process. Two major companies have already committed themselves to biobutanol production as a biofuel additive. Thus, butanol fermentation is on the rise again.**

Key words: **acetone; biobutanol; biofuel; *Clostridium acetobutylicum*; solvents**

Introduction

Butanol is a fermentation product of anaerobic bacteria. In 1862, the famous French microbiologist Louis Pasteur was the first to describe the synthesis of this C_4-alcohol by his "Vibrion butyrique,"[1] probably a mixed culture, containing a similar to or identical with *Clostridium butyricum*.[2] Few bacteria form butanol as a major product. Most of them are clostridia, but *Butyribacterium methylotrophicum* also shows this fermentation pattern.[3] Among the archaea, *Hyperthermus butylicus* produces significant amounts of the alcohol.[4] However, the species most intensively investigated in this respect is *Clostridium acetobutylicum*. Albert Fitz was probably the first microbiologist to obtain a pure culture of such an organism. He called it *Bacillus butylicus* and performed studies on morphology, substrate utilization, and product formation as well as toxicity.[5–8] An even more detailed analysis was carried out by Martinus Beijerinck, who had isolated a similar or even identical bacterium in 1893, which he named *Granulobacter saccharobutyricum*.[9] The name *C. acetobutylicum* was only introduced and validly published in 1926.[10] This organism became famous in both biotechnology and politics. At the beginning of the last century, an approach was launched to produce the precursors of building blocks for synthetic rubber, namely, amyl alcohol, butanol, or acetone via bacterial fermentation. Respective patents were granted to Fernbach and Strange in 1911 and 1912,[11–13] and the process was started in the United Kingdom under supervision of Strange & Graham, Ltd.[14–16] This company also had a collaboration with the chemists Perkins and Weizmann, based at the University of Manchester. The latter ended the cooperation in 1912, but still continued the project. Weizmann succeeded in isolating an organism that produced significantly larger amounts of acetone and butanol than the strains of Fernbach and Strange. This bacterium and the respective fermention process were patented in 1915.[17] At that time, World War I had started and Great Britain was in urgent need for acetone as the essential bulk chemical for production of cordite (smokeless ammunition). Due to its superior product formation, Weizmann's *C. acetobutylicum* became the organism of choice for acetone synthesis and was used in all fermention plants throughout the former British Empire and later also in Britain's ally, the United States. The constant supply of acetone was certainly a decisive factor in winning World War I. Weizmann refused to accept any financial or official acknowledgments by the government, but made clear that he was in favor of a Jewish homeland in Palestine. There is no doubt that this attitude affected the Balfour declaration of 1917, leading eventually to the foundation of the State of Israel. Chaim Weizmann became its first president.

Address for correspondence: Prof. Dr. Peter Dürre, Institut für Mikrobiologie und Biotechnologie, Universität Ulm, 89069 Ulm, Germany. Voice: +49-731-5022710; fax: +49-731-5022719.
peter.duerre@uni-ulm.de

Ann. N.Y. Acad. Sci. 1125: 353–362 (2008). © 2008 New York Academy of Sciences.
doi: 10.1196/annals.1419.009

During the war, the by-product butanol had simply been stored, as there was no ready use for this chemical. This changed shortly after the end of the war when the United States introduced Prohibition in 1920. Besides banning drinkable alcoholic beverages, this also meant a shortage for industrial solvents, such as amyl alcohol, which was also obtained via alcoholic fermentation. On the other hand, automobile production was increased dramatically by Henry Ford's introduction of assembly lines. Thus, there was an urgent need for a solvent to replace amyl acetate (obtained from amyl alcohol) to produce quick-drying lacquers for car manufacturing. Butanol proved to be an excellent basis for synthesizing butyl acetate, and fermentation became the method of choice for its production. Until about 1950, 66% of the butanol used worldwide was produced biotechnologically.[18] The largest plants were located in the United States, for example, Peoria (Illinois) had a fermenter capacity of 96 units with a total volume of 18,168 m^3.[14] A plant with a capacity of 1080 m^3 was operated in Germiston, South Africa, under the apartheid regime until 1982.[19] In the former Soviet Union, at least eight plants were in operation, some of them up to the late 1980s. Dokshukino used four parallel fermenter batteries with a total capacity of approximately 6500 m^3. The processed volume during one cycle of continous fermentation was approximately 2000 m^3.[20] China used butanol fermentation until 2004 and produced approximately 350,000 t/year this way.[21]

The low prices of crude oil at the middle of the last century as well as raising prices for molasses, combined with even lower sugar content, led to the decline of the industrial acetone–butanol fermentation (sometimes called acetone-butanol-ethanol (ABE) fermentation, as ethanol is also produced, but only in minor amounts). Since then, butanol has been almost exclusively produced petrochemically via an oxo reaction from propylene, yielding the intermediate butyraldehyde, which is reduced to butanol by hydrogenation. However, the so-called "oil crisis" in 1973 stimulated scientific research in this field and the subsequent dramatic increase in crude oil prices led to a reintroduction of industrial butanol fermentation (see later in the chapter).

Industrial Importance of Butanol

Butanol (IUPAC nomenclature 1-butanol) is a colorless, flammable liquid with a banana-like odor. It is miscible with all common solvents, but only sparingly soluble in water. In direct contact, it may irritate the eyes and skin. If inhaled, it may cause a narcotic effect. It is an important bulk chemical with a wide range of industrial uses. Almost half of the worldwide production is converted into acrylate and methacrylate esters. Other important derivatives are glycol ethers and butyl acetate. Compounds of minor use are butylamines and amino resins. The esters make use of the double bond of acrylic acid to form homo- and copolymers, which are used for production of surface coatings, adhesives/sealants, elastomers, textiles, superabsorbents, flocculants, fibers, and plastics. Butyl methacrylate is also used in resins, dental products, as an oil additive, as well as for leather and paper finishing. Butyl glycol is a solvent for industrial coatings and hard-surface cleaners. It is also required for electronics. Butyl acetate is used in paints, printing ink, in leather and coatings manufacturing, and as synthetic fruit flavoring in ice cream, cheeses, candy, and baked goods. Butylamine is required for the production of pesticides (thiocarbazides), pharmaceuticals, and emulsifiers. It is also a precursor of butylbenzenesulfonamide, a plasticizer of nylon.

Other applications of butanol and compounds derived from it include paint thinners, hydraulic and brake fluids, perfume manufacturing, safety glass, detergents, flotation aids (e.g., butyl xanthate), deicing fluid, cosmetics (eye makeup, lipsticks, nail-care products, foundations, shaving and personal hygiene products), extraction during the production of drugs as well as alkaloids, antibiotics, camphor, hormones and vitamins, as well as mobile phases in paper and thin-layer chromatography. Butanol has also been recommended for use as a standard odorant by various groups for the training of breath-odor judges.

A new, but very important application came up recently. Bio-based butanol (also called biobutanol) is expected to play a major role in the next generation of biofuels. It can be produced fermentatively from renewable rersources and, thusly, lowers greenhouse gas emissions and dependence on crude oil. It can be blended into standard gasoline similarly to ethanol. The addition of ethanol to gasoline has become quite common meanwhile. In Brazil, 23% is required; in the United States and in part of Europe, 10%. However, butanol has a number of advantages over ethanol. It is less corrosive and has a lower vapor pressure. It does not absorb water, therefore, no phase separation of biofuel/gasoline blends will occur. This means that butanol can be added at the refinery (at any concentration) and be transported and delivered through the existing infrastructure (pipelines, storage tanks, filling stations). Ethanol can only be added shortly prior to use. The energy content of butanol is much higher than that of ethanol, so fuel consumption is similar to that of pure gasoline, whereas in the case of ethanol blends, consumption goes up. Butanol usage does not require modifications to car engines. In 2005, David Ramey drove a 13-year-old Buick across the United States, fueled by pure butanol. Consumption increased by 9% compared to gasoline, but emissions of CO, hydrocarbons, and NO_x were massively reduced. His company, Environmental Energy, Inc. (EEI), is planning to produce ButylFuel™ via a newly developed fermentation process involving two clostridial species.[22] While this is a fairly small enterprise, two major global players, BP and DuPont, also announced the start of fermentative butanol production for use as a biofuel in 2007. In cooperation with British Sugar, an existing ethanol plant in the United Kingdom will be converted into a biotechnological butanol synthesis facility. Starting technology will probably rely on the established processes with solventogenic clostridia, namely, *C. acetobutylicum* or *Clostridium beijerinckii*.[23]

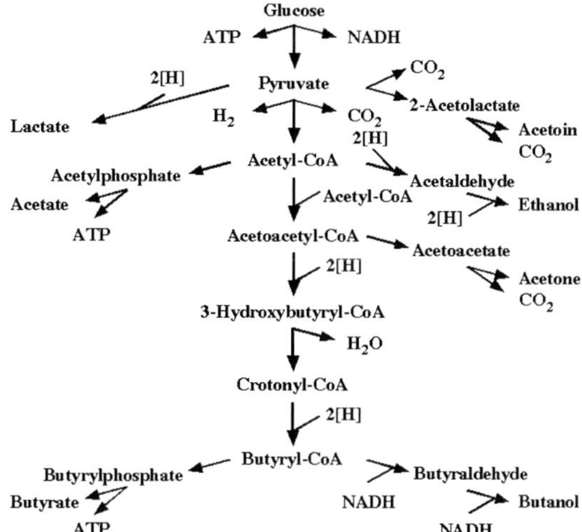

FIGURE 1. Fermentation pathways employed by *Clostridium acetobutylicum*.

Fermentative Metabolism of *Clostridium acetobutylicum*

C. acetobutylicum is the best-studied solventogenic *Clostridium*. Growing on starch or sugars, it first performs a typical butyrate fermentation. Butyrate, acetate, carbon dioxide, and hydrogen are the major products. Ethanol and acetoin are formed in minor amounts, and lactate can be produced under specific conditions. FIGURE 1 shows a scheme of the respective metabolic reactions. At the end of the exponential growth phase, a major metabolic shift takes place. The excreted acids are taken up again and are converted into the neutral products butanol and acetone (in a ratio of typically 2:1). Final solvent concentration at the end of the fermentation is usually approximately 2%. The reason for this shift is the imminent threat of cell death by low pH. Due to acid production, the proton concentration outside the cells is increasing. Since anaerobic bacteria, including *C. acetobutylicum*, are unable to maintain cytoplasmic pH homeostasis, the internal pH will decrease in parallel to the external one, being about 1 unit higher.[24–26] At external pH values of less than 4.5, significant amounts of undissociated acids will diffuse into the cytoplasm and dissociate there, because of the higher pH. This will cause the proton gradient over the cytoplasmic membrane to collapse and thusly lead to cell death. Conversion of butyrate and acetate into solvents increases the external pH again and allows the cells to stay metabolically active for a longer time. This is a distinct advantage over other species that are unable to cope with acid stress. However, the solvents are endangering the cell as well (by inactivating membrane proteins and destroying the membrane), with butanol being the most toxic one. Consequently, a long-time survival program is started by these clostridia simultaneously with the initiation of solventogenesis: formation of endospores. Thus, it is not surprising that the regulatory networks of solventogenesis and sporulation are closely linked (see later in this chapter).

A number of enzymes of *C. acetobutylicum* have been purified and biochemically analyzed.[16] At this point a note of caution must be added. In the past (until the beginning of the 1990s), many different strains were attributed to the species *C. acetobutylicum*. Only after initiation of regular international *Clostridium* meetings (the first one was held in Salisbury, U.K. in 1990) by David Woods (at that time at the University of Cape Town, South Africa), did researchers realize that these strains in fact represent four different species: *C. acetobutylicum*, *C. beijerinckii*, *C. saccharoperbutylacetonicum*, and *C. saccharobutylicum*.[27] This also had major consequences for national culture collections, which had to correct accession numbers for species (e.g., NCIMB 8052 reclassified as *C. beijerinckii*). Thus, it is necessary to check older literature for the strain designation and then refer to the current classification.

Growth on starch requires amylase activity. Two such enzymes have been purified,[28,29] but only one of the respective genes (*amyP*) has been identified unequivocally in the genome sequence. In addition, some other genes are present, encoding proteins, which resemble amylases. Glucose is transported into the cell via a phosphotransferase system.[30] Xylan

TABLE 1. Regulation of *Clostridium acetobutylicum* genes required for solventogenesis

Operon	Relative promoter strength	Upstream OA boxes	Regulatory mechanisms[a]
adc	100	2 + 1 reverse	Induction of transcription Modification of gene product
sol (orfL-adhE-ctfA-ctfB)	1.2	1 reverse	Induction of transcription mRNA processing
bdhB	77	1	Induction of transcription
bdhA	4.5	2	None, constitutive expression under these conditions
adhE2	n.d.[b]	1	Induction of transcription[c]

[a]Switch from acidogenesis to solventogenesis in continuous culture.
[b]Not determined.
[c]Under "alcohologenic" conditions (e.g., substrate combination glucose/glycerol).

can also be used for growth and butanol production,[31,32] and two xylanases and a β-D-xylosidase have been purified and characterized.[33,34] Inside the cell, hexoses are degraded via glycolysis, whereas pentoses are metabolized via the pentose phosphate pathway.[16,35,36]

Solvent formation in *C. acetobutylicum* is catalyzed by acetoacetyl-CoA:acetate/butyrate-coenzyme A transferase (abbreviated as CoA transferase), several butyraldehyde and butanol dehydrogenases, and the acetoacetate decarboxylase. In some strains of *C. beijerinckii*, acetone can be further reduced to 2-propanol (isopropanol) by means of a primary/secondary alcohol dehydrogenase.[37] CoA transferase is required for both butanol and acetone formation, whereas the acetoacetate decarboxylase is specific for acetone synthesis. The situation concerning the aldehyde and alcohol dehydrogenases in *C. acetobutylicum* is more complex. Two different butanol dehydrogenases have been purified and characterized (BdhA, BdhB),[38] and two different bifunctional butyraldehyde/butanol dehydrogenases have been found (AdhE or Aad, AdhE2).[39–41] RNA investigations revealed that the onset of butanol formation coincided with the induction of AdhE, which is only transiently made. Massive butanol production is brought about by BdhB, which is induced later.[42] BdhA is formed constitutively and its conversion of butyraldehyde is just twice as high as that of acetaldehyde. This led to the assumption that this enzyme might be just serving as an electron sink in both low butanol and ethanol production.[16,38,42] AdhE2 is only formed under so-called "alcohologenic" conditions, that is, growth on substrate mixtures with reduced compounds such as glucose/glycerol, which results in the formation of butanol and ethanol, but not acetone.[41,43]

Molecular Biology of *Clostridium acetobutylicum*

The first genes encoding solventogenic enzymes from *C. acetobutylicum* were cloned and sequenced in 1990.[44–46] Meanwhile, the complete genome sequence has been determined. The organism harbors a chromosome of 3.94 Mbp and the megaplasmid pSOL1 of 192 kbp.[36,47,48] The enzymes known to catalyze solventogenic reactions are encoded in five different operons (TABLE 1). The two butanol dehydrogenase genes are chromosomally encoded. *bdhA* and *bdhB* are each organized in monocistronic, adjacent operons.[49] Transcription is initiated at typical σ^A-specific promoters (with 4 and 3 mismatches in the hexamer sequences, respectively) and terminated at rho-independent hairpin structures.[50] The promoter of *bdhB* is 17-fold stronger than that of *bdhA*. Thus, it is in the same range as that of the acetoacetate decarboxylase (*adc*) operon (23% lower activity).[51] Upstream of the promoters two and one, respectively, binding sites for the transcription factor Spo0A have been identified from sequence inspection.[52] When phosphorylated, Spo0A serves as the master regulator of sporulation. This is the regulatory link between sporulation and solventogenesis mentioned before, and such binding motifs are present upstream of the other operons required for solventogenesis as well.[16,52,53] However, it must be noted that according to the RNA analyses, *bdhA* is constitutively expressed, despite the presence of two 0A boxes.[42] This might be either explained by the involvement of additional regulators or it poses a note of caution to the metabolic relevance of *in silico*–detected sequence motifs.

The other three operons are located on the megaplasmid. *adc* is also organized as a monocistronic transcription unit.[54] Three 0A boxes, one of which is in reverse orientation, have been detected upstream of the promoter, which is 1.3-fold more strongly expressed than the *bdhB* promoter.[51,52] However, mutation or even removal of these binding sites only led to decreased induction, but not to constitutive expression.[52,55] This indicated the presence of additional regulators.[16] Two proteins, AdcR and AdcS, that in fact influence *adc* expression have recently been identified.[56] Their exact mode of action is still unknown. In addition to transcriptional regulation, the Adc protein is co- or posttranslationally modified in a still-unknown way.[57] Since the enzyme remains active even in the presence of oxygen, the modification might contribute to stability.

The *sol* operon is located adjacent to *adc*, but with divergent direction of transcription.[39] It contains a small open reading frame *orfL* (function unknown) and the genes *adhE* (encoding a bifunctional butyraldehyde/butanol dehydrogenase), *ctfA*, and *ctfB* (encoding the two subunits of CoA transferase). The operon is controlled by a promoter with an almost perfect σ^A consensus sequence, a reverse 0A box located upstream. Inactivation of the *spo0A* gene did not completely repress *sol* transcription,[58] nor did mutation of the 0A box,[59] indicating that additional transcription factors are involved in control, possibly acting in concert with phosphorylated Spo0A.[16] A second transcription start point identified by primer extension experiments turned out to be an mRNA processing site.[59] It seems that the respective RNase recognizes the secondary structure of the transcript rather than the sequence.[16] A report claiming the upstream gene *orf5* (organized in a monocistronic operon) to encode a transcriptional repressor for the *sol* operon ("SolR")[60] turned out to be wrong, due to erroneously cocloning part of the control region of the promoter of the *sol* operon.[59,61] The composition of the *sol* operon in *C. acetobutylicum* is unique. All other solventogenic clostridia do not carry an *adhE* gene in this unit, but instead a gene encoding a butyraldehyde dehydrogenase (*ald* or *bld*). In addition, being located downstream of *ctfB*, the *adc* gene is part of the *sol* operon.[62,63] A surprising result was the low activity of the promoter of the *sol* operon (1.2% of that of the *adc* promoter).[51] This might be due to the fact that the *sol* operon is only transiently induced and seems to be responsible only for initiation of solventogenesis. Massive butanol production is then catalyzed by BdhB.[42]

A gene required uniquely for butanol production under "alcohologenic" conditions (see earlier in this chapter) is *adhE2*. The effect caused by reduced substrate mixtures, such as glucose/glycerol, can be mimicked by the addition of methyl viologen.[64,65] Such experiments, indeed, in combination with RNA analyses yielded the first indications of an additional AdhE-type enzyme.[66] *C. acetobutylicum* is the first organism known to harbor two such genes.[16] *adhE2* is organized as a monocistronic operon, and two transcription start points have been determined by primer extension. The distal one allows deduction of a typical σ^A-dependent promoter, whereas the proximal one shows only weak homology to the consensus sequence.[67] This very much resembles the situation of the *sol* operon, with one promoter and an mRNA processing site. Sequence analysis showed that a 0A box is located just upstream of the distal transcription start point, further supporting its promoter function.[68]

There is much that is still unknown about the signal(s) causing the onset of solventogenesis. External factors required for reliably initiating acetone and butanol production in fermenters are a surplus of substrate, limiting concentations of phosphate or sulfate, especially high concentrations of butyrate and to some extent acetate, and a pH value of 4.3 or below. Compensation is possible, as, for example, a higher pH will allow solventogenesis if large amounts of butyrate and acetate are added.[69,70] Temperature also affects initiation of solvent formation. All these factors are able to change the topology of DNA. Thus, the degree of supercoiling might serve as a sensor to tell the cell when conditions become so harsh that solventogenesis should be started (for a more detailed review, see Ref. 68). In fact, transcription of the *sol* operon was shown to be directly dependent on relaxation of DNA.[71] As components of an internal signal cascade, the level of NAD(P)H, the ATP/ADP ratio, and the concentration of butyryl-CoA, as well as butyryl phosphate, have been discussed.[25,72–74]

Transcriptional profiling by use of DNA microarrays is a powerful technique for elucidating regulatory networks genomewide. It was introduced into the scientific community about 12 years ago. Since the genome sequence of *C. acetobutylicum* has been known since 2001,[36] this method became applicable as well. The group of Terry Papoutsakis (Northwestern University, Evanston, Illinois) has set up a number of different microarrays, covering part of the genome, as well as all of it. The first report focused on the effects on overexpression of stress proteins on solventogenesis and solvent tolerance,[75] followed by solventogenesis and sporulation, comparing the wild type and a nonsolventogenic, asporogenous strain,[76] *spo0A* overexpression,[77] butanol stress,[78] early sporulation and stationary phase events,[79] and tolerance versus butanol.[80] The data obtained are not yet completely understood. However, these data have been helpful in comparing clostridial sporulation and physiology to that of bacilli,[81] and also yielded a number of unexpected results (e.g., overexpression of solventogenesis genes upon the addition of butanol).[78]

Another approach of genomewide profiling was made by proteome analysis. Comparisons of wild type and an *spo0A* mutant, as well as acidogenic versus solventogenic conditions, have been reported.[57,82] As expected, enzymes, such as Adc, required for solventogenesis were formed under solventogenic conditions, as were stress proteins. Unexpectedly, however, serine biosynthesis enzymes were also induced. This could be confirmed at the transcriptional level.[57,79] The reason for this phenomenon is still unknown, as there are no proteins with a high serine content known to be involved in solventogenesis or sporulation.

Approaches for Improving Fermentation

Three major factors determine the economic competitiveness of the biotechnological butanol production: substrate costs, low product yields versus solvent toxicity, and costs for downstream processing.[83,84] As a substrate, molasses has been used at the peak of industrial butanol fermentation. However, improved extraction procedures and applications led to a lower sugar content and a higher price, rendering this substrate uneconomical. The current situation sees a lot of fluctuation with respect to sugar prices, due to deregulation in a number of countries. Thus, sugar might become an attractive substrate for solvent production and indeed for many fermentation processes. In addition, there are several other options. Since *C. acetobutylicum* is amylolytic, starch is an appropriate substrate as well. Especially in regions with a surplus of starchy, agricultural products (e.g., the Corn Belt of the United States), butanol fermentation might become another attractive business. A problem in using this substrate is that starch gelatinizes upon heat sterilization. Cooling then may lead to formation of oligomers, which are more resistant to enzymatic breakdown.[85] However, in general, the use of food crops is difficult for two reasons: first, there will always be the demand to use it only for nutritional purposes, and second, it will simply not suffice (due to quantity limitations) to fulfill the needs for a biofuel intended, in the end, to completely replace gasoline. Other industrial requirements, however, can be met. Therefore, the search for other substrates was initiated many years ago. Lignocellulose might be the solution. It constitutes a valuable renewable natural resource. Respective hydrolysates had already been used in the past for biotechnological solvent production in the former Soviet Union.[20] However, the process is still too expensive to allow an economic fermentation. In this respect it must be noted that during inspection of the *C. acetobutylicum* genome sequence, a cluster of genes for cellulose degradation was detected.[36] Overexpression of single genes led to functional proteins, a minicellulosome could be detected, but for still not completely known reasons the organism is unable to degrade cellulose.[86–90] Metabolic engineering might help to overcome this obstacle.

Tailor-made strain construction is also currently used to improve solvent tolerance and product yields.[75,78,80] An attempt to improve resistance toward butanol by overexpressing the cyclopropane fatty-acid synthase gene was successful, but also resulted in significantly lower butanol formation.[91] Contrarily, butanol yields have been achieved by targeted mutations that were considered impossible even a decade ago. Topping the list is a strain with an inactivated *orf5* gene and an overexpressed *adhE* gene that produces 238 mM butanol.[92] A future aim certainly will be the construction of a mutant that forms almost no by-products (a homobutanol strain). Acetate and butyrate production can be inhibited by inactivation of the respective kinase genes and acetone by knockout of *adc*. The respective technology has just recently been developed.[93]

Product recovery was traditionally performed by distillation. Since butanol has a higher boiling point than water, this procedure uses up a lot of energy and is therefore very cost-intensive, especially at low butanol concentrations. Other means of recovery are adsorption, gas stripping, liquid–liquid extraction, and membrane-based processes such as perstraction, pervaporation, and reverse osmosis.[83] Recent experiments with a *C. beijerinckii* model fermentation system showed that gas stripping can be successfully applied and was superior to pervaporation.[94] Smaller bubble sizes led to increased foam production, and the necessity of using antifoam reduced productivity.[95] Adsorption using, for example, silicalite, is also promising, as it can be used to concentrate butanol from dilute solutions. Regeneration is performed by heat treatment. The energy requirements for this process are much lower than those of gas stripping and pervaporation.[96]

Typically, the fermentation is carried out using a single production strain. The EEI biomass-to-butanol process (see earlier in this chapter) uses a two-stage, dual-path procedure that relies on two different clostridia. In the first stage, *C. tyrobutyricum* ferments biomass to butyrate, carbon dioxide, and hydrogen, while *C. acetobutylicum* in the second stage converts butyrate into butanol. The technology will be especially appropriate for small biorefineries in farming regions. A totally different approach might be the conversion of butane into butanol by means of a specific monooxygenase.[97]

It is not possible to predict what process(es) eventually will be used in large scale, but it is very clear that new developments and interest in butanol fermentation are increasing, and the process is expected to contribute significantly to our industrial needs in the very near future.

Acknowledgments

Work in my laboratory was supported by grants from the BMBF GenoMik and GenoMikPlus projects (Competence Network Göttingen), and the SysMO project COSMIC (PtJ-BIO/SysMO/P-D-01–06-13), www.sysmo.net

Conflict of Interest

The author declares no conflicts of interest.

References

1. PASTEUR, L. 1862. Quelques résultats nouveaux relatifs aux fermentations acétique et butyrique. Bull. Soc. Chim. Paris May **1862:** 52–53.
2. SAUER, U., R. FISCHER & P. DÜRRE. 1993. Solvent formation and its regulation in strictly anaerobic bacteria. Curr. Top. Mol. Genet. **1:** 337–351.
3. GRETHLEIN, A.J. *et al.* 1991. Evidence for production of *n*-butanol from carbon monoxide by *Butyribacterium methylotrophicum*. J. Ferment. Bioeng. **72:** 58–60.
4. ZILLIG, W., I. HOLZ & S. WUNDERL. 1991. *Hyperthermus butylicus* gen. nov., sp. nov., a hyperthermophilic, anaerobic, peptide-fermenting, facultatively H_2S-generating archaebacterium. Int. J. Syst. Bacteriol. **41:** 169–170.
5. FITZ, A. 1876. Ueber die Gährung des Glycerins. Ber. Dtsch. Chem. Ges. **9:** 1348–1352.
6. FITZ, A. 1877. Ueber Schizomyceten-Gährungen II [Glycerin, Mannit, Stärke, Dextrin]. Ber. Dtsch. Chem. Ges. **10:** 276–283.
7. FITZ, A. 1878. Ueber Schizomyceten-Gährungen III. Ber. Dtsch. Chem. Ges. **11:** 42–55.
8. FITZ, A. 1882. Ueber Spaltpilzgährungen. VII. Mittheilung. Ber. Dtsch. Chem. Ges. **15:** 867–880.
9. BEIJERINCK, M.W. 1893. Ueber die Butylalkoholgährung und das Butylferment. Verhand. Kon. Akad. Wetenschappen Amsterdam 2. Sect., Teil 1, Nr. **10:** 1–51.
10. MCCOY, E. *et al.* 1926. A cultural study of the acetone butyl alcohol organism. J. Infect. Dis. **39:** 457–483.
11. FERNBACH, A. & E.H. STRANGE. 1911. Improvements in the manufacture of products of fermentation. Br. Patent No. 15203.
12. FERNBACH, A. & E.H. STRANGE. 1911. Improvement in the manufacture of higher alcohols. Br. Patent No. 15204.
13. FERNBACH, A. & E.H. STRANGE. 1912. Improvements connected with fermentation processes for the production of acetone, and higher alcohols, from starch, sugars, and other carbohydrate materials. Br. Patent No. 21073.
14. GABRIEL, C.L. 1928. Butanol fermentation process. Ind. Eng. Chem. **28:** 1063–1067.
15. JONES, D.T. & D.R. WOODS. 1986. Acetone-butanol fermentation revisited. Microbiol. Rev. **50:** 484–524.
16. DÜRRE, P. 2005. Formation of solvents in clostridia. *In* Handbook on Clostridia. P. Dürre, Ed.: 671–693. Boca Raton, FL: CRC Press.
17. WEIZMANN, C. 1915. Improvements in the bacterial fermentation of carbohydrates and in bacterial cultures for the same. Br. Patent No. 4845.
18. ROSE, A.H. 1961. Industrial Microbiology. London, U.K.: Butterworths.
19. JONES, D.T. 2001. Applied acetone-butanol fermentation. *In* Clostridia. Biotechnological and Medical Applications. H. Bahl & P. Dürre, Eds.: 125–168. Weinheim, Germany: Wiley-VCH.
20. ZVERLOV, V.V. *et al.* 2006. Bacterial acetone and butanol production by industrial fermentation in the Soviet Union: use of hydrolyzed agricultural waste for biorefinery. Appl. Microbiol. Biotechnol. **71:** 587–597.
21. CHIAO, J.-S. & Z.-H. SUN. 2007. History of the acetone-butanol-ethanol fermentation industry in China: development of continuous production technology. J. Mol. Microbiol. Biotechnol. **13:** 12–14.
22. http://www.butanol.com/
23. http://www2.dupont.com/Biofuels/en_US/
24. DÜRRE, P., H. BAHL & G. GOTTSCHALK. 1988. Membrane processes and product formation in anaerobes. *In* Handbook on Anaerobic Fermentations. L.E. Erickson & D.Y.-C. Fung, Eds.: 187–206. New York: Marcel Dekker.
25. GOTTWALD, M. & G. GOTTSCHALK. 1985. The internal pH of *Clostridium acetobutylicum* and its effect on the shift from acid to solvent formation. Arch. Microbiol. **143:** 42–46.
26. HUANG, L., L.N. GIBBINS & C.W. FORSBERG. 1985. Transmembrane pH gradient and membrane potential in *Clostridium acetobutylicum* during growth under acetogenic and solventogenic conditions. Appl. Environ. Microbiol. **50:** 1043–1047.
27. KEIS, S., R. SHAHEEN & D.T. JONES. 2001. Emended descriptions of *Clostridium acetobutylicum* and *Clostridium beijerinckii*, and descriptions of *Clostridium saccharoperbutylacetonicum* sp. nov. and *Clostridium saccharobutylicum* sp. nov. Int. J. Syst. Evol. Microbiol. **51:** 2095–2103.
28. PAQUET, V. *et al.* 1991. Purification and characterization of the extracellular α-amylase from *Clostridium acetobutylicum* ATCC 824. Appl. Environ. Microbiol. **57:** 212–218.
29. ANNOUS, B.A. & H.P. BLASCHEK. 1994. Isolation and characterization of α-amylase derived from starch-grown *Clostridium acetobutylicum* ATCC 824. J. Ind. Microbiol. **13:** 10–16.

30. TANGNEY, M. & W.J. MITCHELL. 2007. Characterisation of a glucose phosphotransferase system in *Clostridium acetobutylicum* ATCC 824. Appl. Environ. Microbiol. **74:** 398–405.
31. LEMMEL, S.A., R. DATTA & J.R. FRANKIEWICZ. 1986. Fermentation of xylan by *Clostridium acetobutylicum*. Enzyme Microb. Technol. **8:** 217–221.
32. QURESHI, N. *et al.* 2006. Butanol production from corn fiber xylan using *Clostridium acetobutylicum*. Biotechnol. Prog. **22:** 673–680.
33. LEE, S.F., C.W. FORSBERG & L.N. GIBBINS. 1985. Xylanolytic activity of *Clostridium acetobutylicum*. Appl. Environ. Microbiol. **50:** 1068–1076.
34. LEE, S.F., C.W. FORSBERG & J.B. RATTRAY. 1987. Purification and characterization of two endoxylanases from *Clostridium acetobutylicum* ATCC 824. Appl. Environ. Microbiol. **53:** 664–650.
35. DÜRRE, P. & H. BAHL. 1996. Microbial production of acetone/butanol/isopropanol. *In* Biotechnology, Vol. 6, 2nd ed. M. Roehr, Ed.: 229–268. Weinheim, Germany: VCH Verlagsgesellschaft.
36. NÖLLING, J. *et al.* 2001. Genome sequence and comparative analysis of the solvent-producing bacterium *Clostridium acetobutylicum*. J. Bacteriol. **183:** 4823–4838.
37. CHEN, J.-S. 1995. Alcohol dehydrogenase: multiplicity and relatedness in the solvent-producing clostridia. FEMS Microbiol. Rev. **17:** 263–273.
38. WELCH, R.W., F.B. RUDOLPH & E.T. PAPOUTSAKIS. 1989. Purification and characterization of the NADH-dependent butanol dehydrogenase from *Clostridium acetobutylicum* (ATCC 824). Arch. Biochem. Biophys. **273:** 309–318.
39. FISCHER, R.J., J. HELMS & P. DÜRRE. 1993. Cloning, sequencing, and molecular analysis of the *sol* operon of *Clostridium acetobutylicum*, a chromosomal locus involved in solventogenesis. J. Bacteriol. **175:** 6959–6969.
40. NAIR, R.V., G.N. BENNETT & E.T. PAPOUTSAKIS. 1994. Molecular characterization of an aldehyde/alcohol dehydrogenase gene from *Clostridium acetobutylicum* ATCC 824. J. Bacteriol. **176:** 871–885.
41. FONTAINE, L. *et al.* 2002. Molecular characterization and transcriptional analysis of *adhE2*, the gene encoding the NADH-dependent aldehyde/alcohol dehydrogenase responsible for butanol production in alcohologenic cultures of *Clostridium acetobutylicum* ATCC 824. J. Bacteriol. **184:** 821–830.
42. SAUER, U. & P. DÜRRE. 1995. Differential induction of genes related to solvent formation during the shift from acidogenesis to solventogenesis in continuous culture of *Clostridium acetobutylicum*. FEMS Microbiol. Lett. **125:** 115–120.
43. GIRBAL, L. *et al.* 1995. Regulation of metabolic shifts in *Clostridium acetobutylicum* ATCC 824. FEMS Microbiol. Rev. **17:** 287–297.
44. CARY, J.W. *et al.* 1990. Cloning and expression of *Clostridium acetobutylicum* ATCC 824 acetoacetyl-coenzyme A: acetate/butyrate: coenzyme A-transferase in *Escherichia coli*. Appl. Environ. Microbiol. **56:** 1576–1583.
45. GERISCHER, U. & P. DÜRRE. 1990. Cloning, sequencing, and molecular analysis of the acetoacetate decarboxylase gene region from *Clostridium acetobutylicum*. J. Bacteriol. **172:** 6907–6918.
46. PETERSEN, D.J. & G.N. BENNETT. 1990. Purification of acetoacetate decarboxylase from *Clostridium acetobutylicum* ATCC 824 and cloning of the acetoacetate decarboxylase gene in *Escherichia coli*. Appl. Environ. Microbiol. **56:** 3491–3498.
47. CORNILLOT, E. & P. SOUCAILLE. 1996. Solvent-forming genes in clostridia. Nature **380:** 489.
48. CORNILLOT, E. *et al.* 1997. The genes for butanol and acetone formation in *Clostridium acetobutylicum* ATCC 824 reside on a large plasmid whose loss leads to degeneration of the strain. J. Bacteriol. **179:** 5442–5447.
49. PETERSEN, D.J. *et al.* 1991. Molecular cloning of an alcohol (butanol) dehydrogenase gene cluster from *Clostridium acetobutylicum* ATCC 824. J. Bacteriol. **173:** 1831–1834.
50. WALTER, K.A., G.N. BENNETT & E.T. PAPOUTSAKIS. 1992. Molecular characterization of two *Clostridium acetobutylicum* ATCC 824 butanol dehydrogenase isozyme genes. J. Bacteriol. **174:** 7149–7158.
51. FEUSTEL, L., S. NAKOTTE & P. DÜRRE. 2004. Characterization and development of two reporter gene systems for *Clostridium acetobutylicum*. Appl. Environ. Microbiol. **70:** 798–803.
52. RAVAGNANI, A. *et al.* 2000. Spo0A directly controls the switch from acid to solvent production in solvent-forming clostridia. Mol. Microbiol. **37:** 1172–1185.
53. HARRIS, L.M., N.E. WELKER & E.T. PAPOUTSAKIS. 2002. Northern, morphological, and fermentation analysis of *spo0A* inactivation and overexpression in *Clostridium acetobutylicum* ATCC 824. J. Bacteriol. **184:** 3586–3597.
54. GERISCHER, U. & P. DÜRRE. 1992. mRNA analysis of the *adc* gene region of *Clostridium acetobutylicum* during the shift to solventogenesis. J. Bacteriol. **174:** 426–433.
55. BÖHRINGER, M. 2002. Molekularbiologische und enzymatische Untersuchungen zur Regulation des Gens der Acetacetat-Decarboxylase von *Clostridium acetobutylicum*. Ph.D. Thesis, University of Ulm, Ulm, Germany.
56. SCHIEL, B. 2006. Regulation der Lösungsmittelbildung in *Clostridium acetobutylicum* durch DNA-bindende Proteine. Ph. D. Thesis. University of Ulm. Ulm, Germany.
57. SCHAFFER, S. *et al.* 2002. Changes in protein synthesis and identification of proteins specifically induced during solventogenesis in *Clostridium acetobutylicum*. Electrophoresis **23:** 110–121.
58. HARRIS, L.M., N.E. WELKER & E.T. PAPOUTSAKIS. 2002. Northern, morphological, and fermentation analysis of *spo0A* inactivation and overexpression in *Clostridium acetobutylicum* ATCC 824. J. Bacteriol. **184:** 3586–3597.
59. THORMANN, K. *et al.* 2002. Control of butanol formation in *Clostridium acetobutylicum* by transcriptional activation. J. Bacteriol. **184:** 1966–1973.
60. NAIR, R.V. *et al.* 1999. Regulation of the *sol* locus genes for butanol and acetone formation in *Clostridium acetobutylicum* ATCC 824 by a putative transcriptional repressor. J. Bacteriol. **181:** 319–330.
61. THORMANN, K. & P. DÜRRE. 2001. Orf5/SolR: a transcriptional repressor of the *sol* operon of *Clostridium acetobutylicum*? J. Ind. Microbiol. Biotechnol. **27:** 307–313.

62. CHEN, C.-K. & H.P. BLASCHEK. 1999. Effect of acetate on molecular and physiological aspects of *Clostridium beijerinckii* NCIMB 8052 solvent production and strain degeneration. Appl. Environ. Microbiol. **65:** 499–505.
63. KOSAKA, T. *et al.* 2007. Characterization of the *sol* operon in butanol-hyperproducing *Clostridium saccharoperbutylacetonicum* strain N1-4 and its degeneration mechanism. Biosci. Biotechnol. Biochem. **71:** 58–68.
64. RAO, G. & R. MUTHARASAN. 1986. Alcohol production by *Clostridium acetobutylicum* induced by methyl viologen. Biotechnol. Lett. **8:** 893–896.
65. RAO, G. & R. MUTHARASAN. 1987. Altered electron flow in continuous cultures of *Clostridium acetobutylicum* induced by viologen dyes. Appl. Environ. Microbiol. **53:** 1232–1235.
66. DÜRRE, P. *et al.* 1995. Solventogenic enzymes of *Clostridium acetobutylicum*: catalytic properties, genetic organization, and transcriptional regulation. FEMS Microbiol. Rev. **17:** 251–262.
67. FONTAINE, L. *et al.* 2002. Molecular characterization and transcriptional analysis of *adhE2*, the gene encoding the NADH-dependent aldehyde/alcohol dehydrogenase responsible for butanol production in alcohologenic cultures of *Clostridium acetobutylicum* ATCC 824. J. Bacteriol. **184:** 821–830.
68. DÜRRE, P. 2004. Solventogenesis by clostridia. *In* Strict and Facultative Anaerobes: Medical and Environmental Aspects. M.M. Nakano & P. Zuber, Eds.: 329–342. Wymondham, UK: Horizon Bioscience.
69. HOLT, R.A., G.M. STEPHENS & J.G. MORRIS. 1984. Production of solvents by *Clostridium acetobutylicum* cultures maintained at neutral pH. Appl. Environ. Microbiol. **48:** 1166–1170.
70. FREIER-SCHRÖDER, D., J. WIEGEL & G. GOTTSCHALK. 1989. Butanol formation by *Clostridium thermosaccharolyticum* at neutral pH. Biotechnol. Lett. **11:** 831–836.
71. ULLMANN, S., A. KUHN & P. DÜRRE. 1996. DNA topology and gene expression in *Clostridium acetobutylicum*: implications for the regulation of solventogenesis. Biotechnol. Lett. **18:** 1413–1418.
72. GRUPE, H. & G. GOTTSCHALK. 1992. Physiological events in *Clostridium acetobutylicum* during the shift from acidogenesis to solventogenesis in continuous culture and presentation of a model for shift induction. Appl. Environ. Microbiol. **58:** 3896–3907.
73. BOYNTON, Z.L., G.N. BENNETT & F.B. RUDOLPH. 1994. Intracellular concentrations of coenzyme A and its derivatives from *Clostridium acetobutylicum* ATCC 824 and their roles in enzyme regulation. Appl. Environ. Microbiol. **60:** 39–44.
74. ZHAO, Y. *et al.* 2005. Intracellular butyryl phosphate and acetyl phosphate concentrations in *Clostridium acetobutylicum* and their implications for solvent formation. Appl. Environ. Microbiol. **71:** 530–537.
75. TOMAS, C.A., N.E. WELKER & E.T. PAPOUTSAKIS. 2003. Overexpression of *groESL* in *Clostridium acetobutylicum* results in increased solvent production and tolerance, prolonged metabolism, and changes in the cell's transcriptional program. Appl. Environ. Microbiol. **69:** 4951–4965.
76. TOMAS, C.A. *et al.* 2003. DNA array-based transcriptional analysis of asporogenous, nonsolventogenic *Clostridium acetobutylicum* strains SKO1 and M5. J. Bacteriol. **185:** 4539–4547.
77. ALSAKER, K.V., T.R. SPITZER & E.T. PAPOUTSAKIS. 2004. Transcriptional analysis of *spo0A* overexpression in *Clostridium acetobutylicum* and its effect on the cell's response to butanol stress. J. Bacteriol. **186:** 1959–1971.
78. TOMAS, C.A., J.A. BEAMISH & E.T. PAPOUTSAKIS. 2004. Transcriptional analysis of butanol stress and tolerance in *Clostridium acetobutylicum*. J. Bacteriol. **186:** 2006–2018.
79. ALSAKER, K.V. & E.T. PAPOUTSAKIS. 2005. Transcriptional program of early sporulation and stationary-phase events in *Clostridium acetobutylicum*. J. Bacteriol. **187:** 7103–7118.
80. BORDEN, J.R. & E.T. PAPOUTSAKIS. 2007. Dynamics of genomic-library enrichment and identification of solvent tolerance genes for *Clostridium acetobutylicum*. Appl. Environ. Microbiol. **73:** 3061–3068.
81. PAREDES, C.J., K.V. ALSAKER & E.T. PAPOUTSAKIS. 2005. A comparative genomic view of clostridial sporulation and physiology. Nat. Rev. Microbiol. **3:** 969–978.
82. SULLIVAN, L. & G.N. BENNETT. 2006. Proteome analysis and comparison of *Clostridium acetobutylicum* ATCC 824 and Spo0A strain variants. J. Ind. Microbiol. Biotechnol. **33:** 298–308.
83. SANTANGELO, J.D. & P. DÜRRE. 1996. Microbial production of acetone and butanol: Can history be repeated? Chim. oggi/Chem. Today **14:** 29–35.
84. DÜRRE, P. 1998. New insights and novel developments in clostridial acetone/butanol/isopropanol fermentation. Appl. Microbiol. Biotechnol. **49:** 639–648.
85. EZEJI, T.C., N. QURESHI & H.P. BLASCHEK. 2005. Continuous butanol fermentation and feed starch retrogradation: butanol fermentation sustainability using *Clostridium beijerinckii* BA101. J. Biotechnol. **115:** 179–187.
86. SABATHÉ, F., A. BÉLAICH & P. SOUCAILLE. 2002. Characterization of the cellulolytic complex (cellulosome) of *Clostridium acetobutylicum*. FEMS Microbiol. Lett. **217:** 15–22.
87. SABATHÉ, F. & P. SOUCAILLE. 2003. Characterization of the CipA scaffolding protein and in vivo production of a minicellulosome in *Clostridium acetobutylicum*. J. Bacteriol. **185:** 1092–1096.
88. LOPÉZ-CONTRERAS, A.M. *et al.* 2003. Production by *Clostridium acetobutylicum* ATCC 824 of CelG, a cellulosomal glycoside hydrolase belonging to family 9. Appl. Environ. Microbiol. **69:** 869–877.
89. PERRET, S. *et al.* 2004. Production of heterologous and chimeric scaffoldins by *Clostridium acetobutylicum* ATCC 824. J. Bacteriol. **186:** 253–257.
90. SCHWARZ, W.H. 2001. The cellulosome and cellulose degradation by anaerobic bacteria. Appl. Microbiol. Biotechnol. **56:** 634–649.
91. ZHAO, Y. *et al.* 2003. Expression of a cloned cyclopropane fatty acid synthase gene reduces solvent formation in *Clostridium acetobutylicum* ATCC 824. Appl. Environ. Microbiol. **69:** 2831–2841.
92. HARRIS, L.M. *et al.* 2001. Fermentation characterization and flux analysis of recombinant strains of *Clostridium acetobutylicum* with an inactivated *solR* gene. J. Ind. Microbiol. Biotechnol. **27:** 322–328.

93. Heap, J.T. *et al*. 2007. The ClosTron: a universal gene knockout system for the genus *Clostridium*. J. Microbiol. Methods **70:** 452–464.

94. Ezeji, T.C., N. Qureshi & H.P. Blaschek. 2004. Butanol fermentation research: upstream and downstream manipulations. Chem. Rec. **4:** 305–314.

95. Ezeji, T.C. *et al*. 2005. Improving performance of a gas stripping-based recovery system to remove butanol from *Clostridium beijerinckii* fermentation. Bioprocess Biosyst. Eng. **27:** 207–214.

96. Qureshi, N. *et al*. 2005. Energy-efficient recovery of butanol from model solutions and fermentation broth by adsorption. Bioprocess Biosyst. Eng. **27:** 215–222.

97. http://www.igb.fhg.de/WWW/GF/Biokatalyse/en/GFBK_233_Butanol.en.html

Anaerobic Respiration in Engineered Escherichia coli *with an Internal Electron Acceptor to Produce Fuel Ethanol*

JOY DORAN PETERSON[a] AND LONNIE O. INGRAM[b]

[a]*Microbiology Department, University of Georgia, Athens, Georgia, USA*

[b]*Department of Microbiology and Cell Science, University of Florida, Gainesville, Florida, USA*

Environmental concerns and unease with U.S. dependence on foreign oil have renewed interest in converting biomass into fuel ethanol. The volume of plant matter available makes lignocellulose conversion to ethanol desirable, although no one isolated organism has been shown to break bonds in lignocellulose and efficiently metabolize resulting sugars into one product. This work reviews directed engineering coupled with metabolic evolution resulting in microbial biocatalysts that produce up to 45 g L^{-1} ethanol in 48 hours in a simple mineral salts medium and that convert various compounds of lignocellulosic materials to ethanol. Mutations contributing to ethanologenesis are discussed along with adding enzymatic capabilities to existing biocatalysts in order to decrease the commercial enzymes required to reduce plant matter into fermentable sugars.

Key words: ***Escherichia coli*; ethanol; residues; hemicellulose hydrolysate**

Introduction

Escherichia coli is a well-studied facultatively anaerobic bacterium that uses glycolysis via the Embden–Meyerhoff–Parnas (MBP) pathway to convert glucose into pyruvate, requiring adenosine diphosphate (ADP) and oxidized nicotinamide adenine dinucleotide (NAD^+) as cofactors, thus producing adenosine triphosphate (ATP) and reduced NAD (NADH) inside the cell. Electrons from NADH can ultimately be passed to oxygen, but under anaerobic conditions, another electron acceptor is needed. Respiration in an anaerobic environment with an internal electron acceptor is called fermentation. *E. coli* ferments a number of pentose and hexose sugars into a mixture of acids (lactic, acetic, formic, and succinic) and ethanol.[1] The anaerobic bacterium, *Zymomonas mobili*s, and the yeast, *Saccharomyces cerevisiae,* produce ethanol as the sole fermentation product, due to the redox neutral state of ethanol.[2,3] Most ethanol fermentations in the United States today use the yeast, *S. cerevisiae*, to convert starch glucose into ethanol and CO_2, although the anaerobic bacterium, *Z. mobilis,* also metabolizes glucose to ethanol and CO_2 using a different pathway (Entner–Doudoroff). Engineering of the *Z. mobilis* ethanol pathway in hexose and pentose fermenting *E. coli* has been previously reviewed.[4–8] Briefly, pyruvate decarboxylase and alcohol dehydrogenase II genes from *Z. mobilis* were integrated into the chromosome of *E. coli* to generate *E. coli* strain KO11. The *Z. mobilis* K_m [pyruvate decarboxylase (PDC)] is quite low compared with other pyruvate-consuming reactions (FIG. 1), effectively shifting metabolic products to much higher concentrations of ethanol.

Much more ethanol will be needed if we are to focus efforts on decreasing our carbon footprint on the planet by increasing our production and use of renewable energy. Using liquid transportation fuels, such as ethanol, provides an immediate response to environmental concerns and unease with our dependence on foreign oil. The sheer volume of plant matter available makes lignocellulose conversion to ethanol desirable. Producing ethanol from renewable resources in one's home country could help to reverse the trend of ever increasing imported oil; currently over 50% of the oil consumed in the United States is imported (U.S. Energy Information Administration www.eia.doe.gov). The U.S. Energy Policy Act of 2005 (http://www.ferc.gov) requires that 7.5 billion gallons of renewable fuels be incorporated into gasoline over the next 6 years, and ethanol is the most prevalent renewable fuel, with U.S. production

Address for correspondence: Joy Doran Peterson, Microbiology Department, 1000 Cedar Street, 527 BIO SCI, University of Georgia, Athens, GA 30602.

jpeterso@uga.edu

Ann. N.Y. Acad. Sci. 1125: 363–372 (2008). © 2008 New York Academy of Sciences.
doi: 10.1196/annals.1419.020

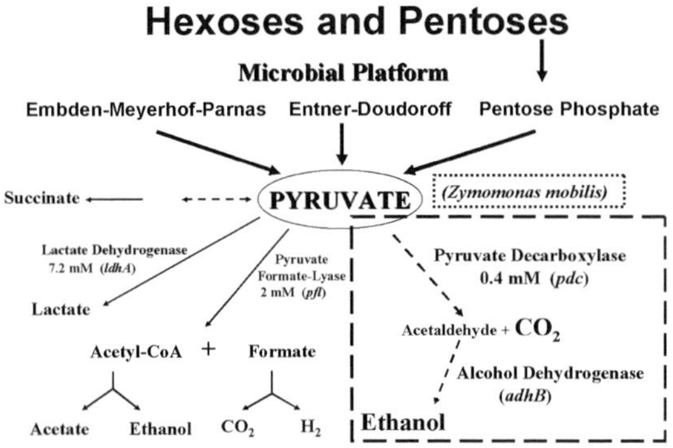

Hexoses and Pentoses
Microbial Platform
Embden-Meyerhof-Parnas Entner-Doudoroff Pentose Phosphate
Succinate PYRUVATE (Zymomonas mobilis)
Lactate Dehydrogenase 7.2 mM (ldhA)
Pyruvate Formate-Lyase 2 mM (pfl)
Pyruvate Decarboxylase 0.4 mM (pdc)
Lactate
Acetaldehyde + CO_2
Acetyl-CoA + Formate
Alcohol Dehydrogenase (adhB)
Acetate Ethanol CO_2 H_2 Ethanol

FIGURE 1. Conversion of hexose and pentose sugars to ethanol by recombinant *E. coli* in conjunction with the *Z. mobilis* ethanol pathway. Native *E. coli* reactions are depicted with a *solid arrow,* while those from *Z. mobilis* are depicted with a *dashed arrow*.

exceeding 4.8 billion gallons in 2006 (FIG. 2). Currently, the majority of ethanol is produced from corn; however, limited supply will force ethanol production from other sources of biomass. The U.S. Department of Agriculture and U.S. Department of Energy published a report describing biomass as feedstock for a bioenergy and bioproducts industry and determined that the United States produces approximately 1.3 billion tons of biomass annually—enough to replace about 30% of our current petroleum usage.[9]

Unlike corn, where the major component is starch, other sources of biomass are composed of 40–50% cellulose, 25–35% hemicellulose, and 15–20% lignin.[10] Most lignocellulosic ethanol process designs are more complex than corn ethanol processes due to the complexity and structural integrity of lignocellulose. Conversion of lignocellulose may include thermochemical and/or mechanical pretreatment to allow enzymatic access, enzymatic degradation to reduce substrates to fermentable sugars, and finally fermentation of those sugars by microorganisms. Another approach is to use agricultural residues that are coproducts of existing harvesting and processing practices. In some cases, the agricultural residues are already collected, partially processed, and require minimal additional treatment before fermenting the sugars to ethanol. This review focuses on recent progress using improved *E. coli* strains for conversion of lignocellulosic sugars into ethanol.

Engineering and Performance of Ethanologenic *E. coli* Strain KO11

Unmodified *E. coli* W is unable to produce ethanol as the major fermentation product; however, the patented KO11 derivative produces ethanol as its major fermentation product from hexose and pentose sugars[11] and the uronic acid constituents of pectin.[12] The broad substrate range, facile genetic system, extensive background of knowledge, and history of industrial use made *E. coli* an obvious choice for metabolic engineering of a microbial biocatalyst for production of ethanol from lignocellulose. In the early ethanologens, pyruvate decarboxylase (*pdc*) and alcohol dehydrogenase (*adh*) II genes from *Z. mobilis* were combined and placed under the control of a single enteric promoter to produce an artificial operon for ethanol production called the *pet* operon. The *pet* operon was first introduced into *E. coli* W on plasmids,[13–16] and thus diverted pyruvate to ethanol as the major fermentation product. For strain KO11, this *Z. mobilis* ethanol pathway, or *pet* operon, was integrated into the chromosome of *E. coli* at the pyruvate formate lyase (*pfl*) locus. The low *Z. mobilis* K_m (PDC) and a mutation in *frd* (-succinate) generated an ethanologen, strain KO11, that produced yields of 95% on glucose in complex media.[17]

Transcriptome analysis was used to investigate the physiological basis for increased growth and glycolytic flux of KO11 relative to *E. coli* B during growth on Luria-Bertani (LB) xylose.[18] KO11 exhibits increased transcript abundance for many xylose metabolic genes and resulting increased enzymatic activity. Increased expression of xylose metabolic enzymes contributes to the observed increase in growth rate and glycolytic flux.

Retaining ethanol production rates and yields over time without antibiotic selection pressure is important to minimizing production costs. Reports regarding the phenotypic stability of KO11 are somewhat mixed. Several groups have replicated our results for high ethanol yields,[19,20] while others have reported lower than expected ethanol yields.[21,22] Still other reports have demonstrated maintenance of KO11 ethanologenicity in continuous stirred tank and fluidized bed reactors for up to 27 days.[23]

Agricultural Residue Fermentations Using Strain KO11

E. coli strain KO11 has been used for production of ethanol from many agricultural residues, including but not limited to rice hulls,[24] sugar-cane bagasse,[25] agricultural residues,[25] corn cobs, hulls and ammonia fiber explosion (AFEX)–pretreated fibers,[26,27] orange peel,[12] pectin-rich beet pulp,[28] sweet whey,[29] brewery waste,[30] and cotton-gin waste.[31] Ethanol production, yield on a total ethanol per gram of sugar basis, and

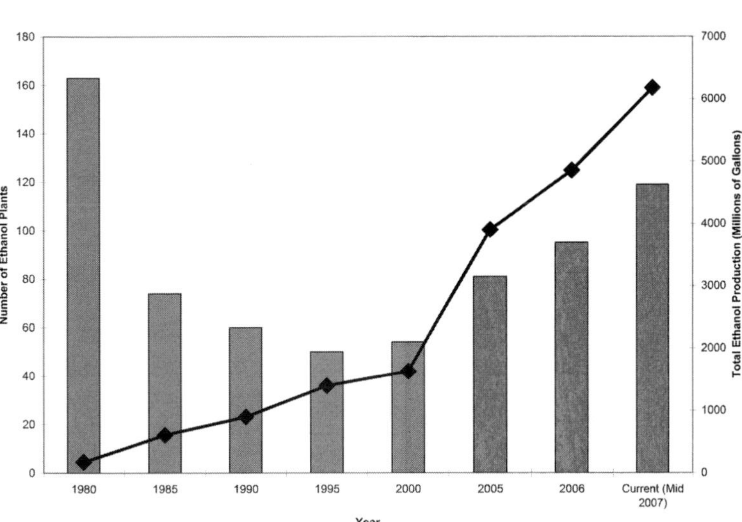

FIGURE 2. Number of ethanol plants (*bars*) and total ethanol production (*line*) from 1980 until present (mid-2007).

TABLE 1. Agricultural and forestry biomass converted to ethanol by *E. coli* strain KO11 or LYO1

Biomass	Biomass pretreatment	Ethanol (g L^{-1})	Fermentation time (h)	gTE gS^{-1a}	Reference
Rice hulls	Dilute acid hydrolysis	46	72	0.46	24
Sugar-cane bagasse	Dilute acid hydrolysis	37	60	0.44–0.51	25
Corn hulls and fibers	Diltue acid hydrolysis	44	72	0.46–0.48	25
Corn stover	Dilute acid hydrolysis	42.4	48	0.45–0.53	25
Beet pulp KO11	Enzymes	40	120	0.32–0.41	28
Corn hulls	Dilute acid hydrolysis	38	48	0.51	26
Pinus sp (softwood)	Dilute acid hydrolysis	35	48	0.48	88
Orange peel	Grinding, enzymes, filtration	28	72	0.41	12
Sweet whey	Proteinase	26	96	0.43	29

[a]gTE gS^{-1} grams of total ethanol divided by grams of sugar.

time to reach maximum ethanol concentration is presented in TABLE 1. Strain KO11 has been the subject of an empirical kinetic model [32] and has been shown to be relatively insensitive to changes in fermentation temperature and pH.[33]

Many of the biomass substrates fermented by KO11 were subjected to dilute-acid hydrolysis. Dilute-acid hydrolysis is an accepted method for depolymerizing plant biomass, although inhibitors of fermentation are generated to varying degrees.[34,35] Recent work has explored mechanisms of toxicity on ethanologenic biocatalysts. Furfural and 5-hydroxymethylfurfural are pentose and hexose sugar derivatives, respectively. Furfural is present in hemicellulose hydrolysate at a concentration of 3–40 mM[36] and can inhibit growth at low concentrations.[37,38] The presence of furfural amplifies the effect of other toxins, as well.[38] Boopathy and colleagues [39] determined the biotransformation of furfural and 5-hydroxymethylfurfural by enteric bacteria, and despite the observed toxicity, ethanologenic *E. coli* KO11 has demonstrated a native ability to transform furfural to less toxic furfuryl alcohol.[40]

Ethanologenic Biocatalyst *E. coli* Strain LYO1

In order to improve the ethanol tolerance of *E. coli* strain KO11, mutant strains with increased ethanol tolerance were isolated using a then novel approach alternating selection for ethanol tolerance in liquid and high antibiotic resistance on solid media.[41] This was important because the antibiotic resistance gene (*cat*) was coupled with the *PET* genes, so increased antibiotic resistance usually correlated with robust ethanol production. The best of these mutants, strain LY01,

TABLE 2. Growth inhibition of *E. coli* LY01 and selected bacteria from exposure to representative hemicellulose hydrolysate toxins

Toxin	Concentration (g/L)	Inhibition (%)	Microorganism	Reference
Furfural	3.4	50	*E. coli* ATCC 1175[a]	39
	2.4	50	*Proteus vulgaris*[a]	39
	2.9	50	*E. coli* LY01[b]	43
5-Hydroxymethylfurfural	2.7	50	*E. coli* ATCC 1175[a]	39
	1.9	50	*P. vulgaris*[a]	39
	3.8	50	*E. coli* LY01[b]	43
Syringaldehyde	0.38	100	*E. coli*[c]	43
	1.2	50	*E. coli* LY01[b]	43
Ferulic acid	0.38	100	*E. coli*[a]	89
	0.7	50	*E. coli* LY01[b]	90
Syringic acid	0.5	50	*Klebsiella pneumoniae*[d]	91
	1.6	50	*E. coli* LY01[b]	90

[a]Anaerobic culture tubes, 10 mL, 37°C, pH = 7.0, inoculum: 10%, 12 h.
[b]Standing tubes, 4 mL, 30°C, initial pH = 7.0, inoculum OD = 0.025, 48 h.
[c]Tubes, 2 mL, 30°C, inoculum: 1×10^7 cells/mL, 24 h.
[d]Shaken tubes, 150 rpm, 10 mL, 30°C, initial pH = 6.5, inoculum 5% v/v, 48 h.

exhibited large colony morphologies due to decreased acid production and tolerated brief exposure to ethanol concentrations as high as 100 g L^{-1}.[41]

The transcriptomes of KO11 and LY01 were compared in LB medium with glucose or xylose and with 0, 1, or 2% w/v ethanol.[42] Forty-nine genes exhibited greater than twofold difference in expression. LY01 exhibited increased glycine degradation and lack of fumarate and nitrate reductase regulator (FNR) function, impacting the availability of pyruvate compared with strain KO11. Increased expression of genes related to betaine synthesis and uptake of protective osmolites were also observed for LY01. *E. coli* strain LY01 is also more resistant than KO11 and many other ethanologens to fermentation inhibitors generated from sugar degradation products.[43] Accordingly, the increased ethanol tolerance of LY01 seems to be a combination of several physiological factors, rather than tolerance conferred by one major mutation in a selected region of the genome.[42,44]

Improved Tolerance to Lignocellulose Hydrolysate Inhibitors by Strain LY01

E. coli KO11 and LY01 were compared for their ability to tolerate representative model aldehydes found in hemicellulose hydrolysate from agricultural and forestry residues.[43] Both organisms have demonstrated a native ability to transform furfural to the less toxic furfuryl alcohol,[40] although LYO1 appeared to be more resistant to furfural and 5-hydroxymethylfurfural during fermentations.[43]

The toxicity of representative alcohol, aldehyde and acid components of hemicellulose hydrolysates were investigated for ethanologenic *E. coli* LY01[38,45,46] (TABLE 2). The toxicity of aldehydes was directly linked to the degree of hydrophobicity.[43] All aldehydes tested inhibited growth; however, only furfural decreased ethanol production.[38] Furfural reductase, purified from LY01 to approximately 50% homogeneity, catalyzes the reduction of furfural with NADPH and appears to be a novel type of alcohol-aldehyde oxido-reductase.[47] Because the furans are toxic themselves and enhance the toxicity of other compounds found in lignocellulose hydrolysates, UV spectra can be used to estimate total furan content,[48] and this value may be used to estimate potential toxicity of a lignocellulosic biomass hydrolysate. Other inhibitors, such as aliphatic acids, appear to inhibit both growth and ethanol production by collapsing ion gradients and increasing the internal anion concentration, rather than by inhibiting essential pathways.[45]

Agricultural Residue Fermentations Using Strain LY01

Sugar-beet pulp was fermented to ethanol using strain LY01, fungal enzymes, and rich nutrient supplementation.[49] Strain LY01 performed well, reaching 40 g ethanol L^{-1}, although increasing substrate concentration to increase ethanol yield above that obtained by strain KO11 was not successful (Peterson, unpublished data). KO11 and LY01 both exhibited reduced ethanol production when solids concentrations

TABLE 3. Comparison of ethanologenesis from xylose using *E. coli* as the biocatalyst

Organism	Xylose (g L^{-1})	Medium	Ethanol (g L^{-1})	Yield (g g^{-1})	Reference
E. coli LY168	90	Min	45.5	0.50	59
E. coli LY168	90	LB	45.3	0.50	59
E. coli KO11	90	LB	43.2	0.48	59
E. coli LY01	90	LB	42.4	0.47	41
E. coli FBR5 (pLOI297)	95	LB	41.5	0.44	63
E. coli KO11	90	Min	26.9	0.30	59
nonrecombinant *E. coli* SE2378	50	LB	20.5	0.41	65

ABBREVIATIONS: LB = yeast extract + tryptone; Min = minerals + 1 mM betaine.

were above 12% w/v. Strain LY01 is currently being used as the parent biocatalyst for further engineering of *E. coli* for pectin-rich biomass fermentations.

Strain LY01 was also selected as the biocatalyst for fermentations using pressurized batch hot water hydrolyzed bermudagrass[50] because this biocatalyst was more resistant to compounds present in grass tissues.[43] Bermudagrass contains lignified cell walls and low-molecular-weight phenolic acids ester-linked to arabinose.[51] Ester-linked 3-(4-hydroxyphenyl)-2-propenoic (*p*-coumaric) and 3-(4-hydroxy-3-methoxyphenyl)prop-2-enoic (ferulic) acids are also found in nonlignified parts of the cell walls.[52] Hot water pretreatment can cause liberation of acetyl groups from hemicellulose, increased depolymerization, and reduction in pH. As the pH is lowered, the hot water pretreatment becomes, in effect, a very dilute-acid hydrolysis.[53] Sugars, furfural, and/or 5-hydroxymethylfurfural were present in minimal concentrations after pretreatment. Phenolic acids, including ferulic and para-coumaric acids were not present in the process water, but were released during fermentations as enzymes liberated these compounds during simultaneous saccharification and fermentation.[50] An ethanol yield of 0.40 g ethanol per g sugar (0.2 g ethanol per g dry weight total pretreated grass) was obtained after pretreatment at 230°C for 2 min with a very low fungal enzyme load.

KO11 and LY01 *E. coli* strains have performed well with high ethanol yields in rich media; however, performance and yields declined in mineral salts media.[54,55] Poor performance in minimal media may be caused by suboptimal partitioning of pyruvate for biosynthesis.[56,57] High NADH and low acetyl-CoA levels inhibit citrate synthase, thus limiting availability of 2-oxoglutarate, required for synthesis of many amino acids. Oxoglutarate is also an important compound for osmotic tolerance. The proposed inhibition of citrate synthase was supported by expression of a NADH-insensitive citrate synthase from *Bacillus* that increased the growth and ethanol production of KO11 substantially.[56] Nuclear magnetic resonance analysis showed that intracellular pools of the osmoprotectants glutamate, trehalose, and betaine are very low in KO11 during anaerobic growth relative to aerobic growth.[58] If these osmolytes are added in the medium, growth and ethanol production of KO11 was increased, suggesting that a lack of osmolytes is causing the decreased performance rather than a specific metabolic demand for glutamate.[58]

Ethanologenic Biocatalyst *E. coli* Strain LY168

In order to improve the economic viability of a lignocellulose-to-ethanol process, strain LY168 was developed to produce high concentrations of ethanol in a simple mineral salts medium.[10] A lactic acid–producing derivative of *E. coli* strain KO11 that grew well and produced large quantities of lactic acid in mineral salts medium, SZ110[59] was reengineered for ethanol production to generate strain LY168.

Conversion of SZ110 from lactic acid production to ethanol production involved several steps beginning with deletion of *ldhA*, the gene for lactic acid production. Acetate accumulation was prevented by elimination of *ackA*, and the native *pfl* gene was restored. To restore ethanol production, the *Z. mobilis pdc*, *adhA*, and *adhB* genes on a promotorless operon were randomly inserted by transposon mutagenesis. Strains were enriched by serial transfers in mineral salts medium and one clone, designated LY160im contained the operon integrated within *rrlE*, a 23S ribosomal RNA subunit, concurrent with the direction of transcription. Two promoters are present to provide expression at high growth rates and basal expression at low growth rates for the *rrlE* gene,[60,61] and, accordingly, a fortuitous site for *PET* integration. To further increase ethanol production in strain LY160im by reducing ethyl-acetate formation, the *Pseudomonas putida* short-chain esterase *estZ* gene was also integrated into the chromosome, resulting in strain LY168. In xylose mineral salts medium

TABLE 4. Derivatives of *E. coli* generated by recombinant DNA technology described in this chapter

E. coli strain	Description	Phenotype	Reference
KO11	*pfl*::Z.mobilis *pet* operon Δ *frd*	Increased ethanol production	4–8, 11, 12, 17, 18
LY01	Directed evolution Multiple changes	Increased ethanol tolerance in rich media	41, 42
SZ110	Δ*Z.mobilis pet* operon Δ*adh*E Δ*ack*A	Growth and lactic acid production in mineral salts media with betaine	59
LY168	Δ*ldh*A +*pfl*B *rrl*E::*Z.mobilis pet* operon + *estZ*	Rapid growth in minimal media, decreased acetate production, increased ethanol production in minimal media	59

with betaine, LY168 produced 0.5 g ethanol per g of xylose, very close to the theoretical maximum of 0.51[59] (TABLE 3). Ethanol production from mineral salts with betaine or rich media (without betaine) (LB broth) were equal, thus alleviating the requirement for costly nutritional supplements. TABLE 4 presents a summary of the major attributes of the *E. coli* strains discussed in this review.

Other Ethanologenic *E. coli* Strains

A series of *E. coli* K12 bacteria were engineered with the same *pet* operon used in engineering of strain KO11. These K-12 derivatives are designated FBR for the Fermentation Biochemistry Research Unit. These strains carry *pfl* (pyruvate formate lyase) and *ldh* (lactate dehyddrogenase) mutations that block pyruvate reduction and limit the NADH,H^+ recycle from glycolysis. Adding the *pet* operon on a plasmid restored fermentative growth, obligating the cell to retain the plasmid or stop growing.[62,63] Strain FBR5 produced ethanol from a variety of substrates at 86–92% of the theoretical yield[63] and has demonstrated stability during long-term growth.[64] However, these strains are dependent on rich media and contain plasmids, and the final ethanol concentration and yields from FBR5 in LB xylose are lower than LY168 in minimal medium.

All *E. coli* strains discussed thus far exhibit recombinant expression of the *Z. mobilis* ethanol pathway; however, ethanol production by a mutant *E. coli* strain lacking foreign genes has been described recently.[65] Because of the limited NAD^+ and maintaining redox balance, wild-type *E. coli* is unable to grow anaerobically in the absence of both *ldhA* and *pflB*.[66] Chemical mutagenesis was used to isolate a ΔldhA/ΔpflB derivative capable of anaerobic growth, designated strain SE2378. SE2378 has a mutation within the pyruvate dehydrogenase operon, and fermented glucose and xylose to ethanol with 82% yield. In native *E. coli*, *pfl* is primarily responsible for production of acetyl-CoA during anaerobic growth, whereas pyruvate dehydrogenase (*pdh*) is reportedly inactive[67] or weakly active.[68] The *pdh* operon mutation produced an additional NADH for each pyruvate, allowing for the balanced production of two moles of ethanol per mole of glucose by a novel pathway not previously known in nature.

Other Improvements

Cheaper Fermentation Media

AM1[69] and NBS mineral salts media[70,71] are two simple mineral salts media developed in our lab to provide a cheaper medium for fermentations (TABLE 5). Both support cell growth and ethanol production at relatively high levels. AM1 is a derivative of NBS, with a 65% reduction in salts. With low total alkali (4.5 mM) and total salts (4.2 g L^{-1}), AM1 was able to support production of ethanol from xylose and lactate from glucose with average productivities of 18–19 mmol L^{-1} h^{-1}. Preparation of crude yeast autolysate and use of this nitrogen source in fermentations using *E. coli* strain KO11 obtained yields similar to those obtained using LB medium.[55]

Relieving Osmolyte Stress

High sugar levels required for high ethanol concentrations create osmotic stress. Osmolytes (see Ref. 72 for review) such as trehalose, betaine, proline, and glutamate help bacteria survive despite changes in extracellular osmolality. In the absence of supplemental protectants added to the medium, trehalose serves as

TABLE 5. Composition of defined minimal media, NBS, and AM1, excluding carbon source[70,92]

Component	Concentration (mmol L^{-1}) NBS + 1 mM betaine	AM1
KH_2PO_4	25.7	0
$K_2H\ PO_4$	28.7	0
$(NH_4)_2H\ PO_4$	26.5	19.9
$NH_4H_2\ PO_4$	0	7.6
KCL	0	2.0
$MgSO_4{\cdot}7H_2O$	1.0	1.5
$CaCl_2{\cdot}2H_2O$	0.1	0
Thiamine·HCl	0.02	0
Betaine·KCl	1.0	1.0
	(μmol L^{-1})	
$FeCl_3{\cdot}6H_2O$	5.9	8.9
$CoCl_2{\cdot}6H_2O$	0.8	1.3
$CuCl_2{\cdot}2H_2O$	0.6	0.9
$ZnCl_2$	1.5	2.2
$Na_2MoO_4{\cdot}2H_2O$	0.8	1.2
H_3BO_3	0.8	1.2
$MnCl_2{\cdot}4H_2O_2$	0	2.5

the major protective osmolyte in *E. coli* K-12 strains.[73] *E. coli* strain W3110 (ATCC 27325), a prototrophic derivative of *E. coli* strain K-12, is an excellent transformation host and was used to examine the importance of trehalose biosynthesis for growth during stress. The native trehalose synthesis pathway activity was increased, resulting in an increase in the growth rate of *E. coli* W3110 in the presence of various osmotic stress agents.[73] Elevated trehalose synthesis coupled with betaine addition increased the tolerance of *E. coli* strain W3110 to xylose, glucose, sodium lactate, and sodium chloride more than either addition alone.[74]

Reducing the Requirement for Fungal Cellulases

The cost of enzymes required for plant cell wall deconstruction is a major consideration in a lignocellulosic ethanol process.[75] In addition, enzymes in the cellulose deconstruction cascade are inhibited by the products generated. To ameliorate this effect, simultaneous saccharification of plant polymers and fermentation to consume these sugars, called SSF, was developed by Gulf Oil Company in 1976.[76,77] In the Gulf SSF process the biomass substrate, enzymes, and yeasts were all placed in one vessel, and the sugars liberated by the enzymes were immediately consumed by the yeast. This could also be achieved by engineering the fermenting organism to help break down the plant material during fermentation. In this fashion, sugars would be liberated by enzymes secreted by the fermenting organism and the resulting sugars fermented to ethanol in one pot. To this end we have been focused on reducing the supplemental cellulase demand by engineering the biocatalysts to produce recombinant cellulase enzymes.

Erwinia chysanthemi endoglucanases, CelZ and CelY, work synergistically to degrade amorphous cellulose and carboxymethyl cellulose.[78] Use of a *Z. mobilis* promoter and addition of the *E. chrysanthemi* out-secretion system (for transport *out* of the cell) resulted in high levels of CelZ expression in *E. coli*.[79,80] Cellobiose is a potent inhibitor of cellulases,[81] and strains able to metabolize cellobiose would be desirable in a biocatalyst for lignocellulosic ethanol production. *Klebsiella oxytoca*, a gram-negative bacterium closely related to *E. coli*, has the native ability to transport and metabolize cellobiose, reducing the initial demand for supplemental β-glucosidase.[82] *E. coli* KO11 has been engineered by adding the *K. oxytoca* cellobiose-utilization operon *casAB*, enabling production of ethanol from cellobiose. With the addition of commercial cellulase, ethanol was also produced from mixed-waste office paper.[82,83]

Using *E. coli* Ethanol Design Scheme for Producing other Commodity Products

E. coli uses many different sugar substrates and produces a wide spectrum of fermentation products. However, redirection of a microorganism's metabolism for efficient production of a single compound usually requires optimizing the expression level of multiple genes, which may not be predictable. Our success in generating microbial biocatalysts capable of producing high titers of chemicals depends on using that organism's natural ability to adapt and evolve. In our studies, the biocatalysts were grown in the desired mineral salts medium with high sugar concentrations long enough for them to evolve in their new environment. This method has resulted in biocatalysts proficient in production of ethanol, lactic acid,[84,85] succinic acid,[86] and acetate.[87]

Conflict of Interest

The authors declare no conflicts of interest.

References

1. CLARK, D.P. 1989. The fermentation pathways of *Escherichia coli*. FEMS Microbiol. Rev. **5:** 223–234.

2. BOTHAST, R.J. & M.A. SCHLICHER. 2005. Biotechnological processes for conversion of corn into ethanol. Appl. Microbiol. Biotechnol. **67:** 19–25.
3. HOFVENDAHL, K. & B. HANS-HAGERDAL. 2000. Factors affecting the fermentative lactic acid production from renewable resources. Enz. Microb. Technol. **26:** 87–107.
4. LIN, Y. & S. TANAKA. 2006. Ethanol fermentation from biomass resources: current state and prospects. Appl. Microbiol. Biotechnol. **69:** 627–642.
5. DIEN, B.S., M.A. COTTA & T.W. JEFFRIES. 2003. Bacteria engineered for fuel ethanol production: current status. Appl. Microbiol. Biotechnol. **63:** 258–266.
6. ZALDIVAR, J., J. NIELSEN & L. OLSSON. 2001. Fuel ethanol production from lignocellulose: a challenge for metabolic engineering and process integration. Appl. Microbiol. Biotechnol. **56:** 17–34.
7. INGRAM, L.O. *et al.* 1999. Enteric bacterial catalysts for fuel ethanol production. Biotechnol. Prog. **15:** 855–866.
8. INGRAM, L.O. *et al.* 1997. Fuel ethanol production from lignocellulose using genetically engineered bacteria. *In* Fuels and Chemicals from Biomass. B. Saha & J. Woodward, Eds.: 57–63. American Chemical Society. Washington, DC.
9. PERLACK, R.D. *et al.* 2005. Biomass as Feedstock for a Bioenergy and Bioproducts Industry: The Technical Feasibility of a Billion-ton Annual Supply: 78 pp. Oak Ridge, TN: Oak Ridge National Laboratory.
10. JARBOE, L.R. *et al.* 2007. Development of ethanologenic bacteria. *In* Advances in Biochemical Engineering/Biotechnology. L. Olsson, Ed., Vol. 108: 350. Berlin: Springer.
11. LIN, E.C.C. 1996. Dissimilatory pathways for sugars, polyols, and carboxylates. *In Escherichia coli* and Salmonella: Cellular and Molecular Biology. F.C. Neidhardt, Ed.: 307–342. Washington, DC: ASM Press.
12. GROHMANN, K. *et al.* 1994. Fermentation of galacturonic acid and other sugars in orange peel hydrolysates by the ethanologenic straw of *Escherichia coli*. Biotechnol. Lett. **16:** 281–286.
13. ALTERTHUM, F. & L.O. INGRAM. 1989. Efficient ethanol-production from glucose, lactose, and xylose by recombinant *Escherichia coli*. Appl. Environ. Microbiol. **55:** 1943–1948.
14. INGRAM, L.O. & T. CONWAY. 1988. Expression of different levels of ethanologenic enzymes from zymomonas-mobilis in recombinant strains of *Escherichia coli*. Appl. Environ. Microbiol. **54:** 397–404.
15. INGRAM, L.O. *et al.* 1987. Genetic-engineering of ethanol-production in *Escherichia coli*. Appl. Environ. Microbiol. **53:** 2420–2425.
16. NEALE, A.D., R.K. SCOPES & J.M. KELLY. 1988. Alcohol production from glucose and xylose using *Escherichia coli* containing zymomonas-mobilis genes. Appl. Microbiol. Biotechnol. **29:** 162–167.
17. OHTA, K. *et al.* 1991. Genetic-improvement of *Escherichia coli* for ethanol-production—Chromosomal integration of zymomonas-mobilis genes encoding pyruvate decarboxylase and alcohol dehydrogenase-II. Appl. Environ. Microbiol. **57:** 893–900.
18. TAO, H. *et al.* 2001. Engineering a homo-ethanol pathway in *Escherichia coli*: increased glycolytic flux and levels of expression of glycolytic genes during xylose fermentation. J. Bacteriol. **183:** 2979–2988.
19. DIEN, B.S., R.B. HESPELL, L.O. INGRAM & R.J. BOTHAST. 1997. Conversion of corn milling fibrous co-products into ethanol by recombinant *Escherichia coli* strains KO11 and SL40. World J. Microbiol. Biotechnol. **13:** 619–625.
20. VON SIVERS M., G. ZACCHI, G. OLSSON & B. HAHN-HAGERDAL. 1994. Cost analysis of ethanol production from willow using recombinant *Escherichia coli*. Biotechnol. Prog. **10:** 555–560.
21. LAWFORD, H.G. & J.D. ROUSSEAU. 1996. Factors contributing to the loss of ethanologenicity of *Escherichia coli* B recombinants pLOI297 and KO11. Appl. Biochem. Biotechnol. **57-8:** 293–305.
22. DUMSDAY, G.J. *et al.* 1999. Comparative stability of ethanol production by *Escherichia coli* KO11 in batch and chemostat culture. J. Ind. Microbiol. Biotechnol. **23:** 701–708.
23. DUMSDAY, G.J. *et al.* 1997. Continuous ethanol production by *Escherichia coli* KO11 in continuous stirred tank and fluidized bed fermenters. Australas. Biotechnol. **7:** 300–303.
24. MONIRUZZAMAN, M. & L.O. INGRAM. 1998. Ethanol production from dilute acid hydrolysate of rice hulls using genetically engineered *Escherichia coli*. Biotechnol. Lett. **20:** 943–947.
25. ASGHARI, A. *et al.* 1996. Ethanol production from hemicellulose hydrolysates of agricultural residues using genetically engineered *Escherichia coli* strain KO11. J. Industrial Microbiol. **16:** 42–47.
26. BEALL, D.S. *et al.* 1992. Conversion of hydrolysates of corn cobs and hulls into ethanol by recombinant *Escherichia coli*-B containing integrated genes for ethanol-production. Biotechnol. Lett. **14:** 857–862.
27. MONIRUZZAMAN, M. *et al.* 1996. Ethanol production from AFEX pretreated corn fiber by recombinant bacteria. Biotechnol. Lett. **18:** 985–990.
28. DORAN, J.B. *et al.* 2000. Fermentations of pectin-rich biomass with recombinant bacteria to produce fuel ethanol. Appl. Biochem. Biotechnol. **84-6:** 141–152.
29. LEITE, A.R. *et al.* 2000. Fermentation of sweet whey by recombinant *Escherichia coli* KO11. Br. J. Microbiol. **31:** 212–215.
30. RAO, K. *et al.* 2007. Enhanced ethanol fermentation of brewery wastewater using the genetically modified strain *E. coli* KO11. Appl. Microbiol. Biotechnol. **74:** 50–60.
31. JEOH, T. & F.A. AGBLEVOR. 2001. Characterization and fermentation of steam exploded cotton gin waste. Biomass Bioenergy **21:** 109–120.
32. OLSSON, L., B. HAHNHAGERDAL & G. ZACCHI. 1995. Kinetics of ethanol-production by recombinant *Escherichia coli* KO11. Biotechnol. Bioeng. **45:** 356–365.
33. MONIRUZZAMAN, M., S.W. YORK & L.O. INGRAM. 1998. Effects of process errors on the production of ethanol by *Escherichia coli* KO11. J. Ind. Microbiol. Biotechnol. **20:** 281–286.
34. DUPREEZ, J.C. 1994. Process parameters and environmental-factors affecting D-xylose fermentation by yeasts. Enzyme Microbial. Technol. **16:** 944–956.

35. McMillan, J.D. 1994. Conservsion of hemicellulose hydrolysates to ethanol. *In* Enzymatic Conversion of Biomass for Fuels Production. M.E. Himmel, J.O. Baker & R.P. Overend, Eds.: 411–437. Washington, DC: American Chemical Society Press.
36. Grohmann, K., R. Torget & M. Himmel. 1985. Optimization of dilute acid pretreatment of biomass. Biotechnol. Bioeng. (Symp.) **15:** 59–80.
37. Beall, D.S., K. Ohta & L.O. Ingram. 1991. Parametric studies of ethanol-production from xylose and other sugars by recombinant *Escherichia coli*. Biotechnol. Bioeng. **38:** 296–303.
38. Zaldivar, J., A. Martinez & L.O. Ingram. 1999. Effect of selected aldehydes on the growth and fermentation of ethanologenic *Escherichia coli*. Biotechnol. Bioeng. **65:** 24–33.
39. Boopathy, R., H. Bokang & L. Daniels. 1993. Biotransformation of furfural and 5-hydroxymethylfurfural by enteric bacteria. J. Ind. Microbiol. **11:** 147–150.
40. Gutierrez, T. *et al.* 2002. Reduction of furfural to furfuryl alcohol by ethanologenic strains of bacteria and its effect on ethanol production from xylose. Appl. Biochem. Biotechnol. **98:** 327–340.
41. Yomano, L.P., S.W. York & L.O. Ingram. 1998. Isolation and characterization of ethanol-tolerant mutants of *Escherichia coli* KO11 for fuel ethanol production. J. Ind. Microbiol. Biotechnol. **20:** 132–138.
42. Gonzalez, R. *et al.* 2003. Gene array-based identification of changes that contribute to ethanol tolerance in ethanologenic *Escherichia coli*: comparison of KO11 (parent) to LY01 (resistant mutant). Biotechnol. Prog. **19:** 612–623.
43. Zaldivar, J., A. Martinez & L.O. Ingram. 1999. Effect of selected aldehydes on the growth and fermentation of ethanologenic *Escherichia coli*. Biotechnol. Bioeng. **65:** 24–33.
44. Gonzalez, R. *et al.* 2002. Global gene expression differences associated with changes in glycolytic flux and growth rate in *Escherichia coli* during the fermentation of glucose and xylose. Biotechnol. Prog. **18:** 6–20.
45. Zaldivar, J. & L.O. Ingram. 1999. Effect of organic acids on the growth and fermentation of ethanologenic *Escherichia coli* LY01. Biotechnol. Bioeng. **66:** 203–210.
46. Zaldivar, J., A. Martinez & L.O. Ingram. 2000. Effect of alcohol compounds found in hemicellulose hydrolysate on the growth and fermentation of ethanologenic *Escherichia coli*. Biotechnol. Bioeng. **68:** 524–530.
47. Gutierrez, T., L.O. Ingram & J.F. Preston. 2006. Purification and characterization of a furfural reductase (FFR) from *Escherichia coli* strain LYO1—An enzyme important in the detoxification of furfural during ethanol production. J. Biotechnol. **121:** 154–164.
48. Martinez, A. *et al.* 2000. Use of UV absorbance to monitor furans in dilute acid hydrolysates of biomass. Biotechnol. Prog. **16:** 637–641.
49. Peterson, J.D. 2006. Ethanol production from agricultural residues. Int. Sugar J. **108:** 178–180.
50. Brandon, S.K., M.A. Eiteman, J.D. Peterson, *et al.* 2007. Hydrolysis of Tifton 85 Bermudagrass in a pressurized batch hot water reactor. J. Chem. Technol. Biotechnol. In Press.
51. Hartley, R. & C. Ford. 1989. Phenolic constituents of plant cell walls and wall degradability. *In* Plant Cell Wall Polymers Biogenesis and Biodegradation. N. Lewis & M. Paice, Eds.: 137–144. Washington, DC: American Chemical Society.
52. Carpita, N.C. 1996. Structure and biogenesis of the cell wall of plants. Annu. Rev. Plant Physiol. Plant Mol. Biol. **47:** 445–476.
53. Hamelinck, C.N., G.V. Hooijdonk & A.P.C. Faaij. 2005. Ethanol from lignocellulosic biomass: techno-economic performance in short-, middle- and long-term. Biomass Bioenergy **28:** 384–410.
54. Martinez, A. *et al.* 1999. Biosynthetic burden and plasmid burden limit expression of chromosomally integrated heterologous genes (pdc, adhB) in *Escherichia coli*. Biotechnol. Prog. **15:** 891–897.
55. York, S.W. & L.O. Ingram. 1996. Ethanol production by recombinant *Escherichia coli* KO11 using crude yeast autolysate as a nutrient supplement. Biotechnol. Lett. **18:** 683–688.
56. Underwood, S.A. *et al.* 2002. Flux through citrate synthase limits the growth of ethanologenic *Escherichia coli* KO11 during xylose fermentation. Appl. Environ. Microbiol. **68:** 1071–1081.
57. Underwood, S.A. *et al.* 2002. Genetic changes to optimize carbon partitioning between ethanol and biosynthesis in ethanologenic *Escherichia coli*. Appl. Environ. Microbiol. **68:** 6263–6272.
58. Underwood, S.A. *et al.* 2004. Lack of protective osmolytes limits final cell density and volumetric productivity of ethanologenic *Escherichia coli* KO11 during xylose fermentation. Appl. Environ. Microbiol. **70:** 2734–2740.
59. Yomano, L. *et al.* 2007. Re-engineering *Escherichia coli* B for ethanol production from xylose in mineral salts medium. Metabolic Eng. In Press.
60. Paul, B.J. *et al.* 2004. rRNA transcription in *Escherichia coli*. Annu. Rev. Genet. **38:** 749–770.
61. Dennis, P.P., M. Ehrenberg & H. Bremer. 2004. Control of rRNA synthesis in *Escherichia coli*: a systems biology approach. Microbiol. Mol. Biol Rev. **68:** 639–668.
62. Hespell, R.B. *et al.* 1996. Stabilization of pet operon plasmids and ethanol production in *Escherichia coli* strains lacking lactate dehydrogenase and pyruvate formate-lyase activities. Appl. Environ. Microbiol. **62:** 4594–4597.
63. Dien, B.S. *et al.* 2000. Development of new ethanologenic *Escherichia coli* strains for fermentation of lignocellulosic biomass. Appl. Biochem. Biotechnol. **84-6:** 181–196.
64. Martin, G.J.O., A.K. Bin Zhou & N.B. Pamment. 2006. Performance and stability of ethanologenic *Escherichia coli* strain FBR5 during continuous culture on xylose and glucose. J. Ind. Microbiol. Biotechnol. **33:** 834–844.
65. Kim, Y., L.O. Ingram & K.T. Shanmugam. 2007. Construction of an *Escherichia coli* K-12 mutant for homoethanologenic fermentation of glucose or xylose without foreign genes. Appl. Environ. Microbiol. **73:** 1766–1771.
66. Matjan, F., K.Y. Alam & D.P. Clark. 1989. Mutants of *Escherichia coli* deficient in the fermentative lactate-dehydrogenase. J. Bacteriol. **171:** 342–348.
67. Cassey, B., J.R. Guest & M.M. Attwood. 1998. Environmental control of pyruvate dehydrogenase complex

expression in *Escherichia coli*. FEMS Microbiol. Lett. **159:** 325–329.

68. DE GRAEF, M.R. *et al.* 1999. The steady-state internal redox state (NADH/NAD) reflects the external redox state and is correlated with catabolic adaptation in *Escherichia coli*. J. Bacteriol. **181:** 2351–2357.
69. MARTINEZ, A. *et al.* 2007. Low salt medium for lactate and ethanol production by recombinant *Escherichia coli* B. Biotechnol. Lett. **29:** 397–404.
70. CAUSEY, T.B., K.T. SHANMUGAM, L.P. YOMANO & L.O. INGRAM. 2004. Engineering of *Escherichia coli* for efficient conversion of glucose to pyruvate. PNAS **101:** 2235–2240.
71. CAUSEY, T.B. *et al.* 2003. Engineering the metabolism of *Escherichia coli* W3110 for the conversion of sugar to redox-neutral and oxidized products: homoacetate production. Proc. Natl. Acad. Sci. USA **100:** 825–832.
72. KEMPF, B. & E. BREMER. 1998. Uptake and synthesis of compatible solutes as microbial stress responses to high-osmolality environments. Arch. Microbiol. **170:** 319–330.
73. PURVIS, J.E., L.P. YOMANO & L.O. INGRAM. 2005. Enhanced trehalose production improves growth of *Escherichia coli* under osmotic stress. Appl. Environ. Microbiol. **71:** 3761–3769.
74. MILLER, E.N. & L.O. INGRAM. 2007. Combined effect of betaine and trehalose on osmotic tolerance of *Escherichia coli* in mineral salts medium. Biotechnol. Lett. **29:** 213–217.
75. INGRAM, L.O. *et al.* 1998. Metabolic engineering of bacteria for ethanol production. Biotechnol. Bioeng. **58:** 204–214.
76. EMERT, G.H. *et al.* 1983. Update on the 50 T/D cellulose to-ethanol plant. J. Appl. Polym. Sci. **37:** 787–795.
77. GAUSS, W.F., S. SUZUKI & M. TAKLAGI. 1976. Manufacture of alcohol from cellulosic materials using plural ferments. U.S. Patent number 3,990,944.
78. ZHOU, S.G. & L.O. INGRAM. 2000. Synergistic hydrolysis of carboxymethyl cellulose and acid-swollen cellulose by two endoglucanases (CelZ and CelY) from Erwinia chrysanthemi. J. Bacteriol. **182:** 5676–5682.
79. ZHOU, S.D. *et al.* 1999. Enhancement of expression and apparent secretion of Erwinia chrysanthemi endoglucanase (encoded by celZ) in *Escherichia coli* B. Appl. Environ. Microbiol. **65:** 2439–2445.
80. ZHOU, S. & L.O. INGRAM. 1999. Engineering endoglucanase-secreting strains of ethanologenic Klebsiella oxytoca P2. J. Ind. Microbiol. Biotechnol. **22:** 600–607.
81. COUGHLAN, M.P. 1992. Enzymatic-hydrolysis of cellulose—An overview. Bioresour. Technol. **39:** 107–115.
82. LAI, X.K. *et al.* 1997. Cloning of cellobiose phosphoenolpyruvate-dependent phosphotransferase genes: functional expression in recombinant *Escherichia coli* and identification of a putative binding region for disaccharides? Appl. Environ. Microbiol. **63:** 355–363.
83. MONIRUZZAMAN, M. *et al.* 1997. Isolation and molecular characterization of high-performance cellobiose-fermenting spontaneous mutants of ethanologenic *Escherichia coli* KO11 containing the Klebsiella oxytoca casAB operon. Appl. Environ. Microbiol. **63:** 4633–4637.
84. ZHOU, S. *et al.* 2003. Production of optically pure D-lactic acid in mineral salts medium by metabolically engineered *Escherichia coli* W3110. Appl. Environ. Microbiol. **69:** 399–407.
85. ZHOU, S., L.P. YOMANO CAUSEY, K.T. SHANMUGAM & L.O. INGRAM. 2005. Fermentation of 10% (w/v) sugar to D(-)-lactate by engineered *Escherichia coli* B. Biotechnol. Lett. **27:** 1891–1896.
86. JARBOE, L.R. *et al.* 2007. Development of ethanologenic Bacteria. *In* Advances in Biochemical Engineering/Biotechnology. T. Scheper, Ed., Vol. 108. Berlin: Springer.
87. CAUSEY, T.B., S. ZHOU, K.T. SHANMUGAM & L.O. INGRAM. 2003. Engineering the metabolism of *Escherichia coli* W3110 for the conversion of sugar to redox-neutral and oxidized products: homoacetate production. PNAS **100:** 825–832.
88. BARBOSA, M.D.S. *et al.* 1992. Efficient fermentation of pinus Sp acid hydrolysates by an ethanologenic strain of *Escherichia coli*. Appl. Environ. Microbiol. **58:** 1382–1384.
89. ZEMEK, J., B. KOSIKOVA, J. AUGUSTIN & D. JONIAR. 1979. Antibiotic properties of lignin components. Folia Microbiol. **25:** 483–486.
90. ZALDIVAR, J. & L.O. INGRAM. 1999. Effect of organic acids on the growth and fermentation of ethanologenic *Escherichia coli* LY01. Biotechnol. Bioeng. **66:** 203–210.
91. NISHIKAWA, N.K., R. SUTCLIFFE & J.N. SADDLER. 1988. The effect of wood-derived inhibitors on 2,3 butanediol production by Klebsiella pneumonia. Biotechnol. Bioeng. **31:** 624–627.
92. MARTINEZ, A. *et al.* 2007. Low salt medium for lactate and ethanol production by recombinant *Escherichia coli* B. Biotechnol. Lett. **29:** 397–404.

Index of Contributors